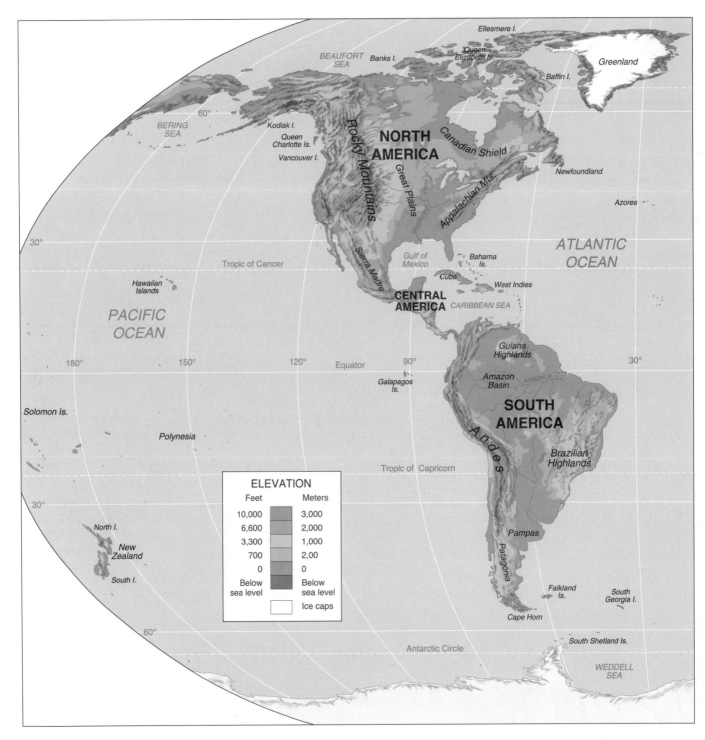

ELEVATION

Feet		Meters
10,000		3,000
6,600		2,000
3,300		1,000
700		2,00
0		0
Below sea level		Below sea level
		Ice caps

Compare this map of Earth's topography with Earth's population distribution shown on the map on the rear endpaper. Humans tend to cluster on flat areas, often near navigable waters.

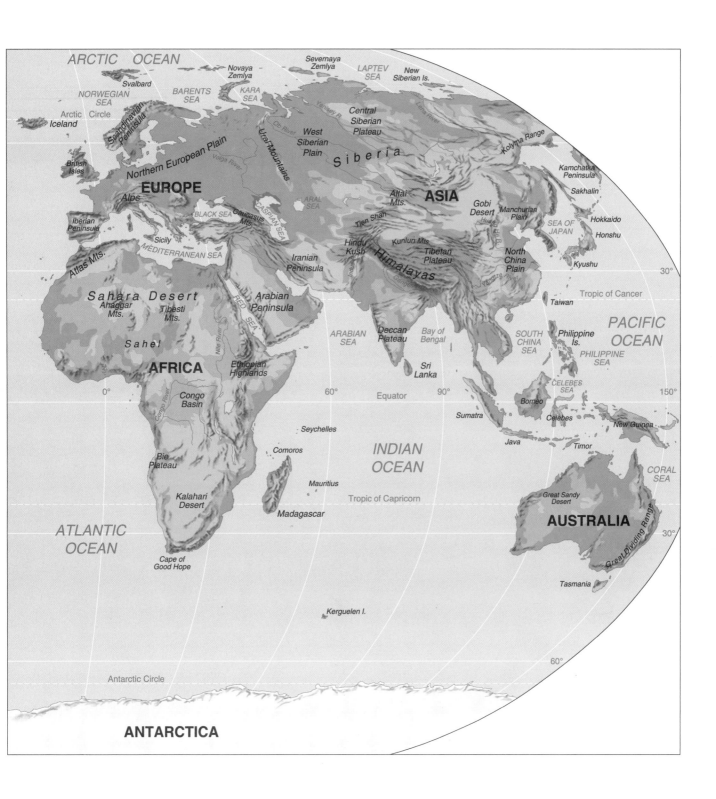

ARCTIC OCEAN

Svalbard

NORWEGIAN SEA

Novaya Zemlya

BARENTS SEA

KARA SEA

Severnaya Zemlya

LAPTEV SEA

New Siberian Is.

Arctic Circle

Iceland

Scandinavian Peninsula

Northern European Plain

Ural Mountains

Ob River

West Siberian Plain

Central Siberian Plateau

Lena River

Siberia

Kolyma Range

British Isles

Volga River

Yenisey R.

Kamchatka Peninsula

EUROPE

Alps

BLACK SEA

ARAL SEA

Altai Mts.

ASIA

Sakhalin

Iberian Peninsula

Caucasus Mts.

CASPIAN SEA

Gobi Desert

Manchurian Plain

SEA OF JAPAN

Hokkaido

Sicily

MEDITERRANEAN SEA

Tien Shan

Huang He (Yellow R.)

Honshu

Atlas Mts.

Iranian Peninsula

Hindu Kush

Kunlun Mts.

Himalayas

Tibetan Plateau

North China Plain

Kyushu

30°

Sahara Desert

Ahaggar Mts.

Tibesti Mts.

RED SEA

Arabian Peninsula

Ganges R.

Yangtze River

Tropic of Cancer

Taiwan

Sahel

Nile River

ARABIAN SEA

Deccan Plateau

Bay of Bengal

SOUTH CHINA SEA

Philippine Is.

PACIFIC OCEAN

AFRICA

Ethiopian Highlands

Sri Lanka

PHILIPPINE SEA

150°

0°

Congo River

Congo Basin

60°

Equator

90°

CELEBES SEA

Seychelles

Sumatra

Borneo

Celebes

New Guinea

Bie Plateau

Comoros

INDIAN OCEAN

Java

Timor

CORAL SEA

Mauritius

Kalahari Desert

Tropic of Capricorn

Great Sandy Desert

Madagascar

ATLANTIC OCEAN

AUSTRALIA

Cape of Good Hope

30°

Tasmania

Great Dividing Range

Kerguelen I.

Antarctic Circle

60°

ANTARCTICA

INTRODUCTION TO
GEOGRAPHY

THIRD EDITION

INTRODUCTION TO GEOGRAPHY
People, Places, and Environment

Edward F. Bergman

Lehman College of the City University of New York

William H. Renwick

Miami University, Oxford, Ohio

Upper Saddle River, NJ 07458

Library of Congress Cataloging-in-Publication Data

BERGMAN, EDWARD F.

    Introduction to geography : people, places, and environment / Edward F. Bergman, William H. Renwick. — 3rd ed.

    Includes bibliographical references (p.   ).

ISBN 0-13-144545-6

    1. Geography.   I. Renwick, William H.   II. Title.

G128.B372 2005

910—dc22   2003027347

Executive Editor: Daniel E. Kaveney
Editor in Chief, Science: John Challice
Associate Editor: Amanda Griffith
Production Editor: Debra A. Wechsler
Executive Managing Editor: Kathleen Schiaparelli
Assistant Managing Editor, Media: Nicole Bush
Vice President of Production and Manufacturing: David W. Riccardi
Media Editor: Chris Rapp
Marketing Manager: Robin Farrar
Manufacturing Buyer: Alan Fischer
Manufacturing Manager: Trudy Pisciotti
Director of Creative Services: Paul Belfanti
Creative Director: Carole Anson
Art Direction: Heather Scott
Cover and Interior Design: Suzanne Behnke
Managing Editor, Audio and Visual Assets: Patricia Burns
AV Production Manager: Ronda Whitson
AV Production Editor: Jessica Einsig
Director, Image Resource Center: Melinda Reo
Manager, Rights and Permissions: Zina Arabia
Interior Image Specialist: Beth Boyd-Brenzel
Cover Image Specialist: Karen Sanatar
Image Permission Coordinator: Carolyn Gauntt
Photo Researcher: Diane Austin
Editorial Assistant: Margaret Ziegler
Cover Image: ©Altitude/Yann Arthus-Bertrand

© 2005, 2003, 2002, 1999 Pearson Education, Inc.
Pearson Prentice Hall
Pearson Education, Inc.
Upper Saddle River, NJ 07458

Pearson Prentice Hall® is a trademark of Pearson Education, Inc.

Printed in the United States of America
10 9 8 7 6 5 4 3 2 1

ISBN 0-13-144545-6

Pearson Education LTD., *London*
Pearson Education Australia PTY, Limited, *Sydney*
Pearson Education Singapore, Pte. Ltd
Pearson Education North Asia Ltd, *Hong Kong*
Pearson Education Canada, Ltd., *Toronto*
Pearson Educación de Mexico, S.A. de C.V.
Pearson Education—Japan, *Tokyo*
Pearson Education Malaysia, Pte. Ltd

This is for all of my geography students who have given me feedback through the years.

EFB

For Debra, with thanks for her patience.

WHR

BRIEF CONTENTS

CONTENTS

6 Cultural Geography 218

7 The Geography of Languages and Religions 266

8 The Human Food Supply 318

9 Earth's Resources and Environmental Protection 358

The Geography of Manufacturing 509

National Economic-Geographic Policies 517

National Trade Policies 528

The Formation of the Global Economy 532

Conclusion: Critical Issues for the Future 540

13 Political Regionalization and Globalization 544

The Collapse of Empires 547

New Unions of States 555

Global Government 571

Protecting the Global Environment 588

Conclusion: Critical Issues for the Future 595

LIST OF MAPS

World Maps

North American Locations

African Locations

Asian and Oceanic Locations (including Russian and the CIS)

European Locations

Central and South American Locations

Cartograms

PREFACE

Many readers of this textbook may not have studied geography since grade school, where geography may have meant simply memorizing place names. Knowing where places are, however, is not all there is to geography. Knowing place names is a tool for studying geography, just as counting is a tool for studying mathematics and reading is a tool for studying literature. In geography as in any other field of study, we begin by gathering some basic information—in the case of geography, this is the *where*.

Once we know the names and locations of environmental features, people, and activities, then geography can proceed to the challenging questions of significance: *What forces* created the physical environment that characterizes various places? *Why* are people and activities located where they are? How do the features and activities at any one place *interact* to make that place unique, and what are the *relationships* among different places? What factors or forces *cause* these distributions of human populations and activities? And how and why are these environmental features and human distributions *changing*? Exploring these questions stretches your mind and imagination. Geography helps you understand current events, and it can provide you with information that is useful in deciding where to live, in seeking or building a career, and in forming your own position on political issues.

This book introduces the principal content of geography, as well as the major tools and techniques of the field. Geography is sometimes subdivided into *physical geography*, which studies the various attributes of the physical environment such as climate, plants and animals, and landforms, and *human geography*, which studies the geography of human groups and activities. Humans and the environment, however, interact; neither can be completely understood without the other.

The contents reflect our sense of logic and our experience of the surest way to accumulate understanding. Instructors may wish to vary the order of the chapters, but many students find it easier to understand certain topics after other topics have already been covered. For example, today many rural people are migrating to cities to escape poverty. Therefore, in order to understand why the world's cities are growing fast, it helps if you have already studied changes in world agriculture and in various national farm policies, so we have placed the discussion of agriculture (Chapter 8) before that of urbanization (Chapter 10).

THREE THEMES IN THIS TEXTBOOK

This textbook emphasizes three themes throughout the study of geography. First, geography examines the interrelationships between humans and their natural environment; second, many basic principles of human geography can be studied and demonstrated in your own hometown; and third, geography is dynamic.

Geography Explores the Interrelationships Between Humankind and the Environment

The study of Earth's climates, soils, vegetation, and physical features is called physical geography, and physical geography sets the stage upon which humans act out their lives. A great deal of human effort is spent wresting a living from the environment, adjusting to it, or altering it.

The first few chapters of this book, therefore, offer an overview of Earth's physical environment. The discussion emphasizes processes operating in the landscape, such as atmospheric circulation, landform change, and vegetation growth. An understanding of these processes is necessary to comprehend the mechanisms whereby humans transform Earth's environments. The theme of human–environmental interaction then weaves through the book. People interpret and evaluate possibilities in their environment, and they alter natural conditions. Any alteration in one of Earth's natural physical and ecological systems, however, will trigger changes throughout the entire system. Because these systems are tremendously complex, many changes may be unexpected or even harmful. For example, people "tame" rivers and build on their floodplains, but that construction can increase damage during a flood. People redistribute plants and animals around Earth, but relocated species can cause unexpected ecological damage. Chapter 2 notes how cities change local climates; these changes affect human health. Pollution threatens the air we breathe, the water we drink, and the soil we till. Chapter 13 emphasizes that environmental protection is one issue that today challenges all humankind, requiring international collaboration for a positive result.

Geography Is Not Restricted to the Study of Exotic Lands and Peoples

The basic principles of geography can be studied in your hometown and even on campus. How do local temperatures and rainfall vary throughout the year? What processes shaped the landforms, and how are these processes changing the land today? What natural hazards affect people in your area? Where have new arrivals in your community come from, and why did they move? Where are local food crops and manufactured goods sold? Are the boundaries of city council districts manipulated to give one group an unfair advantage in elections? How many religions are represented by houses of worship in your town, and are these centers of identifiable residential communities? Can you map the rents on commercial properties in your town, and how do these values reflect accessibility or, perhaps, perception of which neighborhoods are the most elegant?

Study of geography introduces a great range of careers, even though the professionals at work in many fields may not call themselves geographers. People studying or "doing" geography include planners designing new suburbs, scientists working to reduce water pollution, transportation consultants routing new highways, advertisers targeting "junk mail" to zip codes where residents have specific income levels, diplomats negotiating treaties to regulate international fishing, and still more various occupations.

Modern Geography Is Dynamic

It is important to know the current distributions of landforms, people, languages, religions, cities, and economic activities, but none of these patterns is static. Earth's surface is constantly changing: Sometimes it changes slowly, as when erosion wears down a mountain chain over millions of years, and sometimes it changes spectacularly and rapidly as when a new volcano explodes and builds on Earth's surface or a flood prompts a river to change its course. Transportation and communication ties among peoples and regions have multiplied, so social, political, and economic forces constantly redistribute human activities. Maps of these activities reveal only temporary balances between forces for change and forces for stability. What happens *at* places depends more and more on what happens *among* places, and we can understand maps of economic or cultural activity only if we understand the patterns of movement that create them. Modern geography explores the forces at work behind the maps.

Each day the news media report on events in which what happens is directly related to where it happens, and these events trigger changes in geography: A volcano erupts in Mexico; a bountiful harvest in Argentina reduces food prices and thus improves the diet available to Africans; Canadian scientists synthesize a substitute for a mineral previously imported; a new government in Africa redirects international alliances, economic links, and migration streams. American movies and music diffuse U.S. language and culture around the world, while Americans themselves adopt new foods and words such as sushi. Developing countries join the already-developed world in sprouting new industrial sources of air pollution, poisoning their own citizens and changing the chemical composition of Earth's atmosphere sufficiently to change Earth's climate. Protestant Christianity wins converts throughout Latin America; women are accepted into the priesthood, and governments open family planning clinics. Meanwhile, in some African and Asian countries, Islamic fundamentalists win political power and curb women's rights. These events remap world cultural, political, and economic landscapes. Today's dynamic geography doesn't just exist; it *happens*. In every topic convered in this text, it is our goal not only to *describe* distributions and locations, but to *explain* the distributions and locations.

CONTEMPORARY ISSUES IN GEOGRAPHY

What you learn in your reading and in your geography classroom can help you better understand current events and form your own opinions on important questions of the day. Each chapter of this book provides background material for understanding the news. Any number of topics in the book demonstrate this benefit, but here we might mention just two: the treatment of the topic of development, and the treatment of the issue of gender justice.

Development and Environmental Protection

Each one of Earth's billions of people aspires to a high level of material welfare, yet today many people live in conditions of deprivation. The world distribution of wealth and welfare does not coincide with the world distribution of raw material resources. If the possession of raw materials were the key to wealth, then the Republic of Congo and Mexico would count among the richest countries in the world, and Japan and Switzerland would count among the poorest. In fact, the Republic of Congo and Mexico are poor, and Japan and Switzerland are rich. An understanding of the resolution of this paradox is essential to understanding the world today.

This book goes beyond merely describing where there is wealth and where there is not. It weaves together a number of threads of understanding, making clear the *why* of the *where*. Relevant economic principles are individually introduced in the text where appropriate, including adding value to raw materials, sectoral evolution, locational determinants for manufacturing, various nations' economic policies, and the patterns of world trade. New considerations arise virtually every day, such as the discovery of new resources, new technologies, and new governments with new policies. These changes redistribute advantages. Chapter 12 suggests how and why the balance among the factors that support economic development is continuously shifting, and Chapter 13 highlights how these geographic shifts redistribute global sources of pollution. New industries in developing countries generate pollution. Furthermore, those new industries allow rising standards of living, and that in turn allows increased consumption of consumer goods and increased production of waste products.

Economic development is only one aspect of development. Human development includes adequate nutrition, education, political liberty, and the opportunity to live in a healthy, unpolluted environment. Chapter 5 analyzes the geography of health, and Chapter 8 explains the distribution of world food production and trade. Chapter 9 discusses problems and challenges in the consumption of natural resources and the treatment of wastes and disposed items. Chapter 11 maps world education and freedom. In each case, the text suggests what factors might redistribute these activities in the future.

Gender Justice

Almost everywhere, women are worse off than men. They have less power, less autonomy, less money, more work, and more responsibility. The issue of discrimination against women—gender justice—is emerging as a major question of our time. This text does not separate this issue in an isolated chapter, but examines it as it plays a role in each topic through the book.

Chapter 5, on population, notes how discrimination against women can begin even against children in the womb. Some countries record higher rates of abortion for female fetuses than male fetuses. We see that in some countries females get less health care and nutrition; the result is low ratios of females to males in the overall national population. Discrimination against female farmers, discussed in Chapter 8, lowers agricultural productivity and the levels of nutrition in many countries. Chapter 7 examines the differing attitudes toward women taught by various religions. The ramifications of those teachings reach from the issue of priesthood for women into the treatment of women in national laws. Chapter 11 notes the variations in the percentages of males or females in school around the

world, women's role in politics, and other issues of legal rights.

This text readily admits that we do not know the answers to all questions. It invites you to learn what we know and then to join the world of scholarship by increasing the knowledge that we have today. For example, Chapter 10 notes that vast slums are growing in many cities in developing countries largely because of migration from rural areas into the cities. High percentages of people living in cities are uncounted, living in unregistered properties, and performing unrecorded economic activities. The vitality of these urban slums astounds observers and confounds economists, who struggle to understand and measure what is going on. This textbook explains what is happening as we understand it, but acknowledges that we need to devise new methods of measuring and accounting for this activity. Similarly, many of Earth's physical processes are incompletely understood. The process of global warming is frequently in the news today, and yet the number of variables that might cause it and their exact interactions are beyond the range of even the greatest supercomputers. We encourage you to take up the challenge of investigating some of these issues—and, if possible, improving conditions—as possible research topics for your own careers.

MAPS, CARTOGRAMS, AND GIS

A great variety of maps illustrate this book. Some of them include relief shading. You will probably be familiar with traditional maps that illustrate distributions as mosaic patterns of color. On other maps, called flow maps, arrows represent movements of people or of goods, and on these the widths of the arrows often convey the quantity of each flow—the numbers of passengers flying major airline routes across the United States, for example (Figure 1-16).

Several cartograms have also been specially designed for this book. A cartogram is a visual device much like a map, but on a cartogram physical distance is replaced with some other measure in order to convey a visual impression of the magnitude of something. Cartograms of such things as countries' populations, for example (Figure 5-2), or of countries' economies (Figure 12-5), visually convey the relative population or wealth of different countries better than standard land area maps do.

Today maps are drawn and rapidly updated using Geographic Information Systems (GIS) and digital cartographic tools. GIS have become essential and ubiquitous on geographers' desks, and their importance is clear in this book. An expanded discussion of GIS technology has been added in Chapter 1, and we have illustrated how GIS is used in geographic problem-solving throughout the book.

A variety of other visual devices includes tables, bar graphs, and pie graphs, with which you are probably familiar. In each case the captions have been carefully written to help you read these sophisticated images. Each contains a great deal of information.

Special care has been taken in preparing the captions for all illustrations. Each caption guides your eye through or into the figure. Many captions include questions to get you thinking about the information in the image. Many ask you to compare it with information shown on another image in the book in order to develop hypotheses of explanation.

A WORD ABOUT NUMBERS

This book contains many numbers—measurements of populations, economic conditions, production of various commodities, world trade, and more. These measures come from a variety of sources—private organizations, national governments, international organizations—and they are the best available. Such numbers, however, must always be read with two considerations in mind: reliability and date.

The compilation of measures is a tremendously difficult task. For example, the United States is the world's richest country, with many highly skilled government workers, yet the government admits that the national census is probably inaccurate by a factor of 5 to 7 percent. What degree of accuracy, then, can we expect from a poor country that does not have a sophisticated bureaucracy? It is said that the government of India does not have even an accurate count of the number of cities in India. We do not want to promote cynicism about the value or reliability of statistics, but an educated person does exercise judgment about the probable exactitude of any figure.

The second caution is that the measures themselves change. The population of the United States changes every day, yet the United States officially counts its residents only once every ten years—although periodic estimates are more frequently published. It takes a long time to gather and compile statistics, so the measures may seem out of date by the time they are published. This is especially true of international comparative statistics. For example, each year the United Nations Conference on Trade and Development (UNCTAD) publishes a handbook of statistics of world trade, but the book appears three or four years after its date, and many statistics recorded were measured years before the date of the volume. The 2000 volume appeared at the end of 2002 and included statistics that were actually measured as long ago as 1993. The United Nations Demographic Yearbook for 2000 also appeared in 2002, and it contained many "most recent available" statistics dating back to the early 1990s. The figures used to make world maps, as in this text, are sometimes older than

figures given for many individual countries because advanced countries are able to compile numbers faster than poorer countries, and then international organizations need time to compile all of the national figures. Furthermore, governments sometimes change the way they measure things. For example, for many years governments counted and published a statistic called gross national product (GNP), but today that statistic has been replaced by a slightly different measure called the gross national income (GNI). The meaning of GNI is explained in Chapter 12.

The statistics in this textbook are as up-to-date as could possibly be obtained from the most reliable sources as of October 2003. The text notes the direction in which many of these measures are changing, and in many cases we have dared to predict their future direction. The United States population will probably continue to rise, and the percentage of the national labor force working in manufacturing will probably continue to fall. We challenge you to go to the library or to search the World Wide Web to update those measures to today.

FEATURES TO HELP YOU STUDY

Each chapter offers a number of features to help you study:

- **A Look Ahead.** Each chapter begins with a brief outline of the main points to be made in that chapter.
- **Anecdotes.** An anecdote at the beginning of each chapter exemplifies the sort of real-world questions and problems that arise from the material discussed in that chapter.
- **Key Terms.** The key terms in each chapter are printed in **boldface** when each is first introduced. These terms are also listed at the end of each chapter along with the page on which each first appears. All key terms are defined in the Glossary at the end of the book. Some additional terms that may be unfamiliar to some readers but that are less central to the study of geography are printed in *italics*. These terms are defined carefully in the text.
- **Critical Thinking boxes.** Critical Thinking boxes discuss problems or pose questions for which there are no simple or clear answers. These require you to exercise your analytical skills to develop your own viewpoints. Each Critical Thinking box includes questions to start you thinking about the material in the box.
- **Focus On boxes.** Focus On boxes highlight individual problems or case studies of the topics in that chapter.
- **Regional Focus On boxes** (new to this edition). Regional Focus On boxes examine in detail

particular places that exemplify points being made in the text. Many of these places are hotspots in the news.

- **Conclusion: Critical Issues for the Future.** Each chapter concludes with an examination of how contemporary issues discussed in that chapter will continue to challenge us in coming years. Some of these challenges are very real topics of current political debate; some will face humankind all through your lifetime.

- **Summary.** Each chapter closes with a summary of its main points.

- **Questions for Review and Discussion.** At the end of each chapter you will find a list of questions about the subject of that chapter. These review questions test your reading, and their answers can be found in the chapter.

- **Thinking Geographically.** Answers to the questions in this section cannot be found in the text itself. These questions encourage exploration beyond the text—into the library or out into the community. Some students may find these questions good suggestions for papers or research projects.

- **Suggestions for Further Learning.** A short list of up-to-date readings provides sources for further information.

- **Web Work.** This feature contains an annotated list of sites that you may check for updating or for additional information on the topics discussed in the text. These range from the homepages of individual newspapers to the homepages of libraries, government agencies, and of national and international organizations.

- **Three Appendices.** The book also offers three appendices. The first details map projections, their creation, and their uses. The second offers substantial detail of the temperature and rainfall parameters for the various categories of climate defined in the text. The third is a table providing basic statistical information for most of the countries of the world.

THE ATTACKS ON THE UNITED STATES ON SEPTEMBER 11, 2001

On the morning of September 11, 2001, four passenger airliners taking off from U.S. East Coast airports on transcontinental flights were simultaneously hijacked. The hijackers piloted two of those planes into the twin 110-story towers of New York's World Trade Center (WTC). The conflagrations ignited by the planes' full fuel tanks, combined with the shock of the collisions, caused both towers to collapse. A third hijacked plane, also on a scheduled transcontinental flight, was smashed into the headquarters building of the U.S. Department of Defense in Arlington, Virginia (the Pentagon).

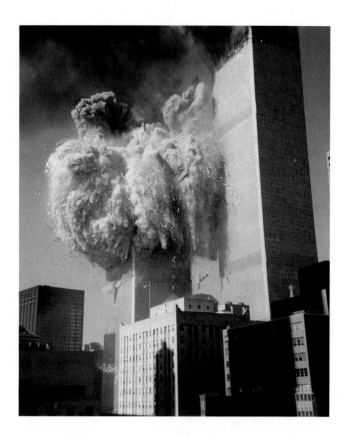

Passengers on the fourth hijacked flight, hearing the fates of the three other flights from conversations with loved ones over their own mobile phones, rallied to retake control of their flight, but, during the fighting on board, that plane crashed in rural Pennsylvania. Hijackers had probably intended to plunge it into the White House or the U.S. Capitol.

The total number of people killed in these hijackings and suicidal crashes was about 3,000, including citizens of the United States, Canada, Pakistan, India, the Netherlands, the United Kingdom, Peru, and many other countries. The dead included individuals of the Christian, Jewish, Muslim, Hindu, and still other religious faiths. This was the greatest massacre on American soil since Japan's surprise attack on the U.S. naval base at Pearl Harbor on December 7, 1941—an event that brought the United States into World War II. The events of September 11 had an instant and profound effect on world politics and world economics.

The United States and allies launched a war against an international terrorist group named Al Qaeda ("the foundation" in Arabic), which had been responsible for earlier terrorist attacks against U.S. and allied governments and peoples. Al Qaeda's leadership was known to be operating from Afghanistan, which was ruled by a fundamentalist Islamic group called the Taliban ("student" in Arabic). (Background to Al Qaeda and the Taliban in Afghanistan is given on page 576). Thus, the overthrow of the Taliban was America's first objective, and that was accomplished within three months.

The United States has since declared a "war against global terrorism," and the ramifications of that war extend into the domestic politics of many countries and throughout international affairs. This textbook incorporates explanatory material in the appropriate chapters and, insofar as possible, reviews the continuing ramifications of these events.

SUPPLEMENTS

The authors and publisher have been pleased to work with a number of talented people to produce an excellent supplements package. This package includes the traditional supplements that students and professors have come to expect from authors and publishers, as well as some new kinds of supplements that involve electronic media.

Online Study Guide: *Introduction to Geography: People, Places, and Environment* Website, by Erick Howenstine of Northeastern Illinois University, gives you the opportunity to explore further topics presented in the book by using the Internet. The site contains numerous review exercises (from which students get immediate feedback), exercises to expand your understanding of geography, and resources for further exploration. In addition to being a valuable tool for student review, the Web site provides an excellent platform from which to start using the Internet for the study of geography. Please visit the site at **http://www.prenhall.com/bergman.**

In addition to the *Online Study Guide*, the book highlights many features intended to assist instructors in preparing and presenting their classes. These instructor-centered supplements include the following:

Transparency Set (0-13-144547-2): More than 125 full-color acetates of illustrations from the text, all enlarged for excellent classroom visibility are available free of charge to qualified adopters.

Instructor Resource Center (0-13-144552-9): Everything you need for lecture preparation, right where you want it. The Prentice Hall Instructor Resource Center CD-ROM makes you more effective, and saves you time and effort. Find all your digital resources in one, well-organized, easy-to-access place. The Instructor Resource Center includes all of the maps and figures from the text, select photographs, PowerPoint presentations to support lectures, test questions, animations, and the instructor's manual, all keyed to the appropriate chapters in the Bergman text. These files are ideal for those professors who use PowerPoint or a comparable presentation software for their classes, or for professors who create text-specific Web sites for their students.

Instructor's Manual (0-13-144550-2): Written by Richard Walasek of the University of Wisconsin-Parkside, the *Instructor's Manual* is intended as a resource for both new and experienced instructors. It includes a variety of lecture outlines, additional source materials, teaching tips, advice about how to integrate visual supplements (including the Web-based resources), and various other ideas for the classroom.

Test Item File (0-13-144546-4): The test item file contains many test questions drawn from the book, available in printed format. It is also available on the *Instructor's Resource Center* CD-ROM.

Test Gen EQ (0-13-144548-0). TestGen EQ is a computerized test generator that lets you view and edit test-bank questions, transfer questions to tests, and print the text in a variety of customized formats. Included in each package is the QuizMaster EQ program that lets you administer tests on a computer network, record student scores, and print diagnostic results.

Course Management: Prentice Hall is proud to partner with many of the leading course-management system providers on the market today. These partnerships enable us to combine our market-leading online content with the powerful course-management tools Blackboard and WebCT. In support of professors who use course management in their classes, Prentice Hall offers all of the test questions that accompany this book in both Blackboard and WebCT format. This allows professors to easily integrate the book's test questions with any pre-existing materials that they have developed for the course.

Rand McNally Atlas of World Geography. If the instructor so chooses, this item is available FREE when packaged with the text. The atlas includes 126 pages of up-to-date, accurate regional maps and 20 pages of illustrated world information tables. Please contact your local Prentice Hall representative for details.

Mapping Workbook (0-13-144964-8). If the instructor so chooses, this workbook is available for FREE when packaged with the text. The mapping workbook contains outline maps for each chapter, based on the maps in the text, ready for student use in identification exercises. Please contact your local Prentice Hall representative for details.

Research Navigator™ is the easiest way for students to start a research assignment or research paper. Complete with extensive help on the research process and four exclusive databases of credible and reliable source material including the EBSCO Academic Journal and Abstract Database, *New York Times* Search by Subject Archive, "Best of the Web" Link Library, and *Financial Times* Article Archive and Company Financials, Research Navigator helps students quickly and efficiently make the most of their research time. Access to Research Navigator is free when purchased with a new Pearson textbook. Please contact your local Prentice Hall representative for details.

ACKNOWLEDGMENTS

Countless colleagues, librarians, and generous individuals both in government and in the private sector helped with information for this text. We wish to thank in a special way Professor Tom L. McKnight of the University of California, Los Angeles, and Professor James M. Rubenstein of Miami University, Ohio, with whom we have worked on earlier textbooks, for their many contributions to our thinking. Any errors in the text are the fault of the authors. We would like to thank the following people for their special help: Emily Nerbun, NAFTA desk of the U.S. Department of Commerce; Marilyn Washington, Japan External Trade Organization; Amy Staples, National Museum of African Art; Laveta Emory, Smithsonian Institution; Priscilla Strain, National Air and Space Museum; Clyde McNair, Agency for International Development; Jim Bohn, Atlanta Regional Commission; Don McMinn, Spriggs and Hollingsworth; Linda Carrico, U.S. Bureau of Mines; Barbara Mathe, American Museum of Natural History; Kerstin Erickson, European Union Information Service; William Usnik, *The New York Times;* Roger Wieck and William Voelkle, the Morgan Library; John Rutter, U.S. Department of Commerce; Gary Cohen; Jerry Hagstrom, *The National Journal;* Debbie Janes, U.S. Environmental Protection Administration; Melissa Starr, World Health Organization; David Sackett, Lehman Brothers; Dan Beard, National Audubon Society; Carey Pickard, the Tubman African American Museum, Macon, Georgia; Ms. Drs. Lydia Van Rietschote, Head of the Department of Sector Policy, Ministry of Education, the Netherlands; Guadalupe Morgan, the World Bank; Professor Francisco Denton, Mexico City; Patricia Dickerson, U.S. Bureau of the Census; Kurt G. Usowski, HUD; Gloria Cuaycong, UN Statistics Division; David Sutherland, NOAA; Louise Genereux, Statistics Canada; Randall Flynn; US Board on Geographic Names; Arthur Lambert; Joseph W. Cherner; David Morgan; Gino Pepoli; plus the many friends and colleagues whose photographs are used and credited in the text. Dr. Ivor Winton of the Department of Geography, Simon Fraser University, British Columbia, Canada, offered careful reading and notes.

We also wish to thank our scholarly colleagues who provided thoughtful suggestions for improving the manuscript during its preparation. These include

Holly R. Barcus, Morehead State University

Daniel Block, Chicago State University

Paul L. Butt, University of Central Arkansas

Joseph M. Cirrincione, University of Maryland

Bryce Decker, University of Hawaii at Manoa

Stanford Demars, Rhode Island College

Leslie Dienes, University of Kansas

Gary Fowler, University of Illinois, Chicago

Michael Fox, Carleton University

Roberto Garza, San Antonio College

Rene J. Hardy, Shoreline College

Erick Howenstine, Northeastern Illinois University

James C. Hughes, Slippery Rock University

Walter Jung, Central Oklahoma University

Tulasi R. Joshi, Fairmont State College

Angelina Kendra, Central Connecticut State University

Rob Kent, University of Akron

Miriam K. Lo, Mankato State University

Ruben A. Mazariegos, University of Texas-Pan American

Ian A. McKay, Wilfrid Laurier University

G. L. "Jerry" Reynolds, University of Central Arkansas

Scott C. Robinson, University of Nebraska Omaha

Lallie Scott, Northeast Oklahoma State University

Christa Smith, Clemson University

James N. Snaden, Central Connecticut State University

David M. Solzman, University of Illinois, Chicago

Robert C. Stinson, Macomb Community College

Christopher J. Sutton, Northwestern State University of Louisiana

Melissa Tollinger, East Carolina University

James Tyner, Kent State University

Thomas B. Walter, Hunter College of the CUNY

Gerald R. Webster, University of Alabama

Kathy Williams, Bronx Community College

John Wright, New Mexico State University

Charles T. Ziehr, Northeastern State University

We owe a debt of gratitude to many people. At Prentice Hall Dan Kaveney, Executive Editor, managed the project from its beginning stages through the journey to production. Ray Mullaney, Editor in Chief of Science and Math Development, was responsible for the development of the book and contributed constructive reviews along the way. We thank Debra Wechsler for her unflagging commitment to quality as she guided us through the production process. Christine Henry served as Marketing Manager for previous editions of this book, and we look forward to working with our new marketing manager, Robin Farrar. Margaret Ziegler, Amanda Griffith, and Chris Rapp provided invaluable editorial assistance with administration, supplements, and digital media respectively. We have enjoyed working with all of these people, and we thank them.

Edward F. Bergman

William H. Renwick

ABOUT THE AUTHORS

Edward F. Bergman stands beside the stone wall that is the very eastern edge of the North American tectonic plate, where it is exposed in Iceland (see Figure 3-5).

Professor Bergman was born in Wisconsin and studied at the University of Wisconsin (Madison), the University of Vienna (Austria), and the University of Washington in Seattle. Today he teaches at Lehman College of the City University of New York and lectures widely across several continents. When not lecturing or writing, he enjoys Manhattan's cultural and social life.

William H. Renwick earned a B.A. from Rhode Island College in 1973 and a Ph.D. in geography from Clark University in 1979. He has taught at the University of California, Los Angeles, and Rutgers University, and is currently Associate Professor of Geography at Miami University. A physical geographer with interests in geomorphology and environmental issues, his research focuses on impacts of land-use change on rivers and lakes, particularly in agricultural landscapes in the Midwest. When time permits, he studies these environments from the seat of a wooden canoe.

INTRODUCTION TO
GEOGRAPHY

A Look Ahead

WHAT IS GEOGRAPHY?

Geography is the study of the interaction of all physical and human phenomena at individual places and of how interactions among places form patterns and organize space. Geography is one of the oldest and most diverse fields of study, so it offers an introduction to a unique breadth of eventual careers.

CONTEMPORARY APPROACHES IN GEOGRAPHY

Geographers employ three principal analytical methods: area analysis, spatial analysis or locational analysis, and geographic systems analysis.

1

INTRODUCTION TO GEOGRAPHY

◄ The eruption of Mount Etna on the Mediterranean island of Sicily in 2002 shown in this spectacular satellite photograph threatened the human settlement around the base of the volcano and polluted the air for hundreds of miles. Mount Etna is the oldest and most active volcano in Europe, but humans live in volcanic regions around the globe, profiting from the rich soils volcanoes often deposit. Humans live in constant interaction with their physical environment, adapting to it, but also changing it for their own use. (AP/Wide World Photos)

DESCRIBING EARTH: COMMUNICATING GEOGRAPHIC INFORMATION

Maps are a characteristic geographic tool, and geographic information systems are new computer tools for storing, mapping, and analyzing geographic data.

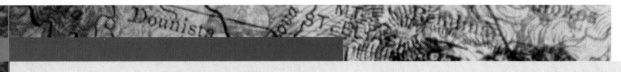

Sailors may soon realize the Northwest Passage across the top of North America between the Atlantic and Pacific Oceans that explorers have sought for 500 years. Technology has delivered stronger icebreakers, and an overall warming of Earth's climate, partly due to human activity, has opened many passages formerly closed by ice and has thinned the ice in others. The United States hopes to sail between the East Coast and Alaska's north slope. Canada objects, however, because Canada claims jurisdiction over all of the islands and water all the way to the North Pole. The United States disputes Canada's territorial claims, because the territory is uninhabited, and the United States continues to explore sea routes. Furthermore, a joint Russian–Canadian study has proposed direct transpolar commercial flights between North America and Asia. During the Cold War between the Communist and the democratic powers, North Americans nervously scanned that airspace for incoming missiles, but today transpolar flights would shave hours off traditional flight times on busy routes. These new routes might also provoke U.S.–Canadian friction over regulation. Thus, changes in the environment, in technology, and in world politics continuously rearrange the relative locations of activities and the transport routes among them, opening new possibilities and occasions for international diplomacy.

WHAT IS GEOGRAPHY?

Many readers of this textbook may not have studied geography since grade school, when geography may have meant simply memorizing place names. Knowing place names is a tool for studying geography, just as counting is a tool for studying mathematics and reading is a tool for studying literature. In geography, as in any other field of study, we begin by gathering some basic information. In the case of geography, this is the *where*. Knowing where places are, however, is not all there is to geography.

Once we know the names and locations of environmental features, people, and activities, then geography can proceed to the challenging questions of significance: *Why* are people and activities located where they are? How do the features and activities at any one place *interact* to make that place unique, and what are the *relationships* among different places? What factors or forces *cause* these distributions of things and activities? How and why are these distributions *changing*? And how, perhaps, can we *alter* those distributions for greater human convenience or profit?

Geography is the study of the interaction of all physical and human phenomena at individual places and of how interactions among places form patterns and organize space. **Physical geography** studies the characteristics of the physical environment. When geography concentrates on topics such as climate, soil, and vegetation, it is a natural science. **Human geography** studies human groups and their activities, such as language, industry, and the building of cities; human geography is a social science. **Cultural geography,** a subfield of human geography, focuses specifically on the role of human cultures. Physical and human geographers share both their approach and a great deal of information, and their analyses of the landscape always weave their understanding together. Thus, geography bridges the physical sciences and the social sciences. **Cartography** (map making) and its computerized extension to **geographic information systems (GIS)** provide tools that help both physical and human geographers store, display, and analyze geographic data.

Geographers investigate the processes underlying all observed distributions and patterns. Technology has accelerated the pace of change, and knowledge of forces of change is critical to understanding and managing global problems that might arise in the future.

Geography offers a way of thinking about problems, and geographers are particularly well equipped to understand interactions among different forces affecting a place. For example, to understand hunger in Africa, geographers examine relationships among climate, soils, agricultural practices, population growth, food prices, environmental degradation, political unrest, and still other relevant factors. To explain unrest in the Middle East, geographers study the interrelationships among the distribution of water and energy resources, differences in religious beliefs, historical settlement patterns, international ties of local groups, and conflicting strategies for achieving economic growth. The interrelationships among factors affecting places help us understand why humans behave as they do. No other scientific discipline takes this approach.

The Development of Geography

Geography in the classical Western world   Every language has a word for the study that we call "geography," but the Greek word that we use, which means

"Earth description," was the title of a book by the Greek scholar Eratosthenes (c. 275–195 B.C.). He was the director of the library in Alexandria, Egypt. The library was the greatest center of learning in the Mediterranean world for hundreds of years. No copies of Eratosthenes' book have survived, but we know from references by other authors that Eratosthenes accepted the idea that Earth is round and that he even calculated its circumference with amazing precision. Eratosthenes also drew fine maps of the world as it was known to him. Hipparchus (180–127 B.C.), a later Greek scholar and director of the library, was the first person to draw a grid of imaginary lines on Earth's surface to locate places precisely.

After the great age of Greek civilization gave way to the Roman era, the Roman Empire produced a number of geographical scholars and compilations of geographical learning. These culminated in the *Guide to Geography* by Ptolemy, a Greek scholar who worked at the library in Alexandria between A.D. 127 and 150. Subsequently, in the long period between the fall of the Roman Empire (A.D. 476) and the European Renaissance of the fifteenth century, Western civilization accumulated little additional geographical knowledge. Much knowledge was actually lost in Western Europe but preserved by Arab scholars.

Geography in the non-European world Outside Europe, geography made considerable advances. The expansion of the religion of Islam (discussed in Chapter 7) fostered travel and allowed research across the wide region from Spain to India and beyond. Islam encourages each believer to complete a pilgrimage to Mecca, in Arabia, at least once in his or her lifetime. This obligation stimulates travel. Muslim scholars, including al-Edrisi (1099?–1154), ibn-Battuta (1304?–1378), and ibn-Khaldun (1332–1406), produced impressive texts describing and analyzing both physical environments and also the customs and lifestyles of different peoples.

China also developed an extensive geographical literature. The oldest known work—*The Tribute of Yu*—dates from the fifth century B.C. and describes the physical geography and natural resources of the various provinces of the Chinese empire. It interprets world geography as a nest of concentric squares of territory. The innermost zone is the imperial domain. The imperial domain is surrounded by the empire, and beyond the empire lies an outermost zone inhabited by "barbarians." The world's oldest map having distances marked with numbers was engraved on copper plate about 325 B.C. The map shows royal tombs in northern Hebei province. In A.D. 267, Phei Hsiu made an elaborate map of the empire, and he is often called the father of Chinese cartography. In the following centuries, Chinese Buddhists wrote occasional accounts of travel to India to visit places sacred to the history of Buddhism, which had spread into China. Chinese maritime trade extended throughout Southeast Asia and

FIGURE 1-1 The Kangnido (detail). When Columbus embarked on his voyage, he knew nothing of Korea, yet the world's finest map was then hanging in the royal palace there. The Kangnido included not only China and Japan but showed India, the countries of the Middle East, and Europe itself. Spain was easy to recognize on this map, which even showed Columbus's native Genoa along the shores of the Mediterranean. The image of Europe was not exact, but no map in Europe had an image of Asia that was as good as this map's image of Europe. (Courtesy of Ryukoku University Library)

the Indian Ocean, and it even reached East Africa, but the Chinese eventually withdrew from more extensive exploration. Few Chinese descriptive geography texts of this activity survive.

The Japanese and the Koreans also engaged in East Asian trade. In fact, to our knowledge, the greatest world map produced before the European voyages of exploration was made in Korea in 1402 (Figure 1-1). This map, the Kangnido, was made by an unknown cartographer, but he or she drew on the combined knowledge of Korea, Japan, and China, including Islamic sources that were known to the Chinese. As a result, this map includes not only East Asia but also India, the countries of the Islamic world, Africa, and even Europe. It demonstrates a far more extensive knowledge of the world than the Europeans had at the time.

The peoples of Africa, of the Western Hemisphere, and of other areas may have had access to geographical accounts of the parts of the world known to them, but either these works have not survived or modern scholars have not yet fully inventoried and studied them.

The revival of European geography Beginning in the fifteenth century, Europe's world exploration and conquest brought so much new geographical knowledge that Europeans were challenged to devise new methods of cataloging or organizing it all.

Focus on
NATIONAL STANDARDS FOR THE STUDY OF GEOGRAPHY

In 1994 the U.S. Congress adopted Goals 2000: The Educate America Act (Public Law 103-227). This act listed geography among the subjects that must be considered fundamental to a national curriculum. Geographical understanding, said Congress, is essential to achieve "productive and responsible citizenship in the global economy." Several academic and scholarly geographical organizations set to work collaborating to produce an agreed core of geographic material and ideas, which was published as *Geography for Life: The National Geography Standards*. These standards specify the geographical subject matter and skills that U.S. students should master.

The goals demonstrate the degree to which geographic knowledge is essential for both understanding and effectively managing environmental and human relations in the twenty-first century. They were established in the hope that all persons educated in the public schools become geographically knowledgeable. These goals were not reached by 2000, but they remain a national target. In this book, we go beyond these standards in the treatment of both subject matter and thinking skills, but the outline of the goals demonstrates the great breadth of the field.

The geographically informed person knows and understands:

The World in Spatial Terms

1. How to use maps and other geographic tools and technologies to acquire, process, and report information from a spatial perspective.
2. How to use mental maps to organize information about people, places, and environments in a spatial context.
3. How to analyze the spatial organization of people, places, and environments on Earth's surface.

Places and Regions

4. The physical and human characteristics of places.
5. That people create regions to interpret Earth's complexity.
6. How culture and experience influence people's perceptions of places and regions.

Physical Systems

7. The physical processes that shape the patterns of Earth's surface.
8. The characteristics and spatial distribution of ecosystems on Earth's surface.

Human Systems

9. The characteristics, distribution, and migration of human populations on Earth's surface.
10. The characteristics, distribution, and complexity of Earth's cultural mosaics.
11. The patterns and networks of economic interdependence on Earth's surface.
12. The processes, patterns, and functions of human settlement.
13. How the forces of cooperation and conflict among people influence and control Earth's surface.

Environment and Society

14. How human actions modify the physical environment.
15. How physical systems affect human systems.
16. The changes that occur in the meaning, use, distribution, and importance of resources.

Uses of Geography

17. How to apply geography to interpret the past.
18. How to apply geography to interpret the present and plan for the future.

One attempt at organizing information gathered by the voyages of discovery is the book *General Geography* (1650) by Bernhard Varen (Varenius in Latin; 1622–1650), a German who taught at the University of Leiden, in Holland. Varen differentiated two approaches to geography: *special geography* and *general geography*. Special geography describes and analyzes places in terms of categories such as local population, customs, politics, economy, and religion. Today we call this approach **regional geography.** Regional geographers usually begin regional studies with a description of the local physical environment—the climate, landforms, soils, and other physical attributes—and then proceed with an inventory much like Varen's, weaving these factors together as they interrelate. The Regional Focus boxes in this textbook will highlight regions that illustrate geographic principles or that are in the news.

TABLE 1-1 Topical Specialties Recognized by the Association of American Geographers

01 Agricultural Geography	19 Gender	37 Physical Geography
02 Applied Geography	20 Geographic Information Systems	38 Planning, Environmental
03 Arid Regions	21 Geographic Theory	39 Planning, Regional
04 Biogeography	22 Geomorphology	40 Planning, Urban
05 Cartography	23 Global Change	41 Political Geography
06 Climatology	24 Hazards	42 Population Geography
07 Cultural Ecology	25 History of Cartography/Historical Cartography	43 Quantitative Methods
08 Cultural Geography		44 Recreational Geography
09 Developmental Studies	26 Historical Geography	45 Regional Geography
10 Earth Science	27 History of Geography	46 Remote Sensing
11 Economic Development	28 Land Use and Conversion	47 Resource Geography
12 Economic Geography	29 Librarianship, Geographical	48 Rural Geography
13 Educational Geography	30 Location Theory	49 Social Geography
14 Energy	31 Marine Resources	50 Soils Geography
15 Environmental Perception	32 Marketing Geography	51 Teaching Techniques
16 Environmental Science	33 Medical Geography	52 Transportation and Communication
17 Environmental Studies (Conservation)	34 Military Geography	53 University/Other Administration
	35 Mountain Environments	54 Urban Geography
18 Field Methods	36 Oceanography	55 Water Resources

What Varen called general geography examines topics of universal application or occurrence. We might study, for example, the geography of climate, water, vegetation, or minerals. This notion corresponds to what we now refer to as **topical** or **systematic geography,** and it is the basic approach taken for the outline of this textbook. Individual geographers concentrate on topics as diverse as the geography of soils (pedology), of life forms (biogeography), of politics (political geography), of economic activities (economic geography), and of cities (urban geography). The Association of American Geographers (AAG), a professional scholarly organization, today recognizes 55 topical specialties (Table 1-1). Regional and topical geography are complementary. A regional study covers all topics in the region under review, whereas a topical study notes how a particular topic varies across regions.

Varen's book was a standard reference for over 100 years. Even Sir Isaac Newton (1642–1727) edited two editions in Latin, one of which was studied by students at Harvard in the eighteenth century.

The human-environment tradition We cannot fully understand the physical environment without understanding humankind's role in altering it, nor can we fully understand human life without understanding the physical environment in which we live.

The German explorer and naturalist Alexander von Humboldt (1769–1859) wove together virtually all knowledge of Earth sciences and anthropology in his great multivolume book *Cosmos.* Von Humboldt's brilliant interrelating of the phenomena of physical nature and the world of humankind had an enormous intellectual impact in the United States (Figure 1-2). This was partly because European Americans had long thought of America as the New World, a veritable Garden of Eden over which they had been given dominion.

George Perkins Marsh (1801–1882), an American scholar and diplomat, expanded on von Humboldt's theme of the interconnections between humankind and the physical environment. While Marsh served as U.S. ambassador to several Mediterranean countries, he was impressed by humankind's destruction of an environment that ancient authors had described as lush and rich. Marsh's book *Man and Nature, or Physical Geography as Modified by Human Action* (1864) was one of the earliest key works in what would become today's environmental movement.

Why study geography? Who is a geographer today? And why should anyone study geography? Today almost all of us are geographers, whether or not we call ourselves that. Geographers include people who plan the need for future schools; develop real estate; manage parklands, forests, or water supplies; plan road tours for rock groups; as well as farmers, who choose which crop to plant on the basis of world markets; and journalists, diplomats, and executives of multinational corporations.

> The information that any citizen needs in order to make an informed decision on an important question of the day is largely geographic.

The information that any citizen needs in order to make an informed decision on an important question of the day is largely geographic. And as world events force rapid change upon us daily, the geography of our world changes too. Almost any topic of the day's news will bring about redistributions—new geographies—of people and activities. Studying geography will not always help you predict the future, but it will help you understand the forces behind these changes. Therefore, when changes come, you may be less surprised. Sometimes you will be able to see changes coming, make changes work for you, or even make the changes yourself.

Geography is one of the most diverse fields of academic study, so it offers an introduction to a unique breadth of eventual careers. A great variety of careers involve studying or "doing" geography, even though the professionals at work may not call themselves geographers (Figure 1-3). As you read through the

FIGURE 1-3 A geographer at work. This geographer is measuring runoff and sediment from a field in England. These factors will help determine whether the land is best devoted to agricultural, commercial, residential or industrial use, or set aside for parkland. (Nigel Cattlin/Photo Researchers, Inc.)

FIGURE 1-2 Alexander von Humboldt. This portrait bust of the explorer and naturalist Alexander von Humboldt was the first statue of a non-U.S. citizen to be erected in New York City's Central Park. It was installed in 1869, marking the one hundredth anniversary of von Humboldt's birth. Today he gazes across the street at the American Museum of Natural History, where statues of the explorers Lewis and Clark gaze back. (EFB photo)

topics in this text, ranging from newspaper distribution to agricultural supplies, from urban planning to the delineation of voting districts, and from multinational marketing of consumer goods to international treaties on environmental preservation, think always of the many ways you might develop a career analyzing or managing any of these activities.

Geographers' studies of *where, what, when, why,* and *why there* necessarily prompt thoughts about *where will* to predict future patterns, and of *how ought.* Geographers' expertise can inform public debates on major issues of the day—not just to help us understand the world but to help improve it.

CONTEMPORARY APPROACHES IN GEOGRAPHY

All geographers collect and analyze geographic information, but they focus on different topics and employ different analytical approaches. Most contemporary geographers employ three analytical methods:

- Area analysis integrates the geographic features of an area or a place.
- Spatial analysis or locational analysis emphasizes interactions among places.
- Geographic systems analysis emphasizes understanding of environmental and human systems and interactions among them.

Let us look briefly at each of these methods.

Area Analysis

Geographers have a long tradition of surveying, describing, and compiling geographic data about places. Each place in the world occupies a unique location and features a unique combination of human behavior and environmental processes that give it a special character. In compiling and analyzing Earth's places, geographers rely on three basic ideas: site, situation, and the concept of the region.

Site   Geographers often speak of specific sites and site characteristics. **Site** describes the exact location of a place, and can be described either in terms of latitude and longitude (discussed below) or in terms of the place's characteristics. Each place has a unique combination of physical and human characteristics. Returning travelers do not describe the latitude and longitude of a vacation spot. Instead, they report the interesting people and customs, different foods, and varied landscapes that made their trip memorable. They may, however, also report that they spoke the same language, used the same credit cards, and saw the same movies advertised as at home. This mix of unique features and common features is what geographers mean when they talk about place.

Relative location and situation   The location of a place relative to other places is called its **situation,** and knowledge of a place's situation helps us understand how it interacts with the rest of the world. **Relative location** influences accessibility, which is indicated by such terms as nearer and farther, easier or more difficult to reach, between, and on the way or out of the way.

Accessibility can be a resource as valuable as mineral deposits or fertile soil.

Relative location is almost constantly changing, as changes in transportation routes and technologies, as well as developments in communication continuously redistribute accessibility. The opening of a new highway, for example, stimulates the development of new housing and new roadside services, such as shopping malls. At the same time, however, the new highway can reroute traffic past older shopping malls, and thus choke them off from customers.

Great cities rise and countries often prosper if they are advantageously situated within national or international patterns of trade, but isolated areas usually fall behind and stagnate. One of history's greatest redefinitions of relative location occurred when Europeans learned how to sail around Africa in order to reach Asia. Much of the commerce that had previously moved by long and difficult caravan journeys overland across Asia and Africa was drawn to the seacoasts. New seacoast cities sprang up, such as Bombay (today Mumbai) in India, Rangoon in Burma (today known as Yangon in Myanmar), and Hong Kong on the coast of China. Major cities in the interior of the Eurasian and African continents were suddenly "on the back road." Some of these, such as Kashgar, a city in western China visited by the explorer Marco Polo in 1275, and Samarkand, a city thousands of years old in today's Uzbekistan, have never recovered their earlier importance (Figure 1-4).

Changes in transportation technology also redistribute accessibility. In the 1950s the Suez Canal was the principal route of oil shipments to Western Europe from

Internal Asian Trade Routes in 1400

FIGURE 1-4 Trabzon, Kashgar, and Merv were once great cities on major trade routes, but when Europeans opened sea routes around Africa and Asia, these and many other cities declined. Few ever recovered their former economic strength.

Regional Focus on
UTICA, NEW YORK

This map shows how new canals, railroads, and highways have continuously rerouted the flows of national economic life in and out of Utica, New York, changing both the relative location of the city within New York State, and, at a smaller local scale of analysis, the relative location of places and land uses within the city.

Utica was founded where the Mohawk River, winding across the top of the map, could be forded. A bridge later spanned the river, and city prospered as a commercial outpost on what was then the American frontier. Utica industrialized when coal could be brought to it from Pennsylvania by the Erie and Chenango Canal (1837), and that canal's route was later paralleled by the tracks of the New York, Ontario & Western Railroad; these tracks can be seen entering the map from the lower left. Industry multiplied when the Erie Canal (cutting across the top of the map) and later the New York Central Railroad (the tracks labeled Conrail cutting across the center) were built. Notice the big buildings and the round storage buildings around Utica Harbor and along the railroad downtown. In the nineteenth century, workers lived close to these factories, and downtown Utica was a busy place. Since World War II, local and interstate highways have provided new accessibility to the suburbs, and virtually all new industrial development has occurred there. New streets were paved, and new buildings can be identified to the North, across the river from the old downtown. Housing has also spread into the suburbs, and shopping and office malls have pulled more activities out of the downtown area. This map combines two topographic maps printed by the United States Geological Survey. Examine a topographic map for your community for signs of changing access routes and land uses.

the Middle East. New oil tankers, however, grew so large that they were incapable of squeezing through the canal, so they had to sail around Africa. This rerouting situated Cape Town "between" the Persian Gulf and Rotterdam, Europe's chief port for importing oil. Cape Town's ship supply and repair facilities won new business.

Each day's news reports the opening of a new transportation or communication facility that will redistribute activities. In 2000, a link was completed across the Oresund Strait between Sweden and Denmark incorporating a series of bridges, a man-made island, and a tunnel (Figure 1-5). The cities of Copenhagen on the

FIGURE 1-5 The Oresund Bridge. Some 55,000 cyclists crossed the Oresund Bridge on June 9, 2000, as part of a three-day festival to dedicate the bridge. Today only cars and trucks are allowed. (Riehle/Laif/Aurora & Quanta Productions, Inc.)

In modern slang, Timbuktoo is a synonym for nowhere, and few Americans even know that there really is such a place. In fact, Tombouctou is a city in Mali at absolute location 16°46′ north latitude, 3° west longitude. It plays no major role in today's world, but in past centuries it was a major urban center. Tombouctou is situated on the Niger River where the river bends farthest north into the Sahara Desert. Thus, Tombouctou marks a cultural border between nomadic Arab peoples to its north and the settled black African peoples to its south. The city was settled by the Taureg people from the north in 1087, and as a major contact point between north and south, it prospered as a market for slaves, gold, and salt and as a point of departure for trans-Saharan caravans to North African coastal cities. From these cities Tombouctou's fame as a center of almost mythical riches spread around the Mediterranean. It was a religious center, and it boasted a great university.

In the fifteenth century, however, Portuguese expeditions sailing around West Africa outflanked the trade routes on which Tombouctou depended. The city declined rapidly. It was captured by the Moroccans in 1591, and by the time the French captured it in 1893, it was little more than an extensive ruin. Today it is a small city of 10,000 people who live surrounded by few remnants of the city's former greatness.

Trans-Saharan Trade Routes in 1400

Trans-Saharan trade routes during the 1400's

Questions

1. Examine the location of Trabzon on the Black Sea in modern-day Turkey (see Figure 1-4). Among what great empires was Trabzon once a major contact and trading point? How important is it today? Do you know of any other once great cities that have declined as trade routes bypassed them? Conversely, what great port cities developed along with new trade routes in the twentieth century?

2. The geographer Ellen Churchill Semple opened her book *American History and Its Geographic Conditions* with the sentence, "The most important geographical fact in the past history of the United States has been their location on the Atlantic opposite Europe; and the most important geographical fact in lending a distinctive character to their future history will probably be their location on the Pacific opposite Asia." This was written in 1903. Have historic developments proved her correct?

Danish side and Malmö on the Swedish side are now joined into one metropolitan area of more than 3 million people. The new facility, with both a highway and railroad tracks, allows a vehicle or train to travel across Europe from northern Norway to southern Spain.

A place's relative location can also change if the territorial scale of the organization of an activity—that is, the extent of territory within which that activity occurs—changes. For example, when trade barriers between countries fall, activities redistribute themselves. Significant redistributions are occurring throughout North America as the economies of the United States, Canada, and Mexico merge through the new trade pacts discussed in Chapter 13. Activities in Europe are readjusting from a national scale of organization to the supranational scale represented by the European Union.

Improvements in transportation and communication links have "shrunk" Earth, so today many activities have expanded their scale of organization to cover the whole globe, a process called **globalization.** An advertisement for one U.S. bank shows a satellite photograph of Earth drifting in space. The headline reads WE DO BUSINESS IN ONLY ONE PLACE. Economic globalization has far outpaced cultural or political integration. Tensions among the different economic, cultural, and political organizing activities generate much of today's international news.

Political Map of the World

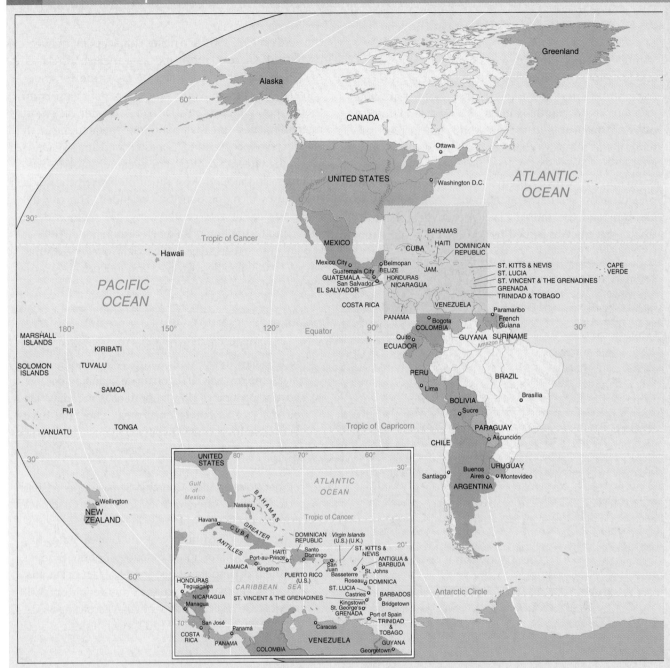

FIGURE 1-6 The world includes almost 200 sovereign states. The number has increased markedly in recent decades, and many of the names of states shown here may be unfamiliar.

Conversely, when a large territory is broken up into smaller individually organized territories, activities will be redistributed and places' accessibility reduced. Until 1991 the U.S.S.R. organized 15 percent of Earth's land surface under one highly unified government before it broke up into 15 independent countries (see Figure 13-5). That political change required reorganizing the scale of all political, economic, and cultural activities throughout the vast territory.

The concept of a region   For purposes of geographic study, we divide Earth into **regions,** which are areas defined by one or more distinctive characteristics or features, such as climate, soil type, language, or economic activity. Geographers define regions and describe and analyze similarities and differences among them.

Probably the most familiar of all maps of regions is the political map (Figure 1-6). People have divided Earth's land surface into almost 200 countries, ranging

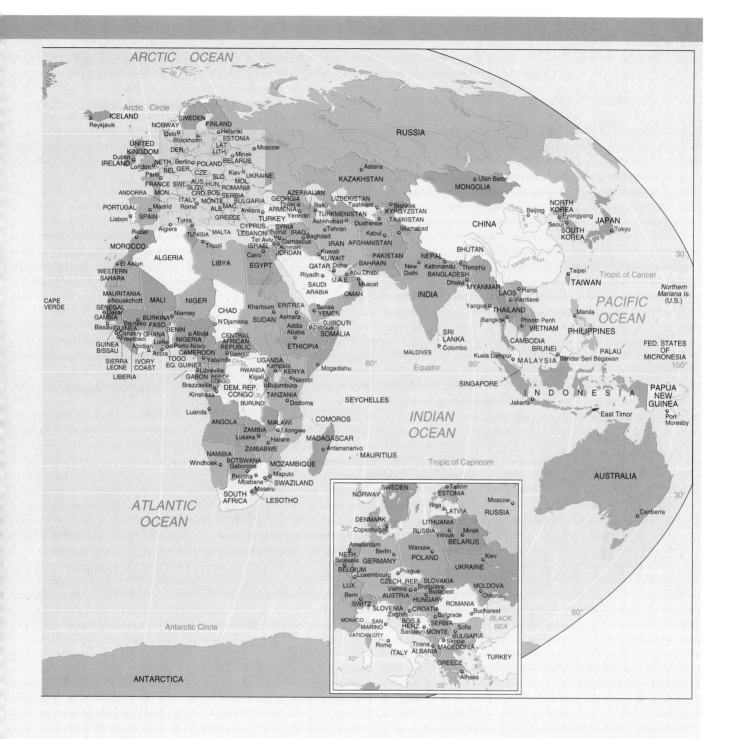

from vast Russia, occupying one-ninth of Earth's land area, to microstates such as Andorra, Monaco, and Singapore. This regionalization of Earth influences almost all other human activities. Geographers study such patterns of how people organize their societies and occupy land.

Of all the features that can be found in any landscape, a geographer chooses only a few specific criteria and then maps those criteria to define a region. The criteria chosen may be either physical or cultural.

Geographers distinguish three kinds of regions. A **formal region** is one that exhibits essential uniformity in one or more physical or cultural features, such as a country or a mountain range. A **functional region,** by contrast, is one defined by interactions among places, such as trade or communication. The city of Chicago, for example, is a formal region whose government covers the legal limits of the city's incorporation. The many commuters and shoppers who circulate daily throughout the city and its suburbs, however, would

Critical Thinking

GROUPING COUNTRIES; NAMING GROUPS

The countries of the world can be grouped into regions in many different ways. Grouping them by continent is fairly easy, as is grouping by membership in certain international organizations such as NATO (North Atlantic Treaty Organization) or OPEC (Organization of Petroleum Exporting Countries). At other times it is useful to refer to groups of countries with broadly similar characteristics, such as large countries, democratic countries, industrial countries, rich countries, and poor countries.

There are enormous differences in levels of wealth among the almost 200 countries of the world today, and this distinction is a particularly important one for geographers. Wealth affects virtually every aspect of a country's human environment, from food availability and disease to pollution and political power. It is useful to make generalizations about geographic patterns of wealth and power, yet generalizations necessarily obscure important differences between countries.

Three major problems confront geographers in identifying groups of countries. One is defining a measurable characteristic of countries, such as wealth. How is wealth measured? In Chapter 12 of this book we will discuss this in more detail. Most often wealth is measured in monetary terms—the average amount of money earned in a country per year. A second problem is the one of deciding how many groups to define.

Should countries be divided into just two groups, or three, five, or 20? One way to avoid this problem is to recognize a spectrum, with poor at one end and rich at the other and many degrees of wealth in between. The third problem is a linguistic one: What should we call the groups, or the endpoints of the spectrum?

One way to refer to these differences is with the words *rich* and *poor*. These terms are fairly simple, descriptive words. They can be misinterpreted, however, because having money is not the only way to be rich. Some people prefer to use the terms *developed* and *developing*, or *more developed* and *less developed*, or *developed* and *underdeveloped*. *Core* and *periphery* are another pair of words sometimes used, as are *North* and *South*.

In this book we will use rich and poor, developed and developing. In doing so, we recognize that such generalizations are never fully accurate and should be interpreted with care.

Questions

1. What does it mean to be rich? Is monetary wealth the most important characteristic to examine in describing and grouping countries?

2. What are the hidden meanings of terms like *developed*, *underdeveloped*, or *developing*?

3. What characteristics do the terms *core* and *periphery* emphasize about rich and poor countries?

---

more readily identify a region larger than Chicago; they would identify the functional region of the Chicago metropolitan area, which includes parts of Illinois, Indiana, and even Wisconsin. The desire of people in northwest Indiana to be on the same time as Chicago confirms that northwest Indiana is part of the Chicago functional region (Figure 1-7).

The third type of region is a **vernacular region.** Vernacular means everyday language, and vernacular regions are defined by widespread popular perception of their existence by people within or outside them (Figure 1-8).

The choice of criteria to define a region takes place in the geographer's mind. Therefore, a region is a concept, an abstract idea. The geographer's concept of a region is similar to the historian's concept of a "period." Time, like space, is continuous, but historians divide time into units of analysis called periods to suit their purposes. Art historians define "styles,"

sociologists define "societies," and many other scholarly fields define their conceptual units of analysis.

Sometimes the features geographers choose to demarcate regions are clear in the landscape. For example, a region defined as a certain forest could be mapped distinctly. Individual countries are usually clearly defined on the landscape by national borders.

In other cases, however, the regions that geographers define are less distinct in the landscape. For example, Chapter 2 maps regions of different climates. On Earth those climatic regions are not sharply defined, but merge into one another imperceptibly. On our map, however, the precise location of the line between one climatic region and its neighbor is determined by the geographer's definition and trained discrimination.

Regions defined by cultural phenomena often merge or overlap as well. For example, people who live between two cities may listen to radio stations located in both cities. If, however, a geographer wants

Metropolitan Chicago

to delineate sharply the market regions of two radio stations broadcasting from the two cities, he or she might choose to draw a line dividing the area where most households listen to the station broadcasting from City A from the area where most households listen to the station broadcasting from City B (Figure 1-9).

Sometimes people confuse regions based on natural phenomena with regions based on cultural criteria. For example, many people refer to Africa as a region. Africa is a continent, but the physical environment is not homogeneous across that vast continent, nor do its many peoples share one culture or historical experience. Historically, the Sahara Desert divided peoples more effectively than the Mediterranean Sea did. Therefore, to the ancient Romans, for example, northern Africa and southern Europe formed a single region focused on the Mediterranean Sea; the word

FIGURE 1-7 The metropolitan region of Chicago includes parts of Wisconsin and Indiana, as well as northeastern Illinois. Commuter trains link Chicago with surrounding communities. While most of Indiana is in the Eastern Time Zone, several counties in northwestern Indiana are in the Central Time Zone to facilitate their links with Chicago, which is on Central time.

Vernacular Regions of North America

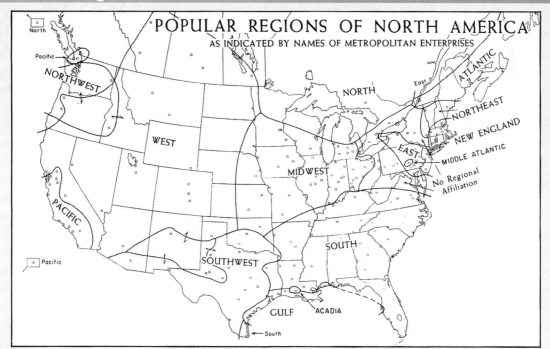

FIGURE 1-8 The regions on this map reflect how local business enterprises in metropolitan areas (the small circles) define themselves in telephone directories. For example, within the region labeled "the South" on this map, businesses most probably identify themselves by that regional designation, such as the Southern Telephone Company. Thus, the regions drawn here are vernacular, defined by popular local everyday language. (After Wilbur Zelinsky, "North America's Vernacular Regions," *Annals, Association of American Geographers,* 70 (March 1980): 14. By permission of the Association of American Geographers)

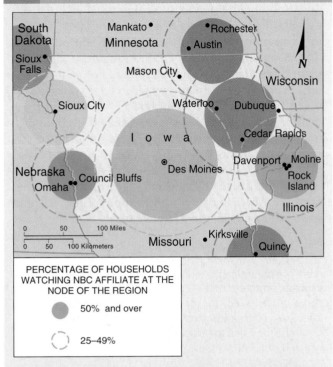

Television Station Viewing Areas Within Iowa

PERCENTAGE OF HOUSEHOLDS
WATCHING NBC AFFILIATE AT THE
NODE OF THE REGION

⬤ 50% and over

◯ 25–49%

FIGURE 1-9 The colored circles on this map are known as areas of dominant influence (ADIs). Most people within any circle watch television programs from the stations in the principal city in that ADI. The construction of new housing, highways, or cable lines may shift an area on the periphery of one city's ADI to another ADI. (After James Rubenstein, *The Cultural Landscape: An Introduction to Human Geography,* updated 7th ed. Upper Saddle River, N.J.: Prentice Hall, 2003)

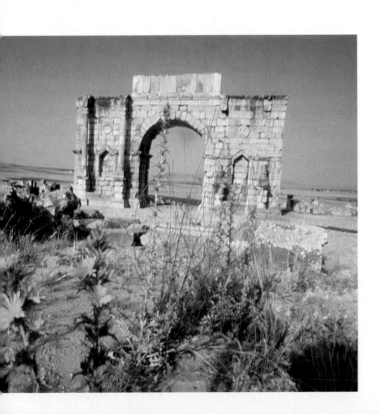

FIGURE 1-10 Roman ruins in Morocco. The remains of ancient Roman buildings haunt the North African desert. These, in Morocco, demonstrate that to the ancient Romans, the lands surrounding the Mediterranean Sea formed one region, and that region was bounded on the south by the Sahara Desert. The African continent was not a region. (Courtesy of the Moroccan National Tourist Office)

mediterranean means "middle of the world." Sub-Saharan Africa was a separate region, distinct and hard to reach (Figure 1-10). Whenever anyone refers to any region, we must be sure that we agree on what area is included and what the criteria of homogeneity are.

Spatial Analysis

A second approach to geographic inquiry, called spatial analysis or locational analysis, looks for patterns in the distribution of human actions and environmental processes and in movements across Earth's surface.

Distribution The **distribution** of a phenomenon means its position, placement, or arrangement throughout space. Geographers recognize distributions and use three terms to define the distribution of any phenomenon under study: density, concentration, and pattern (Figure 1-11).

Density **Density** describes the frequency of occurrence of a phenomenon in relation to geographic area. Density is usually expressed as a number per square kilometer or square mile. Examples include road density (the number of kilometers of roads per square kilometer) and population density (the number of people per square kilometer).

Many statistics are reported by political unit, such as by state or country. In 2000 the population of Russia was 147 million, and the population of Belgium was 10 million, so Russia had almost 15 times as many people as Belgium. Russia, however, is much larger than Belgium, so the population density of Russia was only nine persons per square kilometer (23 per square mile), whereas that of Belgium was 333 per square kilometer (862 per square mile)—37 times that of Russia.

Concentration **Concentration** refers to the distribution of a phenomenon within a given area. If all the occurrences are found in close proximity, we would say the distribution is concentrated, but if they are scattered far from each other, the distribution would be described as dispersed. For example, in many parts of the world, farmers live in villages and travel to their fields in the countryside. The population of such an agricultural landscape is concentrated in villages. In North America, on the other hand, most farmers live in isolated farmhouses located on the land they farm, so the population is dispersed.

Pattern **Pattern** refers to the geometrical arrangement of objects within an area. For example, in most modern

FIGURE 1-11 Density, concentration, and pattern. Assume that the area represented in all figures is one hectare. Density is overall frequency of occurrence of a phenomenon, the number of items per unit of area. The density is six houses per hectare in a_1, whereas in b_1 it is 12 houses per hectare. Concentration refers to the extent of the spread of the feature relative to the size of the area being studied. The houses might be described as dispersed in a_2 and clustered in b_2. Pattern is the geometric arrangement of the distribution of a phenomenon. The pattern of the distribution of houses is linear in a_3 but irregular in b_3.

cities, streets are arranged in a rectangular grid pattern, while in older cities the street layout is more irregular. In areas where rock structures exert strong controls on stream erosion, we sometimes see streams with many right-angle bends; this pattern is called a trellis. Conversely, in areas without structural control, streams and their tributaries form branching or treelike patterns, called dendritic.

Movement

Distance   Interaction among people and places takes place across distance, but there are several ways to measure distance. **Distance** can be measured absolutely, in terms of miles or kilometers, but it can also be measured in other ways. It can, for instance, be measured in time. Most people use time distance measurements in their daily lives without realizing it: "The store is about 20 minutes from here." Sometimes it can be quicker to reach a place that is far away than to reach another place that is, in absolute distance, closer. This may be

the result of rugged **topography** (surface relief) in one direction, direct or winding routes, or different methods of transportation (Figure 1-12).

Geographers also frequently measure distances in terms of the cost required to overcome the distance. Transportation over land is usually more expensive than transportation over water, so two seaports that are far apart in absolute distance may be close in *cost distance.* They may even be closer to each other in cost distance than either is to cities inland from itself. That explains why many political units grew up around the shores of bodies of water. Classical Greece, for example, consisted of cities surrounding the Aegean Sea, and the Roman Empire grew around the Mediterranean Sea.

No matter how we choose to measure distance, we must make some effort to overcome distance when we want to move or transport items. We call that effort or cost the **friction of distance.** The friction of distance usually limits interactions across great distances.

FIGURE 1-12 Measuring distance. When distance is measured by time or by cost, it may seem to be warped when compared to distance measured by absolute units "as the crow flies." For example, drawings (a) and (b) illustrate how the cheapest or fastest distance between two points may not be the shortest. Transportation by water is almost always cheaper than transportation over land, so heavy iron ore will be moved by water if possible. Traveling on a level road is often cheaper and faster than choosing a shorter route crossing rugged land. Drawings (c) and (d) illustrate how the time or cost distance from any point A to a point B may not be the same as from point B back to point A. Most drivers are at one time or another stymied by a situation such as that in drawing (c).

Distance decay The presence or impact of any phenomenon may diminish away from its origin, just as the volume of a sound diminishes with distance. This phenomenon is called the **distance decay** concept. As we travel away from any city, for example, the percentage of the local population that reads that city's newspapers, tunes in to its radio and television stations, or relies on its other services shrinks. For example, we might try to define and map the functional regions of the cities of Los Angeles and San Diego. To the immediate south of Los Angeles, people probably subscribe to Los Angeles newspapers, watch Los Angeles television stations, and rely on service from the Los Angeles airport. As we move southward toward San Diego, the influence of Los Angeles diminishes and yields to that of San Diego.

Physicists have a precise mathematical formula to express the gravitational force that one object exerts on another, and geographers have tried to find analogous mathematical equations to describe patterns of human activities. For example, a geographer might try to write a mathematical equation to define what percentage of households at varying distances from downtown Los Angeles subscribes to Los Angeles newspapers. Perhaps circulation drops 10 percent for every 8 kilometers (5 miles) away from the city center. That

equation could then serve as a model to compare the circulation of newspapers in metropolitan Los Angeles to the circulation of newspapers in other metropolitan areas. A **model** is an idealized, simplified representation of reality that can be used as a standard to compare individual cases in the real world. Comparing the equation model of the distribution of newspapers in metropolitan Los Angeles with patterns in New York, St. Louis, and other metropolitan regions might help us understand how shopping patterns, local professional sports loyalties, or even involvement in metropolitan politics vary from one metropolitan region to another. In some metropolitan regions, the downtown newspapers might circulate farther out into the suburbs, building a greater sense of metropolitan identity.

Diffusion **Diffusion** is the process of an item or feature spreading through time. Any innovation—the use of a tool, a new clothing fashion, or the development of a new technology—originates at a place called a **hearth.** Tracing diffusion of innovation suggests how peoples and cultures interact and influence one another. The simplest image of diffusion suggests concentric waves spreading out from a stone dropped into a pool. Paths of cultural diffusion, however, are seldom so simple.

FIGURE 1-13 Siddhartha Buddha. This richly dressed young man is the Indian prince Siddhartha before he became the Enlightened One (Buddha) and gave up his worldly wealth. The statue exhibits a mix of styles: the curly hair, stocky physique, relaxed stance, and deeply carved drapery are Greek characteristics, but the facial expression is typical of Indian mysticism. The nimbus of light around the prince's head originated with Iranian statues of sun gods. Where could all these cultural influences have come together and this extraordinary figure have been carved? The answer is the region of Peshawar in today's Pakistan, the crossroads of Eurasian travel routes, where the Khyber Pass opens from Central Asia into the plains of the Indian subcontinent. This statue dates from the late second or early third century A.D. (Indian Buddhist sculpture, "Bodhisattva," c. 2nd–3rd century. Schist, chlorite, 45 in. h, 15 in. w, 7 in. d. Eugene Fuller Memorial Collection, Seattle Art Museum. Photo by Paul Macapia)

Geographers define three basic paths of diffusion: relocation diffusion, contiguous or contagious diffusion, and hierarchical diffusion.

Relocation diffusion occurs from widely separated point to point. A nomadic tribe, for example, might wander until it finds a physical environment similar to the one that it left, and then the tribe might settle down. The ancestors of today's Hungarians wandered all the way from Central Asia to the plains of today's Hungary. The ideas and practices they brought with them were spread by relocation diffusion.

Contiguous diffusion occurs from one place directly to a neighboring place. Contiguous diffusion that occurred in the past is often revealed by the dispersion of artistic styles. For example, in the fourth century B.C., Alexander the Great's conquests carried Hellenistic Greek art far into Asia. Central Asian statues of Buddha sculpted during the centuries immediately following these conquests resemble Greek gods (Figure 1-13).

Sometimes diffusion does not occur contiguously across space, but downward or upward in a hierarchy of organization. When such **hierarchical diffusion** is mapped, it shows up as a network of spots rather than as an ink blot spreading across a map. The Roman Catholic Church illustrates both an organizational hierarchy and a geographical hierarchy (Figure 1-14). Each

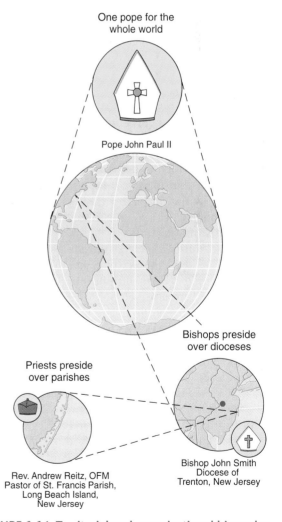

FIGURE 1-14 Territorial and organizational hierarchy. The Roman Catholic Church illustrates both territorial and organizational hierarchy. The pope presides over the entire Church, but the Church is subdivided into dioceses, each presided over by a bishop, and the dioceses are subdivided into parishes.

parish priest is answerable to his bishop, who from his cathedral church (*cathedra* is Latin for "throne") presides over a diocese, which includes many parishes. The bishops, in turn, look up to the Vatican. An announcement from the Vatican diffuses to the cathedrals and then down to the parishes around the world.

As a general rule, more information travels up and down a hierarchy than across any level of the hierarchy. Therefore, news from a parish about an innovation in community service or in a church ceremony might reach the cathedral and even the Vatican before it reaches neighboring parishes. This principle of diffusion up-and-down in a hierarchy is very important, and we will return to it repeatedly. Note that the topmost point in a hierarchy, called the *apex*, may not actually generate innovations. All information, however, does clear through the apex, and most places receive information from the apex.

Advanced societies usually exhibit well-developed hierarchies of cities, and we can study how the hierarchy evolved through time. Once a hierarchy is developed, the diffusion of any specific phenomenon might be traced from the few biggest cities down to the many more smaller cities and then further down to the even more numerous small towns and villages until the landscape is covered (Figure 1-15).

The map of America's most heavily traveled domestic airline routes reflects the national hierarchy of cities (Figure 1-16). To fly from one small city to another small city, a passenger may have to fly first to a nearby big city (which is a movement up the hierarchy), then from that big city to another big city near the ultimate destination (across the hierarchy), and finally from that big city to the destination small city (down the hierarchy again).

When scientists began to study the path of the diffusion of AIDS throughout sub-Saharan Africa, they first assumed that it had diffused contiguously, as is common in poor, underdeveloped regions, where most contact is directly from one village to the next contiguous village. It was discovered, however, that AIDS was being spread across Africa in a hierarchical fashion. Epidemiologists (those who study the spread of disease) wondered how this could be occurring. They found that it was being spread by long-distance truck drivers who travel long distances between large towns and cities and often hire prostitutes in each town along the way. The knowledge of this method of diffusion in Africa caused world health officials to check truck drivers in India, another poor region suffering soaring rates of infection, and to educate them of the necessity for safe-sex techniques. Thus, understanding of the path of diffusion of AIDS in Africa may have saved millions of lives in India and elsewhere.

Many phenomena diffuse hierarchically around the world today. The spread of a disease or of a cultural innovation such as a clothing fashion or a music hit can often be traced among the world's principal

Market Penetration by Television

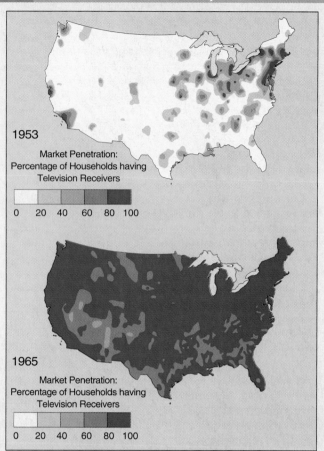

FIGURE 1-15 These maps show how the percentage of U.S. households that had television sets increased from 1953 to 1965. The spots on the first map show that only large cities had television in 1953. The second map shows that by 1965 the innovation had diffused to cover the contiguous 48 states. This occurred before cable television or satellite dishes, so viewers had to receive transmissions from metropolitan centers. How would the pattern of diffusion differ with today's technology? (Reprinted with permission from Brian J. L. Berry, "The Geography of the United States in the Year 2000," *Transactions of the Institute of British Geographers LI,* November 1970, 21–53)

metropolises before it reaches down into the smaller cities in each country. This path of diffusion reveals the world's interconnectedness.

Barriers to diffusion   There are many barriers to cultural diffusion. Oceans and deserts, for example, increase the cost distance or time distance between places. Other topographical features, such as mountains and valleys, historically have blocked human communication and contributed to cultural isolation. In Asia the Himalayan mountains and the deeply cut valleys of the Tongtian, Mekong, Nu, and Brahmaputra rivers make overland travel between China and the Indian subcontinent

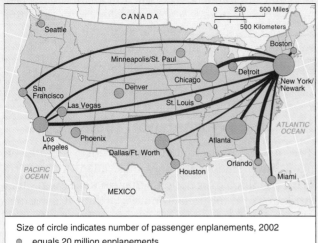

Major Air-Passenger Routes in the United States

Size of circle indicates number of passenger enplanements, 2002

● equals 20 million enplanements

Width of route line indicates total number of passengers on most heavily trafficked individual routes, 2002.

──── equals 1 million passengers

None of the busiest airports or routes in the United States in 2002 included Alaska or Hawaii.

FIGURE 1-16 This map is a good indicator of the urban hierarchy of the United States, although it is a bit distorted by airlines' routing systems. New York is certainly the principal focus of national life. Los Angeles, Dallas, Atlanta, and Chicago are functional "capitals" of individual regions of the country. Minneapolis/St. Paul and Denver have large numbers of total enplanements, but few or no individual routes with more than 2 million passengers. These cities serve as hubs of many local regional routes with small numbers of fliers on each.

extremely difficult, so these two cultures and great concentrations of population are distinctly isolated from one another.

Other barriers to cultural diffusion include political boundaries and even the boundaries between two culture realms. Hostile misunderstanding, distrust, and competition between two culture groups can hinder any communication and exchange between them.

Physical and Human Systems

A third approach in geographic inquiry views Earth as a set of interrelated environmental and human systems. A **system** is an interdependent group of items that interact in a regular way to form a unified whole. Models of systems help geographers see how factors are interrelated—how one activity, force, or event affects others. Individual environmental systems include the climatic processes that produce precipitation, the hydrologic processes that determine what happens to the rain when it reaches Earth's surface, and the character-

istics of the river valleys that receive the water runoff. These individual physical systems are themselves interrelated. For example, an increase in precipitation affects an area's soils, landforms, and vegetation.

Geography also emphasizes interactions between humans and the environment. Human activities are influenced by environmental conditions, but humans in turn modify the physical environment in which they live.

In the following sections we will examine some of the ways in which geographers have studied the interactions among different aspects of the human and physical environment.

Earth's physical systems Geographers study natural processes in terms of four systems: the atmosphere (air), the hydrosphere (water), the lithosphere (Earth's solid rocks), and the biosphere, which encompasses all of Earth's living organisms (Figure 1-17).

The **atmosphere** is a thin layer of gases surrounding Earth to an altitude of less than 480 kilometers (300 miles). Pure, dry air in the lower atmosphere contains about 78 percent nitrogen and 21 percent oxygen by volume. It also includes about 0.9 percent argon (an inert gas) and 0.036 percent carbon dioxide (a crucial percentage, as you will see in upcoming chapters). Air is a mass of gas molecules held to Earth by gravity creating pressure. Variations in air pressure from one place to another cause winds to blow, create storms, and control precipitation patterns.

The **hydrosphere** is the water realm of Earth's surface, including the oceans, surface waters on land

FIGURE 1-17 Earth's physical systems. Geographers regard natural processes as consisting of four open systems, including three composed of nonliving matter (the atmosphere, hydrosphere, and lithosphere), plus the biosphere, which comprises all of Earth's living organisms. The four systems are interrelated and overlap with each other.

(lakes, streams, rivers), groundwater in soil and rock, water vapor in the atmosphere, and ice in glaciers. Water can exist as a vapor, liquid, or ice. Over 97 percent of the world's water is in the oceans in liquid form. The oceans sustain a large quantity and variety of marine life in the form of both plants and animals. Seawater supplies water vapor to the atmosphere, which returns to Earth's surface as rainfall and snowfall. These are the most important sources of fresh water, which is essential for the survival of plants and animals. Water changes temperature very slowly, so oceans moderate seasonal extremes of temperature over much of Earth's surface. Oceans also provide humans with food and a surface for transportation.

The **lithosphere** is the solid Earth, composed of rocks and sediments overlying them. Earth's core is a dense, metallic sphere about 3,500 kilometers (2,200 miles) in radius. Surrounding the core is a mantle about 2,900 kilometers (1,800 miles) thick. A thin, brittle outer shell, the crust is 8 to 40 kilometers (5 to 25 miles) thick. The lithosphere consists of Earth's crust and a portion of upper mantle directly below the crust, extending down to about 70 kilometers (45 miles). Powerful forces deep within Earth bend and break the crust to form mountain chains and shape the crust to form continents and ocean basins. The shape of Earth's crust influences climate. If the surface of Earth were completely smooth, then temperature, winds, and precipitation would form orderly bands at each latitude.

> Human societies must adapt to . . . a local environment as they learn to exploit the local resources. Human societies are not, however, passive.

The **biosphere** consists of all living organisms on Earth. The atmosphere, lithosphere, and hydrosphere function together to create the environment of the biosphere, which extends from the depths of the oceans through the lower layers of atmosphere. On the land surface, the biosphere includes giant redwood trees, which can extend up to 110 meters (360 feet), as well as the microorganisms that live many meters down in the soil, in deep caves, or in rock fractures.

These four spheres of the natural environment interact in many ways. Plants and animals live on the surface of the lithosphere, where they obtain food and shelter. The hydrosphere provides water to drink and physical support for aquatic life. Most life forms depend on breathing air, and birds and people rely on air for transportation. All depend on inputs of solar energy.

Humans also interact with each of these four spheres. We waste away and die if we are without water. We pant if oxygen levels are reduced in the atmosphere, and we cough if the atmosphere contains pollutants. We need heat, but excessive heat or cold is dangerous. We rely on a stable lithosphere for building materials and fuel for energy. We derive our food from the rest of the biosphere.

Plants and animals interact with each other through exchange of matter, energy, and stimuli. An ecological system, or **ecosystem,** is a group of organisms and the nonliving physical and chemical environment with which they interact. **Ecology** is the scientific study of ecosystems. Ecologists study the interrelationships among life forms and their environments in particular ecosystems, as well as among various ecosystems in the biosphere.

Human-Environmental Interaction

The physical environment affects human activities in many ways. Human societies must adapt to local climatic conditions, vegetation, water resources, and other attributes of a local environment as they learn to exploit the local resources. Human societies are not, however, passive. As any human society adapts, it actively and deliberately alters natural conditions. The vegetation that covers much of Earth, for example, is cropland or pastureland that is maintained for human benefit. Thus, interaction between the environment and humankind is reciprocal—the environment affects human life and cultures, and humans alter and transform the environment.

Human culture and cultural landscapes   Geography is one of the social sciences that contributes to our understanding of human **culture.** The word culture is often used to mean only fine paintings or symphonic music, but to social scientists it means everything about the way people live: what sort of clothes they wear (if any); how they gather or raise food; whether they recognize marriages and, if they do, whom (and how many spouses) they think it proper to marry and how they celebrate marriage ceremonies; what sorts of shelters they build for themselves; which languages they speak; which religions they practice; and whether they keep any animals for pets. Human cultures vary significantly in all these aspects. Of all the ways in which the behavior and beliefs of individuals differ, a social scientist chooses a few specific criteria and then defines that bundle of attributes of shared behavior or belief as a *culture.*

People modify their local landscape in the process of making a living, housing themselves, and carrying on their lives. We say that they transform a **natural landscape,** one without evidence of human activity, into a **cultural landscape,** one that reveals the many ways people modify their local environment. Aspects of cultural landscapes include the treatment of the natural environment as well as houses and other parts of the built environment.

Critical Thinking

PARKS, GARDENS, AND PRESERVES

Which landscape in the photograph, the idyllic foreground or the skyscrapers in the background, is a cultural landscape? Both are. Gardens and parks are landscapes that humans have molded according to aesthetic notions. They are cultural landscapes, built on the assumption that in some ways human designs are superior to nature.

Tastes in garden design vary among cultures, and within any culture they change through time. In all cases, however, gardens and parks reflect the prevailing human notions of how best to "improve" nature. The eighteenth-century park designer Lancelot Brown earned his nickname Capability Brown because whenever he

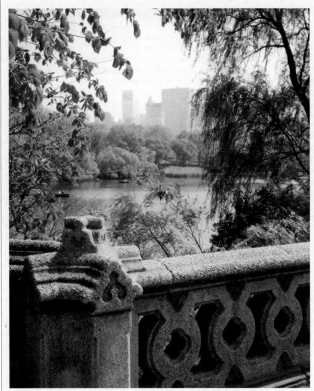

In this photograph, which is artificial: the soil, drainage, topography, and selection of vegetation in the foreground, or the skyscrapers in the background? In fact, all are. New York City's 840-acre Central Park was constructed over the 20 years after 1858. The land was still beyond the fringe of urban development, but a design was approved, the land was blasted and sculpted, and vegetation was planted to achieve an ideal. (EFB photo)

saw a natural landscape, he commented that it had great "capabilities," upon which he would improve. In the nineteenth century Calvert Vaux and Frederick Law Olmsted "built" New York City's Central Park, including its topography and watercourses, and they chose and planted imported species of vegetation. Today naïve visitors marvel that the city has "preserved" this "natural environment" in the midst of skyscrapers, yet it requires an army of engineers and landscape architects to maintain the park. In truth, the park is scarcely less artificial than the surrounding skyscrapers.

Minnesota's huge Mall of America encloses a seven-acre interior landscape that, according to official mall publications, "bring(s) the outdoors indoors . . . inspired by Minnesota's natural habitat—forests and meadows, river banks and marshes." Minnesota's natural conifers, however, cannot survive in a constant temperature of 21° C (70° F). Therefore, the mall's "indoor natural Minnesota forest" is populated by tropical orange jasmines, black olives, oleander, hibiscus, and other species flown in from the South. Few visitors seem to notice, and none complain. The mall is, in fact, a terrarium of exotic species.

Parks and gardens are different from reserves or preserves. These areas are closer to being natural landscapes, because they are set aside to save natural environments, including, usually, the natural wildlife. Most of the properties in the U.S. National Park System, such as Yellowstone National Park, the world's first national park (1872), were originally designated as reserves. These enclaves of nature attract rapidly growing numbers of visitors who demand facilities ranging from hotels to hiking trails to designated scenic viewing spots. The press of their numbers endangers the pristine character of the property. In other words, the people destroy what they come to see. In other cases, properties that were originally reserves are increasingly engineered in order to save them. For example, parts of Niagara Falls are shored up with concrete and steel to prevent them from eroding away and losing their "natural" beauty.

Questions

1. Does the fact that we call our reserves "parks" reflect confusion about what we really want?
2. Should we let our natural reserves become human-made parks?
3. How can we enjoy our reserves and yet save them?

The evidence of human activity is so ubiquitous on Earth that we might be justified in saying that all of Earth is today a cultural landscape. In addition, travelers quickly notice that the cultural landscape is varied. Rural China does not look like rural France or rural Nigeria. One factor in explaining these variations is that the natural environments in these places were different even before humankind set to work on them. The "raw material" that human societies transformed into cultural landscapes affected what humans would do in each.

Another factor in explaining variations in cultural landscapes, however, is that each cultural group creates a distinct cultural landscape, and the better we can learn to interpret a landscape, the more it will reveal to us about the culture that produced it. In the past, human societies developed in greater isolation from one another than today, and the extraordinary diversity of human cultures and cultural landscapes testifies to human ingenuity. Different peoples who live in very similar environments but isolated from one another developed astonishingly different lifestyles. Conversely, some aspects of cultures that have developed in different physical environments are startlingly similar. No direct cause-and-effect relationship between a physical environment and any aspect of culture can be assumed.

DESCRIBING EARTH

Earth is nearly spherical, with a diameter of about 12,735 kilometers (8,000 miles) and a circumference of about 40,000 kilometers (25,000 miles). Its highest point is almost 8,850 meters (29,000 feet) above sea level, and the deepest spot in the oceans is about 10,920 meters (35,800 feet) below sea level. Earth's shape varies slightly from true sphericity due to the effect of centrifugal force. Any rotating body tends to bulge in the "middle" and flatten at the "ends"; thus, the equatorial diameter of Earth is slightly greater (43 kilometers, or 27 miles) than its polar diameter.

The Geographic Grid

Earth rotates continuously about an axis that penetrates Earth's surface at the North Pole and the South Pole. An imaginary plane perpendicular to the axis of rotation that passes through Earth halfway between the poles is called the *plane of the equator*, and this plane intersects Earth's surface at Earth's imaginary midline, or "waist," called the **equator.** We can use the North Pole, the South Pole, and the equator as reference points for defining positions on Earth's surface.

Latitude is angular distance measured north and south of the equator. As shown in Figure 1-18a, we can project a line from any given point on Earth's surface

to the center of Earth. The angle between this line and the equatorial plane is the latitude of that point. Latitude is expressed in degrees, minutes, and seconds. There are 360 degrees (°) in a circle, 60 minutes (′) in a degree, and 60 seconds (″) in a minute. Latitude varies from 0° at the equator to 90° at the North and South Poles.

Lines connecting all points of the same latitude are called **parallels** because they do not intersect. Parallels are imaginary lines, and there can be an infinite number of them—one for each degree of latitude, or for every minute, or for any fraction of a second (see Figure 1-18b).

Longitude is angular distance measured east and west on Earth's surface. Longitude is also measured in degrees, minutes, and seconds. It is represented by imaginary lines, called **meridians,** extending from pole to pole and crossing all parallels at right angles (see Figure 1-18c). Meridians are not parallel to one another. They are farthest apart at the equator, become increasingly close northward and southward, and converge at the poles.

The **prime meridian,** from which longitude is measured, was chosen by an international conference in 1884. It is the meridian passing through the Royal Observatory in Greenwich, England, just east of London. Longitude is measured both east and west of the prime meridian to a maximum of 180° in each direction. Thus, the location of any spot on Earth's surface can be described with great precision by reference to its latitude and longitude. If we say that the dome of the U.S. Capitol in Washington is located at 38°53′23″ north latitude and 77°00′33″ west longitude, we have described its position within about 20 steps.

The same international conference that agreed on the prime meridian also agreed to divide the world into 24 standard time zones (Figure 1-19). When people first began keeping track of time, they defined local time in terms of when the sun rose and set, with noon defined as the time at which the sun is highest in the sky. This timing depends on longitude. The standardization of time from one place to another became necessary when travel between places was fast enough that differences in time from one place to another became important. In the United States, this came with the development of the transcontinental rail network.

When we standardize time across large east-west distances, the usual link between the position of the sun in the sky and time can be stretched considerably. Relative to Earth, the sun travels 360° in 24 hours, or 15° per hour, so time zones that standardize time within each 15°-wide slice of Earth's surface offer a solution. The time at the prime meridian, or 0° longitude, is designated **Greenwich Mean Time (GMT).** Earth rotates eastward, so a clock advances one hour from GMT for each 15° traveled east from the prime meridian. For each 15° you travel west from the prime meridian, you set your clock one hour earlier than GMT.

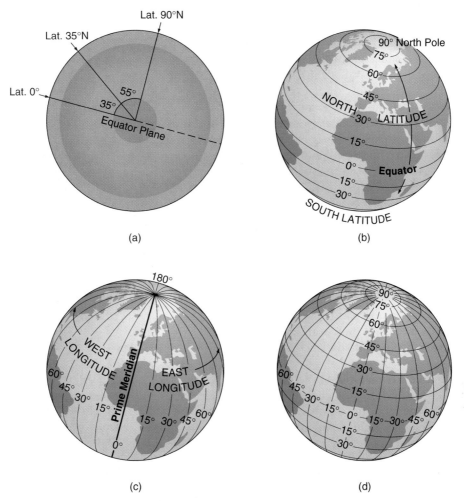

FIGURE 1-18 The geographic grid. These drawings show the development of the geographic grid: (a) the measure of latitude, (b) parallels of latitude, (c) meridians of longitude, (d) the completed grid system.

World Time Zones

FIGURE 1-19 Each of the Earth's 24 time zones represents an average of 15° of longitude or one hour of time. The number in each time zone on this map represents the number of hours by which that zone's standard time differs from Greenwich Mean Time, plus or minus. The colored stripes indicate the standard time zones; gray areas have a fractional deviation. These irregularities in the zones are dictated by political and economic factors.

The meridian for 75° west longitude runs through the eastern United States, so time in the eastern United States is five hours earlier than GMT. (The 75° difference in longitude ÷ by 15° per hour = 5 hours.) When the time is 11 A.M. GMT, the time in the eastern United States is five hours earlier, or 6 A.M. The 48 conterminous U.S. states and the Canadian provinces share four standard time zones, known as Eastern, Central, Mountain, and Pacific. Most of Alaska is in the Alaska Time Zone, which is nine hours earlier than GMT. Canada has an Atlantic Time Zone, which is four hours earlier than GMT, and a Newfoundland Time Zone, which is three and a half hours earlier than GMT.

In Figure 1-19, you can see that the **International Date Line** for the most part follows 180° longitude, although it deviates in several places to avoid dividing land areas. When travelers cross the International Date Line heading east (toward North America), the calendar moves back one day. When travelers cross it going west toward Asia, the calendar moves ahead one day.

Today's world time-zone map does not consist of parallel vertical stripes; it bends for the convenience of individual countries to reduce the number of situations in which a given region is divided among multiple time zones. In many cases it can be much more convenient to be in one time zone than another. In the United States, for example, the Eastern Time Zone is much wider from east to west than longitude alone would dictate: more than 20° instead of the normal 15°. This is because most people in the eastern United States would rather be on the same time as the financial and political centers on the East Coast than an hour different. Most of the state of Indiana is in the Eastern Time Zone. The northwestern part of the state, however, near Chicago, is on Central time because Chicago is on Central time, and Northwest Indiana residents find it more convenient to be on the same time as Chicago.

Global positioning systems   Determining the location of a place on Earth's surface has become much easier with the development of **global positioning systems (GPS).** GPSs are a navigational tool originally developed by the U.S. government for military use, but now available for civilian purposes worldwide. The system consists of a fleet of satellites that orbit Earth, broadcasting digital codes. A portable receiver "listens" to these signals. By measuring very small differences in arrival times of the signals, it determines its own location. Inexpensive receivers are capable of determining locations to within a few meters, while more elaborate systems have accuracies of less than one meter (Figure 1-20). The limitations of the system are relatively few—an antenna must be outside and unobstructed by trees, and the satellite signals may be withdrawn for civilian use at the discretion of the government, so that during

FIGURE 1-20 A GPS device. This tiny handheld global positioning system (GPS) is a spinoff of the satellite navigation system developed for the military. Twenty-four orbiting satellites are constantly transmitting signals that indicate their positions and the time of the transmission. This second-generation GPS device reads four satellites' positions and determines the time elapsed between transmission and reception of the signals. Thus, it can determine not only exactly where you are, but your elevation as well. When such devices are installed in aircraft, yachts, and even cars equipped with computer-generated maps, travelers never need be lost. (Robert Holm & Susan Van Etten/PhotoEdit)

wartime an enemy cannot use them. On May 1, 2000, however, President Clinton signed an Executive Order allowing civilian GPS devices to have the same degree of accuracy as military devices.

GPS is revolutionizing many business operations, especially those related to transportation and mapping. Surveyors, particularly those working in remote or rural areas, determine their locations by GPS instead of using optical devices that are useful only for distances of a mile or two at best, assuming a clear line of sight. Scientists and land managers use GPS to determine their locations when making environmental measurements. Airplanes and ships now navigate by GPS instead of relying on older, ground-based radio systems.

As GPS systems became more widely used, their price dropped rapidly, and this led to further expansion of applications. Many new automobiles now offer GPS-based mapping systems that continually track the vehicle's location and display a moving digital map on the console. Hikers carry GPSs to avoid getting lost, and boaters often carry them to return to their favorite fishing holes or to navigate home in a fog.

Communicating Geographic Information: Maps

One of geographers' most characteristic tools is a map. **Maps** are two-dimensional (flat) representations of some portion of Earth's surface. They are necessarily smaller than the area portrayed, so they cannot show all the things present on the actual surface. Maps show only selected information according to the particular purpose of the map. They are often used to show things that may not be visible on the landscape but are nonetheless important—political boundaries, for example. Maps are simplifications of Earth—models—designed to store and display geographic data. (See also Appendix I.)

Maps take a great variety of forms. Some are just simple drawings intended to show relative location or a path from one place to another. Perhaps the most common map you are familiar with is a road map, showing the locations of towns and cities and the automobile routes that connect them. Airplane pilots use maps that show the elevation of the ground surface, a few things visible from airplanes such as rivers, lakes, urbanized areas, and large highways, areas of the sky in which planes are required to follow certain rules, the locations of towers and navigational beacons, and landing fields. Photos taken from airplanes are maps, as are images collected from satellites and displayed on your TV screen as part of a weather forecast.

Cartography is both a highly technical and a somewhat artistic pursuit, combining the tools of mathematics and engineering with those of the graphic artist. Maps should be accurate, portraying information as it really exists rather than distorted, improperly located, or mislabeled, and they should be visually easy to use, displaying prominently the information a user needs without clutter from unnecessary information. This is why road maps, for example, usually do not show mountains and hills except in the simplest ways. To do so would add many extra lines to a map that is already filled with lines representing roads.

In converting geographic data from their original form on Earth's surface to a simplified form on a map, many decisions must be made about how this information is to be represented. No matter how we draw a map, it cannot possibly show the world exactly as it is in all its detail, nor would we want it to. Scale and projection are two fundamental properties of maps that determine how information is portrayed.

Scale   Maps are smaller than the areas they represent. To know how large an area or how great a distance is represented by a given area or distance on a map, we must know the scale of the map. The **scale** of a map is a quantitative statement of the relative sizes of an object on the map and in reality. Map scales are typically expressed in one of three ways: a written statement (1 inch = 1 mile) or

a representative fraction (1:63,360, meaning that one unit of length on the map corresponds to 63,360 units on the ground), or a graphic scale in which a bar on the map is labeled with ground distances.

Common map scales vary widely. The maps in this textbook showing the entire world, including the oceans, on one page, have a horizontal distance of roughly 25,000 miles (1,584,000,000 inches) on a printed area 6 inches wide, for a scale of 6/1,584,000,000, or about 1:264,000,000. A typical scale used for a road map of a state in the United States might be 1:1,000,000 to 1:2,000,000. A street map of a city might have a scale of 1:10,000, while a surveyor's map of a residential lot might be drawn at 1:100. Maps like the world maps in this book show the land in a very small space, so we call them **small-scale maps.** A map such as a street map that shows a given area in a large space is called a **large-scale map.** A map with a large number in the denominator of a fraction expressing scale (for example, 1/2,000,000) is a small-scale map, whereas a map with a small number in the denominator (for example, 1/100) is a large-scale map (see Figure 1-21).

The amount of detailed information that can be shown about a place is a function of map scale. On a small-scale map there is not enough room to show a lot of detail. If we showed all the numbered highways of the United States on a map the size of a page in this book, the map would be virtually all highway and nothing else. On a large-scale map we have room for more detail about a place, but we cannot show the relations between that place and another that is far away. Figure 1-22 shows Seattle, Washington, viewed at a range of map scales.

Projection   A second fundamental characteristic of maps that governs the way we show information is **projection,** the transferral of locations on Earth's surface to locations on a flat map. For practical reasons, most maps are flat. Flat maps can be printed on paper and folded, or displayed on a video monitor. However, every cartographer faces a problem when drawing a spherical Earth on a flat piece of paper. Some distortion unavoidably results. Cartographers have invented hundreds of projections, but none is completely free of distortion.

So how do we make a flat model of a spherical Earth? One way might be to draw a map on a spherical rubber balloon and then cut the balloon and stretch it in such a way that it could be glued on a page. Or we could draw our map on a transparent globe, place a lightbulb inside it, and project the map onto a screen. This is, in effect, what is done, except that the conversion of the geometry of the sphere to the geometry of the screen is achieved mathematically rather than optically. Regardless of how we make a flat map, the resulting features on the map are different from their equivalents on Earth's surface. The map cannot be an exact scale model of Earth in the same way a globe can.

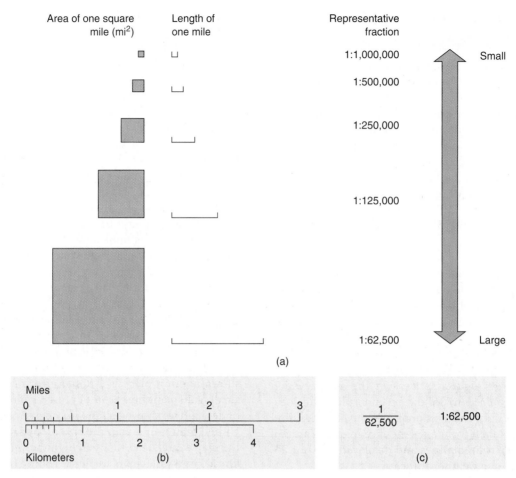

FIGURE 1-21 Map scales. Drawing (a) shows that a small-scale map can portray a large part of Earth's surface, whereas a large-scale map can show only a small part of the surface. Drawing (b) is a graphic scale, and (c) shows fractional scales.

The process of projecting the globe onto the map has introduced distortion.

Several kinds of distortion occur when an image of a round Earth is fit to a flat surface, and as much as we would like to eliminate this distortion, there is no way to do so. When we are mapping a relatively small area of Earth's surface, the difference between a plane and the surface of a sphere is insignificant for most purposes. For large areas, however, and especially for world maps, the distortions are severe. All projections must distort either size (and distance), shape (and orientation), or both. Appendix I describes the methods for making different projections, but here we will examine briefly the advantages and disadvantages of some common methods.

A distortion of size and distance means that the scale in one part of a map is different from that in another part—objects in one area appear larger or smaller in relation to objects in another part of the map than they actually are. It is possible to distort size but preserve the shapes of objects, effectively blowing them up or shrinking them. Maps that distort size but preserve shape are called **conformal maps.**

A distortion of shape and orientation means that in moving from a round surface to a flat one, objects are stretched more in some parts than in others so that their shapes change. A change in shape inevitably means that orientation changes too, such as causing a line that on Earth runs northeast-southwest to instead appear to run east northeast–west southwest. It is possible, however, to make such changes while preserving the sizes of objects relative to each other. Maps that preserve size but distort shape are called **equal-area maps,** or equivalent projections.

One of the most common conformal map projections in use is the Mercator projection (Figure 1-23a). It is named for its Flemish inventor, Gerardus Mercator, who developed it in 1569 for the purpose of navigation. It serves that purpose well, because it preserves shape and thus orientation. On the Mercator projection, lines of latitude and longitude form a perfect rectangular grid. North is always the same direction on the map, and a straight line can be traced on the map by following a constant compass bearing. The rectangular shape of the Mercator map adds another attraction. It fits conveniently on a printed page or

Seattle, Washington, at Different Scales

FIGURE 1-22 These four maps show the city of Seattle, Washington, at a variety of scales. Only major landscape features, cities, and highways are shown in Figure 1-22a. When we zoom in on this map, the amount of information displayed changes, with more detail shown at larger scales. In Figure 1-22b the names of small towns are shown, as are secondary roads. In Figure 1-22c some details of the lakefront shoreline emerge, as do the names of a few major streets. The names of all the individual streets are shown in Figure 1-22d.

on a wall. This, more than its utility in navigation, contributes to the continued widespread use of this projection.

The problem with the Mercator projection is its distortion of size. The distance around the globe at latitude 60° N is half the distance around the globe following the equator, and yet these two distances are the same on a Mercator map. If north-south distances are exaggerated in the same way, as they must be to preserve shape, then the area of a region at the latitude of

the Arctic Circle is shown four times its actual size relative to a region at the equator. The distortion increases poleward. This is why on a Mercator map Greenland (which is centered a little north of 70° N) appears to be similar in size to South America (which straddles the equator), when in reality South America is nine times larger than Greenland. Similarly, the Mercator projection makes North America, Europe, and northern Asia look much larger than they are relative to South America, Africa, and southern Asia. Some people

Conformal and Equal-Area Projections

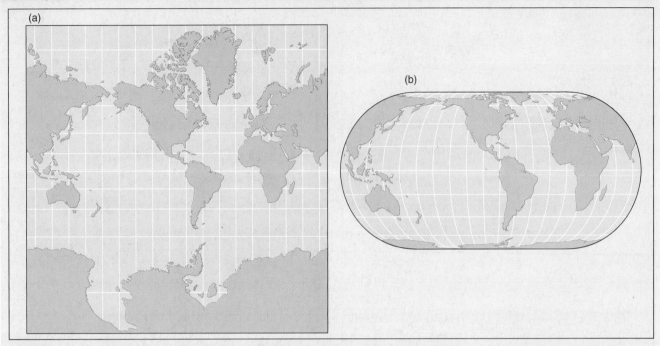

FIGURE 1-23 Flat maps necessarily distort our view of Earth. (a) The Mercator projection, which is a conformal projection, displays the correct shapes but exaggerates sizes in high-latitude areas. (b) An equal-area projection can portray accurate sizes, but shapes are distorted, especially near the edges of the map. For more information about map projections, see Appendix I.

object to the Mercator projection and other similar projections because they exaggerate the size and thus the suggested importance of the wealthier parts of the world relative to the poorer parts.

Size distortions are eliminated in equal-area maps, of which Figure 1-23b is an example. Equal-area maps of the world are much better than Mercator maps for most purposes, because they communicate more accurately the sizes of countries and continents and because in comparing one country with another, size is usually more important than shape. Shape distortions become important, however, because they alter distances. Also, because the polar regions must be drawn much smaller than equatorial regions, there is usually considerable distortion of direction (especially north-south) near the edges of the map.

In addition to shape and size distortions, all flat maps must divide Earth's surface at some point. The surface of a globe is a continuous surface—we can go all the way around without leaving the surface—but a flat map must have edges. If we follow the convention of placing north at the top of the map, then the usual place to cut the surface is in the middle of the Pacific Ocean, which causes the least disruptions of land areas. This is also convenient because the prime meridian can be placed at the center of the map with relatively little separation of land areas. Of course if you live in, say,

New Zealand, this makes it look as though islands in the central Pacific Ocean are very far away and that one would need to travel across Africa and South America to reach them. A New Zealander might prefer a map centered on 180° longitude. Maps produced in the United States often have their edges in central Asia, placing North America at the center.

The choice of projection is fundamental to the way we display information and the way people who use maps receive that information. On the one hand, projections can be as dispassionate as a mathematical equation that describes the alteration of a spherical surface into a planar one. Because maps communicate information, however, the form of the map inevitably influences how and what the map says about a place. Those who use such maps therefore often have very strong opinions about what kinds of distortions are introduced in the projection process.

Insets and other conventions   Other conventions of map projections or printing can lead to misunderstanding. For example, to save space, many maps of the United States tuck a separate map of Alaska into the corner below the U.S. Southwest. This is called an *inset* map. This common practice leads many U.S. schoolchildren to think of Alaska as an island off the coast of California. If the mapmaker also printed the inset map

Focus on
CARTOGRAMS

Standard maps try to depict Earth features in their correct locational and size relationships. Sometimes, however, it may be more important to convey a visual impression of the magnitude of something than to convey a visual impression of exact spatial locations. To do this, cartographers design special images called **cartograms**. All cartograms replace physical distance or size with some other measure. The two main types of cartograms are area cartograms and linear cartograms. On an area cartogram, a region's area is drawn relative to some value other than its land surface area. In Figure 5-2 the value used is population. The greater a country's population, the bigger the country is drawn on the image. When a cartographer makes a conversion from physical space to something else, he or she tries to retain as many spatial attributes of the conventional map as possible. If recognizable shape, proximity, and continuity are preserved, it is easier for the viewer to

compare the cartogram with a standard map. Each of these attributes of space can be retained, however, only by distorting one or more of the others.

Linear cartograms draw our attention to distance. We noted that distance is sometimes measured in time or in cost, and these measures are not the same as actual mileage. A transit map, for example, is a time-distance map. The subway map of Manhattan distorts the island's true shape in order to clarify the subway lines and stops. Subway riders think in terms of stops, not distances. Therefore, on many transit maps, the distance between stops is shown as a uniform distance. Cartograms may look strange to us because the scale depends on something other than the physical units we expect. Cartograms are not, however, lies or tricks. Their purpose is to portray some other important aspect of reality. Area cartograms of such things as countries' populations (Figure 5-2) or economic output (Figure 12-5) visually convey the relative population or wealth of different countries better than standard land area maps do.

at a smaller scale than that of the other states, as used to be common, the confusion about Alaska's location was aggravated by confusion about its size. In fact, Alaska is nearly one-fifth the size of the lower 48 states combined. In January 1990, in order to counter this misrepresentation and the resulting confusion in some students' minds about exactly where Alaska is, the Alaska state legislature passed a resolution "respectfully requesting that all major United States magazines, newspapers, textbook publishers, and map publishers . . . place Alaska in its correct geographical position . . . in the northwest corner."

Several world maps in this book cut out great expanses of the world's oceans so that we can map human activities on the land in greater detail. This shifts the relative positions of the continents and tucks Australia and New Zealand south of Asia, to the West of where they really are. Clear divisions on the maps mark the lines along which the map has been cut, and each portion has markings of latitude and longitude for proper positioning. The first of these maps is Figure 2-30 in Chapter 2. If you want to recheck the continents' relative positions shown more accurately, you can always refer to a globe.

It is conventional to put north at the top of a map, but top and bottom are meaningless in space. In fact, any conventional view conceals a great deal of

information simply because it is conventional; we usually do not notice or consider something we have seen often before.

The inability to understand mapmakers' conventions troubled Mark Twain's fictional character Huckleberry Finn when he and Tom Sawyer were blown east in a balloon. Huck insisted that they were not over Indiana yet because the color of the land below had not changed from the green of Illinois to the pink of Indiana, as shown on his school atlas. "What's a map for?" he demanded. "Ain't it to learn you facts?" Our answer is yes, but a map works for you only if you understand its projection and the explanation of its colors and symbols found in its key, usually at the bottom of the map.

Geographic Information Technology

Computers have transformed many aspects of social, intellectual, and business life in the last few decades, and their impact has been highly significant in geography. Just a few years ago, maps were being drawn with ink pens and rulers, but now they are composed on computers and printed by machine. Location and land-use information was collected by optical surveys and visual inspection of aerial photographs, but now it is collected

Focus on
THE INTERNET

The Internet began in 1969 as a U.S. Defense Department project to allow widely scattered scientists and engineers working on military contracts to share computers and other resources. Data to be transmitted along telephone lines are broken up into packets, and packets can take different routes to their destination. There is no center, but a spiderweb of links, so if one route is destroyed or blocked, a packet simply takes another route. Scientists devised a way of sending messages, called electronic mail, or e-mail, and users invented ways for people to participate in open discussions and created document libraries accessible to all.

Meanwhile, personal computers made computer power affordable, and modems allowed them to hook up over telephone lines to commercial on-line services and bulletin boards—electronic discussion groups and software libraries often set up by enthusiasts. By 1990 the National Science Foundation assumed responsibility for the Internet, first to connect its own supercomputer centers and later to enable links among government and academic researchers. Soon companies were allowed to join.

The year 1993 saw the introduction of the World Wide Web, which offers multimedia capability and hypertext, the ability to link documents to each other by way of hyperlinks. In 1995 the National Science Foundation lifted all remaining curbs on commercial use and began to phase out federal subsidies for the network.

The World Wide Web Consortium (W3C) sets standards for the Internet. It was founded in 1994 by Tim Berners-Lee, the inventor of the World Wide Web, and it has just over 400 members. The Internet Engineering Task Force (IETF) develops agreed technical standards. The nonprofit, privately funded corporation, the Internet Corporation for Assigned Names and Numbers (ICANN), oversees the system of domain names. These bodies are all largely self-created and self-governing. They are open in membership and in argument, and they are consensus-based in their decision-making. They have virtually no powers of enforcement, but the structure has worked smoothly, even during the commercialization of the Internet. As Web use grows, however, and new competitive issues arise (censorship, soaring market values for recognizable or popular domain names, and so forth), the governance may change.

Today the Web is a massive multimedia encyclopedia, unedited (except by the individual contributors), and indexed only by privately created search directories (called *search engines*) and catalogs. Search engines work by periodically visiting every address on the Internet and scanning the titles and usually a portion of the content of any Web pages that may be present. The words found in those pages are then indexed. When you type in a few keywords and ask for a search, you are directed to sites that contain those words. The content of the Internet ranges from the glossy advertising and information pages of multinational corporations to pages of information about governments to thousands

using GPS and digital satellite imagery. Countless hours of work with paper, pencil, and calculator were necessary to perform routine statistical analyses, but now such work is performed in seconds on a desktop computer. Geographic Information Systems (GISs) integrate all of these tasks by using computers to assemble automatically and rapidly collected digital data, analyze such data together with other information that may be on hand, and generate complex and specialized maps for many purposes. GIS technology has revolutionized geographic research and expanded the applications of that work in everyday life.

> GIS technology has revolutionized geographic research and expanded the applications of that work in everyday life.

Automated cartography Anyone who has worked in traditional pen-and-ink technical drawing can attest that such work is tedious and, for most, not particularly exciting. A map is a highly complex technical drawing, and map production using manual techniques is very expensive. When maps had to be updated because of changes in the landscape—new roads, expanded urban areas, changed boundaries and place names—the costs were substantial. For this reason, when *computer-assisted drawing (CAD)* technology emerged, this technology was quickly applied to mapmaking, and sophisticated, specialized digital cartography systems were developed.

of home pages created by people who have assembled some unique information or just have information they want to share. Thus, many Internet users are not just information consumers; they are information producers as well. At very low cost, organizations and individuals anywhere can speak out to the world. The political ramifications of this will be noted in Chapter 11. By 2000 an estimated 500 million people were using the Web.

The Internet was originally free and unfettered, immune to regulation and control by governments. Attempts by oppressive regimes to block information were futile, and it was thought that cyberspace transcended international borders. Optimists saw it as a tool for free exchange of information and even the foundation of a new "brotherhood" among men. But this conventional wisdom is now dead as the freedom of the Internet has been steadily infringed. One battle was a fight over illegal trade in copyrighted material such as pop music. Legislation debated in the U.S. House of Representatives would even have allowed recording corporations to hack into home computers that trade music and video files. This is little better than licensed vigilantism. Threats to the Internet by terrorist organizations have prompted increased electronic surveillance by law enforcement and intelligence agencies. Legislation new in the twenty-first century allows governments to collect and share e-mail traffic and Web-surfing patterns. The Internet has even been militarized. The armed forces of many countries have devoted increased time, money, and effort to offensive cyber-warfare capabilities, including attacks, viruses, and worms that would allow attackers to destroy the systems of their foes. Access to government information on the Web has been restricted, and postings have been limited. Formerly public documents have been removed.

The technologies for restricting access to sites and also, chillingly, for identifying the visitors to individual sites are increasingly successful. For example, *geolocation technology* enables websites to determine the physical locations of users by linking Internet protocol (IP) addresses of users' computers to specific countries, cities, or even postcodes. Individual websites can then modify themselves or be modifed to address the visitor individually. Existing demographic databases describing, for example, what type of people live where you are, can target advertising. Positive aspects of geolocation are that it can help determine the appropriate language in which to present a multilingual website. Electronic commerce vendors, auction houses, pharmaceutical, and financial services industries can prevent the sale of goods that are prohibited or illegal in certain areas. Online casinos can prohibit gamblers where illegal, and rights management policies for video or music broadcasts can be enforced.

Geolocation technology can also be used to find the Internet access point nearest to a particular location. The necessity of doing this arises from the growing popularity of 802.11b or Wi-Fi technology, which provides wireless Internet access to suitably equipped laptops within 100 meters or so of a small base-station.

These systems make it possible to draw new features on maps and edit those features with ease on a video monitor before any paper map is created, in much the same way that word processors make it possible for you to compose a term paper on screen and edit it, correcting typing mistakes and making the needed refinements before you print a paper document. Given the complexity of mapmaking and the specialized skills it requires, this is obviously a big improvement for the cartographer.

Remote sensing satellites Another important dimension of GIS is its ability to make use of digital satellite imagery. The acquisition of data about Earth's surface from a satellite orbiting the planet or from high-flying aircraft is known as **remote sensing.**

Geographic applications of remote sensing include mapping of vegetation and other surface cover; gathering data for large unpopulated areas, such as measuring the extent of the winter ice cover on the oceans; and monitoring changes, such as weather patterns and deforestation.

Landsat satellites, the first of which was launched by the United States in 1972, were the first satellites that were intended primarily for mapping characteristics of Earth's surface. Passive sensors in the Landsat satellites measure the amount of radiation emanating from Earth's surface in particular wavelengths. Sensors primarily measure the amount of radiation of the various colors of visible light, although some measure infrared (heat) energy. In a few cases, active sensors like radar send radiation to Earth and measure the radiation that is reflected back to the satellite. The most recent Landsat Satellite to be launched was Landsat 7, launched in April 1999.

Remote sensing satellites scan Earth's surface much as a television camera scans an image in the thin lines you can see on a television screen. The sensor is moved first across the landscape in a line; it is then moved slightly to scan other lines in succession. At any moment, a sensor is recording the amount of energy from only one place, an area called a picture element, or pixel. A map created by remote sensing is essentially a grid containing many rows of pixels.

The smallest feature on Earth's surface that can be detected by a sensor is determined by the size of the pixel. This is called the resolution of the scanner. Early Landsat sensors had a pixel size of 59 meters by 59 meters (194 feet by 194 feet), compared to 30 meters by 30 meters (98 feet by 98 feet) on later Landsat versions and 10 meters by 10 meters (33 feet by 33 feet) and even 1.5 meters by 1.5 meters on the IKONOS satellite. Even smaller pixel sizes are available on military and intelligence satellites; these data are not, however, available to civilians. The importance of a smaller pixel size is that smaller objects can be detected.

Weather satellites take a broader view and thus have very large pixels, covering several kilometers on a side. This way they can rapidly map a large area such as a continent. Weather forecasters need data about a large area very quickly, because weather systems change rapidly (Figure 1-24).

Dozens of satellites orbit Earth and send back information about its atmosphere, oceans, and land surface, so the amount of information we have available is enormous. Some early GISs were developed to make use of satellite data, and with the growth of powerful desktop microcomputers in the 1980s and 1990s, the use and capabilities of GISs increased exponentially.

Satellites and planes equipped with specialized sensing devices can even read history as it is recorded below the surface. Information available includes changes in plant growth or drainage patterns brought about by human activity and densely compacted soils and erosion associated with agriculture. Archaeologists have also used this kind of information to find long-lost cities.

Human activities cause new and continuing changes in the landscape, and remote sensing is critical to monitoring and evaluating these impacts. For example, accurate assessments of the extent of logging activities are critical to managing forests and the wildlife they support (Figure 1-25). Other important uses of satellite imagery include such diverse applications as estimating nutrient levels in farmland soils, quantifying loss of wetlands in coastal areas, forecasting streamflow based on snow cover, and documenting urban growth (Figure 1-26).

FIGURE 1-24 A satellite image of Hurricane Douglas, west of Mexico, on July 23, 2002. This true-color image was collected by the Moderate Resolution Imaging Spectroradiometer (MODIS) sensor. MODIS sensors are currently deployed on two satellites that together provide data in 36 spectral bands, covering the entire Earth surface every one to two days. (Jacques Descloitres, MODIS Land Rapid Response Team, NASA/GSFC)

In times of war, remote sensing and Global Positioning System devices allow bombardiers to strike with incredible accuracy, as was demonstrated in the war against Iraq in 1991 and in several wars since then. Nevertheless, targeting errors have occurred, and civilians have been killed. Photographs of hospitals or schools mistakenly bombed during the war in Afghanistan in 2001 lost America many friends.

Despite technological advances, therefore, the United States may strengthen agencies that specialize in traditional techniques of gathering information about foreign affairs. The United States might hire more spies, especially language-trained; change the hiring rules and hire double agents and others previously rejected as "undesirable;" improve the professionalism of national information services (when Robert Rubin became Secretary of the Treasury in the Clinton administration, he complained that he had been able to get better information about world affairs when he was a private banker with Goldman Sachs Corporation); and change priorities from high-tech to hiring more information gatherers on the ground. For example, 90 percent of information gathered by U.S. satellites has never been analyzed; representatives on the ground, by contrast, are better able to gather intelligence through overt activities such as simply reading local newspapers, listening to the local radio, speaking with local officials and people on the street, and generally moving about with open eyes and ears. The

July 1984

May 1986

July 1988

September 1995

FIGURE 1-25 A series of Landsat images of part of the Olympic Peninsula in western Washington, showing forest harvest and regrowth. In this false-color image, green tones depict growing vegetation, while reds indicate bare ground. The brightest red patches are recent clear-cuts. Pale reds indicate young vegetation growing after the cut; over time these become pale greens and ultimately dark greens in this imagery. The images show that there was intensive timber harvesting between 1984 and 1986, but that by 1995 much of the harvested land was again covered with vegetation as the forest began to grow back. (National Air Survey Center Corporation)

numbers of people performing these tasks in foreign countries—as well as the number of analysts of foreign affairs stationed in the United States—had been trimmed from the American federal budget through the 1990s, but they are being increased today. New opportunities are opening for individuals trained in geographical analysis.

GIS: A Type of Database Software

A GIS is a special form of database software in which spatial information is an important part of the database. A simple database consists of lists of items—people, commodities, transactions, etc. Each item in the list is associated with multiple bits of information.

FIGURE 1-26 Two Landsat images of Shenzhen, China, taken in 1988 and 1996, showing dramatic urban development in just eight years. Although China's population growth has slowed, many people are moving from rural areas to urban ones as the industrial economy grows. Remote sensing and GIS are essential technologies for documenting and monitoring rapid changes on Earth's surface such as this. (NASA Goodard Space Flight Center, Scientific Visualization Studio)

In a database of land parcels maintained by a city clerk's office, for example, one might expect to find information on the owner (name, address, etc.), the value of the parcel, what kinds of structures exist on the parcel, and so forth. In a GIS, one of the attributes of an item in the database is either its spatial characteristics (boundary information, in the case of a land parcel) or something else (such as an address or name of the district in which it is located) that can be used to locate the item in space.

Using such a database, we can create layers of information, each representing a different characteristic, or attribute, of a place. A single layer can be displayed by itself or combined with other layers to show relations among different kinds of information (Figure 1-27). A GIS allows the user to create many different specialized maps from a single database rapidly.

Types of geographic data in a GIS Geographic data are of many types, and each different type has certain characteristics that determine how the information can be analyzed, interpreted, and displayed. One fundamental distinction is between raster and vector data (Figure 1-28). Raster data are arranged in a rectangular grid of cells, with each cell the same size. A computer display is a raster device. For example, many computer monitors display information in a grid that is 768 rows high by 1024 columns wide, containing 786,432 grid cells. Each of these cells contains specific information—in this case, the color that is to be displayed on the monitor. The pixels in a remotely sensed image usually contain information about the amount of light or energy sensed as coming from a particular place on Earth's surface.

Vector data are based on points with X and Y coordinates that specify location. Vector data are of three basic types: points, lines, and polygons. A point is defined by a single pair of X and Y coordinates. A straight line can be defined by just two points. More complex lines, such as would represent roads or rivers, are defined by a series of points, each defining a straight-line segment of the line. Regions are represented with polygons, which are simply lines that begin and end at the same point. Only four points are needed to define the

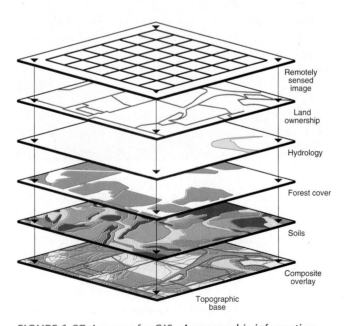

FIGURE 1-27 Layers of a GIS. A geographic information system involves storing information about a location in layers. Each layer represents a different piece of human or environmental information. The layers can be viewed individually or in combination.

Vector-based line

Flat file	
4753456	623412
4753436	623424
4753462	623478
4753432	623482
4753405	623429
4753401	623508
4753462	623555
4753398	623634

Raster-based line

Flat file
0000000000000000
0001100000100000
1010100001010000
1100100001010000
0000100010001000
0000100100000100
0001000100000010
0010000100000001
0111001000000001
0000111000000000
0000000000000000

FIGURE 1-28 Raster and vector data types. Raster data is composed of grids of cells that are coded to indicate a characteristic of a given area, while vector data consists of lines defined by X and Y coordinates of points (vertices) along the line. Remotely sensed images and photographs are typically raster images, while maps showing regional boundaries (such as political maps) are usually stored as vector data. (After Keith Clarke, *Getting Started with Geographic Information Systems,* 4th ed. Upper Saddle River, N.J.: Pearson Education, 2003)

efficient for storing information about regions, such as political units, or lines, such as roads. Modern GISs are able to analyze and display both raster and vector data (including points, lines, and regions) simultaneously (Figure 1-29).

Acquiring digital geographic information A few decades ago all maps were in paper (analog) form, but today nearly all mapping is done digitally using GIS. One of the biggest tasks in the first few decades of GIS technology has been to create the digital geographic database from which maps are made. Creation of this digital resource is done primarily in two ways. One of these is converting existing maps from paper to digital form. To a large extent this has been accomplished for the wealthier parts of the world, although digital data are much less available for developing countries. The other way of creating a digital database is through use of remote sensing satellites, which create digital data directly rather than going through a paper form first.

One way of digitizing paper maps involves manual tracing of lines using a specialized table, following the lines with a device somewhat like a computer mouse. As you might imagine, this is very tedious and time-consuming work. Today most conversion of maps is accomplished using scanners coupled with software that is able to "trace" features on the scanned image digitally and convert them to vectors (lines). While this

border of a rectangular region (Colorado, for example), but many more are needed to define the outline of most geographic regions.

Raster and vector data differ in the kind of information they can portray and the spatial accuracy with which they portray it. In a raster image such as a Landsat image with 30-meter by 30-meter pixels, every place within the pixel is assumed to have the same value for a given characteristic. If the pixel happens to be located in the middle of a lake, this assumption isn't very important—every place within it is water. If, however, the pixel is along the shore of that lake, then only a part of the pixel is water and a part is land. In a satellite image, the value that will be assigned to that pixel will probably fall somewhere between the appropriate value for water and that for land and will represent a surface cover type that does not, in fact, exist at Earth's surface. A vector line, on the other hand, could be created that would show exactly where the shoreline falls, and more accurately portray this feature. As long as pixels are small in relation to the objects being portrayed, raster formats are very useful for storing and manipulating information about complex maps such as a satellite image of a vegetated landscape, with many variations in the colors of vegetation and shapes of vegetation patches. Vectors, on the other hand, are very

FIGURE 1-29 A digital photograph of Burlington, Kansas, and the surrounding landscape, showing both raster and vector data. The photo itself is stored in a raster format with a resolution of about 1 meter. The line showing the boundary between the built-up area of Burlington and the rural land around the town is stored in vector format. (U.S. Geological Survey)

technology has improved dramatically, it still requires considerable expertise as well as expensive hardware and software. This helps explain why creation of a digital database in developing counties is lagging well behind the effort in the United States.

Today a very diverse database is readily available in the United States, much of it freely available on the Internet. For example, the 1:24,000 topographic maps are available in two forms. One of these is simply a scanned image of the map called a DRG or digital raster graphic. This version can be easily displayed on a computer using common image-viewing software. The other version, called digital line graph or DLG, consists of most of the lines from those maps—boundaries, streams, contours, roads, etc.—in vector form. These DLGs are the form that is more often used in GIS analyses. Digital versions of older paper maps are also available at scales of 1:100,000, 1:250,000, and 1:2,000,000.

In addition to these basic map resources, many specialized maps are now available in digital form. For example, the U.S. Geological Survey makes available online land use/land cover maps derived from satellite imagery, maps of hydrologic features and watersheds, and digital elevation models that are used to portray the landscape in 3 dimensions (Figure 1-30). The Fish & Wildlife Service distributes digital versions of its National Wetlands Inventory, and the National Oceanographic and Atmospheric Administration's marine navigational charts are available in digital form. The Census Bureau produces digital maps that show all roads and streets along with the numbering of addresses on them. These data allow you to type an address into a computer and immediately get a map showing the location of the address; such data are also essential for the emergency response systems used by police and other emergency services.

Application of GIS to spatial analysis   As discussed earlier, spatial analysis includes topics such as distribution, density, concentration, pattern, and movement. GIS facilitates quantitative measurement of these geographic properties. Calculation of population density simply involves combining (or overlaying) population data with a map of regions such as counties. Using a GIS, we can determine the total number of people living within the boundary of the county and divide that by the area of the county to determine density. Similarly, a GIS can be used to count the total number of people living within a given distance of some specified location, as part of a measurement of concentration. Changes in the distribution of people, businesses, or forest clearances from one time period to another, as represented in two different layers of a GIS, tell us about movements of people and economic activities over time.

A recent study of the distribution of small ponds illustrates the role of GIS in modern geographic research. An inventory was made of small lakes across the conterminous (48) United States, to determine how many lakes there are and their locations. The analysis began with a satellite-based land cover map produced by the U.S. Geological Survey. Because it was derived from satellite images, this is a raster map, where individual pixels are identified as "row crops", "deciduous forest", "high-density urban", etc. Rivers, lakes, and ocean bays are identified as "water" without further distinction.

The first task was to eliminate from the map all the areas that were clearly not lakes. This was done using map layers that included rivers and the ocean shore. Anything within 1 kilometer of a river or 5 kilometers of the shore was blanked out, so these water areas would not be counted as lakes. Next, lines were drawn around each remaining group of water pixels that touched each other, creating a vector layer in which each lake a single entity. These lakes were then counted—the total is nearly 2.6 million! Figure 1-31 is a map showing their distribution.

The analysis revealed that a very large number of lakes are located in the eastern Great Plains, in areas where natural lakes are rare. In fact, Texas has more lakes than any other state—about 10 percent of the total. Minnesota, which claims "10,000 lakes" on its license plates, in fact ranks seventh among the states in number of lakes according to this study, behind Florida, Oklahoma, Kansas, Missouri, and Mississippi. The good news for Minnesotans is that

FIGURE 1-30 A 3-dimensional perspective view of Mt. Everest. This image was made by "draping" an aerial photograph on a digital elevation model. Similar technology is used in many visualization applications, from television weather forecasts to flight simulation in movies. (National Geographic Image Collection)

The Distribution of Small Lakes in the Conterminous United States

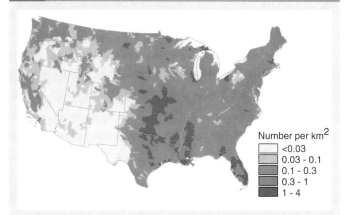

Number per km^2
- <0.03
- 0.03 - 0.1
- 0.1 - 0.3
- 0.3 - 1
- 1 - 4

FIGURE 1-31 The greatest number of lakes are found in the eastern Great Plains, where the vast majority are farm ponds built to store water for farming operations. Large numbers of lakes are also found in areas where they occur naturally, such as the Great Lakes States and Florida. Even in those places many of these lakes were built by humans. (WHR map)

they are being modest—the actual number of lakes there is more like 100,000. While the study does not distinguish between natural and artificial lakes, the great concentrations of lakes in areas where natural lakes are rare confirms that the vast majority of these features are of human origin. Most are small ponds built on farms in the latter half of the twentieth century, many of them as part of government programs. They are most numerous in the eastern Great Plains because this is an area that has enough rainfall to fill the ponds in the spring and early summer, yet it also has a dry season in which this stored water is particularly important.

In addition to the basic task of counting lakes, this study demonstrates the value of GIS for monitoring environmental conditions in a fast-changing world. Each of these farm ponds is so small that, by themselves, they seem insignificant. Taken together, however, they constitute an enormous modification to the environment.. In some areas of the Great Plains, for example, they number more than 3 per square kilometer (one for every 80 acres or so). They store water for cattle, but much water that would otherwise soak into the ground is lost to evaporation, principally during the summer. In some areas the amount of water that is lost from evaporation from farm ponds is enough to reduce the average annual flow in streams by 10 percent—an important amount in a water-stressed area.

This example illustrates many of the approaches of geography we have discussed in this chapter. It demonstrates that answering "where" questions remains an important part of geography. Earth's surface is constantly changing, so even though virtually all of Earth's land surface has been explored and mapped to some level, the maps must continually be redrawn. It illustrates important locational concepts such as distribution and concentration. Spatial interactions are seen in ponds' effects on downstream water availability, the transport of sediment and other substances by streams, and even the migration and distribution of wildlife. The lakes also exemplify the myriad ways in which human activity—in this case especially culture, food production and government policy—and the physical environment interact.

Finally, it is important to note that this study was carried out using data that are available on the Internet, and the data were analyzed on ordinary desktop computers. Anyone with the interest, a modest amount of training in GIS, and the appropriate software could have done this study. It would not, however, have been possible as recently as the mid-1990s. Some of the data were not available or available only to a select few, much expensive computer time would have to have been devoted to the job, and some of the software needed had not yet been written. Recent advances in GIS technology have opened up many such opportunities for geographers to explore the changing face of Earth's surface.

CONCLUSION: CRITICAL ISSUES FOR THE FUTURE

The twenty-first century will present new and difficult challenges to citizens, businesses, and governments. Critical issues include:

- Global-scale environmental change, including deforestation, soil degradation, increased concentrations of atmospheric carbon dioxide, probably global warming, and decreased atmospheric ozone levels.
- Rates of population growth in some regions that strain cultural systems, deplete scarce resources, and produce environmental changes.
- Increasing disparities in wealth and the threat of conflict between rich and poor countries and within individual countries.
- Systems of food production and distribution that do not meet nutritional needs of many people in developing countries.

Confronting problems of this magnitude will require a great deal of creativity and hard work from all segments of society, including geographers. Geography offers several unique perspectives for understanding contemporary issues.

First, geography emphasizes interdisciplinary approaches to contemporary issues. Geographers are trained partly as natural scientists and partly as social scientists, so they have a wide breadth of understanding that makes geography a unique discipline. Geographers work well in interdisciplinary teams analyzing major problems that involve combinations of human and physical factors. They can communicate with specialists in a variety of natural and social sciences, so geographers are often asked to lead group efforts.

Second, geographers have a global perspective. A century ago a famine or war in a distant part of the world had little consequence outside the immediate region, but today local events have worldwide impact. New information technology and transportation systems have created global markets for many commodities. What happens in one place has ramifications in other areas and often throughout the world. Concern over an endangered owl in the U.S. Pacific Northwest causes a rise in lumber prices not only throughout the United States but in Canada and in East Asia as well. A 1986 accident at Chernobyl, a nuclear power plant in Ukraine (then part of the Soviet Union), spread radioactive fallout around the world, heightened fears of nuclear power, and thus contributed to curtailment of nuclear power development worldwide.

Third, geography has a strong component of applied research. Since the earliest days of the discipline, geographers have worked on problems of practical significance, such as finding new routes for commerce and evaluating the natural resources of sparsely inhabited lands. Geographers emphasize studying everyday phenomena about places. They are aware of conditions in the world around them and are accustomed to applying their knowledge to them. GIS has become central to this endeavor, and is likely to remain so.

Chapter Review

SUMMARY

Geography is the study of the interaction of all physical and human phenomena at individual places and of how interactions among places form patterns and organize space. Physical geography studies the characteristics of the physical environment, while human geography studies human groups and activities, such as languages, industries, and cities.

Geographers divide Earth into regions, which are areas defined by one or more distinctive characteristics or features, such as climate, soil type, language, or economic activity. Area analysis is an approach that looks at these features of Earth's surface in a regional context.

Geographers study the distribution of objects across Earth's surface and processes by which human and environmental phenomena move from one place to another. Three properties of spatial distribution include density, concentration, and geometric pattern. Movements of matter, people, and information result in interactions among places along networks through diffusion processes.

The place where a culture originates is called the culture hearth. Various aspects of cultures may spread out and be adopted by other peoples in a process called cultural diffusion. The impact or frequency of any cultural attribute may diminish away from its hearth area; this phenomenon is called distance decay. Some aspects of culture may also develop variations as people who carry that culture wander outward and away from one another. Tracing a course of diffusion may teach us a great deal about how peoples and cultures interact. Diffusion may be by relocation, or it may be contiguous or hierarchical. Barriers to cultural diffusion may be topographic, political, or even cultural. Diffusion does not explain the distribution of all cultural phenomena. Sometimes the same phenomenon occurs spontaneously and independently at two or more places.

Earth's physical environment results from a process of interaction among four systems: the atmosphere, hydrosphere, lithosphere, and biosphere. Cultural systems include customary beliefs, social forms, and material traits. Some geographers regard the distribution of human activities as the result of underlying regularities in cultural systems, while others emphasize the diversity and uniqueness of human behavior in explaining cultural systems. The physical environment influences human actions, although humans can adjust and choose a course of action from many alternatives.

Geographers utilize maps to depict the location of places and to interpret underlying patterns. Remote sensing from satellites and GISs are two recently developed tools that help geographers more fully analyze patterns on Earth's surface. GISs allow rapid integration of many different types of geographic data as well as automated map production.

KEY TERMS

QUESTIONS FOR REVIEW AND DISCUSSION

1. What is geography? Describe differences among physical, human, and regional geography.

2. What contributions did ancient and medieval geographers make to the development of geographic thought?

3. What are the three ways to indicate scale on a map?

4. What are the four types of distortions that can result from map projections?

5. How do remote sensing and GISs contribute to geographic study?

6. What is the difference between a meridian (or longitude) and a parallel (or latitude)?

7. What is the difference between formal regions, functional regions, and vernacular regions? Give examples of each of the three types of regions.

8. What is the difference between density and concentration?

9. What do geographers mean by the concept of spatial interaction?

10. What is the difference between relocation diffusion and contiguous diffusion?

11. What are Earth's four physical systems?

THINKING GEOGRAPHICALLY

1. Cartography is not simply a technical exercise in penmanship and coloring, nor are decisions confined to scale and projection. Mapping is a politically sensitive undertaking. If you were a resident of New Zealand, what changes might you suggest to the world map projections used in this book? Look at how maps in this book distinguish the borders of India with China and Pakistan and of Israel with its neighbors. Can you identify other logical ways to draw these boundaries?

2. Imagine that a transportation device (perhaps the one in Star Trek) would enable all humans to travel instantaneously to any location on Earth. What impact might that invention have on spatial interaction? What if such travel were expensive, so only the rich could afford it?

3. When earthquakes, hurricanes, or other so-called "natural" disasters strike, humans tend to blame nature and see themselves as innocent victims of a harsh and cruel natural world. To what extent do environmental hazards stem from unpredictable nature, and to what extent do they originate from human activity? Should victims blame nature, other people, or themselves for the disaster? Why?

4. Geographic approaches, such as area analysis, spatial analysis, and systems analysis, are supposed to help explain contemporary issues. Find a story in your newspaper that can be explained through application of geographic concepts.

SUGGESTIONS FOR FURTHER LEARNING

Abler, Ronald F., Melvin G. Marcus, and Judy M. Olson, eds. *Geography's Inner Worlds.* New Brunswick, NJ: Rutgers University Press, 1992.

Berleant, Arnold. *Living in the Landscape: Toward an Aesthetics of Environment.* Lawrence, KS: University of Kansas Press, 1997.

Beddoe, D. "GIS Meets Weather Processing Head-On." *GIS World* 10/2 (1997): 52–53.

Brown, Kathryn. "Mapping the future." *Science* 298 (2002): 1874–1875.

Claval, Paul. "The Region as a Geographical, Economic and Cultural Concept." *International Social Science Journal* 39 (May 1987): 159–172.

Cromley, Ellen K., and Sara L. McLafferty. *GIS and Public Health.* New York: Guilford Press, 2002.

de Sherbinin, A., K. Kline, and K. Raustiala. "Remote sensing data: valuable support for environmental treaties." *Environment* 44 (2002): 1: 20–31.

Dodge, Martin, and Rob Kitchin. *Atlas of Cyberspace.* New York: Addison-Wesley, 2001.

Forman, Richard T. *Land Mosaics: The Ecology of Landscapes and Regions.* New York: Cambridge University Press, 1997.

Gaile, Gary L., and Cort J. Willmott, eds. *Geography in America.* New York: Merrill/Macmillan, 1989.

Janelle, Donald G., ed. *Geographical Snapshots of North America.* New York: Guilford, 1992.

Johnston, R. J. *Philosophy and Human Geography,* 2nd ed. London: Edward Arnold, 1986.

———, ed. *The Dictionary of Human Geography,* 3rd ed. Oxford: Basil Blackwell, 1994.

Kirby, Kathleen. *Indifferent Boundaries: Spatial Concepts of Human Subjectivity.* New York: Guildford, 1996.

Livingstone, David. *The Geographical Tradition: Episodes in the History of a Contested Enterprise.* Cambridge, MA: Blackwell, 1992.

Lock, Gary, and Zoran Stancic, eds. *Archaeology and Geographical Information Systems.* London: Taylor & Francis, 1997.

Longley, Paul A., Michael F. Goodchild, David J. Maguire, and David W. Rhind. *Geographic Information Systems and Science.* Chichester, NY: Wiley, 2001.

Mabogunjie, Akin L. "Geography as a Bridge Between Natural and Social Sciences." *Nature and Resources* 20 (1984): 2–6.

MacEachren, Alan M. *How Maps Work.* New York: Guilford, 1995.

Martin, Geoffrey, Preston E. James, and Eileen W. James. *All Possible Worlds: A History of Geographical Ideas,* 3rd ed. New York: John Wiley & Sons, 1992.

Melnick, Alan L. *Introduction to geographic information systems in public health.* Gaithersburg, MD: Aspen Publishers, 2002.

Miller, Harvey J. *Geographic Information Systems for Transportation: Principles and Applications.* New York: Oxford University Press, 2001.

Monmonier, Mark. *How to Lie with Maps,* 2nd ed. Chicago: University of Chicago Press, 1996.

National Geography Standards Project. *Geography for Life: National Geography Standards.* Washington, D.C.: National Geographic Research, 1994.

Raper, Jonathan. *Multidimensional Geographic Information Science.* New York: Taylor & Francis, 2000.

Reed, Michael, ed. *Discovering Past Landscapes.* London: Croom Helm, 1984.

Smith, S. V., W. H. Renwick, J. D. Bartley, and R. W. Buddemeier. "Distribution and significance of small, artificial water bodies across the United States landscape." *Science of the Total Environment* 299 (2002): 2–36.

Strahler, Arthur N. "Systems Theory in Physical Geography." *Physical Geography* 1 (January 1980): 1–27.

Woodward, David, and J. B. Hartley, eds. Volumes are regularly appearing in "The History of Cartography" Project, all published by the University of Chicago Press. *Cartography in Prehistoric, Ancient and Medieval Europe and the Mediterranean* (1987); *Cartography in the Traditional Islamic and South Asian Societies* (1992); *Cartography in the Traditional East and Southeast Asian Societies* (1994).

Zimmerer, Karl S. "Human Geography and the 'New Ecology': The Prospect and Promise of Integration." *Annals of the Association of American Geographers* 84 (January 1994): 108–125.

WEB WORK

The World Wide Web has a tremendous variety of sites of interest to geographers. You might want to visit the home pages of a few geographic organizations in the United States. These include the Association of American Geographers

http://www.aag.org

the National Geographic Society

http://www.nationalgeographic.com/

and the National Council for Geographic Education

http://www.ncge.org

Scholars at the University of Colorado at Boulder maintain the International Network for Learning and Teaching Geography

http://www.colorado.edu/geography/virtdept/resources/contents.htm

Some government agencies that maintain home pages that help introduce visitors to geography include the U.S. Geological Survey

http://www.usgs.gov

and the National Imagery and Mapping Agency

http://www.nima.mil

At the end of 2003, the National Imagery and Mapping Agency is changing its name to the National Geospatial-Intelligence Agency, abbreviated as N.G.A.

The University of Edinburgh, Scotland, lists GIS-related sites

http://www.geo.ed.ac.uk/home/giswww.html

The National Atlas of Canada can be found at

http://atlas.gc.ca/

While that of the United States is at

http://www.nationalatlas.gov/

An excellent site for explaining items of scientific interest in the news is jointly maintained by the University of Wisconsin and the National Science Foundation

http://whyfiles.org

A Look Ahead

ENERGY AND WEATHER

Solar energy drives the circulation of Earth's atmosphere. Energy inputs to the atmosphere vary with latitude and season, and with elevation in the atmosphere. Atmospheric circulation redistributes energy and, in the process, generates weather and storms.

PRECIPITATION

Precipitation may be created when air rises, causing water vapor to condense in the atmosphere. Air may be forced to rise because it is heated at the surface, when passing over mountains, or during storm systems.

2

WEATHER AND CLIMATE

◀ The tanker Prestige, which broke up in high winds off the coast of Spain in November, 2002. The storm that contributed to this incident was linked to a disturbance that began in eastern Asia and traveled more than halfway around the globe. (AP/Wide World Photos)

CIRCULATION PATTERNS

Variations in atmospheric pressure from one place to another cause wind. Air blows from regions of high pressure to regions of low pressure. Pressure differences help explain prevailing wind directions and seasonal differences in winds.

CLIMATE

Climate is the summary of weather that a place experiences. Earth's climates can be grouped into 11 major types.

CLIMATE CHANGE

Climates have changed significantly over the last few million years, including major climate variations within the last 20,000 years. Many different factors contribute to climatic change, including human activities that are raising the level of carbon dioxide in the atmosphere.

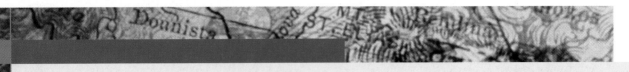

In November of 2002, warm water in the Pacific Ocean around Indonesia provided the energy for a period of heavier than usual rains over that mountanous island nation. The heat released to the atmosphere by those rains created a series of ripples in the jet stream—a band of faster than average winds that circles the globe—near Japan. The ripples, or waves, in the jet stream traveled eastward across the Pacific and spawned a storm system that produced a series of tornados in the United States that killed 34 people. The jet-stream waves continued across the Atlantic, creating a storm off the coast of Spain that caused the breakup of an oil tanker and a massive oil spill on the Spanish coast. The waves also generated floods in the Alps and destructive winds in Germany and Austria—halfway around the world from Indonesia.

The most dramatic weather events in the news are those that cause major damage, like those described above. Floods, mostly caused by heavy rains but some augmented by high sea levels during hurricanes, cause more damage than any other natural hazard. In addition to such extremes, day-to-day weather influences culture and social customs from clothing to agriculture. This chapter explores the mechanisms that cause **weather**—the day-to-day variations in temperature, precipitation, and so forth—and **climate,** the statistical summary of weather over time.

Weather conditions—such as storms, snowfall, clear skies, and warmth or cold—are caused by radiant energy received from the Sun. This solar energy is redistributed from tropical latitudes to the polar regions, from lower levels to upper levels of Earth's atmosphere, and eventually back into space. The movement of energy creates air movements that carry warm air from the tropics to high latitudes, and cold, wintry air from polar regions toward warmer climates. It carries moisture from the oceans over the land and returns drier air to ocean areas.

In the process of moving air and water from one place to another, weather is created. Weather varies from day to day because atmospheric circulation is constantly changing. But variable as it is, circulation tends to follow certain patterns. A zone of rising air is usually found near the equator, causing frequent heavy rains. Over northern Africa the air usually descends, causing clear skies and aridity. In the midlatitudes the air tends to move from west to east most of the time and less often in other directions. In southern and eastern Asia the wind blows from the land to the ocean in winter and from the ocean to the land in summer, causing intense monsoon rains. All these events are linked together in a dynamic atmospheric circulation, as demonstrated by the events of November 2002.

These patterns of weather are one of the fundamental features of Earth's surface that make places different from each other. They make agriculture possible in some places and not in others; they provide abundant water for hydroelectricity in some places and sunshine on beaches elsewhere. More than any other physical attribute of Earth, weather and climate regulate natural systems and constrain the ways in which humans use their environments. Yet even though Earth's circulation is utterly uncontrollable, it is not beyond inadvertent modification by humans, and some believe that intense storms like those of November 2002 will become more common as a result of human effects on the atmosphere. In this chapter we will explore the processes that govern weather, the distribution of climates that results from those processes, and some of the ways in which human activities are modifying weather and climate.

ENERGY AND WEATHER

The movement of energy in the atmosphere can be likened to that in an automobile engine, which uses energy from gasoline to move a vehicle along. An engine releases energy inside cylinders through the burning of gasoline. This concentrated energy drives the vehicle and is dissipated mostly through friction.

The fuel driving the atmosphere is **solar energy,** which heats the air, creating wind and ocean currents. These movements carry heat and moisture from one part of the planet to another before the energy is dissipated into space. In an automobile, motion takes the form of pistons moving back and forth, shafts and gears turning, and wheels spinning. In the atmosphere, air is in motion from one part of the globe to another, sometimes in steady flows and sometimes in intense spinning or vertical movements that produce storms.

The Sun is a large thermonuclear reactor, in which hydrogen atoms combine to produce helium. Most important to us is a byproduct of this reaction: the release of vast amounts of energy. The Sun radiates this energy into space in all directions, and Earth intercepts a tiny fraction of it. Yet this small, steady energy flow is sufficient to power the circulation of the atmosphere and oceans and to support all life on Earth.

As Earth revolves along its elliptical orbit around the Sun each year, it receives this energy across a void averaging about 150 million kilometers (93 million miles). Variations of about 3 percent seasonally in distance between Earth and the Sun cause small variations in the amount of energy reaching Earth, but these are minor in comparison to the effects of latitude and the tilt of Earth's axis.

Incoming Solar Radiation

The amount of solar energy intercepted by a particular area of Earth, or **insolation,** depends on two factors: the intensity of solar radiation, or the amount arriving per unit time, and the number of hours during the day that the solar radiation is striking.

Intensity of solar radiation   The intensity of solar radiation at a particular place on Earth depends largely on the angle at which the Sun's rays hit that place. Daily and seasonal differences in intensity are caused by variations in the **angle of incidence,** which is the angle at which solar radiation strikes a particular place at a point in time. The angle of incidence at a particular place on Earth varies within a day and among seasons.

To illustrate the angle of incidence, consider what happens when a 1-square-meter beam of sunlight strikes a spherical object such as Earth (Figure 2-1). If this beam strikes Earth's surface perpendicularly (that is, at an angle of 90°), the surface it touches also will have an area of 1 square meter. At an angle of less than 90° (that is, an oblique angle), the area illuminated by that beam will be greater than 1 square meter. The intensity of the radiation is reduced, because the beam's energy is spread over the larger area.

As the angle is reduced even further, energy is distributed over a still larger area, and the intensity of radiation becomes less and less. Compared to the level of radiation at a 90° angle, the intensity of energy is about 86 percent at an angle of 60° and 50 percent at an angle of 30°. Thus, the intensity of radiation per unit of surface area is affected greatly by the angle of the Sun's rays, or the position of the Sun in the sky. When the Sun is overhead, sunlight is more intense, and the level of energy received is more concentrated than when the Sun is low in the sky.

On a daily basis the Sun is most intense at noon, when the Sun is highest in the sky, and least intense at sunrise and sunset. Beachgoers are warned to take this daily variation into account: They are more likely to get sunburned around noon than early in the morning or late in the afternoon. Why? Because insolation varies at different times of day.

The angle at which the Sun's energy strikes Earth's surface varies with the season, as well as within a day. Throughout the year, the area of Earth's surface where the Sun is overhead keeps shifting due to Earth's tilt and continual revolution around the Sun (Figure 2-2). As

FIGURE 2-1 The angle of incidence. The angle at which the Sun's radiation strikes Earth's surface, known as the angle of incidence, varies. The angle of incidence affects the concentration of solar radiation (insolation) that a place receives. Radiation is most concentrated when the Sun's rays are perpendicular or vertical to Earth's surface, as occurs when the Sun is directly overhead. As shown, a place receiving more perpendicular insolation has this energy concentrated over a smaller area, making the area warmer. A place receiving more oblique (that is, less perpendicular) insolation has it spread over a larger area, making the area less warm.

Figure 2-2 illustrates, Earth rotates once every 24 hours around its axis of rotation, which is an imaginary line passing through the North and South poles. The axis of rotation is inclined 23.5° away from being perpendicular to the Sun's incoming rays. This axial tilt of 23.5° remains constant, regardless of Earth's point in its orbit or its daily rotation. This axial tilt is a key factor that determines how much insolation any point on Earth receives. Imagine our tilted Earth slowly revolving around the Sun through the four seasons of the year, and how this shifts the overhead point of the Sun. In the Northern Hemisphere summer (June and July), the Sun is higher in the sky with more intense radiation, whereas in the winter the Sun is lower in the sky with less intense radiation. In the Southern Hemisphere, the opposite is the case, and the seasons are reversed there.

The amount of insolation a place receives in a particular season depends on its latitude (Figure 2-3). At noon on the Northern Hemisphere **vernal equinox** (March 20 or 21) and the **autumnal equinox** (September 22 or 23), the perpendicular rays of the Sun strike the

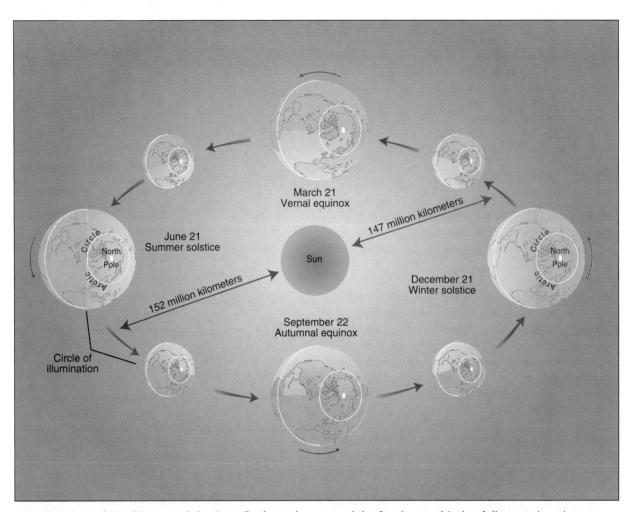

FIGURE 2-2 Earth's orbit around the Sun. Earth revolves around the Sun in an orbit that follows an imaginary plane known as the plane of the ecliptic. The plane of the ecliptic intersects all points on Earth's orbit around the Sun and passes through the Sun as well. Earth's path around the Sun is shaped like an ellipse. As it revolves around the Sun, Earth rotates around the axis of rotation, which is the line passing through the North and South poles perpendicular to the equator. The line passing through the North and South poles is tilted at an angle of 66.5° away from the plane of the ecliptic, or 23.5° away from a line perpendicular to the plane of the ecliptic. This tilt is constant throughout the year, no matter where Earth is in its orbit. The tilt causes the Northern Hemisphere to be more exposed to the Sun's radiation from March to September (Northern Hemisphere spring and summer), while the Southern Hemisphere receives more radiation from September to March (Southern Hemisphere spring and summer).

equator, and the Sun is directly overhead at the equator (Figure 2-3a). At places along the **Tropic of Cancer** (23.5° north latitude) and the **Tropic of Capricorn** (23.5° south latitude), the intensity of radiation is reduced to 92 percent of the level at the equator. At places along latitude 45° north and south, the intensity of solar radiation is 71 percent of the level at the equator, and at places along latitude 60° north and south, the intensity is only 50 percent of the level at the equator. At the North and South poles on the equinoxes, the Sun is seen on the horizon, and horizontal surfaces do not receive any direct radiation.

From the vernal equinox until the autumnal equinox (from late March until late September), the Sun is directly overhead at places along some latitudes in the Northern Hemisphere. The northernmost

latitude receiving direct insolation is the Tropic of Cancer (23.5° north latitude), where the Sun is directly overhead at noon on the **summer solstice** (June 20 or 21). At the September equinox, the Sun is again directly overhead at noon at the equator. From then until the March equinox, places in the southern latitudes receive direct insolation, and the Sun is directly overhead of places along the Tropic of Capricorn (23.5° south latitude) at noon on the Northern Hemisphere **winter solstice** (December 21 or 22).

Day length   The total amount of heat that a particular place on Earth receives in a day is determined by the number of hours during which the Sun's energy strikes the place, as well as the intensity with which it strikes. Places on the equator always receive 12 hours of

FIGURE 2-3 **Variation in angle of solar radiation.** The angle at which solar radiation strikes a particular latitude varies by season. These globes show how Earth receives solar radiation (insolation) on the first day of spring and fall (the equinoxes) (a) and on the first day of Northern Hemisphere summer (b). (From Christopherson, R. W., *Geosystems,* 3rd ed. Upper Saddle River, NJ: Prentice Hall, 1997)

FIGURE 2-4 **Midnight Sun.** On the day of the summer solstice, poleward of 66.5°, the Sun shines the entire 24-hour day, and the Sun appears to travel above the horizon without setting. (Brian Stablyk/Tony Stone Images)

sunlight and 12 hours of night. But in higher latitudes, the amount of daylight varies considerably with the seasons. For example, Winnipeg, Manitoba, Canada, at 50° north latitude, receives about 16¾ hours of daylight at the summer solstice, but only about 7¼ hours of daylight at the winter solstice. In Winnipeg, the Sun is 63.5° above the horizon at noon on the summer solstice, compared to only 16.5° above the horizon at noon on the winter solstice. In a 24-hour day at the summer solstice, Winnipeg receives nearly six times as much solar radiation, measured at the top of the atmosphere, as it does at the winter solstice.

Variations in the length of day from place to place result from the 23.5° tilt of Earth's axis away from the perpendicular. Figure 2-3b illustrates the illumination of the globe at the Northern Hemisphere summer solstice in June. Because the North Pole is located at Earth's axis of rotation, it maintains the same position with respect to the Sun throughout the 24-hour rotation of Earth at the solstice. At the summer solstice, if you stand at the North Pole, you can watch the Sun travel in a circle around you at a constant elevation of 23.5° above the horizon. On the day of the solstice, the North Pole is in full sunlight for the entire 24 hours, a phenomenon called the Midnight Sun (Figure 2-4).

Spatial and seasonal variations in radiation inputs

The amount of solar radiation reaching places at a particular latitude varies through the year because of seasonal changes in the angle of incidence, day length, and distance from the Sun. The tropical areas at low latitudes generally are warm throughout the year because they receive large amounts of insolation in every season, whereas places at high latitudes have strong seasonal contrasts in the amount of sunshine and level of temperature (Figure 2-5).

Because Earth's axis of rotation is tilted, seasonal variations in the amount of incoming solar radiation

are less at places near the equator than at places in higher latitudes. Places at higher latitudes receive somewhat more solar radiation than do places at lower latitudes during the summer, but they receive much less during the winter. The tilt of Earth's axis makes life possible on a seasonal basis at relatively high latitudes; without that tilt, there would not be a warm season at

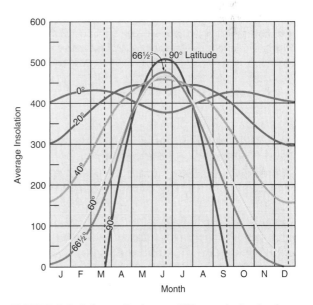

FIGURE 2-5 **Solar radiation at different latitudes by season.** Because Earth's axis of rotation is tilted, seasonal variations in the amount of incoming solar radiation are less at places near the equator than at places in higher latitudes. Places at higher latitudes receive somewhat more solar radiation than do places at lower latitudes during the summer, but they receive much less during the winter.

Average January Temperatures

(a) January

FIGURE 2-6(a) **Global temperatures for January** (convertible to Fahrenheit by means of the scale). In the Southern Hemisphere, the warmest areas are over land, especially Australia. The coldest areas are in continental areas in the Northern Hemisphere, especially Greenland and Siberia. In the Northern Hemisphere, the oceans are warmer than land areas at similar latitudes. (Adapted from National Climatic Data Center, *Monthly Climatic Data for the World* 47, no. 1, January 1994. Prepared in cooperation with the World Meteorological Organization. Washington, D.C.: National Oceanic and Atmospheric Administration. From Christopherson, R. W., *Geosystems*, 5th ed. Upper Saddle River, NJ: Prentice Hall, 2003)

mid and high latitudes, and life would probably be much more concentrated near the equator.

The effects of these variations in energy inputs are clearly visible in world temperature patterns (Figure 2-6).

In general, the highest temperatures occur in low latitudes in both maps. This is because the Sun is highest in the sky in these areas, and therefore the intensity of solar radiation is highest. High solar elevation angles occur throughout the year, and this makes the tropics (the area between the Tropic of Cancer and the Tropic of Capricorn) consistently warm.

High temperatures also occur in high latitudes such as in North America, Europe, and Asia in July. This is because at high latitudes the Sun is high in the sky in summer, and although it is not as high as in the tropics, the day lengths are much greater, so the total

amount of sunshine arriving at Earth's surface in a 24-hour day in midsummer is actually greater at high latitudes than in the tropics. The opposite effect is seen in winter, when high latitudes experience both low Sun angles and short days, causing temperatures to be much lower. Seasonal contrasts in temperature are greatest in high latitudes because seasonal contrasts in incoming solar radiation are greatest there.

Storage of Heat in Land and Water

When heat is absorbed by an object, its temperature rises, and when heat is released, an object cools. The ability of an object to store heat depends on what it is made of. Some materials can absorb or release a large amount of heat with only small changes in temperatures, while others heat and cool quickly with only small

Average July Temperatures

(b) July

FIGURE 2-6(b) Global temperatures for July. The warmest areas in the Northern Hemisphere are over land, especially central Asia and North America. The greatest temperature differences between January and July are over land areas, especially the large landmasses of the Northern Hemisphere. (Adapted from National Climatic Data Center, *Monthly Climatic Data for the World* 47, no. 7, July 1994. Prepared in cooperation with the World Meteorological Organization. Washington, D.C.: National Oceanic and Atmospheric Administration. From Christopherson, R. W., *Geosystems,* 5th ed. Upper Saddle River, NJ: Prentice Hall, 2003)

inputs and releases of energy. Water can absorb and release much larger quantities of heat for a given temperature change than can land. The main reason for this is that a water body such as an ocean can be stirred by the wind and thus carry heat down below its surface to depths of 10 meters (33 feet) or more seasonally. In contrast, land surfaces warm and cool to only about 2 meters (6 feet) depth seasonally. Oceans can store a season's heat in a much larger volume of matter than can land areas, so they do not heat up as much in summer nor do they cool down as much in winter.

The effect on temperature of water's ability to store heat is clearly visible in Figure 2-6. Large landmasses in high latitudes, such as North America and Asia, have very large temperature differences between July and January. Average July temperatures in Siberia reach above 10°C (50°F), while average January

temperatures reach −40°C (−40°F) The difference between July and January temperatures exceeds 50°C (90°F) in eastern Siberia. In contrast, the January and July temperatures in the Aleutian Islands of the north Pacific, at about the same latitude as Siberia, are about 15°C in July and 5°C in January, a range of only about 10°C. The moderate climates of areas near the ocean compared with the climatic extremes of midcontinent regions are a result of this heat storage in water.

Heat Transfer Between the Atmosphere and Earth

Have you ever noticed that on a winter day you feel colder sitting inside near a window or exterior walls than near the building's interior walls, even though air temperatures at both places are about the same? How does moisture on your skin help cool your body? And why does wind blowing on your skin cool it faster than still air? These effects on the temperature of your skin result from processes of heat transfer. Similar processes of heat transfer operating in the atmosphere as well as on your skin are responsible for movement of vast amounts of energy from place to place on Earth.

Radiation   The most important process of heat transfer in the environment is **radiation.** Energy transmitted by electromagnetic waves, including radio, television, light, and heat, is radiation, or radiant energy. You feel heat radiating from a fire. Heat travels at the speed of light from the fire to your skin, which senses it. Radiation can travel through the vacuum of space and through materials, although materials may restrict radiant energy flow. Radiant energy allows us to see the Sun's light energy, feel its heat energy, and listen to its radio energy using electronic machines.

> The most important process of heat transfer in the environment is radiation.

Radiant energy waves have different lengths. The **wavelength** is the distance between successive waves, like waves on a pond. Wavelength affects the behavior of the energy when it strikes matter; some waves are reflected, and some are absorbed.

The Sun's energy comes to Earth as radiant energy, for this is the only way energy can travel through the vacuum of space. Two ranges of wavelengths—called **shortwave** and **longwave**—are most important for understanding how solar energy affects the atmosphere. Most insolation is shortwave energy, with wavelengths between 0.2 and 5 microns (a micron or micrometer is one-millionth of a meter). Wavelengths visible to the human eye account for a small portion of this shortwave energy, from about 0.4 to 0.7 microns. In contrast, most of the energy reradiated by Earth is longwave energy, in wavelengths between 5 and 30 microns.

It is very important that most energy arriving from the Sun is shortwave, while most energy reradiated by Earth is longwave. As energy from the Sun passes through the atmosphere, some wavelengths are absorbed, warming the atmosphere, while others pass through or are reflected, either to be absorbed elsewhere or to travel back into space. The atmosphere is relatively transparent to incoming shortwave radiation, so this shortwave energy from the Sun easily passes through the atmosphere to reach Earth's surface, but when the surface reradiates longwave energy, much of it is blocked and absorbed by the atmosphere. The blockage of outgoing longwave energy causes Earth's atmosphere to

> Of all the gases in the atmosphere, only a few have the property of being relatively transparent to incoming shortwave solar energy but absorbing outgoing longwave radiation.

heat. This heating of the atmosphere is called the **greenhouse effect** because of its similarity to the way glass allows solar energy to enter a greenhouse but limits the loss of heat, causing the temperature inside to rise.

Of all the gases in the atmosphere, only a few have the property of being relatively transparent to incoming shortwave solar energy but absorbing outgoing longwave radiation. Gases with these properties are called **greenhouse gases,** and they are critical to heat exchange in the atmosphere. Among the more important ones are water vapor, **carbon dioxide (CO_2), ozone (O_3),** and methane (CH_4). Although these gases together constitute a small fraction of 1 percent of the atmosphere, they are the most important from the standpoint of atmospheric heating. Water vapor contributes the most to atmospheric heating. Human activities are increasing the amount of some greenhouse gases in the atmosphere, and this is believed to be the chief cause of **global warming.** We will return to this topic later in the chapter.

Latent heat exchange   A second important mechanism of moving energy in the environment is called **latent heat exchange.** It transfers tremendous amounts of energy from low latitudes to high ones, and it is also the most important process causing precipitation.

We can distinguish between two types of heat—sensible and latent. **Sensible heat** is detectable by your sense of touch. It is heat you can feel, from sunshine or a hot pan, and you can measure it with a thermometer. The atmosphere, oceans, rocks, and soil all have sensible heat, for you can feel their relative warmth or coldness. **Latent heat** is "in storage" in water and water vapor. You cannot feel latent heat, but when it is released, it has a powerful effect on its immediate environment.

Latent—which means "hidden"—is a good word to describe the heat that controls the state of water, for it is invisible stored heat. When ice melts, it must absorb heat energy from its surroundings. This is why ice melting in your hand feels so cold—it is absorbing heat from your hand. The heat becomes stored in the meltwater as latent heat.

Latent heat also is stored in water vapor (Figure 2-7). If you ever have had a finger scalded by steam, you know the startling amount of latent heat that was stored in the vapor and conducted from the condensing water into your finger! Latent heat exchange is what happens when water changes state (Figure 2-8).

Here is an analogy that describes the concept of latent heat: Consider a glass containing a mixture of ice

FIGURE 2-7 Mist rising off the warm surface of a lake into cold air. The evaporation of water from the lake removes sensible heat from the lake, cooling the liquid water, and transferring latent heat to the air. (© Leslie & Mark Degner/ CORBIS)

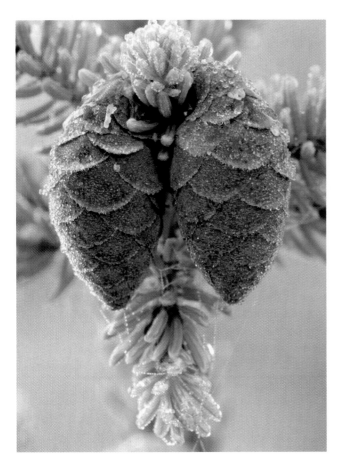

FIGURE 2-8 Dew formation. Longwave radiation from the ground to the atmosphere during the nighttime cools air near the ground. When the relative humidity reaches 100 percent, condensation occurs, and water droplets accumulate on solid surfaces such as on these spruce cones. (Rod Planck/Photo Researchers, Inc.)

and water on a hot summer afternoon. Because the ice water is much cooler than the air and other objects around it, heat energy is absorbed into the water. The heat begins to melt the ice, but as long as the water contains some ice, its overall temperature remains at 0° C (32° F). The water is absorbing enough heat to convert some ice to liquid, but the temperature does not change. The heat the water absorbs to melt the ice is latent heat, because it cannot be directly sensed as temperature.

When water is converted from liquid to vapor, additional heat is required—in fact, much more is needed than is required to melt ice. The latent heat of vaporization—the process that turns liquid to vapor— is about 540 calories per gram, compared to about 80 calories per gram for the latent heat of melting. A calorie is the amount of energy required to raise the temperature of 1 gram of water 1° C.

In the ice-water example, a portion of the heat used to melt the ice came from sensible heat—heat you can sense—expressed as the temperature of the environment around the glass. The remainder came from the latent heat released by water vapor condensing to liquid on the surface of the glass. Thus, the melting of the ice water involved three transfers of heat:

1. Latent heat was converted to sensible heat by condensation of water vapor on the glass.
2. Sensible heat was transferred directly from the warm air to the cool glass.
3. Sensible heat was converted to latent heat in the melting of the ice.

Now let us transfer this knowledge of latent heat to the much larger scale of the atmosphere. The amount of energy involved in latent heat transfers in the atmosphere is vast, especially for major weather systems and hurricanes. Transfer of heat from the land and water surface to the atmosphere by latent heat exchange amounts to about 40 percent of the solar energy absorbed by the surface. Hurricanes gather strength by drawing in warm, humid air and converting the latent heat it contains into sensible heat, which drives the motion of the storm.

Insolation supplies shortwave energy to warm the seawater. As it evaporates, countless molecules of water vapor—each with its own latent heat—hover above the water, available to join the storm as it passes over. As the hurricane accumulates this water vapor, it is also gathering an enormous amount of latent heat energy. The vapor rises in the rotating storm, cools, and condenses. Latent heat is released as sensible heat, a process roughly analogous to adding gasoline to a forest fire.

The moisture in one major hurricane at its peak was estimated to weigh 27 trillion (27,000,000,000,000) metric tons (30 trillion tons). The latent heat released by this condensing water vapor in a single day roughly matched the total amount of energy consumed in the United States for six months. This provides some insight into why hurricanes are dangerous.

Heat Exchange and Atmospheric Circulation

Liquid water changes to vapor primarily at Earth's surface. This water vapor is carried aloft by rising air. In the atmosphere, vapor is converted back to liquid (clouds and rain) or solid (ice clouds and snow) through the formation of precipitation, which we will discuss later in the chapter. This movement of water back and forth between the surface and the atmosphere acts like a great conveyor belt for energy, picking up heat at the surface, carrying it aloft, and releasing it in the atmosphere. The conveyor is driven by convection.

> This movement of water back and forth between the surface and the atmosphere acts like a great conveyor belt for energy, picking up heat at the surface, carrying it aloft, and releasing it in the atmosphere.

Convection is movement in any fluid, caused when part of the fluid (whether gas or liquid) is heated. The heated portion expands and becomes less dense. Therefore, it rises up through the cooler portion. Convection causes the turbulence you see in boiling water and in turbulent clouds overhead. As air is warmed, it expands and becomes less dense. In becoming less dense—that is, weighing less per unit of volume—warm air rises above cooler, denser air, just as a hot-air balloon rises through the cooler air surrounding it.

Think of an island on a sunny day (Figure 2-9). Solar energy passes down through the atmosphere and is absorbed by the sandy surface. As the surface warms, it reradiates longwave energy, some of which is absorbed by the air just above the ground. The water surface warms less than the land does because some of the radiation is used to evaporate water, and some penetrates the water to warm lower depths of the sea. The air therefore grows much warmer over the island than over the water. This warmer air over the island expands, grows less dense, and begins to rise. As the warm air rises, cooler air descends to take its place. The cooler air is then warmed by the surface, and the convection process continues.

Convection also causes horizontal movements of air. Large horizontal transfers of air—and the latent heat contained in water vapor in the air—are major components of Earth's energy system. Horizontal transfer such as this is called **advection.** Heat is advected from tropical areas toward the poles when warm winds blow poleward. Heat in ocean currents moving toward the polar regions is another form of advection.

Let us look at a small-scale example. During one day's time, warm air rising over the land allows cooler

Critical Thinking
CLIMATES IN URBAN AREAS

Climates in major urban areas usually differ markedly from those of surrounding rural landscapes. Localized climate characteristics caused by topographic and land-surface characteristics are called microclimates. Typically urban **microclimates** are warmer, both in winter and in summer. The area of warm microclimates associated with an urban area is called the **urban heat island.** Several factors contribute to these differences. First, cities lack large expanses of vegetation and soil that absorb rainfall and later allow it to evaporate. Evaporation cools the land surface, and without this evaporation the surface remains warmer. Second, the walls of buildings form vertical surfaces that help to trap solar energy during the daytime. Concrete, asphalt, and similar building materials store this heat, and at night these warm surfaces shield the ground between buildings from exposure to a large expanse of relatively cool sky and thus keep it warmer than it would otherwise be. Third, air pollutants may absorb heat and keep the air warmer. Finally, heat released from energy use, such as in heating buildings, adds to the heat available in the city (although this amount is very small in comparison to the amount of energy derived from the Sun).

Questions

1. In your experience, what parts of a city tend to be warmer, and what parts tend to be cooler? For example, in the summer are you cooler in the shade of a skyscraper or in the shade of a tree in a park?

2. What kinds of structures or land uses in cities contribute most to making urban climates different? How could cities be designed differently to reduce excess heating?

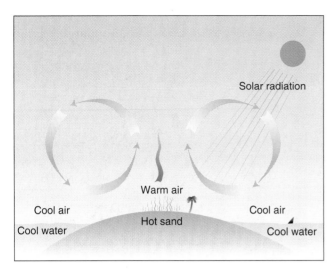

FIGURE 2-9 Convection in the atmosphere. Convection occurs when air is warmed from beneath, expands, and rises. Cooler air descends to replace it. A common example is where land (such as an island, the Florida peninsula, or Cape Cod) is surrounded by water. Convection is strongest in the daytime when the surface is warmed by the Sun. Because of a stronger convection process, the wind tends to be gustier in the daytime than at night.

air from the sea to advect over the land. In other words, wind in some coastal areas blows from the sea toward and over the land. On a large scale, during summertime over Eastern Europe and Central Asia, the season's heating causes winds to blow onto the land from over oceans to the south and east. During winter the land area cools more than the sea, so cold air sinks over the land, warmer air rises over the ocean, and the winds tend to blow from Central Asia toward the warmer Pacific Ocean. Such a circulation pattern is called a monsoon; we will discuss this later in the chapter.

PRECIPITATION

People rely on precipitation to be "normal." Normal rains and snowmelt are necessary for consistent agriculture, to feed Earth's 6.3 billion humans. People who live on lowlands along rivers rely on "normal" precipitation to keep the river within its banks and out of their basements. All plants and animals are adapted to a "normal" amount of moisture for their environment. However, "normal" does not always happen.

From 1998–2002 the southeastern U.S. suffered a severe drought. Reservoirs were drained, domestic water use was restricted, agricultural production was reduced. Much of the U.S. Midwest suffered a severe drought in 1988 that heavily damaged crops and lowered the level of the Mississippi River so much that barges became stranded. In 1990 and 1993 the same regions of the Midwest experienced record wet years, with floods on the Mississippi, Missouri, and other rivers.

Such extremes of precipitation are rare during a typical human lifetime, but they actually occur quite frequently viewed over thousands or millions of years. And at any given time, while most places are experiencing "normal" precipitation, someplace on Earth is experiencing unusual precipitation. Because variations in the amount of rainfall sooner or later affect us all, unusually wet or dry periods make headlines.

Condensation

Precipitation is both a part of energy flows in the atmosphere and a consequence of them. Recall the "conveyor belt" of rising air that carries water vapor, with its latent heat, high enough to become cooled, condense, and cause precipitation. Precipitation is part of the flow of energy, releasing heat in the atmosphere and returning the liquid water back to Earth where it can absorb more heat. To understand this, let us look at the process of **condensation,** the conversion of water from vapor to liquid state.

Air contains water in gaseous or vapor form. Air may hold very little water vapor, as in dry desert air, or it may be filled with water vapor, as in a steamy jungle. The ability of air to hold water-vapor molecules is limited, and this limit varies according to air temperature. We measure the water-vapor content of air by the pressure that the water molecules exert. The maximum water vapor that air can hold is called the **saturation vapor pressure.** Precipitation begins when the pressure of moisture in the air exceeds the saturation amount.

Relative humidity tells us how wet air is. **Relative humidity** is the actual water content of the air expressed as a percentage of how much water the air could hold at a given temperature. For example, if a 30° C (86° F) sample of air contains half the water vapor that it could hold at that temperature, its relative humidity is 50 percent. But if cooled to 22° C (71° F), that same air sample with the same amount of water vapor would be three-fourths saturated, at 75 percent relative humidity. And if cooled to 15° C (60° F), that same air sample would be saturated, at 100 percent relative humidity. If saturated air at 0° C (32° F) is heated to 22° C (71° F), its relative humidity drops to less than 30 percent. This is why indoor air in winter is so dry. Relative humidity tends to fluctuate daily with changing air temperatures, because the amount of water vapor in the air holds fairly steady while temperature rises and falls. Relative humidity typically is lower in the warm afternoon and higher at nighttime.

When air cools, its relative humidity rises, and if the air is cooled enough, it can become saturated. If cooled still further, it becomes supersaturated and contains more water than it can hold in a vapor state. Condensation results from supersaturated air.

Condensation in the atmosphere produces clouds. Clouds form when water-vapor molecules condense around tiny particles of dust, sea salt, pollen, and other particles. The clouds contain small droplets of liquid water or particles of ice—depending on temperature—that are too light to fall. But if these water droplets or ice particles continue to grow as more moisture from the air condenses on them, they eventually become too large to be supported by air currents, and they fall to the surface as rain or snow.

Causes of Precipitation

The movement of air causes precipitation in three ways (Figure 2-10):

- Convection—in which air warmer than its surroundings rises, expands, and cools by this expansion.
- Orographic uplift—in which wind forces air up and over mountains.
- Frontal uplift—in which air is forced up a boundary (front) between cold and warm air masses.

Convectional precipitation   On a warm, humid summer day, the sky is clear in the morning, and the Sun is bright. The Sun warms the ground quickly, and the air temperature rises. Most of the warming of the air takes place close to the ground, because the humid air is a good absorber of longwave radiation, which is being reradiated from the ground. As the air near the ground warms, it expands, becomes less dense, and rises through the surrounding cooler air above, like a hot air balloon. Convection is in progress, conveying humid air higher into the atmosphere.

Because air is a gas, it is compressed by the weight of overlying air. When it rises, air has less weight above it, and the lower pressure allows the air to expand. Compressing a gas causes an increase in temperature, whereas expanding it causes a decrease. The decrease in temperature that results from expansion of rising air is called **adiabatic cooling**; the word *adiabatic* means "without heat being involved."

As air rises, it rapidly cools adiabatically (by expansion) at a rate of about 1°C for each 100 meters (1.8° F per 330 feet) elevation. When the air reaches several hundred meters above the ground, it has cooled to the point where it is saturated, and clouds begin to form (Figure 2-11). Rising warm air may mix with the surrounding cooler air, slowing the convection. If the convection were driven only by heating at the ground surface, it would probably not be strong enough to cause water droplets to grow big enough to fall as precipitation.

But as soon as condensation begins, latent heat—another important source of energy—is released. Latent heat further warms the rising column of air and makes it less dense than the surrounding cool air. The process is self-reinforcing, as the cloud grows rapidly, with strong vertical motion and rapid condensation.

(a) Convectional Precipitation

(b) Orographic Precipitation

(c) Frontal Precipitation

FIGURE 2-10 Causes of precipitation. (a) Convection, (b) orographic uplift, and (c) fronts.

The result is often gusty winds and intense rain: a thunderstorm. Thunderstorms are a common example of intense convectional storms (Figure 2-12).

Convectional storms are responsible for a large portion of the world's precipitation (Figure 2-13). In

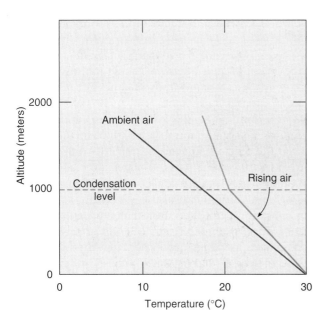

FIGURE 2-11 Adiabatic cooling. As air rises, it cools at a rate of about 1° C per 100 meters (1.8° F per 330 feet). The temperature of the rising parcel of air relative to the temperature of the air around that parcel determines its behavior. If the parcel is warmer than surrounding air, it tends to rise and therefore is unstable. If it is the same temperature or cooler, it resists movement and therefore is stable. In this example, the air parcel is warmer, and thus unstable. Condensation in the air above 1,000 meters (3,300 feet) elevation causes the air to be warmed by the release of latent (or stored) heat, warming the air. This further contributes to the instability of the air. Intense thunderstorms can form under such conditions.

FIGURE 2-12 A convectional storm over the Nevada desert. Storms such as this are most common in summer. (William Belknap Jr./Photo Researchers, Inc.)

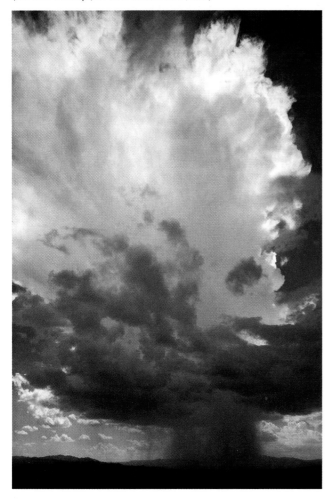

FIGURE 2-13 Clouds revealing convection over the South Florida peninsula. The view is to the south. The land is warmer than the adjacent Atlantic Ocean and Gulf of Mexico, and this warmth stimulates convection and cloud formation. Note the absence of clouds over Lake Okeechobee, the large lake in the center of Florida, because it is cooler, like the ocean. (NASA SS/Photo Researchers, Inc.)

tropical climates, where strong insolation makes temperatures high, all that is needed for intense daily convectional storms is a source of humidity. In midlatitude climates, such storms occur mostly in the summer, because higher temperatures allow the air to hold more moisture, and this means more latent heat can be released, causing strong convection. They are especially common in situations where some other factor is present to favor uplift, the trigger that starts the self-reinforcing growth of the storm. Mountains and fronts, especially cold fronts, often provide the trigger.

Orographic precipitation   Precipitation sometimes occurs when the wind forces air to rise over mountains. This is called **orographic precipitation.** As the air rises, it cools adiabatically (by expansion), the cooling causes condensation, and precipitation results. After air has moved up the windward side of a mountain and over the top, it then descends on the leeward side. As it does so, its relative humidity drops significantly. The leeward side of a mountain range is often much drier than the rainy windward side. A dry region on the leeward side of a mountain range is called a rain shadow. Some of the world's major deserts are arid because they are situated on the leeward side of a mountain range (Figure 2-14).

The western United States provides an excellent example of orographic effects on rainfall (Figure 2-15). Moist air from the Pacific Ocean, the region's major source of moisture, travels from west to east, producing precipitation as it rises first over coastal mountain ranges in Washington, Oregon, and California and then the Sierra Nevada and Cascade ranges. Some of the highest average U.S. rainfall totals occur in the

FIGURE 2-14 The Sierra Nevada mountains in California, seen from the east. Because of orographic effects, the mean annual precipitation at the crest of the Sierras exceeds 100 centimeters (40 inches). But in the foreground, Owens Valley is in a rain shadow and has mean annual precipitation of less than 30 centimeters (12 inches). (Michele Burgess/Corbis/Stock Market)

Sierras and Cascades; but east of the mountain ranges lies the rain shadow, causing dry conditions in eastern Washington, Oregon, Nevada, and Utah. The region is arid because it is cut off from Pacific moisture by the Sierras and Cascades and from Atlantic moisture by the Rocky Mountains to the east.

Orographic precipitation also occurs in the Rocky Mountains, but the amount is less than in the Sierras because the Rockies are more isolated from moisture sources. In the eastern United States and Canada, the Gulf of Mexico and Atlantic Ocean supply ample moisture for precipitation east of the Mississippi Valley. The Appalachians receive more rain than adjacent lowlands because of orographic effects, but the eastern side of the Appalachians does not experience a rain shadow because the region has moisture sources on both sides: the Gulf of Mexico to the southwest and the Atlantic Ocean to the east.

Frontal precipitation   Frontal precipitation forms along a **front,** which is a boundary between two air masses. An air mass is a large region of air—hundreds or thousands of square kilometers—with relatively uniform characteristics of temperature and humidity. An air mass acquires these characteristics from the land or water over which it forms.

In North America, air masses that form over central Canada tend to be cool (because of Canada's relatively high latitude) and dry (because of the region's isolation from moisture sources). This is called a continental polar air mass. In contrast, air masses that form over tropical water, such as over the Gulf of Mexico, tend to be warm and moist and are called maritime tropical air masses.

When a cool air mass, such as continental polar air from Canada, meets a warm air mass, such as maritime tropical air from the Gulf of Mexico, a boundary or front may form between them. Because cool air is relatively dense, it tends to move under less dense warm air, while warm air tends to rise over cool air (Figure 2-16). Fronts are regions characterized by ascending air, cloudiness, windiness, and precipitation.

As air masses migrate across Earth's surface, the fronts move with them. When a cold air mass advances against a warmer one, the boundary is called a **cold front.** The advancing cold, denser air wedges beneath the warm air, forcing it to rise, generating clouds and usually precipitation (Figure 2-16a). A cold front can move quite fast and generate intense thunderstorms if the warm air is sufficiently moist.

The passage of a well-developed cold front in central and eastern North America includes very distinctive weather. Before the front arrives, the air is warm and moist, with the wind typically from the south or southwest. As the front arrives, intense precipitation falls, and in the summer thunderstorms usually form. As the front passes, the wind shifts to the west or northwest and the temperature falls quickly. Then the sky

(a)   (b)

FIGURE 2-15 Topography and precipitation in Oregon. (a) A shaded relief map showing the north-south-trending Coast Ranges and Cascades separated by the Willamette Valley in western Oregon, and plateaus of eastern Oregon. (b) Mean annual precipitation. Note that precipitation is highest in the Coast Ranges and Cascades, somewhat lower in the Willamette Valley, and quite low in eastern Oregon. The high rainfall in the mountains is due to orographic lifting of air as it moves eastward off the Pacific Ocean. By the time this air reaches the plateau east of the Cascade Range most of the moisture has been removed, and rainfall there is a fraction of that received along the coast. (*Atlas of Oregon,* 2nd ed., Copyright 2001, University of Oregon Press)

FIGURE 2-16 Fronts are boundaries separating warm and cold air. A cold front (a) generates precipitation as the cold air drives under the warm air like a wedge, lifting the warm air. This rapid vertical motion of air produces the tall, puffy cumulous clouds characteristic of a cold front. In a warm front (b), warm air rides up over cold air, causing precipitation to fall from broad, flat layers of stratus clouds. (From Christopherson, R. W., *Geosystems,* 5th ed. Upper Saddle River, NJ: Prentice Hall, 2003)

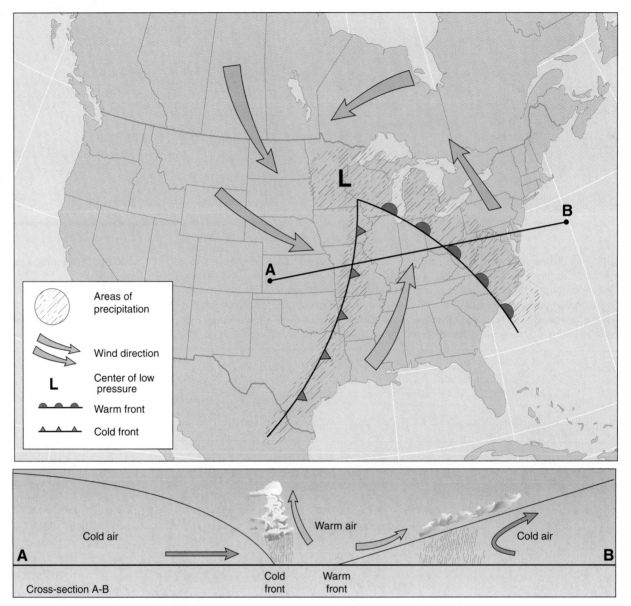

FIGURE 2-17 Patterns of winds, clouds, and precipitation around an idealized large midlatitude storm. The map view shows the position of the fronts at the ground surface. The cross section shows the locations of the fronts aloft.

clears, and cool dry air arrives, with blue skies and bright sunshine.

When, on the other hand, a warm air mass advances against cooler air, the boundary is a **warm front.** The warm air rides up over the cool air as though the cool air were a gentle ramp (Figure 2-16b). Precipitation along a warm front is less intense than along a cold front, and a warm front often passes without significant precipitation. In winter, a warm front sometimes causes freezing rain, as rain formed when the warm air aloft falls through colder air below, freezing as it reaches the ground. Air masses and the front between them may become stalled and not move, creating a stationary front. Note the cold and warm fronts depicted on the weather map (Figure 2-17).

Because a cold front normally involves rapid vertical motion of warm and relatively humid air, the clouds that form along it are typically cumulus clouds, which are tall, puffy clouds with billowy tops. In warm fronts, the warm air rises gradually over the cold air, and the stable layering of warm air over cold produces broad, flat layers of clouds known as stratus clouds.

CIRCULATION PATTERNS

Air is made of molecules, so air has mass. Earth's gravity attracts this mass of air to the surface, giving it weight. We think of air as a lightweight substance, but the thickness of air over your head has considerable weight,

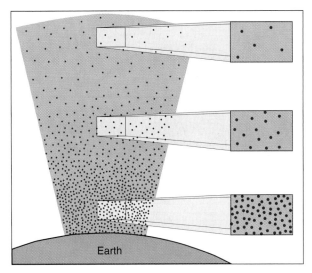

FIGURE 2-18 **Density and pressure of air.** Visualizing the density of air (greater near Earth's surface) and atmospheric pressure. (From Christopherson, R. W., *Geosystems,* 5th ed. Upper Saddle River, NJ: Prentice Hall, 2003)

pressing on you and Earth's surface at an average of about 1 kilogram per square centimeter (14.7 pounds per square inch). This means that directly above each square centimeter of Earth is a column of air over 480 kilometers (300 miles) tall that weighs on average about 1 kilogram (Figure 2-18).

Atmospheric pressure varies with altitude because the higher you go, the less air exists above you. Thus, atmospheric pressure is greater at sea level than in "mile-high" Denver or atop Mount Everest (8.8 kilometers or 5.5 miles high). Because sea level is a surface that is at virtually the same height worldwide, scientists use the average atmospheric pressure at sea level as a world standard. Atmospheric pressure is measured with a *barometer.* The average atmospheric pressure at sea level read on a barometer is 1,013.2 millibars (a force that supports a column of mercury 29.92 inches tall in a mercury barometer).

Pressure and Winds

At any location, atmospheric pressure varies with conditions. For example, the air above the island in Figure 2-9 is warmer than the air around it, and therefore it is less dense. Because warm air is less dense, it weighs less, and the atmospheric pressure over the island is lower than the pressure over the cool sea. As the warm, lighter air over the island rises, higher pressure over the cool sea forces air horizontally toward the island to replace the rising warm air. The difference in pressure between the two places is the *pressure gradient.*

Differences in pressure produce wind. In Figure 2-19, if pressure at the surface is high, air descends, and winds blow away (diverge) from the area of high pressure. But if pressure at the surface is low,

air ascends vertically, and horizontal winds converge toward the area of low pressure.

Coriolis effect If Earth did not rotate, winds would simply blow in a straight line from areas of high pressure to areas of low pressure. However, on our real spinning planet, winds follow an indirect, curving path. This deflection of wind (and any other object moving above Earth's rotating surface) is called the **Coriolis effect.**

To understand Coriolis effect, imagine two people throwing a ball back and forth on the back of a flatbed truck (Figure 2-20a). Both people and the ball are moving at the same speed and in the same direction, so they do not change position relative to one another. The ball game is the same as if the truck were not moving.

On the merry-go-round (Figure 2-20b), when the ball is thrown in a straight line, the turning platform moves the other person away from where the ball arrives. As the two people throw the ball back and forth, the ball follows its expected straight-line path, but the receiver has been moved by the merry-go-round. To the players, it appears that the ball has followed a curving path. Similarly, as Earth rotates beneath moving air, the air appears to move in a curve. This deflection of motion is only apparent to those on the spinning surface. Viewed from above, the path would be straight.

Near the equator, Earth is like the flatbed truck in Figure 2-20a; the Coriolis effect is zero at the equator. But toward the poles, rotating Earth is more like a spinning disk, and the Coriolis effect is strongest in the polar regions. This is shown in Figure 2-20c, where a plane flies from the North Pole in a straight line toward the equator but arrives far to the west of its intended destination because Earth rotated eastward during the hours that the plane was in flight.

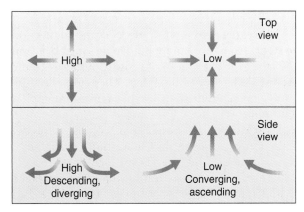

FIGURE 2-19 **Wind movement is partly determined by pressure.** Air descends in regions of high pressure. Air ascends in regions of low pressure. Wind blows from areas of high pressure to areas of low pressure. (From Christopherson, R. W., *Geosystems,* 5th ed. Upper Saddle River, NJ: Prentice Hall, 2003)

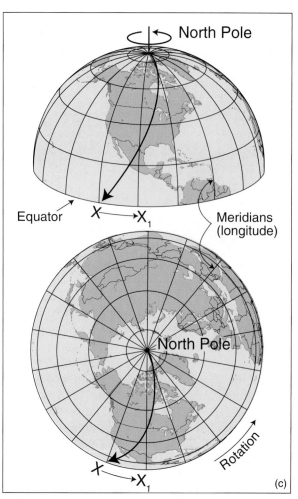

FIGURE 2-20 Coriolis effect. (a) When two people on a moving flatbed truck throw a ball to each other, the people and the ball are moving at the same speed and in the same direction, so they do not change position relative to one another. (b) When two people on a merry-go-round throw a ball to each other, the ball follows a straight-line path, but the people have moved. To the people the ball appears to be following a curve. This apparent curving of the ball demonstrates the Coriolis effect. (c) When an airplane flies in a straight line from the North Pole to the equator, it would actually land west of its intended destination because Earth would rotate eastward during the flight. (From Tarbuck, E. J., and Lutgens, F. K., *Earth Science,* 7th ed. New York: Macmillan, 1994)

The net result of the Coriolis effect is a deflection to the right (from the standpoint of a moving object) in the Northern Hemisphere. In the Southern Hemisphere the deflection is to the left. The effect is significant only when considerable distance is involved: A pilot must consider it in setting the flight path of a several-hour flight across the ocean, but not for a model airplane across a backyard. The Coriolis effect helps weather forecasters understand the behavior of storms.

Figure 2-21 shows the result when the Coriolis effect is added to the vertical air movements in high- and low-pressure areas. It puts a spin on the winds, adding a spiral motion to create cells that rotate clockwise and counterclockwise. In the Northern Hemisphere (Figure 2-21a), high-pressure cells are rotating clockwise, and low-pressure cells are rotating counterclockwise. The opposite patterns hold in the Southern

Hemisphere (Figure 2-21b). To summarize, differences in air pressure create winds, and Earth's spin curves their flow.

Global Atmospheric Circulation

Understanding global circulation patterns is very important. These patterns control our weather, climate, and agriculture. At the global scale, atmospheric circulation operates like the convection system over the warm island we described previously. Beginning at the equator, we can identify four zones of circulation patterns:

1. Intertropical convergence zone
2. Subtropical high-pressure zones
3. Midlatitude low-pressure zones
4. Polar high-pressure zones

Circulation Patterns **63**

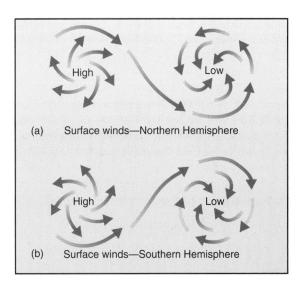

FIGURE 2-21 Wind movement with Coriolis effect added.
(a) The Coriolis effect, caused by Earth's rotation, deflects
wind to the right of its expected path in the Northern
Hemisphere. This causes spiraling circulation around
high- and low-pressure centers, in the directions shown.
(b) The same pattern reversed for the Southern Hemisphere.
Compare with Figure 2-19. (From Christopherson, R. W.,
Geosystems, 5th ed. Upper Saddle River, NJ: Prentice Hall,
2003)

Intertropical convergence zone In the tropics, de-
pendable year-round inputs of solar energy heat the air,
expanding it and creating low pressure. As a result,
convectional rising of air occurs daily above Earth's
equator. This forms the **intertropical convergence
zone (ITCZ),** so called because it is a zone between the
Tropics of Cancer and Capricorn where surface winds
converge (Figure 2-22).

As air rises in the ITCZ, convectional precipitation occurs, usually as afternoon thunderstorms. Air
converges toward the equator at the surface, replacing
the rising air. The Coriolis effect deflects this moving
air to the right in the Northern Hemisphere to form
the Northeast Trade Winds and to the left in the Southern Hemisphere to form the Southeast Trade Winds.
The **trade winds** are so-called because they were very
important to sailing ships. For example, the northeast
trades in the Atlantic carried European ships to the
Caribbean. Aloft, air circulates away from the ITCZ
both northward and southward, toward subtropical latitudes. This air has lost most of its moisture in the daily
rainfalls, and it is now warm and dry.

Subtropical high-pressure zones The warm, dry air
that spreads poleward from the ITCZ descends at about
25° north and south latitudes. This creates zones of
high pressure that are especially strong over the oceans.
These **subtropical high-pressure (STH) zones** are areas
of dry air, bright sunshine, and little precipitation.

This descending dry air associated with subtropical high-pressure cells creates an arid climate, so most

of the world's major desert regions are on land in this
zone, at about 25° north and south latitudes. Further,
most are on the western edges of continents, including
deserts in northern Mexico and southwestern United
States, the Atacama in Peru and Chile, the Sahara in
northern Africa, the Namib in southern Africa, and the
Australian desert. The eastern sides of these continents
are under the same high-pressure cells, but they tend to
be more humid because the circulation brings in warm,
humid tropical air.

On the poleward sides of the subtropical high-pressure cells, circulation is toward the poles. But these winds
are deflected by the Coriolis effect, so winds prevail from
the southwest in the Northern Hemisphere and from the
northwest in the Southern Hemisphere. The prevailing
westerly wind makes an eastbound transatlantic flight in
the midlatitudes normally about an hour shorter than a
westbound flight.

Midlatitude low-pressure zones Poleward of the
subtropical high-pressure zones are the **midlatitude low-
pressure zones.** These lower-pressure areas experience
convergence of warm air blowing from subtropical
latitudes and cold air blowing from polar regions.
The warm and cold air masses collide in swirling low-
pressure cells that move along the boundary between
the two air masses, which is known as the **polar front.**
Winds in these regions generally blow from west to east
and are therefore called westerlies. (Winds are named
for the direction from which they blow.)

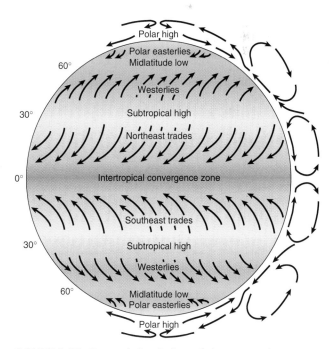

FIGURE 2-22 General circulation of the atmosphere.
This simplified illustration of average atmospheric
circulation patterns shows the major circulation zones by
latitude: intertropical convergence zone (ITCZ), subtropical
high-pressure (STH) zones, midlatitude low-pressure zones,
and polar high zones.

Polar high-pressure zones   In the polar regions, the intense cold caused by meager insolation creates dense air and high pressure. Because the air is so cold, it contains very little moisture, and convection and precipitation are limited. The snow and ice around the poles comes from snow accumulating over many thousands of years.

Seasonal Variations in Global Circulation

Global circulation patterns vary with the seasons. We can illustrate typical conditions during January and July (Figure 2-23).

Global circulation in January   During January, Earth's average atmospheric pressure and winds are arranged into broad zones according to latitude (Figure 2-23a). Conditions in the Northern and Southern hemispheres roughly mirror each other. The ITCZ is generally at about 5° to 10° south latitude.

High-pressure cells predominate in the subtropical regions just north and south of the ITCZ, especially over the subtropical oceans. The Southern Hemisphere, where it is summer in January, displays the

clearest tendency for high pressure over the oceans and low pressure over land. The Northern Hemisphere has a high pressure region in eastern Asia, as well as in the eastern Pacific and Atlantic oceans.

In the midlatitudes, less continuous low-pressure regions in the Northern Hemisphere appear most clearly over oceans. The midlatitude low-pressure zone is much more consistent in the Southern Hemisphere, because of the absence of land between 40° and 70°. A monthly average map obscures the presence of storms, but it is useful to show the general condition of the atmosphere during Northern Hemisphere winter.

Global circulation in July   By July, January's major pressure and wind zones have moved northward. In some cases they have moved from being over land to water areas, or vice versa (Figure 2-23b). The ITCZ, which was south of the equator in January, is almost entirely in the Northern Hemisphere during July, as far north as 30° north latitude in southern Asia. The subtropical highs are still present over the oceans, although they are strengthened in the Northern Hemisphere and weakened in the Southern.

The most notable change in the subtropics and midlatitudes is the replacement of high pressure over

(a) January

FIGURE 2-23 Worldwide average atmospheric pressures (in millibars) and winds in a Northern Hemisphere winter in January and summer in July. (a) In January, the intertropical convergence zone (ITCZ) is generally south of the equator, especially over land areas. In the Northern Hemisphere, the subtropical high-pressure cells are weak or absent, and a large and intense high-pressure cell dominates interior Asia, with deep low-pressure areas over the North Pacific and North Atlantic.

Asia with low pressure. In fact, this region has the world's highest average pressures during January and the world's lowest average pressure during July. As a result, wind directions in the vicinity are reversed. This seasonal reversal of pressure and wind produces a **monsoon circulation,** in which winter winds from the Asian interior produce extremely dry winters in most of south and east Asia, while summer winds blowing inland from the Indian and Pacific oceans result in wet summers. These wet summers are the famous monsoon season of heavy rains in southern Asia.

Seasonal changes also occur in the midlatitudes. The extreme low-pressure regions over the northern oceans in January are replaced in July with generally weak low pressure and inconsistent winds, although storms with strong winds may periodically pass through the regions. In the Southern Hemisphere winter (July), the low-pressure system is especially deep and consistent around the globe because of the absence of large landmasses to break it up as in the Northern Hemisphere. The south polar high is similarly strong in July.

Remember that with the shift in seasons between January and July, solar radiation also changes. During July in the Northern Hemisphere, solar radiation inputs are highest north of the equator, while in January,

insolation is greater south of the equator. Circulation patterns also shift seasonally, bringing strong seasonal contrasts in weather as wind direction and zones of frequent storms move north and south.

Ocean Circulation Patterns

When wind blows over the ocean, it exerts a frictional drag on the sea surface, creating waves and currents (Figure 2-24). Continuing wind adds energy to the waves, building them into larger ones that may travel thousands of kilometers until they break onto a distant shore. The continuing drag of prevailing winds also causes broad currents in the ocean's surface layers. Contributing to these currents are differences in seawater temperature and salinity. These give water different densities, promoting movement of currents from areas of greater density to those of less density. This is similar to the way a high-pressure area in the atmosphere causes wind currents to blow toward a low-pressure area. Both these currents—oceanic and atmospheric—are important in redistributing heat around Earth, a role played by the Gulf Stream, for example (Figure 2-25).

Gyres are prominent features of oceanic circulation. These wind-driven circular flows mirror the

(b) July

FIGURE 2-23 *(continued)* (b) In July, the ITCZ has shifted to the north, and low pressure has replaced high pressure over interior Asia. Weak low pressure also is seen over North America, and the subtropical highs are prominent. In the Southern Hemisphere, the midlatitude low-pressure belt along latitudes 40–60° south is very strong.

FIGURE 2-24 Distribution of global wave heights. A global map of mean wave heights for the period September 25 to October 4, 1992, made from radar altimetry data collected by the *TOPEX/Poseidon* satellite. Magenta areas have wave heights of less than 1 meter (3 feet); red areas have wave heights of 7–8 meters (23–26 feet). Wave heights indicate wind speed. The area of large waves in the southern Indian Ocean results from intense storms at the time these data were collected. Another area of storms is visible in the North Pacific. The calmest areas are in protected waters, such as the Gulf of Thailand and the western Caribbean Sea. (NASA/Science Photo Library/Photo Researchers, Inc.)

movement of prevailing winds (Figure 2-26); you can see this by comparing the oceanic circulation figure with Figure 2-23. Gyres form beneath tropical high-pressure cells. The Gulf Stream forms the western limb of the gyre in the North Atlantic.

Where ocean currents circulate warm water from low equatorial latitudes to higher latitudes, they are carrying heat poleward by advection. Such flows are balanced by cool currents traveling equatorward, most notably along the west coasts of subtropical land areas. Such cold currents cool the lower portions of the atmosphere above them, reducing convection. With less convection occurring, adjacent landmasses may be very dry. Some of the driest areas on Earth owe their aridity partly to this effect, most notably the Atacama desert of Peru and Chile. Occasional shifts in ocean circulation such as **El Niño** (see the focus box on El Niño) can cause significant variations in weather.

Storms: Regional-Scale Circulation Patterns

Variations in circulation like El Niño (ENSO) and the Asian monsoons are examples of recurring changes in the atmosphere. These range in scale from global phenomena like ENSO, to a continent-size land-water convection pattern such as the Asian monsoon, to regional-scale systems such as hurricanes, down to the scale of a local thunderstorm.

Storms are especially important because they are areas of concentrated convection that bring precipitation and sometimes strong winds. On a regional scale, large storms can affect areas hundreds to thousands of kilometers across. Such storms, which are called **cyclones,** are large low-pressure areas in which winds converge in a counterclockwise swirl in the Northern Hemisphere and clockwise in the Southern Hemisphere. We will look at two types of cyclones: tropical cyclones (hurricanes or typhoons) and midlatitude cyclones.

Tropical cyclones are intense, rotating convectional systems that develop over warm ocean areas in the tropics and subtropics, primarily during the warm season. When such storms have wind velocities exceeding 119 kilometers per hour (74 miles per hour), we call them **hurricanes** in North America, **typhoons** in the western Pacific, and *cyclones* in the Indian Ocean.

These storms typically develop in the eastern portion of an ocean within the trade-wind belt. They begin as areas of low pressure (rising air) and converging winds, drawing in warm, moisture-laden air. This humid tropical air contains a great deal of energy, both as sensible

FIGURE 2-25 The Gulf Stream. In this composite of satellite images of the North American East Coast from late May 1996, brown shades indicate warm water, and blue indicates cool water. The strong flow of warm water from southwest to northeast is plainly visible, including swirling eddies in the flow where the warm current moves past cooler water. (Microstock/Photri-Microstock, Inc.)

Oceanic Circulation, Showing Major Currents

FIGURE 2-26 Note how the ocean circulation is clockwise in the Northern Hemisphere and counterclockwise in the Southern Hemisphere, driven mainly by circulation around the subtropical high-pressure zones. (From Christopherson, R. W., *Geosystems*, 5th ed. Upper Saddle River, NJ: Prentice Hall, 2003)

Focus on
EL NIÑO/LA NIÑA

Ocean currents generally are much more consistent than atmospheric winds, but oceanic flows do vary in speed and direction. One of the most important of these variations is called El Niño. The term is Spanish for "the (male) Child," a reference to the Christ child, because the phenomenon occurs around Christmastime. El Niño is a circulation change in the eastern tropical Pacific Ocean that occurs every few years. In this change, the usual cool flow from South America westward is slowed and sometimes reversed, replaced by a warm-water flow from the central Pacific eastward. The counterpart of El Niño, in which especially cool waters are found in this region, is known as La Niña (the female child). The change in ocean currents is linked to a change in atmospheric circulation patterns in the Pacific Ocean known as the Southern Oscillation; the two phenomena together are referred to as El Niño-Southern Oscillation (ENSO).

Peruvian fishermen named El Niño, and they knew its effect: The reversed flow of water deprived fish populations in the eastern Pacific of their food and contributed to the collapse of the Peruvian anchovy indus-

try in the 1970s. ENSO events are far reaching because a modification of circulation in one part of the globe may cause circulation patterns to change elsewhere in North and South America and the Pacific region. For example, El Niño events are linked to flooding in the U.S. Southwest, droughts in Australia, and reduced rainfall in India. La Niña events often bring wet weather in south and Southeast Asia, and dry conditions in the southern United States.

In 1997 a particularly strong El Niño event developed, with ocean temperatures in the eastern Pacific about 5° C (9° F) above non-El Niño levels. This led to greatly increased convection in the eastern Pacific, and rains in normally dry coastal Peru. At the same time, the western equatorial Pacific was cooler than normal, causing drought in Southeast Asia. Forest fires raged out of control in Indonesia, causing severe air-pollution problems in Kuala Lumpur, Malaysia, and intense storms pounded coastal California. By 2000 the pattern had reversed, and La Niña conditions dominated. This, along with warm water in the western Pacific, was a major contributor to the 1998–2002 drought in the southern United States.

Focus on
TORNADOES

Tornadoes are an extreme form of weather created when energy conditions in the atmosphere are such that extremely intense convection occurs. They generate the highest surface wind velocities known to occur on Earth, and the destruction they cause can be near total, though fortunately very localized.

Tornadoes are intense columns of rising air, usually associated with thunderstorms. As the air rises, it creates a partial vacuum (an area of very low pressure) that draws air in toward it. As this air is drawn in, it creates a swirling vortex, much like the swirling motion of water as it drains out of a sink. As the air swirls in, it gains tremendous speed, sometimes exceeding 300 kilometers per hour (185 miles per hour). This vortex usually moves horizontally, sometimes touching the ground and sometimes rising above it, leaving an erratic and usually very narrow path of destruction.

Most tornadoes occur in one of two circulation patterns: in association with intense thunderstorms embedded in hurricanes, and along especially strong cold fronts in the midlatitudes. The south-central United States has the greatest frequency of tornadoes of any place in the world. In the United States they occur mostly in spring and early summer, when contrasts between cold, dry continental air and moist, humid air from the Gulf of Mexico are greatest. Such conditions

Tornado damage in Grandview, Missouri, May 5, 2003. A series of strong storms in the first two weeks of May 2003 generated over 300 tornadoes in the midwestern United States, killing 38 people. (© Reuters News Media Inc./ CORBIS)

create cells of rapidly rising air, including very intense thunderstorms and sometimes tornadoes.

heat and latent heat in the water vapor, but especially latent heat. As this energy is drawn into the developing storm, condensation and the resulting release of latent heat intensifies the convection. The center of low pressure grows more intense, and as wind speed grows, the Coriolis effect causes a spiraling circulation to develop, counterclockwise in the Northern Hemisphere and clockwise in the Southern Hemisphere.

These tropical storms move with the general circulation over the subtropical Atlantic, Pacific, and Indian oceans, from east to west. As they travel across the warm ocean surface, they often intensify. By the time they reach the western portion of the ocean, their atmospheric pressures may be as much as 10 percent lower than average sea-level pressure, with winds exceeding 150 kilometers per hour (93 miles per hour).

Hurricanes thrive on warm, moist air, so they are most intense over ocean areas during the warm season. They lose intensity over land, because they lose

their source of energy. Furthermore, the smooth ocean surface favors development of high winds. In contrast, the hills and trees of land areas slow the wind by friction.

Hurricanes' greatest threat to humans is therefore in tropical and subtropical coastal areas, on the eastern margins of the continents and in southern Asia, where monsoon circulation draws air northward from the Indian Ocean. When a hurricane strikes land, the combination of intense wind and extremely low pressure causes a **storm surge,** an area of elevated sea level in the center of the storm that may be several meters high. The surge carries large waves crashing inland onto low-lying coastal areas, with devastating results. In fact, the majority of hurricane deaths and damage result from storm surges. Adding to this hazard, tornadoes commonly are spawned as a hurricane comes ashore.

Perhaps the coastline most vulnerable to tropical cyclones is on the Bay of Bengal, including Bangladesh and West Bengal in eastern India. The land in this area

Tornado Risk in the United States

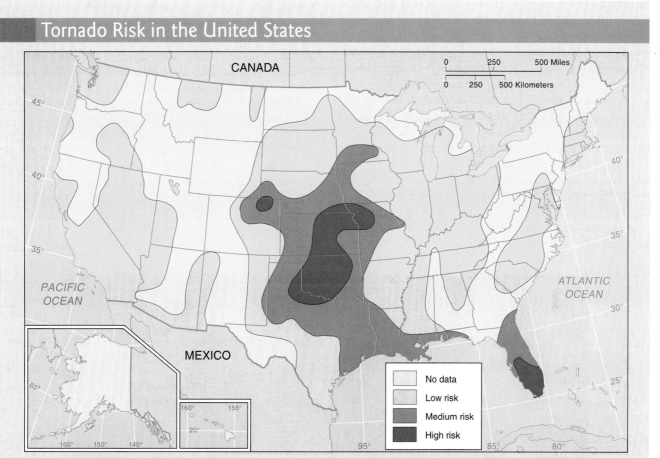

Tornado risk in the United States is greatest in the southern plains, the Southeast (especially central Florida), and the Midwest. (Alaska and Hawaii not to scale.)

is low (elevation less than 2 meters or 6 feet), consisting mostly of the Ganges River delta, and it is densely populated. Bangladesh has millions of people and very limited resources for coping with periodic storms. Although satellite-based warning systems help evacuate people in advance of approaching storms, transportation is difficult, and people often are unable to reach high ground. In recent decades, several severe storms struck the area, including one in 1970 that killed about 300,000 people.

In midlatitude areas another type of storm dominates the weather: the midlatitude cyclone. The midlatitude cyclone has little in common with the tropical cyclone (hurricane), for it occurs in cooler, less humid latitudes that cannot provide the immense volume of heat and water vapor needed to generate a hurricane. **Midlatitude cyclones** are centers of low pressure that develop along the polar front. They move from west to east along that front, following the general circulation in the midlatitudes. Midlatitude cyclones are usually much less intense than hurricanes but much more common.

In a midlatitude cyclone, air is drawn toward the center of low pressure from both the warm and the cold sides of the polar front. Where warm air is drawn toward cold, typically on the eastern side of the storm, a warm front develops. On the western side, the spiraling motion causes cold air to drive under the warm air, forming a cold front. As the center of low pressure moves eastward, these fronts move with it, bringing precipitation to areas over which they pass. The passage of a front at the surface usually is marked by a significant change of temperature, precipitation, and shifting winds. The repeated passage of such storms creates highly variable weather conditions, with alternating cold and warm air as the polar front moves back and forth across the land. Occasionally, especially in North America, strong cold fronts associated with midlatitude cyclones produce tornadoes (see the focus box on Tornadoes).

The Weather on August 25, 1998

Global and regional circulation patterns just described are visible in satellite images of Earth, such as the one shown in Figure 2-27. These patterns may not be obvious at first glance, but a trained eye can identify wind directions, fronts, and precipitation patterns even in a satellite image. Here they are shown with appropriate weather-map symbols on the image. The image shown in Figure 2-27 shows most of the Western Hemisphere in late summer. In this view of Earth, from about 22,000 miles out in space, we see a little less than half the globe. The South and North Poles are not visible, but the view does extend to almost 80° latitude, north and south.

The intertropical convergence is visible as a broken line of cloudy areas extending across the North Pacific at about 10° N, continuing east across the Caribbean Sea. The ITCZ consists mainly of convective storms, some of which are scattered and some of which are concentrated in areas of low pressure. On the northern margins of the ITCZ we see three tropical cyclones. One is in the early stages of formation in the tropical Atlantic. Another, named Hurricane Bonnie, is slamming into the North Carolina coastline. A third is visible near the left-hand edge of the image in the tropical North Pacific.

In the midlatitudes we see sweeping curves of clouds, many of which show the locations of fronts. In August it is winter in the Southern Hemisphere, and two particularly well-developed storms are visible there. A third can be seen in the North Atlantic, approaching Europe.

The atmosphere is continually in motion, and on any given day the atmosphere contains many storms, some intense and some mild. This circulation occurs in the vertical dimension as well as horizontally; areas of rising air are cloudy and may produce precipitation, while areas of sinking air are cloud-free. The locations of storms and high-pressure cells change from day to day, producing weather variability from one day to the next.

FIGURE 2-27 Weather pattern on Earth. Full-disk image of Earth, August 25, 1998. See the text for description of patterns shown in this image. (National Climate Data Center/National Oceanic and Atmospheric Administration/Seattle)

CLIMATE

The atmospheric circulation patterns we see on any given day are the result of energy distribution patterns occurring at that time, and these change rapidly from hour to hour and day to day. Geographers also study temperature, precipitation, and other atmospheric phenomena that persist for much longer periods than a thunderstorm or a heat wave.

Climate is the summary of weather conditions over several decades or more—a place's weather pattern over time. The vegetation, natural resources, and human activities that characterize a particular region of Earth are heavily influenced by its climate. At first glance, describing the climate of a place may appear to be a straightforward task of calculating annual averages of temperature and precipitation. In reality, climate is extremely difficult to describe. Air temperature and precipitation change not only day to day but year to year, and other factors such as windiness, humidity, and occurrence of storms must also be considered. We will discuss this in detail later in the chapter.

> Climate is the summary of weather conditions over several decades or more—a place's weather pattern over time.

Climate changes over time. Earth has alternated between warming periods and cooling periods. Glaciers have covered large areas during cool periods, and they have melted during warm periods. The human population has survived by adapting to the alternating periods of warming and cooling. Ironically, human activity now may be causing climatic changes that will fundamentally alter the character of the planet.

People are affected by various elements of climate. The number of hours of sunshine in a day and the degree of cloudiness are important to anyone who is active out of doors—gardeners, farmers, hunters, builders, vacationers at the beach—and residents of homes heated by solar energy. Amounts of rainfall and snowfall make great differences in how we build our homes and roads from place to place. The windiness of a place is important in designing structures and harnessing wind power. Pollutants in the air can have both short-term and long-term effects on human and animal health. Counts of pollen and mold spores are important to those who suffer allergies.

Air Temperature

To understand a region's distinctive climate, the two most important measures are air temperature and precipitation. In everyday speech, we refer to "hot or cold climates" and "wet or dry climates." The most obvious differences in air temperature on Earth are those between the tropics and the poles and those between winter and summer. These variations are caused by latitudinal and seasonal variations in solar energy inputs.

Air temperature also varies with elevation. This variation occurs because the atmosphere is warmed by longwave radiation from Earth's surface and because of the adiabatic cooling of rising air described earlier. A very general rule of thumb is that temperature drops about 6.4° C per 1,000 meters (3.5° F per 1,000 feet). This is why mountain regions are cooler than adjacent lowlands (Figure 2-28). Air temperature also varies considerably within a few meters of the ground surface. In a desert the ground surface temperature can reach 70° C (160° F) during the daytime, while air that lies 2 meters above the ground rarely exceeds 55° C (130° F). At night the desert ground may be significantly cooler than the air above.

To minimize the effects of the ground surface on temperature records, climatologists usually measure air temperature at a height of about 1.4 meters (55 inches) inside a ventilated shelter that shades the thermometer and other instruments from direct sunlight. Climatologists at these weather stations usually report high and low temperatures for each day and calculate the average daily temperature as the average of the two readings.

Temperature patterns also show the effects of topography, proximity to the ocean, and moisture availability (Figure 2-29).

In both summer and winter, the higher elevations of the Rockies and Appalachians are cooler than adjacent lowlands. At very high elevations temperatures are so low that year-round snow cover is possible. Permanent snow occurs at an elevation above approximately 4,000 meters (13,100 feet) in the southern Rockies and as low as 1,000 meters (3,300 feet) in the Northern Rockies and the mountains of Alaska.

The smallest seasonal differences in temperatures in North America are found in areas near the ocean, especially along the Pacific Coast. Westerly winds bring

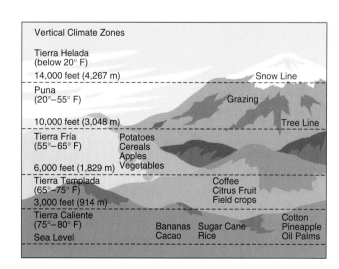

FIGURE 2-28 Vertical climate zones. Throughout much of Latin America, elevation defines these vertical climate zones. Distinct crops characterize each zone.

Average January Temperatures

(a)

FIGURE 2-29 Average temperatures (°C) in North America during January and July, based on data collected from 1931–1960. (a) In January, the lowest temperatures occur in the interior of the continent and in high-altitude areas. Higher temperatures occur in the South and along the West Coast. (b) In July, the highest temperatures occur in the interior of the continent, especially in the desert southwest. Temperature differences between northern and southern

air from ocean areas to coastal regions in the west, while on the Atlantic Coast the airflow is normally from the continent, so seasonal temperature variations are greater. The effects of moisture availability and sunshine on temperature are visible in the deserts of southwestern United States and northwestern Mexico, where generally clear skies allow ample sunshine, but water is not available to dissipate that heat through evaporation. The highest recorded temperatures in North America occur in this region.

Precipitation

The amount of precipitation is extremely variable between places and over time. A thunderstorm can unleash heavy rain in one place and a light sprinkling just a short distance away. Worldwide, precipitation generally ranges from virtually none in some desert areas to more than 300 centimeters (120 inches) in tropical areas. A few tropical mountain areas have recorded more than 10 meters (33 feet) of rainfall per year.

Precipitation usually is measured by collecting rain or snow in a cylindrical container that is marked in millimeters or hundredths of an inch. Snowfall normally is recorded as an amount of liquid water rather than the depth of snow accumulated on the ground, because the amount of water in snow varies widely with the snow's texture. Weather stations typically report daily precipitation, although the instruments are capable of recording the amount for other periods of time, such as a minute or an hour.

Figure 2-30 shows worldwide average precipitation. Note how the levels correspond to global circulation patterns: the rainy ITCZ around the equator, the midlatitude high-pressure areas with their dry air, the mountain rain shadow areas, and the dry polar regions.

Average rainfall amounts tell us much about the climate of a region, but there is also much they do not tell. The timing and reliability of rainfall are equally important. In midlatitude areas precipitation frequently takes the form of gentle rainfall, whereas in tropical regions it may come in torrential and destructive downpours. In northern Nigeria, for example, 90 percent of all rain falls in storms of more than 25 millimeters (1 inch) per hour. That is, by way of comparison, half the average monthly rainfall of London,

> Average rainfall amounts tell us much about the climate of a region, but there is also much they do not tell.

Average July Temperatures

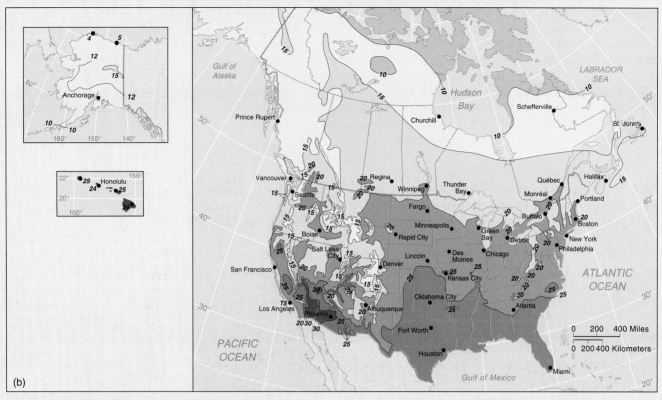

(b)

FIGURE 2-29 *(continued)*
portions of the United States are less than in January, and storm systems traveling along the polar front are less intense as a result. Be careful when trying to interpolate between temperature lines, because large changes may occur over short distances, depending on altitude, slope, soil type, vegetation, and urban development. (Alaska and Hawaii not to scale.)

Average Annual Precipitation Worldwide

cm	in.
200 and over	80 and over
105–199	60–79
100–149	40–59
50–99	20–39
25–49	10–19
Under 25	Under 10

MODIFIED GOODE'S HOMOLOSINE EQUAL-AREA PROJECTION

FIGURE 2-30 Generally, climates with higher temperatures require greater rainfall to maintain vegetation. (From Christopherson, R. W., *Geosystems*, 5th ed. Upper Saddle River, NJ: Prentice Hall, 2003)

England. In Ghana, cloudbursts regularly occur at a rate of 200 millimeters (8 inches) per hour, four times London's monthly total. In Java, an Indonesian island, a quarter of the annual rainfall comes in downpours of 60 millimeters (2.4 inches) per hour, more than Berlin, Germany, gets in an average month.

In sudden, intense storms a greater portion of the rainfall runs off the land rather than soaking in, sometimes increasing soil erosion. Studies in the West African nation of Burkina Faso found that in one year nearly 90 percent of all erosion took place in just six hours.

The regularity of rainfall is also important. In Western Europe and eastern North America, rainfall does not vary much from month to month. New York's rainiest month receives only one and a half times as much rain as its driest month, London's rainiest month is two times greater, and Berlin's is two and a half times greater. Delhi receives about the same total annual rainfall as London, but Delhi, by contrast, receives only 10 millimeters (0.2 inch) of rain in November and more than 175 millimeters (7 inches) in both July and August. Zungeru, in central Nigeria, gets 54 times as much rain in the wettest month as in the driest.

Soil can absorb and hold only a limited amount of water for crops to use. Think of this as a bank account deposited in the wet season and drawn on in the dry season. Although many tropical countries get what seems to be adequate annual rainfall, they get it in a lump sum, like a huge win at a casino. Most is squandered, and the little that is left in the soil bank is used up within a few months. For example, in Agra, India, where the Taj Mahal is located, rainfall exceeds the current needs of vegetation only in July and August. As the rains decrease in September, the water stored in the soil is used up in just three weeks, leaving nine months of the year in which water supplies are inadequate.

Human societies can adjust to variations in rainfall if these variations are regular or predictable. Irrigation can help even out the supply of water over the growing season. In the Sahel, south of the Sahara, farmers grow fast-maturing crops such as sorghum and millet that shoot from seed to ripened ear in three or four months.

It is much more difficult, however, to cope with rains that come irregularly from year to year. In Europe and wetter portions of North America, the amount of annual rainfall varies by less than 15 percent per year on average. In many of the tropical and subtropical regions, though, it fluctuates from 15 to 20 percent, and in the semiarid and arid lands, by up to 40 percent or more. In the lean years, crops fail. If the rains are late, the growing season is cut short and yields are greatly diminished. If imported food is not available, famine can follow. That is why on the Indian subcontinent the monsoon is awaited with such hope and trepidation, and a delay of a week or two causes panic.

Water availability   Evaluating water availability involves much more than just measuring rainfall. Plants demand water and thus play a key role in what happens to the precipitation. Plants take water up through their roots and evaporate it through their leaves, releasing it into the atmosphere as water vapor, a process called **transpiration.** Therefore, most climate classification systems consider precipitation in relation to what a region's vegetation needs. Plants consume very large amounts of water, totaling about two-thirds of all precipitation that falls on Earth's land areas. A single tree may remove 100 or more liters of water from the soil in a single day. The amount of water that could evaporate from a damp soil or be transpired by plants if all of their needed moisture were available is called potential evapotranspiration (POTET; see Chapter 4). POTET indicates the moisture demand that plants make on their environment. Evapotranspiration is a fundamental part of the hydrologic cycle, which will be discussed in detail in Chapter 4.

Because heat energy must be available to evaporate water, warmer climates make possible a greater POTET. In tropical climates, where temperatures typically exceed 20° C (68° F) throughout the year, annual POTET can exceed 150 centimeters (60 inches). In contrast, in midlatitude climates, annual POTET totals 50 to 90 centimeters (20 to 35 inches), and annual POTET is only a few centimeters in polar regions.

In distinguishing between arid and humid climates, measurement of potential evapotranspiration is important. Parts of tropical East Africa and central Minnesota both receive the same average annual rainfall of about 80 centimeters (30 inches). Yet trees are sparse in East Africa and abundant in Minnesota. The reason is that East Africa is hotter and therefore has a much greater potential evapotranspiration than central Minnesota. East Africa's POTET exceeds its annual precipitation, so it cannot support as many trees.

CLASSIFYING CLIMATE

Classifying climate types is useful for many purposes. Knowing which places have certain climates allows analysis and planning by climatologists, geographers, geologists, social scientists, government, and industry. Describing and naming specific climate types gives us vocabulary to use in describing places and communicating such information to others.

Most climate classifications combine information about temperatures and precipitation to take into account the effects of temperature on water availability for vegetation. One example of such a classification is one devised in 1918 by German geographer Wladimir Köppen (Table 2-1 and Figure 2-31; see also Appendix II). Köppen used the distribution of plants to help him draw boundaries between climate regions. He identified five basic climate types and subdivided each to

TABLE 2-1 Major Climate Types and Their Köppen Equivalents

Climate Type	Climate Characteristics
Humid tropical	
Af	Tropical, constantly warm and humid, with no dry season
Am	Tropical, constantly warm and humid, but with a short dry season
Seasonally humid tropical	
Aw	Tropical, constantly warm and humid, but with a pronounced dry low-Sun season and wet high-Sun season
Desert	
BWh	Hot desert climate
BWk	Cool desert climate
Semiarid	
BSh	Hot semiarid (steppe) climate
BSk	Cool semiarid (steppe) climate
Humid subtropical	
Cfa	Humid, warm subtropical climate, with hot summers and no dry season
Cw	Humid, warm subtropical climate, with hot summers and dry winters
Marine west coast	
Cfb	Marine west coast climate, with warm summers and no dry season
Cfc	Marine west coast climate, with cool summers and no dry season
Mediterranean	
Cs	Mediterranean climate, with dry, warm summers and cool, wet winters
Humid continental	
Dfa	Humid continental climate, with hot summers, cold winters, and no dry season
Dwa	Humid continental climate, with hot summers and dry, cold winters
Dfb	Humid continental climate, with warm summers, cold winters, and no dry season
Dwb	Humid continental climate, with warm summers and dry, cold winters
Subarctic	
Dfc	Moist subarctic climate, with cool summers, very cold winters, and no dry season
Dwc	Moist subarctic climate, with cool summers and very cold, dry winters
Dfd	Moist subarctic climate, with cool summers, frigid winters, and no dry season
Dwd	Moist subarctic climate, with cool summers and frigid, dry winters
Tundra	
ET	Tundra climate, with very cool, short summers and frigid winters
Icecap and ice sheets	
EF	Ice-cap climate, with temperatures consistently below freezing

reveal important distinctions. A vegetation-based approach has the advantage of defining climates in terms that have more meaning for other environmental features rather than just selecting an arbitrary temperature or precipitation level as the boundary between two climates. Climatologists have modified Köppen's classification many times but have retained its essential elements. It remains the most widely used classification system.

Köppen's scheme uses uppercase letters A, B, C, D, and E to define the major climate types. These letters are assigned in relation to the latitudes at which climates are found, with A climates nearest the equator and E climates toward the poles. A, C, D, and E climates are distinguished primarily by temperature, with A being the warmest, but B climates are distinguished primarily by precipitation. Tropical (A) climates are warm all year. Dry (B) climates have limited moisture. Warm midlatitude climates (C) have cool winters and warm summers. Cold midlatitude climates (D) have cold winters and mild summers. Polar climates (E) are cold all year. In mountainous areas large differences in climate occur over short distances, causing detailed climatic patterns that cannot be shown on a world map. These areas are mapped as H, or highlands.

World Climates

A HUMID LOW-LATITUDE CLIMATES

Af
Am   Humid tropical

Aw   Seasonally-humid tropical

B DRY CLIMATES

BWh
BWk   Desert

BSh
BSk   Semiarid

FIGURE 2-31 This world map of climate patterns is largely based on the Köppen system of climate classification.

EARTH'S CLIMATE REGIONS

Take a moment to become acquainted with Figure 2-31, because the discussion in this entire section will refer to it. Climate zones form rough horizontal bands around Earth, following lines of latitude. Tropical (A) climates

predominate in the very low latitudes, whereas polar (E) climates predominate in the higher latitudes. These horizontal bands show the powerful influence of temperature on climate. As explained earlier, higher latitudes receive less insolation than do lower latitudes. Observe the striking correspondence between this map

C WARM MID-LATITUDE CLIMATES

- Cfa / Cw — Humid subtropical
- Cfb / Cfc — Marine west coast
- Cs — Mediterranean

D COLD MID-LATITUDE CLIMATES

- Dfa / Dwa / Dfb / Dwb — Humid continental
- Dfc / Dwc / Dfd / Dwd — Subarctic

E POLAR CLIMATES

- ET — Tundra climate
- EF — Icecap and ice sheets

H HIGHLAND

Denotes highland areas in which climates are cold due to elevation, or local climate variations are too great to show on this map

MODIFIED GOODE'S HOMOLOSINE EQUAL-AREA PROJECTION

and the patterns shown in the world maps of temperature (Figure 2-6), precipitation (Figure 2-30), and atmospheric circulation (Figure 2-23). In particular, clear correspondence exists between dry (B) climates and subtropical high-pressure cells. Dry climates occur on the western sides of continents in the latitudes between 20° and 35°, in both Northern and Southern hemispheres.

You also can see the influence on climate caused by mountain ranges. The North American Rockies, the Andes of South America, and the Tibetan Plateau in Asia are mapped as H, or highland, climates. These areas are generally colder than lowland areas because of their high altitude, but they are also areas of great spatial variability in climate. East of the Rockies and Andes are rain shadow deserts. Africa's Sahara is a **desert,** or B, **climate** because it is dominated by the subtropical high-pressure zone of the eastern Atlantic. This desert extends into interior Asia, because that region is isolated from oceanic moisture sources, both by mountains and distance from the sea.

In the following descriptions, we will focus on the core characteristics of climates, which are briefly summarized in Table 2-1 and described in detail in Appendix II. Graphs called *climographs* depict the monthly average temperature and precipitation for a typical site in each climate zone. You should recognize, however, that enormous variability occurs within each climatic region and that broad transitions, not sharp boundaries, exist between them. In high mountain regions, the high altitude causes cold conditions most of the year. Most mountain regions have much variation in elevation, with both snowcapped peaks and vegetated valleys.

Humid Low-Latitude Tropical Climates (A)

Humid tropical climates (Af, Am) Humid tropical **climates** (Af and Am climates in Figure 2-31) lie mostly within 10° north and south of the equator, but can extend to 20° north and south. These areas are under the rainy ITCZ and include the world's tropical rain forests—its steamy jungles. Tropical climates share important characteristics: Throughout the year, they are warm and humid. Because tropical areas are warm throughout the year, the variation in temperature in a single day is greater than the difference between the warmest and coolest months of the year. (This is the opposite of what most midlatitude residents experience. There, temperature differences between winter and summer are far greater than between day and night.) In the tropics the air can hold a large amount of moisture because of the high temperatures, and precipitation usually is heavy.

High temperatures and ample rainfall mean that tropical climates support abundant plant growth and productive agriculture; if water supply and soil fertility are adequate, two or three harvests per year are possible. The high temperatures mean that a great deal of energy is available to evaporate water, so POTET is very high. Because precipitation is greater than evapotranspiration during most of the year, the tropics generally are humid. But many tropical areas experience distinct wet and dry seasons, as will be discussed below.

The most extensive areas of humid tropical climate are in the Amazon River basin of South America, equatorial Africa, and the islands of southeast Asia. The small country of Singapore (Figure 2-32) typifies a humid tropical climate. Temperatures are warm

Af: Singapore
Precipitation: 241.3 cm
Temperature Range: 2°C
Lat/long: 1°10'N, 103°51'E

FIGURE 2-32 Locator map for Singapore and climograph.

throughout the year, averaging between 26°C and 28°C (79–82°F). Annual precipitation totals about 240 centimeters (94 inches), and every month brings at least 150 millimeters (6 inches). Annual precipitation substantially exceeds evapotranspiration, so the area is very humid and features lush plant growth.

Humid tropical climates are influenced by the ITCZ most of the year, so rainfall in these areas primarily is convectional (Figure 2-33). Atmospheric instability results from intense daily insolation, and the high moisture content of the atmosphere favors development of severe convectional storms. Rain may fall in the afternoon or evening nearly every day, with mornings generally clear. Under these conditions, the annual totals of rainfall exceed 200 centimeters (80 inches) in many areas, and POTET is generally 120 to 170 centimeters (48 to 68 inches) per year. In oceanic areas of this climate, tropical cyclones are generated, forming hurricanes or typhoons. Southeastern Asia, the Pacific islands of Oceania, and the Caribbean are especially prone to these tropical storms. Some humid tropical climates experience a brief dry season. These are designated "Am."

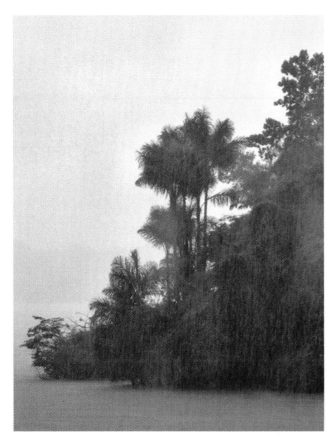

FIGURE 2-33 A thunderstorm in the Amazon Basin. Frequent intense rainfall is characteristic of this tropical humid (Af) climate. (Francois Gohier/Photo Researchers, Inc.)

Aw: Managua
Precipitation: 120.4 cm
Temperature Range: 2°C
Lat/long: 1°0'N, 86°20'W

FIGURE 2-34 Locator map for Managua and climograph.

Seasonally humid tropical climates (Aw)

In many areas of the humid tropics, rainfall is concentrated in part of the year, allowing for a distinct dry season. This is caused by seasonal shifts in the location of the ITCZ or by monsoonal circulation patterns (in southern and southeastern Asia). The ITCZ moves north in the Northern Hemisphere summer (May–October) and south between November and April, and rainfall shifts with it.

Central and South America and Africa have substantial areas of seasonally wet climates caused by this shift in the ITCZ. In Managua, Nicaragua, for example (Figure 2-34), rainfall is heavily concentrated in the six-month period from May through October, with little rain falling in the other six months. Hurricanes, which occur during the high-Sun period, also contribute to high rainfall averages. Annual rainfall totals about 120 centimeters (48 inches), roughly half that of Singapore. Average monthly temperatures vary between 25° and 29° C (77–84° F). The seasonal contrasts are caused by movement of the ITCZ. The lack of moisture in the dry season contributes to strong seasonality of vegetation growth (Figure 2-35).

In southern and southeastern Asia precipitation is also seasonal, because of the monsoonal circulation pattern. In the summer the Asian landmass heats and

FIGURE 2-35 Savanna vegetation in eastern Africa. This type of vegetation is characteristic of seasonally humid tropical climates. In the background, Mt. Kilimanjaro, Tanzania, at 5,895 meters (19,340 feet) elevation, has an ice-cap climate even though it is located just 3° south of the equator. Global warming has contributed to rapid shrinkage of the ice cover on Mt. Kilimanjaro; the ice may be gone by 2015. (Michele Burgess/Corbis/Stock Market)

develops an extensive low-pressure region that draws air in from all directions. The low-pressure cell draws over the land a steady flow of moist air from the Indian and Pacific oceans, and the rainfall is intense. Monsoonal rains are critical to agriculture in Asia. The timing of this circulation change varies from year to year, and if the rain arrives too early or too late, crop yields may be severely affected.

Some of the highest average rainfall totals in the world (5–10 meters or 16–33 feet or even more) occur on the southern slopes of the Himalayas, where mountains cause orographic lifting. In winter, however, intense high pressure develops over central Asia, and the wind in southern and eastern Asia blows from the north and northwest. The air warms as it descends from the Tibetan Plateau, and extremely dry conditions result to the south of the Himalayas.

Dry Climates (BW and BS)

Dry climates cover 35 percent of Earth's land area, although fewer than 15 percent of the world's inhabitants live in such areas. Drylands are generally located in bands immediately to the north and south of the low-latitude humid climates. The most extensive region of dry lands extends from North Africa to Central Asia; the African portion is known as the Sahara (the Arabic word for "desert"). Sand dunes cover only a small portion of the Sahara and other drylands; more common is a cover of stone or gravelly soil.

Dry climates are distinguished by a formula that relates average annual temperature to average annual rainfall. Higher rainfall totals are necessary to classify a place as humid (A, C, or D) under warm conditions than under cold conditions. Dry climates can be subdivided into warm types (BWh, BSh) and cool types (BWk, BSk).

Desert climates (BWh, BWk)   The most extensive regions of warm, dry climate are found in the subtropics on the western sides of continents. It is interesting to study this in Figure 2-31. The deserts (BWh) of northwestern Mexico and the Atacama Desert of coastal Peru border the Pacific; the Sahara and Namib/Kalahari deserts of northern and southern Africa border the North and South Atlantic; and the Australian Desert borders the Indian Ocean. These BWh climate regions result from air descending in the subtropical high-pressure zones, over the eastern parts of the oceans (in other words, off the west coasts of the continents).

Cold ocean currents can also increase the aridity of these areas. Ocean currents generally circulate clockwise in the Northern Hemisphere and counterclockwise in the Southern Hemisphere. Thus, on the eastern side of the ocean, the currents flow from high latitude to low latitude, bringing cooler water toward the tropics. The cold ocean surface cools the lower

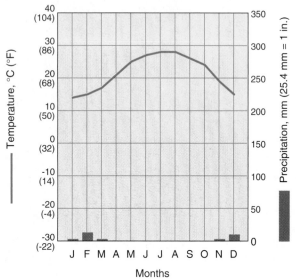

BWh: Cairo
Precipitation: 3.0 cm
Temperature Range: 14°C
Lat/long: 30°1'N, 31°14'E

FIGURE 2-36 Locator map for Cairo and climograph.

layers of the atmosphere, inhibiting evaporation and convection and therefore reducing precipitation.

In warm desert climates like that of Cairo, Egypt, high temperatures exceeding 40° C (104° F) are common (Figure 2-36). The high temperatures combined with low rainfall produce the greatest differences between annual precipitation and POTET. Annual POTET exceeds 150 centimeters (60 inches) in many BW regions; but as precipitation totals less than 20 centimeters (8 inches), the demand greatly exceeds the supply. In these conditions only plants with specialized water-collecting and water-holding characteristics can survive (Figure 2-37).

Semiarid climates (BSh, BSk)   Not all the world's drylands are barren of vegetation; in many areas sufficient rain falls to support extensive seasonal or perennial plant cover. A climate with enough moisture to support such vegetation is semiarid, and typically regions of **semiarid climate** lie in transitional areas between deserts and more humid regions. Semiarid climates are designated BS in the Köppen scheme. Annual precipitation is only 20–50 centimeters (8–20 inches). Most of the world's grasslands, or **steppes,** are in this climate

FIGURE 2-37 A desert environment in Tunisia. Potential evapotranspiration in this warm desert is much greater than precipitation, and only vegetation adapted to severe moisture stress can survive. (Richard Steedman/Corbis/Stock Market)

type, which supports extensive grazing activities but only limited agriculture (Figure 2-38).

Some semiarid climates are caused by rain shadow effects that isolate the continental interiors from moisture sources. In Asia, for example, a mountain belt stretches east-west from the Caucasus to China, blocking Indian Ocean moisture from the interior. The North American Rockies and South American Andes similarly block Pacific moisture.

Seasonal temperature contrasts are characteristic of midlatitude BSk climates. Long summer days and bright sunshine warm the ground, and because of lack of water, this heat is not dissipated through evaporation. Unlike tropical deserts, however, midlatitude dry climates are cool part of the year, so that plants require less moisture to survive (Figure 2-39). In some areas, such as interior basins of the western United States or western China, agriculture is possible with water imported from wetter mountain areas.

Warm Midlatitude Climates (C)

In the midlatitudes, seasonal variations of insolation profoundly influence temperature. In summer these latitudes receive more radiation per day than equatorial areas. In winter, however, insolation is cut to half or even a third of summer levels. Therefore, the midlatitudes experience a distinct cool season, commonly known as a winter. In subtropical locations, this winter season may be a month or two in which frost can occur. Winter causes life to become restricted to organisms that can tolerate at least occasional freezing conditions.

Lower temperatures mean that annual precipitation and evapotranspiration are generally less in midlatitude climates than in the tropics. Like the tropics,

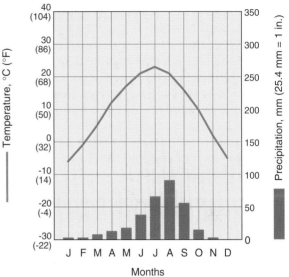

BSk: Lanzhou
Precipitation: 31.2 cm
Temperature Range: 29°C
Lat/long: 36°4'N, 103°44'E

FIGURE 2-38 Locator map for Lanzhou and climograph.

FIGURE 2-39 Caucasus Mountains. Semiarid climates have enough moisture for significant plant growth at least part of the year, but moisture shortages are common. Grasses, such as those in this area in the Caucasus Mountains of Georgia, southwestern Asia, are particularly well adapted to semiarid climates. (D. Thomas/Photo Researchers, Inc.)

however, midlatitude climates include a wide range of precipitation patterns, including continuous humidity and distinctly seasonal rainfall. Precipitation in the midlatitudes is heavily influenced by the polar front—the boundary between warm tropical air and cold polar air along which midlatitude cyclones form and travel. The significance of these storm systems is that daily weather tends to be more variable in the midlatitudes than in the tropics. This climatic zone experiences rainy spells separated by dry spells, and rainfall can occur at any time of day rather than concentrated in the afternoon and evening.

In many midlatitude environments—in the central United States, for example—seasonal precipitation patterns also are influenced by temperature. Because warmer air holds more moisture, greater rainfall occurs in summer than in winter. In the southeastern United States, midlatitude cyclones are less frequent in summer than in winter. But summer rainfall is more intense because more water is present in the air, so winter and summer receive similar amounts of rainfall.

Humid subtropical climates (Cfa, Cw)
Humid subtropical climates occur in latitudes between about 25° and 40° on the eastern sides of continents and between about 35° and 50° on the western sides. These climates are relatively warm most of the year but have at least occasional freezing temperatures during the winter. Frost occurs, so many plants adapted to tropical humid climates cannot survive, and most humid subtropical climates have deciduous species that lose their leaves in autumn and become dormant in winter. Eastern China, the southeastern United States, and parts of Brazil and Argentina are the largest areas of humid subtropical climates.

The subtropical high-pressure zones that create deserts on the western margins of continents (the eastern margins of oceans) create humid conditions on the eastern sides of continents. The circulation around these high-pressure cells brings warm, moist air from low latitudes to the eastern margins of the continents (clockwise in the Northern Hemisphere and counterclockwise in the Southern Hemisphere). This moisture supplies energy to midlatitude cyclones that generate plentiful rainfall in these areas.

Precipitation may occur all year or be concentrated in summer. Summer rainfall may result from greater moisture in the warmer air, as in New Orleans (Figure 2-40), or from monsoonal effects, as in southeastern China. Cool winter temperatures bring occasional frost in low latitudes, and snowfall occurs poleward, but the ground is not frozen or snow-covered for long. The warm, humid conditions of Cfa climates support a long growing season with forest vegetation under natural conditions (Figure 2-41) and high crop yields in agricultural regions, sometimes with two harvests in a year.

In parts of southern and Southeast Asia, the monsoon circulation causes a distinct winter dry spell,

Cfa: New Orleans
Precipitation: 161.5 cm
Temperature Range: 15°C
Lat/long: 30°0'N, 90°5'W

FIGURE 2-40 Locator map for New Orleans and climograph.

FIGURE 2-41 Wetlands in Louisiana. Humid subtropical climates support lush vegetation, such as this wetland forest in Louisiana. Many of the world's most productive agricultural regions are in this climate. (Randy O'Rourke/Corbis/Stock Market)

producing a subtropical winter-dry climate (Cw). The reduced precipitation falls at a time of year when POTET is low, and it does not have as much of an impact on agriculture or water resources as it would if it occurred in the summer.

Marine west coast climates (Cfb, Cfc)

On continental west coasts between about 35° and 65° are mild climates with small temperature variations and plentiful moisture year-round. Major regions include western North America between northern California and coastal Alaska, southern Chile, northwestern Europe, southeastern Australia, and New Zealand. These areas are cooler than humid subtropical climates (Cfa), especially in summer. They are subdivided into cooler types (Cfc) and warmer types (Cfb).

Marine climates are moderated by ocean temperatures. Typical summer temperatures are 15°C to 25°C (60–77°F), rather than the 25°C to 35°C (77–95°F) inland at these latitudes. Frost-tolerant plants stay green all year in the mild winters of −5° to +15°C (23–59°F), even if snow falls. Maritime influence on temperature is so great that winter temperatures normally associated with subtropical latitudes are found as far poleward as 55° (Figure 2-42). Palm trees survive in coastal areas of

FIGURE 2-43 Normandy, France. Normandy has mild winters despite its relatively high latitude (49° N). (David Frazier/Corbis/Stock Market)

southwest England, at the same latitude as Winnipeg, Manitoba, in south-central Canada. Kodiak, Alaska, in a maritime location at latitude 57° N, has the same January average temperature as Richmond, Virginia, which is under continental climate influence at latitude 37° N. Clearly a location on the West Coast in midlatitudes significantly raises winter temperatures relative to East Coast or interior locations.

The North Atlantic and the northeastern and southern Pacific are known for stormy weather, and these storms bring plenty of moisture to the adjacent land areas (Figure 2-43). Midlatitude cyclones are frequent, especially in winter. The low temperatures mean that air contains less moisture than at lower latitudes, so drizzle is common. Annual rainfall is about 70 to 120 centimeters (28 to 47 inches). Where mountains are near the coast, as in western North America, Chile, and Norway, orographic effects contribute to high rainfall, 2 to 3 meters (6 to 9 feet) per year. The current highest average annual rainfall in the mainland United States is in a marine climate—the west coast of the Olympic Peninsula in Washington, where annual rainfall totals 3 to 4 meters (10 to 13 feet).

Mediterranean climates (Cs)

On the western margins of continents, another distinctively seasonal climate occurs. The dry-summer **Mediterranean climate** envelops the Mediterranean region of Europe and parts of northern Africa, lending this climate type its familiar name. A similar climate occurs in extreme southern Africa, southwestern Australia, California, and central coastal Chile. Precipitation is seasonal and caused by northward and southward movement of the subtropical high-pressure zones. In summer, these zones move poleward and bring aridity; in winter they move toward the equator and are replaced by more frequent storms of the midlatitude low-pressure zone.

In Mediterranean climates, cool, rainy winters bring occasional frost or snow, with frequent midlatitude

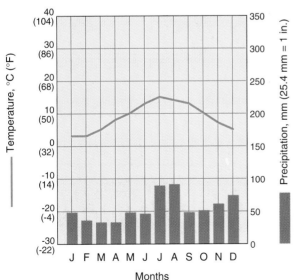

Cfb: Edinburgh
Precipitation: 65.8 cm
Temperature Range: 12°C
Lat/long: 55°57'N, 3°12'W

FIGURE 2-42 Locator map for Edinburgh and climograph.

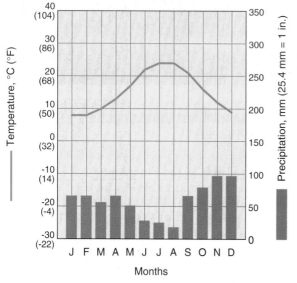

Cs: Rome
Precipitation: 71.4 cm
Temperature Range: 16°C
Lat/long: 41°54'N, 12°30'E

FIGURE 2-44 Locator map for Rome and climograph.

cyclones. Warm, dry summers have temperatures of 20°C to 35°C (68–95°F) (Figure 2-44). In Rome, Italy, for example, little or no rain may fall between May and October.

Mediterranean climates have significant water-availability problems. Precipitation occurs in winter, when human, animal, and vegetative demand for water is low. However, very little rainfall occurs in the summer, when POTET is highest. Plants with roots that can reach deeper soil moisture stay green during the summer, when shallow-rooted grasses dry out. Because water is scarce in the summer, many plants grow mostly in the winter and spring, even though there is less sunlight then (Figure 2-45). Agriculture in Mediterranean climate regions requires storage of winter precipitation in reservoirs for irrigation in summer.

Cold Midlatitude Climates (D)

Humid continental climates (Dfa, Dwa, Dfb, Dwb)

Like Mediterranean climates, **humid continental climates** have strong seasonal contrasts. In the latter case, however, the contrasts are of temperature rather than moisture availability. Continental climates are so named because they are remote from the ocean and therefore deprived of the sea's input of moisture and moderating influence on temperature. These climates occur between about 35° and 60° latitude in the interior and eastern portions of Northern Hemisphere continents. Many eastern coastal areas at high latitudes also have strong seasonal contrasts because the prevailing westerly winds extend this humid continental

FIGURE 2-45 **Variation in water availability.** Seasonal variations in water availability change the appearance of the landscape in Mediterranean climate regions, such as coastal California. The wet season (left) is typically winter and spring; the dry season (right) is summer and autumn. (Connie Coleman/Getty Images Inc.—Stone Allstock)

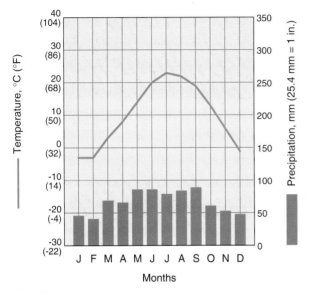

Dfa: Chicago
Precipitation: 80.8 cm
Temperature Range: 26°C
Lat/long: 41°45'N, 87°40'W

FIGURE 2-46 Locator map for Chicago and climograph.

climate to them. D climates occur exclusively in the Northern Hemisphere because very little landmass exists in the Southern Hemisphere at these latitudes.

Summer temperatures are warm because long days and high solar angles cause more radiation in a day than tropical locations receive. In winter, days are short and the Sun is low in the sky, so temperatures often fall below freezing. In more southerly humid continental areas like Chicago, Illinois (lat. 42° N), January temperatures may average below freezing, but rain still occurs in winter (Figure 2-46). In more northerly places like Moscow, Russia (lat. 56° N), winter brings many weeks of subfreezing temperatures.

In most humid continental climates, precipitation occurs throughout the year from frequent midlatitude cyclones. Typical annual rainfalls are 60 to 150 centimeters (24 to 60 inches), usually ample for plants, with an annual evapotranspiration of 50 to 90 centimeters (20 to 35 inches). In winter, plants need little water, but in summer they demand every drop that falls and sometimes more (Figure 2-47). During summer dry spells, farm fields may require supplemental irrigation.

Subarctic climates (Dfc, Dwc, Dfd, Dwd) At the northern edge of the humid continental climates, winter temperatures are cold enough and the growing season short enough that agriculture is generally not possible (Figure 2-48). The only plants that survive in these areas are those able to weather extreme cold. No abrupt temperature or precipitation boundary exists between the humid continental climate and the **subarctic climate**; the difference literally is one of degrees. Conifers dominate; the broad expanse of spruce, fir, and larch covering much of northern Canada and Siberia is called *boreal* (meaning northern) *forest*, lending the name boreal forest climate to the subarctic

FIGURE 2-47 Birch woodland. Humid continental climates have strong seasonal contrasts, as seen in these two views of a birch woodland in winter and summer. (Gregory K. Scott/Photo Researchers, Inc.)

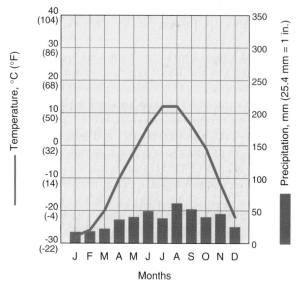

Dfc: Churchill
Precipitation: 44.3 cm
Temperature Range: 40°C
Lat/long: 58°45'N, 94°5'W

FIGURE 2-48 Locator map for Churchill and climograph.

FIGURE 2-49 Winter near Hudson Bay, Canada, a subarctic climate region. (Stephen J. Krasemann/Photo Researchers, Inc.)

climate (Figure 2-49). Like the humid continental climate, this climate is essentially limited to the Northern Hemisphere, because so little landmass exists at these latitudes in the Southern Hemisphere.

Extremely cold −20° C to −10° C (−4° F to +14° F) subarctic winters feature snow cover and months of temperatures well below freezing. As days lengthen in spring, temperatures rise rapidly, and 15- to 20-hour summer days bring mild temperatures approaching those of much warmer climates. The growing season is short, generally June to September, in a cool summer (10° C to 20° C, 50° F to 68° F), with warm spells above 30° C (86° F). Temperatures warm enough for plant growth occur as early as April, but frosts may come as late as June. Similarly, the first frost is often in September, even though temperatures to 20° C (68° F) occur into early October.

Annual rainfall is low, generally 20 to 50 centimeters (8 to 20 inches), but ample for plants because virtually no evapotranspiration occurs for six months of the year. Winter precipitation usually is locked up as snow and ice until the spring melt releases this water to

plants. Summer water shortages are rare, even though evapotranspiration may exceed precipitation during part of the summer.

Polar Climates (E)

High-latitude climates are characterized by two important features: low average temperatures and extreme seasonal variability. These climates occur at and around the poles and are differentiated by temperature alone, because temperature determines whether water is present in liquid form. These climates result from scant insolation, varying from zero during winter at the poles to among the highest daily totals on Earth for a brief period at midsummer.

Tundra climates (ET) Tundra temperatures are low all the year, rising above freezing only for a brief, cool summer (Figure 2-50). The annual average temperature is below freezing. Winters below −20° C (−4° F) are common, typically staying below freezing for three to five months. Midsummer daytime highs rarely exceed 10° C to 15° C (50° F to 59° F), even though the Sun can shine 20 to 24 hours a day. The soil 1.5 to 2 meters (5 to 6.5 feet) below the surface is insulated from seasonal temperature changes, so at this depth soil temperatures are close to the annual average. The condition of permanently frozen ground is known as **permafrost.**

Annual precipitation rarely exceeds 30 centimeters (12 inches). Most falls as snow, covering the ground for months except where the ground is bared by the wind. Although annual rainfall is slight, water is plentiful because evapotranspiration rates are low. Most of the year water is present as ice rather than liquid. In the brief summer when the ground thaws, the water may not be able to drain from the soil because of

ET: Barrow
Precipitation: 11.0 cm
Temperature Range: 32°C
Lat/long: 71°28'N, 156°0'W

FIGURE 2-50 Locator map for Barrow and climograph.

FIGURE 2-51 Ellesmere Island. Flowers bloom during the brief summer growing season in Ellesmere Island, northern Canada, located in a tundra climate region. (Stephen J. Krasemann/Photo Researchers, Inc.)

permafrost below the surface; saturated soils are therefore common.

At its northern limit, the boreal forest gives way to a treeless landscape in which vegetation survives the long winter by staying close to the ground, where it is protected from winter's extremes (Figure 2-51). This low-lying hardy vegetation, called tundra, gives the **tundra climate** its name.

Ice-cap climates (EF) Near the poles, and high in some mountains at lower latitudes, are climates in which even the warmest month averages below freezing (Figure 2-52). Virtually all of Antarctica and most of the Arctic have this **ice-cap climate.** Permafrost is extensive and thick beneath a year-round cover of snow or ice. The extremely low temperatures prevent the air from holding much water. Annual precipitation usually is less than 10 centimeters (4 inches), falling as snow and compacting into a thick mass of glacial ice that flows to the sea rather than melting in place (Figure 2-53). Cores of ice extracted in Antarctica indicate average annual precipitation of at most a few centimeters. Due

EF: McMurdo
Precipitation: 0 cm
Temperature Range: 21°C
Lat/long: 77°0'S, 170°0'E

FIGURE 2-52 Locator map for McMurdo and climograph.

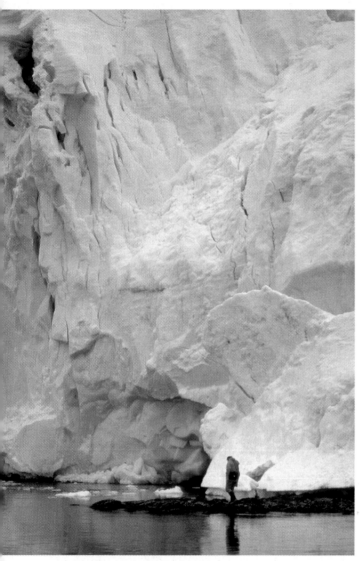

FIGURE 2-53 The edge of the Antarctic ice cap. In ice-cap climates, temperatures are only rarely above freezing, and glaciers form from the accumulation of snow. Virtually all of Antarctica has an ice-cap climate, with little glacial melting. The glaciers flow to the sea, forming icebergs. (Robert W. Hernandez/Photo Researchers, Inc.)

to low precipitation and windy conditions, Antarctica contains desertlike dry valleys that are free from snow or ice cover.

CLIMATE CHANGE

So far we have seen that climates vary from place to place. Climates also vary over time. The world climate map is based on averages over periods of a few decades in the mid-twentieth century. However, the world climate map has not always looked like this. Considerable change has taken place through the years and is likely to continue in the future.

Humans have long been able to modify environments, sometimes radically, by building settlements, clearing forests for agriculture, domesticating animals, manufacturing goods, damming rivers, extracting resources, and spreading waste throughout the air and seas and across the land. Earth's land surface is profoundly different today from what it was only 200 years ago. For example, the forests of the eastern United States were almost completely removed between 1600 and 1900, and an entire ecological community—the tall-grass prairie of the Great Plains—has been replaced by agricultural crops since 1800. Only recently have we come to realize that humankind, in addition to modifying Earth's surface, has set in motion global-scale climatic changes. Debate continues about the nature of these changes.

To interpret the significance of future climate change, we must understand patterns of past climate changes. These can give us valuable clues in determining whether recent changes are similar to those of the past or whether they are unique because of human actions.

Climatic Change over Geologic Time

Viewed over geologic time, which is about 4.6 billion years, the climate of the last 2 million years has been quite exceptional. This period, which includes our present time, is known to geologists as the **Quaternary Period.** Earth has experienced more climatic variability during the past 2 million years than in most of the previous 200 million years. Periods of similar variability probably occurred prior to 200 million years ago, but the record of the first 4 billion years of Earth's history is not very clear. In any case, it is clear that the Quaternary glaciations are an exceptional episode in recent Earth history.

Climate during the Quaternary Period has included intervals in which global average temperature was as much as 10° C (18° F) cooler than the present and warm intervals in which the climate was as warm or warmer than today. In the last 1 million years there have been about 10 of these cold intervals, occurring fairly regularly about once per 100,000 years (Figure 2-54). During the cooler periods, great continental ice sheets, like those covering much of Antarctica and Greenland today, extended over much of North America, northern Europe, and northern Asia. Periods of glaciation were also periods of lowered sea level, because water was taken out of the sea and stored on land in the form of glacial ice. The portion of the Quaternary Period in which these glaciations occurred is called the **Pleistocene Epoch.** In everyday language we describe these times of cold climate as the Ice Ages. Even areas not covered by ice were much different in the past. Areas that today are covered with deciduous forest were frozen tundra, and lakes existed in places that now are deserts. Subtropical climates extended well into the midlatitudes.

Earth's climate shifted between warm and cool periods between 10 and 30 times during the

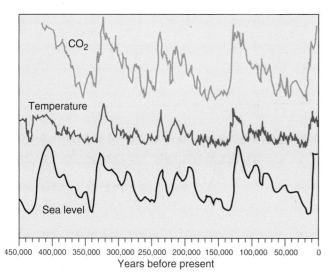

Years before present

FIGURE 2-54 Atmospheric carbon dioxide, temperature, and sea level over the last 450,000 years. Carbon dioxide and temperature data are based on samples from the Vostok ice core, Antarctica. Sea levels are based on deep-sea sediment cores. The strong similarities in the patterns demonstrate that atmospheric carbon dioxide is very closely linked to climate, particularly temperature and glacial cycles. The timing of the major glaciations at an interval of about 100,000 years suggests that variations in the geometry of Earth's orbit around the Sun are key, although interactions among the biosphere, oceans, and atmosphere are also important. (Redrawn from Sigman, D. M., and Boyle, E. A., in *Nature* 407 (2000): 859)

Pleistocene Epoch; the exact number of cold periods is not known. The last cold period reached its maximum, as indicated by the extent of glacial ice, only about 18,000 years ago—within the period of the human archaeological record. At that time global average temperatures were about 5°C (9°F) cooler than the present. The melting of the ice back to its present extent was completed only about 9,000 years ago, or about 7000 B.C. We now appear to be within a period of relatively warm climate.

Within the past 1,000 years climate has varied, though less dramatically than in the distant past. Recent climate changes have been important in European history. For example, between A.D. 800 and 1000, seafaring pirates and adventurers from present-day Scandinavia, called the Vikings, extensively explored the North Atlantic Ocean and established settlements in Greenland and North America. Lack of sea ice during those centuries as a result of especially warm temperatures aided the Vikings' exploration.

Cooling occurred in the Middle Ages, beginning about A.D. 1200. The period from about 1500 to 1750, when temperatures were especially cool, is known as the **Little Ice Age.** Glaciers advanced in Europe, North America, and Asia. Since the early 1800s and especially after 1900, climates have warmed relatively steadily. Most of Earth's glaciers have been shrinking since the early 1800s. Short cooling intervals occurred in the 1800s, including one in 1884–1887 caused by the erup-

tion of the volcanic island of Krakatau in Indonesia. Krakatau spewed so much ash and sulfur dioxide into the atmosphere that it prevented some of the short-wave insolation from reaching Earth's surface. The 1930s and 1940s were relatively warm, then cooling occurred from about 1945 to 1970. Climate has warmed dramatically since 1975. It is important that these climatic fluctuations have occurred well within the period of human occupation of Earth. As noted, glacial melting from the most recent Ice Age was particularly rapid about 15,000 to 10,000 years ago. Archaeologists have found that agriculture and cities probably developed about 10,000 to 8,000 years ago. Many scholars of early human culture believe climate change was fundamental to the development of civilizations, influencing the availability of water for crops and perhaps driving major migrations.

The period of glacial melting and warming between 15,000 and 5,000 years ago coincided with expansion of human settlement in northern Europe and North America. Settlement extended gradually northward on the European mainland and became well established throughout much of the British Isles by 2000 B.C. Meltwater from the glaciers raised sea level, flooding low-lying areas and separating land areas. Water filled the English Channel, isolating the British Isles from Europe, and the land bridge that once connected North America and Asia went underwater, halting migration from Asia into North America. Rising sea level also led to rapid filling of the Black Sea about 7,500 years ago.

Modern human inventions such as heating, housing, transportation, and industry have made us less dependent on day-to-day conditions. But we remain strongly influenced by the dynamics of the atmosphere, both short-term (weather) and long-term (climate). Many believe that as a result of human actions, climate will change much more rapidly in the future.

Possible Causes of Climatic Variation

Why does climate change over time? Many potential causes exist, but it is difficult to say which are important and which we can ignore. Understanding the causes of climatic change is of critical importance, because if we learn that climate is affected by human activity, such as burning fossil fuels (coal, oil, and natural gas), then we have the opportunity to take action to limit such effects. On the other hand, if climatic change were entirely governed by natural processes, then there would be no need for concern about potential human impacts on the atmosphere. We will summarize three of the main possible causes of climatic change: astronomical factors, geologic processes, and human modification of Earth's surface and atmosphere.

Astronomical hypotheses Inputs of solar energy drive atmospheric circulation and thus climate. Changes in insolation could alter climate. Humans know little about variations in the amount of solar radiation reaching Earth, because it is difficult to observe the Sun through the filter of the atmosphere. Recent studies based on satellite observations of the Sun are beginning to suggest that variations in solar output could be responsible for some of the climatic variations observed in the last few hundred years, causing some (but not all) of the warming of the twentieth century.

In the long run—over tens of thousands of years—we know that the geometry of Earth's revolution around the Sun fluctuates (Figure 2-55). Earth spins like a top, but a slightly unbalanced one, and because of this imbalance it wobbles slightly and very slowly as it spins. The wobble causes the tilt of Earth's axis to vary between 22° and 24° instead of staying constant at around 23½°. Other variations in the geometry of Earth's orbit around the Sun also occur, with periods of tens of thousands of years. It appears that the 100,000-year timing of major cold and warm periods over the past 1–2 million years corresponds to these variations in Earth-Sun geometry, although the mechanisms of this linkage are controversial. These variations do not explain shorter-term climatic change, however.

One short-term astronomical factor that may cause variation in the Sun's radiation output is sunspots. Sunspots are relatively cool regions on the surface of the Sun that vary in number and appear and disappear over a cycle lasting 11 years. Sunspot numbers also vary over longer periods. Sunspots affect the output of solar energy. Some climate variations appear to have periods close to the 11-year sunspot cycle. The sunspot cycle also affects concentrations of ozone in Earth's upper atmosphere.

Geologic hypotheses   Geologic factors also may cause short-term and long-term climatic change. One long-time-scale mechanism is continental drift, or plate tectonics. This remarkable phenomenon will be discussed further in Chapter 3, but briefly, continental drift is the movement of great crustal plates that make up Earth's surface at rates of up to a few centimeters per year. After millions of years continents might move great distances, opening up oceans or draining them. In addition to affecting ocean shapes and currents, the locations of continents and oceans directly affect many features of atmospheric circulation, including the ITCZ, the monsoon circulation of Asia, midlatitude cyclones, and the stormy zone around Antarctica.

Movements of continental plates have also caused the formation of major mountain ranges, such as the Himalayas, Andes, and Rocky Mountains. Continental movements are ponderously slow, so they cannot explain variations within the Quaternary Period. But they might help explain why no major glaciations occurred for more than 200 million years before then.

Volcanic eruptions can influence climate for a few years by injecting large amounts of dust and gases—especially sulfur dioxide—into the upper atmosphere. These gases reduce the amount of solar radiation filtering through the atmosphere to Earth. Past periods of more frequent volcanic eruptions may have lowered temperatures at other times.

Human causes   Processes like sunspot cycles and continental drift are beyond human control. However, humans are active participants in the changing climate. Two important ways that humans influence climate are altering the atmosphere and removing vegetation.

The carbon dioxide content of the atmosphere has increased dramatically since the start of the Industrial Revolution in the late eighteenth century (Figure 2-56). The elevated CO_2 levels cause warming, because CO_2 is a greenhouse gas. Analyses of air trapped in Antarctic ice reveal that the concentration of carbon dioxide was higher during past warm periods and lower during glacial periods (Figure 2-54). However, some past changes in CO_2 concentrations appear to be a result of changes in temperature, not a cause. Dead plant matter lying in swamps and in the soil decays during periods of warm climate, releasing more carbon dioxide to the atmosphere. During cool periods carbon accumulates on Earth's surface, reducing atmospheric CO_2. Also, the oceans may store more carbon when they are cold and less when they are warm. We can see a very close link between climate and the biosphere.

(a)

(b) Earth's wobble   (c) Variation of tilt

FIGURE 2-55 Variations in Earth's orbital geometry.
Climate may vary with long-term variations in the geometry of Earth's orbit around the Sun. These include (a) the annual variation of distance between Earth and the Sun, (b) the orientation of the Earth's axis, and (c) the degree of tilt of that axis relative to our orbit around the Sun.

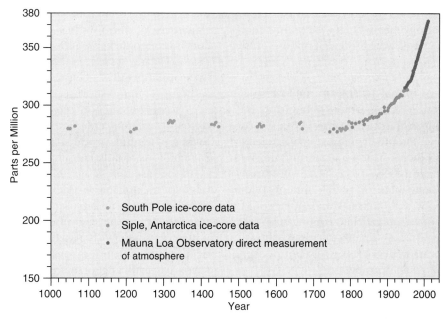

FIGURE 2-56 **Concentration of carbon dioxide in the atmosphere.** Carbon dioxide concentrations are now about one-third higher than two centuries ago, as a result of burning fossil fuels in factories, cars, and homes.

Global Warming

Forecasting weather beyond five days is very difficult, so prediction of future climate is an immense challenge. At the present state of the art, we cannot plan for the next 10 or 20 years with confidence. Keep this in mind when trying to understand the dilemma that scientists and government officials face when considering actions to limit global warming.

During the twentieth century, Earth's temperature increased slightly less than 1° C (1.8° F) (Figure 2-57). Throughout the 1990s the body of scientific evidence linking this temperature rise to emissions of CO_2 accumulated rapidly, so that today few scientists doubt that increased CO_2 is the principal cause of global warming.

The future of global warming is always uncertain, but the consensus is that unless output of CO_2 slows dramatically, world average temperature could rise by a few degrees celsius in the next century.

Despite this scientific consensus, there are many uncertainties. One is that we do not know how rapidly atmospheric CO_2 content will increase. That will depend on whether we continue to expand our use of fossil fuels as we have in the past. Another uncertainty is that we are not sure how water in the atmosphere, which plays a major role in regulating climate, will be affected by the increased levels of greenhouse gases.

Thus, we can only guess how global warming will affect specific regions. Will storm tracks shift? How will a place's current levels of rainfall, snowfall, and

FIGURE 2-57 **World surface temperature history of the last 146 years, based on records of measured temperatures.** Temperatures are shown relative to the average for 1961–1990. The trend shows three distinct periods of warming, from about 1855–1880, 1910–1945, and 1975–2003. The total warming over the period of record is about 1° C (1.8° F).

temperature change? Will the seasons change? How will plants and animals be affected? How large an effect will this have on sea level, worldwide energy use, and food supply?

The consequences of global warming   One serious effect of global warming that is widely believed to be possible is a worldwide rise in sea level of perhaps 1 to 5 meters (3 to 16 feet). People living near coasts would face danger from rising seas. The danger would not be from constant inundation, because sea level would rise very gradually over years, allowing people time to relocate, raise structures, or build dikes. The danger would be from occasional severe storms that would cause sudden flooding farther inland. The Dutch have shown that well-built dikes can hold back the sea, but poorer countries cannot afford such protection.

Another possibility is that climate change could reduce water supply in some regions. Consider a mid-latitude environment such as eastern Nebraska that averages 60 centimeters (24 inches) precipitation and 55 centimeters (22 inches) evapotranspiration. The 5 centimeters (2 inches) of precipitation that is not transpired by plants flow into streams and rivers. A small increase in evapotranspiration, due to a warmer climate, could sharply reduce water flow to the region's streams and rivers. Semiarid regions and densely populated subhumid areas depend on river flow for irrigation, drinking water, and waste removal. These areas might suffer severe water shortages if warming increases evapotranspiration and decreases river flow. However, if this warming brings greater precipitation, agriculture may be helped rather than harmed. Agricultural production might be especially helped in areas currently receiving little precipitation. Warming could also lengthen the growing season in high-latitude areas and make agriculture possible in areas of Canada and Siberia where today it is not. There is some evidence to suggest that the warming in the western Pacific that is tied to droughts in the southeastern U.S. and elsewhere (as well as the November 2002 storms discussed at the beginning of the chapter) may be related to global warming, but as yet we are not certain of these effects.

A third possibility is that warming will increase storminess. More heat in the atmosphere can mean more humidity, and atmospheric humidity is the energy source driving hurricanes and other intense storms. While we do not know whether global warming is a factor in the increased frequency of weather-related disasters, there is statistical evidence to suggest as much, and certainly the possibility of more frequent weather disasters is alarming.

Should we attempt to halt global warming?
Although global warming is in progress and humans are significant contributors to the problem, we have not reached global consensus on how we should try to stop it or if we should stop it at all. Those who argue for

immediate action emphasize the potentially severe consequences of warming in some areas. But it is hard to convince people to spend money to prevent an event that does not have dramatic, obvious effects on our lives and one that occurs over long periods of time rather than in the short term.

Reducing carbon dioxide concentrations will be difficult because we depend on fossil fuels in our daily lives, and energy producers employ many people and earn billions of dollars each year. Significantly reducing fossil-fuel use is possible only if we consume less energy or shift to alternative energy sources. An alternative to reducing fossil-fuel use is to try to trap and store ("sequester," in the scientific jargon) CO_2. Several alternatives are being discussed, from storing CO_2 in the deep ocean or underground to increasing photosynthesis by fertilizing plants in the ocean. Any of these alternatives would be expensive. Would people prefer to cut back on their use of coal-generated electricity or spend money on new ways to produce electricity? Either alternative is expensive and inconvenient.

Another reason for not acting to curb global warming is the belief that it is easier to adapt to climatic change than to prevent it. People already adjust to changing weather, commodity prices, and technology from year to year, so why shouldn't we be able to adjust to climate change, too? The United States has refused to join the Kyoto Protocol, an international treaty that would limit CO_2 emissions (see Chapter 13) on the grounds that a strategy of adapting to global warming would be less costly and more effective than attempting to reduce it.

We must remember, though, that humans are not the only life on the planet. We are only one form of life and one part of many interacting ecosystems. Some animals and plants could not adapt to a climate change, and thus *our* human-environment interaction might be responsible for the extinction of other species. Is the possibility that we can adapt to climate change sufficient reason not to halt our contribution to global warming?

CONCLUSION: CRITICAL ISSUES FOR THE FUTURE

While we are reasonably certain that global warming is caused at least in part by human activity, there are many uncertainties. Foremost among the uncertainties is how the change in global average temperature will be reflected at the regional and local level. How will prevailing circulation patterns change? How will precipitation amounts be affected? What areas will become wetter, and what ones dryer? Will storms and other extreme weather become more frequent or more intense?

Our understanding of the dynamics of Earth's climate is improving rapidly, and within the next several years answers to many of these questions will be available. When this happens we will have to decide what actions, if any, should be taken in response to climate change. We may want to prevent further impacts, or we may decide that the costs of living with them are less than the costs of prevention. The first of these challenges primarily concerns atmospheric scientists, while the second is more of a social and political problem. Both are key geographical issues.

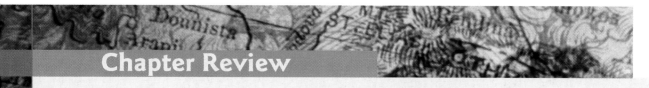

Chapter Review

SUMMARY

Solar radiation received by places at a particular latitude varies with season and by day length. The annual total radiation is highest and varies least over the seasons in low latitudes, while day length and angle of incidence cause greater seasonal variations at high latitudes. Energy arrives from the Sun as shortwave energy and is reradiated by Earth as longwave radiation. Insolation absorbed by Earth's surface and atmosphere is eventually returned to space. Human activities may be leading to global warming, especially through emission into the atmosphere of greenhouse gases such as carbon dioxide.

Precipitation forms when humid air rises, is cooled, and the water vapor in it condenses. Three processes create most precipitation: convection, orographic uplift over mountain ranges, and interaction between cold and warm air masses along fronts. Convectional precipitation dominates the tropics, while the main form of precipitation in the midlatitudes is frontal.

Wind blows from areas of high pressure to areas of low pressure, influenced by the Coriolis effect, which is a result of the rotational force of Earth. Global circulation patterns include bands of low pressure in the tropics, high pressure feeding trade winds in the subtropics, low pressure in the midlatitudes, and high pressure at the poles.

Climate is weather and its seasonal variations averaged over time. The most important variables in defining climate are temperature and precipitation. The Köppen scheme classifies climate into five main categories: A, B, C, D, and E. Humid tropical (A) climates occur in the tropical low-pressure zone and are dominated by the intertropical convergence. Dry climates (B) predominate in the subtropics, generally on the western sides of continents, and in continental areas isolated from moisture sources. Warm midlatitude climates (C) occur in subtropical areas on the eastern sides of continents and on west coasts at higher latitudes. Cool midlatitude climates (D) occur in continental areas, mostly in the Northern Hemisphere. Polar climates (E) occur at high latitudes. A sixth category (H) is mapped in mountain areas.

Earth's climate has varied greatly within the past 2 million years. Average global temperatures have been both higher and much lower than today. Glaciers covered much of Earth's surface during cold periods, the last of which ended about 12,000 years ago. Several possible causes for these variations have been proposed, including changes in the geometry of Earth's orbit around the Sun, geologic factors such as plate tectonics and volcanic eruptions, and changes in the composition of the atmosphere, some human-caused. Human-induced global warming, caused primarily by burning fossil fuels that release carbon dioxide into the atmosphere, is underway, demonstrating the key role humans play in Earth's physical processes.

KEY TERMS

adiabatic cooling p. 56
advection p. 54
angle of incidence p. 47
autumnal equinox p. 47
carbon dioxide p. 52
climate p. 46
cold front p. 58
condensation p. 55
convection p. 54
Coriolis effect p. 61
cyclone p. 66
desert climate p. 78
El Niño p. 66
front p. 58
global warming p. 52

greenhouse effect p. 52
greenhouse gases p. 52
gyre p. 65
humid continental climate p. 84
humid subtropical climate p. 82
humid tropical climate p. 78
hurricane p. 66
ice-cap climate p. 87
insolation p. 47
intertropical convergence
   zone (ITCZ) p. 63
latent heat p. 52
latent heat exchange p. 52
Little Ice Age p. 89
longwave energy p. 52

marine west coast climate p. 83
Mediterranean climate p. 83
microclimate p. 54
midlatitude cyclone p. 69
midlatitude low-pressure zones p. 63
monsoon circulation p. 65
orographic precipitation p. 58
ozone p. 52
permafrost p. 86
Pleistocene Epoch p. 88
polar front p. 63
polar high-pressure zones p. 64
Quaternary Period p. 88
radiation p. 52
relative humidity p. 55

QUESTIONS FOR REVIEW AND DISCUSSION

1. When do minimum and maximum daily amounts of solar radiation arriving at the top of the atmosphere occur at your latitude? How do angle of incidence and day length affect these seasonal differences?

2. What processes are responsible for the major energy exchanges between Earth's surface and the atmosphere?

3. Why, in terms of energy availability and convectional processes, do hurricanes happen (a) mostly over oceans and (b) in late summer and autumn?

4. How do climatic conditions in coastal areas differ from those in continental interiors? Why do these differences occur?

5. What are the major features of a midlatitude cyclone? What sequence of weather would one expect to observe as a midlatitude cyclone passes?

6. In what way does temperature affect the definition of a climate as humid or arid? Why?

7. For each of the 11 climate types described in this chapter, describe how the characteristics of the climate relate to the typical circulation patterns in the atmosphere.

8. Describe major variations of global average temperature that have occurred during (a) the past 1,000 years, (b) the past 10,000 years, (c) the past 100,000 years.

9. What are the major causes of global warming? What are some of its likely effects on human activities?

THINKING GEOGRAPHICALLY

1. For a two-week period, keep a daily journal of the weather, including such things as air temperature, wind direction, cloudiness, and precipitation. For the same period, clip the weather map from a daily newspaper. Then compare the two records.

2. In what ways do topography and/or land cover affect the weather where you live?

3. The following is a good group research activity. From your library or an on-line source such as the National Climatic Data Center (**www.ncdc.noaa.gov**), obtain a list of daily record-high and -low temperatures for your location, since record-keeping began, if possible. Enter the data into a spreadsheet or program that will perform sorting, and sort the records by year. What percentage of record-high temperatures occur in the second half of the period of record? What percentage of record-low temperatures occur in the second half of the period of record? Can you spot any other trends?

4. How might an increase in average annual temperature of 5° C (9° F) affect your day-to-day life?

5. Compare the map of climates (Figure 2-31) with the world population map (on the rear endpaper). In what climate regions are the greatest concentration of people found? Why?

SUGGESTIONS FOR FURTHER LEARNING

Alley, R. B., and others. "Abrupt climate change." *Science* 299 (2003): 2005–2010.

Akinremi, O. O., and others. "Precipitation trends on the Canadian prairies." *Journal of Climate* 12, no. 10 (1999): 2996–3003.

Arnell, N. W. "Climate change and global water resources: Global environmental change, part A." *Human and Policy Dimensions* 9 (1999): S31–S49.

Bonan, G. B. "Effects of land use on the climate of the United States." *Climate Change* 37, no. 3 (1997): 449–486.

Broecker, Wallace S. "Chaotic climate." *Scientific American* 273 (November 1995): 62–68.

Epstein, P. R. "Climate and health." *Science* 285, no. 5426 (1999): 347–348.

Grimm, Alice M., and others. "Climate variability in southern South America associated with El Niño and La Niña events." *Journal of Climate* 13, no. 1 (2000): 35–58.

Hansen, J., and others. "A common-sense climate index: Is climate changing noticeably?" *Proc. Natl. Acad. Sci.* 95, no. 8 (1998): 4113–4120.

Karl, Thomas R., Nicholls, Neville, and Jonathan Gregory. "The coming climate." *Scientific American* 276 (May 1997): 78–83.

Kerr, R. A. "A warmer Arctic means change for all." *Science* 297 (2002): 1490–1492.

Lettenmaier, D. P., and others. "Water resources implications of global warming: A U.S. regional perspective." *Climatic Change* 43, no. 3 (1999): 537–579.

Leung, L. R., and others. "Potential climate change impacts on mountain watersheds in the Pacific Northwest." *Journal of the American Water Resources Association* 35, no. 6 (1999): 1463–1472.

Meehl, G. A. "Pacific region climate change." *Ocean & Coastal Management* 37, no. 1 (1997): 137–147.

Melillo, J. M., and others. "Soil warming and carbon-cycle feedbacks to the climate system." *Science* 298 (2002): 2173–2176.

Monon, S., J. Hansen, L. Nazarenko, and Y. Luo. "Climate effects of black carbon aerosols in China and India." *Science* 297 (2002): 2250–2253.

Parry, M. L., T. R. Carter, and M. Hulme. "What is a dangerous climate change?" *Global Environmental Change* 6, no. 1 (1996): 1–6.

Rosenberg, N. J., and others. "Possible impacts of global warming on the hydrology of the Ogallala Aquifer region." *Climatic Change* 42, no. 1 (1999): 677–692.

Schindler, D. "From acid rain to toxic snow." *Ambio* 28, no. 4 (1999): 350–355.

Schoonmaker, David. "El Niño and unemployment." *American Scientist* 85, no. 4 (July 1997): 319–321.

Stenseth, N. C., and others. "Ecological effects of climate fluctuations." *Science* 297 (2002): 1292–1296.

Stix, Gary. "Green Policies. (Insurers Cope with Global Warming)." *Scientific American* 274 (February 1996): 27–28.

Sumner, G. "Precipitation Weather." *Geography* 81, no. 4 (1996): 327–345.

Sutton, R. T., and M. R. Allen. "Decadal predictability of North Atlantic sea surface temperature and climate." *Nature* 388, no. 6642 (1997): 563–567.

Tayanc, M., and H. Toros. "Urbanization effects on regional climate change in the case of four large cities of Turkey." *Climatic Change* 35, no. 4 (1997): 501–524.

Thompson, L. G., and others. "Kilimanjaro ice core records: Evidence of Holocene climate change in tropical Africa." *Science* 298 (2002): 589–593.

Vargas, A. B. "Changes in the general circulation and its influence on precipitation trends in Central America: Costa Rica." *Ambio* 23, no. 1 (1994): 87–90.

Wilbanks, T. J., and others. "Possible responses to global climate change: Integrating mitigation and adaptation." *Environment* 454, no. 5 (2003): 28–38.

Yarnal, B., and others. "Severe convective storms, flash floods and global warming in Pennsylvania." *Weather* 54, no. 1 (1999): 19–24.

WEB WORK

Weather data are widely available on the World Wide Web from a great many sources. Here are just a few of the good ones. Intellicast is one of many commercial weather-information providers. Its Web site contains an excellent array of satellite imagery, in addition to up-to-date forecasts:

http://www.intellicast.com/

The U.S. government's atmospheric science programs are housed in the National Oceanographic and Atmospheric Administration. Its home page has links to the National Weather Service, which carries out forecasting and warning programs, and to the National Climatic Data Center, which stores and distributes historic climate data, including both statistics and satellite images.

http://www.noaa.gov/

Environment Canada maintains an excellent Web site with a wealth of Canadian environmental information, including weather and climate data:

http://www.msc.ec.gc.ca/

At its home page, the Carbon Dioxide Information Analysis Center of the Oak Ridge National Laboratory has done an excellent job of assembling data related to global warming and other global atmospheric pollution problems.

http://cdiac.esd.ornl.gov/

Two sites contain a wealth of information on global climate policy and climate change. One is Weathervane, a Digital Forum on Global Climate Policy, which is maintained by Resources for the Future:

http://www.weathervane.rff.org/

The other is the U.N. Intergovernmental Panel on Climate Change:

http://www.ipcc.ch/

A Look Ahead

PLATE TECTONICS

Earth's surface is in motion, both horizontally and vertically. Earth's crust consists of several moving pieces, known as tectonic plates. The movement of these pieces causes earthquakes, volcanic eruptions, and formation of mountain ranges at plate boundaries.

SLOPES AND STREAMS

Water, wind, ice, and gravity shape Earth's surface. Rocks are broken down into smaller pieces through exposure to air and water. The pieces then are moved downhill by gravity or are carried by streams, wind, or ice to other locations. Streams play a major role in shaping landforms. They erode material from some places, transport it downhill, and deposit it elsewhere.

3

LANDFORMS

◄ Plumes of sediment entering the Gulf of Mexico, as seen in a satellite image taken in March 2001. Delivery of sediment from the central United States to the mouth of the Mississippi has been significantly altered by a variety of human actions ranging from soil erosion in the Corn Belt to levee construction in Louisiana. (NASA Headquarters)

ICE, WIND, AND WAVES

Glaciers—rivers of ice—carve landforms in regions under continuous snow cover, such as mountaintops and the poles. In deserts, where vegetation is scarce, the wind shapes landforms. Along coastlines, waves caused by wind blowing across the ocean surface cause intensive erosion, which rapidly changes landforms in these areas.

THE DYNAMIC EARTH

Earth's surface is continually changing. Change on Earth's surface is rapid enough to affect human settlements and natural resources. In particularly dynamic environments, natural hazards like floods, earthquakes, and landslides are a major problem.

Slowly but surely, New Orleans is sinking into the sea. Along with it, Louisiana is losing more than 6,000 hectares of coastal wetlands each year, about 40 percent of all the coastal wetland that is lost in the United States annually. New Orleans itself is sinking at about 1 meter per century. The city on average is 3 meters below sea level. The reason the land is dry rather than covered with seawater is that pumps run continuously to remove water from the city. The prospect of increased sea-level rise caused by global warming only aggravates the problem.

Primarily, New Orleans is sinking because it is located on the Mississippi Delta. For millions of years the Mississippi has delivered sediments eroded from the North American interior to the Gulf of Mexico, where they are deposited on the seafloor. As these sediments accumulate, they are compacted under their own weight, and slowly the land sinks. Levees adjacent to the channel have been built higher in order to protect the city from flooding, but at the same time this prevents new sediment from being deposited. The loss of wetlands is more than a curiosity; they are a habitat for migratory birds, including nearly 20 percent of the U.S. winter population of ducks and geese. Wetlands are also critical to maintaining the productivity of sport and commercial fisheries and hunting lands that together bring $35 million to Louisiana's economy each year.

Some changes on Earth's surface take place gradually, at rates of a few millimeters or less per year. Others, like the earthquake that shook Algeria in May of 2003 (Figure 3-1), are sudden and dramatic. Whether gradual or catastrophic, changes in the Earth occur everywhere. In this chapter we will learn about the causes and rates of such movements and how they shape mountains, hills, and valleys. This chapter looks at Earth's rocks, soil, and surface landforms, which comprise the lithosphere, the outermost part of one of the four "spheres." (Look back at Fig. 1-17). **Geomorphology** is the study of landforms and the processes that create them.

FIGURE 3-1 Earthquake damage in Algeria. The damage shown here was caused by a severe earthquake that struck northern Algeria in May 2003, killing 1,700 people and injuring 7,000. (AP/Wide World Photos)

Earthquakes and gradual sinking of the land are newsworthy mostly because of their effects on humans. They also are notable because most of the time Earth's lithosphere appears to be fixed and unchanging relative to the other three spheres—the atmosphere, the biosphere, and the hydrosphere—which change rapidly. The atmosphere changes daily—hot to cold, wet to dry, windy to still. We can see seasonal changes in the biosphere—plants and animals appear, grow, reproduce, and die. The hydrosphere is visibly dynamic—rivers flood and later dry up. But the lithosphere seems static. Landforms like plains, hills, and valleys do not change noticeably in a lifetime; soil is soil; rock outcrops seem permanent as monuments.

Landforms, in their remarkable variety, did not just appear as we see them at some distant time. Landforms constantly change, although often imperceptibly slowly. As happens in the atmosphere, the lithosphere is driven by continual transfers of energy and matter, and it interacts with the other three spheres. Some of the energy that modifies the lithosphere comes from deep within Earth. Heat contained in Earth's core drives movement on the planet's surface.

Occasionally, Earth's surface changes quickly. This reminds us that the lithosphere is quite dynamic—earthquakes shake the land, volcanoes spew forth hot gases and molten rock, and ocean waves pound the shore. Humans insist on getting in the way of these dramatic changes in landforms by building settlements in places where earthquakes, volcanoes, and hurricanes are likely to occur. Because of the rarity of these events, people discount their importance and thus become vulnerable when they do strike.

Our actions have profoundly changed the lithosphere, as they have changed the other spheres.

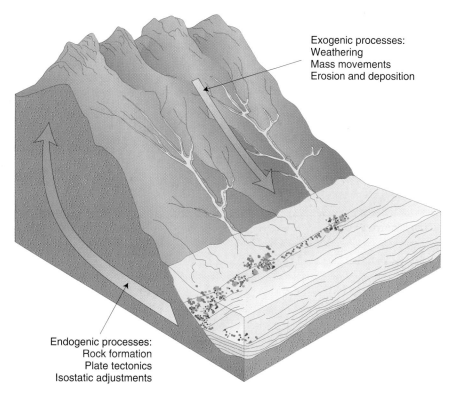

Exogenic processes:
Weathering
Mass movements
Erosion and deposition

Endogenic processes:
Rock formation
Plate tectonics
Isostatic adjustments

FIGURE 3-2 Endogenic and exogenic processes. The landforms around you are the product of interaction among endogenic and exogenic processes. Endogenic processes involve movement of Earth's crust through tectonic action. Crustal uplift may be caused by plate tectonics, or by isostatic adjustments, which cause the land to rise when rock mass erodes away. Exogenic processes wear down rocks once they reach the surface.

Agricultural practices are depleting soil fertility. In semiarid areas, overgrazing by animals has produced desertlike conditions. In many agricultural areas, soil is being eroded at rates 10 to 100 times greater than natural rates of erosion. Some estimates of human-caused earth moving suggest that more earth is moved by humans than by all natural processes combined.

Geographers studying the shape of Earth's surface—its topography—recognize that it includes many features that seem to have distinctive characteristics (Look ahead to map on the front endpaper). Elements of Earth's surface that have such identifiable form—its mountains, valleys, hills, and depressions— are called **landforms.** They are built through a combination of endogenic (internal) and exogenic (external) processes. **Endogenic processes** are forces that cause movements beneath or at Earth's surface, such as mountain building and earthquakes. These internal mechanisms move portions of Earth's surface horizontally and vertically, raising some parts and lowering others. Even as these internal forces are building Earth's features, these features are simultaneously attacked by **exogenic processes,** which are forces of erosion, such as running water, wind, and chemical action.

Endogenic and exogenic forces continually move and shape Earth's crust (Figure 3-2). Endogenic processes form rocks and move them to produce mountain ranges, ocean basins, and other topographic features. As these rocks become exposed, exogenic activities go to work. They erode materials, move them down hillslopes, and deposit them in lakes, oceans, and other low-lying areas. We will examine all these processes.

PLATE TECTONICS

If you were to view Earth from a great distance, as we sometimes view other planets through telescopes, the most distinctive topographic features would be the enormous mountain ranges arranged in linear patterns that extend for thousands of kilometers. Three especially prominent mountain ranges on Earth's land surface include the Rockies of North America, connecting through Central America to the Andes of South America; the Himalayas, extending across Asia; and the north-south system of mountains in Eastern Africa. Large as these highly visible mountain ranges are, none rank as the world's longest. That title belongs to a mountain system beneath the oceans, the interconnecting mid-ocean ridges that are more than 64,000 kilometers (40,000 miles) long.

For millennia, people believed that Earth's continents and oceans were fixed in place for all time and that Earth was only a few thousand years old. This

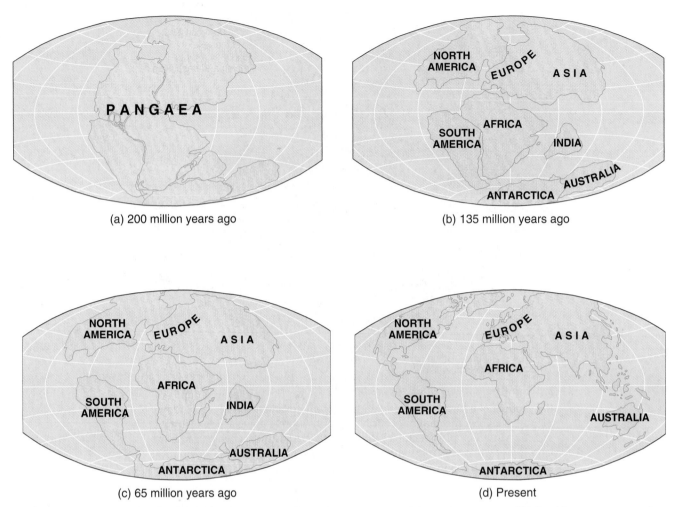

(a) 200 million years ago

(b) 135 million years ago

(c) 65 million years ago

(d) Present

FIGURE 3-3 **Past plate movements.** Two hundred million years ago (a) Earth's continents were all joined in one supercontinent known as Pangaea. This continent gradually broke apart, beginning with the opening of the Atlantic Ocean from south to north and the movement of India away from Antarctica toward Asia (b and c). The world of today is shown in (d).

notion of a "fixed Earth" was challenged early in the twentieth century by German earth scientist Alfred Wegener. He argued that Earth's land areas once had been joined in a single "supercontinent," now known as Pangaea, and that over thousands of years the continents had moved apart (Figure 3-3). Because Wegener could not explain why the continents moved, his ideas were rejected at the time. But researchers in the 1960s vindicated Wegener by working out an explanation, called **plate tectonics theory.** Since that time, plate tectonics theory has provided us with explanations of the origins of Earth's great mountain chains, volcanoes, and many other important phenomena.

> Thousands of earthquakes—sudden movements of Earth's crust—occur every day.

Earth's Moving Crust

Earth resembles an egg with a cracked shell. Earth's crust is thin and rigid, averaging 45 kilometers (28 miles)

thickness. Beneath this rigid crust the rock is like a very thick fluid and is slowly deformed by movements within Earth. While far from the free-flowing substances we know as liquids, the rock just beneath the crust, known as the **mantle,** is fluid enough to move slowly along in convection currents, driven by heat within Earth's core. These currents are analogous to winds in the atmosphere, which carry heat away from Earth's surface. Geologists believe that this motion of the mantle causes pieces of Earth's rigid crust to move; those pieces are called **tectonic plates.** This is the plate tectonics theory. Movement of the plates causes earthquakes to rumble, volcanoes to erupt, and mountains to be built (Figure 3-4).

Earthquakes Thousands of **earthquakes**—sudden movements of Earth's crust—occur every day. Figure 3-5 shows major earthquakes zones around the world,

Volcanic archipelago   Trench   Midocean ridge   Trench

Ocean

Upwelling

Mantle

Convection cells

FIGURE 3-4 Mechanisms of plate movement. Plate tectonic theory explains the occurrence of moving continents, mid-ocean ridges, deep-sea trenches, and earthquakes. At center, rock magma melted in Earth's interior rises along mid-ocean ridges, emerging to chill into new oceanic crust. Along this ridge, plates of oceanic crust are spreading apart. Where crustal plates collide (at left and right), crust is forced downward (subducted) and recycled (melted) back into the interior. This generates new magma, which migrates toward the surface, sometimes emerging as a volcano. Less dense continental crust (the continents) rides on moving plates of basaltic crust. This plate tectonic process is believed to be driven by convection currents in Earth's mantle, shown by arrows. (From E. J. Tarbuck and F. K. Lutgens, *The Earth,* 4th ed. New York: Macmillan, 1993)

notably clustered where two plates meet. The place where Earth's crust actually moves is the **focus** of an earthquake. The focus is generally near the surface but can be as deep as 600 kilometers (372 miles) below Earth's surface. The point on the surface directly above the focus is the **epicenter** (Figure 3-6). The tremendous energy released at the focus travels worldwide in all directions, at various speeds through different layers of rock.

Most earthquakes are too small for people to feel, and they are detectable only with a **seismograph**, a device that records the quake's **seismic waves**, or vibrations. Earthquake intensity is measured on a 0-to-9 logarithmic scale developed by Charles F. Richter in 1935. Magnitude 3 to 4 earthquakes on the Richter scale are minor; magnitude 5 to 6 quakes can break windows and topple weak buildings, and magnitude 7 to 8 quakes are devastating killers if they affect populated areas.

Several factors in addition to an earthquake's intensity determine the damage it causes. Generally, damage is greater at places closer to the epicenter and at places built on ground subject to landsliding or collapse. Earthquake damage is also greater where buildings are not designed to absorb the shaking. To appreciate this, let us contrast two major earthquakes: one in California and one in Armenia. In 1989 an earthquake struck the San Francisco Bay area. The Bay Bridge buckled, freeways crumbled, and 67 people died. Some 100,000 buildings were damaged or destroyed, many in the Marina District, a neighborhood built on unstable fill in the San Francisco Bay. The death and destruction was limited, however, considering the quake's magnitude of 7.1 on the Richter scale. In contrast, in 1988 an earthquake in Northwestern Armenia killed nearly 55,000 people, injured 15,000, and left at least 400,000 homeless. Although registering 6.9 on the Richter scale—lower than the California quake—and in a city much smaller than San Francisco, this earthquake trapped many inside collapsing buildings not designed to withstand earthquakes. People in less economically developed societies cannot finance the cost of earthquake-proofing their structures, as is routinely done with new buildings in quake-prone communities in wealthy societies like the United States.

Earthquake prediction is unreliable. Potential movement zones are closely monitored, and computer models attempt to replicate conditions along plate boundaries. But predicting an earthquake is like trying to forecast the weather several months from now. It just cannot be done accurately with current technology.

Volcanoes   Like earthquakes, volcanoes are clustered along boundaries between tectonic plates (Figure 3-5).

Earth's Tectonic Plates

FIGURE 3-5 (a) The major plates of Earth's crust move relative to one another, generally at rates of a few centimeters per year. (b) These motions cause earthquakes that are concentrated along plate boundaries. Ridges with rift valleys at their centers are formed where plates are moving away from each other, generally in ocean areas. Mountain ranges are created where plates converge, sometimes with deep-ocean trenches on the seaward side of the convergence area.

Movement within Earth and between the plates generates **magma** (molten rock). Being less dense than the surrounding rock, magma migrates toward the surface. Some reaches the surface and erupts, and it is then called **lava.** A **volcano** is the surface vent where lava emerges. The magma may flow over the surface,

forming a plain of volcanic rock, or it may build up to form a mountain. The chemistry of the magma/lava determines its texture and therefore the type of landform it builds.

Shield volcanoes erupt runny lava that cools to form a rock called basalt. They are called shields because

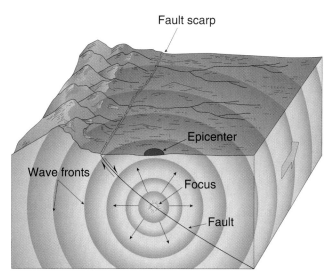

FIGURE 3-6 Earthquake focus and epicenter. Where Earth's crust actually moves is the focus of an earthquake. The point on the surface directly above is the epicenter. Displacement of the surface creates a fault scarp.

clouds of sulfurous gas high into the atmosphere. It may also pour lethal gas clouds and dangerous mudflows down the volcano's slopes. Repeated eruptions build a cone-shaped mountain, made up of a mixture of lava and ash layers.

Eruptions of composite cone volcanoes have killed tens of thousands of people at a time, but such disasters are much less frequent than severe earthquakes. One of the greatest volcanic explosions in recorded history was the 1883 eruption of the island of Krakatau in present-day Indonesia. Two-thirds of the island was destroyed, and the event killed about 36,000 people, most of whom died in a flood triggered by the eruption. Ash discharged into the atmosphere by Krakatau significantly blocked sunlight and caused noticeable cooling of Earth's climate for a couple of years.

The deadliest volcanic eruption in the twentieth century, in 1902, was Mt. Pelée on the island of Martinique in the Caribbean. The eruption produced hot ash flows that wiped out the city of St. Pierre in minutes, leaving 29,000 people dead and only one survivor, a man protected from the holocaust in his prison cell. This catastrophe exemplifies some of the difficulties in avoiding known natural hazards. Mt. Pelée steamed and rumbled, giving months of warning before erupting. The fearful citizens of St. Pierre wanted to leave, but the politicians convinced them to stay in the city to vote in an imminent election—one that did not happen. Ninety-five years later, in 1997, a similar hot-ash eruption in the Soufrière Hills on the nearby island of Montserrat destroyed the principal city on the island, Plymouth, but in this case the city was evacuated in advance, and the loss of life was minor in comparison to that on Martinique 95 years earlier.

Thousands of volcanoes stand dormant (inactive, but with the potential to erupt) around the world.

of their shape (Figure 3-7). Each of the Hawaiian Islands is a large shield volcano, although the only currently active one is Mauna Loa, on the island of Hawaii (the "Big Island"). These generally sedate volcanoes make news on the rare occasions when they grow more active, and flows of lava threaten settlements. The mid-ocean ridges are formed of similar basaltic lava.

Explosive volcanoes that cause death and destruction are more likely to be **composite cone volcanoes** (Figure 3-8). Composite cones are made up of a mixture of lava and ash. Their magma is thick and gassy, and it may erupt explosively through a vent. The eruption sends ash, glassy cinders (called pyroclasts), and

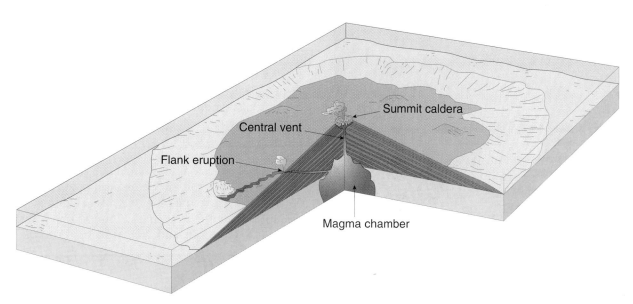

FIGURE 3-7 Shield volcano. Gentle shield volcanoes, exemplified by the Hawaiian Islands, are the largest volcanoes on Earth. Lava may erupt on the flank of the volcano or from the craterlike caldera at the top of the mountain. (From E. J. Tarbuck and F. K. Lutgens, *Earth Science,* 7th ed. New York: Macmillan, 1994)

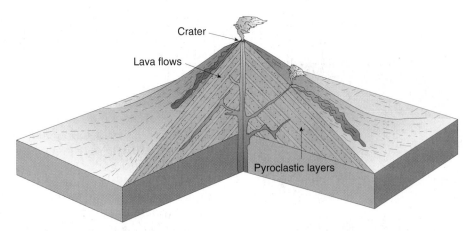

Crater

Lava flows

Pyroclastic layers

FIGURE 3-8 Composite cone volcano. Composite cone volcanoes, composed of alternating lava and ash, are relatively explosive. (From E. J. Tarbuck and F. K. Lutgens, *Earth Science,* 7th ed. New York: Macmillan, 1994)

About 600 are actively spewing lava, ash, and gas—some daily—but they rarely cause damage. Others, however, have not erupted in hundreds of years, and people have settled nearby (Figure 3-9). When warnings are not available, the danger can be great. The 1985 eruption of Nevado del Ruiz, in Colombia, triggered giant mudslides that buried most of a town, killing 23,000 people. In general, predicting volcanic eruptions is more accurate than predicting earthquakes, because volcanoes give many warnings before erupting.

Types of Boundaries Between Plates

Three types of boundaries form between moving plates of Earth's crust. The type of boundary depends on whether the plates are spreading apart, pushing into each other, or grinding past each other.

FIGURE 3-9 Popocatepetl eruption, January 2001. Mexico's Mt. Popocatepetl, just 60 km (40 miles) from Mexico City, has been active since 1996 after a long period of dormancy. Rapid growth of the Mexico City metropolitan area has increased the number of people threatened by such eruptions. (Pablo Spencer/AP/Wide World Photos)

Divergent plate boundaries A boundary where plates are spreading apart is a **divergent plate boundary.** People seldom are aware of plates spreading apart, first because it happens at a rate of only a few centimeters per year (1 inch = 2.54 centimeters) and, second, because it happens mostly deep in mid ocean. The Mid-Atlantic ridge is a well-known example. Divergent plate boundaries occur on land, too. The rift valleys of East Africa are an example, visible in Figure 3-5. These valleys are hundreds of meters deep and extend thousands of kilometers from Mozambique in the south, to the Red Sea in the north.

Where two plates are diverging on the seafloor, a phenomenon called **seafloor spreading,** lava continually erupts. Rapidly chilled by seawater, it solidifies to form a new seafloor crust. This is how much of the seafloor crust forms, very slowly spreading from mid-ocean ridges.

Convergent plate boundaries A boundary where plates push together is a **convergent plate boundary.** Material from one plate—the subducting plate—is slowly forced downward by the collision, back into the mantle. Seafloor crust is denser than continental crust. Because of this difference in density, when a plate of continental crust collides with a plate of oceanic crust, the denser oceanic plate sinks beneath the lighter continental crust. The oceanic plate is carried into Earth's mantle, where some of it is remelted. This magma then migrates toward the surface, causing volcanic eruptions at sites above the plunging plate. This occurs, for example, along the west coast of South America where the South American and Nazca plates converge (see Figure 3-5).

Transform plate boundaries A boundary where the plates neither converge nor diverge, but grind past each other, is a **transform plate boundary.** California's

San Andreas Fault is an example. Along this fault, the Pacific plate is moving northwest relative to the North American plate (Figure 3-5a). The boundary between these plates is not a smooth one, and ridges and mountains are built as the two plates grind against one another. The plates bind for long periods and then abruptly slip, causing the earthquakes that frequently strike California.

Vertical movements of Earth's crust   Parts of the crust move vertically as well as horizontally. As two plates collide, material may be forced downward into Earth's interior or upward to form mountains. Over millions of years, vertical movements along plate boundaries produce mountain ranges thousands of meters high.

Vertical movement of crust also occurs because the crust "floats" on the underlying mantle, like a boat floating in water. If material is added to the crust, it sinks, and if material is removed, it rises. Deposition of sediment or accumulation of ice in glaciers can cause the crust to sink. These vertical movements caused by loading or unloading the crust are called **isostatic adjustments.** Removal of material by erosion or melting of glaciers allows the crust to adjust isostatically, or "rebound." In some places, crust that was buried under continental ice sheets has risen vertically over 100 meters (330 feet) in the last 15,000 years because of glacial melting.

> In some places, crust that was buried under continental ice sheets has risen vertically over 100 meters (330 feet) in the last 15,000 years because of glacial melting.

Rock Formation

Although by human standards Earth's surface moves very slowly—by at most a few centimeters per year—this movement produces Earth's great diversity of rocks. As Earth's crust moves, its materials are eroded and deposited, heated and cooled, buried and exposed.

Types of rocks   Rocks can be grouped into three basic categories that reflect how they form:

Igneous rocks are formed when molten crustal material cools and solidifies. The name derives from the Greek word for fire—the same root as the English word ignite. Examples of igneous rocks are basalt, which is common in volcanic areas, including much of the ocean floor, and granite, which is common in continental areas.

Sedimentary rocks result when rocks eroded from higher elevations (mountains, hills, plains) accumulate at lower elevations (like swamps and ocean bottoms). When subjected to high pressure and the presence of cementing materials to bind their grains together, rocks like sandstone, shale, conglomerate, and limestone are formed.

Metamorphic rocks are created when rocks are exposed to great pressure and heat, altering them into more compact, crystalline rocks. In Greek the name means "to change form." Examples include marble (which metamorphosed from limestone) and slate (which metamorphosed from shale).

Minerals   Minerals are natural substances that comprise rocks. Each type of mineral has specific chemical and crystalline properties. Earth's rocks are diverse in part because the crust contains thousands of minerals. The density of rocks depends on the kinds of materials they contain. Denser rocks are dominated by compounds of silicon, magnesium, and iron minerals; they are called **sima** (for *si*licon-*ma*gnesium). Less dense rocks are dominated by compounds of silicon and aluminum minerals; they are called **sial** (for *si*licon-*al*uminum).

Denser sima rocks make up much of the oceanic crust. Less dense sial rocks make up much of the continental crust. The lower density and greater thickness of sial rocks cause the continents to have higher surface elevations than the oceanic crust, just as a less dense dry log will float higher in water than a denser wet one.

The formation and distribution of many minerals is caused by the movements of Earth's crust. Vast areas of the continental crust, known as **shields,** have not been significantly eroded or changed for millions of years (Figure 3-10).

Shield areas often contain rich concentrations of minerals, such as metal ores and fossil fuels. Shields are located in the core of large continents, such as Africa, Asia, and North America. Many of the world's mining districts operate where these continental shields are exposed at the surface.

Stress on rocks   Crustal movements along plate boundaries exert tremendous stress on rocks. Despite their rigidity, the rocks bend and fold. When stressed far enough, they fracture along cracks called **faults.** The fractured pieces may then be transported to new locations. Fracturing takes place in different ways, depending on the type of boundary. Near a divergent plate boundary, rocks break apart because they are stretched; the fracture is called a *normal fault* (Figure 3-11). Near a convergent plate boundary, rocks fracture because they are compressed; such fractures are called *reverse faults* (because they are the opposite of normal faults) (Figure 3-12), or *thrust faults* if there is large horizontal movement. Alternatively, the crust may rumple like a rug, creating folds. The Appalachian Mountains, the western edge of the Wasatch Range in Utah (Figure 3-13), and the Himalayas are examples of mountain ranges created by faulting and folding along

Continental Shields

FIGURE 3-10 The ancient cores of the continents are exposed in several areas of Earth's surface. The rocks exposed in these areas are typically rich in mineral ores.

FIGURE 3-11 A normal fault. The diagrams show cutaway views of what happens beneath Earth's surface along a normal fault. (a) Rocks break apart because they are stretched. (b) How erosion would alter the block that has been uplifted along a normal fault. (From E. J. Tarbuck and F. K. Lutgens, *The Earth,* 4th ed. New York: Macmillan, 1993)

FIGURE 3-12 A reverse fault. (a) Rocks break when stressed by forces that compress them. (b) How erosion alters the surface of rocks that have been uplifted along a reverse fault. Later erosion shapes the surface, and may obscure the actual orientation of the fault. (From E. J. Tarbuck and F. K. Lutgens, *The Earth,* 4th ed. New York: Macmillan, 1993)

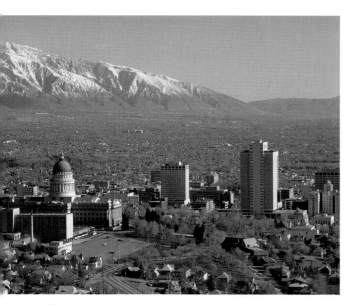

FIGURE 3-13 The Wasatch Fault. The Wasatch Mountains in Utah rise abruptly from the Salt Lake valley along the Wasatch Fault, an active normal fault just east of Salt Lake City. (Tom McHugh/Photo Researchers, Inc.)

convergent plate boundaries. Faulting also occurs along transform boundaries. Rock movement near a transform boundary is mostly parallel to a plate boundary rather than perpendicular, as in the other two types of boundaries.

Rocks and landforms   Differences in geologic structures and rock types from one place to another are a critical part of the geographic variability of Earth's

FIGURE 3-14 Satellite image of the central Appalachians in Pennsylvania. The winding ridges are formed by resistant layers in folded sedimentary rocks, exposed by erosion of weaker layers that now occur in valleys. (Photo Researchers, Inc.)

surface. These geologic features influence the surface in three different ways. (1) Movement of the crust such as that along faults creates landforms like those illustrated in Figures 3-11, 3-12, and 3-13. (2) Variations in the resistance of rocks to exogenic processes cause weak rocks to be removed more rapidly while resistant rocks remain in place. This creates **structural landforms,** in which the shape of the land reflects the underlying rock structures. The ridges and valleys of the central Appalachians are formed in this way (Figure 3-14). The ridges are formed by resistant layers, often sandstone and conglomerate, while the valleys are underlain by weaker materials such as shale. (3) Differences in the minerals contained in rocks affect the characteristics of soils that form on them. Soils overlying limestone, for example, tend to be rich in calcium, a major component of limestone. The following section describes the exogenic processes that help to expose these geologic influences on the landscape.

SLOPES AND STREAMS

Plate tectonics theory helps us to understand the causes of mountain-building motions in Earth's crust. But landforms created through endogenic processes are attacked by exogenic processes as they are being formed (Figure 3-15). The wearing down of Earth's crust through exogenic processes reshapes Earth's crust into new landforms. Exogenic processes shape Earth's surface in two principal steps. Rocks are first broken down into smaller pieces through weathering. Then they are carried by gravity down the slopes of hills or are transported by water, wind, or ice from one place to another.

FIGURE 3-15 Faulted landforms. Death Valley, California, was formed by faulting that raised the land in the foreground and the distance, while lowering the valley floor. The valley is filled with sediments eroded from the adjacent mountains. (Deborah Davis/PhotoEdit)

Weathering

Weathering is the process of breaking rocks into pieces ranging in size from boulders to pebbles, sand grains, and silt, down to microscopic clay particles and dissolved solids. Without weathering, the force of gravity and the agents of water, wind, and ice would have nothing to move. Rocks begin to break down the moment they are exposed to the weather at Earth's surface. They are attacked by water, oxygen, carbon dioxide, and temperature fluctuations. Weathering is the first step in the formation of soil, which will be discussed in more detail in Chapter 4. Weathering takes place in two ways: chemical weathering and mechanical weathering.

Chemical weathering　Rocks may be broken down as a result of **chemical weathering,** which is a change in the elements that compose rocks when they are exposed to air and water (Figure 3-16). Chemical weathering occurs faster in places with warm temperatures and abundant water, such as in humid tropical environments. Acids released by decaying vegetation also chemically weather rocks. Some of the dissolved products of chemical weathering are carried away by water seeping through soil and rocks, a process called *leaching.* The water eventually may carry these materials to rivers and thence to the sea. This is the source of the salinity (dissolved salt) of the oceans.

One example of chemical weathering is oxidation. Iron is a common element in rocks, and it combines with oxygen in the air to form iron oxide, or rust. Iron oxide has very different properties from the original iron—it is physically weaker and more easily eroded. You can see the effects of oxidation on iron or steel surfaces exposed to the weather; the rusty oxide easily flakes away.

Another example of chemical weathering is the decomposition of calcium carbonate. It is a major component of limestone and other sedimentary rocks. Calcium carbonate dissolves in water, separating into ions of calcium and carbonate that are carried by streams into the sea. In some areas of limestone bedrock, underground water may remove large quantities of rock by chemical weathering beneath the surface. Water flowing underground dissolves passageways and even large caverns in the limestone. If the caverns collapse, they create depressions called sinkholes at the surface. Such underground erosion produces a distinctive form of topography called **karst,** named after a region in Croatia with this type of landscape. Karst topography is found in many parts of the world, including the Caribbean (especially Puerto Rico, Cuba, and Jamaica), several parts of the southeastern United States (especially Florida, Kentucky, and Missouri), and southeastern China.

Mechanical weathering　Rocks are also broken down by physical force. This process is called **mechanical weathering.** Rocks expand and contract with frequent changes in temperature, and this action causes them to break apart. Highway potholes in the northern United States and Canada illustrate these processes. Rainwater seeps into roadway cracks and freezes into ice crystals when the temperature turns colder. The water, which expands about 9 percent when frozen, pushes apart the pavement, and opens up the potholes in a phenomenon called frost-wedging. Plant roots growing in cracks between rocks also contribute to mechanical weathering; you probably have observed sidewalks that have been heaved by tree roots.

Mechanical and chemical weathering work together to break down rocks. Often, mechanical forces open cracks, and water seeps in to weather the rock chemically.

Moving Weathered Material

Once rocks are weathered, they may be carried from one place to another. Material most commonly moves downhill by gravity. This happens in two ways: by mass movement or by surface erosion. In **mass movement,** rocks roll, slide, or freefall downhill under the steady pull of gravity. In **surface erosion,** water—which flows downhill because of gravity—carries solid rock particles with it. Surface erosion may also result when the wind or ice carries material from one place to another.

Material moves faster down steeper hills than down gentler ones. The steepness of a hill is measured through slope, which is the difference in the elevation

FIGURE 3-16 Weathering. Granite exposed along the coastline of Nova Scotia reveals the effects of weathering on rocks. Water has penetrated the rock and weakened the bonds between grains, allowing the rock to be easily broken open. Freezing and thawing along with salt spray from the sea have contributed to the weathering. (Norm R. Catto)

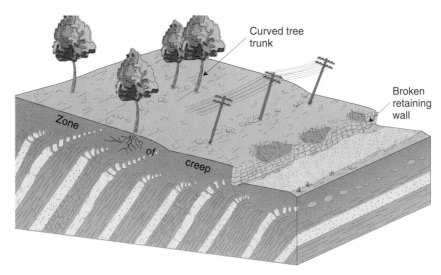

FIGURE 3-17 Soil creep. Soil creep is the most common form of mass movement. Creep, as the name suggests, is a very slow downslope movement of the uppermost part of the soil. (From E. J. Tarbuck and F. K. Lutgens, *Earth,* 7th ed. Upper Saddle River, NJ: Prentice Hall, 2002)

between two points (known as the *relief* or *rise*) divided by the horizontal distance between the two points (known as the *run*). The greater the rise and shorter the run between the two points, the faster the movement of materials down the hill. Wherever slopes occur, gravity is available to move material. Even the gentlest slope provides the potential energy necessary to move at least some material downward, either through mass movement or surface erosion. Erosion is usually much more rapid, however, on steeply sloping land than on gentle slopes.

Mass movement The most common form of mass movement is **soil creep.** As the name suggests, creep is a very slow, gradual movement of material down the slope of a hill (Figure 3-17). A tiny movement can cause creep—a rodent digging, a worm burrowing, an insect pushing aside soil. Creep occurs near the surface, in the top 1 to 3 meters (3 to 10 feet) of soil.

More dangerous and dramatic mass movements, such as rock slides and mudflows, can occur on steep slopes, especially during wet conditions. Steep slopes are prone to rock slides because the force of gravity pushing down on the rocks is likely to exceed the strength of the rocks. Landslides on steep slopes can follow intense rains, because material with a high water content is heavier, weaker, and less able to resist the force of gravity. The sliding material may break down into fluid mud, which flows downhill (Figure 3-18). Houses built on very steep slopes, such as along the west coast of North America, risk damage from landslides and mudflows.

Surface erosion The most common form of surface erosion is caused by rainfall. Intense rain sometimes

falls faster than soil can absorb it. Water that cannot infiltrate, or soak into the ground must run off the surface. As it runs off the surface, water picks up soil particles and carries them down the slope. With enough runoff, water can carve channels into the landforms.

The smallest channels eroded by the flow of water—only a few centimeters deep—are called *rills*. Rills are so small that soil creep or a farmer's plow can obliterate them. If channels gather enough water, however, they become larger and permanent carriers of water. As these stream channels deepen, they gather water and eroded soil from adjacent slopes. When the

FIGURE 3-18 Landslide. This landslide occurred on a steep slope on Santa Cruz Island, California, following heavy rain. The heavy rain weakened the rocks, triggering the landslide. Small scars near the top of the landslide reveal fractured surfaces in the rock. Once the weathered rock began to slide downhill, it was jostled and broken apart by the motion and turned to mud. By the time it reached the bottom of the slope, the landslide had become a mudflow. (WHR photo)

Critical Thinking

WEALTH AND NATURAL HAZARDS

When we hear about natural hazards such as wildfire, earthquake, landslides, and coastal erosion, one place that often appears in the headlines is Los Angeles. Los Angeles seems to have more than the usual share of such calamities. Dramatic scenes of houses toppled by landslides splash across the TV screens, and often the homes are very expensive ones. Some of the most hazardous neighborhoods in Los Angeles are also among the most desired. The hills overlooking Hollywood and Malibu and the Palos Verdes Peninsula, with its spectacular ocean views, are examples. The steep slopes scattered on mountainsides around Los Angeles offer isolation from city traffic and spectacular views, on clear days at least. The steep slopes are also the places most vulnerable to important hazards such as wildfire and landsliding caused by winter rains. In contrast, the poorest neighborhoods in Los Angeles are generally flat and far removed from such hazards. South Central Los Angeles is an example. People who live in these neighborhoods have other hazards to deal with, and they are just as vulnerable to earthquakes as anyone in the area. But it does seem that in Los Angeles the rich people have chosen to live in the more hazardous environments. This may cause an increase in the property damages that occur when disasters strike, because rich people's houses cost more than poor people's apartments.

Los Angeles is just one city, and a unique one at that. What about other cities? In some Latin American cities, the wealthier neighborhoods tend to be closer to the city center rather than on the outskirts. In the case of coastal cities like Rio de Janeiro, Brazil, the wealthy tend to live on lowlands, while many poor people live in shantytowns on steep hillsides. When landslides happen there, the poor may be more vulnerable than the rich.

Because every place is unique, it is difficult to make generalizations about the relation between wealth and vulnerability to hazards. Most people don't think about such hazards when they choose a place to live; other considerations are usually much more important. When we hear about disasters triggered by earthquakes, floods, or other natural hazards, however, the outcomes in poor countries are often very different from outcomes of similar events in rich countries.

Questions

1. Scan the news reports of the last few months for reports of major natural disasters. Who suffered the most? Who was affected the least? Did wealth help protect people from the danger? Did it make them more vulnerable? Or was wealth irrelevant?

2. What environments in your community are more vulnerable to natural hazards, and which ones are less vulnerable? Who lives in the vulnerable places?

---

streams gather enough water, they form ravines, valleys, and canyons.

Surface erosion by water is relatively slow in most natural environments because the ground is covered by grass and trees. But on large parts of Earth's surface, humans have removed vegetation by clearing forests and plowing fields. Once the vegetative ground cover is removed, slow surface erosion can suddenly become severe erosion (Figure 3-19). Ground where vegetation has been removed can suffer more surface erosion in a few months than it experienced during the previous several thousand years. The eroded soil contributes to water pollution downstream, and the remaining soil may be less productive for agriculture.

> Streams shape their channels by alternately eroding and depositing material on their beds and banks.

Stream drainage Streams collect water from two sources: groundwater and overland flow (Figure 3-20). When rain falls on the land surface, most of it soaks into the soil and accumulates as **groundwater.** Groundwater migrates slowly through the soil and underlying rocks. During dry periods, most of the water flowing into streams is supplied not by rainfall but by groundwater. If rain falls intensely, the soil may not be able to absorb it as fast as it falls. Water then may run directly to streams through **overland flow**.

A stream drains groundwater and overland flow from an area called its **drainage basin.** Drainage basins may be a small as a farm field, or as large as a major portion of a continent. (Smaller basins are nested

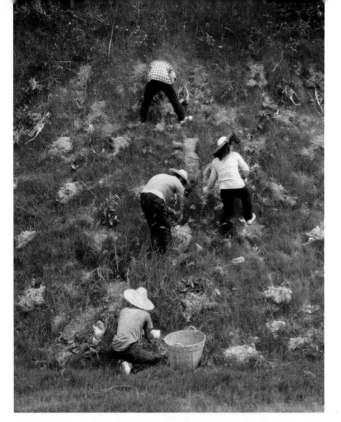

FIGURE 3-19 Agricultural workers plant crops on a steep slope in Hong Kong. Land scarcity requires cultivation on steep slopes in many parts of the world. (© Martin Jones/ CORBIS)

within larger basins). The greater the area of its drainage basin, the more water a stream must carry. Rivers in dry areas may be exceptions, because they may lose water to evaporation. A basin with plentiful runoff from groundwater and overland flow may carve a complex network of many channels to remove the water and sediment. Small rills deliver water and material to larger streams, which join others to form still larger rivers, which flow to the sea (Figure 3-21).

The volume of water that a stream carries per unit of time is its **discharge.** Discharge ranges from a few cubic centimeters per second in rills to over 200,000 cubic meters per second at the mouth of the Amazon River, the world's largest. Discharge of any stream usually increases after storms and decreases during dry spells.

Drainage density is the combined length of all of the stream channels in a basin, divided by the area of the drainage basin. A basin that has soil capable of absorbing and storing most of the rain that falls will usually have a low drainage density. Landscapes with soils that cannot absorb rainfall very rapidly are more easily eroded to form channels, and they tend to have higher drainage densities.

Streams shape their channels by alternately eroding and depositing material on their beds and banks.

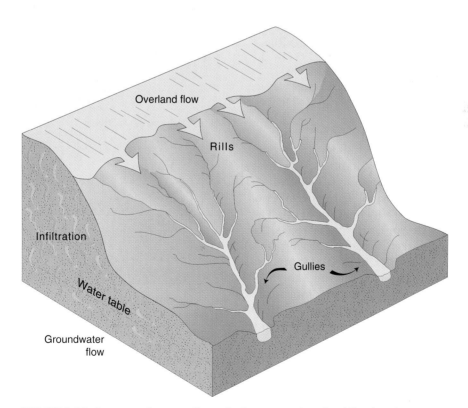

FIGURE 3-20 Sources of streamflow. Surface water (overland flow) and groundwater (water below the water table) are the two sources of water for streams. Under the pull of gravity, rainwater soaks into the soil. The water in the ground slowly migrates through soil and rock. The surface flow follows the lowest path available, eventually running into streams. Streams merge into other streams and rivers, so most water ultimately ends up in the ocean. If rainfall is intense and the soil cannot absorb it all, overland flow can become great enough to cause flooding.

FIGURE 3-21 A MODIS view of the mouth of the Yangtze River, China. The Yangtze enters from the west (left). The Yangtze carries an enormous sediment load, and construction of the Three Gorges Dam will reduce this flow of sediment to the sea, with unknown consequences for the coastal region. (Jacques Descloitres, MODIS Land Science Team/NASA Headquarters)

FIGURE 3-22 Floodplain of the Missouri River, Missouri. This image shows elevations in the Missouri River floodplain, as measured by satellite-borne radar. The floodplain, in shades of blue and purple, was hit hard by a flood in 1993. Upland areas are shown in shades of yellow. The yellow patches within the floodplain are areas of forest. The radar detects the height of the treetops and maps these areas as higher elevations than the surrounding fields even though the ground elevations are about the same. (NASA Headquarters)

The turbulent, swirling motion of the water erodes particles from the channel. The water transports these particles, along with loose sediment in the channel and minerals dissolved in the water. This movement of material in a stream is called **sediment transport.** The amount of sediment a stream carries increases as discharge increases, and larger streams typically carry more sediment than do smaller ones. Transport also increases greatly after a heavy rain. During periods of lower flow, less sediment is transported downstream and a channel may partially fill with sediment.

A stream is also responsible for shaping its **floodplain,** which is a nearly level surface at the bottom of the valley through which the stream is flowing (Figure 3-22). The surface of the floodplain is formed from deposits of sediment when the stream periodically floods. Flowing water tends to **meander,** or change direction from side to side, and this contributes to widening the channel. The channel continually shifts from side to side as the stream erodes material from one side of the channel, where current is swifter, and deposits it on the other side, where the slower current has less energy (Figure 3-23).

By continually eroding and depositing material in channels and floodplains, streams tend toward a stable condition, known as **grade.** A graded stream transports exactly as much sediment as it has collected

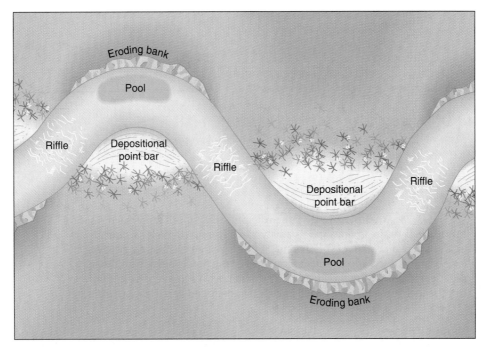

FIGURE 3-23 Features of a meandering channel. Erosion occurs primarily on the outsides of bends while deposition occurs on the insides. Depth usually alternates between deep portions or pools and shallows or riffles.

(Figure 3-24). Streams rarely operate at a condition of grade for long, because daily changes in weather and disturbances from erosion and human activities continually upset the balance. As the stream's stable condition is upset and the transport of sediment increases or decreases, the shape of the stream channel may change. An especially heavy flow following a storm may cause the channel to shift. When increased erosion upstream generates more sediment than a stream can carry, the excess is deposited in the channel or on the floodplain (Figure 3-25).

Most sediment carried by a stream does not move to its final resting place in a single step.

The deposition of excess sediment slowly raises the elevation of a stream which in turn reduces the difference in elevation between places upstream and downstream, and also reduces the slope. Lowering the slope reduces the amount of sediment arriving from upstream.

Most sediment carried by a stream does not move to its final resting place in a single step. Typically, sediment becomes temporarily stored in a floodplain, then eroded, transported, and deposited a second time, then a third time, and so on many times along its

FIGURE 3-24 An idealized profile of a stream and its numerous tributaries. Note the concave-upward shape of the profile, with steeper gradient toward the heads of streams and gentler gradient downstream. As the stream flows downhill, it gathers more and more water, and its erosive power is greatly increased. The gentler gradient downstream slows the flow, balancing the erosive power of the stream with the amount of sediment it must carry. (From E. J. Tarbuck and F. K. Lutgens, *Earth Science*, 7th ed. New York: Macmillan, 1994)

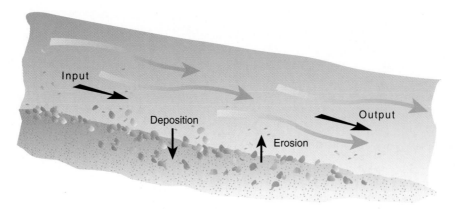

Input

Deposition

Erosion

Output

FIGURE 3-25 Stream erosion. Streams erode and deposit material in response to variations in streamflow and its ability to transport sediment. Sediment flows in and out of a portion of the channel. If more sediment enters than leaves, the excess is deposited and the channel shrinks or the floodplain accumulates sediment. On the other hand, if more sediment leaves than enters this portion of the channel, it means that the channel is being eroded and therefore enlarged.

journey. Eventually, most sediment reaches the sea. Where a river enters the sea, the water velocity drops abruptly, and the sediment may form a large area of deposited sediment called a **delta.**

Rainfall also is an important erosion agent in dry areas, even though it is rare. Because deserts lack vegetation, the occasional rainfalls cause rapid erosion. Dry soils are much more prone to erosion than are vegetated ones—look at any construction site after a heavy storm. Thus, a dry landscape often is covered with gullies and dry stream channels, even though only a few centimeters of rain may fall per year (Figure 3-26).

A prominent landform in many deserts is the **alluvial fan.** Alluvial fans are broad, gently sloping deposits formed from sand and gravel where a fast-

FIGURE 3-26 Erosion by water in the desert. Erosion by the San Juan River has exposed folded sedimentary rocks in southeastern Utah. (Tom Bean/Corbis/Stock Market)

moving stream emerges from a narrow canyon onto a broad valley floor (see Figure 3-15, where large alluvial fans are shown on the far side of the valley). When the stream emerges, the water quickly spreads out and loses velocity. Given the limited amount and frequency of rainfall, a stream in a desert is unlikely to carry sediment very far. Instead, sediment that the stream can no longer carry is deposited in an alluvial fan. Occasional floods crossing the alluvial fan can disturb the pattern and encourage wind erosion. Large depositional areas in desert valleys form important sources of wind-eroded sediment.

Increased erosion from human activity In addition to shaping landforms over millions of years, erosional forces are important in human time scales because human modification of Earth's surface is usually much more rapid than that occurring over geologic time. Human activities sharply increase the amount of sediment being eroded into streams. Major contributors include deforestation, agriculture, and urban development.

People clear forests because they want to use trees for fuel, lumber, and paper, or they want to use the land for another purpose, especially agriculture and urban development. Erosion has increased as a result of elimination of the vegetation cover for both reasons, but it is particularly severe where agriculture has replaced the forest.

For much of Earth's agricultural land, erosion of soil into streams is a major loss. To meet the needs of a growing population, food production expands in two principal ways—by opening up new land for agriculture and by using existing farmland more intensively. But both strategies can result in erosion of the rich soil necessary for productive agriculture.

Regional Focus on

MANAGEMENT OF THE LOWER MISSISSIPPI RIVER

Despite its location in the flat, tectonically stable interior of North America, the Mississippi River is a remarkably dynamic environment. The river drains nearly 3 million square kilometers (1.25 million square miles), about a third of the United States. Its vast floodplain, 100 km (60 miles) wide in places, is the resting place of sediment eroded from the Rocky Mountains in the west, the Appalachians in the east, and most of the land between. At the mouth of the river, deposition of sediment each year extends the delta further out into the Gulf of Mexico.

This extension of the delta seaward has an important effect on the river: It increases the horizontal distance the water has to travel to reach sea level. The vertical distance to the sea remains the same, however, so the channel slope decreases through time (slope = rise/run). The delta today reaches far out into the Gulf of Mexico, about 300 kilometers (190 miles) past Baton Rouge, following the river to the southeast. And yet due south of Baton Rouge the Gulf is only about 100 kilometers (62 miles) away. If the river could find a shorter path to the ocean, the water would quickly be diverted to this steeper path, and the longer, gentler channel would be abandoned.

In the geologic past, the Mississippi has repeatedly extended its delta well out into the gulf and then suddenly switched to a new, shorter route to sea level. The channel has switched locations several times in the last 10,000 years, with the mouth of the Mississippi at times reaching 300 kilometers to the west, near the Texas-Louisiana border. The switches happen suddenly, usually triggered by a large flood that overflows the channel and begins to cut a new, steeper, faster channel. If the channel were left to natural processes, such a switch would be imminent.

In 1973, one of the largest floods in recent years struck the lower Mississippi and very nearly did cause the channel to abandon its present course and follow a different path, known as the Atchafalaya River, to the Gulf. Engineers regulating the river have been aware of this possibility for many years, and a structure called the Old River structure for regulating the flow was built at the place where the diversion is most likely. Had this structure been washed away in the 1973 flood, as nearly happened, then the seaport facilities along the river from Baton Rouge to New Orleans and beyond would suddenly have found themselves without a river! Such a disaster was narrowly averted in 1973, but the root cause of the problem—the extension of the delta into the Gulf—remains. We can be sure that the engineers in charge will be vigilant.

Lower Mississippi River

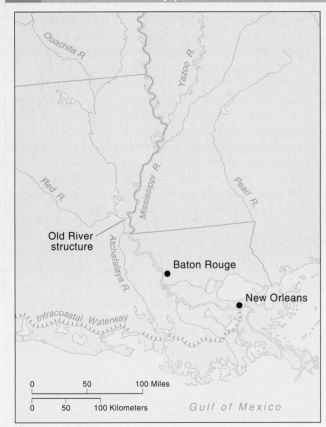

The lower Mississippi River enters the Gulf of Mexico at the tip of a long extension of the delta. A shift in the channel to the west would dramatically shorten the river's path to the Gulf.

In places with rapid population growth, new lands may be opened for agriculture that may not be suited for intensive farming. For example, farmers in East Africa are clearing, tilling, and planting extremely steep mountain slopes. The rate of surface lowering by erosion in such areas can reach a few millimeters per year. Within a few years the rich topsoil may be completely eroded in such areas. The consequences of this erosion for food production will be discussed in Chapter 8.

Opening up new land for agriculture was a major contributor to increased erosion in the United States during the eighteenth and nineteenth centuries. More than 2 million square kilometers (800,000 square miles) of forest were cleared and replaced with plowed fields and pastures in the eastern and midwestern United States. This deforestation probably increased the rate of soil erosion by 10 to 100 times.

Erosion of agricultural land is further increasing as farmers use existing fields more intensively. The soil on about one-fifth of U.S. cropland is being lost faster than it can be replaced by natural soil formation. Also, to obtain higher yields, farmers plant profitable crops like corn, soybeans, and wheat every year, instead of periodically planting cover crops such as clover or alfalfa which restore nutrients to the soil. By failing to restore nutrients to the soil, farmers can exhaust its productivity.

Soil erosion lessened in much of the eastern United States during the twentieth century because farmland was abandoned. Its soil was depleted of nutrients, and more productive land was available in the Midwest. Much of the former farmland in the eastern United States has returned to forest, and erosion is less severe today than it was in the past. When the prairies west of the Mississippi were brought under the plow in the late 1800s, erosion became a problem there, reaching crisis proportions in the 1920s. Since the 1930s, government-sponsored soil conservation measures have significantly reduced erosion on U.S. farms, but the problem remains serious in some areas.

A heavy rainfall or rapid snowmelt may dump an overload of eroded sediment into a stream, and it may be deposited on the floodplain downstream. Such sediment can bury nutrient-filled soils already in the floodplain, as well as roads and buildings constructed on the floodplain. Flooding may increase if a sediment-choked stream overflows its banks.

Even after the rate of soil erosion into a stream declines, a large quantity of sediment from the valley bottom may continue to be sent downstream. Many U.S. rivers have high sediment loads in part because they are currently excavating from their floodplains sediment deposited during a past time of severe agricultural erosion.

Urban development also increases erosion. Land is cleared of vegetation for building new houses, factories, and shops. Erosion rates during the construction phase often are hundreds of times greater than for undisturbed land. Evidence of this erosion is usually clear right after a storm—small channels that deepen downslope and mud washed onto adjoining areas, for example. If construction proceeds slowly, the land may be subject to high rates of erosion for several years.

Once land has been developed, surfaces once covered with vegetation are covered instead with roofs, streets, and parking lots. Because rain does not soak into these nonporous surfaces, it must run off into stream channels. Sewers collect storm water in cities and often discharge it into a stream at a rapid rate. If the sewers are unable to handle the flow during a heavy rainfall, low-lying areas in the city may flood. To handle increased discharge, streams in urban areas may enlarge their channels through erosion of their banks, damaging adjacent property.

ICE, WIND, AND WAVES

In some parts of the world, running water is a less powerful erosional agent than other forces. In some cold places, for example, the land is covered with ice that flows downhill, grinding away at rocks. The wind can cause surface erosion in places with sparse vegetation, especially in the world's extensive deserts. Wind is a much less powerful erosion agent than running water, for wind can carry only smaller particles up to sand sized, and it can carry these only short distances. In coastal areas, a combination of water and wind can cause surface erosion. Waves—which are driven by winds—pound the shore with turbulent water. Even the strongest rocks and vegetation are eventually broken down by the continual rushing forth and back of the waves.

Glaciers

Greenland, the South Pole, and many high mountain areas are currently covered with thick layers of moving ice, called **glaciers. Alpine glaciers** form on the peaks of individual mountains, wherever snow accumulates year after year without melting. **Continental glaciers** over 3 kilometers (almost 2 miles) thick cover vast areas of Greenland and Antarctica.

Glaciers are rivers of ice that flow from places where snow accumulates year to year to warmer places where the ice melts. Water enters the head of the glacier as snow (Figure 3-27). The glacier flows downhill until the ice eventually melts and leaves the glacier at its terminus as meltwater or icebergs.

Glaciers flow very slowly, usually at rates of a few meters to a few hundred meters per year. Glaciers may change size from one year to the next, depending on variations in weather and climate. They grow when they receive more snowfall at their head, and they shrink if warm temperatures increase the melting at the terminus. For the past 200 years most of the glaciers of the world have been shrinking as a result of climatic warming over that time period (Figure 3-28). Glaciers

FIGURE 3-27 The Rhône Glacier, Switzerland. (above) Crevasses (cracks) on the glacier's surface are caused by flow. For most of the past 200 years, this alpine glacier has been shrinking, with melt exceeding accumulation. As a result, the glacier is much smaller than it was in the 1700s, probably as a result of climatic warming in the 1800s and 1900s. (Gabe Palmer/Corbis/Stock Market)

FIGURE 3-28 Shrinking glacier. (right) The Qori Kalis Glacier, Peru, in 1978 (top) and the same view in 2000. Most of the world's glaciers have been shrinking rapidly in recent decades as a result of climatic warming. (Lonnie Thompson/Byrd Polar Research Center)

can also act erratically. For many years a glacier may move only a few meters per year, then for a few years it may suddenly surge forward at several hundred meters per year, before slowing again.

A glacier is like a conveyor belt because it picks up sediment from areas of erosion and drops it in depositional areas. As ice accumulates and begins to flow, the glacier picks up material, and where it melts, the sediment is deposited. Glacial deposits play an especially important role in shaping landforms in places with rapid melting, because large quantities of material are dropped in these places, forming **moraines** (Figure 3-29). A **terminal moraine** is a ridge of material dumped at the end of the glacier.

A glacier is like a conveyor belt because it picks up sediment from areas of erosion and drops it in depositional areas.

Meltwater leaving a glacier deposits some of the debris close to the glacier in a broad, gently sloping plain, known as an **outwash plain.** The outwash plain contains a thick layer of rocks deposited close to the glacier in a layer of sand and gravel that can exceed a thickness of 100 meters (330 feet). The finer silt and clay materials are usually carried much farther and may be deposited in lakes, seas, or distant valleys.

Impact of Past Glaciations

Only 20,000 years ago—a very short time in Earth's 4.6-billion-year history—glaciers covered much of North America, Europe, and northern Asia (Figure 3-30).

These glaciers shaped landforms as they advanced. When Earth's atmosphere warmed, these rivers of ice melted back toward the poles or upslope to cold areas, but they left behind debris that shapes the landforms we see today in many regions.

As they advanced and retreated, alpine glaciers created distinctive landforms. In an unglaciated mountainous area, a stream may carve a V-shaped valley (Figure 3-31a, p. 120).

As an alpine glacier flows through a V-shaped valley, it scours away the rock and rounds the valley bottom into a U-shape (Figure 3-31b). When the ice melts, a U-shaped valley remains, surrounded by knife-edged ridges (Figure 3-31c).

The advance and retreat of continental glaciers have influenced the distribution of many human activities. The following are examples of how glaciation has

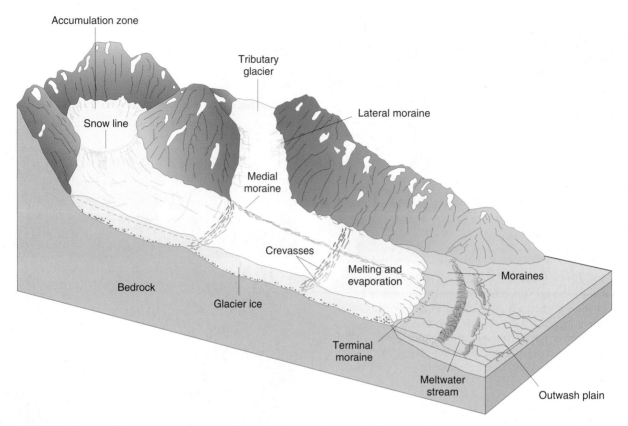

FIGURE 3-29 Landforms of alpine glaciation. Alpine glaciers are fed by snow in a zone of accumulation and flow downhill. They melt at lower elevations, leaving debris to accumulate at the end (terminal moraine), side (lateral moraine), or middle (medial moraine) of the glacier or be carried away in meltwater streams. (From R. W. Christopherson, *Geosystems,* 5th ed. Upper Saddle River, NJ: Prentice Hall, 2003)

affected three important natural resources in North America: soils, drinking-water supplies, and transport routes.

Soils The advance and retreat of glaciers in southern Ontario and in the U.S. Midwest broke up bedrock, ground it into sand and silt, and left a great layer of debris covering the land. This rock contains important nutrients such as calcium and magnesium that are released by chemical weathering. The soils that have resulted are highly fertile and help make this region one of the most productive agricultural areas in the world. Other factors also contribute to the productivity of this region, and we will see in Chapter 8 how soil and climate interact with human inputs to determine regions' agricultural productivity.

Water supply Retreating glaciers left sand and gravel deposits capable of yielding a large supply of high-quality groundwater (Figure 3-32). Long Island, New York, which is more than 100 miles long and about 20 miles wide, is the top of a terminal moraine deposited about 20,000 years ago, marking the southern extent of the most recent glaciation. Much of Long Island consists of outwash deposits of sand and gravel that absorb precipitation readily. This groundwater is

the primary source of water for 2.5 million inhabitants of Long Island.

With rapid population growth in recent years, however, demand for water has exceeded the rate at which the groundwater is being recharged. As more of the island is paved or built over, precipitation is blocked from entering the ground. As a result, the level of groundwater has lowered, and saltwater from the Atlantic Ocean has seeped in to replace the depleted fresh groundwater in some places. The quality of the groundwater has also suffered by pollution from landfills, industry, and residential septic systems. Similar problems are occurring in other urbanized areas of the northeastern United States and Europe that depend on groundwater left behind by retreating glaciers.

Transportation routes Before the last continental ice sheet formed tens of thousands of years ago, the Missouri River flowed northward into Hudson Bay in Canada instead of southward to the Gulf of Mexico. The Ohio River did not exist; another river to the north carried water from the western slope of the Appalachians to the Mississippi. But as the glaciers melted and began to recede, these rivers were blocked by the glacier, and their water could not take the more northerly routes to the sea. Instead, these rivers, plus the

Extent of Continental Glaciation During the Last Ice Age

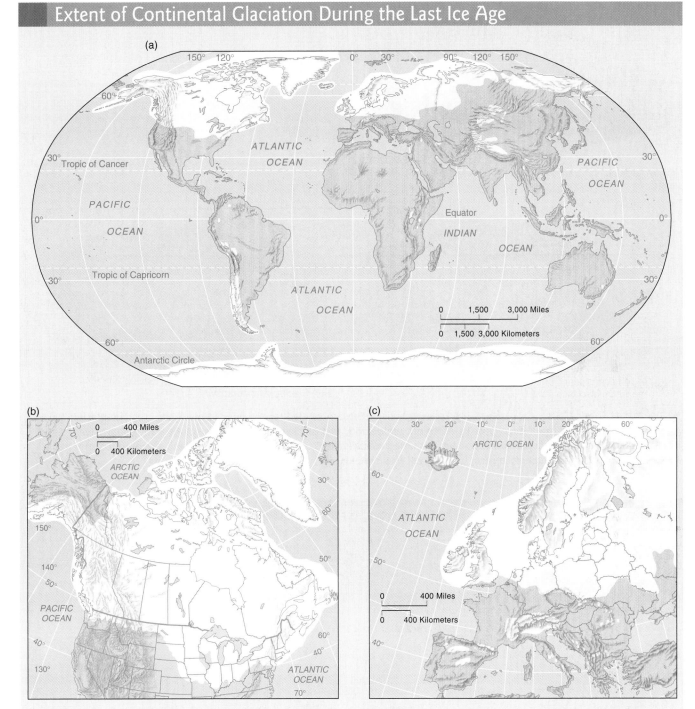

FIGURE 3-30 Much of the Northern Hemisphere was ice-covered 20,000 years ago. These maps show the extent of glaciation (a) worldwide, (b) North America, (c) Europe.

meltwater from the glaciers, were forced to flow along the edge of the glacier until they emptied into the Mississippi River. The lower Missouri River and the Ohio were thus created as meltwater channels following the edge of the glaciers, and when the glaciers melted, they remained in this new position. The receding glaciers also carved the Great Lakes, which explorers eventually used to reach the interior of the continent.

European exploration of the North American continent during the seventeenth and eighteenth centuries and eventual settlement of the North American interior were heavily influenced by water transport. This exploration and settlement followed the St. Lawrence River to the Great Lakes and the major inland rivers feeding into the Mississippi, such as the Ohio, Missouri, and Illinois. Travel from the U.S. East Coast to the interior

(a) Unglaciated topography

(b) Region during period of
    maximum glaciation

(c) Glaciated topography

FIGURE 3-31 Typical features of glaciated landscapes. (a) Stream-eroded terrain prior to glaciation. (b) The glaciers have their way with the land, gouging, scraping, scouring, bulldozing, and plucking rock and soil. (c) After the climate warms and the glacier melts, distinctive landforms remain: U-shaped glacial valleys, new lakes and streams, and sharp-edged mountain ridges. (From E. J. Tarbuck and F. K. Lutgens, *Earth,* 4th ed. New York: Macmillan, 1993)

Glacial Deposits in Long Island, New York

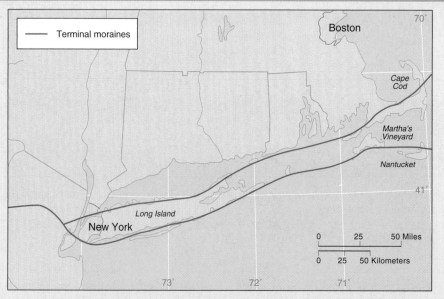

FIGURE 3-32 Long Island is covered by glacial deposits. The island is actually the top of a system of terminal moraines that also includes Martha's Vineyard, Nantucket, and Cape Cod, Massachusetts. The sandy soils of Long Island absorb water readily, and groundwater resources once were abundant. Today, however, these resources are threatened by overuse, intrusion of saltwater from the ocean, and pollution.

Glacial Effects on North American Drainage

FIGURE 3-33 Glacial meltwater channels profoundly altered the drainage of North America and created paths that were exploited by early explorers and later in the construction of canals. A series of major lakes were scoured by glaciers at the margins of the Canadian Shield, including the five Great Lakes, Lake of the Woods, Lake Winnipeg, Reindeer Lake, and others. Several of these are connected by channels formed by glacial meltwater draining from one lake to another, including the Red River-Minnesota River system, which at one time drained meltwater from Lake Winnipeg into the Mississippi River system. The Ohio and Missouri Rivers were created by meltwater flowing around the margins of the continental ice sheet (compare this map with Figure 3-30b). Some of these lakes and rivers are connected by canals, often constructed along glacial meltwater channels. Such canals include (1) the Erie Canal, and (2) the Illinois Barge Canal.

was difficult, because no convenient direct water route existed from the Atlantic to the Great Lakes and the Mississippi River system. After the United States gained its independence from Britain at the end of the eighteenth century, many canals were built in **meltwater channels** left behind by receding glaciers (Figure 3-33).

The Champlain Canal follows a meltwater channel to connect the Hudson River (which flows into the Atlantic at New York City) with Lake Champlain (which empties into the St. Lawrence River). The Erie Canal follows a meltwater channel for part of its route across New York State to connect the Mohawk River with Lake Ontario. The Illinois Barge Canal connects the Mississippi River with Lake Michigan at Chicago by following a meltwater spillway occupied by the Illinois River.

Effects of Wind on Landforms

Wind is an important shaper of Earth's landforms, especially in the dry regions. Wind erosion is also significant wherever the soil is not well covered with vegetation. This includes deserts, farmland, and coastal areas where beaches are kept free of vegetation by waves washing the shore.

Wind is not capable of moving large particles like gravel, but it can carry great amounts of fine-grained sediment such as sand. Where wind velocities are lower, or where topography encourages deposition, the sand accumulates in **dunes.** These accumulations of shifting sand are difficult places for vegetation to become established.

When the fine sand is eroded from the soil surface, larger rocks, pebbles, and gravel are left behind. The landforms in the desert can form a hard, armored surface, called **desert pavement.** The most popular image of a desert landform is a sand dune, but about 90 percent of Earth's desert areas are covered with desert pavement rather than sand. Wind erosion is minimized once desert pavement has formed, because the remaining rocks are too heavy to be moved by the wind.

Wind also can affect landforms in humid regions, because very fine-grained sediment eroded by the wind can be carried some distance before being deposited (Figure 3-34). Thick layers of windblown silt, called **loess,** blanket many areas, including central China and the Mississippi River valley of the United States. Originally, this material was carried by meltwater from northern glaciers to nearby valley bottoms. Wind then carried the sediment to adjoining areas, forming loess, a fine agricultural soil.

> A coast is an especially active area of erosion because an enormous amount of energy is concentrated on the shorelines from pounding waves.

Running water subsequently has carved deep gullies and ravines in these loess deposits.

Windblown sand is common in coastal zones as well as in dry climates because of the lack of vegetation. Along coasts, however, vegetation is usually scarce because of the pounding of waves, rather than because of the climate.

Coastal Erosion

A coast is an especially active area of erosion because an enormous amount of energy is concentrated on the shorelines from pounding waves. Land may be lost to erosion or gained through deposition at rates up to several meters per year.

Waves   Winds blow across the sea surface, transferring their energy to the water by generating waves. Waves are a form of energy, traveling horizontally along the boundary between water and air. As the wind blows harder and longer, it transfers more energy and generates bigger waves. Waves also grow larger with an increase in the expanse of water across which the wind blows.

The speed at which a wave travels is affected by its **wavelength.** A small ripple travels very slowly, whereas an ocean swell wave may travel 10 to 50 kilometers per hour (6 to 30 miles per hour). A **tsunami,** which is an extremely long wave created by an underwater earthquake, may travel hundreds of kilometers per hour (Figure 3-35).

A wave can travel thousands of kilometers across deep ocean water relatively unchanged, but when it nears the shore, the shallow bottom restricts water motion and distorts the wave's shape. The wave is slowed, causing the top part of it to rush forward and break (Figure 3-36).

The energy of a wave is released as a tremendous erosive force of rushing water on the beach. Beach pebbles and granules of sand are rolled back and forth, constantly being ground ever finer. The finest particles are carried into deep water and settle on the seafloor. The larger sand- and gravel-size particles are left behind on the shore to form a **beach,** which is a surface on which waves break and constantly move sand up and down.

A beach reflects the characteristics of the waves that form it. A coastal area pounded by larger waves will likely form a beach of very coarse material, because the waves carry the finer sand offshore and deposit it in quieter waters. The shape and size of the beach can vary if storms hit during some seasons, while other seasons are quiet.

FIGURE 3-34 Wind erosion. Satellite view of a dust storm in southwest Asia. This image, taken May 2, 2003, shows dust blowing on the Arabian peninsula (left), Iran, Pakistan, and Afghanistan. Plumes of sediment laden water are also visible entering the Arabian Sea from India (right). (Jeff Schmaltz, MODIS Rapid Response Team/NASA Headquarters)

Longshore current   When you watch waves break on a beach, the most obvious motion of the water is perpendicular to the shoreline—the waves move up the

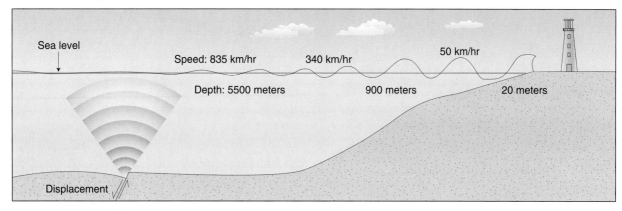

FIGURE 3-35 Tsunami. Earthquakes beneath the sea cause *tsunamis* (the Japanese word for "harbor wave"). A seafloor fault slips, violently displacing the water above, creating a very low wave at the surface that moves at a very high speed. As the wave comes ashore, it can build to heights of tens of meters. Historically, tsunamis have killed thousands in coastal settlements. After losing 150 people in a 1946 tsunami, Hawaii established the Pacific Tsunami Warning Center, which alerts nations around the Pacific when a tsunami is detected. (From E. J. Tarbuck and F. K. Lutgens, *The Earth,* 4th ed. New York: Macmillan, 1993)

beach and then recede. When the waves break, the energy gives a push to the water in a direction parallel to the shore, and the repeated breaking of many waves generates a **longshore current** traveling parallel to the shore (Figure 3-37). The longshore current is like a river, carrying sediment through **longshore transport** from areas where it is eroded by waves and depositing it where breaking waves lose the energy to carry it—usually in deep water. Longshore currents can carry enormous amounts of sediment great distances.

> Like rivers, landforms along shorelines are shaped by the balance between sediment arriving in a portion of the shore and being removed from it.

Like rivers, landforms along shorelines are shaped by the balance between sediment arriving in a portion of the shore and being removed from it. Distinctive landforms develop, such as the beaches, barrier islands, and spits illustrated in Figure 3-38. If more sediment is removed than arrives, the coast is eroded—which is the condition of most of the world's shorelines (Figure 3-39). But in some areas, more sediment arrives than is removed, and the land area grows.

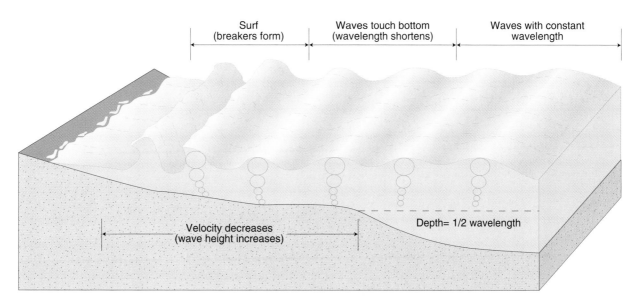

FIGURE 3-36 Breaking waves. As waves reach shallow water, they change shape, with long, low waves becoming short and tall, eventually breaking and releasing their energy on a beach. (From E. J. Tarbuck and F. K. Lutgens, *Earth,* 7th ed. Upper Saddle River, NJ: Prentice Hall, 2002)

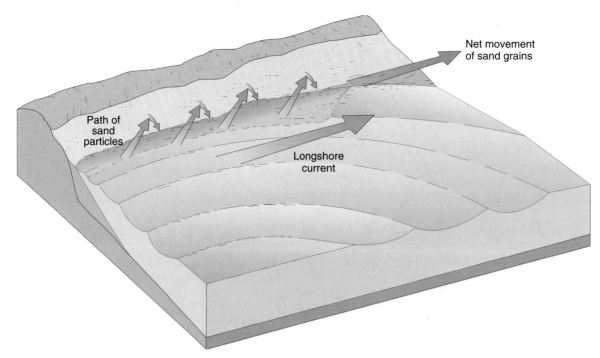

FIGURE 3-37 Longshore current. The longshore current is driven by waves breaking at an angle to the coastline. The current flows parallel to the shore. Breaking waves stir beach sand causing it to be carried along with the current. (From E. J. Tarbuck and F. K. Lutgens, *Earth,* 7th ed. Upper Saddle River, NJ: Prentice Hall, 2002)

FIGURE 3-38 Coastal deposits. Deposition along coastlines creates beaches in small embayments (a), spits across the mouths of some bays (b), and barrier islands on gently sloping coastlines (c).

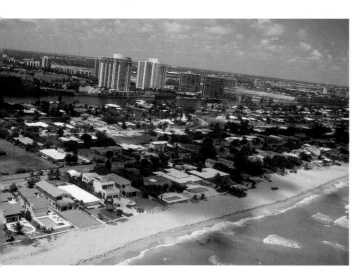

FIGURE 3-39 Beachfront homes in Fort Lauderdale, FL. The land in the foreground is a barrier island, built by longshore transport of sand along the shoreline. Most of the beaches in this part of Florida are eroding, and sand must be imported to replace what is lost to the waves. If sea level rises, erosion rates will increase. (Alese and Mort Pechter/Corbis/Stock Market)

Sea-level change The edge of the land—the shoreline—is defined by the elevation of the sea, or **sea level.** We think of sea level as being fixed, but on most shorelines it is not constant; it continually rises or falls relative to the adjacent land. In the short term, the sea can rise or fall several meters because of tides and storms. But two long-term factors can cause sea level to rise or fall: climate change and movements of Earth's crust.

Over the past few hundred years, sea level has risen on Earth as a whole, at about 1 millimeter per year along coasts that are otherwise stable. Sea level has risen because the volume of seawater has increased. The increase results from the melting of glaciers, due to the overall warming trend since the eighteenth century, as well as from warming of the sea itself. This change in sea level is small compared to the sea-level rise of about 85 meters (280 feet) that occurred at the end of the most recent Ice Age.

Sea-level changes are significant at two different time scales. In the short term—a few decades—the direction of sea-level change affects how a shoreline erodes. If sea level rises, the water offshore becomes deeper and waves break closer to the land, causing more erosion. If sea level falls, the shallower water causes waves to break further offshore, dissipating their energy and reducing shoreline erosion. Because many shorelines have gentle slopes, a minor increase in sea level can translate into a much larger landward migration of the shoreline (Figure 3-40).

Over thousands of years, large sea-level change can reshape shorelines. During continental glaciation, the sea was substantially lower because more of the world's water was frozen in glacial ice on the land. With sea level lower, rivers in coastal areas cut deep valleys as they approached the sea. When the glaciers melted, sea level rose worldwide, about 15,000 to 20,000 years ago, drowning the river valleys and creating large bays such as the lower St. Lawrence River, Chesapeake Bay, and Delaware Bay.

Sea level has fallen rather than risen relative to the adjacent land in some places. This has left inland soils, animals, and vegetation that are typical of beaches. For example, much of the U.S. West Coast has been tectonically uplifted in the last few million years as the North American and Pacific plates have ground together. As a result, the region lacks the many deep river mouths of the U.S. East Coast, but it does have former shorelines that are now above sea level, forming **marine terraces** (Figure 3-41).

Human impact on coastal processes The sea has been important for thousands of years for trade, communication, and food. This long tradition has established most of the world's densest populations

FIGURE 3-40 Sea-level and coastal erosion. Sea-level rise and shoreline position are directly related. On gently sloping shorelines, such as coastal plains, relatively low rates of sea-level rise can translate into significant rates of shoreline movement. For example, if the land rises only 1 meter for each kilometer inland (a ratio of 1:1,000), a 1-millimeter-per-year rate of sea-level rise could mean horizontal shoreline displacement of 1 meter per year. Such rates of shoreline erosion usually result in major problems in developed coastal areas.

FIGURE 3-41 Marine terrace. The broad green area in this photo of California's Big Sur coast is a marine terrace, a former beach raised above the sea by tectonic uplift. Such terraces are common on the West Coast of the United States. (Renee Lynn/Photo Researchers, Inc.)

near the sea. The shoreline is an attractive place to build houses and recreational facilities, so coastal areas worldwide have become focal points for settlement and investment.

A drawback of coastal living is erosion from shifting shorelines. In wealthy societies like the United States, the typical response is to try to control coastal processes by modifying the coastline. These structures achieve the desired erosion control and stabilize the shoreline—but only temporarily. For example, people build groins perpendicular to the shore to slow longshore transport of sediment (Figure 3-42). A groin constructed in one location along the shore to interrupt the movement of sand will cause less sand to be deposited on a beach farther along the shore, worsening the erosion problem there. People also build sea walls parallel to the shore, but the constant pounding of waves removes sand from around the sea walls and ultimately undermines them.

The shoreline is one of the most dynamic features of Earth's surface. On an eroding barrier island of the eastern United States, erosion over the past few decades could average between 10 centimeters and 2 meters (between 4 inches and 7 feet) per year. At that rate, someone who takes out a 30-year mortgage to buy a house that is 30 meters (100 feet) from the beach may

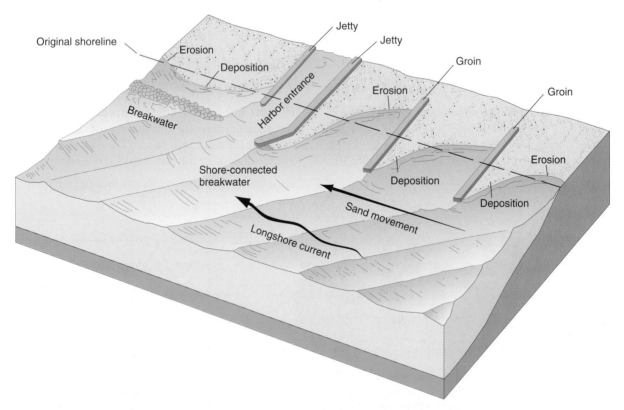

FIGURE 3-42 Human modification of coastal processes. Attempts to modify the original shoreline only change the pattern and cause problems elsewhere. (Adapted from R. W. Christopherson, *Geosystems,* 5th ed. Upper Saddle River, NJ: Prentice Hall, 2003)

see waves lapping at its edge by the time the last payment is made. And if a hurricane hits, the house may be washed away even sooner.

Homebuilders may not know about the threat of erosion when they build new houses on an attractive beach. A coastal development is usually well established before its residents discover the dynamic nature of the coast. By then it is too late to abandon the substantial investment, and more money is spent to protect the structures already built. Most of the erosion and damage to these communities takes place during major storms such as hurricanes. After such a disaster, sympathy for the victims induces the government to help rebuild communities at considerable public expense. Over the long run, though, buying up the properties and relocating the families would probably be cheaper than repeatedly rebuilding shorefront communities.

THE DYNAMIC EARTH
Rates of Landform Change

We can now see how past geologic events have shaped the world we live in and how, occasionally, dramatic Earth movements occur. Most landforms change so slowly that they provide metaphors for permanence, such as everlasting hills and rock of ages. Rates of horizontal movement of continents relative to one another are typically millimeters per year to centimeters per

year. Vertical movements of the land are somewhat slower—generally less than a few millimeters per year. Rates of landscape lowering by weathering and erosion are even slower. The surface of a mountain may erode at 5 to 50 millimeters per century, and a gentle slope at only 0.1 to 1 millimeter per century.

Change may be somewhat faster where processes are strong and materials are weak. For example, massive granite usually erodes very slowly, but a steep slope of fractured shale may erode rapidly. Similarly, waves erode a shoreline of glacial sand and gravel much faster than one of solid rock. Much of the Massachusetts shoreline, which consists of glacial deposits, is eroding horizontally at a rate of 0.5 to 2 meters per year (2 to 7 feet). But in Maine, resistant rocky coasts are eroding at less than 10 millimeters a year (0.4 inch).

One way to assess the significance of a rate of change is to determine how long it takes to create a landform that reflects the prevailing environmental conditions. For example, how long does it take for a river to create a channel of suitable size to carry its runoff? How long does it take for waves to shape a coastline into a smooth beach? How long does it take for volcanic rock to develop into mature soil? The time required to shape the land varies greatly, depending on the processes and the materials involved (Figure 3-43).

In virtually every landscape, features that take a long time to develop can retain the marks of past events, such as climate and sea-level changes tens of thousands of years ago. These same landscapes also

	Hours	Days	Weeks	Years	10's of years	100's of years	1,000's of years	10,000's of years	100,000's of years	Millions of years
Fluvial landform										
River channel				▬▬▬▬▬▬						
Valley fill						▬▬▬▬▬				
Drainage density				▬▬▬▬▬▬						
Slope forms										
Soil profile						▬▬▬▬▬▬				
Slope profile					▬▬▬▬▬▬▬▬▬▬					
Coastal forms										
Beach face profile	▬▬▬▬▬									
Beach dune profile		▬▬▬▬▬▬▬▬▬								
Coastline plan form				▬▬▬▬▬▬						

FIGURE 3-43 Length of time to form selected landscape features. The length of time needed to create a landform is short where processes are strong and materials weak. But if large amounts of material must be moved, landforms can take very long periods to be shaped by weak processes.

contain more dynamic features that respond to the weather or changes in land use and so reflect the environmental conditions of the present or very recent past.

Another way to view rates of change is to see their effects on human activity in particularly sensitive areas, such as along coastlines. In the Mississippi Delta, land is sinking relative to the sea at about 10 millimeters per year (four tenths of an inch). The city of New Orleans was founded in 1718 at a site close to sea level. If the land sinks at 10 millimeters per year for 200 years, the total subsidence is 2 meters (7 feet), a significant amount for a low-lying city. This subsidence also complicates management of floods in the lower Mississippi Valley.

Environmental managers must understand natural rates of landform change so they can predict and manage human impacts on the land. To maintain naturally productive soils, for example, we cannot allow soil to erode severely and then expect it to recover in a few years. On the other hand, if we pave an area and thus alter the amount of water entering a stream, we should not be surprised that the stream quickly floods and causes increased erosion.

Environmental Hazards

Some environmental processes, such as volcanic eruption, earthquakes, landslides, tornadoes, and hurricanes, occur so rapidly and violently that they cause widespread death and destruction. From a human perspective, these changes are momentous, because at any given location they are rare in human time scales. From a geologic perspective, however, they are routine.

In some areas, rapid change is frequent enough for people to recognize the threat and avoid it. But a coastal lowland may experience a severe hurricane only once in several decades, so people feel they can safely build homes there. Because many environmental hazards are once-in-a-lifetime events, people often rebuild after a disaster rather than move to a safer area. They believe that another similar event is unlikely in the near future, and the risk is too small to justify giving up a livelihood or a home. Rebuilding often is accompanied by public pressure to build new structures to protect the community, such as a levee or sea wall.

Unfortunately, structures such as sea walls on the shoreline or levees along rivers cannot protect people and property against the most extreme events in a given environment. And in many cases, these structures worsen the problem. For example, levees confine floods within a narrow channel instead of letting the water spread out across a floodplain, so the water height during a flood increases. Protective structures also worsen the problem by giving people a false sense of security, leading them to make greater investments, thus increasing their vulnerability. When the next flood, storm, or wildfire strikes, the damage is even greater.

As was discussed in Chapter 1, humans adapt to the natural environment, and they also modify it. Geographers studying these problems recognize that natural disasters are usually not purely natural. Rather, they are caused by a combination of natural events and human vulnerability. In a flood, a house is destroyed for a simple reason. Someone chose to build where a flood can occur. In most cases, the best approach is not to control the threat but instead to change human behavior.

In the wake of the 1993 floods on the Missouri and Mississippi rivers, some levees were neither repaired nor rebuilt. Instead, flood-prone land was bought or leased by the government so that future floods can inundate this land instead of damaging valuable properties again. Allowing land to be inundated instead of keeping out floodwaters has multiple benefits: Damage is reduced to structures in the path of the flood, and downstream communities are protected. Approaches to hazard management that focus on avoiding hazards rather than confronting them are usually much cheaper in the long run.

CONCLUSION: CRITICAL ISSUES FOR THE FUTURE

Earth's surface includes some features that have remained much the same for millions of years, and others that have changed dramatically in the last few hundred years. Where landforms are stable and unchanging, humans usually do not have significant impacts on the land. Human activity can destabilize landforms, however, meaning danger for people living in that environment. A destabilized drainage basin might now have more frequent floods that erode riverbanks. An unstable coastline might now experience severe erosion. A deforested hillslope now might be prone to catastrophic landsliding.

As Earth's population grows and more land is developed for human use, people must learn to live with instability rather than try to control it. When Earth is moving, either in the sudden jolts of an earthquake or through gradual erosion, human structures are likely to fail. Future urban growth and land development must be sensitive to Earth's dynamic nature by avoiding hazardous areas. Floodplains and other low-lying areas should be left undeveloped or used for parks. Construction of housing on landslide-prone slopes should be discouraged, and soils that are easily eroded should be protected with conservation measures.

Geographers' training and experience in both the natural and social sciences place them in a unique position to help solve problems of human occupancy of hazardous environments. Geographers have been leaders in identifying natural hazards and developing strategies for coping with them. Such skills are increasingly important on our densely populated and dynamic Earth.

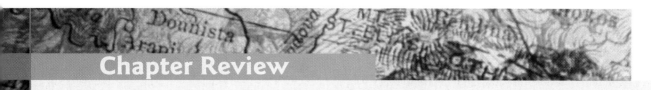

Chapter Review

SUMMARY

Major landforms of the world are created by a combination of endogenic (internal) and exogenic (external) landforming processes. Endogenic mechanisms are forces that cause movement beneath Earth's surface, raising some portions and lowering others. The most significant of these are movements associated with plate tectonics. This motion can cause earthquakes, volcanoes, and mountains, depending on whether the boundaries between plates are convergent, divergent, or transform.

Exogenic processes are forces of erosion that wear down Earth's crust and reshape it into new landforms. Rocks are first broken down into smaller pieces through chemical and mechanical weathering. Then they are carried by gravity down the slopes of hills or by wind from one place to another. Streams collect water from groundwater and overland flow and transport the water down to the sea. Streams are conveyor belts for sediment from hillsides, and they play a major role in shaping landforms.

In many mountain and poleward environments, glaciers flow across the land, eroding rock and depositing it where higher temperatures melt the ice. Wind plays a major role in moving material in deserts, and the absence of vegetation allows even infrequent rainfall to shape the land. Along coastlines, waves caused by wind blowing across the ocean surface cause intensive erosion and rapidly change landforms.

Most change on Earth's surface is slow in human terms, usually taking thousands of years to significantly reshape the land. But in some areas, change may be dramatic. Geographers have learned to identify areas that are subject to rapid or sudden change and to understand better how natural processes occur and how they affect human settlements. In general, we have learned that in dynamic environments it is better to live with nature and avoid hazards than to attempt to control them.

KEY TERMS

alluvial fan p. 114
alpine glacier p. 116
beach p. 122
chemical weathering p. 108
composite cone volcano p. 103
continental glacier p. 116
convergent plate boundary p. 104
delta p. 114
desert pavement p. 122
discharge p. 111
divergent plate boundary p. 104
drainage basin p. 110
drainage density p. 111
dune p. 122
earthquake p. 100
endogenic processes p. 99
epicenter p. 101
exogenic processes p. 99
fault p. 105
floodplain p. 112
focus (of an earthquake) p. 101
geomorphology p. 98

glacier p. 116
grade p. 112
groundwater p. 110
igneous rocks p. 105
isostatic adjustment p. 105
karst p. 108
landforms p. 99
lava p. 102
loess p. 122
longshore current p. 123
longshore transport p. 123
magma p. 102
mantle p. 100
marine terrace p. 125
mass movement p. 108
meander p. 112
mechanical weathering p. 108
meltwater channel p. 121
metamorphic rocks p. 105
moraine p. 117
outwash plain p. 117
overland flow p. 110

plate tectonics theory p. 100
seafloor spreading p. 104
sea level p. 125
sediment transport p. 112
sedimentary rocks p. 105
seismic waves p. 101
seismograph p. 101
shield p. 105
shield volcano p. 102
sial p. 105
sima p. 105
soil creep p. 109
structural landform p. 107
surface erosion p. 108
tectonic plate p. 100
terminal moraine p. 117
transform plate boundary p. 104
tsunami p. 122
volcano p. 102
wavelength p. 122
weathering p. 108

QUESTIONS FOR REVIEW AND DISCUSSION

1. What are endogenic and exogenic processes? Make a list of mechanisms that affect the shape of Earth's surface, and classify each as endogenic or exogenic.

2. What three types of relative motion occur at plate boundaries? What are examples of boundaries where each of these three types of motion is occurring?

3. What is isostatic adjustment, and what are its causes?

4. What environments favor rapid rates of mass movements?

5. What is a drainage basin? What measurable characteristics of drainage basins are important to understanding river processes?

6. What landforms provide evidence of sea-level rise? Sea-level fall?

7. How has soil erosion from agricultural lands modified sediment transport, erosion, and deposition in rivers?

8. What are some examples of landforms whose shape can be understood in terms of the processes acting on them? How long does it take for these landforms to develop?

THINKING GEOGRAPHICALLY

1. What were the environmental conditions where you live (or go to school) at the time of the last glacial maximum, 20,000 years ago? Were glaciers present? If not, how was the climate different? How was the vegetation different? What is the evidence that reveals these differences?

2. In your library, obtain a copy of a Soil Survey for your area. Soil Surveys in the United States are published by the Natural Resource Conservation Service, and typically available at most libraries. Read the sections pertaining to the development of soils in the area. Select two different sites that exemplify different kinds of landforms or deposits and visit these sites, comparing what you see with what is written in the soil survey.

3. Look at a topographic map of an area that includes a river. Can you identify the meanders of the channel?

Measure the width of the river. Now measure a distance along the channel that is long enough to include several bends of the river. Count the number of right-hand bends in the measured portion of the river, and divide into the distance. The result is the meander wavelength, normally 10 to 15 times the river width. Try these measurements on other maps to see how consistent the relation is.

4. Mount Saint Helens in Washington State erupted violently in 1980. The Loma Prieta earthquake in California occurred in 1989. How far apart are these two points on a map? Was there a possible connection between these two events? List reasons why there might be a connection, and why there might not.

SUGGESTIONS FOR FURTHER LEARNING

Abrahams, Athol D., and R. A. Marston. "Drainage basin sediment budgets: An introduction." *Physical Geography* 14 (May–June 1993): 221–224.

Boardman, John. "Periglacial geomorphology." *Progress in Physical Geography* 15, no. 1 (1991): 77–82.

Collins, M. J., and J. C. Knox. "Historical changes in upper Mississippi River water areas and islands." *Journal of the American Water Resources Association* 39, no. 2 (April 2003): 487–500.

Cooke, R. U., A. Warren, and A. Goudie. *Desert Geomorphology*. London: UCL Press, 1993.

Cooke, R. U., and J. C. Doornkamp. *Geomorphology in Environmental Management: A New Introduction*. Oxford: Clarendon, 1990.

Galgano, F. A., and others. "Beach erosion, tidal inlets and politics: The Fire Island story." *Shore & Beach* 67, no. 2–3 (1999): 26–32.

Goudie, Andrew. "Human influence in geomorphology." *Geomorphology* 7 (July 1993): 37–60.

Higgitt, D. L. "Soil erosion and soil problems." *Progress in Physical Geography* 17, no. 4 (1993): 461–472.

Lindstrom, M. J., D. A. Lobb, and T. E. Schumacher. "Tillage erosion: An overview." *Annals of Arid Zone* 40, no. 3 (September 2001): 337–349.

Karlen, W., and others. "Glacier fluctations on Mount Kenya since ~6000 cal. years b.p.: Implications for Holocene climatic change in Africa." *Ambio* 28, no. 5 (1999): 409–418.

McHugh, M., T. Harrod, and R. Morgan. "The extent of soil erosion in upland England and Wales." *Earth Surface Processes and Landforms* 27, no. 1 (January 2002): 99–107.

Nash, D. J. "Arid geomorphology." *Progress in Physical Geography* 27, no. 2 (June 2003): 284–303.

Nuhfer, Edward B. *The Citizen's Guide to Geologic Hazards*. Arvada, CO: American Institute of Professional Geologists, 1993.

Parshall, T., and D. R. Foster. "Fire on the New England landscape: Regional and temporal variation, cultural and environmental controls." *Journal of Biogeography* 29, no. 10-11 (October–November 2002): 1305–1317.

Phillips, J. D. "Sources of nonlinearity and complexity in geomorphic systems." *Progress in Physical Geography* 27, no. 1 (March 2003): 1–23.

Ritchie, J. C., D. E. Walling, and J. Peters. "Application of geographic information systems and remote sensing for quantifying patterns of erosion and water quality." *Hydrological Processes* 17, no. 5 (April 15, 2003): 885–886.

Rondeau, Bernard, and others. "Budget and sources of suspended sediment transported in the St. Lawrence River, Canada." *Hydrol Processes 2000* (Chichester, England: John Wiley & Sons LD) 14, no. 1: 21–36.

Scull, P., J. Franklin, O. A. Chadwick, and D. McArthur. "Predictive soil mapping: A review." *Progress in Physical Geography* 27, no. 2 (June 2003): 171–197.

Solomon, S. M., and others. "Coastal hazards and associated management issues on South Pacific Islands." *Ocean & Coastal Management* 42, no. 6–7 (1999): 523–554.

Stethem, C., B. Jamieson, P. Schaerer, D. Liverman, D. Germain, and S. Walker. "Snow avalanche hazard in Canada—a review." *Natural Hazards* 28, no. 2-3 (March 2003): 487–515.

Summerfield, Michael. *Global Geomorphology.* London: Longman, 1991.

Trenhaile, Alan S. *The Geomorphology of Canada: An Introduction.* Toronto: Oxford, 1990.

Viles, H. A., and A. S. Goudie. "Interannual, decadal and multidecadal scale climatic variability and geomorphology." *Earth-Science Reviews* 61, no. 1-2 (April 2003): 105–131.

Walker, J. "Application of geomorphology to the management of river-bank erosion." *Journal of Institute of Water & Environmental Management* 13, no. 4 (1999): 297–300.

Wilby, R. L. "Greenhouse Hydrology." *Progress in Physical Geography* 19, no. 3 (1995): 351–369.

Xiubin, He, and others. "The 1998 flood and soil erosion in Yangtze River." *Water Policy* 1, no. 6 (1998): 653–658.

WEB WORK

The U.S. Geological survey is the U.S. government agency responsible for studying and monitoring Earth's surface, and it provides a wide range of resources concerning earthquakes, geological resources, water resources, and landforms.

http://www.usgs.gov/

The Geological Survey of Canada has a web page that describes many of its activities and, of course, it is bilingual!

http://www.nrcan.gc.ca/gsc/

A site at Michigan Technical University will take you to a map of nearly all active volcanoes in the world.

http://www.geo.mtu.edu/volcanoes/

Volcano World, sponsored by the U.S. National Aeronautics and Space Administration, also has excellent imagery.

http://volcano.und.edu/

The Geography Department at the University of Edinburgh has a mapping program that will produce maps of recent earthquakes.

http://www.geo.ed.ac.uk/quakes/quakes.html

A Look Ahead

BIOGEOCHEMICAL CYCLES

The hydrologic cycle is a continual worldwide flow of water among the four spheres: Water in the atmosphere condenses as clouds, falls as precipitation, runs off into the ocean or is stored in the ground, and returns to the atmosphere through transpiration by plants or evaporation. Humans play major roles in all major biogeochemical cycles.

CARBON, OXYGEN, AND NUTRIENT FLOWS IN THE BIOSPHERE

The carbon cycle is a fundamental link among the atmosphere, oceans, biosphere, and soil. Atmospheric carbon is absorbed by the biosphere during photosynthesis, stored in living organisms and in the soil, and released through respiration. Fossil-fuel combustion is a major source of carbon to the atmosphere.

SOIL

Soil is the uppermost part of the solid Earth. Soil is formed through a combination of physical, chemical, and biological processes. Soil characteristics vary with climate, and the world map of soils has patterns similar to the world climate map. In many areas, people have reduced the ability of soil to support plant life by removing nutrients and exposing soil to erosion.

ECOSYSTEMS

Earth's surface includes many ecosystems, in which plants, animals, and their physical environments interact. Water, nutrients, organic matter, and energy

4

BIOGEOCHEMICAL CYCLES AND THE BIOSPHERE

◄ This tower in an oak savanna woodland is instrumented for measuring exchanges of carbon dioxide and water vapor between the biosphere and the atmosphere. (Dennis Baldocchi)

move among living things and the soil, biosphere, atmosphere, and hydrosphere. Photosynthesis is the basis of food chains and supports animal life. Plants are adapted to varying availability of light and water and have developed forms that enable them to compete and to survive environmental stress. Ecological communities may change over time through succession, especially following disturbance.

BIOMES: GLOBAL PATTERNS IN THE BIOSPHERE

Earth's ecosystems are grouped into biomes characterized by particular plant and animal types and usually named for climates or dominant vegetation types. World maps of soils, vegetation, and climate show strong similarities because climate exerts a powerful control on soil formation and plant growth. Humid tropical environments have diverse forest vegetation and nutrient-poor soils. Arid and semiarid regions have cacti, grasses, and shrubs that are adapted to low-moisture stress. Humid midlatitudes are dominated by forests: winter-deciduous forests in warmer areas and coniferous forests in colder, drier areas. High-latitude areas have either sparse vegetation or low herbaceous tundra that can survive extreme cold in winter. Throughout the populated world, vegetation and soils show extensive human impact. The biosphere is a system in which humans play a crucial role.

Global warming is under way, and the warming is making it hard for the polar bears around Churchill, Manitoba, to find enough to eat. Churchill lies on the southwest shore of Hudson Bay and is a port for oceangoing vessels that carry grain produced on the Canadian Prairies to foreign destinations. The bay is frozen much of the year, so the port is only open for a brief period from midsummer to autumn. While the ice is a barrier to shipping, it is essential for the polar bears that inhabit the region. They feed largely on seal pups, which they catch in snow lairs that the seals create over breathing holes in the sea ice. When the bay is ice-free, the seals spend most of their time in the water, safe from the bears. The summer is thus the hungry season for the bears.

Over the last few decades, temperatures in central and northern Canada have been gradually rising. It is expected that global warming will be felt most strongly in high latitudes, and there is considerable evidence that these effects are occurring now in Churchill. Today the ice melts in Hudson Bay about three weeks earlier in the summer than it did 25 years ago, and it is freezing later in the fall. That means that the hungry season for the bears is lasting several weeks longer, and the bears are having a harder time making it through the summer. It is expected that the warming will ultimately force the bears to migrate northward, and the tundra vegetation around Churchill will be replaced with species currently found much farther south.

Traditionally, studies of diverse topics like vegetation, weather, human population, and industrial activity have been carried out in isolation from each other. But scientists have increasingly recognized the intimate interaction among all Earth's systems, including natural and human-dominated phenomena. **Desertification,** a process by which semiarid vegetation and soil become more desert-like as a result of human use, has occurred in semiarid lands around the world. This is an example of environmental change caused by a combination of natural and human processes operating at Earth's surface. Land degradation in various forms is occurring on a significant portion of the world's agricultural land. It reduces the fertility of the soil and necessitates greater investment in agricultural development to feed Earth's growing population.

In this chapter, we examine biogeochemical cycles and the biosphere, including the hydrologic, or water, cycle. Water moves between the hydrosphere, biosphere, lithosphere, and atmosphere. It plays a critical role in regulating environmental processes, determining spatial patterns of vegetation and soil on Earth's surface, and linking all of Earth's subsystems. The biosphere is the thin layer of living things on Earth's surface, of which we are an inseparable part.

> Because of these interactions among climate, soil, living things, and rocks, patterns of plant and animal life correspond closely with patterns of climate, topography, and rock types.

The biosphere is intimately connected to Earth, ocean, and air. Plants depend on solar energy for **photosynthesis,** which is the formation of food in green plants as a result of exposure to light, and on atmospheric circulation for temperature regulation, carbon supply, and water supply that make growth possible. At the same time, plants and animals absorb and release oxygen and carbon dioxide to the atmosphere, and in so doing, powerfully influence atmospheric composition and climate. Plants and animals help convert rocks to soil, the dynamic medium that stores water and nutrients and supports life, as well as regulating nutrient supplies in the soil. Because of these interactions among climate, soil, living things, and rocks, patterns of plant and animal life correspond closely with patterns of climate, topography, and rock types.

The flow of water between the lithosphere, biosphere, and atmosphere is an example of a biogeochemical cycle. **Biogeochemical cycles** are recycling processes that supply essential substances such as carbon, nitrogen, and other nutrients to the biosphere. Like movements of water, biogeochemical cycles also connect Earth's subsystems.

In previous chapters, we described environmental changes that occurred over the past few million years. These changes have been driven mostly by natural climatic variability. But during the last 10,000 years (since the end of the Pleistocene Epoch), and especially the last 1,000 years, the human population has expanded so rapidly that we now profoundly influence global environmental patterns as much as the most dramatic climatic variations did in the Pleistocene Epoch. Just a few thousand years ago the human species had only a minor environmental impact. Today, however, our large numbers, our enormous consumption of natural resources, and our widespread construction projects and other environmental modifications give us a significant role in global biospheric processes like energy exchange, food webs, and movements of materials on the surface of Earth. As a consequence, we can understand Earth's environments today only by recognizing how humans modify and regulate them.

> We can understand Earth's environments today only by recognizing how humans modify and regulate them.

THE BIOGEOCHEMICAL CYCLES

Processes operating in the biosphere—the growth and decay of plants and animals, indeed all life processes—depend on exchanges of energy and matter. Earth receives a constant supply of energy in the form of light from the Sun. Matter, including the essential substances water and carbon, is available

> Biogeochemical cycles are the pathways by which energy and matter are transformed and recycled in Earth systems.

in the atmosphere, hydrosphere, and lithosphere, as well as in the biosphere itself.

The law of conservation of energy states that energy cannot be created or destroyed under ordinary conditions, but it may be changed from one form to another. Similarly, the law of conservation of matter states that matter may be changed from one form to another under ordinary conditions but cannot be created or destroyed. The exception to "ordinary conditions" is nuclear reactions, in which matter is converted to energy.

While energy and matter are not created or destroyed, they are constantly being transformed. Biogeochemical cycles are the pathways by which energy and matter are transformed and recycled in Earth systems (Figure 4-1). The dynamic earth processes that we have discussed so far (plate tectonics, mountain building, erosion) all involve the exchange of energy and matter. Processes operating on the living organisms of the biosphere also depend on exchanges of energy and matter.

This section focuses on one of the two most important biogeochemical cycles, the **hydrologic,** or water, **cycle.** The next section examines another important cycle, the carbon cycle. Water is essential for all life to survive on Earth. Carbon is the most important element in biological processes on Earth.

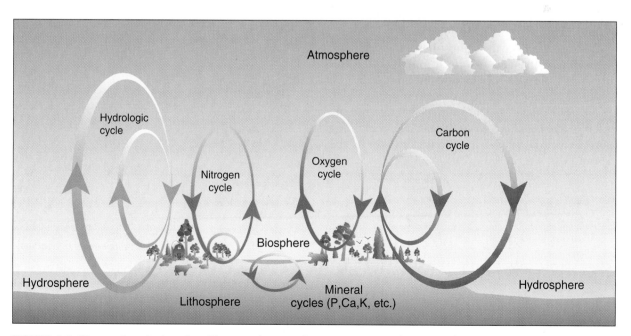

FIGURE 4-1 Biogeochemical cycles. These cycles transfer matter among the atmosphere, biosphere, hydrosphere, and lithosphere. The cycles represented here are shown in greatly simplified form. Important minerals cycling in the environment include phosphorus (P), calcium (Ca), potassium (K), and others.

The Hydrologic Cycle

Water is central to every part of the biosphere. Some of the distinctive features of water that make it so important include the following:

- Earth's temperatures allow water to exist in all three states—solid, liquid, and gas (vapor)—and to change readily from one state to another. Water can permeate rock as groundwater, exist in frozen reservoirs we call glaciers, flow worldwide in the oceans, and drift worldwide through the atmosphere. Water is the only common substance that exists in all three of these states at normal environmental temperatures.

- Relatively large amounts of heat energy are involved in water's changes among solid, liquid, and gaseous states.

- Water is an excellent solvent, readily dissolving many substances, making them more mobile and available for chemical reactions with whatever water contacts.

- Most important, all living things are primarily water (roughly 70 percent in the case of humans).

- As we saw in Chapter 2, geographical variation in water availability is the key to understanding many environmental variations at the global scale.

Water cycles through the atmosphere, lithosphere, and hydrosphere by means of evaporation, condensation, precipitation, and runoff. Water falls from the atmosphere to the ground and ocean through condensation and precipitation. **Runoff** carries water from the land to the sea. Evaporation converts liquid water in lakes and oceans into vapor, returning it into the atmosphere. This flow is the hydrologic cycle.

Water is stored in the atmosphere, hydrosphere, and lithosphere in solid, gaseous, and liquid forms (Figure 4-2). The salty oceans are by far the largest reservoirs of water—about 96.5 percent of all water on Earth. Nearly 2 percent is stored in glacial ice, and another 0.9 percent is saline groundwater. The remaining 0.8 percent is available as fresh water, mostly in the

ground. Only about 0.014 percent of all water is in rivers and lakes. Although the atmosphere is powerful in controlling weather and climate, it contains a mere 0.001 percent of all the world's water.

The total quantities of water present in various forms change little from year to year for Earth as a whole. But over hundreds or thousands of years, climate substantially alters the size of glaciers, and the volume of water in the oceans also changes. As explained in Chapter 3, about 20,000 years ago the amount of water stored in glacial ice was much larger, and as a result, sea level was substantially lower than today.

Water Budgets

A **water budget** is an accounting of all the inflows and outflows of water in a given system over some time period. Water budgets are essential tools for water-resource managers because they determine the water available for human use. For example, river water is used for domestic and commercial supplies, as cooling water for power plants and industries, hydroelectric power generation, irrigation, fish and wildlife habitats, navigation, recreation, and waste removal. Knowing the amount available for these purposes is important, yet in many areas we have only limited information on streamflows. Measuring precipitation and determining evapotranspiration allows one to predict average river flows (Figure 4-3).

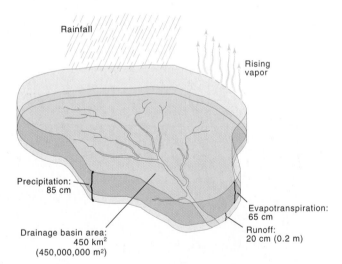

Volume of runoff: 0.2 m x 450,000,000 m²
= 90,000,000 m³ per year
= 2.85 m³ per second (average)

FIGURE 4-3 Calculating streamflow. Average streamflow can be calculated from the water budget for a drainage basin, showing the close connection between water resources and climate. The average annual evapotranspiration is subtracted from the annual precipitation to determine the average annual runoff, expressed as a depth of water on the land. This value, when multiplied by the drainage basin area (in comparable units) gives the volume of runoff carried by the stream in a year.

FIGURE 4-2 Quantities of water in storage. The oceans hold most water; glaciers hold the next largest amount, and the remainder is in streams, lakes, groundwater, and the atmosphere. (Data from Jones, 1997)

FIGURE 4-4 The hydrologic cycle: Flows of water among land, sea, and air. The quantities of water transferred in an entire year are shown. (Modified from R. W. Christopherson, *Geosystems,* 5th ed. Upper Saddle River, NJ: Prentice Hall, 2003)

Water budgets help us understand why river flows are so much greater in humid regions than arid regions. For example, Canada's Mackenzie River has a drainage area of 1.8 million square kilometers and an average flow of 9,600 cubic meters per second. The Nile River in Egypt has a much larger drainage area of 3 million square kilometers but discharges only 950 cubic meters per second because so much of the basin is arid. The Mackenzie's annual flow is equivalent to a depth of about 17 centimeters of water spread over the entire basin area; the Nile's annual flow is equivalent to an average depth of only 1 centimeter.

Estimates of the global average rates of evaporation, condensation, precipitation, and runoff are shown in the global water budget (Figure 4-4). Each process varies geographically with the amount of water available. Excess water evaporated into the atmosphere over the oceans is carried by wind over land areas, where it condenses into clouds and falls as precipitation. About two-thirds of the water that falls on land areas evaporates there, and the remaining one-third drains into rivers, which return the excess to the sea as runoff (Figure 4-5). A significant amount of river flow (an average of about 5 percent worldwide) is returned to the atmosphere via plant-water use on irrigated lands rather than flowing to the ocean as liquid water.

The water budget shown in Figure 4-4 does not show the variability of water from place to place and season to season. Evaporation from the sea depends on

the insolation available to heat water. Precipitation and evaporation occur in both land and ocean areas, but not in equal amounts; over the oceans, more evaporation than precipitation occurs.

FIGURE 4-5 The Amazon River. This satellite image shows the confluence of the Rio Solimoes and the Rio Negro near Manaus, Brazil. The Rio Negro, entering from the upper left, originates in the Amazonian lowland and carries very little sediment. The Rio Solimoes is muddy with sediment derived from the Peruvian Andes. (GSFC/JPL, MISR Team/NASA Headquarters)

Evapotranspiration and local water budgets When we focus on local water budgets, we recognize the important role of plants in removing water from the soil. In well-vegetated areas, much of the conversion of water from liquid to vapor takes place in the leaves of plants. This water is replaced by water drawn from the soil through plant roots. This plant mechanism is called transpiration, and when combined with evaporation we call the process **evapotranspiration (ET).**

Evapotranspiration is low in winter and high in summer, reflecting active transpiration by plants. In the winter, when water use is low, precipitation exceeds evapotranspiration. The surplus water either runs off or becomes stored as groundwater. In springtime, plants develop leaves and water use rises rapidly, so that by early summer it exceeds precipitation. Initially, this excess water demand is met by water stored in the soil, but by late summer, soil moisture is exhausted and water use is limited. At this time shallow-rooted grasses may wither, while deep-rooted shrubs and trees remain green. In autumn, evapotranspiration falls with the temperature, and once again precipitation exceeds evapotranspiration. At this time, excess precipitation restores soil moisture to higher levels, and runoff increases again.

A local water budget compares precipitation and evapotranspiration. For a humid midlatitude site such as Chicago, Illinois, a graph of a water budget reveals that precipitation occurs fairly consistently throughout the year (Figure 4-6).

Water budgets vary tremendously from one climate to another (Figure 4-7). In some humid climates,

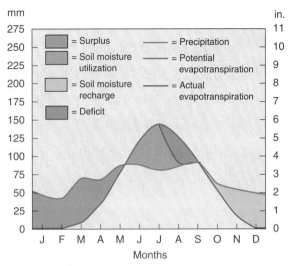

FIGURE 4-6 Water budget diagram for Chicago, Illinois, a typical humid midlatitude site. Precipitation is shown as a blue line, and potential evapotranspiration is shown in red. In the cool months precipitation is greater than potential evapotranspiration, and the excess becomes runoff (dark blue shaded area). In the warm season potential evapotranspiration is greater than precipitation. Part of this excess demand is made up by soil-moisture use (green shading), but not all. The difference between potential evapotranspiration and actual evapotranspiration is a deficit (brown shading). The soil water that was used in early summer is recharged in autumn (light blue shading). The orange area at the bottom of the graph is water that falls as precipitation and is part of actual evapotranspiration.

Focus on
POTENTIAL AND ACTUAL EVAPOTRANSPIRATION

Evapotranspiration occurs whenever water is available for plants to transpire and the air is not saturated with humidity. But ET rates vary tremendously over time and from place to place. Energy is necessary to evaporate water. When water vaporizes from a liquid to a gas, energy is absorbed (it becomes latent heat in the water vapor), which cools the plant leaves. Thus, the rate of evapotranspiration depends mostly on energy availability. ET occurs fastest under warm conditions and it virtually halts below freezing. Atmospheric humidity and wind speed also are significant: ET is faster on dry, windy days than on humid, calm days.

In warm weather, conditions may favor high ET, but the soil may be too dry to supply this demand. Therefore, we distinguish between potential ET and actual ET. **Potential evapotranspiration (POTET)** is the amount of water that would be evaporated if it were available. **Actual evapotranspiration (ACTET)** is the amount that actually is evaporated under existing conditions. If water is plentiful, then ACTET equals POTET. But if water is in short supply, ACTET is less than POTET. ACTET never exceeds POTET.

Comparing POTET with precipitation is a useful way to describe water availability in various climates. For example, if precipitation always is greater than POTET, plants have plenty of water, and the climate is quite humid. If precipitation is less than POTET most of the time, then plants never get as much water as they need, the natural vegetation is adapted to dryness, and the climate is arid. Farmers using irrigation to supply water to plants can calculate the difference between POTET and precipitation and thereby know how much water they must apply to crops. POTET can be calculated from mean monthly temperature data.

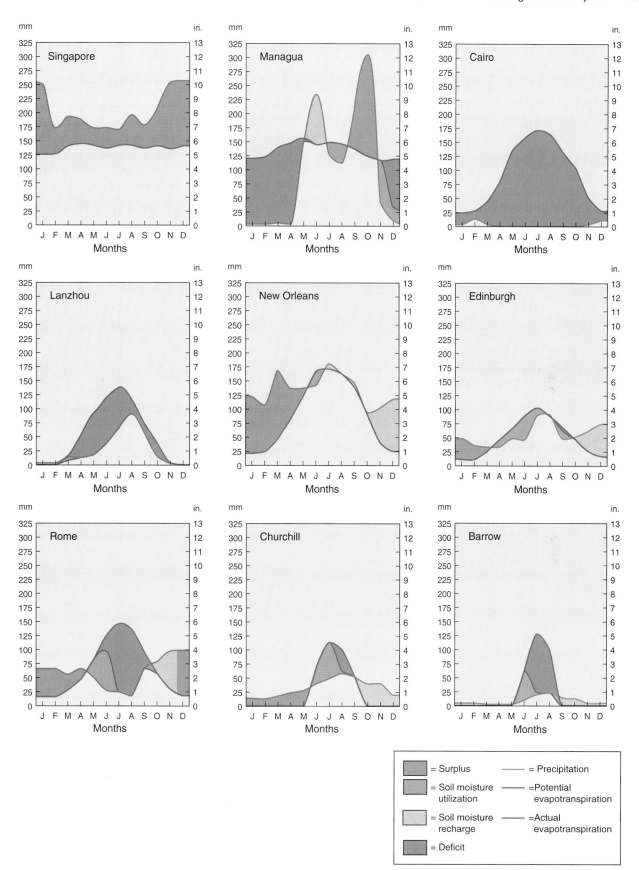

FIGURE 4-7 Water budget diagrams for selected stations. Compare with the climatic data and descriptions in pages 78 to 88 in Chapter 2.

precipitation almost always meets demand, and stored soil moisture provides the small additional amount of water needed during brief dry spells. In semiarid and arid climates, the demand for water considerably exceeds precipitation during the year, so soil moisture is never fully replenished, and plants must withstand severe moisture deficits.

Soil's role in the water budget   Most of Earth's land area is covered with a layer of rock debris and organic matter known as *soil*. Soil is critical in the water budget, for it stores water and makes it available for evapotranspiration; soil is a water "bank." How effectively soil does this depends on how fast it can absorb precipitation, how much water it can store in the root zone, and the ability of local vegetation to transpire water into the atmosphere.

Soil's **infiltration capacity** determines whether rainfall infiltrates (soaks into) the soil or runs off. Infiltration is good, and most precipitation will soak in when dense vegetation exists at ground level, a layer of organic litter covers the soil, and the soil is kept porous by worms and insects. If the soil is bare or compacted, infiltration is poor, less water is stored, and runoff is greater.

Soil texture—the size of mineral particles in the soil—affects how much water soil can hold. Sandy soils drain quickly and hold little water (pour water on sand and watch how quickly it soaks in). Soils with high contents of finer particles—silt and clay—drain slowly (pour water on modeling clay and see how quickly it runs off). Deep-rooted plants (several meters) have a longer opportunity to absorb water as it drains through the soil; shallow-rooted plants (half a meter or less) must rely on moisture contained in the upper soil. Thus, shallow-rooted grasses and herbs generally use less water (and potentially allow more runoff) than deep-rooted trees. Soil and vegetation management is critical for maintaining adequate water resources for agriculture and other human needs.

Vegetation and the Hydrologic Cycle

Forests demonstrate the close association between vegetation and the water budget. Forests occur where ample moisture is available for most of the year. Trees demand large volumes of water. A single large tree can transpire 1,000 liters per day in warm weather. Trees gather water through extensive and usually deep root systems and transpire through leaves that often are tens of meters above the ground surface. Trees use so much water that they are the dominant medium for returning rainwater to the atmosphere in large forested regions like eastern North America and the Amazon River basin.

This role of forests has water-management implications. In the southeastern United States, areas formerly covered with slow-growing deciduous trees (that is, trees that shed their leaves in autumn) have been replanted with faster-growing evergreens to feed the timber industry. Because evergreens do not lose all their leaves in the winter, they keep growing and transpiring water year-round when deciduous trees cannot. Evergreen needleleaved trees also catch more precipitation on their leaves than broadleaved deciduous trees do, and this water is lost to evaporation. Studies in an experimental watershed in North Carolina showed that this measurably increased evapotranspiration and correspondingly decreased streamflow.

In contrast to trees, grasses are relatively shallow rooted, so they experience significant variations in moisture availability. When soil water is plentiful, grasses grow quickly and transpire at rates similar to trees. But during periods of limited soil moisture, grasses become dormant, and transpiration virtually ceases. This helps grasses survive dry seasons, as in semiarid climates, where most trees cannot grow.

Deforestation—clear-cutting of forest—in the Amazon River basin is proceeding at a rate of about 30,000 square kilometers (11,600 square miles) per year, or about 0.6 percent of the basin's area per year. By 2000 a total of over 650,000 square kilometers (250,000 square miles) had been deforested, an area about the size of Texas (Figure 4-8). Deforestation may be critical to the water balance of this region. In this tropical rain forest, rainfall averages 200 to 300 centimeters (80 to 120 inches) per year, and evapotranspiration is about 110 to 120 centimeters (44 to 48 inches) per year. The excess precipitation that is not evapotranspired runs off to the Atlantic Ocean via the Amazon River, which has the world's greatest discharge. The Atlantic is the original source of water that falls as precipitation on the Amazon basin. This recycling presents a fine example of the hydrologic cycle at work.

The Amazon basin's interior, however, is more than 2,000 kilometers (1,200 miles) from the Atlantic, and most of the atmospheric water falls closer to the coast. This precipitated water infiltrates the soil, trees transpire it back into the air, and easterly winds carry it further inland. This cycle repeats, moving water step by step westward. Thus, rainfall in the Amazon basin's interior may have been precipitated and transpired several times in its westward journey.

> Rainfall in the Amazon basin's interior may have been precipitated and transpired several times in its westward journey.

Forests clearly are integral to this process. If the forests continue to be cut over very large areas, and if the grasses that replace the trees transpire much less than the trees would, this could reduce precipitation in the interior. Such precipitation changes

FIGURE 4-8 Deforestation in Santa Cruz, Bolivia. This map, based on satellite imagery, shows land-use change over a quarter century. Gray areas are areas that remained in forest and woodland as of 1998. Other areas are colored according to the year in which a site was converted to agriculture. Land use types are delineated with lines. Solid white lines show planned colonies, dashed white lines indicate spontaneous colonies, and dotted white lines show Mennonite colonies. Other regions of development are non-Mennonite soybean farms. Natural vegetation formations in this image are delineated with black lines. (Marc Steininger/NASA Headquarters)

are today only predictions in computer models, but they help to show the important role that vegetation plays in the hydrologic cycle.

CARBON, OXYGEN, AND NUTRIENT FLOWS IN THE BIOSPHERE

Carbon is not the most abundant element on Earth, but in combination with hydrogen, it is the most important for sustaining life. Compounds of carbon and hydrogen are the major component of the foods that plants produce and animals consume, and of the fossil fuels (coal, oil, natural gas) that are our most important sources of power. Living things constantly exchange carbon with the environment, by photosynthesis, respiration, eating, and disposal of waste.

Exchanges of carbon among the biosphere, atmosphere, oceans, and rocks are collectively called the carbon cycle (Figure 4-9). An **oxygen cycle** is inextricably linked with the carbon cycle, and we will consider them together. We now look at these cycles.

The Carbon and Oxygen Cycles

In the **carbon cycle,** carbon in the form of atmospheric carbon dioxide (CO_2) is incorporated into carbohydrates in plant tissues through photosynthesis. Animals consume plants and use the plant carbohydrates for their own life processes. Through respiration, which occurs in almost all living organisms, carbon is returned to the atmosphere (as CO_2). Carbon dioxide in also added to the atmosphere by the combustion of fossil fuels. Thus, these three processes—photosynthesis, respiration, and combustion—cycle carbon and oxygen back and forth between living things and the environment.

Photosynthesis can be shown by the following equation:

$$\text{carbon dioxide} + \text{water} + \text{energy} \rightarrow \text{carbohydrates} + \text{oxygen}$$

For this reaction, land plants obtain carbon dioxide from the air, water from the soil, and energy from solar radiation. They store carbohydrates in tissue for later use and release oxygen to the atmosphere. Plants are the source of atmospheric oxygen, without which animals could not exist.

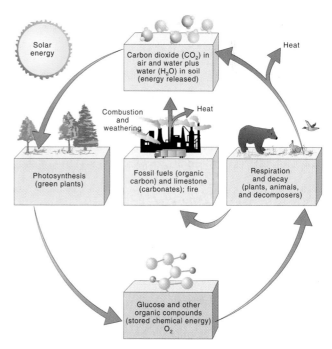

FIGURE 4-9 The carbon cycle. Carbon is taken from the atmosphere and stored in biomass through photosynthesis. It is returned to the atmosphere through respiration. Combustion of fossil fuels releases vast quantities of carbon to the atmosphere from long-term storage in rocks. (From R. W. Christopherson, *Geosystems,* 3rd ed. Upper Saddle River, NJ: Prentice Hall, 1997)

Focus on

GEOGRAPHY, GEOGRAPHIC INFORMATION SYSTEMS, AND THE GLOBAL CARBON BUDGET

Concern about the accumulation of carbon dioxide in the atmosphere is mounting, and countries are moving toward implementation of an international treaty to control carbon dioxide emissions. Therefore, the need for accurate information on carbon dioxide release and uptake is increasing. One of the most important parts of the puzzle is to compile detailed knowledge of the amounts of carbon released and taken up by the biosphere. Geography and geographic information systems (GIS) are critical to that problem.

Many different flows must be estimated in calculating the carbon dioxide budget of the atmosphere, hydrosphere, biosphere, and lithosphere. Some of these are easier to estimate than others. We know fairly closely, for example, how much carbon is emitted by fossil-fuel combustion because we know approximately how much coal, oil, and natural gas each country burns. Another large number in the budget—the amount that goes from the atmosphere to the oceans each year—can be calculated based on knowledge of the physical processes that control the exchange of carbon dioxide between water and air. On the other hand, estimating the amount of carbon released by deforestation or taken up by forest growth is much more difficult. It is highly variable from place to place. Other key processes, too, such as the accumulation and decay of organic matter in the bottoms of reservoirs, the release

of carbon dioxide from decaying peat bogs, or the buildup and loss of organic matter from agricultural soils, are very poorly known. These processes, however, are the very ones that can be managed directly by humans, so they must be accounted for in any international agreements.

Geography and GIS provide the tools for extending a few isolated measurements to take into account processes spread over large areas, so geographers are playing a big role in studies of the carbon budget. For example, we can directly measure rates of carbon uptake by photosynthesis or release through deforestation for small areas like a few square meters or a few hectares. Such measurements involve tasks like collecting, weighing, and analyzing vegetation, or carefully sampling and analyzing carbon dioxide in air samples as shown in the chapter-opening photo. We could not possibly do these things for every hectare of land in the United States, but we can do this at many individual sites representing different land-cover types—coniferous forest, deciduous forest, lakes, wetlands, farmlands, and so forth. Then, using GIS, we can make detailed maps of land cover (often based on satellite imagery) showing the spatial extent of each different kind of surface, and we can assign rates of carbon exchange for each of these types of surface based on our field measurements. The GIS then does the counting for us to estimate total carbon movements over large areas. Not only is the GIS useful in that accounting process, but, in combination with satellite imagery, it can help monitor and account for changing land-cover characteristics over time (see, for example, Figure 1-26). Geography and geographers will continue to play a central role in this work.

Respiration involves the opposite reaction:

$$\text{carbohydrates} + \text{oxygen} \rightarrow \text{carbon dioxide} + \text{water} + \text{energy (heat)}$$

In respiration, carbohydrates are broken down by combination with atmospheric oxygen to CO_2 and water. Energy is released in the process. Some of this energy in lost as heat and some is stored in chemical compounds for later use in other life processes.

The lithosphere is a major storehouse of carbon. Through geologic time, carbon enters the lithosphere slowly through rock formation, principally from oceanic sediments such as limestone but also through creation of fossil fuels such as coal.

The spatial and temporal distribution of photosynthesis is determined mostly by climate (Figure 4-10).

Seasonal cycles of solar radiation are reflected directly in the CO_2 content of the atmosphere. The air we breathe has been monitored for carbon dioxide concentration since the late 1950s at the Mauna Loa observatory in Hawaii (Figure 4-11):

- As expected, CO_2 varies annually with the seasons by a few parts per million.

- The greatest concentrations in the Northern Hemisphere are during spring, because during winter, photosynthesis is reduced (or ceases completely in cold climates) while respiration continues, returning CO_2 to the atmosphere.

- The lowest concentrations are during autumn. CO_2 concentration drops during the summer because plants are actively photosynthesizing,

Seasonal Variations in Plant Growth

(a)

(b)

FIGURE 4-10 A global map of plant growth in July 1987 (a) and January 1998 (b), from satellite data. On land areas, dense vegetation is shown in dark green with sparse vegetation and desert shown in pale green and tan. Note that some high-latitude areas in the Northern Hemisphere, where it is summer, are shown in the same color as the tropical forests, while southern Africa and South America, where it is winter in July, are shown in the same colors as deserts. (NOAA)

removing CO_2 from the atmosphere and storing it in the biosphere.

■ The Mauna Loa data also clearly reveal the steady increase in atmospheric CO_2 levels that results from fossil-fuel consumption.

The extraction and combustion of coal, oil, and natural gas since the Industrial Revolution that began in the late 1700s is a new process affecting the carbon cycle (Figure 4-12). Rates of carbon released into the atmosphere through this process have increased steadily over

FIGURE 4-11 **Atmospheric carbon dioxide concentrations measured at Mauna Loa, Hawaii.** Mauna Loa, located at 19.5° north latitude in the mid-Pacific Ocean. Data collected there reflect general Northern Hemisphere conditions. The annual increase and decrease of carbon dioxide concentrations corresponds to seasonal patterns of plant growth and decay, with atmospheric concentrations decreasing in the Northern Hemisphere summer and increasing during the winter. (Data courtesy of Carbon Dioxide Information and Analysis Center, Oak Ridge National Laboratory)

FIGURE 4-12 **Harvesting peat for fuel in Ireland.** The Irish climate provides ample moisture for plant growth, but because of the cool temperatures, organic matter does not decay rapidly. Plant remains accumulate to form peat. Substantial amounts of carbon are stored in high-latitude peat deposits like these. (Tom Bean/Corbis/Stock Market)

the past 200 years, and they are expected to continue to increase at least well into the twenty-first century. The long-term carbon dioxide increase so evident in Figures 2-54 and 2-56 illustrate this. Uncertainty about future fossil-fuel use makes it difficult to predict future atmospheric carbon dioxide concentrations.

Deforestation

Land-use change, especially clearing of forests, is important to these biosphere-atmosphere exchanges. In many parts of the world, most notably in tropical forests, forests are being cleared to harvest trees for lumber or fuel and to make way for agriculture. In some areas

trees are growing back, but at a global level, more trees are being cut than are growing. This is important to global carbon flows because relatively large amounts of carbon can be stored in **biomass** (living and formerly living matter) in forests, and if trees are cut and burned, the carbon that they contain is released to the atmosphere. If the wood is used as lumber to build houses, the carbon is stored as long as the houses stand.

The most rapid deforestation, and hence the largest release of carbon, is in tropical Central and South America, West Africa, and Southeast Asia (Figure 4-13). Deforestation rates in most midlatitudes are much lower, and in many areas, such as the eastern United States, forests are regrowing and are thus a *sink*, or net storage facility, for carbon. In arctic regions

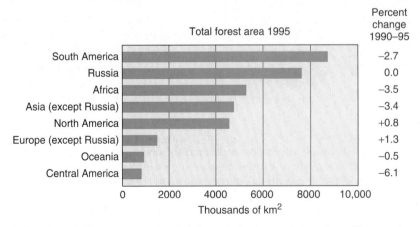

FIGURE 4-13 **Forest extent and deforestation rates by region.** The most extensive forests in the world are in Siberia and the Americas, but major forest areas also exist in Africa and southern Asia. Deforestation is most rapid in the tropics; forest area is increasing in Europe and North America. (Data courtesy of World Resources Institute)

large amounts of carbon are stored in soils in the form of peat deposits—accumulations of plant matter in swamps and bogs that are the first step in the formation of coal. How these areas might change if climate warms is a point of concern. Warming would increase both respiration and photosynthesis, but we do not know whether the net change would store more carbon in biomass or in the atmosphere.

SOIL

The interface between the lithosphere and the biosphere is the soil. **Soil** is a dynamic, porous layer of mineral and organic matter, in which plants grow. Soil is the uppermost part of the lithosphere, and at the same time is a key portion of the biosphere. Soil is a storage site for water, carbon, and plant nutrients.

Soil properties are attributable to five major factors:

1. *Climate* regulates both water movements and biological activity.
2. *Parent material* is the mineral matter from which soil is formed.
3. *Biological activity*—plants and animals—moves minerals and adds organic matter to the soil.
4. *Topography* affects water movement and erosion rates.
5. All these factors work over *time*, typically requiring many thousands of years to create a mature soil.

Soil Formation

The first step in soil formation is weathering, or the breakdown of rock into smaller particles and new chemical forms. Chapter 3 explained the two ways that rocks weather: mechanical weathering (such as ice expansion or tree roots growing in the cracks between rocks) and chemical weathering (such as dissolving limestone in water in the soil). The parent material from which soil is formed is important because it influences the chemical and physical characteristics of the soil, especially in young soils.

Water plays an important role in rock weathering and soil formation. A water budget indicates how much water moves through the soil in a place and measures the significance of water in local weathering. For example, if mean annual precipitation is 100 centimeters and evapotranspiration is 70 centimeters, then 30 centimeters of water would seep down through the soil in an average year. This water would be a powerful weathering agent. However, if an area has 70 centimeters of precipitation and the atmosphere is capable of evaporating 150 centimeters of water, virtually all water is evapotranspired by plants, so little would percolate down through the soil to help weather the rocks.

In a very humid climate, much water passes through the soil and leaches out soluble minerals on its way. Because of this, soils in humid climates generally have lower amounts of soluble minerals such as sodium and calcium compared to soils in dry climates. In semiarid areas, water enters the soil and picks up soluble minerals, which are drawn toward the surface as water is evapotranspired; soils of semiarid and arid climates often have a layer rich in calcium near the surface.

Plant and animal activities are also critical to soil formation. Plants produce organic matter that accumulates on the soil surface, and animals redistribute this organic matter through the soil. Plants and animals also play a role in weathering processes. Topography also affects the amount of water present in the soil, largely through controlling drainage and erosion. Steeply sloping areas generally have better drainage than flat or low-lying areas, and they are often more eroded. Finally, soil formation is a slow process that takes place very gradually over thousands of years. Soils that have only been forming for a few hundred or even a few thousand years have very different characteristics from those that have been modified by chemical and biological processes for tens of thousands of years.

Soil Horizons

Soil is a complex medium, containing six principal components:

1. *Rocks* and *rock particles,* which constitute the greatest portion of the soil and which may weather, releasing nutrients needed for plant growth.
2. *Humus,* which is dead and decaying plant and animal matter that holds water, supports soil organisms, and supplies nutrients.
3. *Dissolved substances,* including phosphorus, potassium, calcium, and other nutrients needed for plant growth.
4. *Organisms,* including animals such as insects and worms and many microorganisms, including bacteria, and fungi.
5. *Water from rainfall,* which is necessary for plant growth and which helps to distribute other substances through the soil.
6. *Air,* which shares soil pore spaces with water and which is necessary for respiration by plant roots and soil organisms.

These substances are not uniformly distributed in soils, but found in layers called soil **horizons** (Figure 4-14). Soil horizons are formed through the movement of water, minerals, and organic matter vertically in the soil and by variations in biological and chemical activity at different depths in the soil. In this way they are different from layers in sediments, which are deposited in sequence from the bottom up.

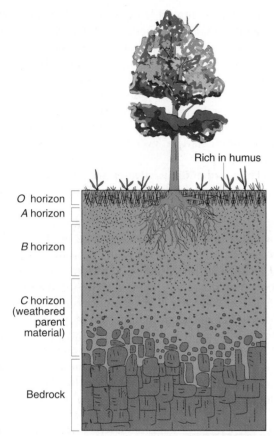

Rich in humus

O horizon
A horizon

B horizon

C horizon
(weathered
parent
material)

Bedrock

FIGURE 4-14 A soil profile. The sequence from the bottom up shows the typical steps of soil formation. Solid rock at the bottom weathers into large pieces, which weather further into ever smaller particles. At top is the organic layer, which contains organic matter from decayed plants and animals. Designations like A are given to horizons in well-developed soils. Poorly developed soils lack some of these layers. (From E. J. Tarbuck and F. K. Lutgens, *Earth Science,* 7th ed. New York: Macmillan, 1994)

For example, litter—leaves, twigs, dead insects, and other organic matter—accumulates to form a horizon at the surface known as the *O* (organic) horizon. As this litter decays, organisms such as insects, worms, and bacteria consume it and carry it underground, where it helps form the *A* horizon. Waste from these burrowing animals as well as their dead carcasses add more organic matter to the *A* horizon. In many soils the *A* horizon contains much of the nutrients that support plant life.

Water may erode materials from the soil surface, or carry some substances down from the *A* horizon to the *B* horizon. Clay minerals formed from chemical weathering often accumulate in the *B* horizon. In some environments organisms play an important role in moving materials between the *A* and *B* horizons. In dry regions soluble minerals such as calcium accumulate in the *B* horizon. The *C* horizon, beneath the *B* horizon, contains weathered parent materials that have not been

altered as completely by soil-forming processes as materials above it have.

Thousands of Soils

Soils are divided into 11 broad clusters, called **soil orders** (Figure 4-15 and Figure 4-16, p. 150). However, these 11 orders are divided into 47 suborders, about 230 great groups, 1,200 subgroups, 6,000 families, and thousands of soil series. In other words, Earth's dynamic lithosphere, hydrosphere, atmosphere, and biosphere have worked as a team to create an enormous diversity of soil. There is nothing simple about "ordinary dirt!"

Detailed maps developed by the U.S. Department of Agriculture's Natural Resource Conservation Service are used by farmers and land-use planners because they show the great local variability of soils. Often, two or three different great groups of soils occur within a few kilometers of each other, and these groups are subdivided into the subgroups, families, and series of more specific characteristics. Because of this great local variability, generalizations about soils must be made with caution.

Climate, Vegetation, Soil, and the Landscape

At a global scale, distinctive environmental regions derive from intimate connections among climate, vegetation, and soil:

- *Humid tropical* and *subtropical soils* such as oxisols and ultisols are highly weathered. They have lost soluble minerals due to heavy precipitation, which may deplete soils of essential nutrients, so fertilization is necessary to increase crop yields. Tropical soils are usually oxidized and red in color.

- *Arid region soils* such as aridisols are high in soluble minerals, because little leaching occurs with the scant rainfall. These soils are low in organic matter, but they can be very productive if sufficiently irrigated.

- *Midlatitude humid soils* such as alfisols and spodosols are moderately leached where they are dominated by deciduous forest vegetation and heavily leached where they are developed beneath the acidic litter of coniferous forests.

- *Midlatitude subhumid soils* such as mollisols are fertile and are the resource base for many important grain-producing regions.

These climatic soil regions are interrupted by major mountain chains of the world. Poorly developed soils such as inceptisols predominate because erosion removes soil as rapidly as it is formed.

In Figure 4-17 (p. 151) we "zoom in" on a portion of these world maps to examine vegetation and soil at a larger map scale. A close relation among climate, vegetation, and soil also exists at the scale of local ecosystems. You can see that vegetation and soils are far more diverse than is revealed in the large-scale map. This semiarid area of northwestern Colorado is in a sandstone and shale upland, incised with narrow valleys. It is an open woodland dominated by pinyon pine and Utah juniper trees and a shrub-grass community that is mostly sagebrush (Figure 4-18, p. 151). Within a few square kilometers, you may see forest of one or more types, disturbed areas, shrublands, and grasslands. The caption to Figure 4-17 describes the soil-vegetation relations here.

Soil Problems

In modern commercial agriculture and in many traditional systems, nutrients are added artificially to the soil. These nutrients may be delivered in the form of animal manure, by plowing into the soil a soil-enriching crop, or as manufactured inorganic fertilizers. The more intensively a soil is worked, the more important such additions become. Unfortunately, increasingly intensive agriculture has meant that nutrients, especially organic matter, are not being replaced as rapidly as they are withdrawn. Soil fertility usually declines over decades, so the problem is less apparent than if it occurred over just a few years. Variations in yield from year to year caused by weather, insects, plant diseases, and changing technology mask the effects of long-term soil degradation. Nevertheless, a long-term decline in soil quality is in progress in many parts of the world.

To this problem, add erosion, accelerated by cultivation. Plowing a field rips apart the soil-holding system of plant roots and lays bare the soil to attack by rain and wind. Erosion removes the uppermost part of the soil—the O, or organic horizon, usually its most fertile part—and with it go both nutrients and the ability to store them. In many agricultural regions, topsoil loss has ranged from several centimeters (1 to 2 inches) to tens of centimeters (4 inches to a foot or more). Often, the depth of erosion amounts to a significant part of the entire A horizon. In semiarid areas the process can so degrade the soil that only plants adapted to lower soil moisture and nutrient availability, as in deserts, can survive. This is the process of desertification mentioned at the beginning of the chapter. In some cases, desertification may be irreversible, at least in short periods of time. In such cases, grassland and other relatively productive vegetation types can be replaced with only sparse grasses—shrubs forming clumps of vegetation separated by bare ground (Figure 4-19, p. 152). Soil organic matter is reduced, and when rain falls, it is more likely to run off the surface, increasing erosion and decreasing the soil's ability to support vegetation. As a result, many fewer animals can be supported by the available forage. Even though the climate today is not fundamentally altered, the land appears more desertlike than it did before excessive exploitation by humans.

Soil Fertility: Natural and Synthetic

Soil fertility refers to the ability of a soil to support plant growth through making nutrients available. Under a given set of conditions, more fertile soils produce higher crop yields than less fertile soils. Fertility, however, is a difficult thing to measure, because other factors that affect yields are also very important—such as the way the soil is plowed and the way crops are planted. Also, some soils may be relatively infertile with one set of management techniques but highly fertile when managed differently.

Plants need sunlight, carbon dioxide, oxygen, and water. But they also require nutrients for growth, including nitrogen, phosphorus, potassium, calcium, magnesium, sulfur, and others. In general, fertile soils are those that store and release sufficient quantities of nutrients and water for optimal plant growth. Nutrients are present in many different forms—dissolved in soil water, as part of organic matter, attached to the surfaces of mineral grains, and in the crystalline structure of minerals. Some forms are more available to plants than others. Organic matter, both living and dead, stores nutrients and makes them available to plants. In most soils, depletion of organic matter reduces fertility. Addition of manure to the soil helps replace lost organic matter. Most available nutrients occur in the uppermost soil horizons: the O, A, and upper B horizons. Cultivation of the soil and harvesting agricultural crops accelerates the breakdown of organic matter and losses of nutrients stored in the soil. Sustaining soil fertility requires maintaining soil nutrients and maintaining the soil's ability to store and release them to plants.

Farmers long have known that measures must be taken to restore soil fertility. In traditional agricultural systems, this has meant a *fallow period*, a season when a field is allowed to rest and nothing is harvested from it. During this rest period the natural processes of plant growth, organic matter accumulation, mineral weathering, and biological activity can restore fertility. Earth's burgeoning population demands more agricultural products, so fallow periods have been reduced or eliminated. Many areas that once were fallowed are now cultivated every year, reducing or eliminating the opportunity to restore natural soil fertility.

The low cost, availability, and easy application of inorganic fertilizers has led to replacing organic (natural) fertilizers with manufactured inorganic fertilizers. Manufactured fertilizers contain valuable nutrients, but only

World Soils

FIGURE 4-15

Oxisol—a deeply weathered, heavily oxidized soil of the humid tropics, red in color.
Aridisol—dry desert soil with limited organic matter, limited chemical weathering, and accumulations of soluble minerals.
Mollisol—grassland soil of subhumid-semiarid land, dark with plentiful organic material. Mollisols are an important agricultural soil in U.S. Midwest.
Alfisol—a moderately leached soil of humid subtropical and midlatitude forests, with clayish *B* horizon.
Ultisol—red- to yellow-colored, well-weathered soil with clayish *B* horizon typically found in warm, humid climates.
Spodosol—acidic soil formed under coniferous forest, with a light-colored, sandy *A* horizon and red-brown *B* horizon in which iron accumulates.

MODIFIED GOODE'S HOMOLOSINE EQUAL-AREA PROJECTION

Entisol—poorly developed soil usually as a result of low weathering rates, with characteristics dominated by parent material.

Inceptisol—weakly developed soils often found in mountainous areas where erosion limits soil development.

Vertisol—soils with high content of clay minerals that swell and shrink on wetting and drying. Deep cracks open up in the dry season.

Histosol—organic soils found in peat bogs and other areas of organic-matter accumulation.

Andisol—young volcanic soils developing on ash and lava deposits.

Mountain soils and **ice fields** are also mapped. Areas mapped as mountain soils have too much local variability to show individual soil types at this map scale. Ice fields have no soil.

FIGURE 4-16 Six soil orders. (a) This oxisol in Rwanda has a reddish color caused by oxidation of iron. (b) A mollisol in Nebraska has a dark, organic-rich and fertile *A* horizon. (c) This spodosol in New York is identified by the white, sandy surface horizon. (d) An inceptiosol in Massachusetts shows only weak horizon development because the soil has been subjected to erosion. (e) Areas of volcanic activity often have andisols, with distinct ash layers, like this one in Japan. (f) This entisol in Connecticut has not had much time to develop, and so does not have distinct horizons. Compare with soil descriptions on pages 148–149. (Loyal A. Quandt/National Soil Survey Center, Natural Resources Conservation Service, U.S.D.A.)

organic matter can help maintain the ability of the soil to store those nutrients. In the short run (a single crop season), nutrients can be added to soil to supply plants' needs. But in the long run, the ability of the soil to support plant growth is diminished.

Farmers and resource managers around the world increasingly worry about the impact of soil degradation on the world's food-producing capacity. About one-fifth of U.S. cropland is being eroded faster than it can be replaced. American agricultural policy is oriented toward helping farmers better conserve the soil. The challenge is to develop and adopt agricultural practices that sustain the land, not only in wealthy nations but in developing countries as well.

> Farmers and resource managers around the world increasingly worry about the impact of soil degradation on the world's food-producing capacity.

The study of landscape patterns at local and regional levels, using vegetation and soil maps, almost always reveals human disturbance. As settlement increases, environmental managers worry about human impact on ecosystems. Although global generalizations such as reports of the amount of forest area lost in the tropics per year are popular, impacts are best understood at a local scale.

ECOSYSTEMS

An **ecosystem** includes all living organisms in area and the physical environment with which they interact (Figure 4-20). An ecosystem can cover a small area,

FIGURE 4-17 A soil map for part of Rio Blanco County, Colorado. This map shows the close correspondence between landforms, soils, and vegetation. Sagebrush prefers deep, well-drained soils (36); pinyon-juniper woodland prefers thin, rocky soils (73). On gently sloping uplands, moderately thick soils have developed from sandstone bedrock; vegetation is dense-to-open woodland of pinyon pine and juniper, with scattered sagebrush and grasses (70); small patches of drier, windblown material show a distinctive fine-textured soil with sagebrush (33, 64, 66). (Map and photo from U.S. Department of Agriculture, Soil Conservation Service)

such as a field or a pond. On any scale of analysis, large or small, certain fundamental elements exist:

- *Producers*: green plants and other organisms that produce food for themselves and for consumers that eat them.
- *Consumers*: organisms that eat producers, other consumers, or both.

FIGURE 4-18 The landscape of Rio Blanco County, Colorado. View is taken from lower center of map in Figure 4-17, looking north (toward the top of the page). (WHR photo)

- *Decomposers*: small organisms that digest and recycle dead organisms (decomposers include bacteria, fungi, insects, and worms).
- *Materials and energy necessary for production and consumption to occur*: water, mineral nutrients, gases such as oxygen and carbon dioxide, and energy (light and heat).

Ecosystem Processes

Energy from the Sun is the starting point for understanding the operation of an ecosystem. Sunlight makes photosynthesis possible, and green plants produce food in the form of carbohydrates. The food is distributed through an ecosystem by way of a **food chain.**

Plant-eating animals, known as **herbivores,** begin the food chain by consuming plants (Figure 4-21). **Carnivores,** which are meat-eating animals, eat herbivores and may in turn be eaten by other carnivores or by **omnivores,** such as humans, who eat both plants and animals. Most of the food that animals consume is used to keep their bodies functioning. Animals excrete some of the food, and when they die, their bodies are rich in

FIGURE 4-19 Desertification. Cattle grazing on sparse grass cover near Chaco Canyon, New Mexico. Only 300 years ago, large herds of native grazers and newly introduced cattle could be supported on extensive grasslands. But today only a few animals can be supported in this semiarid environment. (Bernard Boutrit/Woodfin Camp & Associates)

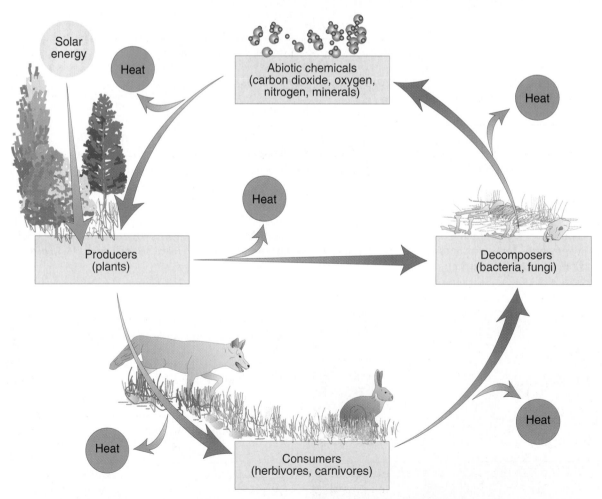

FIGURE 4-20 Ecosystems. Ecosystems consist of plants, animals, decomposers, and the physical environment with which they interact. Energy is supplied to an ecosystem from the Sun, passed through it via a food chain, and dissipated as heat.

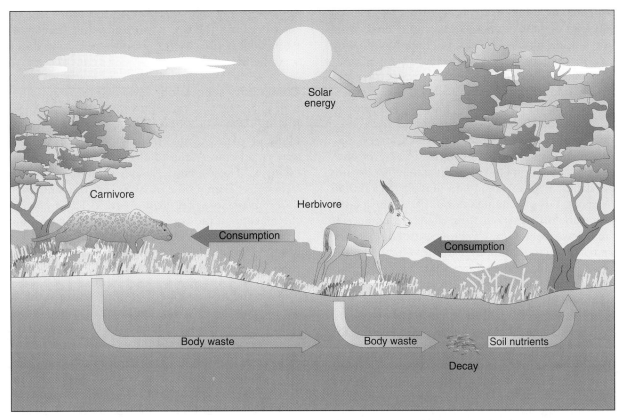

FIGURE 4-21 A food chain. Green plants are the primary producers of food, which is eaten by the primary consumers: herbivores and omnivores. Carnivores, or secondary consumers, derive their energy from other animals rather than directly from plants. At each step of the food chain, energy is lost as heat, and waste materials are broken down by decomposers. (From T. L. McKnight, *Physical Geography: A Landscape Appreciation.* 4th ed. Englewood Cliffs, NJ: Prentice Hall, 1993)

stored-up nutrients. Decomposers attack animal excretions and dead bodies. These small organisms return chemical nutrients to the soil and the atmosphere. Some of these provide nutrients to new plant growth, completing the cycle.

A step in the food chain is called a **trophic level**. At each trophic level, food is passed from one level to the next, but most of the energy is lost. As a rule of thumb, about 10 percent of the energy consumed as food at a given trophic level is converted to new biomass, and the remaining 90 percent is burned off and dissipated as heat. Because of this loss, the amount of biomass decreases as we go from the first trophic level—the green plants—to higher levels. This is why we find large numbers of herbivores such as mice and rabbits in natural systems, but very few large carnivores like wolves and lions.

Some chemicals in the environment, especially persistent pesticides such as DDT, create problems passing through a food chain. These chemicals do not break down easily in the environment, so they accumulate in animal tissues. If a pesticide accumulates in an animal rather than being excreted, then at each trophic level the concentration of that pesticide increases, a process called **biomagnification.** Consider

this example: If a falcon weighing 1 kilogram consumes 50 smaller birds, each weighing ⅕ of a kilogram, and if each of these smaller birds consumes in its lifetime 2 kilograms of plant-eating insects, then the falcon indirectly consumes 100 kilograms of insects. If each of these insects consumed, during its lifetime, 10 times its weight in plants, then the falcon has indirectly consumed 1,000 kilograms of plants! Therefore, even a very small concentration of pesticides at a low trophic level, such as a plant, can become high concentrations in large carnivores. In some birds of prey such as eagles and falcons, this causes eggs to fracture, killing chicks as they incubate. Because of these problems, DDT and similar pesticides have been banned or severely restricted in much of the world, and the pesticides used today break down relatively rapidly to reduce this problem.

Plant and animal success in ecosystems   Within any particular ecosystem, living things compete for resources such as nutrients and water. The more successful plants and animals in this competition will dominate that environment.

Success of a species means that it survives, thrives, or even comes to dominate an ecosystem for a long period. Humans are an example: surviving for the first 150,000 years of our existence as a species, then thriving over the last few thousand years, and today dominating Earth (with dramatic consequences for other species).

The success of one species over another results from competition. In the case of plants, this competition is for light, water, nutrients, and space. Although plants require all these factors to grow, in any ecosystem one factor usually is restricted, which forces competition and adaptation. For example, in an arid environment, plants compete for scant water but do not need to compete for the abundant sunlight. In a humid environment, water is abundant, but plants must compete for sunlight. An area with adequate water and light may have poor soils, so plants must compete for nutrients. Through evolution, plants that have adapted life forms, physiological characteristics, and reproductive mechanisms that allow them to succeed in particular environments have survived. The plants that compete best in an environment dominate the cover there.

What adaptations are effective in competing for space and nutrients? In a humid area height is effective: Plants that are able to grow tall capture the most light and shade out their competitors. An ecosystem with extreme cold, drought, fire, or short growing seasons favors organisms that have adapted to survive or even benefit from the stress. For example, some species have adapted by developing the ability to take advantage of occasional fires. After a fire has felled other vegetation and heated the soil, long-dormant seeds respond to the flash heating and germinate quickly. They occupy the burned site, completing their life cycle

and spreading new seed before the other vegetation recovers from the fire. Once the other vegetation regains dominance, the fire-resistant seeds lie dormant, awaiting the next fire opportunity.

Community succession   As plants grow and multiply, they alter the environment, sometimes changing it so that it becomes more suitable for the growth of other kinds of plants. Thus, the plants of the community may change over time until a stable plant population emerges. These plants remain until something, perhaps a fire, changes the physical environment and then different plants again begin to grow. These changes in the makeup of the community are called **succession.** A common example of community succession occurs when agricultural land is abandoned in eastern North America (Figure 4-22).

- In the first couple of growing seasons, the fields sprout fast-growing herbaceous weeds and other pioneer species. These spread seeds widely and tolerate bright sunshine.

- Over the next few growing seasons, the weeds and pioneer species are succeeded by slower-growing perennial shrubs and trees, such as pines, which prefer bright sunshine and gradually shade out the smaller plants beneath them.

- Hardwoods that tolerate shade grow beneath the pines, eventually overtaking them in height. As the pines die away, a forest of shade-tolerant species dominates the canopy. They are replaced by their own seedlings as they die, thus leading to

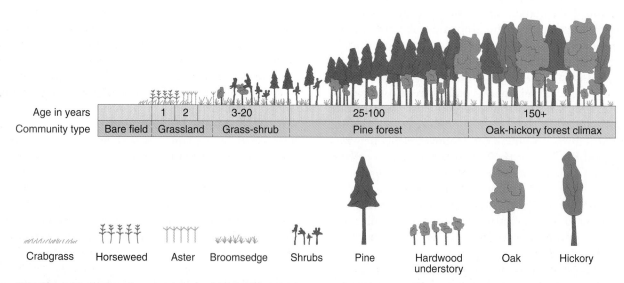

Age in years	1	2	3-20	25-100	150+
Community type	Bare field	Grassland	Grass-shrub	Pine forest	Oak-hickory forest climax

Crabgrass   Horseweed   Aster   Broomsedge   Shrubs   Pine   Hardwood understory   Oak   Hickory

FIGURE 4-22 Succession sequence. Clearing a field sets the stage for the succession sequence shown here, from left to right. This example is typical of conditions in the southeastern United States. In each successional stage, modification of the local microenvironment allows a new assemblage of plants and animals to become established. The sequence culminates in the climax hickory forest. (From R. W. Christopherson, *Geosystems,* 3rd ed. Upper Saddle River, NJ: Prentice Hall, 1997. Adapted from E. P. Odum, *Fundamentals of Ecology,* 3rd ed., Figure 9-4, p. 261 © 1971 Saunders College Publishing. Adapted by permission.)

a stable association of plants in equilibrium with local soil and climate, called a **climax community.**

The length of time necessary for succession to take place varies, depending on the growth rates and lifespans of plants, rates of seed introduction and establishment, and whether soil development proceeds along with plant growth. If the seeds of the late-successional species are present at the time succession begins and growth rates are fast, a well-developed forest may be established in a few decades. The illustration shows a more common situation, where a mature forest is achieved in 150–200 years. But if succession also involves soil development through weathering and accumulation of organic matter, it may take thousands of years for a site to be converted from bare rock to a mature forest.

In environments subject to frequent disturbance, such as fire, windstorm, disease outbreaks, floods, and volcanic eruptions, it may take a long time to establish a climax community. Vegetation may actually change continually, never reaching climax.

Earth's recent history has been one of dramatic change, including dramatic climate changes and human transformation of vast areas. Even environments that have not been dramatically altered by humans are subject to natural disturbances such as storms, fire, or outbreaks of plant or animal diseases. These environmental changes have profoundly influenced vegetation distributions in many areas, favoring plants that are adapted to dynamic conditions. Very little of Earth's surface can be regarded as stable or undisturbed.

Biodiversity

Biodiversity is the diversity of species present in any environment. Biodiversity is important, for it brings multiple food options to living things, improving each one's odds for survival. Thus, biodiversity can improve the stability of an entire community.

Earth's biodiversity is vast: There are an estimated 10 million plant and animal species. Each of these has specific needs. For plants, critical needs include water, light, and nutrient availability, the absence of soil disturbance, effective seed dispersal, and good germination conditions. For animals, the most critical requirement is food supply, meaning a sufficient quantity of specific plants or animals to feed upon. In a world of rapidly changing environments, the critical needs of many species increasingly are not met.

Habitat loss is so widespread that a significant portion of the world's species may be threatened with extinction. The largest single cause of extinction is land-use change, resulting from increasing human settlement worldwide. About 39 percent of Earth's land area (excluding Antarctica) now is in cropland or permanent pasture. Between 1960 and 2000, the land area in such agricultural uses increased about 10 percent, whereas the world's forest area decreased about 5 percent. Hunting, species introductions (such as new predators that out-compete indigenous species), and pollution also contribute to extinction. In Europe and North America, radical changes in the landscape have taken place over centuries of human habitation. While a few extinctions were noted, little attention was paid to the broader, unknown impacts of that landscape change. Rapid change now is occurring in the Amazon basin at the same time that scientists are beginning to explore tropical environments. Biologists working in the Amazon discover new species virtually every time they look, raising concern that deforestation is causing extinctions of species even before we discover them.

Biosphere reserves To reduce the impact of land-use change on species diversity, protected areas are being established by national governments and by international agencies. One important international effort is the United Nations Biosphere Reserve Program. In South America, for example, about 7.4 percent of the entire continent (490,000 square kilometers or 190,000 square miles) now is nationally protected, including 29 biosphere reserves (some overlap nationally protected areas) that cover about 508,000 square kilometers (196,000 square miles). Approximately 6.4 percent of all land area is nationally protected, and 337 biosphere reserves cover about 3.7 percent of Earth's surface. The degree of protection (restrictions on land use and disturbance) and level of enforcement, however, vary considerably.

The size of individual protected areas is important. A single animal may range over only a few square kilometers, but that small area cannot ensure species survival. A reserve must be large enough to sustain enough individuals to allow genetic diversity in the breeding population. It is difficult to decide how large a protected area should be, and this disagreement can become part of the political tug-of-war among conflicting potential land users.

Today conservationists are beginning to recognize that setting aside a few patches of relatively "natural" landscape will not be sufficient to preserve Earth's remaining biodiversity. Rather, we must also consider ways of promoting biodiversity in the managed landscape—as on farms and in commercial forests and even urban areas. One reason is the sheer size of human-dominated land areas in comparison to the small patches that have been protected. The land available for protection is insufficient to retain biodiversity across the landscape. Another reason is the belief that biodiversity in itself provides important goods for humans, perhaps through enhancing biological productivity or promoting ecological stability. If biodiversity contributes these services, then we need to maintain that biodiversity where we live and work—not just in remote preserves.

Biodiversity is a barometer of the consequences of human modification of the environment. A commitment to biodiversity means a commitment to a broad range of environmental policies—not just preserving wilderness areas, but maintaining diversity everywhere.

BIOMES: GLOBAL PATTERNS IN THE BIOSPHERE

Earth's ecosystems are grouped into **biomes** characterized by particular plant and animal types, usually named for a region's climate or dominant vegetation type. Biomes typically contain many ecosystems. A terrestrial biome's nature reflects two especially visible features: climate and vegetation. Underlying a biome's label is a diverse community of characteristic plants and animals.

The global distribution of terrestrial biomes (Figure 4-23) imitates very closely the distribution of climate regions described in Chapter 2 (Figure 2-31). This correspondence reflects a close link between climate and vegetation. To grow, plants depend on the two most important elements of a climate: energy and water.

We will now examine the vegetation and climates that distinguish major biomes: forest, savanna, woodland, scrubland, grassland, desert, and tundra.

The Major Terrestrial Biomes

FIGURE 4-23 This map shows vegetation that would exist if humans did not disturb the vegetation, as in agricultural regions. Map generalization at this scale obscures many local variations in vegetation types, especially in mountains and populated areas. (From R. W. Christopherson, *Geosystems,* 5th ed. Upper Saddle River, NJ: Prentice Hall, 2003)

Tropical rain forest
Midlatitude broadleaf deciduous forest
Needleleaf or boreal forest
Temperate rain forest
Tropical savanna
Tropical seasonal forest and scrub   } Savanna, scrubland, and open woodlands
Mediterranean shrubland (chaparral)
Tall grass
Short grass   } Midlatitude grasslands
Warm desert and semidesert (DBW)
Cold desert and semidesert (DBC)   } Desert
Arctic tundra
Alpine tundra   } Tundra
Ice

Figure 4-24 is a matrix showing some of the associations among temperature, water availability, and vegetation characteristics. The distribution of animal species is strongly influenced by the distribution of vegetation; thus, our descriptions of biomes will focus on plants.

The global distribution of terrestrial biomes imitates very closely the distribution of climate regions described in Chapter 2.

years) with spreading canopies. Forests occur over a wide range of humid and subhumid climates, where ample water is available most of the year. We group forests into four broad categories: *tropical rain forest*, *midlatitude broadleaf deciduous forest*, *needleleaf* or *boreal forest*, and *temperate rain forest*.

Forest Biomes

Forest biomes are dominated by trees: tall, woody-stemmed perennials (plants that live for at least several

Tropical rain forest biome Tropical forests and woodlands dominate the humid equatorial environments of Central America, the Amazon River basin, equatorial Africa, and Southeast Asia (Figure 4-25). Tall, broad-leaved trees retain their leaves

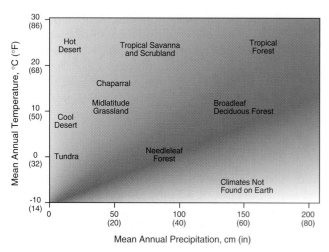

FIGURE 4-24 Generalized relation between vegetation type and climate. The amount of precipitation necessary to support a given amount of vegetation (such as forest relative to grassland) is greater in warm climates, where evapotranspiration is higher, than in cold climates. Cold climates with high rainfall amounts are not found on Earth.

FIGURE 4-25 Tropical rain forest. Rainforest vegetation is generally dense and multistoried, as exemplified by this scene from the tropical island of Borneo. (Courtesy of Julia Nicole Paley)

all year. A **tropical rain forest** has a top layer, or canopy, and two more layers beneath that (Figure 4-26). Each layer has different dominant species and associated animal communities.

The highest canopy consists of very tall trees—heights of 40 to 60 meters (130 to 200 feet)—spaced some distance apart so that sunlight reaches the next lower canopy. Underneath is the dense middle canopy. Most photosynthesis occurs in the two upper layers, where sunlight is most intense. The middle canopy blocks most light, keeping the forest floor in twilight, so plants at the lowest levels must tolerate deep shade to survive. Climbing vines are common in many tropical rain forests, as are plants that grow on the surfaces of other plants to gain height advantage without having long trunks or stems reaching to the ground.

Tropical rain forests are noted for biodiversity, with hundreds of tree species within a single hectare (about 2.5 acres). In contrast, midlatitude forests may have only five to 15 tree species per hectare, and high-latitude coniferous forests may have only three to five. Tropical diversity of trees is paralleled by a great variety of animals. The diverse vertical structure adds a diversity of habitats and species. A walk through a zoo illustrates the vast number of large animal species of the world that are native to tropical climates. Non-human primates are found almost exclusively in the tropics, especially in tropical forests. The wide diversity of life in the tropical rain forest places this biome at the center of controversy over deforestation, species extinctions, and biodiversity.

In the rainforest biome, much nutrient storage is in living biomass rather than in the soil. Organic matter decays rapidly in warm, wet conditions, and nutrients are not easily retained in the soil. When leaves fall on the forest floor and begin to decompose, the nutrients released are quickly absorbed and recycled by living plants. This is yet another reason why deforestation is a major concern. If forests are removed, their store of nutrients may be lost quickly to leaching and runoff, leaving only the impoverished soil.

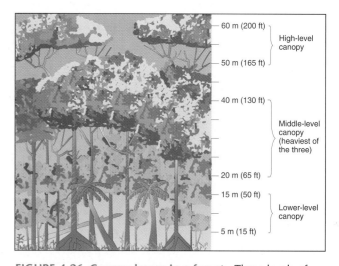

FIGURE 4-26 Canopy layers in a forest. Three levels of forest canopy can be distinguished in the tropical forest. At the top is the high-level canopy, in which the tallest trees emerge above the rest. The densest canopy layer is in the middle levels. A third, lower-level canopy is composed of shade-tolerant species. (From R. W. Christopherson, *Geosystems*, 5th ed. Upper Saddle River, NJ: Prentice Hall, 2003)

Midlatitude broadleaf deciduous forest biome   The **broadleaf deciduous forest** exists in subtropical and midlatitude humid environments where seasonally cold conditions limit plant growth (Figure 4-27). The eastern United States and much of China share this biome, which explains why some Chinese ornamental species are popular on lawns in the eastern United States. On the map you can see that this biome occurs mostly in the Northern Hemisphere.

During the summer growing season, long days and a high solar angle promote rapid growth, so that annual plant growth may attain 60 to 75 percent of that in the tropics, even though the growing season is only five to seven months. These plants have evolved wide, flat leaves that capture as much sunlight as possible. Broadleaf deciduous forests are much less diverse than are tropical rain forests, but these forests still support many species. Because these forests are located in the midlatitudes, many animal species that make their home there either hibernate in winter or migrate to warmer locations.

Soils of this region vary with the underlying geology, but they generally are leached less than are soils of the humid tropics. This is because annual precipitation usually is lower than in the rain forest and because lower temperatures slow chemical weathering. Lower temperatures also contribute to reduced decay rates of organic matter. Consequently, undisturbed soils typically have a substantial *O* horizon, ranging in thickness from a few centimeters (1 to 2 inches) to over 10 centimeters (4 or more inches). When these forests are cleared and the soil is brought under cultivation, the soils generally are quite fertile, but that fertility declines once the organic matter decays and is not replaced.

Needleleaf or boreal forest biome   In poleward portions of the midlatitudes, a needleleaf (coniferous) evergreen forest flourishes. The name "boreal forest"

FIGURE 4-27 **Forest in Michigan.** A forest in the transition zone between midlatitude deciduous forest and northern coniferous forest in northern Michigan. (Stan Osolinski/Corbis/ Stock Market)

comes from the northern location of this vegetation (Latin *borealis* = "northern"). The **boreal forest** thrives in cold, continental midlatitude climates. There is little land in the Southern Hemisphere in latitudes 50°–70°, and thus extensive boreal forests are restricted to the Northern Hemisphere. The most extensive boreal forest regions are in central Canada and Siberia. Needleleaf forests also occur in some areas of seasonally dry soils.

During cold winters, low humidity and frozen ground cause moisture stress, but needleleaved trees survive because the leaves have low surface area in relation to their volume, and they are covered with a waxy coating to reduce water loss. This adaptation allows them to survive in colder climates that would kill broadleaved trees. Because they are evergreen, these trees photosynthesize whenever temperatures are warm enough, instead of having to wait until after the last frost to produce leaves. Most needleleaved trees are evergreen, but not all. The larch, for example, is a needleleaved deciduous tree.

Coniferous forests also occur in the southeastern United States, where sandy soils drain quickly and have limited water storage. During warm summers, these soils dry out, and the moisture-retaining properties of coniferous trees allow them to compete well against broadleaved trees. The ability to sprout from roots and grow rapidly following forest fires is an added advantage of many coniferous species in areas of summer aridity.

Needleleaves accumulate beneath coniferous forests, and as they decay, they produce strong acids in water that percolate down through the soil. This acidic water leaches the upper soil of all but the most insoluble minerals, leaving a light-colored, sandy upper soil layer. Some of the iron and aluminum leached from the upper soil accumulates lower in the soil profile. This produces a distinctive soil associated with coniferous forests called a *spodosol* (see Figure 4-15).

Temperate rainforest biome   On the west coasts of North and South America, lush evergreen coniferous forests grow. These temperate rain forests are found in areas of marine west coast climate, where moderate temperatures and ample rainfall allow year-round plant growth. The evergreen trees grow faster in summer, but they can photosynthesize in winter despite occasional freezing temperatures. The temperate rain forest is much less diverse than the tropical rain forest; it is dominated by relatively few species.

Savanna, Scrubland, and Open Woodland Biomes

Savannas and open woodlands have trees spread widely enough for sunlight to support dense grasses and shrubs beneath them. This vegetation is common in

Critical Thinking
HUMAN-DOMINATED SYSTEMS

In recent years, quantitative study of biogeochemical cycles at the global scale has determined the amounts of materials that are processed in various environments per unit time, in part to understand the relative importance of different components of those cycles. One of the conclusions of this research is that a considerable amount of material processing—conversion of substances from one form to another—is carried out by humans. In some cases humans do more processing than nature does.

Consider some examples. We have already seen that humans are playing such a major role in global carbon cycle that the quantity of CO_2 in the atmosphere is steadily increasing, despite the vast amounts of carbon processed each year by the biosphere. The nitrogen cycle is another important cycle, in which nitrogen is taken from the atmosphere and converted to forms that can be used by plants and animals to build proteins and other molecules. This bioavailable nitrogen is then passed through food chains before being broken down and released back to the atmosphere. Humans *fix* nitrogen (convert it to biologically available forms) through fertilizer manufacture, and today more nitrogen is fixed by humans than by natural processes. Food production through photosynthesis is a critical biogeochemical process in which humans play a major role.

> A considerable amount of material processing—conversion of substances from one form to another—is carried out by humans. In some cases humans do more processing than nature does.

About 8 percent of global photosynthesis takes place in agricultural lands. Humans play an even larger role in consuming plant matter, either directly as food for themselves and their domestic animals and as fiber such as lumber, or indirectly by controlling its characteristics or fate, such as grass grown on golf courses or crop residues left in the field. Estimates of the amount of biological production that is "appropriated" by humans range from about 3 to 40 percent, depending on what human uses are included in the estimate.

These conclusions are of great importance, not only for scientific and practical purposes but in describing the physical relationship between humans and the environment. This, in turn, has implications for the way we think about nature and our role in it. In short, it goes to the very heart of the human-environment relationship.

Questions

1. How important is it that we separate human systems from natural systems in our environmental management policies? Should we be investing in wilderness preservation or improvement of managed landscapes?

2. Are humans distinct from nature or a part of it? Should nature, as something over which humans have little or no influence, be preserved?

climates that have a pronounced dry season. The term **savanna** refers especially to vegetation characteristic of large seasonally dry areas in tropical Africa and South America. The rainy season brings green, lush grass, but during the dry season the grass dries out and only deeper-rooted trees continue to photosynthesize. If the dry season is pronounced, even the trees may lose their leaves, and in some areas dry-season deciduous trees are common.

Fire is common in the savanna, aided by the seasonal aridity and probably often caused by humans as well as lightning. Fire does not destroy most grasses, which grow back quickly. Slower-growing trees might in time become dominant, shading out the grasses, but they are more susceptible to fire damage, so the savannas remain. Whether the savannas of Africa and South America are the result of human activity is often debated. Some expanses of African savanna occur in climates that support forests elsewhere, supporting the hypothesis that human-caused fires have created the extensive African savannas.

Soils of tropical savannas usually are deeply weathered, as in the humid tropics, but they may be less leached of nutrients as a result of the reduced rainfall. High temperatures and reduced plant growth combine to lower the organic content. Many soils of these regions are quite productive if sufficient irrigation water is available.

In more arid margins of tropical savannas, a seasonal forest or thorn-scrub vegetation occurs. Drought-resistant trees and shrubs dominate, as well as cacti, and grasses are less important. Savanna-like open woodlands (also called parklands) also are found in some semiarid midlatitude areas.

Mediterranean climates promote a distinctive shrubland vegetation type called *chaparral.* This is a shrub woodland dominated by hard-leaved trees and shrubs that withstand the severe summer aridity. Fire is common in this area because of summer aridity and flammable waxy-leaf coatings. Many plant species in the chaparral are adapted to frequent fire, and some even require fire for their continued presence in the landscape (see the Critical Thinking box, p. 160).

Midlatitude Grassland Biome

In Figure 4-23, extensive midlatitude grasslands are shown in deep red and brown. Major areas include interior Asia, the Great Plains of Canada and the United States, and central Argentina (Figure 4-28). The midlatitude semiarid climate of these areas features hot summers, cold winters, and moderate rainfall. In moister portions, the grass is usually taller (up to 2 meters or 6 to 7 feet), and the biome is called **prairie.** In drier areas, such as Asia, the biome is called short-grass prairie, or **steppe.**

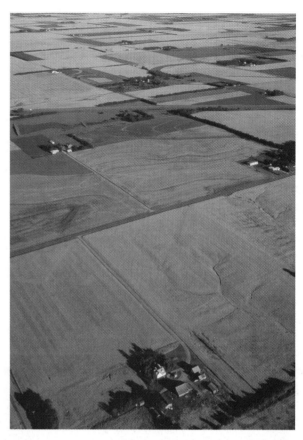

FIGURE 4-29 Former tall grass prairie. Today most of the tall grass prairie has been converted to row crops. (Russell Munson/Corbis/Stock Market)

FIGURE 4-28 Tall grass prairie in Kansas. A small remnant of formerly extensive tall-grass prairie in the Flint Hills of Kansas. (Frank Siteman/StockBoston)

Grasses are well suited to this climate because they grow rapidly in the short season when temperature and moisture are favorable (generally spring and early summer). During dry or cold periods, aboveground parts of these plants die back, but the roots become dormant and survive. This trait also allows grasses to survive fire and grow back rapidly, using available moisture at the expense of trees or shrubs that might invade. Many experts believe that occasional fires, either natural or deliberately set by humans, are at least partly responsible for establishing and maintaining extensive grasslands.

The soils of these midlatitude grasslands have a dark upper horizon with high nutrient content and are very fertile. Because of this fertility, many of the world's midlatitude grasslands are "breadbaskets"—primary production areas for wheat and other cereal grains (Figure 4-29). In North America, virtually all original *tall-grass prairies* are farmed to produce wheat, corn, soybeans, and other small grains. Much of the *short-grass prairie* present at the time of European settlement remains, although it has been heavily grazed.

Desert Biome

In **deserts,** moisture is so scarce that large areas of bare ground exist, and the sparse vegetation is entirely

Critical Thinking

FIRE AND CHAPARRAL IN CALIFORNIA

In the chaparral woodlands of California, fire is an important natural part of ecosystems. Unfortunately, people have built homes and businesses close to the chaparral. Conflagrations like the 2003 fires near Los Angeles destroy entire communities, causing vast property damage and occasionally taking lives. When a fire starts near a populated area, the natural reaction is to extinguish it at once. Paradoxically, this action to halt fire actually makes conditions worse. In the chaparral, plant growth produces flammable biomass that does not decay as rapidly as it is produced, so it accumulates over time.

Fire threatens homes in suburban Los Angeles. (Nate Rawer/AP/Wide World Photos)

The longer it accumulates, the more fuel is available for severe fires. If a fire starts in a recently burned area, little fuel is available, and the fire burns slowly at cooler temperatures, and it does not spread widely. But if a fire starts where one has not burned for a long time, a hot, dangerous fire is likely.

Consequently, many areas that routinely would burn have not burned, and fuel accumulations—dry chaparral brush—are unusually large. When a fire starts during dry, windy conditions, it becomes very dangerous and uncontrollable. Its high temperatures make it much more destructive than fires during moist conditions.

Decades of firefighting in coastal California and the western United States have not eliminated fires; they have only allowed fuel to accumulate. Normal, "natural" fires, low temperature and slow burning, have been eliminated and replaced by infrequent but catastrophic conflagrations. Fire managers now use controlled burns in some areas to reduce fuel accumulations.

Questions

1. If you lived in a fire-prone neighborhood, would you support the use of controlled burns even if there were a small risk that the fire would get out of control?

2. Can you think of other examples, perhaps where you live, of situations where poor environmental decisions in the past have locked us into protecting structures and other investments at costs that may be unreasonably large?

adapted to moisture stress. Plants with such adaptations are called **xerophytes.** Some desert plants are drought-tolerant varieties of those common in more humid areas, such as grasses. Others, such as cacti, are almost exclusive to deserts. Most desert plants are adapted to limit evaporative losses, with thick, wax-coated leaves or with needles rather than leaves, protecting cacti from being consumed for the water they contain.

Another common adaptation is water-collecting root systems. These are extensive shallow roots to gather occasional rainwater or long tap roots to reach deep groundwater. Many desert plants survive long periods without moisture through dormancy, either as a mature plant or as a seed. When occasional rains do occur, these dormant plants spring to life and flower in a few weeks before becoming dormant again. Animal life in

the desert is dispersed because of the reduced rates of plant production, but it is still diverse. Many desert animals are nocturnal, foraging and hunting at night to avoid dehydration during the hot daytime hours.

Scant moisture and consequent slow chemical activity leave many desert soils poorly developed. Leaching of soluble nutrients is limited, and usually the horizons in desert soils are developed through vertical movements of soluble minerals, controlled by rainfall and evaporation. When water enters the soil, it may dissolve some minerals, but as evaporation draws this water back to the soil surface, these minerals are redeposited. As a result, high concentrations of soluble minerals near the surface (especially salts) are common. Desert soils may be fertile if irrigated, but irrigation usually introduces more minerals to the soil, and the soil can become salty.

Tundra Biome

In cold, high-latitude environments or in high mountains above the treeline, freezing and short growing seasons severely limit plant growth. During winter, water is locked up in the soil as ice. Dehydration and abrasion by blowing snow damage exposed upper parts of plants. Therefore **tundra** is dominated by low, tender-stemmed plants and low, woody shrubs. These survive the cold by lying dormant below the wind, often buried in snow, growing only in the short, cool summer. Animal life in this biome is very limited in winter, but in summer the tundra comes alive with insects, migratory birds, and grazers such as reindeer and caribou.

As in deserts, tundra soils often are poorly developed because of slow chemical activity. Permafrost occurs where mean annual temperature is below 0° C (32° F). Soil horizons are weakly developed and often disrupted by frost action. In poorly drained areas, organic matter often accumulates rather than decays, and peat is formed. As was mentioned earlier, peat is an accumulation of plant debris and an early stage in the formation of coal.

Because of slow plant growth in this environment, recovery from disturbances can be very slow. This, combined with the disruption of permafrost caused by construction activities and vegetation disturbance, has led to concern about the long-term effects of oil exploration and related activities in Alaska, Siberia, and similar areas.

Natural and Human Effects on the Biosphere

Plants, soils, climate, and human activity each reflect the influence of the other. In the absence of humans, climate would be the strongest control on vegetation. Climate controls soil formation and plant growth so strongly that, with the exception of agricultural areas, world vegetation and soil patterns correspond closely to world climates. Humans, however, have had profound influences on ecosystems in most of the world's land areas, including the 37 percent of land area (excluding Antarctica) that is cropland or permanent pasture (Figure 4-30). Locally, topography and geology also exert strong influence on vegetation and soils.

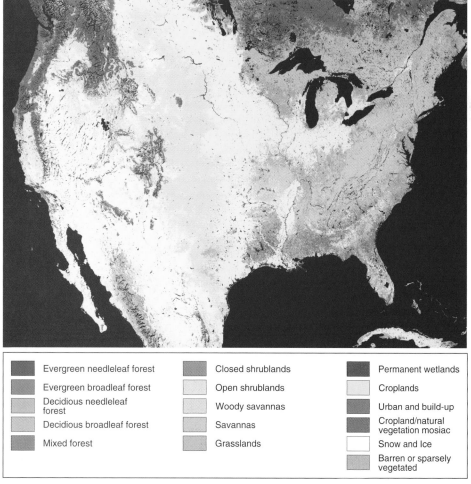

Evergreen needleleaf forest	Closed shrublands	Permanent wetlands
Evergreen broadleaf forest	Open shrublands	Croplands
Deciduous needleleaf forest	Woody savannas	Urban and build-up
Deciduous broadleaf forest	Savannas	Cropland/natural vegetation mosiac
Mixed forest	Grasslands	Snow and Ice
		Barren or sparsely vegetated

FIGURE 4-30 Land cover of part of North America, based on satellite imagery. Compare the extent of agricultural land in with the distribution of natural vegetation types as shown in Fig. 4-23. (MODIS/NASA Headquarters)

Regionally and globally, climate is in turn affected by ecological processes and soil-vegetation linkages:

- Plants regulate evapotranspiration from land surfaces and thus moderate the hydrologic cycle. If vegetation is cut back, reducing the number of plants to draw water from the soil and transpire it to the atmosphere, moisture flow to the atmosphere may be reduced. The excess water runs off to the sea as river flow. In this manner, vegetation could modify climate by affecting humidity.

- Soil eroded by wind from semiarid and arid areas may cloud the atmosphere with dust and modify energy exchanges there, thus modifying climate.

- The biosphere stores and regulates carbon, helping to control the global carbon budget. If vegetation grows more quickly and absorbs atmospheric carbon dioxide faster than decaying organic matter can return carbon dioxide to the air, there is net storage of carbon in biomass. If the reverse is true, there is net release of carbon as carbon dioxide to the atmosphere. Thus, vegetation partly controls the level of this important greenhouse gas in the atmosphere.

Desertification is a good example of the impact of interactions among human and natural processes. Economic and social forces, especially population growth and demand for food, cause us to populate semiarid areas with domesticated animals or to replace native vegetation with plant crops. The reduction in vegetation cover causes increases in runoff and wind erosion, which reduce the amount of water and nutrients present in the soil, further limiting vegetation growth. The loss of vegetation may also affect the atmosphere by increasing amounts of dust it contains or by reducing the amount of solar energy absorbed by the ground surface. Changes in dust or sunlight absorption are hypothesized to affect temperature and precipitation amounts, although such effects have not been confirmed by observation. These intimate relations among the soil, water, air, and living things lead some scientists to view Earth as a single interconnected dynamic system—almost a functioning organism—rather than just a set of cause-and-effect linkages. We can see all these subsystems interacting with each other in much the same way that different parts of a single living organism interact.

CONCLUSION: CRITICAL ISSUES FOR THE FUTURE

Beginning thousands of years ago but especially within the past two centuries, humans have disrupted ecosystems worldwide, largely through spreading population increase and increasing demands on resources. By disturbing environments, we have unbalanced them, and they probably now are much more dynamic than they are naturally. We have increased biomass stores in some areas and reduced them elsewhere; we have altered the distribution of biomes and nearly eliminated many ecosystems; we have changed the sediment and nutrient flows of rivers and lakes.

Only a few thousand years ago, Earth was regulated by non-human processes, and people were only negligible players, no more significant than any other common species. Today we are major participants in Earth's environmental processes, often the dominant ones. Environments in which human impacts are minor are increasingly scarce, and the future of Earth's environmental systems depends on how we manipulate biological and physical processes at the surface of Earth.

The scientific study of human interactions with the biosphere is still young, and there is much to be learned. Until recently most studies of human-environment relations have been carried out at relatively local scales. But today we recognize that human impacts are truly global, and we need to increase our understanding of Earth's systems at that scale.

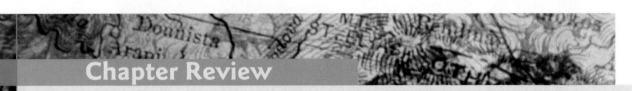

Chapter Review

SUMMARY

Climate is the dominant control on local environmental processes, primarily through its influence on water availability and movement. Knowledge of the water budget is essential to understanding both physical and biological processes; the water budget is a critical regulator of ecosystem activity. This influence extends to human use of the landscape, through determining the natural vegetation cover to agricultural potential.

Carbon is the basis of life on Earth. It cycles through the atmosphere, oceans, biosphere, and soil. Photosynthesis transfers carbon from the atmosphere to the biosphere, and respiration returns it to the atmosphere. Plant growth is highest in warm, humid climates where water and energy are plentiful. Plant growth is less in dry or cool climates. Large amounts of carbon are exchanged between the atmosphere and the oceans, and the oceans are a major sink for carbon dioxide added to the atmosphere by fossil-fuel combustion.

The outermost layer of Earth's surface in land areas is soil, a mixture of mineral and organic matter formed by physical, chemical, and biological processes. Climate is a major regulator of soil development, through its control on water movement. Plant and animal activity in the soil produces organic matter and mixes the upper soil layers. In humid regions, water moves downward through the soil, carrying dissolved substances lower in the profile or removing them altogether. In arid regions, these substances are not so easily removed. Soil characteristics reflect these processes, and there is close correspondence between the world climate map and the world soil map.

Photosynthesis is the basis of food chains and ecological systems. Plants compete for water, sunlight, and nutrients; and plants that are best adapted to compete for limiting factors will dominate a given environment. Such adaptations help explain the world distribution of major vegetation types. Disturbances modify ecological communities, and succession following disturbance is a fundamental process of landscape change. Very large portions of the world's land surface have been modified by human activity, and humans are major players in most of the world's ecosystems.

The world vegetation map closely mirrors the world climate map. Ecologically diverse and complex forests occupy humid environments, storing most nutrients in their biomass. In arid and semiarid regions, sparse vegetation is adapted to moisture stress. Forests adapted to winter cold are in humid midlatitude climates, with broadleaf forests in warmer areas and coniferous forests in subarctic latitudes. In cold, high-latitude climates, cold-tolerant short vegetation occupies areas that have a mild summer season. Vegetation is absent in ice-bound polar climates.

Overall, climate is the strongest control on *natural* vegetation, but humans have had profound influences on ecosystems in most of the world's land areas. Plants regulate evapotranspiration from land surfaces and thus moderate the hydrologic cycle. The biosphere stores and regulates carbon, helping to control the global carbon budget. Only a few thousand years ago, Earth was regulated by nonhuman processes, and people were only negligible players. Today, however, humans are globally significant and locally dominant players in Earth's biogeochemical cycles. Environments in which human impacts are minor are increasingly scarce.

KEY TERMS

actual evapotranspiration (ACTET) p. 138
biodiversity p. 155
biogeochemical cycle p. 134
biomagnification p. 153
biomass p. 144
biome p. 156
boreal forest p. 159
broadleaf deciduous forest p. 159
carbon cycle p. 141
carnivore p. 152
climax community p. 155
desert p. 161
desertification p. 134

ecosystem p. 150
evapotranspiration (ET) p. 138
food chain p. 152
herbivore p. 152
horizon p. 145
hydrologic cycle p. 135
infiltration capacity p. 140
omnivore p. 152
oxygen cycle p. 141
parent material p. 145
photosynthesis p. 134
potential evapotranspiration (POTET) p. 138
prairie p. 161

respiration p. 142
runoff p. 136
savanna p. 160
soil p. 145
soil fertility p. 147
soil order p. 146
steppe p. 161
succession p. 154
trophic level p. 153
tropical rain forest p. 158
tundra p. 163
water budget p. 136
xerophytes p. 162

QUESTIONS FOR REVIEW AND DISCUSSION

1. Diagram the hydrologic cycle, using boxes and arrows. Proportion box sizes to indicate the relative amounts of water stored at each point in the cycle. Label the arrows linking the boxes with the appropriate hydrologic process (evapotranspiration, condensation, precipitation, runoff, and so on).

2. Examine the water-budget diagram in Figure 4-7 for a climate similar to the one in which you live. Explain the seasonal patterns of precipitation, potential evapotranspiration, actual evapotranspiration, soil-moisture use and recharge, runoff, and moisture deficit.

3. What are photosynthesis and respiration? How are they related? How do they vary seasonally in relation to temperature and water availability? How do they vary in relation to major climate types?

4. What is soil texture? How does it affect the movement and storage of water in the soil? How are soil-forming processes affected by water availability?

5. For each major climate type described in Chapter 2, describe the major vegetation type generally associated with that climate.

THINKING GEOGRAPHICALLY

1. Go to the U.S. Geological Survey's surface water data web site (**http://waterdata.usgs.gov/nwis/sw**) and locate monthly flow data for a river near you. Download the data and make a bar graph showing mean monthly flow. Explain the variations in river flow in terms of seasonal variations in precipitation and evapotranspiration.

2. In your library, find the soil survey for a place near you where excavation is occurring, such as for a highway or building foundation. Discover the soil characteristics in the area. Then visit the excavation site and view the soils in the field, comparing what you see with the descriptions in the soil survey.

3. Consider a typical water budget for a midlatitude continental site in which soil-moisture use is important for part of the year. How is actual evapotranspiration affected by the amount of water stored in the soil, and the rooting depth of the vegetation?

4. World maps of climate and vegetation have many similarities. If climate were to change significantly, do you think the vegetation map would also change? Why or why not?

5. How has the vegetation cover where you live changed in the past 200 years? How do you think this change may have affected the local water budget?

SUGGESTIONS FOR FURTHER LEARNING

Abrams, Marc D. "Fire and the Development of Oak Forests." *Bioscience* 42 (May 1992): 346–353.

Bonnie, R., and others. "Counting the cost of deforestation." *Science* 288, no. 5472 (2000): 1763–1764.

Box, E. O. "Plant functional types and climate at the global scale." *Journal of Vegetation Science* 7, no. 3 (1996): 309–320.

Bryce, S. A., and others. "Ecoregions: A geographic framework to guide risk characterization and ecosystem management." *Environmental Practice* 1, no. 3 (1999): 141–155.

Cincotta, R. P., and others. "Human population in the biodiversity hotspots." *Nature* 404, no. 6781 (2000): 990–992.

Cox, C. Barry. *Biogeography: An Ecological and Evolutionary Approach*, 5th ed. Oxford: Blackwell, 1993.

Crifasi, R. R. "The political ecology of water use and development." *Water International* 27, no. 4 (December 2002): 492–503.

Davies, H. R. J., and S. M. A. H. Alredaisy. "Desertification: Or perhaps not? White Nile, Sudan, 1980–92." *GeoJournal* 37, no. 4 (1995): 479–482.

DellaSala, D. A., and others. "A global perspective on the biodiversity of the Klamath-Siskiyou ecoregion." *Natural Areas Journal* 19, no. 4 (1999): 300–319.

Desaigues, Brigitte, and others. "Estimation of the social benefits of preserving biodiversity." *International Journal of Environment and Pollution* (Inderscience Enterprises Ltd.) 12, no. 4 (1999): 400–413.

Ferraro, P. J., and A. Kiss. "Direct payments to conserve biodiversity." *Science* 298 (2002): 1718–1719.

Flannagan, M. D., and C. E. Van Wagner. "Climate change and wildfire in Canada." *Canadian Journal of Forest Research* 21, no. 1 (1991): 66–72.

Global Envionmental Change Programmes. "Global change and the Earth system: A planet under pressure." *Science* (Stockholm: IGBP) no. 4 (2001).

Goudie, Andrew. *The Human Impact on the Natural Environment*. New York: Blackwell, 1993.

Gresswell, R. E. "Fire and aquatic ecosystems in forested biomes of North America." *Transactions of the American Fisheries Society* (American Fisheries Society) 128, no. 2 (1999): 193–221.

Groombridge, B., and M. D. Jenkins. *World Atlas of Biodiversity*. Berkeley: University of California Press, 2002.

Hancock, J. "Biodiversity superstores." *Biologist* 45, no. 5 (1998): 207–208.

He, Hong S., and others. "Integration of GIS data and classified satellite imagery for regional forest assessment." *Ecological Applications* 8, no. 4 (1998): 1072–1083.

Jackson, J. B. C., and others. "Historical overfishing and the recent collapse of coastal ecosystems." *Science* 293 (2001): 629–638.

Jones, J. A. A. *Global Hydrology*. Edinburgh Gate, England: Addison Wesley Longman, 1997.

Kelly, Robert C. "The carbon conundrum: Global warming and energy policy in the third millennium." *CountryWatch* (2002).

Lackner, K. S. "A guide to CO_2 sequestration." *Science* 300 (2003): 1677–1678.

Lewandrowski, J., and others. "Estimating costs of protecting global ecosystem diversity." *Ecological Economics* 29, no. 1 (1999): 111–125.

Liu, J., and others. "Protecting China's biodiversity." *Science* 300 (2003): 1240–1241.

McGregor, G., G. E. Petts, A. M. Gurnell, and A. M. Milner. "Sensitivity of alpine stream ecosystems to climate change and human impacts." *Aquatic Conservation* 5, no. 3 (1995): 233–247.

Myers, Norman. "Tropical forests: The main deforestation fronts." *Environmental Conservation* 20 (Spring 1993): 9–16.

Pimm, Stuart L., and J. L. Gittlemen. "Biological diversity: Where is it?" *Science* 255 (February 21, 1992): 940.

Post, Wilfrid M., T. -H. Peng, W. R. Emanual, A. W. King, V. H. Dale, and D. L. DeAngeles. "The global carbon cycle." *American Scientist* 78 (July–August 1990): 310–326.

Postel, Sandra L., G. C. Daily, and P. R. Ehrlich. "Human appropriation of renewable fresh water." *Science* 271 (1996): 785–778.

Reid, Walter V., and M. C. Trexler. *Drowning the National Heritage: Climate Change and U.S. Coastal Biodiversity.* Washington, D.C.: World Resources Institute, 1991.

Schlesinger, William H., J. F. Reynolds, G. L. Cunningham, L. F. Huenneke, W. M. Jarrell, R. A. Virginia, and W. G. Whitford. "Biological feedbacks in global desertification." *Science* 247 (March 2, 1990): 1043–1048.

Senay, G. B. "Combining AVHRR-NDVI and land-use data to describe temporal and spatial dynamics of vegetation." *Forest Ecology and Management* 128, no. 1, 2 (2000): 83–91.

Sioli, Harald. "The effects of deforestation in Amazonia." *The Ecologist* 17, no. 4, 5 (1987): 134–138.

Turner, Billie Lee. *The Earth as Transformed by Human Action.* New York: Cambridge University Press, 1990.

Vitousek, Peter M., P. R. Ehrlich, A. H. Ehrlich, and P. A. Matson. "Human appropriation of the products of photo-synthesis." *BioScience* 36 (June 1986): 368–373.

Williams, J. R. "Addressing global warming and biodiversity through forest restoration and coastal wetlands creation." *Science of the Total Environment* 240, no. 1–3 (1999): 1–9.

Williams, K., and others. "Sea-level rise and coastal forest retreat on the west coast of Florida, USA." *Ecology* 80, no. 6 (1999): 2045–2063.

Williams, Michael. *Deforesting the Earth.* Chicago: University of Chicago Press, 2003.

Wilson, E.O. *The Future of Life.* New York: Knopf, 2002.

Zhang, Peichang, and others. "China's forest policy for the 21st century." American Association for the Advancement of Science, *Science* 288, no. 5474 (2000): 2135–2136.

WEB WORK

The World Wide Web is an especially rich source of information relating to the global environment. Among international governmental agencies, the United Nations Environment Program (UNEP) site provides a useful gateway to United Nations activities:

http://www.grida.no/

Data gathered by UNEP and many other governmental and nongovernmental agencies is available through CIESIN, a non-profit organization that provides information to help scientists, decision-makers, and the public better understand their changing world:

http://www.ciesin.org/

For information on programs aimed at protecting the biosphere, the International Union for Conservation of Nature is a key data-collection and dissemination organization. In particular, its Biodiversity Conservation Information System (BCIS) helps coordinate conservation efforts among diverse governmental and non-governmental organizations:

http://iucn.org/

In the United States, the United States Geological Survey (USGS) is home to two key environmental research groups. The Water Resources Division collects and disseminates water resource information in the United States, and the Biological Survey is responsible for research on the nation's biological resources. You can find the USGS at:

http://www.usgs.gov/

The Natural Resources Conservation Service, formerly the Soil Conservation Service, is the key to information about soils in the United States as well as climate and water resource information of interest to the agricultural community:

http://nrcs.usda.gov/

Environment Canada provides a central point of access to a wealth of environmental information pertinent to Canada.

http://www.ec.gc.ca/envhome.html

A Look Ahead

THE DISTRIBUTION AND DENSITY OF HUMAN SETTLEMENT

About 90 percent of Earth's 6.3 billion people occupy only 20 percent of Earth's surface. The distribution and density of settlement is due largely to attributes of the physical environment, to factors of history, and to varying rates of population growth.

WORLD POPULATION GROWTH

Earth's total population is still rising, but the rate of increase is slowing. Most population increase is occurring in the poor countries, and some people fear that Earth may be overpopulated or facing potential overpopulation.

OTHER SIGNIFICANT DEMOGRAPHIC PATTERNS

Sex ratios vary greatly among national populations, and these variations may be explained as the result of both economic and cultural factors. The median age of the human population overall is rising, and this presents distinctly different challenges to the rich countries and the poor countries.

POPULATION, POPULATION INCREASE, AND MIGRATION

◀ Such abandoned farmhouses are a common sight throughout the United States and Canadian prairie provinces. World population is increasing, but at a decelerating rate, and humankind is concentrating in urban areas. (© Annie Griffiths Belt/CORBIS)

MIGRATION

Most geneticists believe that all the world's peoples descended from a single stock that migrated away from Africa as long as 200,000 years ago. Since 1500, the great migrations that have made our world include migrations of Europeans to the Western Hemisphere, Asia, Oceania, and Africa; of Africans, mostly to the Western Hemisphere; of Indians; and of Chinese.

MIGRATION TODAY

Migration of people from poor countries to rich countries, of people from rural areas to cities (discussed in Chapter 10), and of people seeking greater political freedom continues today. Pressures of migration have become a political issue in many rich countries.

On May 7, 2003, all of France's major news-papers carried an unprecedented open letter addressed to France's working men and women by Prime Minister Jean-Pierre Raffarin. "Our retirement system," he wrote, "conceived more than 50 years ago, no longer corresponds to the current and future demographic reality." In France's system, today's workers pay the pensions currently enjoyed by the previous generation. As each generation lives longer, however, and low national birth rates foretell a falling total population, the financial burden on each worker has grown. In 1960, four workers financed each pensioner, but in 2000 it was only two; and if present statistical trends continue, by 2020 each worker will have to support a retired worker. The pension system will collapse. The workers responded to the Prime Minister's letter by calling a national strike. In most other rich countries, including, to some degree, the United States, comparable demographic realities apply. All solutions proposed to the pension cost-squeeze are politically controversial, including raising taxes, raising the birth rate, welcoming more immigrants, requiring people to work longer, or reducing pension benefits.

Most poor countries face a different demographic challenge: Birth rates are also falling, but they remain relatively high. The numbers of children born every year challenge the societies' ability to feed them, clothe them, educate them, and eventually provide jobs for them.

This chapter examines the world's **population geography**—that is, the distribution of humankind across our planet. The relative distribution of humankind is constantly changing, and this is because of two factors. One is the fact that different countries have different internal population dynamics, such as the numbers of births and deaths and the consequent rates of increase or decrease. The second factor is human **emigration** out of some places and **immigration** into others. Migration has affected the distribution of the world's peoples in the past, and continuing migration affects world affairs today.

This chapter will occasionally include ideas and information about **demography,** which is the study of individual populations in terms of specific group characteristics. These characteristics may include the distribution of ages within the group, the relative numbers of males and females, income levels, or any other characteristic. Demography means "describing people."

THE DISTRIBUTION AND DENSITY OF HUMAN SETTLEMENT

In 2003 the population of Earth was about 6.3 billion, but people are not evenly distributed across the landscape (see the map on the rear endpaper). The map shows that large areas of Earth's surface are surprisingly uninhabited. About 75 percent of the total human population lives in the Northern Hemisphere between 20° and 60° North latitude, and even great expanses of that area are sparsely populated. About 90 percent of the population is concentrated on less than 20 percent of the land area.

The map reveals three major concentrations of population: (1) East Asia, where eastern China, the Koreas, and Japan total almost 1.6 billion people; (2) South Asia, where India, Pakistan, and Bangladesh together account for another 1.4 billion; and (3) Europe from the Atlantic to the Ural Mountains, with almost another billion. Southeast Asia forms a secondary concentration, with about 600 million people; the eastern United States and Canada are home to almost 275 million.

These five population concentrations, plus parts of West Africa, Mexico, and areas along the eastern and western coasts of South America, are densely populated, with more than 25 people per square kilometer (60 per square mile). More than half of Earth's land area, by contrast, has fewer than 1 person per square kilometer (2.6 per square mile) (Figure 5-1). A surprising portion of Earth's land surface is virtually uninhabited: central and northern Asia, northern and western North America, and the vast interiors of South America, Africa, and Australia. The countries that occupy these spaces are enormous in area, but they contain relatively few people. The three African countries of Chad, Niger, and Mali, for example, together cover 39 times the area of South Korea, but their combined population is half that of South Korea.

Figure 5-2 is a cartogram of world population. The size of each country reflects its population, not its land-surface area. The most populous countries are shown as the biggest, and the least populous as the smallest, no matter how big or small their land-surface areas. The single country of India, therefore, is drawn considerably larger than the entire continents of South America or Africa.

A number of factors combine to explain the distribution of the human population. In this section we

FIGURE 5-1 Low population density. The enormous island of Greenland (2.2 million square kilometers or 840,000 square miles—more than three times the size of Texas) is inhabited by only 59,000 people. The fact that any people live here is a tribute to how human culture and technology allow us to adapt to all environments. (George Holton/Photo Researchers, Inc.)

will consider the impact of the physical environment and history on population patterns. In the next section we will take a close look at rates and distributions of world population increase. We will examine these factors as hypotheses of explanation for the population distribution.

Population Density

It is often thought that the population density of a place reflects the productivity of the local environment. **Arithmetic density** is the number of people per unit of area, and in this measure the countries of the world range from Mongolia's two people per square kilometer (five per square mile) to Bangladesh's 882 per square kilometer (2,284 per square mile) to the city-state of Singapore (Figure 5-3, p. 174), with its 5,699 people per square kilometer (14,760 per square mile). The map on the rear endpaper reveals, however, that arithmetic density varies greatly even within individual countries. In Egypt, for example, the arithmetic density is nearly 2,000 people per square kilometer (5,200 per square mile) in the delta and valley of the Nile River, but it is only three persons per square kilometer (eight per square mile) in the rest of the country, which is desert.

Some scholars believe that a country's **physiological density,** which is the density of population per unit of cropland, tells us more about its ability to support its population. In this measure, the countries range from Australia's 2.89 hectares (7 acres) per

capita to Singapore's total lack of any cropland to feed its 4.2 million people.

Do these statistics tell us anything about a country's standard of living? No. In the past, physiological density may have influenced the well-being of primitive societies that were entirely dependent on local agriculture. Today, however, there are several reasons why measures of population density do not reveal national welfare. We will study world agriculture in greater detail in Chapter 8 and economic development in Chapter 12, but we should note here that few countries actually farm all their potential cropland, and scholars differ about how much potential cropland each nation has. The amount of potential cropland can be increased by discovering crops that can thrive under new conditions, by treating soils in new ways, and by expanding irrigation. Chapter 4 noted, however, that the amount of potential cropland can be decreased by misusing and thus degrading soils or by expanding urban or industrial land uses.

Furthermore, the amount of land people have to sustain themselves is less important than the productivity of that land. A fertile region, for example, could presumably support a higher population density than an infertile land; in other words, the fertile area has a higher **carrying capacity.** Productivity also depends on the level of local technology and other inputs, including labor. Some countries enjoy technologically sophisticated and highly productive agriculture—even on infertile soils. Other countries suffer technological backwardness and low productivity—even on fertile soils.

Finally, no countries are fully dependent upon local agriculture anyway. Trade and circulation free societies from the constraints of their local environments and allow them to draw resources from around the world. Physiological density is irrelevant to the welfare of a rich city-state such as Singapore, which imports food in exchange for manufactured goods and services, or to any densely populated but industrial country such as Japan or Belgium. Neither arithmetic density nor physiological density fully explains how the people at any place feed themselves today, or how they support themselves. Some of the most densely populated areas on Earth are rich, while others are poor. Conversely, some sparsely populated regions are rich, while others are poor. No clear correlation between density and welfare can be drawn.

Climate The clearest relationship between population distribution and environmental factors is that population densities are low in most of the world's cold areas and dry areas. (Compare the map on the rear endpaper with Figures 2-30 and 2-31.) The major exceptions to this rule are places where rivers flowing through dry areas provide water for irrigation, such as India, Pakistan, and Egypt (Figure 5-4, p. 174).

Earth's coldest areas are inhabited only where mines or other special resources make it worthwhile to

A Cartogram of World Population

FIGURE 5-2 On this population cartogram, the size of each country reflects its population, not its actual land area. On such a population cartogram the whole African continent is smaller than India, because it is less populous. South America is also relatively small.

support populations at work there, as in parts of central Asia or high in the mountains of South America. People will settle in harsh areas if it is profitable to do so.

Some warm and wet equatorial regions are also sparsely populated, such as the Amazon basin in South America, the Congo basin in Africa, and the island of New Guinea. Despite the lush vegetation of these re-

gions, conventional agriculture has not proved successful. Some soils are poor in nutrients. In other cases, agricultural crops cannot compete with other plants. A warm, moist climate helps a wide variety of plant and animal forms to flourish, but many of these are hostile to humans and their agriculture. Insects thrive, as do the diseases they carry. In addition, these environments support countless forms of parasites, microbes, and fungi that weaken and kill humans, wilt and blight their plants, eat crops alive in the fields, or quietly feast on them in granaries or storerooms.

Heat also affects humans' ability to perform physical labor. Work generates heat, and the body has to lose excess heat to work efficiently. The productivity of manual workers decreases by as much as one half when the temperature is raised to about 35° C (95° F), as is quite common in the tropics. The invention of the air conditioner has made it more comfortable for people to settle in the warm regions of the United States, and most of the country's population growth since 1970 has taken place in the southern Sunbelt. U.S. citizens have

FIGURE 5-3 Singapore. Modern skyscrapers encroach on historic colonial buildings in Singapore, a city that is an independent country. (© Macduff Everton/CORBIS)

FIGURE 5-4 The Nile delta. This photograph illustrates why the ancient Greek historian Herodotus called Egypt "the gift of the Nile." A thin strip of cultivable land winds northward through uninhabitable desert and then fans into a delta where it meets the Mediterranean Sea. Each year the Nile River floods and recedes, leaving a narrow plain of rich black mud, although construction of the High Dam at Aswan has significantly reduced this annual input of natural fertilizer. (Jeff Schmaltz, MODIS Rapid Response Team/NASA Headquarters)

spent more than $1 trillion on air conditioners and an additional $30 billion each year on electricity to keep cool. Cultural and technical innovation can often overcome many environmental restraints, but it may be expensive to do so.

Tropical Asia is home to several great concentrations of people, but these concentrations are mostly in seasonal regions that are not wet year-round. Some of these, such as the island of Java or the Deccan Plateau of India, offer rich volcanic soils, and others are well adapted to the construction of flooded fields for growing rice, called *sawahs* (Figure 5-5).

Rice yields the highest number of calories per acre of any known crop. Sawah flooding plus intensive supervision and care limit the growth of competing vegetation, and algae in the water supply nitrogen as a plant nutrient. It seems almost always possible to increase the yield of a rice terrace by working it just a little bit harder, so sawahs can support increasing numbers of cultivators on a given unit of cultivated land. In addition, rice terraces create a stable ecosystem.

In conclusion, most of the world's population is concentrated in areas of seasonal environments that are not too wet, too hot, too dry, or too cold. In fact, most of Earth's inhabitants live in warm mid-latitude climatic regions (designated C in the Köppen climate classification system in Figure 2-31) or in tropical areas that experience distinct seasons (categories Am and Aw).

Topography and soils   Topography often affects population distribution, although its effect is less prominent than that of climate. People tend to settle on flatlands because of the ease of cultivation, construction, and transportation. Thus, most of the densest concentrations of population in the world are found on level terrain, often near navigable waters (compare the

FIGURE 5-5 Rice terraces. The construction and intensive cultivation of rice terraces such as these on the Indian Ocean island of Bali supports a very high human population density. (SuperStock, Inc.)

map on the rear endpaper with the map on the front endpaper).

Flatness alone, however, is insufficient to attract a large population. Usually there is an association of other environmental attributes, such as fertile soil, available water supply, and moderate climate, that helps to explain high population density. Flatlands in central Siberia, western Australia, and central Brazil are nearly unpopulated because other factors, such as the climate or soils, diminish the area's productivity.

Conversely, sloping land (hills and mountains) usually has a sparse population density, although this is not always the case. High density is found in many mountainous portions of South America, Japan, New Guinea, Southeast Asia, Central Europe, and elsewhere. Specialized attractions, such as mineral resources, often help explain such concentrations. Moreover, in most mountainous areas, the settlements actually are concentrated in valley bottoms, even though they may be small and constricted. Thus the connection between topography and population density is too inconsistent to validate a direct relationship.

People usually settle areas of fertile and potentially productive soil unless there are powerful negative factors (compare the map on the rear endpaper with Figure 4-15). For example, floodplain soils are typically fine-textured and full of nutrients, so they can be used intensively for agriculture. Thus, most river floodplains are densely populated unless they are located in extremely cold or dry areas. People do not necessarily avoid areas of poor soils, however, because some poor soils can be enriched with fertilizers.

Overall, the complexity of human history and the variety of human adaptations teach that environmental conditions alone can never explain population densities.

History   History helps explain the pattern of human settlement. The populations of China and the Indian subcontinent achieved productive agriculture and relative political stability thousands of years ago. These peoples domesticated many plants and animals very early. **Domestication** is the process of taming and training animals and of sowing, caring for, and harvesting plants for human uses—largely for food. Intensive cropping and irrigation yielded generous food supplies, which supported rising populations. The Western European population multiplied when world exploration and conquest brought Europe new food crops—notably the potato. Later, during the Industrial and Agricultural Revolutions, European productivity multiplied. Migrants from Europe settled and brought

> Most of the world's population is concentrated in areas of seasonal environments that are not too wet, too hot, too dry, or too cold. In fact, most of Earth's inhabitants live in warm mid-latitude climatic regions or in tropical areas that experience distinct seasons.

European technology to more sparsely populated areas, so some of today's secondary population concentrations grew as extensions of Europe. These include Eastern North America, coastal South America, South Africa, and Australia and New Zealand. It has been estimated that Europeans and their descendants increased from 22 percent of the world's population in 1800 to 35 percent by 1930.

Human population distribution can also be affected by the demarcation of cultural territories, particularly political territories. Two governments on opposite sides of an international border may have different environmental policies, and these can drastically alter the carrying capacity of the environment. Furthermore, a country may manipulate the distribution of its population. It might subsidize or command the settlement of harsh territories in order to occupy them effectively, or it might establish settlements along its borders for defense.

WORLD POPULATION GROWTH

Earth's population reached 1 billion around 1800, 2 billion by 1930, 4 billion by 1975, and stood at 6.3 billion by 2003. It continues to increase by about 77 million each year (Figure 5-6). The rate of increase, however, is decelerating. In the period 1965–1970, the average annual population increase was 2.06 percent, but it fell to 1.73 percent for the period 1985–1990, to 1.35 percent from 1995–2000, and is registering 1.22 percent from 2000–2005.

The rate of population increase represents a balance among several demographic statistics. First we will

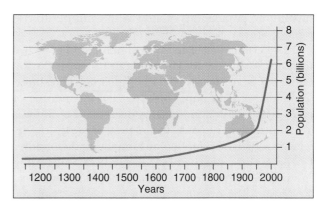

FIGURE 5-6 The increase of human population. The human population grew rapidly from about 1550 until about 1950, and then its growth accelerated to a still higher rate of increase after 1950. The total number is still increasing greatly today, but the rate of increase is slowing.

Focus on
NATIONAL CENSUSES

Counting a country's population might seem to be a fairly easy task, but in fact throughout history—and in many countries today—it has been difficult or impossible. Most countries counted their people only to record how many were on the tax rolls—or ought to have been. The Canadian census of the late seventeenth century was probably the world's first census undertaken solely to count the population. The United States was the first country to require a regular census for the purpose of reapportioning seats in the representative legislature (see Figure 11-28). Such a census has been taken every 10 years since 1790.

The United Nations annually publishes the *Demographic Yearbook*, but some figures in it are only estimates. Some countries lack the administrative apparatus to carry out a full census effectively. Portions of the country may be inaccessible, and the population may be moving within the country, as from the countryside into the cities. Portions of the population may avoid being counted out of superstition, distrust of the central government, or the wish to avoid taxation. In April 2000 one-third of American households failed to return their census forms, so census takers had to visit those homes. The difficulty of taking an accurate census even in the United States should warn us not to assume that all census numbers are exact. The government of Turkey ordered everyone to stay home from 5 A.M. to 7 P.M. on October 22, 2000, in order to be counted. China's national census taken in November 2000 is the greatest effort any nation has ever made. Over 6 million census takers were employed, and they gathered a number of social and economic statistics as well as population numbers.

In the middle of 2003, the world's ten most populous countries contain almost two-thirds of all humankind: China (1,300 million people), India (1,100), the United States (290), Indonesia (235), Brazil (182), Pakistan (150), Russia (145), Bangladesh (140), Nigeria (135), and Japan (127).

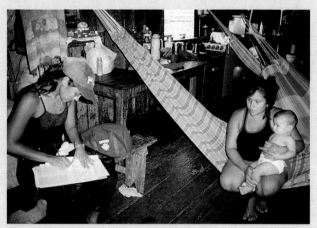

Trying to count everyone. This census taker was one of 180,000 searching out Brazil's population in October 2000. (Larry Rohter/NYT Pictures)

define each of these statistics, and then we will explain through the next several pages how changing relationships among these statistics cause Earth's total population and the populations of individual countries to rise or fall.

The **crude birth rate** is the annual number of live births per 1,000 people. The **crude death rate** is the annual number of deaths per 1,000 people. The difference between the number of births and the number of deaths is the **natural increase** or **decrease.** For individual countries or regions, the natural increase or decrease can then be modified by subtracting the number of people who emigrated out of that area and adding the number of people who immigrated into that area. The result is the overall population growth or population decrease, and that figure can be expressed as a percentage of the total population. For example, Mexico currently has a birth rate of 21.92 births per 1,000 people, and a death rate of 4.97 deaths per 1,000 people, for a rate of natural annual increase of 1.695 percent. When the net rate of emigration of 2.65 people per 1,000 of population is taken into consideration, however, the growth rate falls to 1.43 percent.

Geographers also investigate each country's **fertility rate,** which is the number of children born per year per 1,000 females in a population. A great deal of attention is paid to each country's **total fertility rate,** which is the average number of children that would be born to each woman in a given society if, during her childbearing years (15–49), she bore children at the current year's rate for women that age. This statistic is thought to carry great predictive value. A total fertility rate of about 2.1 stabilizes a population, so this value of 2.1 children is called the **replacement rate.** If a country's total fertility rate falls below 2.1 and the country does not experience immigration, that country's population will ultimately decrease.

Population Projections

A **population projection** is a prediction of the future, assuming that current population trends remain the same or else change in defined ways. Projecting population is an uncertain task because the smallest rise or fall in the percentage of increase today would increase or decrease the total population in the next century by hundreds of millions of people. Even if worldwide crude birth rates and total fertility rates continue to fall, the total number of people on Earth will continue to grow. This is because the number of young women reaching childbearing age is larger than ever before.

Earth's population will go on rising, but the sooner the world's total fertility rate falls to the replacement rate, the lower the total population will be at which the population eventually stabilizes. Many observers insist that the sooner Earth reaches a constant population— **zero population growth** overall—the better. When will this happen? At what figure?

The United Nations offers a range of projections in answer to this question. Today the world total fertility rate is 2.7. If it drops to the replacement rate as early as 2015, then the world's population will stabilize at about 8 billion. If, however, the total fertility rate does not fall to replacement level until 2055, then the population will not stabilize until it has reached 13 billion at the end of the twenty-first century. In 2003, the United Nations estimated that Earth's population will reach 7 billion in 2013, 8 billion in 2028, and 8.9 billion in 2050. As recently as 2000 the United Nations had estimated that Earth's population would reach 9.3 billion in 2050. The reduction of the 2050 estimate by 400 million people was attributed one-half to rising

death rates in some countries (due to AIDS, discussed below) and one-half to falling fertility rates.

Projected totals are falling, but many people still fear that it might be difficult to sustain the world's economy and political order with billions more people, especially if it remains true that some of the most dense concentrations are in some of the poorest areas.

Rates of Population Increase Vary

The overall population projection conceals wide variations in the rate of population increase from region to region and country to country. For example, the annual growth for the period 1995–2000 was 2.4 percent in Africa, 1.6 percent in Latin America, and 1.4 percent in Asia, but only 0.8 percent in North America. Europe reported no population growth at all. Figure 5-7 shows how the rates of population increase vary from country to country. These different growth rates suggest that there will be significant shifts in the geographical distribution of the world population. By 2050, Africa's share of Earth's population is expected to increase from 13 to 20 percent and Latin America's share from 8.5 to 9.1 percent. Asia's share, however, is projected to fall from 61 to 59 percent, North America's from 5.1 to 4.4 percent, and Europe's from 12.2 to 7 percent.

Some scholars compute the **doubling time** for each country's population—that is, the number of years it would take the country's population to double at present rates of increase. A number compounding at

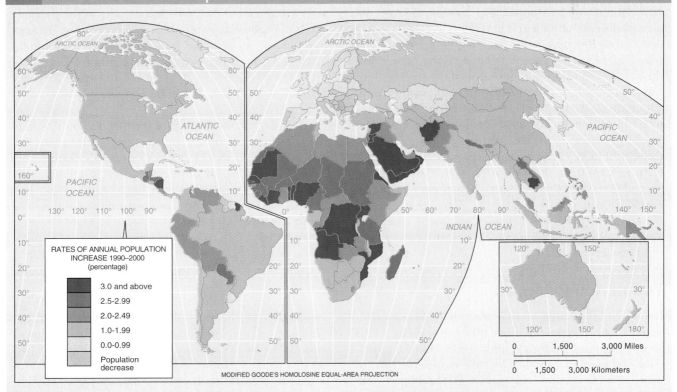

Rates of Annual Population Increase

RATES OF ANNUAL POPULATION INCREASE 1990–2000 (percentage)

- 3.0 and above
- 2.5-2.99
- 2.0-2.49
- 1.0-1.99
- 0.0-0.99
- Population decrease

MODIFIED GOODE'S HOMOLOSINE EQUAL-AREA PROJECTION

FIGURE 5-7 Most countries with the highest rates of population increase are in Africa and the Near East. (Data U.S. Census Bureau)

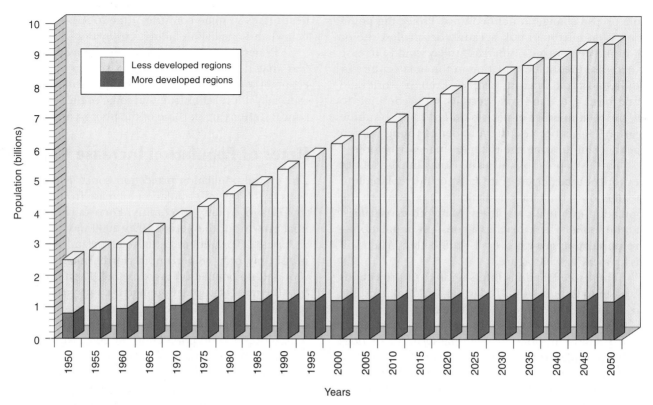

FIGURE 5-8 World population, 1950 to 2050. By one projection, the population of today's developed countries is expected to hold steady, while the populations of today's developing countries will continue to increase. (Data courtesy of the United Nations)

3 percent annually (the rate of population increase in Uganda) doubles in only 23 years, but a population compounding at only 1 percent annually (as in the United States) takes 70 years to double.

Overall rates of population increase vary significantly between the rich and poor countries. Ninety-five percent of the population increase projected to the year 2050 will be in countries that today are poor, those least able to support a larger population (Figure 5-8). The share of the world population living in the countries that are poor today will increase from 80 percent today to 90 percent by 2050. In these countries, economic growth is in a race with population growth. In order to achieve rising per capita (per person) incomes in these countries, the rate of economic growth must exceed the rate of population growth. If the rate of economic growth equals the rate of population growth, per capita incomes will remain static; but if economic growth rates fall below population growth rates, per capita incomes will fall.

Population Pyramids

Demographic analysis reveals that in countries with high fertility rates, the populations are young. This is often represented by a graphic device called a **population pyramid** (Figure 5-9), which represents two aspects of a population: age and gender. Young people are indicated at the base of the pyramid, so a wide base

indicates large numbers of young people. As the individuals of the same age (called a *cohort*) grow older together through time, the horizontal bar that represents that cohort moves toward the top of the pyramid. Males and females are shown on separate sides of the pyramid to show their relative numbers in each cohort. Demographers can project the future structure of a population based on this knowledge. For example, if we know how many people are 20 years old today, we can estimate how many people will be 50 years old 30 years in the future.

Today the median age of the world's population is 26.4, but the various countries and regions demonstrate great differences in the ages of their populations (Figure 5-10). Where birth rates are high, median ages are generally lower than where birth rates are low. The median ages of the populations in many countries in Latin America, Africa, and the Near East are between 18 and 25 years. In contrast, the median ages of populations of wealthier countries are higher. In Western Europe, for example, the median age is 38; in the United States it is 36.

In the countries with low median ages, a high percentage of the total population is drawing on national resources, but it has not yet reached its most productive working years. These countries are challenged to feed, clothe, and educate youngsters. As the youngsters reach their mature working age, the national economies must be able to provide productive

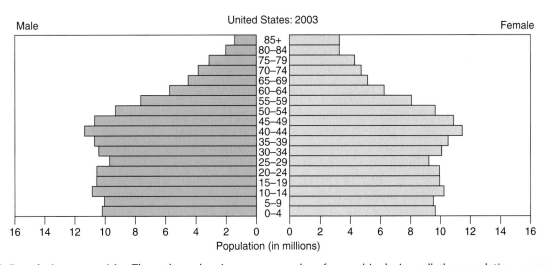

FIGURE 5-9 Population pyramids. These three drawings are examples of a graphic device called a population pyramid, which shows the age and sex structure of a country's population. The pyramid for Uganda (a) shows a broad base—indicating many children—tapering to a narrow top of fewer older people. Birth rates are high, but life expectancies are limited. This pyramid is typical of those of poor countries. The "pyramid" for Germany, by contrast (b), shrinks at the base, indicating a birth rate that has been falling steadily since about 1970. Life expectancy is long. Germany's relative lack of elderly men is the continuing evidence of losses in World War II (1939–1945). The population pyramid of the United States (c) shows a bulge of people between 30 and 50 years of age—the "baby boom" of 1945–1965. Below this group the figure shrinks to reveal the "baby bust," but then, when "boomers" began to reach their childbearing years, the number of births rose slowly in a phenomenon called the "echo of the baby boom." (U.S. Census Bureau)

Median Ages of National Populations

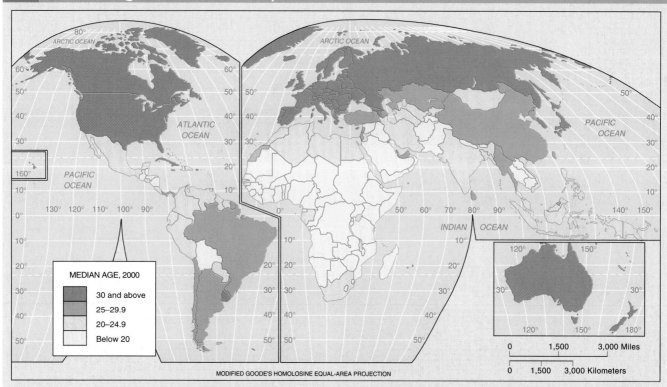

MEDIAN AGE, 2000

- 30 and above
- 25–29.9
- 20–24.9
- Below 20

MODIFIED GOODE'S HOMOLOSINE EQUAL-AREA PROJECTION

FIGURE 5-10 Comparison of this figure with Figure 5-7 shows that populations are young where rates of population increase are high. (U.S. Census Bureau)

jobs for them. If economic growth has not provided enough jobs, their frustration may break out in civil disorder. Economic development is urgently needed.

Those countries with higher median ages, by contrast, must provide health care and services to populations that are increasingly elderly. These countries' labor forces shrink, but they must be able to support increasing numbers of retired people.

The **dependency ratio** of a country suggests what proportion of its people are in their most productive years. It is defined as the ratio of the combined population of children less than 15 years old and elderly people over 64 years old to the population of those between 15 and 64 years of age. The larger the percentage of dependents, the greater the financial burden on those who are working to support those too young or too old to work. A dependency ratio will be high if a country has either a high percentage of young people (which is typical of the poor countries) or else a high percentage of elderly people (which is typical of the rich countries). If a country is poor, a highly dependent population compounds its financial difficulties. In Ethiopia, for example, which is a poor country, the dependency ratio is 1.08, and most of the dependents are youngsters. In the United States, by contrast, the ratio is only 0.54, and a growing share of the dependent population is made up of elderly people.

The Demographic Transition

Scholars who have studied population growth in the countries that today are rich have defined a model to describe these countries' historical experience. That model is called the **demographic transition.** The demographic transition model defines a pattern of growth that exhibits three distinct stages.

(1) In *Stage One*, both the crude birth rate and the crude death rate are high, so the population does not increase rapidly. All countries experienced this stage in the past, when the human population was a fragile number in constant danger of significant reduction, locally if not globally, by periodic epidemics.

(2) As incomes increase and medical science advances (which is usually, at first, simply an understanding of hygiene), crude death rates drop dramatically (Figure 5-11). The **infant mortality rate,** which is the number of infants per thousand who die before reaching 1 year of age, falls almost immediately (Figure 5-12). Another factor lowering death rates is the improvement in the quantity and quality of food that resulted from the Agricultural Revolution (discussed in Chapter 8). Crude birth rates, however, remain high. Thus, in *Stage Two* of the demographic transition, crude death rates are falling, but crude birth rates remain high, so the rate of natural increase is high.

Crude Death Rates

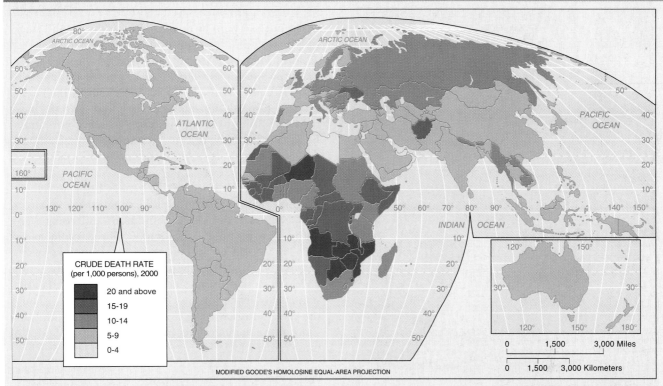

CRUDE DEATH RATE
(per 1,000 persons), 2000

- 20 and above
- 15-19
- 10-14
- 5-9
- 0-4

MODIFIED GOODE'S HOMOLOSINE EQUAL-AREA PROJECTION

FIGURE 5-11 Death rates are high in very poor countries (sub-Saharan Africa and South Asia), moderately high in wealthy countries with aging populations (Europe), and low in countries with very young populations and at least modest health-care availability (Latin America and China). Crude death rates are falling almost everywhere, but environmental disasters, local famine, epidemics, or war can still substantially increase a country's crude death rate. (U.S. Census Bureau)

Infant Mortality Rates

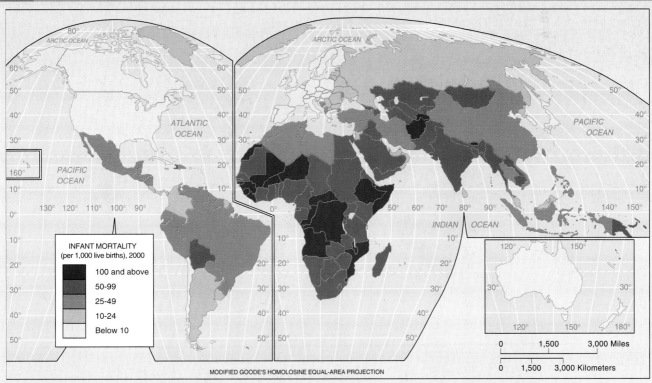

INFANT MORTALITY
(per 1,000 live births), 2000

- 100 and above
- 50-99
- 25-49
- 10-24
- Below 10

MODIFIED GOODE'S HOMOLOSINE EQUAL-AREA PROJECTION

FIGURE 5-12 This map shows the same contrast as Figure 5-11. Infant mortality rates are low in the rich countries, but they remain tragically high in poor countries. (U.S. Census Bureau)

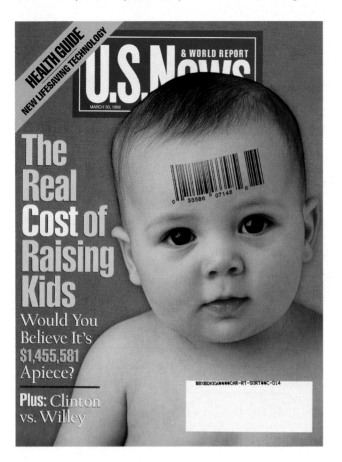

Several theories have been offered to explain the persistence of high birth rates during the second stage of the demographic transition. In traditional societies, children may be economic assets. They provide more hands to help in the fields. Also, people traditionally expect their children to look after them in their own old age. If infant mortality rates are high, parents have many offspring to ensure that some survive into adulthood. Not until one or two generations have passed will adults accept the reality of lower infant mortality rates and start having fewer children.

(3) Eventually, crude birth rates begin to fall. In Europe, this decline in childbearing occurred along with economic growth, urbanization, and rising standards of living and education. Children came to be seen as expenses rather than as economic assets (Figure 5-13), and many parents began to have fewer children so that they could provide a higher material standard of living for the children they did have. Thus, the rich countries of the world today demonstrate *Stage Three* of the demographic transition. Their crude death rates remain low, their crude birth rates are also low,

FIGURE 5-13 The cost of children. Traditional societies often think of children as economic assets; modern societies, however, calculate their costs. (© 1998 U. S. News & World Report, L.P. Reprinted with permission)

Falling Total Fertility Rates

(a)

FIGURE 5-14 The "population explosion' that many observers feared in the 1950s and 1960s (a) is fizzling. Total fertility rates remain high in many poor areas, but today (b) women virtually everywhere are bearing fewer children. (U.S. Census Bureau) and total fertility rates are at or below the replacement level. Without substantial immigration, many rich countries' populations are expected to fall. Italy, for example, has a total fertility rate of only 1.2 and a median age of 40.2 (the world's highest). Its population is projected to fall 28 percent between 2000 and 2050, from 57 million to 41 million. Reasons cited by Italians for having fewer children typify the social changes in the rich countries: Women are pursuing careers, daycare services are limited, urban housing is short, contraceptives are increasingly available, abortion has been legalized, and people choose to enjoy material goods rather than to bear and raise children. As this demographic transition model has been described in the rich countries, the transition took place over centuries.

In the period from 1950 to 1980, scholars studying world population believed that the demographic transition model would eventually apply to all countries. They noted that birth rates and total fertility rates were still high in the poor countries, and they described these countries as being in the second stage of the demographic transition. Scholars concluded that when these countries achieved economic development, they too would enter the third stage. Therefore, the 1974 World Population Conference in Bucharest, Romania, concluded, "Economic development is the best contraceptive."

Several characteristics of the demographic transition model do still apply in today's poor countries. Crude birth rates and total fertility rates are falling, but remain relatively high (Figure 5-14). The average total fertility rate for the countries of West Africa during the period 1995–2000, for example, was 6.1. In many poor countries, it remains economically rational for the poor to have more children—for all the same reasons as it was in the past for the populations of today's rich countries (Figure 5-15). It can even be argued that some African countries face labor shortages in agriculture, and because they lack capital for agricultural machinery, increases in rural labor forces would increase food output.

Furthermore, in even the world's poorest countries, infant mortality rates are dropping because antibiotics and immunization today help many babies to survive who would have died in past years (Figure 5-16). Modern medicine also keeps people alive longer, and the combined effect of lowering infant mortalities and keeping people alive longer expands **life expectancy,** which is the average number of years that a newborn baby within a given population can expect to live (Figure 5-17). Today life expectancies are lengthening fast in most of the world. The combined effect of lowering infant mortalities and lengthening life expectancies is to increase total populations.

(b)

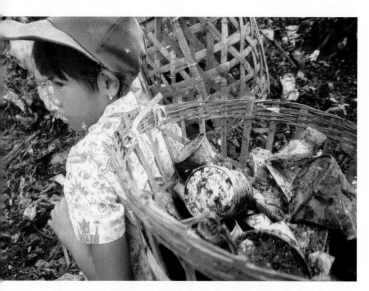

FIGURE 5-15 Indonesian child worker. This young girl is helping support her family by collecting potentially valuable pieces of scrap. Children can be an economic asset if they are put to work in fields or factories. Chapter 12 will explain how international treaties are trying to stop the abuse of child labor by prohibiting products of their labor in international trade. (AP/Wide World Photos)

Therefore, some circumstances in today's poor countries are comparable to the second stage of the demographic transition as it was experienced in the countries that are today rich.

Is the Demographic Transition Model Still Relevant Today?

Statistical trends noted through the 1990s, however, triggered doubts in some scholars' minds about whether the demographic transition model really does apply to today's poor countries. Already by the time of the opening of the Third U.N. Conference on Population and Development, in Cairo, Egypt, in 1994, experts identified two trends occurring in the poor countries:

(1) Total fertility rates are dropping much faster than they did in the countries that today are rich. From 1975 to 2000, the world total fertility rate fell from 4.5 to 2.7. In East Asia, the decline was from 5.1 to 2.6; in Africa, the decline was from 6.6 to 5.1; and in Latin America, the decline was from 5 to 2.7.

(2) Falling total fertility rates are not necessarily associated with improving incomes and standards of living. Rates are falling in countries where the economy is stagnating or even declining, as in Bangladesh, Syria, and Turkey. Per capita incomes in many African economies have fallen, but total fertility rates have fallen, too. These observations demolish the model's assumption that economic growth is necessary to trigger falling total fertility rates.

These statistical trends convinced many scholars that today's decreases in total fertility rates in the poor countries are fundamentally different from the fall of fertility rates that occurred in today's rich countries. Therefore, new causes must be found to explain them. Some scholars are abandoning the demographic transition model. Theories of explanation focus on three new factors: the growing role of family-planning programs, new contraceptive technologies, and the power of mass media.

(1) Family-planning programs   During the demographic transition in today's rich countries, the concept of family planning was not quickly accepted. Early leaders in the population-control movement in Europe and the United States, such as Margaret Sanger (1883–1966), were even arrested for immorality or creating public disturbances. Today, by contrast, lowering the crude birth rate has become the focus of most countries' efforts to reduce overall population increases. Over 90 percent of the people of developing countries live under governments committed to reducing the rate of population growth (Figure 5-18). Today national public health services seek clients and have removed or lowered many of the barriers to the availability of contraception.

Empowering women is key to lowering fertility rates. The 1994 Cairo Conference, the 1995 Fourth World Conference on Women (held in Beijing, China),

FIGURE 5-16 Lowering infant mortality. Bill Gates, the world's richest man, founder and chairman of Microsoft Corporation, administers a dose of oral polio vaccine at a health clinic in India. The Bill and Melinda Gates Foundation has dedicated billions of dollars to medical care in the world's poor countries. The United States has been polio-free since 1993, but polio still exists in seven countries: Afghanistan, Egypt, India, Niger, Nigeria, Pakistan, and Somalia. (AP/Wide World Photos)

Life Expectancies

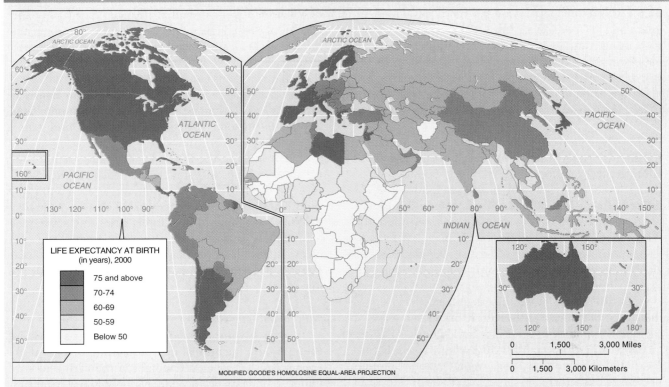

FIGURE 5-17 People in North America, Europe, and the richest countries in Japan and Oceania can expect to live much longer lives than people in the poor countries. Life expectancies are lengthening everywhere, but much of the human population still cannot be expected to reach what we would call "old age." Compare this map with Figures 5-11 and 5-12. Countries with high death rates have low life expectancies. (U.S. Census Bureau)

FIGURE 5-18 Family planning. This Indonesian government billboard suggests that smaller families can be healthier, happier, and better off. It is part of the government's family-planning initiative. (Courtesy of U.S. Agency for International Development)

and follow-up conferences held at U.N. headquarters in New York in 2000 and 2002 all emphasized that key factors in successful family planning include universal access to education and health and relative equality in status, employment opportunities, and full political rights for women. The Beijing Conference declared that "women have the right to decide freely and responsibly on matters related to their sexuality" and must be able to do so "without coercion, discrimination and violence." International delegates called for strict punishment of domestic violence, including domestic rape, which is not recognized as a crime in many countries.

(2) Contraceptive technology The demographic transition occurred in Europe without modern contraceptive methods. Today, however, technology has provided new means of birth control. The U.N. has replaced the slogan that it proclaimed in Bucharest in 1974 ("Economic development is the best contraceptive") with a new slogan: "A contraceptive is the best contraceptive."

Sterilization is the most widespread form of birth control in the poor countries. In Brazil, for example, an estimated 50 percent of all married women of reproductive age have been sterilized. New drugs and devices range from intrauterine devices to pills and injections, so that by 2000, 58 percent of currently married women in the world (70 percent in the more developed regions and 55 percent in the less developed regions) were using contraceptives. These techniques reduce reliance on abortion, but about 60 million abortions are still performed annually, and improperly performed abortions kill about 500,000 women each year. Efforts to develop a male contraceptive pill continue.

(3) The role of the mass media   During the demographic transition in today's rich countries, information about contraceptive methods diffused slowly because educational levels were low and mass communication was limited. Today, however, communication and the mass media in developing countries have accelerated the diffusion of ideas about family planning. Many governments, including those of Ghana, Nigeria, Gambia, Zimbabwe, and Kenya, broadcast family-planning messages on both radio and television.

Furthermore, public health officials have noted that even commercial television programs have a tremendous impact in lowering fertility rates. They detach sexuality from childbearing. Whereas sex is procreational in a traditional society, modern media often present sex as a recreational activity. Popular television programs transmit images, attitudes, and values of a modern, urban, middle-class existence in which families are small, affluent, and consumer-oriented. Popular television has been credited with lowering fertility rates in India, Pakistan, the Philippines, Kenya, Brazil, and China.

Obstacles to population control   Despite technological advances, family-planning programs face obstacles. The manufacture and distribution of contraceptive devices and instructions for their use is expensive. Two-thirds of the users of condoms, diaphragms, and contraceptive sponges live in the industrialized world, and attitudes about family planning in the rich countries determine whether these countries offer assistance to the poor countries—even for those poor countries that want it.

Religion can block birth-control programs. The Roman Catholic Church officially opposes all forms of birth control, and some other religions also preach against specific forms of birth control. Many religious groups adamantly oppose abortion.

Another obstacle to birth control is the low status of women. In numerous societies women still lack political and economic rights, have limited access to education, and in general exercise little control over their own lives.

In areas where male children are preferred to females, birth rates remain high for several reasons. If a couple's first child is female, they will continue having children until a male child is born. In these traditional societies, a girl is often regarded as just another mouth to feed, a temporary family member who will eventually leave to serve her husband's relatives. Also, in many societies a woman's parents must provide her with a dowry, which is a substantial financial payment at the time of her marriage. A son, on the other hand, means more muscle for the farmwork and someone to care for aged parents. In several traditional Asian and African religions, only sons can burn the necessary ritual offerings to ancestors. A son carries on the family name. If women win the right to work outside the home, as is happening in some societies, the financial preference for male children is reduced.

Throughout the world, fertility rates are lower in urban areas than in rural areas. Moreover, the larger the city, the more likely women are to use contraception. Family-planning services may be easier to reach. Children cost more to raise in big cities, and space is at a premium. Today half of the world's people live in cities, and that percentage is increasing (see Chapter 10). This development will probably continue to lower total fertility rates.

Birth-control programs: Some examples   The most dramatic and significant change in fertility has been in China, where the annual rate of natural population increase dropped from 2.9 percent in the early 1960s to under 1 percent by the end of the century. China is home to about 21 percent of the human population, so that decline alone accounts for much of the change in world trends. In 1979 China launched a one-family/one-child policy, offering incentives to reduce childbearing, such as financial rewards and special privileges for small families, and penalties for exceeding the targets. Unfortunately, this policy perpetuated traditional prejudice in favor of males. Some women may have been coerced into having abortions, and there were even reports of *female infanticide*, the killing of female children. China ended the one-family/one-child policy in 2002 and introduced a new program in which those who want more children are required to pay a "social compensation fee." Chinese statistics suggest that the former policy never worked, anyway: In 2000, 80 percent of Chinese children under 14 lived in families with siblings. The policies were enforced more strictly in the cities than in the countryside, where births are often concealed. China is determined to continue to reduce its birth rate. It has manged to lower its total fertility rate to only 1.8, but nearly a quarter of the country's people are under 15 years of age, so total numbers of births will probably remain high for many years.

India's population grew from 342 million at independence in 1947 to 1 billion in 2000. If its growth rate continues at 1.6 percent per year, it will probably overtake China as the world's most populous country before 2050. Today the government sponsors family-planning programs, and radio and television advertising and information emphasize that small families are healthier and happier. A new social security program assures people that they will be taken care of by the government in their old age. Abortions are legal, but sterilization accounts for about 90 percent of India's program. Total fertility fell from 6 in 1951 to 3.1 in 2000. Programs in India's individual states also discourage large families. In Rajasthan, for example, village council members lose their seats if they have more than three children.

Iran's population counted only 35 million in 1979, but in the 1980s the government encouraged Iranians to have more children as a sign of national strength. Population growth soared to a 4 percent increase annually. The government soon realized that a rising population would inevitably be a problem for the future, so a 1993 law sponsored birth control, terminated subsidized health insurance and food coupons for children after a couple's third, and provided for free birth-control pills and condoms. Today Iran's population is 70 million, and its rate of annual increase has fallen to 1.24 percent.

The Mexican population quintupled from 19.6 million in 1940 to 103 million in 2000, but the National Population Council, established in 1974, has developed a network of family-planning clinics. The total fertility rate fell from 7 in 1965 to 2.5 today. The total population will nevertheless continue to grow by 1 million per year for at least 30 years.

Brazil's total fertility rate has fallen from 5.75 in 1979 to 2.2 today. The fall has been attributed to urbanization, women going to work outside the home, literacy, and television.

Changes in World Death Rates

A decrease in the crude birth rate is not the only event that can reduce the rate of human population increase. An increase in the crude death rate has the same effect. Natural disasters and wars claim tens of thousands of lives around the world each year. Death rates have risen in Russia since about 1990; this has been attributed to the physical and mental stresses of chaotic economic and political circumstances since the fall of the communist government and the dissolution of the U.S.S.R. (see Chapter 13). Russian income has fallen more than 10 percent, use of tobacco and misuse of alcohol has spread, and infectious diseases have multiplied. Life expectancy has fallen. At the same time, fertility rates have fallen, so in the early twenty-first century Russia's population was decreasing by about one million per year.

Infectious diseases   The study of the incidence, distribution, and control of disease is called **epidemiology.** As countries develop, the leading causes of death change from infectious and parasitic diseases, which are caused by disease-causing organisms called **pathogens** entering and multiplying within the body, to degenerative diseases, which result from deterioration

or degeneration of the body. This change in the leading causes of death is called the **epidemiological transition.**

As recently as the 1970s, scientists tended to see infectious diseases as a series of problems to conquer, ticking off victories like notches on a gun. Smallpox was the greatest cause of death by infection in human history, but all but a few smallpox viruses kept for scientific study were supposedly destroyed by 1995. Some countries, however, kept smallpox viruses and other pathogens for purposes of war. The Biological Weapons Convention of 1972 pledges the countries that signed it to destroy their existing stocks of biological weapons. Virtually all nations signed it, but the United States, Russia, North Korea, and possibly others held back stocks of weapons-quality pathogens. Therefore, beginning in 2001, as a defensive measure, the U.S. government started production of smallpox vaccine. Vaccinations began with military personnel, health workers, and government officials, including President George W. Bush. The wisdom and necessity of widespread vaccination has, however, been questioned, because the vaccination itself endangers some who receive it.

Shortly after the terrorist attacks on the United States of September 11, 2001, and through the following months, spores of inhalation anthrax, a rare and particularly virulent disease bacterium, were sent through the U.S. mails. Most of the recipients and sufferers were working in offices of political or media power, so government officials drew the conclusion that the United States was facing the spread of a disease deliberately introduced, that is, an attack of biological warfare. As of late 2003, no evidence had been uncovered that the anthrax attacks were carried out by the people responsible for the attacks of September 11 or by others acting together with them. The mailings may have been the work of another foreign organization, of domestic U.S. terrorist organizations, or of an American scientist hoping to spur U.S. anti-biological weapons defense. Anthrax is not easy to spread, although it could be spread widely with disastrous health results if a method could be achieved of converting it into an aerosol. So far, it is not believed that anyone has.

New drug-resistant strains of several pathogens are appearing around the world, and scientists are challenged to develop medicines to combat them. Some pathogens evolve and mutate naturally. Also, patients sometimes fail to complete the full portion of prescriptions of antibiotic medicines they receive; this allows pathogens to develop drug resistance. The use of antibiotics in animal feed and sprayed on fruits and vegetables during food processing also increases opportunities for resistant organisms to evolve and thrive.

As much as 70 percent of all antibiotics produced in the United States are fed to healthy livestock to promote growth. This allows the emergence of drug-resistant bacteria, which are today widespread in commercial meats and poultry and can already be found in consumers' intestines. For example, the Bayer

Corporation makes a drug called Baytril, which is used in large-scale poultry production worldwide. The widespread use of Baytril in chickens has been shown to reduce the effectiveness of a similar drug made for humans, Cipro, for some types of infections. European governments ban the nontherapeutic use of antibiotics in animals. Many disease-causing pathogens are passed back and forth between humans and other animals with which we live in close proximity. Many influenza strains, for example, originate in Asia because great numbers of ducks and pigs—common animal hosts for viruses—live there in close proximity to human beings. The Great Flu Pandemic which swept the world in 1918–1919 killing as many as 40 million people originated in China, as did the pandemics of Asian flu in 1957 and the Hong Kong flu in 1968.

In November 2002 a new viral infection, called SARS (severe acute respiratory syndrome), appeared in southern China. The Chinese government withheld information about the disease to avoid embarrassment, and this action abetted the disease's global spread. Only when the Chinese government came to realize that the global panic was damaging Chinese credibility—and thus, economic development, paralyzing business travel and tourism—did the government take emergency measures. By then, however, the disease had spread around

the world, and thousands were suffering. The epidemic subsided, but the discovery of the SARS virus in civet cats and other animal species that are both kept as pets and to be eaten suggests how the virus may have been transmitted to humans. It also raises the possibility that we may never eradicate the virus if it is harbored in wild animals.

The single leading communicable cause of death in the world today is the human immunodeficiency virus (HIV), which causes acquired immune-deficiency syndrome (AIDS). AIDS destroys the body's ability to fight infections, and it was responsible for 3.2 million deaths in 2002, almost 5 percent of all deaths. It is believed that the AIDS virus emerged in Africa in the 1950s; by 2002 about 42 million people worldwide had contracted the AIDS virus (Figure 5-19). As with SARS, denial among both social and political leaders in some countries provided the epidemic with an ideal environment for continued spread. India was probably the country with the greatest number of infected people (about 3.5 million), but the disease was spreading fast in East Europe, Russia, and Central Asia. Two-thirds of all people infected people were in sub-Saharan Africa. In several African countries, AIDS-related illnesses were the leading cause of adult death. AIDS has shortened life expectancy by 17 to 20 years in Botswana, Zimbabwe,

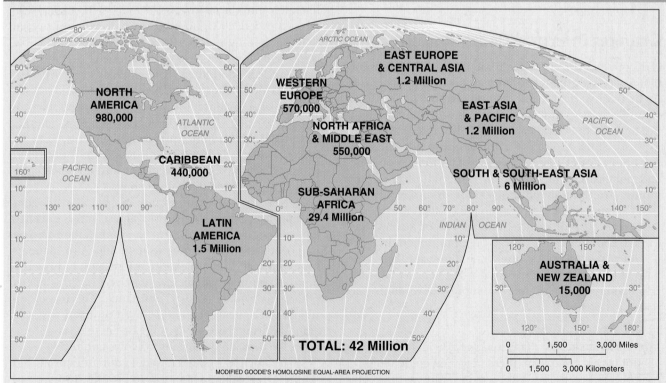

Adults and Children Estimated to Be Living with HIV/AIDS at the End of 2002

TOTAL: 42 Million

MODIFIED GOODE'S HOMOLOSINE EQUAL-AREA PROJECTION

FIGURE 5-19 Today scientists think that the human immunodeficiency virus (HIV) evolved in Africa and spread from other primates to humans there. It continues to spread around the world, but Africa is particularly afflicted. (The World Health Organization)

and Zambia. AIDS causes massive human suffering, and it also has a pernicious economic impact. It weakens and kills adults in their prime years, thus destroying national economies. New techniques of treating the disease offer promise, but their success is not yet proven. The high cost of these drugs has been lowered through international cooperation among pharmaceutical companies and national governments, but many countries lack the health systems necessary to distribute these drugs effectively.

The second leading communicable cause of death is diarrheal diseases, responsible for about 4 percent of all deaths. The most important of these is cholera. New strains of cholera appear regularly, and they are spread through contaminated water and poor sanitation. Cholera periodically surges out of Asia.

Tuberculosis is the third leading communicable cause of death, causing about 3 percent of the total. Approximately one-third of the human population carries TB bacilli in their bodies, but in the vast majority of cases the disease is latent and cannot be transmitted. New strains of drug-resistant tuberculosis, however, are multiplying; and the war against this disease continues.

Malaria is responsible for about 2 percent of human deaths. The mosquitoes that carry the disease live in a wide variety of climates, so malaria occurs widely; but about 80 percent of the deaths occur in sub-Saharan Africa. Drugs are normally used to treat malaria, but drug-resistant strains are appearing in Southeast Asia and spreading to Africa. In addition, mosquitoes are developing resistance to insecticides. Thus, health officials fear that the appearance of "supermosquitoes" armed with drug-resistant "supermalaria" could increase the number of deaths from malaria in coming years.

Threatening new viruses continue to emerge. These include not only HIV and SARS but the Ebola virus (first identified in the Sudan in 1976), the Sabiá virus (identified in Brazil in 1990), the Hanta virus (identified in the U.S. Southwest in 1993), and the Nipah virus (first identified in Malyasia in 1998). These are all fatal to humans, but, luckily, most are difficult to transmit. The Ebola virus, for example, can be transmitted only through body secretions, and it kills its host so fast that there is little time for transmission to new victims.

Pathogens may spread through new medical practices such as blood transfusions and organ transplants, through sexual practices, or through needle sharing among drug addicts. Also, human alterations of the environment can create breeding grounds for new viruses and increase the number of pathways viruses can take to new populations. These include deforestation and urbanization, which bring virus-carrying rodents and people together; international trade, which unwittingly transports virus-bearing insects from country to country; and farming methods that allow viruses from birds and mammals to mingle. Global warming is increasing the portion of Earth hospitable to the mosquitoes that carry malaria; they are appearing at previously cooler higher latitudes and higher elevations. Global warming may redistribute other pathogens and diseases in ways we cannot foresee. Global travel also triggers international outbreaks of infectious diseases. For example, the West Nile virus has spread across North America since it was first brought from the Mideast in 1999. Epidemiologists note that the number of transfers of pathogens from animals to humans seems to be increasing: AIDS, SARS, Ebola, West Nile virus, Lyme disease, and still others have won headlines in recent years. Epidemiologists are not certain why this increased transfer is happening.

Human environmental alteration may, on the other hand, eliminate a pathogen's habitat, eradicate the pathogen, and prevent future epidemics. For example, the completion of Egypt's Aswan Dam in 1971 destroyed the floodwater habitat of the *Aedes aegypti* mosquitoes, carriers of Rift Valley fever virus. By 1980, Rift Valley fever had virtually disappeared from Egypt.

Degenerative diseases Most degenerative diseases are difficult to treat because treatment requires changes in the behavior of the sufferers, who are often adults. Degenerative diseases are leading causes of death in rich countries, but they are increasing in several developing countries. The World Health Organization (WHO—a United Nations Agency) estimated that Earth's number of overweight people equaled the number of undernourished by 2000—about one billion of each. This is a turning point in the history of humankind. Some 300 million people were obese, that is, in serious medical danger, and that number was rising fast. In the rich countries, causes include the switch to restaurant and take-out meals, which usually have higher calories than home-cooked meals, plus increasingly sedentary lives lacking exercise. In the developing countries, increased dietary fat—largely attributable to the spread of cheap, high-quality cooking oil—has helped to raise daily caloric intake by an average of 400 calories per capita since 1980. In these countries, too, urbanization and sedentary lives take their toll. Overweight and obesity may trigger adult-onset diabetes, from which some 30 million people in East and South Asia alone were already suffering by 2003, plus heart diseases.

Tobacco products are predicted to account for an increase of 50 percent in global cases of cancer by 2025. Tobacco now accounts for about 4.9 million annual deaths worldwide through lung cancer and many other debilitating diseases, and one-fifth of the people now living in the developed world—more than 250 million people—may die prematurely of smoking-related illnesses. WHO projects that tobacco may account for over 10 million deaths a year—10 percent of all deaths—by 2030. In 2003, 192 nations signed a treaty regulating tobacco crops, advertising, trade, and other aspects of marketing. Ratification is proceeding, but the United States defends the "free speech" rights of the tobacco industry.

Is Earth Overpopulated?

Many people fear that Earth may become overpopulated, or that it is already. People disagree, however, on the meaning of the word "overpopulation." Some people complain that their neighborhood is overcrowded or that new construction and crowding are everywhere lowering the quality of life. Studies of rats reveal that crowded rats suffer mental and physical breakdown. Does this mirror human behavior? We do not know.

Discussions of world population growth inevitably invoke the ideas of Thomas Robert Malthus (1766–1834), who asked whether humankind would always be able to feed itself. Malthus's statement of the relationship between population and food supply still demands our attention. Malthus was a professor of political economy and also a clergyman, and in 1798 he published, anonymously, his *Essay on Population*. His essay, stated most simply, argued that food production increases arithmetically: 1-2-3-4-5 . . . units of wheat. Population, however, increases geometrically: 2-4-8-16-32 . . . people. Therefore, the amount of food available per person must decrease as the population increases.

The human population can be kept in balance with food supplies only through checks on population increase. Malthus defined two types: positive checks and preventive checks. *Positive checks* refer to premature deaths of all types, such as those caused by war, famine, and disease. *Preventive checks,* by contrast, are human actions designed to limit births, such as a decision by a young couple to delay marriage and childbearing. According to Malthus, couples should have a sense of responsibility for the economic welfare of the children they might produce; at the least, couples ought to fear their own inevitable economic decline if they produce too many children. "There are perhaps few actions that tend so directly to diminish the general happiness as to marry without the means of supporting children," he thundered. "He who commits this act, therefore, clearly offends against the will of God."

Malthus did not believe that preventive checks could effectively control population growth. He pointed to high birth rates among the poor as evidence. Therefore, he predicted that the future of humankind would consist of endless cycles of war, pestilence, and famine. This is the pessimistic conclusion of the **Malthusian theory.** People who share Malthus's pessimism are often today called *neo-Malthusians* (new Malthusians) or *Catastrophians*.

Malthus was replying to the French philosopher Jean Antoine Condorcet (1743–1794). Condorcet had argued that increases in productivity and improvements in technical ingenuity would keep pace with the needs of Earth's rising population, so per capita material welfare could actually rise. Condorcet also suggested that, given education and prosperity, most people would voluntarily limit their family size. People who share Condorcet's optimism are often referred to as *Cornucopians;* a cornucopia is a horn of plenty.

So far, Condorcet's predictions have proven more accurate than those of Malthus. The human population has sextupled since Malthus's time, yet the widespread starvation that he feared has not happened. Many people are hungry, but this deprivation does not result from insufficient food on Earth. People are hungry because the food is not priced within their reach and distributed to all. The problems are political and economic, not technological. Will these political problems ever be solved? We do not know. Chapter 8 details how humankind has managed to multiply the food supply. Could mounting numbers of people ever place demands on food and materials so great that per capita supplies will necessarily fall? That question will be answered by future developments in technology, and we cannot possibly foresee what those can be.

OTHER SIGNIFICANT DEMOGRAPHIC PATTERNS

We have seen how variations in birth rates and death rates affect both overall world population growth and the growth rates in different parts of the world. Two other demographic statistics vary so significantly around the world that they demand geographical analysis. One is variations in the sex ratios among national populations, and the other is the aging of the human population.

Sex Ratios in National Populations

About 105 male children are naturally born for every 100 females. This ratio occurs everywhere in the world except in countries where ultrasound, amniocentesis (medical procedures that provide information on the state and sex of a fetus developing in the womb), and abortion are available, and where there is a strong preference for male children. The demographic impact is dramatic. In South Korea, for example, where fetal testing to determine sex is common, the birth ratio is 114 to 100; in China it is 117 to 100; and in Taiwan it is 110 to 100. It must be assumed that female fetuses are aborted at a higher rate than male fetuses. These dramatic imbalances in sex ratios will cause problems in forming new families 20 years from today. Taiwan and some states of India have limited or banned the use of ultrasound scanners and amniocentesis.

Although it is natural for more males to be born than females, women tend to live longer than men when both receive similar medical care. Therefore, a population should naturally have an approximately

Male-to-Female Population Ratios

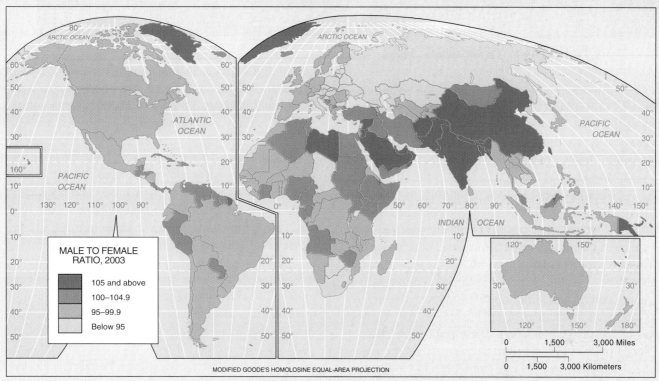

FIGURE 5-20 The numbers of males per 100 females in national populations, as shown here, vary substantially around the world. What factors could explain these variations? (U.S. Census Bureau)

equal number of males and females, or perhaps slightly more females. In the world in the year 2003, there were 101 males for each 100 females, but Figure 5-20 shows how sex ratios vary widely.

What determines national sex ratios? The ratios of females to males are high in the rich countries of North America, in Western Europe, and in Japan. This observation suggests the hypothesis that sex ratios vary with national wealth. Wealth might be expected to reduce female mortality for two reasons: (1) Birth rates are generally lower in richer countries, thus reducing maternal mortality. (2) Improved medical care is generally available in richer countries during pregnancy and birth.

Sex ratios in the poor countries, however, do not show a consistent pattern. In some poor countries men outnumber women, but in others, in tropical Africa, for example, women outnumber men.

In fact, variations in sex ratios can be explained only by a combination of economic and cultural factors. In the countries of North America and Europe and in Japan, women may suffer many kinds of discrimination, but they are not generally discriminated against in access to medical care. Moreover, in these areas social differences tend to increase mortality among men. Men more often fall victim to violence, for example, and to workplace accidents. In most of Asia and Africa, however, women do not get the medical

care, food, or social services that men get. As a result, fewer women survive than would be the case if they had equal care. The rise of death rates in the formerly communist countries of Eastern Europe afflicts mainly men, so male-to-female population ratios are falling in Russia, Belarus, Ukraine, Armenia, Latvia, Romania, Bulgaria, and other countries in that region.

Great variations exist within each world region. In Asia, Japan has approximately the same male-female ratio as European countries, and in some countries of Southeast Asia women outnumber men. In South Asia, however, particularly India and Pakistan, the ratios of men to women are high. The high ratio of men to women in China may be attributable to the preference for boys.

Sharp contrasts exist even within individual countries. In the northern Indian states of Punjab and Haryana, for instance, which are among the country's richest, there are only 86 females per 100 males, whereas in the state of Kerala in southwestern India there are 103 females per 100 males, a ratio similar to that in Europe, North America, and Japan.

Women's positions are everywhere more favorable if they can earn an income outside the home. That reduces boys' advantage as potential supporters in the parents' old age, and females suffer less relative deprivation from birth. Later in life, working women can rely on their own resources.

Critical Thinking
HOW WILL YOU RETIRE?

As the American population ages and the balance between workers and retirees shifts, you might start planning for your own retirement, even if it is still many years away. The three traditional sources of income for American retirees were pensions, personal savings, and Social Security.

Corporate pension plans multiplied through the twentieth century. Most were "defined benefit plans" in which a worker, after attaining seniority, received a set payment for life, based on a combination of salary and years of work. Clause 401 (k) of The Tax Reform Act of 1978, however, triggered a new kind of pension plan, in which control and responsibility rested with the worker, who saved for retirement in an account that received favorable tax consideration. These plans were cheaper for corporations even if corporations matched some portion of the workers' savings; from 1978 to 1998, when the great majority of total savings were in such plans, employers' contributions to pensions fell 22 percent. In 2003, about one-half of American workers in the private sector still had no employer-sponsored retirement plan of any kind, but the stock market decline early in the twenty-first century devastated many workers' savings. Furthermore, analyses revealed that many corporate pension accounts were substantially underfunded.

Many Americans find it hard to accumulate individual savings in a culture so rich in attractive consumer goods. The personal savings rate in the United States has fallen from about 8 percent of household income in the early 1990s to 3.5 percent in 2003. Other industrialized nations average about 7–15 percent savings rates.

Social Security payments were never intended to allow retirees to live comfortably, just to prevent starvation, and today Social Security is widely acknowledged to be fragile, as "Boomers" reach retirement (see Figure 5-9). Suggestions to lower benefits meet strong resistance. Therefore, the U.S. is raising Social Security taxes by raising the base amount of a person's income that is taxed, although most Americans already pay more in Social Security taxes than in income taxes. The government is also steadily increasing the age at which workers may retire and receive benefits. Law current in 2003 has the retirement age for Social Security and medicare slowly rising to 67 (from the current 65) by the year 2027.

Questions

1. What will be the required minimum age for retirement to collect Social Security when you hope to retire?

2. Try to project what the U.S. population pyramid (Figure 5-9) will look like when you retire.

3. If you have a job, do you now pay more income tax or Social Security tax?

Six of the ten countries with the highest ratio of males to females are Arab countries, where women's rights are not always guaranteed. Some demographers think this suggests that societies that limit women's rights do not count all women in the census. If females are actually undercounted in these countries, then the reported higher ratios of males may not be accurate. Certainly women are better off in countries that acknowledge female deprivation and seek to remedy it. Efforts to achieve this can be influenced by providing women with education and the right to vote.

The Aging Human Population

Through the twentieth century, improvements in health and hygiene triggered a rise in the median age of Earth's population for the first time in history. The median age increased from 23.5 years in 1950 to 26.4 in 2002, and it is projected to reach 37.8 by 2050. The percentage of people age 60 or older is expected to rise from the current 10 percent to 22 percent by 2050.

In the rich countries, around 20 percent of the populations are already age 60 or older; that figure may rise to nearly 33 percent by 2050. These aging populations challenge the possibility of sustaining economic growth and equitably distributing wealth among generations. Higher percentages of national budgets are devoted to senior citizen centers than to elementary schools. Funds for medical research are dedicated to degenerative diseases such as cancer, heart disease, and Alzheimer's, when enormous sums of money buy relatively short advances in longevity. To support elderly populations, young workers may have to increase their contributions to pension funds, or else pension benefits may have to fall. Alternatively, workers may have to work more years before retiring. Most rich countries are already raising the age at which people may collect pensions.

Today's poor countries face different problems. These countries still exhibit high birth rates, challenging them to care for and educate children, but their birth rates are falling. Therefore, their median ages will soon rise fast. About 8 percent of the populations of today's poor countries are over 60 years of age, but that percentage will rise to 20 percent by 2050. The median age of Mexico's population, for example, was 22 years in 2000, but it is projected to be 34 years by 2050. In many poor countries the elderly are today cared for by their families. As the ratios of old to young rise, the countries will have to build up their national incomes and devise national pension systems. They have perhaps two generations' time in which to do this.

MIGRATION

Human beings do not stay put; they never have. Wanderings and migrations of people have distributed and redistributed populations throughout history and even prehistory, and significant redistributions continue.

Geographers who analyze human movements divide the causes for those movements into **push factors** and **pull factors.** Push factors drive people away from wherever they are. Push factors include starvation and political and religious persecution. Pull factors attract people to new destinations. Pull factors include economic opportunity and the promise of religious and political liberty. Physical geography can also serve as a push or pull factor. For example, we can say that migrants from northern cold regions have found the warmth of the U.S. Sunbelt attractive and that environmental disruption in the African Sahel, by contrast, has been found intolerable by emigrants from that region. The environment itself never actually pushes or pulls; people decide to migrate after considering environmental factors.

People choose a particular destination because of what they think they will find there, not because of the reality of conditions at that place. Therefore, migrants can be surprised by the reality of what they find at their new homes. Sometimes migrants find success beyond their expectations, but in other cases they are disappointed (Figure 5-21).

Migration has not always been voluntary. Some migrations have been forced; and millions of people suffered the tragedy of migrating in slavery, regardless of push or pull considerations of their own. Others have fled persecution in their homelands, and millions of others, called *internally displaced persons* (*IDPs*), have been forced to flee from one region to another within their own homeland. For example, millions of Angolans have fled the civil war that has raged in their country since the 1970s; many have fled abroad, and many more are IDPs.

Many migrants complete whatever documentation is required for them formally to settle in their host countries, but today great numbers of people are crossing borders without completing legal papers. These migrants are called **undocumented, irregular,** or **illegal immigrants.**

The thriving City of Eden as it appeared on Paper   *The thriving City of Eden as it appeared in Fact*

FIGURE 5-21 Disappointed migrants. Charles Dickens's comic novel *Martin Chuzzlewit* (1844) recounts these gentlemen's dismay when they migrate to the American frontier and find that it is much more primitive than land developers had suggested. Many of Dickens's American readers took offense, but in fact such misrepresentations were common. Many other migrants to the United States, by contrast, have found success beyond their original hopes. (Courtesy of the Library of Congress)

Movements of people can often be explained by studying flows of trade or by mapping the most convenient transport routes away from places people want to leave. People also follow information. Successful migrants write home, and they pave the way for new arrivals by providing them with employment and financial assistance. In this way, linkages called **migration chains** are forged. Information often travels between formerly imperial powers and their former colonies, as between France and several African countries or between the United States and the Philippines. In these cases, potential migrants absorb some of the culture of their destination even before they leave home. Information flows can be astonishingly place-specific. An estimated 80 percent of the Hispanics living in metropolitan Washington, D.C., are from the tiny Central American country of El Salvador—many of them undocumented. Nobody knows exactly how word first spread throughout El Salvador to trigger this migration.

Many migrants intend to stay in their new location only until they can save enough capital to return home to a higher standard of living. These **sojourners** are usually men who are either unmarried or who have left their families in their home country. Many sojourners eventually decide to stay, at which time they send for their wives and families. The United States has always attracted great numbers of sojourners. In various years between 1890 and 1910 the number of Italians leaving for Italy was as high as 75 percent of the number of Italians arriving in the United States. Even today many Greeks, Irish, Caribbean peoples, and Africans shuttle back and forth between their homelands and the United States.

The ease of international travel and the number of sojourners today has created a new condition in which villages in the poor countries form building blocks of immigrant urban neighborhoods in rich countries. For example, when in 1982 a terrible fire killed 24 people in a tenement building in Los Angeles, authorities were astonished to learn that several hundred residents of the building were all neighbors from one small village, El Salitre, in the Mexican state of Zacatecas. Committees in the rich country not only send money home but they decide on what expenditures will be made in their hometown—whether to build a new school or a new water-treatment plant, for example.

Some migrations have numbered in the millions, but some small migrations have been disproportionately significant because the migrants played key roles in transforming the government, language, economics, or social customs in their new homelands. As we shall see, the impact of the international migrations of Indians and Chinese, for example, or of Caribbean blacks into the United States, are not fully realized by their numbers alone.

In several cases the impact of international migrants in their new homes has harmed the welfare of the natives, or **indigenous peoples** of that region. The

United Nations estimates that today there are 300 million indigenous people in 70 countries, defined as "descendants of the original inhabitants of a land who were subjugated by another people coming after them." These include many Native American groups, the Bushmen of South Africa, and the Lumad of the Philippines. The United Nations hopes to ensure some degree of self-determination for these peoples. Self-determination does not necessarily mean political separatism, but the right to pursue and protect their own culture.

Prehistoric Human Migrations

Human beings are one single species, and all its members can successfully interbreed. Some geneticists believe that they have sketched a family tree for all humankind through studies of DNA, deoxyribonucleic acid. DNA is a component of genes, which are the hereditary units that determine characteristics of organisms. These scientists believe all humans descended from a single stock that lived in Africa about 150,000 to 200,000 years ago (Figure 5-22). The people who are closest to the root of our common DNA are the Vasikela Kung of the northwest Kalahari Desert region. Some members of our common stock stayed "home," while others migrated away to Asia (probably across the Arabian Peninsula) between 56,000 and 73,000 years ago and to Europe between 39,000 and 51,000 years ago. Scientists are not certain about the first peopling of the Americas. Both archaeological and DNA evidence suggest several different occurrences as long as 16,000 years ago, some on foot across a land bridge that once linked what is now Siberia and Alaska, but perhaps other occurrences by boat from Asia or Europe.

Humankind's earliest migrations are deep in prehistory, but research and formulation of hypotheses about the origins and diffusion of humankind continue in geography, biology, anthropology, archaeology, and, as we shall see, linguistics. This is an exciting frontier of human self-knowledge.

The Migrations of Peoples since 1500

For purposes of understanding contemporary geography, we will begin our study of human migrations 500 years ago, before the European voyages of exploration and conquest. In 1500, Earth's many peoples and cultures were relatively isolated from one another. The world we know is the product of their migrations and mingling. The most significant population transfers during these years have been migrations of Europeans to carve enclaves of settlement around the world, migrations of blacks out of Africa, and migrations of other groups instigated by European expansion.

European migration to the Americas The replacement of natives by newcomers was the greatest in the Western Hemisphere (Figure 5-23). The Native

Prehistoric Human Migrations

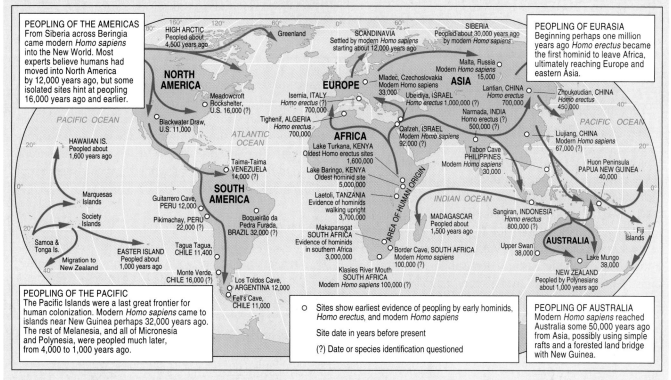

FIGURE 5-22 This map illustrates the latest hypotheses about migrations of humans out of Africa to settle around Earth.

European Migration in the Nineteenth Century

FIGURE 5-23 The nineteenth century saw the greatest migrations of Europeans. Europeans regarded the Western Hemisphere as a New World open to European settlement, and their massive migrations forever changed hemispheric demographics.

Critical Thinking

THE EAST-WEST EXCHANGE OF DISEASE

When the Europeans first came to the Western Hemisphere, they brought diseases that were new to the native populations. The Native Americans had no immunity to these diseases, so the diseases decimated them. Did the Europeans take new diseases back to Europe that decimated the European populations? No. Why not? Why was the direction of this terrible trade so one way?

One possible reason for the virulence of Eastern Hemisphere diseases in the Western Hemisphere is that the peoples of the Eastern Hemisphere had, through history, domesticated many more animals than the peoples of the Americas had. Many disease-causing pathogens are passed back and forth between humans and other animals with which we live in close proximity,

and we and the other animal sufferers develop resistance to these pathogens. The peoples of the Americas had domesticated very few animals, so they had developed little resistance to any diseases. That helps explain why they fell victim to the new diseases arriving from the East. The only disease that, it seems, the Americans gave to the Europeans was syphilis, which a member of Columbus's crew took back to Europe.

Questions

1. Trace the origins and diffusions of some modern epidemics around the world. You might study the great Spanish influenza epidemic of 1918.

2. What factors of world trade, travel, and tourism would probably accelerate the diffusion of a new disease today, and what routes might it take?

American population fell victim to slavery, warfare, and most important, to European diseases, including smallpox, influenza, measles, and typhus.

We have only estimates of the Native American populations at the time of the European conquests, and they range from 8 million to 112 million. The numbers of deaths were appalling. There may have been 25 million Native Americans in Mexico before the arrival of the Spaniards; by 1600 that number had declined to about 1.2 million. The population of the Inka Empire in South America plummeted from about 13 million in 1492 to 2 million by 1600. (You may be familiar with the spelling "Inca," which is Spanish, whereas the spelling "Inka" is Native American Quechua. For reasons discussed in Chapter 7, Inka is preferred today.) The estimated 5 million Native Americans in Brazil shrank to only 220,000 today. In what is today the United States and Canada, the Native American population fell from about 2 million in 1492 to a low of 530,000 in 1900.

North America received some 40 million Europeans before 1950. Today they and their descendants make up the majority of the population. The number of self-identified Native Americans in the United States tripled between 1960 and 2000 to 2.5 million, perhaps reflecting growing cultural pride, but they still constitute less than 1 percent of the national population. Canada's 976,000 self-identified "First Nations" individuals constitute just over 3 percent of Canada's population.

Both the United States and Canada have mixed policies of both assimilating Native Americans and also

segregating them (sometimes, historically, against their will) on reservations, where native cultures may be preserved. In both countries Native Americans retain preferred hunting and fishing rights over substantial areas in addition to their reservations, over which they exercise more substantial, but never complete, control. Canada has granted the Inuit people (Eskimos) political domain over roughly one-fifth of the country, the eastern half of the former Northwest Territories, now the territory of Nunavut (see Figure 11-23). Continuing litigation of other extensive Native claims embroils national politics. In the United States, Native American lands represent 2.2 percent of the total land area.

About 4.5 million Europeans, mostly from the Mediterranean area, migrated to what came to be called Latin America, and that region developed a complex racial structure. Ibero-American (descended from inhabitants of Spain and Portugal) societies were composed as a pyramid of varying proportions, with a large base of Native Americans and blacks; a lesser number of people of mixed race (a person of mixed white-black ancestry is usually called a *mulatto*, one of mixed black-Native American a *zambo*, and one of mixed white-Native American a *mestizo*); and a minority of whites on top. Several Latin American colonies and later the successor independent states institutionalized racial status rankings. Native Americans lost their rights in their own homelands, even though they may have constituted the majority of the population. Only since the mid-1980s have some South American countries, including Peru, Venezuela, Colombia, and Brazil, set aside lands for Native Americans.

To varying degrees, racial animosities linger throughout Latin America. Through much of 2002, for example, political events in Venezuela captured the world's attention. There hostility lingers between the *pardos,* brown-skinned people of Amerindian or African ancestry, and the *mantuanos,* people of European appearance and pretensions. Hugo Chávez, an army officer who had launched an unsuccessful coup d'etat in 1992, rode this distinction to political power, winning election to the presidency in 1998 while openly referring to himself as "an Indian" against the "rotten white elites." After taking power, he suspended the Congress and Supreme Court. The mantuano minority launched a coup d'etat against him in April 2002. Chávez was ousted but eventually reinstated.

Native Americans still make up one-third to one half or more of the populations of Mexico, Guatemala, Ecuador, Bolivia, and Peru (Figure 5-24). In Mexico the Native Americans have played the most conspicuous political role, but Native Americans are waking to power in Peru, Bolivia, and Ecuador. When Alejandro Toledo won election as Peru's first Native American President in 2001, he staged an inauguration ceremony at the Inka city of Machu Picchu presided over by a native holy man.

Among the southernmost midlatitude states, Chile claims to be 95 percent mestizo, and the Native Americans were practically exterminated in Uruguay and Argentina. In Central America most people are mestizos. Costa Rica is the only country in which whites form the majority.

The African diaspora Black peoples have migrated out of Africa, either freely or in slavery, for centuries.

FIGURE 5-24 Sacsayhuaman. These Native Americans' ancestors built this fortress in Peru in the fifteenth century. Today the ruins are a tourist attraction. (Robert Frerck/Corbis/Stock Market)

These migrations are known collectively as the **African diaspora,** from the Greek word for "scattering," used first by the Jews to describe their worldwide migrations.

Modern racial slavery originated in medieval Islamic societies. Muhammad himself owned slaves, and Arabs and Persians invented the long-distance slave trade that transported millions of sub-Saharan captives out to a realm stretching from Islamic Spain across to India. The Islamic slave system lasted over a period of 12 centuries, and in total numbers it probably surpassed the African slave trade to the Western Hemisphere (Figure 5-25). The trade diminished slowly, lingering longest in the Arab countries. (It was finally banned in Saudi Arabia in 1962 and in Mauritania in 1980.) The absence of a large black population across this enormous region today can be explained by the high mortality rate, by assimilation, and by the practice of castrating slaves. In central India, however, there are still communities of black descendants of African slaves.

International agencies insist that slavery persists in Mauritania and that Arabs from northern Sudan continue to enslave the blacks of the southern Sudan.

The transport of slaves to the Western Hemisphere European transportation of black slaves to the Western Hemisphere had a dramatic impact on population geography. In the fifteenth century Europeans took a few hundred slaves from West Africa to Europe and the Atlantic Islands, and in the 1520s Europeans began transporting slaves to the Western Hemisphere (Figure 5-26). Black Africans served as a labor supply to replace the dying Native Americans.

From 1526 until transatlantic slave trade came to an end in 1870, about 10 to 12 million slaves reached the Western Hemisphere from Africa. The sources of the slaves reflected patterns of European colonialism in Africa. For example, the Portuguese African colonies of Angola and Mozambique provided the slaves of Portuguese-ruled Brazil, whereas West Africa supplied most of the slaves for North America. DNA samples from the Western Hemisphere black populations are today providing evidence of their ancestors' African homelands.

The pattern of resettlement in the Western Hemisphere reflected the uses of slave labor. Slaves were brought to tropical and semitropical plantations in the U.S. Southeast, to the islands of the West Indies, and throughout Latin America, to the *tierra caliente* (hot land) at elevations of less than 914 meters (3,000 feet). Mexico was an exception because there the slaves were used in the cities and mines. Mexico never had a significant rural black population, and eventually Mexico's black population died out or mixed with the other groups almost entirely. Mexico abolished slavery at independence in 1821.

Blacks in the Americas today Descendants of slaves now constitute the majority of the population on most

The Movements of Slaves out of Africa

FIGURE 5-25 Arab slave dealers first transported black slaves North and East, throughout the entire region from Spain to India. Transport of slaves to the Western Hemisphere started later and stopped earlier. Only about 5 percent of the slaves brought to the Western Hemisphere were brought to the area that became the United States.

West Indian islands and in Belize. In South America, blacks still occupy, for the most part, the lowland areas. After centuries of intermarriage, tropical South American countries still have significant minority black, zambo, or mulatto populations.

In the United States, about 12 percent of the population is black—35 million people—giving the country one of the world's largest black national populations. The black population was originally concentrated in the Southeast, but blacks began to migrate to the cities of the North just before the time of World War I. Between 1910 and 1970 some 6.5 million African Americans headed northward, and the percentage living in the South fell from 70 percent to 50 percent. The civil rights struggles of the 1960s created new opportunities for blacks in the South, and partly for that reason and partly for reasons not fully understood, African Americans have been migrating back to the South since the mid-1970s. About 52 percent of African Americans lived in the South in 1980; some 55 percent in 2000 (Figure 5-27).

The twentieth century has also witnessed a steady migration of West Indian and Caribbean blacks into the United States—almost 3 million between 1960 and 2000 alone. A high percentage of these people brought capital or job skills, and they and their descendants

have played a role in twentieth-century U.S. politics and culture disproportionate to their numbers. For example, the first black U.S. Secretary of State, Colin Powell, was the son of immigrants from Jamaica. Statistical descriptions of America's foreign-born blacks and of their children reveal high educational attainment, high income levels, substantial representation in professional and managerial occupations, and low levels of unemployment. The flow of migrants and sojourners to the United States, combined with the flow of U.S. tourists to the Caribbean, has caused a degree of acculturation sometimes referred to as the Americanization of the Caribbean.

The cultural and economic might of black Americans exerts a tremendous attraction in black Africa. Black America's literature and consumer products blanket the continent, and largely because of America's black population and its many achievements, American culture has diffused throughout Africa to a surprising degree. It is displacing the cultural influence of Africa's European former rulers, despite the fact that the United States never had any African colonies.

European migration to Asia and Oceania The Spanish, French, Dutch, Germans, Portuguese, British, and Americans all created empires in Asia, but few

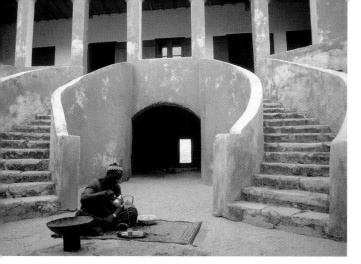

FIGURE 5-26 A slave market. Tens of thousands of Africans passed through this old slave market on Goré Island off the coast of today's Senegal on their way to the Western Hemisphere. In 1992, Pope John Paul II stood in the doorway in the background, known as the Door of Tears, and blessed them and their descendants. Today the Senegalese government is restoring the island's historic fortifications as a museum and is building an international conference center. Furthermore, African countries are sending delegations to the Americas to perform religious ceremonies at slave burial grounds, asking for forgiveness for having sold them into slavery. (Brian Seed/Getty Images Inc.—Stone Allstock)

people from these countries settled permanently in the Asian colonies. In no cases did any of these peoples come to make up significant proportions of mainland Asian populations. The most important European settlers in Asia and Oceania can conveniently be divided into those who migrated by land—the Russians—and those who migrated by sea—settlers from the British Isles who went to Australia and New Zealand.

Russian expansion   After the Russians rose up against their Mongol rulers in the sixteenth century, they sent a steady stream of settlers to the East, over the Ural Mountains into Asia. Between 1860 and 1914 almost 1 million Russians resettled in the East, and this flow continued within the political framework of the U.S.S.R. (look ahead to Figure 13-5). In 1990 the population of Kazakhstan, which was ostensibly a republic of the Kazak people, was 38 percent ethnic Russian. Russians migrated particularly into the cities in the non-Russian republics. In 1990 Tashkent, then the capital of the Uzbek Republic and today the capital of independent Uzbekistan, was primarily Russian.

When the U.S.S.R. dissolved in 1991, some of the newly independent republics imposed rigid language and citizenship requirements, triggering a migration of some 6 million Russians into Russia. Many of these families had not lived in Russia for generations. The position of Russians who remain in the newly independent Republics is still being defined.

Australia and New Zealand   More than 2 million Europeans, mostly from the British Isles, moved to Australia and New Zealand in the nineteenth century, and another wave of 3 million Europeans (half British) arrived between 1945 and 1973. They all but exterminated the native peoples. Latest censuses count only 150,000 Aborigines in Australia (of a total population of 19 million) and 400,000 Maoris among New Zealand's 3.8 million people. Extensive litigation in both countries has recently returned lands to the native peoples and awarded them substantial payments.

In both Australia and New Zealand, commercial farming and ranching thrived from the beginning of white settlement. Wool, wheat, and later frozen meat were sent to markets within the British Empire. In Australia, white settlement in the dry areas was hazardous and only intermittent until spectacular mineral discoveries attracted some people inland. In general, however, the white populations of both Australia and New Zealand have always been highly urbanized along or near the coasts (Figure 5-28, p. 202).

About 88 percent of Australia's 19 million people are of British and European ancestry, but the country abandoned an explicit "white Australia" policy in 1973. Today 40 percent of Australia's 100,000 annual immigrants come from Asia, and 7 percent of the national population is of Asian ancestry. A full 22 percent of Australians in 2002 were foreign-born, the highest percentage of any country.

European migration to Africa   Europeans migrated to Africa and settled in substantial numbers only where environmental conditions made possible a Mediterranean or European style of life and commercial agriculture.

East Africa, Northwest Africa, and West Africa   The East African highlands attracted white settlement after the opening of the Suez Canal in 1867. Europeans claimed the fertile lands in today's Kenya, Tanzania, and Uganda, and much of the black population ended up as workers on the white-owned farms or in white enterprises. Tens of thousands of whites remain scattered throughout East Africa, but the black-majority governments rescinded their legally privileged position.

In Northwest Africa, too, European colonial governments reserved the best lands for European settlers. During French rule of Algeria from 1830 to 1962, some 1.2 million Europeans settled there—about as many as in all 26 countries of tropical Africa combined. When Algeria, Morocco, Tunisia, and Libya gained their independence, most white-owned enterprises were nationalized or confiscated. Few Europeans remain in these countries.

West Africa never experienced substantial white settlement. Europeans were discouraged by the heat and humidity. Today whites still come or serve as business executives, technicians, or even government advisers, but they make up insignificant percentages of the populations.

Southern Africa   Southern Africa includes large areas that Europeans found well suited to settlement and

Distributions and Migrations of African Americans, 1950

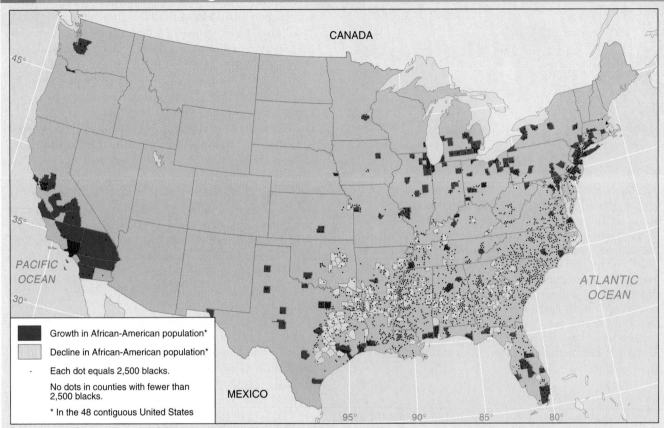

FIGURE 5-27 Each dot on these maps represents 2,500 African Americans, and the colors indicate whether each county gained or lost black population in the previous decade. The counties that contain fewer than 2,500 African Americans have no dots, and in large cities the density of black residents becomes so great that it is hard to distinguish the number of dots.

The map for 1950 reveals that during the decade of World War II (the 1940s), black populations declined in many southern rural counties, whereas the black populations of many northern cities increased rapidly. Also, many blacks moved

creation of a European lifestyle. Several colonies did achieve considerable prosperity—for the whites—but always based on repression and exploitation of the native black majority. As these colonies won independence, white racist regimes yielded to black-dominated governments (Figure 5-29).

The Republic of South Africa is the leading state in southern Africa. In 1671 the Dutch established Capetown colony as a staging point for voyages around Africa to Asia, and numerous Dutch farmers, called Boers, came to settle. When Britain seized that strategic colony in 1806, many Boers fled north, but the British eventually conquered all of South Africa and set up the Union of South Africa in 1910. That state was ruled by its white minority. International opposition to the government's policy of strict racial segregation, called **apartheid**, caused the government to declare itself a republic, totally free of any allegiance to Britain, in 1961.

The government divided the population into four groups: whites (British and Boers combined, totaling

13 percent of the total population in 1996), Asians (3 percent), people of mixed race (or "Colored," 9 percent), and blacks (75 percent). Each group held a fixed status. The Colored are mulatto products of early mixing between Boers and Khoikhoi, nomadic native peoples of southwest Africa. Most of the Asians came to the eastern province of Natal to work on the sugar plantations, but many eventually rose into the middle class or even to substantial wealth. Nevertheless, to restrict their political rights, the South African government maintained the legal fiction that they would someday return to Asia. Their residency was recognized as permanent only in 1963.

Parliament had three chambers one white, one Asian, and one Colored. Blacks were excluded from participation, for the government planned to move them into segregated *Bantustans* (homelands), on the basis of ethnic roots (Figure 5-30). The Bantustans covered only 13 percent of the territory, areas of low agricultural potential and without other resources. The

Distributions and Migrations of African Americans, 1990

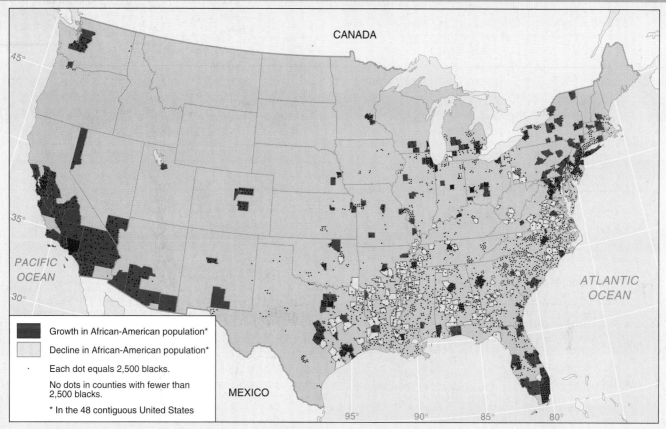

to California and Washington, where airplane manufacturing and other industries grew. The second map indicates that between 1980 and 1990 the migrations continued. The rural South continued to lose black residents, while the black populations of the cities in almost all regions increased. The black populations of many western states remain negligible. Returns from the census of 2000 reveal that these patterns of distribution and migration have persisted. (Data from U.S. Census Bureau)

blacks obviously would never be able to support themselves in these lands, but would have to sell their labor within South Africa, where they would be immigrant workers without political rights.

By the late 1980s the system of apartheid had begun to collapse under the pressures of international sanctions and internal rebellion. A new governmental system was devised, and free elections open to people of all races were held in April 1994. Nelson Mandela, a black leader who had been imprisoned for 27 years during the struggle for civil rights, won the presidency.

The government faces tremendous pressures. Blacks understandably want the life of conspicuous affluence that many South African whites enjoyed, but South Africa does not have the wealth to provide that for all its people. Furthermore, South Africa is considerably richer than its neighbors, so an estimated 2–8 million Africans from neighboring countries have illegally migrated into South Africa. In 2000, South

Africa's Home Affairs Minister Mangosuthu Buthelezi complained that "the presence of illegal aliens impacts on housing, health services, education, crime, drugs, transmissible diseases—need I go on?"

The histories of some other states of southern Africa echo that of the Republic of South Africa. The Portuguese colonies of Angola and Mozambique did not have many white settlers when they became independent in 1975, and most of those left. Botswana also had few white inhabitants when it gained independence in 1966; and when Namibia gained independence from South Africa in 1990, whites made up only 5 percent of its population. Zambia won independence in 1964 with a black-dominated government, which in 1975 nationalized all private landholdings and other enterprises. Today few whites remain. Whites struggled to retain political power in Zimbabwe (previously known as the colony of Southern Rhodesia), as in South Africa, but power was transferred to blacks in 1980, and today few whites remain. Whites long

Population Distribution in Australia and New Zealand

FIGURE 5-28 Vast stretches of inner Australia are virtually uninhabited.

Southern Africa

FIGURE 5-29 Substantial numbers of Europeans settled in the midlatitude lands of southern Africa when these were colonies, and today's independent countries are still working to achieve rising standards of living for all.

The New South Africa

FIGURE 5-30 South Africa's new 1994 constitution abolished the Bantustans and established one central government and nine provincial legislatures.

retained title to much of the best agricultural land, but that land was seized, and today land redistribution is virtually complete.

The migration of Indians When the British ruled the Indian subcontinent, they encouraged migration to other parts of their empire. In many places these Indian migrants went, they rose to positions that were socially and economically above the natives, as noted already in South Africa.

In British East Africa, Indians generally made up less than 5 percent of the populations, but when Uganda, Kenya, Malawi, and Tanzania achieved independence in the 1960s, Indians dominated the new countries' economies. Indians' high visibility made them targets of the majority blacks' ethnic hostility, and Uganda even expelled more than 80,000 Indians. Several East African countries have recently tried to entice Indians to return, and Indians are regaining economic strength in Uganda and Kenya.

Indians settled in Myanmar, and several hundred thousand remain, but they have faced discrimination since Myanmar achieved independence in 1948. Singapore is about 6 percent Indian. Indians still form key elements of the populations in several West Indian countries. Belize is about 2 percent Indian and Jamaica about 3 percent, but Trinidad and Tobago is about 40 percent Indian, and Guyana on the South American mainland is almost 50 percent Indian. The population of the Pacific Ocean island of Fiji is almost equally split between Indians and indigenous Fijians.

Today the 20 million nonresident Indians make up one of the world's most widespread diasporas, and in many places they have enjoyed great success. Annual Indian migration to the United States has reached almost 50,000. Indians have won about one-half of all visas set aside for skilled workers, and in the 1990s Indian entrepreneurs established, by one estimate, over 800 high-tech corporations in California alone. Links are so strong between the U.S. computer industry and the industry in India that President Bill Clinton, while on a visit, referred to the Indian city of Bangalore as "Silicon Valley East," after California's concentration of high-tech industries (see Figure 12-12). First-generation Indian-Americans enjoy the highest median family income of any foreign-born group.

Several reasons have been cited for Indians' international success. One is that they have been able to take advantage of the network of commercial outposts established by the former British Empire. A second reason is their facility with English, which is increasingly the world language of business. India has more English speakers than any country except the United States. Some Indians have themselves cited the cosmopolitan character of Indian culture as a third explanatory factor in their international success. The population of India is so diverse in terms of skin colors, languages, religions, and other cultural attributes that many Indians

learn during childhood how to cooperate with people of all backgrounds. A fourth factor is that Indians in many countries help one another. If an Indian cannot borrow money from a local commercial bank to start a new business, local fellow Indians often raise the necessary capital. Hard work and frugality are additional factors often cited by Indian entrepreneurs themselves, but these characteristics often describe groups that still do not enjoy the success that the Indians have.

Nonresident Indians constitute a valuable resource for India itself, because they have been able to transfer investment funds, technology, and marketing and financial expertise to India. The Indian government grants them privileges not available to other foreigners. For example, they may own land in India and repatriate funds elsewhere.

The Overseas Chinese From early in the nineteenth century until about 1930, waves of South Chinese left their homeland to work in the British, French, and Dutch Asian empires (Figure 5-31). Their success mirrored that of the Indians. At the time the nations of Southeast Asia gained their independence, their economies were firmly in the hands of small Chinese minorities, who were often linked across international borders by bonds of family, clan, and common home

FIGURE 5-31 The number after each country is the percentage of its population that is Chinese. Even where the Overseas Chinese make up small minorities of the population, they exercise financial influence far beyond their numbers. Overseas Chinese control the economies of most of these countries.

province. The Chinese commercial supremacy and their international links still survive, so that today these 50 million **Overseas Chinese** constitute a formidable economic power throughout Southeast Asia.

Just as the migrant Indians faced native ethnic hostility, so have the Chinese. For example, ethnic resentments dominate the history of Malaysia. When the British ruled Malaysia as a colony, they founded the cities of Singapore and Kuala Lumpur. The populations of both cities came to be overwhelmingly Chinese, although the majority of the population of the entire colony was Malay. Malaysia was granted independence in 1957, but animosity was so great between Chinese and Malays that in 1965 Singapore broke off as an independent Chinese-dominated city-state. Even in Malaysia, however, the Chinese minority still controls the economy, although the Malay-dominated government blocks Chinese and Indians from advancement in government positions and encourages Malay entrepreneurship. The population of Malaysia today is about 59 percent Malay, 32 percent Chinese, and 9 percent Indian.

In Indonesia the 6 million Chinese control an estimated 80 percent of the country's private industry. Indonesian corporations owned by ethnic Chinese count among the world's largest, and the Indonesian government requires them to invest in businesses owned by other ethnic groups. A 1965 communist coup attempt, supposedly backed by China, ignited anti-Chinese sentiment, and as many as 500,000 ethnic Chinese were slaughtered. Anti-Chinese violence flared again during Indonesia's political and financial instability in the late 1990s into the twenty-first century, and the future of the Chinese in Indonesia is unclear. The loss of their skills and capital would devastate the Indonesian economy.

Overseas Chinese play an elite role in the Philippines, and Thailand's commercial infrastructure is almost entirely Chinese. Thailand's largest corporation, Charoen Pokphand, was founded by two brothers from South China in 1921. Today it owns farms, food-processing facilities, and a variety of factories around the world.

Ambitious Chinese also established themselves in New South Wales, Australia; along the west coast of the Western Hemisphere from British Columbia to Peru; and in small but significant numbers in the Caribbean.

Tens of thousands of Chinese migrated to California in the mid-nineteenth century, and a new wave of Chinese migration to the United States began in 1965. Today Chinese Americans make up 1 percent of the U.S. population, up from 0.4 percent in 1980.

Tens of thousands of Hong Kong Chinese emigrated before that city reverted to Chinese rule in 1997. Many of them headed for Australia, and wealthy migrants made Vancouver, Canada—today almost 20 percent ethnic Chinese—a major financial center. Significant numbers of Chinese have also settled in metropolitan Toronto.

The Overseas Chinese have proved to be a valuable resource for their homeland, just as the nonresident Indians have. Overseas Chinese have raised investment capital, invested in homeland industries, introduced new technologies to China, and distributed Chinese products worldwide.

MIGRATION TODAY

Human migration has by no means come to an end. The United Nations estimates that in 2002 175 million people were living in countries other than the ones in which they were born. Large-scale migrations still make daily news. The United Nations' Universal Declaration of Human Rights affirms anyone's right to leave his or her homeland to seek a better life elsewhere, but it cannot guarantee that there will be anyplace willing to take anyone. As in the past, the major push and pull factors behind contemporary migration are economic and political. People are trying to move from the poor countries to the rich countries and from politically repressive countries to more democratic countries. In addition, millions of people

> Pressures for migration are growing, and in coming years they may constitute the world's greatest political and economic problem.

are fleeing civil and international warfare. Pressures for migration are growing, and in coming years they may constitute the world's greatest political and economic problems.

Many migrants may be poor, uneducated, or unskilled, but usually they are also enterprising, in good health, and of working age. They seek opportunity, and it takes courage, stamina, and determination to pull up roots and head away to a foreign and perhaps hostile land. In 1993, Mexico's President Carlos Salinas de Gortari said, "I don't want any more migration of Mexicans to the U.S., especially those migrants who are very courageous and take the impressive risk of going to the U.S. looking for jobs. That is precisely the kind of person I want here."

Other migrants are highly educated and skilled, and even the richest countries compete to attract them. Many were born in poor countries, and they received fine educations either at elite institutions in their homelands or in rich countries. The poor countries want these people to return.

The first half of this chapter noted that populations are declining in many of today's rich countries, and some observers may ask why migrants from today's poor countries cannot simply migrate to fill "vacancies" in these countries. New migrants would provide the young labor force that rich countries need in order to sustain the pension benefits of aging populations. The United Nations has estimated the numbers of immigrants each rich country would have to accept each year until 2050 in order for its total population to stabilize rather than fall. These numbers range from 29,000 for France to some 350,000 for Germany. The barriers to such mass migration, however, are cultural, economic, and political.

Refugees

Some migrants are defined in international law as being **refugees,** and these people have special rights. A refugee, as defined by a 1951 Geneva Convention, is someone with "a well-founded fear of being persecuted in his country of origin for reasons of race, religion, nationality, membership of a particular social group or political opinion." Signatories to the Convention are obligated to accept refugees and grant them safety, called **asylum.** Through recent years, civil wars and international wars displaced hundreds of thousands of people and sent many streaming over international borders. The United Nations High Commissioner for Refugees defined 19.8 million people as "of concern" on January 1, 2002 (Figure 5-32).

Some countries have broadened their definition of a "social group" to which they will grant asylum. For example, the United States, Canada, and several European countries have recognized persecution due to sexual orientation as a claim to refugee status. The European Parliament has accepted the legal principle that in some countries women are customarily treated so badly that they constitute a persecuted social group. Both the United States and Canada have also offered asylum to women from countries where, in the Canadian government's words, customs place women "in a more vulnerable position than men." Such countries include several Arab and African countries and Israel.

While many migrants are genuine refugees, most migrants are seeking a new home where they can work and enjoy a higher material standard of living. If they are fleeing poverty in their homelands, they are sometimes loosely called *economic refugees.* In just the five years from 1995–2000, the more-developed regions of the world received 12 million migrants from the less-developed. Many economic refugees claim to be genuine refugees, and it is difficult for host countries to ascertain who is a genuine political refugee. Claims to asylum have been abused, and many rich countries have been tightening their scrutiny of claims.

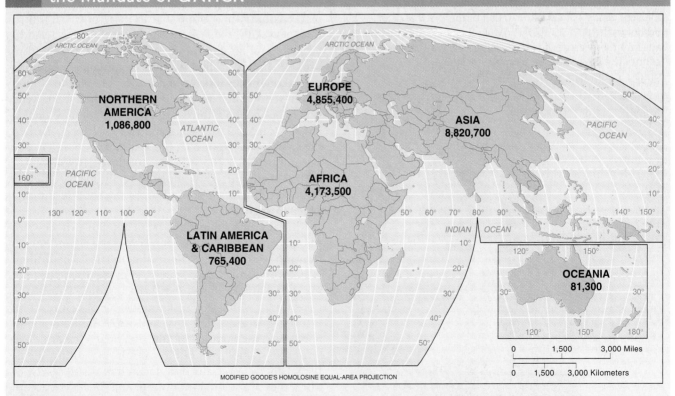

Estimated Number of Persons of Concern Who Fall Under the Mandate of UNHCR

FIGURE 5-32 This map shows the world's refugee population on January 1, 2002. It does not include the millions of people who are IDPs. (Data from U.N. High Commissioner for Refugees)

The Impact of International Migration

The impact of international migration is greater than the numbers might suggest. Immigrants are often greeted with apprehension, and immigration has become an explosive political issue in many receiving countries. This is for several reasons. (1) Official statistics substantially underestimate actual numbers; undocumented migration is rising everywhere. (2) Migrants are usually in the peak years of fertility. Therefore, they are playing an increasing role in the total population growth in the rich target countries. (3) Migrant settlements are generally concentrated in a few places within a country. This usually adds to the immigrants' visibility and increases the perception of cultural differences. (4) Many migrants today seem to be assimilating to the cultures and habits of their new homelands only slowly—if at all; this may be either because they are discriminated against or because they cling to their native traditions.

The peoples of the rich and democratic countries are beginning to exhibit "compassion fatigue," and doors are closing all around the world (Figure 5-33). Moreover, whenever unemployment rises, immigrants who were once grudgingly tolerated because they were willing to work in low-wage or dirty jobs suddenly come to be seen as competitors and become targets of wrath. The European countries are growing more selective; they welcome educated and skilled immigrants, but not poor or unskilled immigrants.

Every demographic impetus toward migration will be multiplied over the next two or three decades as populations rise in the poor countries. Either substantial flows of investment capital and technology will be directed to the poor countries and create new job opportunities in them, or else migration from the poor countries will be stimulated by the combination of large increases in population, excess labor supply, rising social and political turbulence, and persistent or worsening inequalities between rich and poor countries. Furthermore, the pull of the rich countries is being strengthened by global communications media and social networks spreading of images and information about them. Many poor people are growing desperate, and tragic stories of attempted migration can be found in the rich countries' newspapers almost every day: would-be immigrants drowned when flimsy boats capsize in the Mediterranean Sea or Straits of Florida, or found suffocated in cargo containers arriving on rich countries' highways or at airports. Altogether an estimated 200 million people are today under the sway or in the hands of traffickers in human beings who abuse them in sexual exploitation, economic slavery, forced labor, debt enslavement, or other forms of exploitation.

Some countries may be relieved to see people emigrate. The emigrants may be politically troublesome, or the country may hope that its emigrants will return with new skills, make investments in their homelands, or just send money home. Money sent home from workers abroad, called **workers' remittances,** supports several countries today. Mexican workers in the United States send home as much as $10 billion each year, which is Mexico's third-largest source of foreign exchange, after oil and tourism. Egypt, Nigeria, Turkey, the Dominican Republic, the Philippines, and Greece are among other countries that each receive billions of dollars each year. The estimated total value of international remittances is several times the total of international foreign aid.

Many rich countries accept foreign workers, but they do not grant them citizenship or the rights that citizenship would bring. The host countries accept them only as sojourners, and it may be unclear whether the migrant workers themselves hope to stay. The conditions under which the host countries grant citizenship, or even guarantee full civil and legal rights to aliens, have become matters of international concern. Unscrupulous employers often profit by exploiting aliens. Illegal aliens especially have faced hazards in search of work, they are seldom unionized, and they often accept lower wages and worse working conditions than legal residents or nationals will.

FIGURE 5-33 **Keeping them back.** This high fence, seen here between San Diego, California, and Tijuana, Mexico, winds its way across much of the U.S.-Mexican border in the U.S. Southwest. While the fence may be an extreme measure, it is seen as the only way to control the flow of eager migrants into the United States. (Jack Kurtz/The Image Works)

FIGURE 5-34 An immigrant neighborhood. This Arab butcher shop in Brussels illustrates how North African immigrants offer a variety of services and goods to their compatriots. (Courtesy of Robert Mordant)

Migration to Europe

About 16 million non-European legal immigrants live in the 15 countries of the European Union in 2003 (see Chapter 13), representing from 7 to 10 percent of the host countries' populations (Figure 5-34), but Europe has not always easily absorbed immigrants. The European countries have not historically thought of themselves as multicultural, and even though many immigrants come from former colonies, many Europeans fear the arrival of increasing numbers of Asians and Africans. In a 1999 poll, 66 percent of EU citizens described themselves as "a little racist;" and an additional 33 percent described themselves as "very racist" or "quite racist."

Compared to the Europeans around them, immigrant groups have often achieved lower levels of education, occupy substandard housing, and receive inferior services. Furthermore, many immigrants from the Near East and North Africa are Muslim; and, as we shall discuss in Chapter 7, their religion sometimes clashes with traditional European Christian culture. Generalizations about Europe's immigrant Muslims need many qualifications. Variations among their cultural and national backgrounds, and among the countries to which they migrate, may play a bigger part in their ability to integrate than their religion does. Overall, however, Europe's Muslims are much more likely than non-Muslims to be jobless and ill-educated, harassed by police, and discontented. If they feel equally estranged both from the Muslim lands and also from their new homes (or, in the case of the second generation, from the European land of their birth), they may turn to Islam as an identity. European secularism sharpens that tendency. Thus, some immigrants do not want to be acculturated to traditional European cultures. This contrasts strongly to the many millions of immigrants to the United States, who have traditionally seen the United States as admirable and the place of their future prosperity.

Most Western European Muslim communities condemned the 9/11 attacks on the United States, but many hesitated to endorse U.S. retaliation on Muslim countries. Several of the perpetrators of the attacks on the United States had actually lived in Western Europe for years. In February 2003, a German court convicted Mounir El Motassadeq, a 28-year-old Moroccan student, to 15 years in prison for his role as "accessory to murder" in the attacks; this ended the first trial anywhere of anyone connected with the attacks. Cells of al-Quaeda and other terrorist groups have been uncovered in Paris, London, Berlin, and other leading European cities. Some have committed terrorist acts, but the extent of others' involvement is under investigation. After murderous attacks on German and French tourists and workers in North Africa and in other Muslim lands early in 2003, a recording from al-Quaeda leader Ayman al-Zawahiri warned, "The young holy warriors have already sent messages to Germany and France. However, if these dosages are not enough, we are prepared with the help of God to increase the dosage." These discoveries have compounded many Europeans' fears of their Muslim neighbors. It has been asked whether Western Europe may contain enclosed societies in which hate can be preached and treachery plotted while all around, non-Muslims remain utterly unaware.

Nevertheless, as immigrant populations rise, Europeans are beginning to recognize immigrants' contributions and to lower barriers at least to immigrants with skills. The count of 650,000 self-employed immigrants in the European Union in 2000 revealed that many immigrants were professionals or business owners. Many French people began to accept the idea of France as multicultural when the star of France's World Soccer Cup-winning team in 1998 was a French-born son of Algerian immigrants, Zinedine Zidane.

Europeans are sharply curtailing the granting of asylum. Most European countries have decreed, for example, that people who had passed through one or more countries on their way from their homelands are no longer refugees, but are merely "country shopping" for the most favorable new site to settle. Therefore, they will not accept anyone claiming refugee status who came overland through a third country. European states have also lengthened the lists of countries from which visitors' visas are necessary, begun fining airlines for bringing people without visas, and even sending people back to the country they were last in. The reduction of legal migration to Europe has increased illegal migration, which may have reached 500,000 in the year 2000.

European states have also redefined the criteria for citizenship. For example, France traditionally accepted the idea that an individual's citizenship is based on his or her place of birth, a concept called *jus soli* (law of soil). The United States accepts *jus soli*. France, however, has made it more difficult for children born in France of foreign parents to become French citizens. Some other countries, including Germany, have traditionally defined citizenship on the basis of the citizenship of one's ancestors. This concept is called *jus sanguinis* (law of blood). Germany has tightened conditions for claims to German citizenship.

A fleet made up of ships from five Western European nations has begun patrolling the Mediterranean to turn back migrants sailing across the sea, but sealing borders will not solve the problem of existing immigrants and their European-born children and grandchildren. European governments are directing new attention to what the Dutch call *inburgering*, that is, the forming of citizens. Governments are requiring classes in language, history, culture and traditions, and the rights and duties of citizenship. Western Europeans are also welcoming the extension of the European Union to the East and immigrants from Eastern Europe as an alternative to immigrants from Africa or the Near East (see Chapter 13).

Migrations of Asians

An estimated 4 to 5 million Asians work abroad, and in many cases it is doubtful whether many will ever return to their homelands. Meanwhile, their remittances to their homelands are an important element in the economies of their home countries.

The absence of these workers lowers the local unemployment rates in their homelands, but it also constitutes a loss of important skills needed back home. Although the home country might not have the resources to reward these workers adequately, it frequently needs their contributions. A young Filipino woman, for example, may be needed in her hometown as a schoolteacher, but staying there she would earn $40 per month. If she is earning $150 per month as a maid in Singapore, she is better off, but the Philippines suffers. An estimated 60,000 Filipino women—many college graduates—work as maids in Singapore.

The oil wealth of some Arab states has drawn into them many Asian workers who seem little disposed ever to go home. In 2000, the population of the United Arab Emirates was less than 20 percent native-born; that of Kuwait only 35 percent native-born. The expatriate workers in these rich countries often come from poorer Muslim lands, and they are generally more susceptible to radical interpretations of Islam. Some

Arab Gulf states have adopted *indigenation* policies, reserving jobs once monopolized by foreigners for Arabs, but as long as the Middle East remains an international hot spot, and as long as Arab oil supplies are important to world prosperity, any destabilizing migration into Arab oil-rich countries is of world concern.

Japan, of all the world's richest countries, limits immigration the most strictly. Between 1945 and 2000, fewer than 250,000 foreigners acquired Japanese citizenship, including those who married Japanese. The most significant minority in Japan is the Japanese of Korean ancestry—about 0.5 percent of the total population. These people are a reminder that Japan ruled Korea from 1910 until 1945, and they suffer discrimination. With a total fertility rate of only 1.3 and a median age of 40.2, Japan's population is projected to fall 17 percent by 2050. Japan would need to import 600,000 workers per year to keep its working population stable between 2000 and 2050, and by that time the population would be one-third foreign-born. The government has reduced the cost of children—by expanding child-care facilties, for example—and in 2000 the Ministry of Justice broadened the category of jobs that foreigners in Japan (today about 1.1 million) could hold. He urged Japanese "aggressively to carry out the smooth acceptance of foreigners."

Migration to the United States and Canada

The single largest migration flow for the past 150 years has been migration to the United States. In recent years the United States has accepted 600,000 to 800,000 immigrants per year. Some 7 to 9 million illegal immigrants may also live in the United States.

The 2000 census found 10 percent of the U.S. population to be foreign-born, up from 7.9 percent in 1990. The nation's overall birth rate is low, so immigration is responsible for about one-third of U.S. population growth. If immigration continues at current levels, the foreign-born percentage of the population may account for nearly two-thirds of population growth projected to the year 2050.

The source areas of U.S. immigrants have changed throughout history, thus changing the makeup of the national population. Immigration was totally unrestricted until the late nineteenth century, when the government enacted rules explicitly designed to keep out Chinese and other Asians. Later, in the 1920s, the government issued quotas for various nationality groups. The first quotas favored British, Irish, and Germans, who had formed, along with African Americans, the bulk of the population in 1890. By the 1920s, however, Southern and Eastern Europeans had already established a strong presence. After the 1920s a maximum quota was set on immigrants from the Eastern Hemisphere (in practice, Europe), but there was in theory no ceiling on immigrants from the Western Hemisphere. In 1965 the United States

U.S. CENSUS BUREAU CATEGORIES

All categorizations of people in the U.S. Census are self-descriptions. People are asked "what" they are, and they choose among given categories as they wish. The choices, however, have changed several times through recent decades, reflecting changing attitudes about categorizing people.

In 1990 the U.S. Census Bureau offered five choices "White;" "Black;" "American Indian, Eskimo and Aleut;" "Asian or Pacific Islander;" and "Other." This selection, however, triggered widespread criticism. The Bureau classified Native Hawaiians, for example, as "Asian or Pacific Islanders," but many Native Hawaiians would rather be classified as "American Indians." They do share certain historical experiences; also, if Native Hawaiians were classed as "American Indians," they would enjoy access to some affirmative action programs that are closed to them as "Asians."

In 1997, therefore, the federal government began to allow people to classify themselves as being of more than one race by checking two boxes. To avoid the diminution of the minority voice in certain government programs, however, anyone who counted himself or herself as being both of a minority race and also being white, would be counted in the totals for minorities. Some observers found this reminiscent of long-repudiated segregationist racist categorizations called "one drop" systems. In these, anyone having just "one drop" of black blood was considered black and suffered segregation.

For the 2000 census, the Census Bureau addressed Native Hawaiian concerns by separating them from Asians and categorizing them with other Pacific Islanders. Thus the six categories of "race" used in 2000 were: "White;" "Black or African American;" "American Indian and Alaska Native;" "Asian;" "Native Hawaiian and Other Pacific Islander;" and "Some Other Race."

In the 2000 census, 6.8 million people (2.4 percent of the total) chose to identify themselves as being of more than one race. People under 18 years of age were twice as likely to be identified as being multiracial than were people 18 or older. This reflects increasing ethnic mixing in the United States, but also, perhaps, that younger people, whose heroes may include proudly multiracial Mariah Carey, Derek Jeter, and Tiger Woods, feel less bound by traditional ethnicities.

The U.S. government has also tried to count Spanish-speakers living in the United States. The history of this question and this category illustrates again the difficulties of describing and categorizing the population. In the 1950 census the Bureau asked how many people used Spanish as their mother tongue, and in 1960 it tried asking how many people had Spanish surnames. During the 1970s, several government agencies adopted a new category: "Hispanic." The exact meaning of the term is not clear, but it would seem to include people who speak Spanish or who have ancestry in some Spanish-speaking country. It would be foolish to suggest that all such people share a common "culture" (see Chapter 6) beyond sharing a language background, and some people who consider themselves Hispanic do not speak Spanish. This category "Hispanic" was nevertheless adopted by the Census Bureau in 1980, 1990, and 2000. People were asked to identify themselves in one of the racial categories, and then a second question asked: Are you Hispanic? A person could claim to be of any one of the six "races" in the first question, and then also claim to be Hispanic.

For some people, their identity as an Hispanic is more important to them than any racial categorization. Forcing them to identify themselves "racially" in the first question, they insist, forces them to be race-conscious in a way that Hispanic society traditionally is not, and that they do not want to be. Partly in response to this feeling, some analyses divide the U.S. population into the categories "White," "Black," "American Indian," "Asian," and "Hispanic." Whenever you see statistics reported in this way, however, you must remember that the category "Hispanic" has been created by pulling out from each of the six racial groups those people who reported themselves as "Hispanic" in the second question.

The word "Hispanic" excludes Brazilians, whose native language is Portuguese (see the box on page 236). Throughout Latin America the word "Hispanic" is reserved to refer to things Spanish, or from Spain. People throughout Latin America do use the words "Latinoamericanos" or "Latinos" to refer to themselves—as in the name of the baseball stadium in Havana, Estadio Latinamericano. Therefore, today, increasingly, U.S. "Hispanics" prefer to be called "Latinos."

An increasing number and share of people in the United States are growing to resent any form of racial or ethnic classification, and as mixing continues, racial and ethnic categories may be meaningless within a few decades. In 2003, however, Californians voted down a Racial Privacy Initiative that would have protected people from disclosure of racial identity to government.

TABLE 5-1 Difference in Population by Race and Hispanic or Latino Origin, for the United States: 1990 to 2000

Subject	1990 CENSUS		CENSUS 2000		DIFFERENCE BETWEEN 1990 AND 2000	
			Using race alone for Census 2000		Using race alone or in combination for Census 2000	
	Number	Percent of total Population	Race alone	Race alone or in combination	Percent difference (based on 1990)	Percent difference (based on 1990)
	(1)	(2)	(3)	(4)	(5)	(6)
Race						
Total population	**248 709 873**	**100.0**	**281 421 906**	**281 421 906**	**13.2**	**13.2**
White	199 686 070	80.3	211 460 626	216 930 975	5.9	8.6
Black or African American	29 986 060	12.1	34 658 190	36 419 434	15.6	21.5
American Indian and Alaska Native	1 959 234	0.8	2 475 956	4 119 301	26.4	110.3
Asian	6 908 638	2.8	10 242 998	11 898 828	48.3	72.2
Native Hawaiian and Other Pacific Islander	365 024	0.1	398 835	874 414	9.3	139.5
Some other race	9 804 847	3.9	15 359 073	18 521 486	56.6	88.9
Hispanic or Latino and Race Total population	**248 709 873**	**100.0**	**281 421 906**	**281 421 906**	**13.2**	**13.2**
Hispanic or Latino (of any race)	22 354 059	9.0	35 305 818	35 305 818	57.9	57.9
Not Hispanic or Latino	226 355 814	91.0	246 116 088	246 116 088	8.7	8.7
White	188 128 296	75.61	94 552 774	198 177 900	3.4	5.3
Black or African American	29 216 293	11.7	33 947 837	35 383 751	16.2	21.1
American Indian and Alaska Native	1 793 773	0.7	2 068 883	3 444 700	15.3	92.0
Asian	6 642 481	2.7	10 123 169	11 579 494	52.4	74.3
Native Hawaiian and Other Pacific Islander	325 878	0.1	353 509	748 149	8.5	129.6
Some other race	249 093	0.1	467 770	1 770 645	87.8	610.8

Source: U.S. Census Bureau

changed the rules again, in a way that tended to favor Latin Americans and Asians (Table 5-1 and Figure 5-35). In 1965 European immigrants outnumbered non-Europeans 9 to 1, but within 20 years that proportion was reversed.

The latest changes in the source areas of immigrants will again bring long-term consequences to national demography. In 1990 three-quarters of all Americans were non-Hispanic whites; by 2050 that percentage may fall below half. Immigration is the principal factor in the increase in the number and percentage of the total population that is Asian (11.9 million, 4 percent of the total). The increase in the number and percentage that is Hispanic, however, to 35.3 million, 12.5 percent of the total, is due to two processes. (1) Hispanics form the largest share of all immigrants. Through recent years about 20 percent of all immigrants have in fact come from just one country, Mexico, and Mexicans today make up about 30 percent of all foreign-born U.S. residents. (2) Hispanic birth rates are the highest among U.S. groups. Furthermore, they have the lowest median age of any U.S. group (26.6), so Hispanics would make up an increasing share of the national population even if immigration were ended in 2000. In the 2000 census, 35.3 million people claimed to be Hispanic, and 34.7 claimed to be black. Thus Hispanic Americans may be said to outnumber black Americans, although 36.4 million people claimed to be of mixed black-plus-some-other race. Already by 2002, however, Hispanics numbered 38.8 million, and black plus mixed-black only 38.3 million. If statistics include the 3.5 million U.S. citizens in Puerto Rico, the number of U.S. Hispanics surpassed that of African Americans about 1992.

In 2000, two-thirds of all foreign born lived in just six states (California, New York, Florida, Texas, New Jersey, and Illinois). Figure 5-36 shows a demographic

(a)

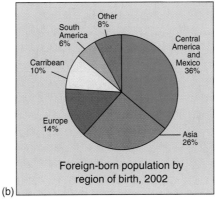

(b)

FIGURE 5-35 (a) Sources of immigrants to the United States and (b) foreign-born population by region of birth. The graphs in part (a) show how the source areas of immigrants to the United States have changed through history. The percentages of immigrants from Northern and Western Europe have dropped steadily, as the percentages from Latin America and Asia have risen. These trends have affected the composition of the national population. Graph (b) shows what percentage of the total foreign born population in 2002 had come from each region. (U.S. Census Bureau)

analysis of the population of each state. About two-thirds of new immigrants settled in just 10 metropolitan regions (New York, Los Angeles, San Francisco, Miami, Chicago, Washington, D.C., Houston, Dallas, San Diego, or Boston). In these areas, they often replace native-born residents, so the populations of New York and Los Angeles, for example, are today 40 percent foreign-born. Some plains states that are losing population are trying to lure foreign-born settlers. Iowa, for example, has a lower population than it did 20 years ago, and today it publicizes itself as an "immigration enterprise zone" in order to attract new residents.

Other features of current U.S. immigration legislation give preference to immigrants who bring either skills or capital for investment and make it easier for foreign students already in the United States to stay.

Canada welcomed from 175,000 to 220,000 new immigrants each year through the late 1990s, so foreign-born Canadians numbered 5.4 million in 2001, a full 18.4 percent of the total population. Immigration now represents over half of total Canadian population growth. Canada has also experienced a recent shift in the source area of its immigrants. In the 1950s more than 80 percent of Canadian immigrants were drawn

The United States Population in 2000

KEY

USA	281.4	← Total population in millions
*White	69.1	
*Black	12.1	*Percentage of those who claimed to be of only one race.
*Asian	3.6	
**Hispanic	12.5	**Of any race.

State data (T = total population in millions; W = White %; B = Black %; A = Asian %; H = Hispanic %):

State	T	W	B	A	H
ME	1.3	96.9	0.5	0.7	0.7
AK	0.6	69.3	3.5	4.0	4.1
MN	5.0	89.4	3.5	2.9	2.9
MI	9.9	80.2	14.2	1.8	3.3
VT	0.6	96.8	0.5	0.9	0.9
NH	1.2	96.0	0.7	1.3	1.7
WA	5.9	81.8	3.2	5.5	7.5
ID	1.3	91.0	0.4	0.9	7.9
MT	0.9	90.6	0.3	0.5	2.0
ND	0.6	92.4	0.6	0.6	1.2
WI	5.4	88.9	5.7	1.7	3.6
KY	4.0	90.1	7.3	0.7	1.5
MA	6.3	84.5	5.4	3.8	6.8
RI	1.0	85.0	4.5	2.3	8.7
OR	3.4	86.6	1.6	3.0	8.0
UT	2.2	89.2	0.8	1.7	9.0
WY	0.5	92.1	0.8	0.6	6.4
SD	0.8	88.7	0.6	0.6	1.4
IA	2.9	93.9	2.1	1.3	2.8
IN	6.1	87.5	8.4	1.0	3.5
OH	11.4	85.0	11.5	1.2	1.9
NY	19.0	67.9	15.9	5.5	15.1
CT	3.4	81.6	9.1	2.4	9.4
CA	33.9	59.5	6.7	10.9	32.4
AZ	6.1	75.5	3.1	1.8	25.3
CO	4.3	82.8	3.8	2.2	17.1
NE	1.7	89.6	4.0	1.3	5.5
MO	5.6	84.9	11.2	1.1	2.1
IL	12.4	73.5	15.1	3.4	12.3
WV	1.8	95.0	3.2	0.5	0.7
PA	12.3	85.4	10.0	1.8	3.2
NJ	8.4	72.6	13.6	5.7	13.3
NV	2.0	75.2	6.8	4.5	19.7
NM	1.8	66.8	1.9	1.1	42.1
KS	2.7	86.1	5.7	1.7	7.0
AR	2.7	80.0	15.7	0.8	3.2
TN	5.7	80.2	16.4	1.0	2.2
VA	7.1	72.3	19.6	3.7	4.7
MD	5.3	64.0	27.9	4.0	4.3
DE	0.8	74.6	19.2	2.1	4.8
HI	1.2	24.3	1.8	41.6	7.2
OK	3.5	76.2	7.6	1.4	5.2
LA	4.5	63.9	32.5	1.2	2.4
MS	2.8	61.4	36.3	0.7	1.4
AL	4.4	71.1	26.0	0.7	1.7
NC	8.0	72.1	21.6	1.4	4.7
D.C.	0.6	30.8	60.0	2.7	7.9
TX	20.9	71.0	11.5	2.7	32.0
GA	8.2	65.1	28.7	2.1	5.3
SC	4.0	67.2	29.5	0.9	2.4
FL	16.0	78.0	14.6	1.7	16.8

FIGURE 5-36 This image shows the percentage of the population in each state that was of each of four groups in 2000.

from Europe, but today that figure is less than 25 percent. About 60 percent of Canada's new citizens now arrive from Asia. In 1999 an ethnic Chinese from Hong Kong became Canada's governor general, the country's head of state. Canada has set a goal of increasing immigration gradually until inflows reach 1 percent of its population, up from about 0.85 percent in 2002. In Canada, the concentration of the foreign-born population is even greater than it is in the United States. In 2001, 94 percent of all foreign-born lived in Canada's 27 metropolitan areas, with Metropolitan Toronto, Vancouver, and Montreal alone accounting for three-quarters of them. Thus North American metropolitan areas have become increasingly "exotic" enclaves of ethnic diversity markedly different from the countries' non-metropolitan populations.

Canada selects its immigrants according to a system in which applicants are assigned points on the basis of characteristics such as their ability to speak English or French, workforce skills, family ties, and ethnic diversity. How these factors are to be weighed is a political issue in Canada, but Canada is today giving more weight to skills and less to reuniting families.

New arguments over immigration Many people in the United States are beginning to question continuing immigration. Some argue that immigrants cost Americans money for public services. Others note that immigrants contribute to American economic growth. The National Academy of Sciences concluded in 1997 that immigration produces substantial economic benefits for the United States as a whole. A great many

Critical Thinking
IS IMMIGRATION A SUBSTITUTE FOR EDUCATION?

Many critics have long argued that the labor force provided by immigration has allowed the United States to slight the education and development of the nation's native workforce. They have suggested that the country invest in educating and training its own workers rather than rely on immigrant labor. In his 1895 speech "Let Down Your Bucket," the black educator Booker T. Washington (1856–1915) encouraged southern industrialists to seek their labor force around them in the willing blacks who had so long "tilled your fields, cleared your forests, built your railroads and cities," rather than "to look to the incoming of those of foreign birth and strange tongue and habits . . . to buy your surplus land, make blossom the waste places in your fields, and run your factories." This would require "helping and encouraging them . . . and education."

Today, more than 100 years later, some U.S. industrialists insist that the national labor force still has not been properly trained, so skilled immigrant labor is still needed. The president of the National Association of Manufacturers said in 1995, "The shortage of available expert workers is a growing concern of American business. It is important that American industry continue to have access to skilled foreign workers who are in short supply."

In 1990 the U.S. government created the H-1B visa to permit foreigners with a college degree or higher to work in the United States for a renewable three-year term for employers who petition on their behalf. The number of such visas granted annually has risen to 195,000, and many people think there should be no limit. Another type of visa, called L-1, was designed for intercompany transfers; over 325,000 foreigners worked in the United States in mid-2003 on L-1 visas.

In May 2000, Gene Sperling, the president's economic policy adviser, said, "The best way to increase the supply of skilled workers is to improve the education and training of United States citizens and other people already in the country. However, at times, U.S. businesses need additional access to the international labor market to maintain and enhance our global competitiveness, particularly in high-growth new technology industries and particularly in tight labor markets." Some critics suggest that importing such workers is only a way to restrain wages in such fields.

Questions

1. Which is cheaper, importing a trained labor force or educating native-born workers? What cost may be associated with failing to educate and train Americans?

2. Can you think of reasons why native-born citizens may not be receiving the education that equips them for job openings?

immigrants hold arduous low-paying jobs, and this slightly reduces the wages and job opportunities of low-skilled Americans. Some scholars argue that this adverse impact on U.S. poor is immoral. While unemployment is low, however, few people protest immigration. America's labor and union organizations have traditionally stood against immigration, but in 2000 the AFL-CIO announced in favor of an amnesty for illegal immigrants and a repeal of most sanctions against their employers. It has been suggested that the unions hope to organize the new workers.

Some people argue that the United States is importing poverty. The poverty rate is significantly higher for immigrants than for the native-born (18 percent versus 12.1 percent in 2000), and while immigrants and their U.S.-born children make up only 20.5 percent of the U.S. population, they make up 26 percent of the U.S. poor population. The newcomers and their children, however, will make up the labor force to support the growing number of pensioners in the near future.

Substantial costs of immigration must be borne by local governments in the areas where immigrants concentrate, and these costs exasperate local taxpayers. In California, for example, 59 percent of voters approved a 1994 ballot initiative to deny public health and education to illegal aliens. Courts, however, ruled that step unconstitutional, so today Los Angeles spends about $2 billion per year educating illegal immigrants and an additional $350 million per year for their health care.

Most arguments over immigration are often based in economics, but immigration policy has political and cultural aspects as well. For example, should the United States absorb immigrants from Mexico deliberately to preclude potential political turmoil there? And is there an "American culture," and can that culture be sustained as the racial and ethnic makeup of the population changes? The United States was dubbed the

melting pot in a 1914 novel of that title, but various groups have maintained their cultural identities, and the traditional melting pot image cannot be sustained as demands rise for appreciation of U.S. "multiculturalism." Perhaps the United States will redefine itself as a **cultural mosaic,** which is what Canada proudly calls itself. As yet, however, many Americans seem blind to the changing composition of the national population. Of the 1,477 characters on the four major television networks' prime-time series in the 1999–2000 season, only 3 percent were Latino.

The United States must decide how widely to open its borders, and on what criteria of admission. Even the question of the definition of citizenship emerged in national politics in 1996, when the Republican Party platform called for a constitutional amendment "declaring that children born in the United States of parents who are not legally present in the United States or who are not long-term residents are not automatically citizens." Perhaps the United States will follow France and thus discard its historic *jus soli* definition of citizenship. In any event, immigration policy will continue to be a political issue in the foreseeable future.

CONCLUSION: CRITICAL ISSUES FOR THE FUTURE

Many poor countries are losing their best educated people and most skilled workers through emigration. This migration from less developed to more developed countries has been called the **brain drain.** In some cases a country's schools are qualifying students for professional careers faster than the economy can absorb them. In other cases the skilled or educated seek political freedom, and in still other cases the country's best students go abroad for training to help their own people, but they stay abroad and never return home.

The United States profits most from the drain of skilled labor from other countries, and in fact, the nation's scientific establishment would suffer without them. About one-third of U.S. Nobel prizewinners in the sciences have been foreign born. Thousands of senior scientists, engineers, and other professionals migrate to the United States each year, and about half of the young foreigners who come for education never return home. America's high-tech industries draw on the world's best minds. Almost half of the scientists and engineers working in California's Silicon Valley are foreign-born.

The drain of scientists, engineers, and physicians from poor countries undermines the ability of these areas to develop. Any advanced country is reluctant to expel these skilled immigrants. That country would suffer its own brain drain, and the skilled would surely just go to another advanced country anyway. How to keep these skilled professionals in their homelands, or to attract them back, is a problem that must be solved to avoid an ever widening gap between the world's rich and poor countries.

Chapter Review

SUMMARY

The population of Earth is about 6.3 billion people, and 90 percent of the population is concentrated on less than 20 percent of the land area. Major population concentrations are in East Asia, in South Asia, and in Europe. Secondary concentrations are in Southeast Asia and eastern North America. Much of Earth's land surface is virtually uninhabited.

Factors that explain this distribution include the physical environment, technology, history, and local differences in rates of population increase. Most of the population is concentrated in areas of mild midlatitude or seasonal humid tropical climates, but people will settle in harsh areas if it is profitable. In some places local population densities may be high, but trade and circulation free societies from the constraints of their local environments.

The populations of China and the Indian subcontinent long ago achieved productive agriculture and relative political stability. Europeans multiplied when they gained material wealth and improved their food supplies as the result of world exploration and conquest, and later again during the Industrial and Agricultural Revolutions. European migrants took technology to more sparsely populated areas.

The world population is increasing, but the rate of growth is slowing. The rate of increase in each country represents a balance among the crude birth rate, the crude death rate, the fertility rate, the total fertility rate, and migration. Most of the increase is occurring in the poor countries. The history of population growth in today's rich countries traces a demographic transition. In the first stage both the crude birth rate and the crude death rate were high. In a second stage medical science improved, so crude death rates dropped. Therefore, rates of population increase rose. In the third stage crude birth rates dropped, so today rates of natural population increase are low. Many rich countries' populations are not even reproducing themselves.

In today's poor countries, crude birth rates are dropping fast, even without significant increases in material well-being. Theories of explanation focus on the growing role of family-planning programs, new contraceptive technologies, and the educational power of mass media. Family-planning efforts face obstacles in traditional cultures, religions, and in some cases the cost of devices.

The total population will continue to grow because the number of young women reaching their childbearing years is larger than ever before, but the sooner the world's total fertility rate falls to the replacement rate, the lower the total population will be at which the projected numbers level off. The current most probable estimate is that Earth's population will reach 8.9 billion by 2050.

Natural population decreases can result either from an increase in the death rate, which could result from disease or war, or by a lowering of the birth rate. Today almost all poor countries are committed to family planning. A preference for large families is traditional in some areas, and having children may be economically rational for some families in poor countries, but many people in poor countries want small families. The low status of women inhibits family planning in some societies. Urbanization usually lowers total fertility rates.

The Malthusian theory argues that the rate of growth in the human population will always outpace the rate of growth in food supplies, triggering cycles of famine, disease, and war.

The natural ratio between females and males in a human population should be one to one, but, worldwide, men outnumber women. The sex ratio varies greatly in different countries. This may be due to a combination of economic and cultural factors.

The median age of humankind as a whole began to rise early in this century. In the rich countries, questions arise about the possibility of sustaining economic growth and about the equitable distribution of wealth among different generations. Many poor countries must quickly build up their national incomes and devise national welfare or social security programs.

Human migrations have redistributed populations throughout history, and significant migrations continue. Consideration of push factors drives people away from wherever they are, and consideration of pull factors attracts them to new destinations, but some migrations have been forced. Sojourners intend to stay in their new location only temporarily.

The most significant population transfers during the past 500 years have been migrations of Europeans around the world, migrations of blacks out of Africa, and migrations of other groups instigated by European expansion.

It is sometimes difficult to differentiate political refugees, who are fleeing politically repressive countries, from economic refugees, who are trying to move from the poor countries to the rich countries. The brain drain of educated people and skilled workers from poor countries undermines their ability to develop.

KEY TERMS

African diaspora p. 197
apartheid p. 200
arithmetic density p. 171
asylum p. 205
brain drain p. 214
carrying capacity p. 171
crude birth rate p. 176
crude death rate p. 176
cultural mosaic p. 214
demographic transition p. 180
demography p. 170
dependency ratio p. 180
domestication p. 175
doubling time p. 177

emigration p. 170
epidemiological transition p. 187
epidemiology p. 187
fertility rate p. 176
immigration p. 170
indigenous peoples p. 194
infant mortality rate p. 180
life expectancy p. 183
Malthusian theory p. 190
melting pot p. 214
migration chains p. 194
natural increase or decrease p. 176
Overseas Chinese p. 204
pathogens p. 187

physiological density p. 171
population geography p. 170
population projection p. 176
population pyramid p. 178
pull factors p. 193
push factors p. 193
refugee p. 205
replacement rate p. 176
sojourners p. 194
total fertility rate p. 176
undocumented, irregular, or illegal immigrants p. 193
workers' remittances p. 206
zero population growth p. 177

QUESTIONS FOR REVIEW AND DISCUSSION

1. Briefly describe the distribution of the human population.

2. Explain the demographic transition model, then explain the new developments in demographics that challenge the continuing validity of that model.

3. What current developments may increase the world's death rate?

4. What have been the major currents of the African diaspora?

5. Who and where are the "Overseas Chinese"? Why are these people of such importance in the world today?

6. Differentiate a refugee from an economic refugee.

7. What is the Malthusian theory?

THINKING GEOGRAPHICALLY

1. How has physical geography affected the distribution of people in any given continent or region?

2. Investigate any country's family-planning efforts and experience.

3. Do you know any elderly people who have retired and are now collecting pensions? Where did they spend their working years? Where are they living now?

4. Do you know how many generations of your family have moved either to this country or within this country? What push and pull factors motivated them?

5. Has your local community seen net in-migration or net out-migration in recent years? If in, where from? If out, where to? What push and pull factors did they consider?

6. Compare the cartogram in Figure 5-2, countries drawn according to their population size, with Figure 12-5, countries drawn according to their economic output, with a standard area world map. Do you think that one or the other of these criteria—population, output, or area—ought to determine the amount of time you devote to each country in your studies? Which ought to determine how much time on the evening news is devoted to each country? Can you suggest any other measures of each country's "importance"?

7. Are any major refugee streams reported in this week's headlines? If so, what are the push factors?

SUGGESTIONS FOR FURTHER LEARNING

Ager, Alastair. *Refugees: Perspectives on the Experience of Forced Migration.* New York: Cassell Academic, 1999.

Alba, Richard, and Victor Nee. *Remaking the American Mainstream: Assimilation and Contemporary Immigration.* Cambridge, MA: Harvard University Press, 2003.

Boyle, P. J., Keith Halfacree, and Vaughan Robinson. *Exploring Contemporary Migration.* New York: Addison-Wesley, 1998.

Bradshaw, York, and Stephen Ndegwa, eds. *The Uncertain Promise of Southern Africa.* Bloomington: Indiana University Press, 2000.

Brettell, Caroline, James Hollifield, and Barry Chiswick, eds. *Migration Theory: Talking Across Disciplines.* New York: Routledge, 2000.

Chaliand, Gérard, and Jean-Pierre Rageau. *The Penguin Atlas of Diasporas.* New York: Viking Penguin, 1995.

Cohen, Roberta, and Francis M. Deng, eds. *The Forsaken People: Case Studies of the Internally Displaced.* Washington, D.C.: Brookings Institution, 1998.

Diamond, Jared. *Guns, Germs and Steel: The Fates of Human Societies.* New York: W.W. Norton & Co., 1999.

Dummett, Michael. *On Immigration and Refugees.* New York: Routledge, 2001.

Garrett, Laurie. *Betrayal of Trust: The Collapse of Global Public Health.* New York: Hyperion Press, 2000.

Goldin, Liliana, ed. *Identities on the Move: Transnational Processes in North America and the Caribbean Basin.* Austin, TX: University of Texas Press, 2000.

Hayter, Teresa. *Open Borders: The Case Against Immigration Controls.* New York: Pluto Press, 2001.

Honari, Morteza, and Thomas Boleyn, eds. *Health Ecology: Health, Culture, and Human-Environment Interaction.* New York: Routledge, 1999.

Hyndman, Jennifer. *Managing Displacement: Refugees and the Politics of Humanitarianism.* Minneapolis: University of Minnesota Press, 2000.

Livi-Bacci, Massimo. *A Concise History of World Population: An Introduction to Population Processes, 3rd ed.* New York: Blackwell, 2001.

Lutz, W., B. C. O'Neill, and S. Scherbov. "Europe's population at a turning point." *Science* 299 (2003): 1991–1992.

Malcolmson, Scott. *One Drop of Blood: The American Misadventure of Race.* New York: Farrar, Straus & Giroux, 2000.

Menjivar, Cecilia. *Fragmented Ties: Salvadoran Immigrant Networks in America.* Berkeley: University of California Press, 2000.

National Academy of Sciences, National Research Council. *The New Americans: Economic, Demographic, and Fiscal Effects of Immigration.* Washington, D.C.: The Academy, 1997.

Newman, James L. *The Peopling of Africa: A Geographic Interpretation.* New Haven: Yale University Press, 1995.

Pagden, Anthony. *Peoples and Empires.* New York: Modern Library, 2003.

Population Reference Bureau. *Population Bulletin.* Washington, D.C., quarterly.

Preston, Samuel, Patrick Heuveline, and Michel Guillot. *Demography: Measuring and Modeling Population Processes.* Malden, MA: Blackwell, 2001.

Rumbaut, Ruben, and Alejandro Portes, eds. *Ethnicities: Children of Immigrants in America.* Berkeley: University of California Press, 2001.

Salgado, Sebastiao. *Migrations: Humanity in Transition.* New York: Aperture, 2000.

Segal, Ronald. *Islam's Black Slaves.* New York: Farrar, Strauss and Giroux, 2001.

Shinagawa, Larry, and Michael Jang, eds. *Atlas of American Diversity.* Walnut Creek, CA: Altamira Press/Sage Publications, 1998.

Simon, Julian, ed. *The State of Humanity.* Cambridge, MA: Blackwell, 1995.

Sowell, Thomas. *Migrations and Cultures: A World View.* New York: Basic Books, 1996.

U.S. Committee for Refugees. American Council for Nationalities Service. *World Refugee Survey*. Washington, D.C., annually.

Weeks, John R. *Population: An Introduction to Concepts and Issues, 7th ed.* New York: Wadsworth, 1998.

Weinstein, Jay A., and Vijiyan Pillai. *Demography: The Science of Population*. Boston: Allyn & Bacon, 2000.

Whitman, Jim. *Migrants, Citizens and the State in Southern Africa*. New York: St. Martin's Press, 2000.

World Health Report. Geneva: World Health Organization, annually.

World Population Prospects. New York: United Nations Population Division, periodic.

WEB WORK

A good place to start looking for information about population, health, and migration is at the United Nations central page for population studies:

http://www.un.org/esa/population/unpop.htm

The U.S. Census Bureau is at

http://www.census.gov/

And it also maintains a special page for its International Programs Center

http://www.census.gov/ftp/pub/ipc/www/

Two private organizations maintain informative home pages with links to other interesting sites. They are the Population Reference Bureau

http://www.prb.org/

And the Population Connection

http://www.populationconnection.org/

The World Health Organization, a U.N. agency, can be found at

http://www.who.int/en/

World diseases and epidemics can be followed at the Web page that the U.S. Centers for Disease Control and Prevention maintains for travel warnings

http://www.cdc.gov/travel/

The International Organization for Migration, an intergovernmental body that tracks migration figures, can be found at

http://www.iom.int/

The Migration Information Service is at

http://www.migrationinformation.org/

The home page for the United Nations High Commissioner for Refugees can be found at

http://www.unhcr.ch/cgi-bin/texis/vtx/home

For immigration to the United States, the Bureau of Citizenship and Immigration Services is now part of the U.S. Department of Homeland Security

http://www.bcis.gov/graphics/index.htm

And for Canada, always try Statistics Canada at

http://www.statcan.ca/start.html

A Look Ahead

CULTURAL EVOLUTION CONTRASTS WITH CULTURAL DIFFUSION

Evolutionism studies the sources of cultural change that are within individual cultures themselves, whereas diffusionism emphasizes how cultures spread out from the places they originate and are adopted by other peoples. Folk cultures preserve traditions, but popular cultures embrace changing norms.

IDENTITY AND BEHAVIORAL GEOGRAPHY

The behavior of any individual is partly unique to that person, but much of any person's behavior is specific to his or her cultural group. People often group themselves and interact with other groups defined by culture, race, ethnicity, or identity, even though it may be difficult for an observer to identify these

CULTURAL GEOGRAPHY

◀ These Americans of German ancestry are marching in New York City's annual Von Steuben Day Parade, commemorating the contributions of German-Americans to the United States. General von Steuben was a hero of the U.S. War of Independence. Cultures change, but Americans of many backgrounds cling to historic cultural traditions even in the country known as the "melting pot." Studying the evolution and diffusion of cultures is key to understanding human geography. (AP/Wide World Photos)

groupings "objectively." Geographers study how individuals and groups perceive and respond to their environments.

CULTURE REALMS

The entire region throughout which a culture prevails is called a culture realm. Culture realms may be defined on the basis of religion, language, diet, customs, economic development, or still other criteria. Some aspects of culture realms are visible in the landscape.

THE GLOBAL DIFFUSION OF WESTERN CULTURE

Much of today's world cultural geography is the legacy of the diffusion of European culture that accompanied European political and economic global supremacy from about 1500 until about 1950. The attacks on the United States that occurred on September 11, 2001, shocked Americans and other global citizens into reflection on America's role in the world today.

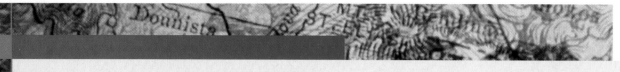

Imagine that deep in the Brazilian rain forest there is a human community that is completely isolated. The people carry on their lives in total ignorance of the rest of us here on the planet, knowing nothing of international trade, of international political affairs, or even of the government of Brazil, which exercises no effective jurisdiction over them. Certainly they are affected by the complicated affairs of the rest of us. They see our airplanes in the sky, and their local environment is changing because the rest of us are polluting the atmosphere, but they explain these things either as natural phenomena or as the work of spirits. These isolated people—let's call them the "Solitary" people—believe that they are the only humans on the Earth (Figure 6-1).

The Solitary people would have developed their own way of life, including a language to communicate among themselves, a religion, and Solitary ways of organizing their families and society. They must depend entirely on their local environment for all of their needs, including food, clothing, and shelter. Thus, for example, the Solitary people would have learned over time which local plants and animals can be eaten. If the Solitary people receive no imports from the outside world, neither do they produce anything for others. Everything produced locally is consumed locally, and any surplus of food or goods is stored for emergencies or else just wastes away. Activities are probably closely related to the seasons.

As the Solitary people make use of their local environment, they can also transform it. They can clear forests or plant new ones, drain swamps or dig irrigation systems, fertilize the soil or deplete and waste it through poor agricultural practices (Figure 6-2). If they burn local vegetation, they make their own contribution to world atmospheric pollution. They can even redesign local landforms, as many societies have reduced steep mountainsides to stepped terraces for agriculture. People are never entirely passive; rather, they interact with their local environment.

FIGURE 6-1 A discovery of an isolated people. This haunting photograph records the first contact between these formerly isolated people and the rest of humankind. On August 4, 1938, a U.S. scientific expedition entered the supposedly uninhabited Grand Valley of the Balim River in western New Guinea. They found over 50,000 people living in Stone Age circumstances totally unaware of the existence of anyone else on Earth. We cannot say for certain whether any totally isolated groups live on Earth today, but it is less probable each year. (Courtesy of the Department of Library Services, American Museum of Natural History)

FIGURE 6-2 A cultural landscape. This aerial photograph of the Grand Valley of the Balim was taken a few weeks before the expedition actually contacted the valley's inhabitants. It must have been a surprise to look down into a supposedly uninhabited valley and see these neat clearings and settlements outlined by irrigation ditches. This was clearly a cultural landscape, one modified by considerable human effort. (Courtesy of the Department of Library Services, American Museum of Natural History)

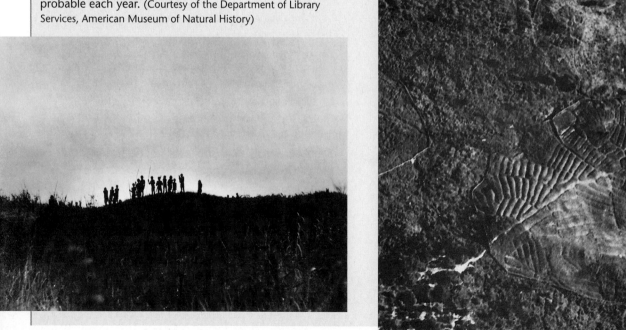

These imaginary Solitary people will have developed in isolation their own distinctive culture, and they will have constructed their own distinctive cultural landscape, as both of these terms were defined in Chapter 1. This chapter explores how cultures originate, develop, and spread.

CULTURAL EVOLUTION CONTRASTS WITH CULTURAL DIFFUSION

The word culture was introduced in Chapter 1 to describe everything about the way a people live: their clothes, diet, articles of use, and customs. Geographers study the origin, diffusion, and extent of all aspects of cultures. An object of material culture is called an *artifact,* which means literally "a thing made by skill." Culture, however, includes patterns of behavior as well as possessions. The way a group does things—its interpersonal arrangements, family structure, educational methods, and so forth—are part of its culture, and so are the facts of a group's mental or imaginative life—its poetry, music, language, and religion.

Human cultures are never static; they are constantly changing, as people learn new techniques and develop new cultural traits. The forces that cause these changes may be divided into two general categories. Some evolve within the society itself, but others are triggered by contact with other societies. The Solitary people, for example, were isolated, so their culture would change only through internal evolution. They might discover new uses for local plants through the years, or they might devise new religious ceremonies. **Evolutionism** is the point of view that the most important sources of cultural change are embedded in cultures, and change is internally determined. **Diffusionism,** by contrast, emphasizes how various aspects of cultures spread out from the places they originate and are adopted by other peoples. The process of spreading is called **cultural diffusion,** and the process of adopting some aspect of another culture is called **acculturation.** Cultural diffusion can be actively imposed, as when an outside power conquers a region and imposes its way of life, or it can be freely chosen, as when one group discovers and adopts some aspect of a different culture that it considers superior to its own.

This chapter will begin with an examination of two classic evolutionist theories. Both emphasize the role that technological and economic evolution play in the development of cultures and in the interaction between societies and the environment. Both theories have been used to suggest that certain developments in the past were inevitable, or even that certain events are inevitable in the future.

The second part of this chapter will then analyze diffusionistic approaches to the study of culture.

Through history, the diffusion of goods, ideas, and techniques among peoples has increased, and the original individual characteristics of world cultures have come to be overlain or commingled with new shared characteristics. This mingling has created new and original cultural combinations.

The last section of this chapter analyzes the worldwide diffusion of European culture that accompanied European global conquest and economic dominance. European political dominance has retreated, but it left behind cultural, political, and economic legacies. The United States inherited Europe's leadership, and today many people are reevaluating America's global role.

Theories of Cultural Evolution

Humankind has, for the most part, adapted itself to varying environmental conditions through cultural and technological evolution, not through biological evolution. For example, humans survive Minnesota winters by inventing furnaces and warm clothes, not by evolving furry bodies. Our cultural or technological tools allow us to survive anywhere. Scientific research stations can be maintained even in the frozen wilderness of Antarctica or in the driest desert, although supplies may have to be brought to the local residents. Thus, humans exist in every biome.

The theory of human stages   One theory of cultural evolution argues that all cultures evolve through certain stages of development, and it defines these stages by the way in which the culture exploits the environment. This theory was first articulated by a Roman general, Marcus Tarentius Varro (116–27 B.C.), and it remained almost unchallenged until the nineteenth century.

Varro argued that humankind originally derived its food from things that people hunted or harvested naturally. He referred to humans in this stage of evolution as **hunter-gatherers.** Varro stated that humans then domesticated animals and moved into the evolutionary stage of **pastoral nomadism.** Pastoral nomads have no fixed residences, but drive their flocks from one place to another to find grazing lands and water. If their movements are regular and seasonal—as, for example, between mountain and lowland pastures—their movements are called *transhumance.* The evolutionary stage of pastoral nomadism was in turn followed by settled agriculture. Agriculture was at first *subsistence agriculture,* which means that people raised food only for themselves, but subsistence agriculture slowly evolved into *commercial agriculture,* which is raising crops to sell (see Chapter 8). The final stages of social and economic evolution were urbanization and industry.

Varro's theory was generally accepted until the German naturalist and explorer Alexander von Humboldt (1769–1859), whom we met in Chapter 1, pointed out that the native peoples of South America had never experienced a stage of pastoral nomadism.

Therefore, Varro's theory could not be true of all societies everywhere. The different ways of life described by Varro might be "categories" of human societies at different times and places, but they are not "stages" in one only possible path of social evolution.

Today there are still human groups living in each category described in this theory of cultural evolution, and we will look at them in greater detail in Chapter 8. Varro's theory that all human societies advance through the same series of stages, however, carries a danger. According to this unilinear theory, societies cannot just be "different" from each other; some must be "ahead" and others "behind." Such presumptions have been used by "advanced" peoples to justify replacing "more primitive" peoples. For example, in the nineteenth century many Americans argued that it was inevitable that the Native North Americans' way of life (pastoral nomadism and subsistence agriculture) would be destroyed and replaced by American farmers. Thus, any activity causing this replacement was morally neutral. Other thinkers, however, argued that human behavior is based on conscious choices. Therefore, the results of human actions can never be scientifically inevitable and morally neutral.

Historical materialism

Historical materialism is a school of thought that tries to write a plot for human history based on the idea that human technology has increased humankind's control over the environment. Karl Marx (1818–1895) was the founder of historical materialism.

Marx said that humankind has progressively conquered the physical environment in order to improve material welfare. A contemporary ruler may or may not be wiser than one in the past, but a modern tractor can do more work than a horse dragging a wooden plow. Therefore, historical materialists assert, history's plot is the advance of technology. Marx insisted that all social evolution is rooted in technological evolution. This is because technology determines any society's economic system, and the economic system, in turn, determines the society's political and social life (Figure 6-3). Thus, as technology advances, technological change triggers changes in all these other aspects of society. Marx's argument that all aspects of a society are rooted in its technology and economic system has always aroused great controversy.

Marx was an optimist, and he agreed with Condorcet (see Chapter 5) that technology would eventually produce material abundance for all people. Goods would no longer be scarce, so people would no longer dispute their allocation through economic and political structures. Goods would be created and distributed "from each according to his abilities; to each according to his needs."

Some of what Marx wrote has turned out to be misguided guesses, and some of it is simply false, but much of what Marx wrote is generally accepted. Today no

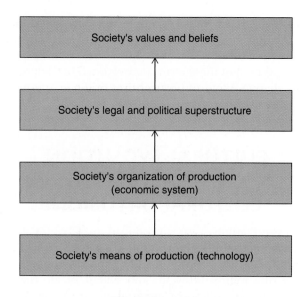

The Environment
(from which humankind must wrest material welfare)

FIGURE 6-3 The theory of historical materialism. Historical materialists believe that any society's productive technology determines its economic system and that its economic system determines its political system. Ultimately, even a society's values and beliefs rest on its technological foundation. Technology is always improving, so all the other aspects of any society are constantly being dragged along and forced to change, too. Technology is the truly revolutionary force in society.

one doubts that science progresses almost inexorably, and that scientific progress often upsets society's economic, political, and even philosophical assumptions. The cloning of a lamb and the birth of another with human genes in 1997, for example, challenges our laws and even our ideas of life. Many of the political movements that have used Marx's name, however, disgrace his memory. The communist parties that seized power in Russia in 1917 and were imposed on other peoples in the twentieth century bear no relationship to the humanitarian society that Marx envisioned material abundance would bring.

Historical materialism contrasts with Malthusianism

Recently some economists and ecologists have attacked historical materialism from a new direction. Historical materialism notes that in the past, technological progress offered rising standards of living to human populations, and historical materialists assume that this rate of technological advance will continue. This optimism characterizes some of our leading scientists and industrialists today, who would be classified among the Cornucopians discussed in Chapter 5.

Today many concerned scientists and critics, however, belong to the opposing neo-Malthusian camp. They warn that there is no guarantee that technology will continue to provide rising standards of living for the increasing human population. Confidence that technology will always solve problems of scarcity might lead humankind to disaster. Marx himself never imagined the explosive growth of the human population that has occurred since his death. At that time the human population was only about 1.5 billion, and it was growing at an annual rate of less than 1 percent. The human population today is over 6 billion, and it is rising at an annual rate of just over 1 percent. The neo-Malthusians ask whether increase of the human population is, in fact, outpacing technological development or whether it might do so in the future.

These are some of the most controversial arguments of our time, and you will hear echoes of both the neo-Malthusian and the Cornucopian positions throughout this book. People of different political and economic philosophies offer different answers, and by the time you finish reading this textbook, you will have more information on which to form ideas of your own.

> The choices and constraints involved in utilizing the natural environment are often as much cultural, economic, political, and social as they are technological.

Cultures and Environments

Marx wrote that the key to understanding a society is the degree of control that society has over its environment. This degree of control is more important than the nature of the environment itself. Some writers, however, argue that variations in the physical environment itself explain even the variations among human cultures. The simplistic belief that human events can be explained entirely as the result of the effects of the physical environment is called **environmental determinism.** The study of the ways societies adapt to environments, by contrast, is called **cultural ecology.**

Many writers have emphasized climate as a major control on human affairs. The Greek physician Hippocrates (460?–377? B.C.) argued in his book *Airs, Waters, Places* that civilization flourishes only under certain hospitable climatic conditions. Historian Arnold Toynbee (1889–1975) turned Hippocrates' idea that a hospitable climate nourishes civilization upside down with his own *challenge-response theory.* Toynbee argued that people need the challenge of a difficult environment to put forth their best effort and to build a civilization. A rich environment encourages only sluggishness. Today whenever anyone expresses a preference for an environment of seasonal change, as in the northern United States, over the almost tediously fine weather of southern California, that person is suggesting the challenge-response theory.

Other aspects of the natural environment have been nominated as the key factor determining a people's life. West Virginia's state motto, "Mountain people are always free," echoes a belief that rugged topography creates rugged individualists. Some political scientists have argued that societies in environments that are

either particularly wet or particularly dry have necessarily developed totalitarian governments to organize and care for waterworks. Many environmentally deterministic theories are intriguing, but exceptions can always be found. For example, the Dutch have managed their waterworks democratically for hundreds of years (Figure 6-4).

The ability to find exceptions to any environmentalist theory should warn us that no simple theory completely explains the relationship between the environment and human societies. Human affairs are not simple, and when we examine any world situation or historic event carefully, environmental determinism, or any other single-factor explanation, proves insufficient. Nineteenth-century French geographers proposed the theory of **possibilism** as an antidote to environmental determinism. Possibilism insists that the physical environment itself will neither suggest nor determine what people will attempt, but it may limit what people can profitably achieve. The choices and constraints involved in utilizing the natural environment are often as much cultural, economic, political, and social as they are technological. For example, Canadians have the technological capability to overcome environmental limitations and to grow bananas in greenhouses in Canada, but they choose not to do so. It is cheaper for them to buy bananas grown in Costa Rica.

Environmental determinism today Assumptions about the physical environment's influence in human affairs still affect economic and political debates. For example, today many of Earth's tropical regions are poor, and there are three main hypotheses about why they are poor.

1. One theory emphasizes that the tropical environment handicaps human societies. Tropical heat multiplies the activity of organisms that are inimical to people and to agriculture, and it reduces human work efficiency. Tropical storms frequently are violent, and many tropical soils are poor.

2. Another school of thought, however, points out that many tropical areas were long held as colonies by countries in the temperate latitudes. This colonial experience left physical, cultural, and economic legacies that, combined with the continuing patterns and terms of world trade (discussed in Chapter 12), best explain tropical regions' current poverty.

3. A third point of view insists that there is something about the cultures of the peoples of the tropical regions that retards their economic growth. Perhaps their cultures do not encourage economic growth or these countries are poorly governed.

The Netherlands vs. the Sea

FIGURE 6-4 More than half of the Netherlands lies below sea level, so most of the country today would be under water if it were not for massive projects to modify the environment. Since the thirteenth century the Dutch have reclaimed more than 4,500 square kilometers (1,800 square miles) of *polders,* which are pieces of land created by draining water. Most polders are reserved for agriculture, although some are used for housing, and one contains Schiphol, Europe's busiest airport. The second distinctive modification of the environment is the construction of massive *dikes* to prevent the North Sea from flooding much of the country. The Dutch have built dikes in two major locations, the Zuider Zee project in the north and the Delta Plan project in the southwest. With these massive projects finished, attitudes toward the environment have changed in the Netherlands. A plan adopted in 1990 calls for returning 26 square kilometers of farms on polders to wetlands or forests. In the infrared satellite photograph showing the northern area of the map (b), the vegetation shows as red, more developed areas as green. ([a]: From W. H. Renwick and J. M. Rubenstein, *An Introduction to Geography.* Upper Saddle River, NJ: Prentice Hall, 1995; [b] Science Photo Library/Earth Satellite Corporation/Photo Researchers, Inc.)

Each side in this debate may be partly correct. Which side you take will determine how you allocate responsibility for the current poverty. The first theory faults uncontrollable environmental factors; the second faults the rich, formerly colonial powers; and the third faults the peoples and governments of the tropics themselves. Furthermore, your assumptions will determine what solutions you suggest to the problem of poverty in the tropics. If the environment is to blame, nothing can be done. If the former colonial powers are to blame, perhaps they should pay reparations. If the tropical peoples themselves are to blame, they should change their habits or work harder.

All forms of environmentalist assumptions remain common in popular discussion and debate. Often they seem to be harmless figures of speech, but they are intellectually dangerous. They predispose conclusions and close the mind to alternative explanations. Watch carefully for these simplistic traps.

Cultural Diffusion

The isolation experienced by the imaginary Solitary people is very different from the way most of us live today. Through history, global communication and transportation have increased, and trade and other cultural exchanges have multiplied. Most people no longer exploit their local environments and develop their cultures in isolation. They are interconnected by transportation and communication of goods, people, ideas, and capital. More people eat imported foods and combine them in imported recipes. They wear imported clothes in imported styles, and they fashion their built environments out of imported materials in styles that originated among distant peoples. Your daily life undoubtedly exemplifies this interconnectedness. Many articles that you use and other aspects of your activities draw on materials and ideas from around the world. To describe all this movement, we use the term *circulation.*

FIGURE 6-5 Manufactured clothes. This little boy wears a traditional robe, as is legally required in Bhutan, the world's most isolated country. Bhutan struggles to preserve its traditional culture, and it allows only a limited number of foreigners to visit each year. The collar of a modern blue shirt, however, peeks out from underneath the robe. The shirt, the doll, and other aspects of his attire and possessions are clearly Western imports. It may be that traditional Bhutanese culture has become only a façade. (Courtesy of Brett Cobb)

Almost everywhere today, cultural diffusion is more important than cultural evolution. Anthropologist Clifford Geertz wrote in 1995, "The very notion of isolation lacks, these days, much application. There are very few places—there may not be any—where the noises of the all-over present are not heard, and most anthropologists work by now in places where such noises all but drown out local harmonies." What Geertz calls "the noises of the all-over present" is cultural diffusion, and those "noises" overwhelm what he calls "local harmonies," which is local cultural evolution. Find a picture of the most isolated people you can, and chances are high that they will be wearing or carrying something that they did not make themselves (Figure 6-5). What happens *at* places depends more and more on what happens *among* places, and mapped patterns of

> **What happens *at* places depends more and more on what happens *among* places. . . . Today geography doesn't just *exist*; it *happens*.**

economic or cultural factors can be understood only if we understand the patterns of movement that create them and continuously rearrange them. Today geography doesn't just *exist*; it *happens*.

The American anthropologist Clark Wissler (1870–1947) called the places where cultures are developed *geographical culture centers*. He stated a principle, called the *age-area principle*, that if traits diffuse outward from a single geographical culture center, the traits found farthest away from that center must be the oldest traits. This description of cultural diffusion suggests concentric waves spreading out from a stone dropped into a pool. Geographer Carl O. Sauer (1889–1975) elaborated geographical studies of cultural diffusion. Chapter 1 explained that today we call the place where a distinctive culture originates the hearth area of that culture, and Chapter 1 outlined the paths of cultural diffusion. Geographers can investigate the distributions of any cultural attributes; they are limited only by their imagination and curiosity.

Diffusion does not explain the distribution of all phenomena. Sometimes the same phenomenon occurs spontaneously and independently at two or more places. In the history of mathematics, for example, the idea of the zero and its use as the basis of a numerical place system was conceived in two places: among the Maya, a Native Central American people, and among the ancient Hindus of India or perhaps the Babylonians before them. Diffusionists once argued that Phoenicians must have sailed across the Atlantic from their homeland in the eastern Mediterranean thousands of years ago and taught the Maya how to use the zero. There is, however, no evidence for this assumption, and it underestimates the ingenuity of the Maya. To demonstrate diffusion, we must be able to illustrate its path from one culture to another.

Folk culture   Today's rapid pace of cultural innovation and diffusion requires us to differentiate folk culture from popular culture. The term **folk culture** refers to a culture that preserves traditions. Folk groups are often bound by a distinctive religion, national background, or language, and folk cultures are conservative and resistant to change. Most folk-culture groups are rural, and relative isolation helps these groups maintain their integrity. Folk-culture groups, however, also include urban neighborhoods of immigrants struggling to preserve their native cultures in their new homes. The term folk culture suggests that any culture identified by the term is a lingering remnant of something that it is embattled by the tide of modern change.

Cultural geographers have identified culture hearths and routes of diffusion of a surprising number of folk cultures across the United States. Geographers

Regional Focus on
LAHIC

The village of Lahic is one of the world's most extraordinary examples of an isolated community that has created and maintained a distinctive local culture. Lahic is a village of some 2,000 people, perched high on a ragged cliff surrounded by the towering Caucasus Mountains in the Republic of Azerbaijan. Nobody knows when this village was first settled or who the first settlers were, but the unique local language provides a clue. It is a form of Persian (Iranian) that died out in Iran hundreds of years ago. Perhaps some ancient Iranian emperor sent settlers to Lahic to defend the empire's borders against the peoples to the North.

Still today Lahic is connected to the outside world by only a rugged dirt path that winds its way around precipitous gorges and is closed by rain and snow much of the year. Wires from Baku, the capital of Azerbaijan some 128 kilometers (80 miles) away, brought electricity and telephone service to the village in the 1980s, but services are provided only a few hours a week.

The desolate landscape of Lahic allows the people to raise livestock, but they cannot feed themselves, so long ago they developed a tradition of exchanging craftwork for food. In their isolation, the people of Lahic developed artisanry skills that have made their handcrafted products famous and prized around the world. Virtually every kind of handcrafted object has been made here and exported for hundreds of years: objects in copper and brass, carpets, handcarved and wrought iron, guns, cutlery, clothing, and leather goods. Every home is a workshop. Lahic crafts have long been treasured by collectors throughout the Middle East and Europe. Traders discovered them many centuries ago and sold them for high prices in bazaars in Baghdad, Shiraz, and other great Middle Eastern cities. A display of copperware made in Lahic won a gold medal at the Paris World Exposition in 1878, and today Lahic crafts are on display in museums from London and Paris to Moscow and Istanbul.

A 1997 U.N. report on Azerbaijan noted that "Lahic . . . has kept almost perfectly intact a microcosm of cultural features which has preserved its communal harmony and social cohesion across the centuries. The spirit of the Middle Ages still lingers there."

These days, more young people are going off to school or to find new job opportunities in Baku and thereby abandoning their village's ancient crafts. How much longer can this time capsule survive?

Traditional artisan workmanship. This artisan in Lahic, Azerbaijan, demonstrates the skills that have earned renown for products from his isolated village and kept its economy thriving for hundreds of years. (Stanton R. Winter/New York Times Pictures)

who have investigated the architecture of houses, for example, have differentiated three types and traced the diffusion paths of these styles westward with various immigrant groups (Figure 6-6). Barns and other structures are also built in distinct architectural styles that reveal the origins of their builders. Folk geographic studies in the United States range from studies of folk songs, folk foods, folk medicine, and folklore to objects of folk material culture as diverse as locally produced pottery, clothing, tombstones, farm fencing, and even knives and guns.

In North America the Amish provide an example of a folk culture. The Amish are a sect within the Mennonite Protestant religious denomination formed in Switzerland in the sixteenth century. Early Amish immigrants to North America settled in Pennsylvania and spread across the Midwest and into Canada. The Amish stand out because they wear plain clothing and shun modern education and technology. They prosper by specializing their farm production and marketing their produce, but they severely curtail the choice of goods that they will accept in return (Figure 6-7).

Many African Americans have long felt that their traditional folk cultures were taken from them during the years they suffered in slavery, yet careful investigation has yielded a rich variety of cultural elements in some African-American communities today that are traceable to specific African origins. These range from

Paths of Diffusion of House Types

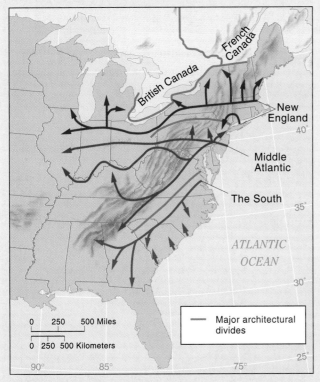

FIGURE 6-6 Geographers have identified three main styles of houses that originated on the U.S. Eastern Seaboard and diffused westward along with pioneer settlers. (From Rooney, Zelinsky, and Louder, eds., *This Remarkable Continent*. College Station, TX: Texas A & M Press, 1982.)

nate anywhere, and it tends to diffuse rapidly, especially wherever people have time, money, and inclination to indulge in it.

Popular material culture usually means mass culture—that is, items such as clothing, processed food, books, CDs, and household goods that are mass produced for mass distribution. Whereas folk culture is often produced or done by the people at-large (folk singing and dancing, cooking, costumes, woodcarving, etc.), popular culture, by contrast, is usually produced by corporations and purchased. It is, therefore, often related to social class, as defined by income and education.

Mass manufacturing lowers the cost of items, but in order for it to succeed, consumer taste must be homogeneous. Such "mass taste" necessarily requires some sacrifice of individuality and cultural identity. The United States has long been the world's largest relatively homogeneous consumer market. This has lowered the cost of items and raised Americans' material standard of living. The substantial unity of the Canadian and U.S. markets has raised Canadian standards of living, too,

FIGURE 6-7 Popular culture mingles with a folk culture. The straw hat, suspenders, and plain black boots identify this Pennsylvania boy as Amish, but his in-line skates demonstrate how most traditional folk cultures allow adoption of some aspects of modern popular culture. (Keith Myers/New York Times Pictures)

musical styles to the meticulous sweeping of front yards (a tradition traced from villages of coastal Nigeria to rural Georgia) to the patterns of speech discussed in Chapter 7.

The United States is relatively homogeneous in culture, modern and technologically advanced. Cultural diffusion and change churn across the country rapidly and repeatedly. In such a country the identification and study of folk cultures help us appreciate and cherish the richness of some of our remaining folk traditions. Chapter 5 noted that Canada, by contrast, more carefully nurtures and preserves its diversity as a "cultural mosaic."

Some newly independent countries are today emphasizing or even resurrecting traditional folk cultures as a way of enhancing national identity and community. When the nation of Eritrea won independence in 1993, for example, one of the new government's first acts was to open a national university mandated to research and define folk literature, folk music, and other distinct Eritrean traditions.

Popular culture   Popular culture, by contrast, is the culture of people who embrace innovation and conform to changing norms. Popular culture may origi-

Focus on

THE DIFFUSION OF ANGLO-AMERICAN RELIGIOUS FOLK SONGS

The map of the diffusion of white spiritual songs from eighteenth-century New England exemplifies folk cultural geography. These songs almost seem to have "slid" down the Appalachian Mountains into the Upland South and then into the Lowland South, where many of them are still popular. Some individual migrants carried the songs westward with them, but for the most part the acceptance of these songs diffused from community to community. They have a simple melodic structure, which reveals their folk quality, and their first person narrative reveals their Protestant Christian faith (for example, "Don't you see my Jesus coming? He'll embrace me in His arms"). These songs did not diffuse in other directions, because of cultural barriers. Further to the south, the songs' Protestant religious sentiment hindered their diffusion into largely French Roman Catholic Louisiana and Hispanic Roman Catholic southern Texas. To the north, both religious and political barriers prevented their diffusion. Quebec in Canada is also largely Roman Catholic.

At the same time as these songs were spreading southwestward, they largely disappeared from their culture hearth in the Northeast. This was probably because that region urbanized rapidly and its folk culture was replaced by a popular culture.

The Diffusion of Protestant Religious Folk Songs

Protestant religious folk songs diffused from New England toward the southwest until stopped by cultural barriers. (Reproduced by permission of the American Folklore Society from *Journal of American Folklore* 65:258 [Oct.–Dec. 1952]. Not for further reproduction.)

but some Canadians believe that it has threatened Canadian cultural identity.

The popular culture of the United States exhibits a great degree of national homogeneity, but individual consumer goods win varying degrees of acceptance (called *market penetration*) in different regions of the country. The regions are often the vernacular regions mapped as Figure 1-8. The national marketing managers of corporations might call themselves "applied cultural geographers" who specialize in this type of analysis. The soft drink Dr Pepper, for example, originated in the U.S. South, and it still enjoys its greatest acceptance there. The U.S. South is today less culturally distinct than it was decades ago, but a cultural heritage, much of which lingers from earlier folk traditions, still exists (Figure 6-8).

Geographers investigate the origins and diffusion paths of popular material culture and also of popular social culture. For example, people devise new ways of living, working, and playing, and they innovate in education and in employer-employee relations. Sport is an important part of popular culture, and popular culture regions can be defined by sporting preferences (Figure 6-9). Regions differ in their popular entertainments—a movie that is a success in Seattle, for example, may flop in Houston. Cultural geographers want to know why. The radio stations in different regions across the United States play varying mixes of country, gospel, rock, classical, and other popular kinds of music; DJs base their decisions on market research such as that shown in Figure 6-10 (p. 231). Geographers have investigated these and many more attributes of popular culture.

The marketing of popular culture can overwhelm folk culture. How many of the characters and tales of traditional U.S. folk culture—or even of genuine historical heroes and heroines—have survived the onslaught of mass-produced and mass-marketed merchandise related to comic-strip characters? Who is most famous in the United States today: Paul Bunyan

Coke vs. Pepsi

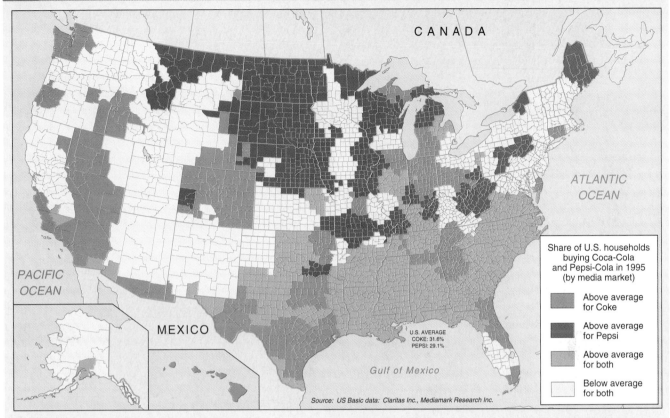

FIGURE 6-8 Both Coca-Cola and Pepsi were born in the U.S. Southeast in the nineteenth century—Coca-Cola in Atlanta, Georgia, and Pepsi-Cola in New Bern, North Carolina—and each is consumed in the region at a rate above the national average. Perhaps the warm climate explains that fact. Nationwide, Coke drinkers tend to be urban and financially better off. Pepsi drinkers are somewhat older, suburban, and rural. Alaska and Hawaii are not to the same scale. (*U.S. News & World Report,* August 5, 1996)

(a mythical folk hero), Johnny Appleseed (a real person, John Chapman, 1774–1845), or the Lion King, a cartoon product of popular culture? The answer is probably the Lion King.

IDENTITY AND BEHAVIORAL GEOGRAPHY

At any time a great many disputes disturb the world's peace. Widespread international wars, civil wars, urban riots, insurrections, and street fighting persist. The press often refers to the parties to the disputes as being of different culture groups, races, ethnic groups, or other identities, but outsiders to the disputes find it difficult to understand the bases for animosity. To understand the background of many contemporary news items, it is important to understand what those categories mean and how they influence people's behavior. When geographers study identities, perception,

and patterns of behavior, geographers' work borrows ideas from the study of psychology but also makes fresh contributions.

Grouping Humans by Culture, Race, Ethnicity, and Identity

When we define human groups, we must be sure that we know and agree on exactly what the criteria of inclusion are and whom the group includes. Unclear or vague references to groups may cause misunderstanding or even insults.

Culture groups   The definition of a culture may include a great number of characteristics or just a few. For example, all social scientists agree that language is an important attribute of human behavior. Two people who share a language share something very important. If, however, those two people hold different religious beliefs, feel patriotism for two different countries, and

Television Baseball Viewing

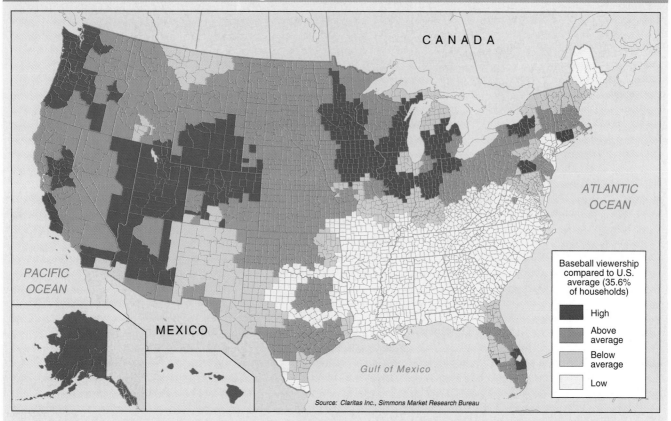

FIGURE 6-9 Several hypotheses have been offered to explain this pattern of television baseball viewership. The game's relative unpopularity in the South and in metropolitan areas may be partly explained by baseball's weak appeal among blacks, which can be attributed to the sport's historically racist policies. Furthermore, in metropolitan regions having two or more teams, such as Chicago and New York, fans may be going out to the ballpark rather than watching games on television. Alaska and Hawaii are not to the same scale. (From Michael J. Weiss, *Latitudes & Attitudes,* Boston: Little, Brown, 1994, p. 55)

eat different diets, social scientists may insist that although those two people share one attribute of culture, they do not share one culture. The great number of English speakers around the world today share few attributes other than language, so we would not say that they all share one common culture. If, however, two people do share a language, religious beliefs, political affiliation, and dietary preferences, then social scientists would agree that those two people share a culture.

A *subculture* is a smaller bundle of attributes shared among a smaller group within a larger, more generalized culture group. For example, Italian Americans, Chinese Americans, and African Americans share subsets of cultural attributes within the larger American culture. Sometimes even one single attribute—shared loyalty for a sports team, for example—can bind individuals so strongly that that single attribute is termed a subculture.

Ultimately, cultural affiliation may be a matter of the feelings or the preferences of the individuals. Two people may share so many cultural attributes that observers insist that they share a culture, yet they may

hate each other. They may even kill each other because, *to them,* their differences are crucial. Each day's newspaper proves this statement. In some cases people who share religious beliefs kill each other over other issues. Psychologist Sigmund Freud (1856–1939) described what he called "the narcissism of the minor differences," pointing out that the most vicious and irreconcilable quarrels often arise between peoples who are to most outward appearances nearly identical. Our study of political geography in Chapter 11 will cite examples of countries that outsiders regarded as homogeneous that have nevertheless broken out in civil war. Outsiders did not see the cultural fault lines within the countries. In other cases, people whom observers would describe as very different from each other feel strong bonds. What they share is more important to them than their differences. We must always investigate peoples' feelings in order to understand their loyalties and their animosities.

Races Human beings are one single species, and humans evolved too recently to have accumulated any

Tastes in Music

FIGURE 6-10 Purchasers of classical music compact discs are generally urban, upper-income, and highly educated, whereas a much wider social spectrum and geographic distribution of people buy country music. Alaska and Hawaii are not to the same scale. (From Michael J. Weiss, *Latitudes & Attitudes*, Boston: Little, Brown, 1994, p. 55)

significant amount of genetic change or mutations. Nevertheless, anthropologists have traditionally used secondary biological characteristics to divide all humankind into **races.** The traditional criteria for these subdivisions have been external features such as eyefolds or skin color. The crudest classification system based on these criteria divides humankind into three races: Caucasoid ("white"), Mongoloid ("yellow"), and Negroid ("black").

A second classification system is based on the analysis of blood types. Classifications of people by blood type yields a different set of categories ("races") from classification of people by skin color or other external features. Individuals cross over categories, and the fact that they do casts doubt on the utility of either system.

Geneticists are currently working on a third system of racial classification: differences in people's inherited genetic makeup. These researchers reach conclusions that do not agree with those of any other classification system. Geneticists place the Japanese, Koreans, and Inuits, for example, close to the Caucasoids, although those groups have traditionally been included among Mongoloids.

In fact, it is impossible to make unambiguous distinctions among races. How many races there are, which groups of people are in each, and even the very meaning of the concept of race are all open to question.

Attempts at racial classification have nevertheless often been used to buttress racist ideologies. **Racism** is a belief in the inherent superiority of one race over another and the linking of human ability, potential, and behavior to racial inheritance. Racism is wrong, immoral, and injurious. Racists' search for "racial purity" is nonsense in a world in which people have intermingled as they have throughout human history.

Ethnic groups The concept of an **ethnic group** is frequently confused with the concept of a cultural group. The word *ethnic* comes from the Greek for "people," and the definition of an ethnic group may depend upon almost any attribute of biology, culture, allegiance, or historic background. The word has historically been used in a pejorative sense. Its meanings have included "alien," "pagan," and often "primitive." Some social scientists nevertheless define ethnic groups and study the groups' characteristics or attributes. Ethnomusicology, for example, is the study of ethnic groups'

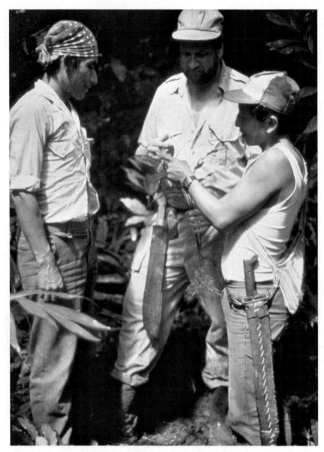

FIGURE 6-11 An ethnobotanist. Dr. Michael Balick of the New York Botanical Garden (center) learns plant uses from these traditional healers of the Kekchi Maya people in Belize. Dr. Balick is researching potential anti-AIDS and anticancer agents for the National Cancer Institute. The healers' clothing reflects how fast these people are being acculturated and their traditional knowledge is being lost. (Photo by S. Matola, courtesy of Michael Balick and The New York Botanical Garden)

music, and ethnobotany is the study of ethnic groups' knowledge of the uses of plants (Figure 6-11). The migrations discussed in Chapter 5 are sometimes described as migrations of ethnic groups.

Ethnocentrism is the term given to the tendency to judge other cultures by the standards and practices of one's own, and usually to judge them unfavorably. Practices in other cultures that may seem strange to us, however, may in fact be sensible and rational. Conversely, some aspects of our own culture may seem strange or even offensive to others. Most Americans, for example, assume that a man should have one wife and a woman one husband at a time, but almost any number of spouses are allowed in a series. This would shock many people. Americans, in turn, may be shocked to learn that Tibetans assume that a woman is married to all sibling brothers at once. This is called *fraternal polyandry.* Through thousands of years, fraternal polyandry has prevented unsustainable population increases and the fragmentation of landholdings in Tibet's poor

mountain valleys. The social ramifications of fraternal polyandry confound most Americans. An American might ask how to identify the father of any given child. To a Tibetan, however, it makes no difference, and a Tibetan might consider the question prurient.

Any geography book will contain examples of ways of life that contrast with your own. None is necessarily "right" or the best for everybody. All people have to overcome the initial assumption that "different from" the way we do things ourselves is "worse than" our way, and people everywhere can learn to appreciate and respect the integrity of other peoples' behaviors. Shrewd people even learn from others whenever they can.

How people identify themselves   Culture groups, races, and ethnic groups may also be differentiated from the concept of personal identity, or self-identification. Many people insist that their identity—who they are—is shaped by their membership in a group or groups. Affiliation with an identity group is not only a question of how much people share, but of what that means to them.

Some identities have the names of traditional cultures (Italian, Jewish, Polish), but diversity has many bases other than cultural. Some identities correspond to self-perceived races; some correspond to religions. Some identities are regional (Southern, Western, Puerto Rican), whereas still others are new groups modeled on traditional cultures (Hispanic, Asian American). Some new identities are demanding social recognition (women, gay people, the disabled). *Tribe* is still another word that is used loosely to describe a group of people who may share biological background, race, religion, or political community, but for whom sharing no single one of these attributes is essential. Nationality is still another powerful identity, and that will be discussed in Chapter 11. Today people remark on the "culture" of such groups, but in fact it is not obvious that a distinct culture corresponds to each identity.

Furthermore, any individual has several identities. A middle-class, Baptist, Republican, disabled, black mother of West Indian ancestry and U.S. citizenship may feel different allegiances, and at different times her behavior may be determined by any one of those identities.

Ethnicity is largely a matter of self-identification. For example, new migrants to the United States often struggle to retain their inherited culture for a generation or two, but the generations born in the new homeland may quickly acculturate and even disdain their parents' ethnic self-identification. Sometimes people are aware of their ethnicity only when someone of their group wins celebrity and they want to bask in reflected glory (an Olympic athlete, for example) or hide from the shame (a criminal), when they are discriminated against, or when politicians appeal to ethnicity. In New York State, for example, the plurality of the population was of Italian background in 1982, but that fact may not have been important to the people until Mario Cuomo emphasized that he was the first Italian American to run for the governorship; he appealed for Italian Americans to vote for him on the basis of ethnic solidarity.

Sometimes identity is not chosen by the individual, but ascribed to the individual by someone else, and people who find themselves labeled together may develop a shared identity in reaction to the labeling. The many peoples of Asia, for example, may feel that they share very little, but when they come to the United States and are viewed by others as "Asian Americans," they may band together, form a community, and begin to act in concert.

Therefore, identity involves at least three points of view: our own choice, the members of the group to which we ascribe ourselves (who may accept us or reject us as a member), and the general society around us (which may accept or deny our self-identification). For example, Jews lived in Germany for hundreds of years and made important contributions to German art, science, and politics. They thought of themselves as Jewish Germans. Nevertheless, throughout the Nazi period the Germans rejected Jews as Germans.

In other circumstances, a person might be included in a group against his or her own wishes. For example, during World War II, the British felt that anyone born in Germany must be presumed to owe loyalty to Germany, so the British interned anyone who had been born in Germany. Many of the people interned, however, including Jews and anti-Fascists, actually felt no loyalty to Germany. The United States interned Americans of Japanese ancestry, although none was ever proved to be traitorous, and many Japanese Americans fought valiantly in the war.

In still other circumstances, a group might claim an individual as a member, even though that individual repudiates their claim. Serbia is one of several countries that occasionally draft unlucky tourists who are second- or even third-generation descendants of emigrants. To the Serbian government, these people are still Serbs. These people may deny their Serbian identity in Serbia, but after they return to the United States, they may attend the local Serbian Orthodox church and participate in a Serbian street festival in their hometown.

To understand many of the disputes in today's news, we must examine, in each case, the interplays of the identities of the participants. We cannot understand people's behavior unless we understand how they view the world, and this includes their own identity and the identities of others.

Behavioral Geography

Behavioral geography is a subfield of cultural geography that studies our perception of the world around us and how our perception influences our behavior. The political commentator Walter Lippmann (1889–1974)

differentiated "the world outside" from "the pictures in our heads." The world outside is the way things really are, but the pictures in our heads may be based on preconceptions, misperceptions, or incomplete understanding. Geographers call these pictures in our heads **mental maps.** Geographers borrow from psychology the theory of **cognitive behavioralism,** which argues that people react to their environment as they perceive it. In other words, people make decisions on the basis of mental maps. Thus, we cannot understand why people make the decisions they do or act as they do by studying only the real environment in which they acted. We must discover what was in their heads when they made their decisions.

For example, in 1898 Senator George Hoar of Massachusetts cast a key vote for the annexation of Hawaii because, according to his diary, he drew a line on a map from Alaska to southern California, saw that Hawaii fell inside that line, and concluded that its possession was necessary for the country's defense. In fact, Hawaii does not lie inside such a line. Check its position on a globe. No one has been able to discover what sort of map Senator Hoar was looking at (perhaps one with an inset map of Hawaii?), but his misperception affected history. We make choices based on our perceptions of the world, but we must act in the real world. If there is a difference between what we think about the world and the way the world really is, then we may get into difficulty. We might try to do something impossible or fail to take advantage of some opportunity.

Where we choose to live, shop, or visit depends on our feelings about places—whether we think certain places are good or bad, beautiful or ugly, safe or dangerous (Figure 6-12). Geographers have investigated how we perceive environmental threats, either natural (earthquakes) or human-made (chemical or nuclear installations). How we perceive these threats affects our actions or reactions toward them.

Our individual notions of usefulness or of aesthetic value affect our evaluation of landscapes. Most urbanites, for example, agree that the Painted Desert in Arizona is beautiful, and it is a major tourist attraction. A farmer, however, might be appalled at such a landscape (Figure 6-13).

Studies that have grouped men, women, children, the elderly, the disabled, and individuals categorized in many other ways have taught us that each of these groups perceives the environment in significantly different ways. Therefore, they may almost be said to live in different environments. This knowledge is useful in the design of new environments, such as housing. Housing for the hearing impaired, for example, should have open interior sight lines. Studies launched jointly in 2003 by the American Institute of Architects, neuroscientists at the National Institutes of Health, and the U.S. General Services Administration are seeking to discover how workers are affected by their environments by monitoring heart rates and other signals of physical

(a)

(b)

FIGURE 6-12 These are the mental maps of places' residential desirability in the heads of students at the University of California at Berkeley (a) and the University of Alabama (b). High numbers indicate a positive image in the students' minds. Students at both places seemed to like where they were. Berkeley students gave coastal California a score of 100; Alabama students rated Alabama 100. The Alabamans, however, rated the region of Berkeley (60–70) much higher than the Californians rated Alabama (0.0). Both seemed to have positive images of Colorado and low images of the Dakotas. These preferences are not necessarily based on personal experience. We all have opinions of places we have never visited, and a visit might surprise us. (From Peter R. Gould, "On Mental Maps," in Michigan Interuniversity Community of Mathematical Geographers, Discussion Paper #9, 1966. Reprinted with permission.)

and mental stress levels. Results may help architects design more efficient and healthy work environments.

Differences in perception and behavior may be studied among different cultures. Human behavior is rooted in biology and physiology, but it is filtered through culture. One should never assume that someone else will do what you would do if you were in his or her position. Other people have different life experiences, different values, and different perceptions. "Their heads," we say, "are in a different place." To

avoid controversial topics, we sometimes say, "Don't go there."

Cross-cultural research into human mental functioning and behavior suggests significant differences between European and Asian cultures. Compared to European Americans, for example, East Asians tend to think more holistically, to pay greater attention to context and relationship, to rely more on experience-based knowledge than abstract logic, and to show more tolerance for contradiction. Westerners, by contrast, tend to

FIGURE 6-13 Farmland in Iowa. This black soil photographed near Martinsburg, Iowa, is so rich that a farmer might find this a beautiful view. What is your aesthetic reaction? (Courtesy of John G. Daniels)

be more analytic, to detach objects from their context, to avoid contradictions, and to rely more on formal logic. For example, when asked to decide which two—of a panda, a monkey, and a banana—go together, Japanese tend to select the monkey and the banana; monkeys eat bananas. Europeans, by contrast, tend to select the panda and the monkey, because these both belong in the category *animals*. The differences in observed behavior are not genetic, for Asian Americans tend to think as European Americans do. Different methods of thinking observed in different cultures may ramify into every aspect of human behavior.

Proxemics is the cross-cultural study of the use of space. People of different cultures balance the way their mind relies on input from their eyes, ears, and noses differently, even if they have the same physical sensory capacity. They do not agree, for instance, on crowding. In a room in which most Americans would feel uncomfortably crowded, most Middle Easterners would not. In designing a library, it helps to know that North Americans are more distracted by sounds than by sights, whereas Japanese are more distracted by sights than by sounds. Therefore, people talk in Japanese libraries, but the libraries have study cubicles. Proxemics affects human behavior at all levels from polite interpersonal behavior to the design of buildings or even whole cities.

Some scientists believe that humans exhibit **territoriality.** Many animals lay claim to territory and defend it against members of their own species. Most aspects of territorial behavior among animals have to do with spacing, protecting against overexploitation of that part of the local environment on which the species depends for its living. Applying these studies to human behavior is complicated by the interplay between human biology and culture. People defend their standing space in a crowd (they bump back), urban gangs defend their turf, and nations defend their land; but it remains unclear whether this behavior is biologically determined human territoriality.

CULTURE REALMS

The entire region throughout which a culture prevails is called a **culture realm** or **culture region.** Any aspect of culture may be used to define a culture realm. Religion is an important aspect of culture, so religion is often chosen as a criterion. The use of the phrase "the Christian World," for instance, implies that the prevalence of Christianity across a large region unites the peoples of that region. Furthermore, this region might significantly be contrasted with, for instance, "the Islamic World." Each is a great culture realm, and Chapter 7 examines how the prevalence of each of those religions encourages other similarities across those realms. Other criteria that may be chosen to define culture realms include language, diet, customs, or economic development. These topics will be discussed individually in the following chapters.

Problems in Defining Great Culture Regions

Many geographers and other social scientists have drawn maps dividing the world into great culture regions. Scholars will never agree upon one single pattern of world culture regions, however, because different scholars have different purposes, so they choose different criteria. The use of different criteria yields different regions.

Our delineation of culture regions is important because our pattern predisposes us to stereotype the people within each region as homogeneous, rather than to recognize the differences that may exist within each region. A region, by definition, should display a certain homogeneity, but that homogeneity may exist only in the eyes of an outsider. Insiders may see the differences among themselves as greater than their similarities. For example, Westerners may label Asia as a culture region. A Korean, however, might insist that the many cultural traits that differentiate him from a Burmese are more important than anything they share. Asia does in fact exhibit greater cultural diversity within it than, for example, Europe does. Asia does not exist as a homogeneous place in the mind of a Burmese, a Japanese, a Vietnamese, or a Cambodian. It exists only in the mind of an outsider. In Chapter 12 we will see how the term the Third World lumps countries that are very different and obscures understanding of any of them.

Critical Thinking

IS LATIN AMERICA A REGION? HOW DID IT GET ITS NAME?

Some criteria of homogeneity justify labeling Central and South America "Latin America." "Latin" suggests that certain important aspects of the culture spring from the cultural tradition of ancient Rome (*Latium*), and, in fact, most of Latin America was long ruled by either Spain or Portugal, both of which are Latin countries. Most Latin Americans today speak Spanish or Portuguese, both languages descended from ancient Latin, and most of the peoples there belong to the Roman Catholic Church. Italy itself, however, played no historic role in Latin America. Why, then, don't we call the area Luso-Hispanic America? ("Luso" meaning Portuguese, because today's Portugal was the ancient Lusitania) or Ibero-America? The more one thinks about the term Latin America, the more mysterious its origin becomes.

The truth is that the term Latin America originated as political propaganda. During the U.S. Civil War,

France's Emperor Napoleon III thought he saw an opportunity to take over Mexico. France had never had any colonial interests there, but Napoleon III invaded on the pretext that France, a Latin nation, was avenging military humiliations suffered by the Spanish and was therefore "defending Latin honor." On July 3, 1862, Napoleon published a letter in the French newspaper *Le Moniteur* introducing the term Latin America to justify what was really French imperialist aggression. Eventually the French were defeated and retreated, but Napoleon's sly propagandistic term has long survived him.

Questions

1. What similarities and differences are there among the countries from Mexico to Argentina? List differences and similarities among the countries in sub-Saharan Africa, in Asia, and in Europe.

2. Which do you think are more important in each case, the similarities or the differences?

Two questions must be asked of any pattern of culture regions. (1) Has the cartographer drawn the region correctly according to the criteria that he or she chose? For example, when mapping language regions of South America, did the cartographer remember that Brazilians speak Portuguese, not Spanish? (2) The second question to ask is, Are the criteria chosen meaningful? For example, does the fact that Brazilians speak Portuguese tell us anything important about Brazilians? Does it provoke questions about how that affects Brazil's interaction with Spanish-speaking South Americans? Any regionalization scheme is useful if it is both accurate and meaningful.

If a cartographer does not apply the same criterion consistently on a single map, the map is meaningless and confusing. For example, on Figure 6-14 the World Bank divided the world into eight "demographic regions," but there are no criteria of homogeneity within any individual region, nor are there consistent criteria of heterogenity differentiating one region from another. Some regions are individual countries; other regions are defined by their national economic systems (past or present); still other regions, such as "Middle Eastern crescent," seem to be defined on the basis of religion; and the region "Other Asian Countries" is made up of countries "left over!" Each of these regions contain such great differences within it that any statistic reported to describe that region or to compare it with other regions is of little value.

Visual Clues to Culture Realms

Culture realms can reveal themselves through visual clues. These include the language of posted signs, the clothing the local people are wearing, and the goods available in local shops. In the built environment, building materials, architecture, and settlement patterns are all visible manifestations of cultures. People usually rely on local materials for building, so in one place stone may be the traditional building material, in another place brick, and in another wood (Figures 6-15 and 6-16). Innovations in the uses of these materials may diffuse across cultures.

Styles of architecture often represent adaptations to climatic conditions, so a particular style may be adopted in different regions with similar climates (Figure 6-17, p. 239). Cultural preferences are sometimes so powerful, however, that an architectural style may diffuse beyond the limits of where its building materials can be found and even beyond the range where that architecture is comfortable. The style of the Italian Renaissance architect Andrea Palladio (1508–1580), for example, spread to England because of aesthetic preferences, despite the fact that Palladian buildings are uncomfortable in England's damp, cool climate. From England the preference diffused to America, even to areas that lacked both the appropriate climate and the necessary building materials. As Europe and the United States came to dominate other regions,

Focus on

REGIONALISM IN THE UNITED STATES

Interest in American regionalism has swung in and out of popularity among scholars throughout history, and today, despite a commercial landscape of Starbucks, Gaps, and McDonalds that look the same from coast to coast, the study of American regions and regionalism is winning new interest. In 1976 the University of Nebraska opened its Center for Great Plains Studies, and this Center was followed by The Center for the American West at the University of Colorado at Boulder, the Appalachian Regional Studies Center at Radford University in Virginia, the Southwest Center at the University of Arizona, the Center for the Study of Southern Culture at the University of Mississippi, the New England Studies Program at the University of Southern Maine, the Southern California Studies Center at the University of Southern California, and still more.

The University of North Carolina Press published *The Encyclopedia of Southern Culture* in 1989, and that volume was followed by similar volumes for other regions. Even New York City has its own *Encyclopedia of New York,* published in 1995. The National Endowment for the Humanities may establish 10 regional centers to support relevant studies, including folk dancing and other regional activities.

Many critics doubt whether genuine folk cultural regions still exist in the United States. They ask whether we are trying to preserve something or whether we are only re-creating something because we fear homogenization. Some regional characteristics were based on characteristic livelihoods. Can we restore whaling in New England, or Western cowboys?

World Bank "Demographic Regions"

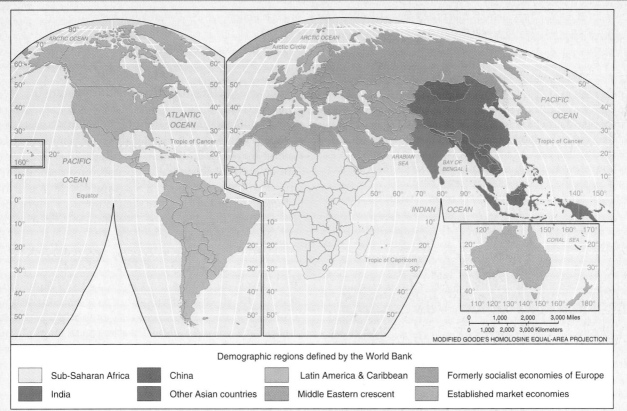

Demographic regions defined by the World Bank

- Sub-Saharan Africa
- India
- China
- Other Asian countries
- Latin America & Caribbean
- Middle Eastern crescent
- Formerly socialist economies of Europe
- Established market economies

FIGURE 6-14 What internal homogeneity does each of these "demographic regions" exhibit? Virtually none. Therefore, any information provided about each of the regions on this map is not very useful because the cartographer did not define clear criteria for the regionalization scheme. (From World Development Bank, *World Development Report 1993*)

Traditional Building Materials

FIGURE 6-15 The materials that are locally available—stone, brick, bamboo, or wood—are used everywhere in traditional architecture, called vernacular architecture.

FIGURE 6-16 The mosque at San, in the West African country of Mali. This building demonstrates that when people have mud and very little wood, they can still build extraordinary structures. It is in the distinctive Dyula architectural style, named for a trading people who diffused the style as they moved around West Africa. It is quite different from the Middle Eastern architectural styles that many Americans expect mosques to reflect. Compare it with Figure 7-23. (James Stanfield/National Geographic Image Collection)

Palladian architecture diffused throughout the world (Figure 6-18).

Public statuary and monuments may reveal local cultural values, but they may also proclaim that a place is dominated by a cultural or political outsider. For example, several Eastern European countries fell under the domination of the U.S.S.R. after World War II (see Chapter 13). Statues of Russian "liberators" were often set up in their public places. Since these countries have regained their freedom, however, the local people have replaced these statues with statues of genuine local heroes. Budapest, Hungary, for example, boasts new statues of ancient Hungarian royalty and Christian saints. Many cities in the U.S. South erected monuments to commemorate the Confederate cause, but as African Americans gained political power there, they insisted the monuments be taken down.

Settlement patterns   The designs of settlements reflect cultural differences, so a trained observer can see in the look and layout of whole towns and cities the cultural backgrounds of their builders (Figure 6-19). City planning will be discussed in Chapter 10, but rural societies can be differentiated by the way that some cluster housing settlements, whereas others isolate settlements in individual farmsteads.

FIGURE 6-17 **Simla, India.** The building in the left foreground is a bungalow, which is a Hindi word for a low-sweeping single-story house with a roof extending out over a veranda. Such houses were first built in the mountain foothills of northern India, but the style was copied throughout the British Empire and eventually the United States. Today in English the word *bungalow* means almost any sort of small house. Simla was the summer capital of the British Indian Empire between 1864 and 1947. It was cooler up at Simla (2,200 meters/7,100 feet; notice the vegetation) than down at New Delhi in the plains about 275 kilometers (170 miles) to the south. The medieval-style cathedral in the background seems an odd presence in northern India. It is not, however, a medieval building. It was built in the nineteenth century in the medieval style that was popular for Christian churches then (and still is today). Wherever the British went, they took their culture and traditions, including architectural styles. (Courtesy of the Library of Congress)

FIGURE 6-18 **A Palladian building in the United States.** This house, called the Morris-Jumel Mansion, was built in 1765 in Manhattan 10 miles north of Wall Street. The choice of the Palladian architectural style is highly symbolic. Standing virtually on the frontier of Western civilization, this style boldly—almost arrogantly—announced European conquest and the coming of European civilization to the New World. Symbolism and aesthetic taste can prevail even over comfort, for although this style is appropriate for the Italian climate, in New York it would have been impossible to keep warm. Furthermore, in Italy it would have been built of stone. Here it was built entirely of wood masquerading as stone. (Courtesy of Robert F. West/Morris-Jumel Mansion)

FIGURE 6-19 **Settlement patterns.** An expert on African cultures would recognize the pattern and spacing of the buildings in this view as typical of the Kisti people of northwest Kenya. (M & E Bernheim/Woodfin Camp & Associates)

In those societies that cluster housing, farmers may choose to live together in clusters ranging from a few dozen homes up to thousands. There are no dwellings in the surrounding farmland, but the farmers journey out to work in the fields each day. The farm buildings are usually concentrated together with the human settlement. There may be no economic reason for the clustering; it is a cultural choice.

Such compact villages may be found in many forms: irregular; wandering along a principal street, river, or canal; clustered about a village common; or checkerboard (Figure 6-20). Clustering may reveal the need for common security against bandits or invaders, family or religious bonds, or communal land ownership. The government may deliberately cluster the population to supervise it or provide education or health care. Clustering may also have environmental reasons— people may cluster at water sources, for example, or on dry places when the surrounding land is swampy.

A. Muang Nan, Laos

B. Ban Mae Sakud, Thailand

0   15   30 Meters

0   50   100 Feet

N

Houses   Sheds and farm buildings   Direction of sleeping

FIGURE 6-20 Village settlement patterns in Laos and Thailand. Arrangement of houses in two Southeast Asian communities. (a) The front gables of Lao houses, such as those in Muang Nan, Laos, face one another across a path, and the backs face each other at the rear. Ridgepoles are set perpendicular to the path and parallel to a stream if one is nearby. Inside the house, the head of a sleeping person is in the opposite direction to that of a sleeping neighbor, so that heads and feet are always together. (b) In Ban Mae Sakud, Thailand, houses are not set in a straight line, because of a local belief that evil spirits move in straight lines. Ridgepoles are set parallel to the path, and the heads of all sleeping persons face east. (From James M. Rubenstein, *The Cultural Landscape: An Introduction to Human Geography,* Updated 7th ed. Upper Saddle River, NJ: Prentice Hall, 2003.)

Clustering is more common among farmers than among livestock ranchers, except in Africa, but it generally characterizes settlement across much of Europe, Latin America, Asia, Africa, and the Middle East.

Isolated farmsteads, by contrast, characterize those areas of Anglo-America, Australia, New Zealand, and South Africa that were settled by Europeans, and also some parts of Japan and India. The conditions for isolated farmstead settlement usually include peace and security in the countryside; agricultural colonization of the region by individual pioneer families rather than by socially cohesive groups; agricultural private enterprise, as opposed to communalism; unit-block farms in which all of a farmer's land is in a block rather than in scattered parcels; and well-watered but well-drained land.

The history of the United States provides examples of both patterns of settlement. In the southern colonies in the seventeenth and eighteenth centuries, settlement was characterized by widely dispersed, relatively self-sufficient plantations. New England settlement, by contrast, reflects the social cohesion of the

settlers' society. The people of New England were tightly bound in religious communities, so they advanced westward in tiers of adjacent, well-planned towns. The reason for the differences in the settlement patterns was not economic; it was a cultural choice.

Forces that Stabilize the Pattern of Culture Realms

Despite the force of diffusion, a number of factors tend to fix the geography of culture realms. Culture leaves its mark on the landscape. The fixed pattern of activities, land uses, transport routes, and even individual buildings guides, restricts, or predisposes future patterns and activities. The construction of a factory, for example, represents a great investment of money, and once the factory is operating, it relies on a local workforce and develops ties to local suppliers. An industrial complex such as this cannot easily be picked up and moved. **Inertia** is the term for the force that keeps things stable. All of a people's fixed assets in place—railroads, pipelines, highways, airports, housing, and more—are called the **infrastructure.**

Historical geography is the subfield within geography that studies the geography of the past and how geographic distributions have changed. Historical geographers can sometimes read landscapes as if through time the landscape has been overlain with layer after layer of peoples using the land in different ways and organizing it for different purposes. The landscape is like an old manuscript on which a reader can discern earlier erased writings (Figure 6-21). Scholarly re-creations of past land uses were defined as *sequent occupance* studies by the geographer Derwent Whittlesey in 1929. An alternative approach to historical geography focuses on how the transformation, use, and organization of the landscape is continuously changing.

Culture includes a set of values and ways of doing things, and culture groups seldom get displaced or eliminated entirely. Culture is learned behavior, and cultural norms are handed down through generations. Groups may be quick to take up new techniques or products, but tradition is a powerful force in human life, and imported ideas or lifestyles only slowly transform a people's entire cultural inheritance. This inertia is reinforced by the existence of territorially

sovereign states and the enormous power that governments exert over their citizens to teach and to enforce norms of behavior. This power includes, as we have seen, the power to resurrect and promote folk traditions. Lee Kuan Yew, Senior Minister of Singapore, emphasized to an interviewer the role of culture in various Asian countries: "Culture is very deep-rooted; it's not tangible, but it's very real: the values and perceptions, attitudes, reference points, a map up here [he tapped his head], in the mind."

Each local culture is today a unique mix of what originated locally and what has been imported, and each unique culture is a local resource, just as surely as the minerals under the soil and the crops in the fields.

Most peoples value their culture, and they usually try to preserve key aspects of it. This influences the way they interact with other peoples. A peoples' self-consciousness as a culture may be codified in their religion, as it is with the Jews, or in their sense of their own history. This is called **historical consciousness.** Many Americans have difficulty understanding other people's historical consciousness, because one aspect of American culture is an optimistic denial that history can shackle future opportunity. Other peoples, however, nurture their traditional cultures and their historical consciousnesses. Many people act the way they do because that is the way their ancestors acted, or to right wrongs or to account for deeds or misdeeds that took place long ago. Several peoples fighting in the Balkans in the 1990s, for example, felt that they were avenging medieval battles.

One major theme in geography is the tension between forces of change and forces for stability. Cultures evolve, cultures diffuse, and peoples can transform themselves and their behavior, but cultures and culture realms also have elements of stability. Any cultural pattern or distribution maps the current balance between those forces.

Trade and Cultural Diffusion

Cultural isolation is usually accompanied by economic self-sufficiency, but trade diminishes peoples' cultural isolation at the same time as it expands their economic possibilities. Trade breaks down cultural isolation and triggers cultural diffusion. Every item in trade is a product of the culture that originates it, so economic exchange is one of the most important forms of cultural diffusion. Choosing goods is an act of self-definition, of social and cultural identity (Figure 6-22). Trade, economics, and culture are intertwined.

The study of how various peoples make their living, how economies develop, and what peoples trade is **economic geography.** Trade releases people from dependence on their local environment. It allows them to draw resources from around the world. Fewer and

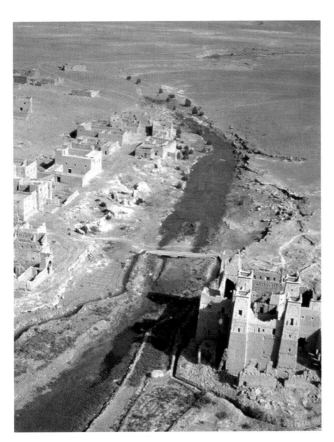

FIGURE 6-21 An abandoned city. An abandoned city prompts us to ask, "Who was here before us? What did they do? What happened to them?" This ruin is in Morocco. (Courtesy of Moroccan National Tourist Office)

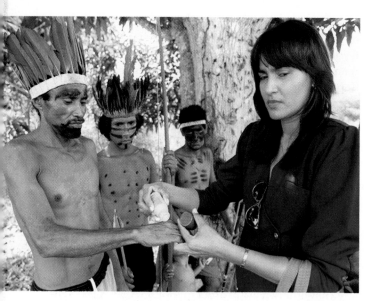

FIGURE 6-22 New consumer goods change cultures. An Avon Lady demonstrates a product to Tembe tribesmen in Brazil's Amazon region. The tribesmen obviously have their own cultural tradition about makeup, but the introduction of new cosmetic products will change these people's culture in ways that cannot be foreseen. (John Maier, Jr./The Image Works)

fewer people anywhere rely on their local environment for all of their needs, so the link between the resources of any local environment and the well-being of the people who live there has been weakened. The Japanese, for example, live in an environment that is poor in natural resources, yet the Japanese have grown rich through trade.

As peoples come into contact with others and begin to trade, they usually first export only whatever they have in surplus and have no use for. They usually view imports as luxuries unnecessary to their way of life. Trade, however, triggers far-reaching cultural and economic changes.

For example, imagine three isolated communities in three different environments (Figure 6-23). Village A is located along a river plain. The people there catch fish, which they fry; they grow rice, which they eat as rice cakes; and they distill rice into sake (rice wine). Village B lies on the slope of nearby hills. The people there have domesticated grapes, which they have learned to distill into wine, and oats, which they make into oat bread. They have also domesticated goats, which clamber about the rugged hillsides. They milk the goats for dairy products, and they roast goat meat. The people in village C, up on a nearby plateau, grow corn for cornbread and to be distilled into whiskey and have domesticated cattle, which they eat as broiled steaks. Each village has developed other aspects of a unique culture, too, including its own language, religion, and customs, but let us focus on their diets as representative of their cultures.

The construction of a road and the commencement of trade among the three villages will probably trigger at least four significant changes.

1. Each village will have access to the products of each of the other villages. The people of village A, for example, will taste broiled steaks, and the people of village B will be introduced to fried fish. We might call this the simple *addition of cultural possibilities.*

2. New cultural combinations, or *cultural permutations*, will appear: fried steaks, broiled fish, roast beef, and more. A good example of cultural permutation is the appearance of pita fajitas on menus in southern California. A pita fajita is a piece of pita bread, an item brought to southern California by immigrants from the Near East, stuffed with fajita ingredients, usually beef or chicken, which were introduced to Mexico by the Spaniards long ago and now to southern California by Mexican immigrants. Nobody knows who first devised this permutation. People often find surprising new uses for objects created by other cultures and societies, even when they do not grasp the fundamental values or technologies that created the objects in the first place.

All aspects of the cultures of all three villages will experience these two results—cultural addition plus cultural permutation. Residents of one village may convert to the religion of another, or perhaps the religions will blend. The languages, styles of architecture, music, games, clothing and other customs of all three villages may add up and also permutate. Those who profit from a new trading system and exchange with other cultures often challenge and overturn local politics and traditions.

FIGURE 6-23 Isolated cultures yield to interaction. The inhabitants of three isolated villages in three different environments learn to use local resources, and they develop three distinct cultures. When a new road connects those villages, however, the lifestyles of all three villages may be transformed. Their cultural possibilities will multiply, and their economies will evolve.

3. The residents of all three villages will see how imports can raise their standard of living. In order to pay for the imports they want, they will dedicate more effort to *producing items for trade—that is, for markets.* The change from self-sufficiency to production for markets is one of the greatest transitions in history. The existence of trade and markets gives each village, for the first time, an incentive to produce surpluses of its local goods. Agronomists (economists who specialize in agriculture) say that, in general, "The market produces the surplus." In other words, if farmers have a market where they can sell their surplus for a profit, they are likely to produce a surplus. Furthermore, when people specialize in producing an item, they become more efficient, and the quantities they produce increase. Therefore, the total amount of food—of all goods—produced in all three villages will probably increase.

Production for the market affects cultures in another subtle way: The people may produce less of those local goods that cannot be exported, but more of those local goods that win broader markets. Traditional local items may be altered in order to increase exports. For example, many Native American tribes once made traditional craft items such as baskets, jewelry, and clothing for themselves only. Then they began to produce these items for tourists, and today they concentrate on producing those items and styles that tourists favor. They even alter traditional designs if requested to do so by customers. Thus, people may slowly surrender their own traditional culture.

As trade multiplies, more of what people produce anywhere is consumed elsewhere. Conversely, a growing percentage of the things people use are produced elsewhere. Eventually, if the terms of trade are favorable, people dedicate most of their efforts to producing export products, and they rely on imports even for their necessities. They have surrendered their self-sufficiency and become dependent on trade, and they surrender their cultural isolation and experience cultural change.

4. Village B will probably develop a market larger than those in village A or village C. This is because of its situation between the two other villages. Village B is the most convenient place in the pattern or network of exchange, called the *central place,* so it will probably grow to be the largest.

World Trade and Cultural Diffusion Today

Chapter 12 examines world trade in detail, but we must note here that virtually all peoples are today experiencing this evolution from self-sufficiency and cultural isolation to trade and cultural exchange. The share of any country's territory that is devoted to export production may be small, but the shares of the national population and the national income that are involved are rising everywhere. Even within individual countries, growing cities create markets for food from surrounding rural regions. This draws the rural population out of subsistence agriculture into commercial agriculture.

Regions or peoples have not always freely chosen to enter into the system of production and exchange. Some areas were forced into specialized production by colonialism. Peoples in other regions were forced to buy goods, and they had to develop exports in order to pay for these goods. The British actually went to war to force the Chinese to buy opium (1839–1842). Opium is an addicting drug, but all kinds of new goods—electronic equipment, leisure activities, or styles of clothing—may trigger our desire to enjoy them. This process has been called the creation of new **felt needs**—things people begin to think they need. As people begin to depend on the availability of desirable goods, they become more deeply enmeshed in the web of trade and circulation.

Trade leads to specialization of production and to greater production, but it does not benefit all regions equally. Some regions prosper, and others fall behind. Why international trade causes this to happen is a principal subject of research in economic geography. Many different reasons will be suggested throughout this book.

We noted earlier that items of popular culture achieve differing market penetration throughout the United States, suggesting differing regional cultures. One can study market penetration of individual consumer items around the world. In fact, the international marketing divisions of large corporations spend millions of dollars each year in order to explore why some products have worldwide appeal, whereas others are successful only in geographically restricted markets. Salespeople strive to break down cultural differences, but the integrity of each culture realm resists complete homogenization. Therefore, cross-cultural advertising and marketing of consumer goods present fascinating cultural-geographic questions. For example, Domino's Pizza has spread around the world, but the corporation has learned that Germans prefer smaller pies than North Americans do, and Japanese like pies topped with squid and sweet mayonnaise. A few consumer products have already achieved almost global diffusion: Coca-Cola, Sony, and Kodak. Why do people around the world dress differently, eat different foods, and speak different languages, but almost everywhere emulate the Marlboro Man? Is there anyplace where people refuse cola drinks (Figure 6-24)? Yes, Utah. Why there? Many of Utah's devout Mormons avoid the caffeine of the original colas and have not taken to the decaffeinated brands.

Consumer goods are only one aspect of culture. In some places people wear American blue jeans, but

FIGURE 6-24 A global product. Some consumer goods practically blanket the Earth. (Courtesy of Julia Nicole Paley)

they wear them to political demonstrations against the United States. They like blue jeans, but that does not mean that they like everything about U.S. culture. In fact, when popular culture from one country overwhelms the folk culture of another, this process may be viewed as offensive cultural aggression. We will examine below how this process feeds some peoples' animosity against the United States.

The Acceleration of Diffusion

In the past, travel and transportation were more difficult and expensive than they are today, whether we measure the cost in time, money, or in any other unit. The friction of distance was so high that only a few things moved far, and those things moved slowly. Over the past 200 years, however, technology has reduced the friction of distance and accelerated the diffusion of cultural elements (Table 6-1). We often hear that the world has "shrunk." As the cost and time of moving almost anything—people, food, energy, raw materials, finished goods, capital, information—have steadily fallen, things

Focus on

THE DEVELOPMENT OF "THE GLOBAL CHILD"

Previous to the year 2001, the Mattel Corporation manufactured for Asian markets a version of the popular "Barbie" doll with Asian facial features. When, however, Mattel realized that Asian children were happy with a "Caucasian" Barbie, Mattel stopped making an "Asian" model. The corporation estimates that by 2004 a full 80 percent of its goods will be global, and only 20 percent are geared to local audiences. Matthew Bousquette, the President of Mattel Brands, has said, "The general overriding belief was that kids are different around the world, but we discovered that wasn't the case. Kids are more alike than they are different." He meant, presumably, kids from families with enough money to buy manufactured toys and who are, therefore, probably reasonably well-to-do and aware of consumer marketing and advertising. Thus Mattel, Hasbro, the Lego Company, and other multinational giant toy manufacturers that used to sell toys and gear in a variety of local styles are today increasingly designing and marketing one version worldwide.

Several influences mold the "global child," including the following: (1) The rapid worldwide expansion of cable and satellite TV channels, which, along with movies and the Internet exposes millions of children to the same manufactured icons. In 2003 the Disney Corporation, for example, operated 24 Disney-branded cable and satellite channels in 67 countries outside the United States. (2) The international reach of retail giants such as Wal-Mart, Toys 'R' Us, and Carrefour is widening. These retailers enter into agreements with toy companies for global promotions. In October 2002, for example, Mattel introduced a "Rapunzel Barbie" simultaneously in 59 countries. Wal-Mart Worldwide featured the doll in stores and in television ads in 35 languages. A video and DVD was released and played in theaters worldwide the same day. Tie-ins include Barbie laptop computers, Barbie kitchen sets, a travel van, and more. The Barbie website is maintained in 8 languages. Similarly, in 2001 the movie *Harry Potter and the Sorcerer's Stone* was released in 3,700 theatres in the United States and 10,000 more in 100 countries elsewhere. Toys and other tie-ins launched simultaneously accounted for over $1 billion in sales. Some variations in regional toy markets still exist. Action figures, for example, make up 6 percent of the toy market in the United Kingdom and 5 percent in the United States, but only 1 percent in Germany. Cheerleader-themed goods are meaningless outside the United States.

TABLE 6-1 Transport and Communication Costs 1920–1990, Expressed in 1990 Dollars

Year	Average ocean freight and port charges per short ton of import and export cargo	Average air transport revenue per passenger mile	Cost of a 3-minute telephone call New York to London
1920	$95	NA	NA
1930	60	$0.68	$244.65
1940	63	0.46	188.51
1950	34	0.30	53.20
1960	27	0.24	45.86
1970	27	0.16	31.58
1980	24	0.10	4.80
1990	29	0.11	3.32

The Internet makes the transmission of information even cheaper than any of these measures. The United Nations estimated in mid-2000 that it cost $75 to send a document from Madagascar to the Ivory Coast by courier, $45 to fax it, and 20 cents to e-mail it.

NA—Not applicable

Courtesy Institute for International Economics

are not necessarily so fixed in place as they were in the past. Many activities have been significantly released from the constraints of any given location. If an activity can move or relocate freely, we call it **footloose.**

Originally, information could move only as fast as a person could carry it, but electronics disengaged communication from transportation. When Samuel Morse demonstrated the first intercity telegraph line in 1844, a Baltimore paper wrote, "This is indeed the annihilation of space." The telephone, invented in 1879, furthered the annihilation of space by allowing a person, figuratively, to be in two places at the same time.

The annihilation of space has continued with electronic mail, facsimile machines, computer modems, and other electronic devices collectively known as the **electronic highway.** More people with personal electronic devices are plugged in everywhere all the time (Figure 6-25). Schoolchildren with personal computers can tap into networks of information that were unavailable to the world's leading scholars 25 years ago.

The activities of communicating with other people through an electronic network or even of playing a game alone with a computer create a new mental world—a new "place" where that activity is occurring, called *virtual reality*. That extension of reality through global electronic means of communication is **cyberspace,** a word coined by William Gibson in his 1985 science fiction thriller *Neuromancer.* The term may be applied widely. Many office workers, for example, collaborate through electronic networks without sitting down together in one office. One worker may be at home tending children, a second in a car on the road, and a third in an airport waiting room. The electronic highway connects them in a virtual office in cyberspace, so office work is increasingly footloose.

The compression of space compresses time. The latest music heard on radio stations in Los Angeles

FIGURE 6-25 **Where are you?** People talking on the telephone are, functionally, two places at once: where their bodies are, and the place with which they are communicating. Can these people be described as "absent-minded"? (Ed Quinn/CORBIS)

and New York will be heard in Nairobi, Kenya, and Montevideo, Uruguay, before the week is over.

New means of communication grant access to news and cultural elements from all parts of the world, but we continue to be selective in what we pay attention to (Figure 6-26). Our own backgrounds; our education,

Relative World Coverage in One Newspaper

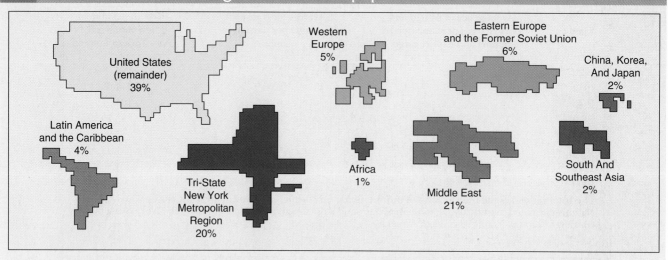

FIGURE 6-26 The relative sizes of world regions on this illustration correspond to the coverage devoted to each on the front page of *The New York Times* for two months in 2003. New York has a large and influential Jewish population, and this is reflected in the large amount of coverage of news from the Middle East. These proportions were measured before the war against Iraq focused American attention on the Middle East. Miami has a large and influential Latin American population, and Miami has important trade links with Latin America. Considering those facts, how do you think "the view from Miami" would differ? How about Los Angeles? How about the city in which you live? Measure your local newspaper and see.

perceptions, and prejudices; and the media to which we are exposed affect our understanding of other peoples and places and how we rank their relative importance. For a great variety of reasons, we are more knowledgeable about or more interested in some places than in others. These reasons are not necessarily related to the size or population of a particular region.

The fluidity of world economic and political groupings today has stimulated many theories of what new groupings might form in the future. The political scientist Samuel Huntington predicted that "the fundamental source of conflict [in coming decades] will not be primarily ideological or primarily economic. The great divisions among humankind and the dominating source of conflict will be cultural." He warned of a coming "clash of civilizations." He defines a civilization as "the highest cultural grouping of people and the broadest level of cultural identity people have short of that which distinguishes humans from other species," and his list of major contemporary civilizations includes "Western, Confucian, Japanese, Islamic, Hindu, Slavic-Orthodox, Latin American and possibly African civilization." With few exceptions, these are the major religious groupings and regions discussed in Chapter 7. Professor Huntington suggests that these realms are the cultural equivalents of the world's geological tectonic plates, discussed in Chapter 3. If that is true, then the areas where they meet might experience virtually unending strife. Geographers have traditionally referred to such areas as *shatterbelts*. This book will explore a few, including Bosnia (Muslims caught between Eastern and Western Christians), Sudan (an Arab

Muslim north battling the Black Christian and animist south), and Kashmir (between Pakistan and India).

It remains unclear, however, whether the peoples within each of Professor Huntington's "civilizations" share any real affinity, or why these affinities might trigger hostilities against other groups. The Muslim extremist Osama bin Laden hoped that the attacks on the United States on 9/11 would trigger a war between Islam and the West, and thus ignite a "clash of civilizations." As the news proves each day, however, such rhetoric simplifies profound matters of identity. Few people rely on their religion as their exclusive identity, but blend that with their nationality and even their common humanity. People around the world refuse to identify themselves exclusively according to typologies held either by Islamic extremists or by U.S. scholars. Hypotheses of global culture wars simplify at least the definition of identity, the impact of migration (see Chapter 5), and global economic and political forces.

The Challenge of Change

Maps of culture realms, of regions of economic specialization, and of political jurisdictions reveal current distributions of specific human activities. Human activities, however, are not static. They are dynamic, continuously organizing and reorganizing, forming and reforming. Therefore, no pattern of the organization and use of territory is stable. Distributions and patterns are disrupted and reshaped repeatedly, just as the patterns in a kaleidoscope are. A map of human activities at any time is comparable to a weather map. We

Critical Thinking
CAN CULTURES BE PRESERVED?

In the 1930's coastal Rio de Janeiro was the capital of Brazil, and the country had scacely begun the task of exploring its great interior expanses, let alone developing them. From 1941 to 1961 a series of expeditions, known together as the Roncador-Xingu Project, opened up 1,500 km (930 miles) of trails into the forest, explored 1000 km (621 miles) of rivers, carved airstrips, and opened over 40 new towns. Along the Xingu River the Project met 14 indigenous nations that had seldom seen outsiders. At that time, native peoples were considered scarcely human and were, in fact, hunted as animals. As a result of the Project, the government created Xingu National Park, 26,000 square kilometers (10,040 square miles) where previously warring tribes learned to live together. The first director of the Park,

Orlando Villas Boas (1914–2002), insisted that his task was to bring in health care workers, but to keep other Brazilians and tourists out, to refrain from imposing white man's logic or meddling in village affairs, and to keep the peoples' ethnobotanical knowledge out of the hands of pharmaceutical companies.

Questions

1. Can the well-meaning goal of "protecting" other peoples become in fact a policy of denying them access to the world's ideas and goods, thus keeping them isolated in a "human zoo"?

2. Who should make the decision whether or not people should be allowed to choose for themselves whether they want the world's ideas and goods?

3. Do you agree with all of Villas Boas's goals?

---

cannot understand the activities if we do not know what forces are at work. For geographers to answer the question "Why there?" we must study activities and forces.

In nature, it is variations in air pressure that cause the winds to blow from points of high pressure to low pressure. Gravity causes water to flow to the sea. In studies of human activities, variations in local job opportunities or political liberties spur people to migrate. Variations in interest rates or in profitability distribute investment capital, and the places that attract investment enjoy economic development. The force of religious belief drives people on pilgrimages to holy sites. Religious conversion triggers the flow of ideas. Variations in places' natural endowments initiate trade flows. Produce flows to markets, raw materials to manufacturing sites, tourists to attractions or to beaches. Variations in transport capability or freight rates direct flows of traffic. All these forces overcome geographic inertia.

The geography of global manufacturing, which we will examine in detail in Chapter 12, provides a good example of an activity that is redistributed continuously. The factors that redistribute manufacturing include new products; new technologies; new raw materials (the replacement of metals with plastics, for example); new sources of traditional raw materials; new technologies of manufacture; new governments with new policies regarding investment, taxation, or environmental preservation; growing and shrinking labor supplies; and the opening and development of new markets. When the executives of global corporations decide where in the world to build new factories, they balance these and still more factors, and each of these factors changes every day!

This textbook will detail many redistributions: Some religions are winning new converts, expanding geographically and assuming new influence in world affairs, whereas others are withering away. Some languages are demonstrating the flexibility to adapt to new communication technology and are therefore gaining users at the expense of others. The relative rise and fall of nations' influence in world political or economic affairs wins or loses new adherents to those nations' products, cultural artifacts, or lifestyles. If the Solitary people were to be discovered today, people in other nations might adopt some aspect of Solitary culture—perhaps Solitary music or a food or medicinal crop. At the same time, the Solitary way of life would be transformed by the infusion of products and ideas from the outside world.

It is important to be familiar with the current distributions of human activities, but if you learn the reasons why things get distributed the way they do, then you will have learned something that will be useful for the rest of your life. Former French president François Mitterand often said, "History has accelerated." Geographic redistributions have, too.

THE GLOBAL DIFFUSION OF EUROPEAN CULTURE

Despite the rich variety of indigenous local cultures around the globe, the world is increasingly coming to look like one place. In consumer goods, architecture, industrial technology, education, and housing, the Western model is pervasive. To ethnocentric Westerners who presume the superiority of their own culture, or to local people who have accepted Western culture, this may seem natural. To them this may be "modern" life, or "progress," or "development."

A dispassionate observer, however, might expect more diversity, more styles and models of development. Why are so many people around the world imitating Western examples and adopting aspects of Western culture? In much of the world, acculturation to the Western way of life is rapidly replacing both the positive and negative features of other cultures. It is the most pervasive example of cultural diffusion in world history. It illustrates all types, paths, and processes of diffusion that we have studied, and an understanding of European cultural diffusion over the past 500 years is essential to understanding the world we live in today.

Europe's Voyages of Contact

Europeans came to play a central role in world history and world geography because it was they who paved the way for the modern system of global interconnectedness. In the fifteenth century the great cultural centers of the world—the Inka and Aztec empires in the Western Hemisphere, the Mali and Songhai in Africa, the Mughal in India, Safavid Persia, the Ottoman Turks,

the Chinese, and all of the lesser empires and culture realms—were still largely isolated from one another.

The European voyages of exploration and conquest connected the world (Figure 6-27). It was not inevitable that the Europeans would be the ones to do this. The Chinese were actually richer and more powerful than the Europeans, and earlier in the fifteenth century they had launched fleets of exploration that had reached the east coast of Africa. We can hardly imagine how different world history and geography would be if the Chinese had continued their initiatives and gone on to explore and conquer Africa and the Western Hemisphere. But they did not. Instead, the Chinese government focused on internal affairs, and the Europeans continued with world exploration and conquest.

The first European initiative was Prince Henry of Portugal's conquest of the city of Ceuta on the north coast of Africa in 1415. He learned about trans-Saharan caravan routes and the riches of West Africa, and he was inspired to finance studies of naval skills. Improvements in sailing technology allowed the Portuguese to sail down the west coast of Africa to reach beyond the Sahara Desert. Prince Henry, known as "The Navigator," launched the era of European seaborne colonial empires. Soon Spain, Holland, France, and England were racing to secure colonies. In contrast to these

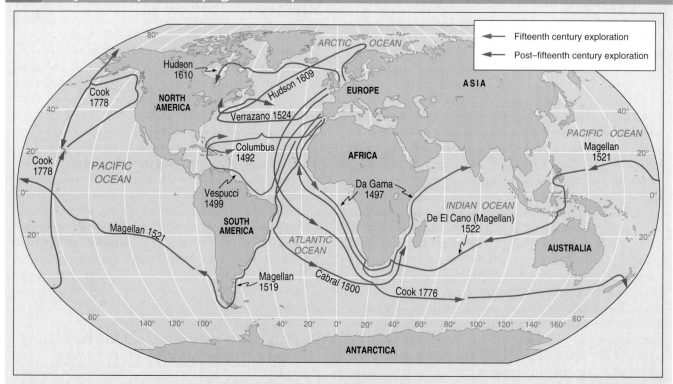

Major European Voyages of Exploration

FIGURE 6-27 Each of the European voyages of exploration was a daring enterprise. The Portuguese first established a string of bases around the coasts of Africa and Asia all across to Japan, where they established a trading post in 1543. Other European nations followed, and the race for discovery of new riches, of new converts to Christianity, and for help against Christianity's powerful Islamic foe was launched.

seaborne empires, Russia at the same time forged eastward overland from the Russian homeland in Europe west of the Ural Mountains. The Russian Empire pushed across Siberia to the Bering Strait, crossed over into Alaska, and eventually established colonies down the North American West Coast as far south as today's California.

This European outreach triggered the **Commercial Revolution,** between about 1650 and 1750. The development of the first oceangoing freighters that could carry heavy payloads over long distances allowed a tremendous expansion of trade. The evolution of superior ships was paralleled by the evolution of superior naval gunnery and of additional useful technologies such as clocks, which were perfected in order to determine longitude at sea. This first era of outreach ended with the death of the British explorer Captain James Cook in 1779. By then Europeans were the first people in world history who could draw a fairly accurate outline map of all the world's continents and major islands. Eventually the United States would inherit European culture and share its power.

Expansion and cultural diffusion   From the time of exploration on, Europe did not actually originate every "modern" idea and then impose it on the rest of the world. Europeans learned from others, too, and then transplanted around the world the ideas that they had adopted from elsewhere. "Knowledge is power," said English philosopher Sir Francis Bacon (1561–1626), and because it was the Europeans and not the Chinese or the Inka or any other group who had contacted all other civilizations, Europe became the clearinghouse of world information and products. Global diffusion was fixed hierarchically, with Europe as the apex. Europe became cosmopolitan—that is, familiar with many parts of the world. Other peoples, no matter how great their native civilizations were, remained more localized in the world Europe was creating.

Chapter 5 described the migrations of peoples triggered by European expansion. Consider, too, what Europeans did with agricultural products. They took sugarcane from Asia and planted it in the Caribbean region. They replanted bananas from Southeast Asia to South America, cocoa from Mexico to Africa, rubber from South America to Southeast Asia, and coffee from Arabia to Central America. All these foods remain major products of international trade and important factors in the export economies of many countries.

Europeans not only relocated the production of many goods around the world, they also introduced many products into world trade, both for the European home market and for other overseas markets that they pioneered. Europeans introduced many Indian goods, for instance, into China, and South American products into Africa and Asia. An Englishman introduced tea plants from China into India, leading to the development of a tea industry in India and Ceylon (today

known as Sri Lanka). Europeans created world markets, and they profited by controlling every stage: production, transportation, and marketing.

The story of Coca-Cola, one of the world's most familiar consumer products, exemplifies cultural blending. The drink was formulated in Atlanta, Georgia, and today symbolizes westernization so powerfully that the large-scale infusion of Western products into the non-Western world is often referred to as the "Coca-colonization of the world." The two original ingredients from which Coca-Cola takes its name, however, are coca, a Native South American Quechua word for the tree whose leaves supply a stimulating drug (used today to make cocaine and crack), and cola, a word in the language of the West African Mandingo people for the nut that supplied the other original stimulating ingredient. Westerners borrowed the knowledge of both of these ingredients from their far-flung hearth areas, combined them, and marketed the drink worldwide.

Thus, modern world culture is not exclusively a Western product. It is presumptuous for Westerners to think that it is. Even non-Westerners, however, often fail to see the many contributions of non-Western peoples because the route of global diffusion was through the Western powers.

Economic Growth Increased Europe's Power

Europe pulled ahead of the rest of the world economically as it underwent the tremendous transformation of the **Industrial Revolution.** Between about 1750 and 1850, Europe evolved from an agricultural and commercial society to an industrial society relying on inanimate power and complex machinery. We cannot fully explain why Europe experienced this transformation before any other part of the world, but we can list several factors that enabled Europe to industrialize. The voyages of discovery and conquest resulted in an influx of precious metals and other sources of wealth, stimulating industry and a money economy. The expansion of trade encouraged the rise of new institutions of finance and credit. In the mid-sixteenth century the joint stock company was developed, allowing many investors to share both potential profit and risk in new enterprises. The creation of stock markets where stocks could readily be bought and sold granted capital new liquidity, or easy conversion from one form of asset to another. This created, in the words of English writer Daniel Defoe (1660–1731), "strange unheard-of Engines of Interest, Discounts, Transfers, Tallies, Debentures, Shares, Projects."

In 1769, James Watt designed the steam engine, which multiplied the energy available to do work. Subsequent inventions and technical innovations in manufacturing, applied first to textiles and then across

FIGURE 6-28 The first iron bridge. This is the world's first iron bridge, built over England's Severn River in 1779. It demonstrated iron's strength, and the bridge's bold design invited other uses for the new material. Its builder, John Wilkinson, launched the first iron boat in 1787, and when he died he had himself buried in an iron coffin. (George Fischer/Bilderberg/Aurora & Quanta Productions, Inc.)

a broad spectrum of goods, dramatically increased productivity. Factories and industrial towns sprang up, canals and roads were built, and later the railway and the steamship expanded the capacity both to transport raw materials and to send manufactured goods to markets (Figure 6-28). New methods of manufacturing steel, chemicals, and machines played important parts in the vast changes. These innovations occurred first in Great Britain and subsequently spread to continental Europe and to North America.

Beginning in the eighteenth century, Europe also first experienced an **Agricultural Revolution.** This development, to be examined in detail in Chapter 8, both increased food production and released agricultural workers from the land, thereby creating a supply of labor for industry.

As a result of the Industrial and Agricultural revolutions, Europe and European settlements around the world drew far ahead of any other places and peoples in productive capacity. As recently as 1800, the per capita incomes of the various regions of the world were close. If we index the Western European per capita income in the year 1800 as 100 units of wealth, then estimated per capita incomes in North America were 125, in China 107, and in the rest of the non-European world 94. By 1900, however, European and North American incomes were several times those of non-Western peoples.

Commercial contacts and economies   At the beginning of the age of the European voyages, European demand for foreign products such as spices, sugar, fruits, and North American furs grew rapidly. Soon

the Europeans were no longer content to trade with native peoples for these goods, and the Europeans themselves established overseas estates and plantations and applied large-scale techniques to specialized production.

European commercial plantations at first concentrated along the coasts, but in the nineteenth century the railroad allowed penetration of the continental interiors to superior agricultural lands or, later, as Europe industrialized, to mineral deposits. The world's railway network expanded from 200,000 kilometers (125,000 miles) in 1870 to over 1 million kilometers (625,000 miles) by 1900. European treaty ports (ports that by treaty had to be kept open for trade) and coastal footholds became inland empires (Figure 6-29). The development of the steamship also facilitated the transport of minerals, and increased quantities of minerals supplied Europe's multiplying factories. Between 1840 and 1870 the world's merchant shipping rose from 10 million tons to 16 million tons, and then it doubled in the next 40 years.

New cities emerged in the non-European world as coordinating centers for these commercial activities; new ports sprang up along the seacoasts. The major seaport cities from the Straits of Gibraltar around Africa, across the Indian Ocean, and all throughout South Asia are the products of European contact (Figure 6-30). The same is true for most of the port cities of the Western Hemisphere. The railroads and associated commercial economies at first affected only a small percentage of the population, but over time an increasing share of the population and territory were drawn into the emerging global economy. In some countries, however, such as India, the modern commercial economy still overlies a traditional subsistence economy. There may be little exchange between the two economies unless refugees from the collapse of the traditional economy flee to the slums of the modern cities.

Political conquest   In two waves of exploration and conquest—the first extending from 1415 to 1779 and the second occurring at the end of the nineteenth century—Europe (and, in the second period, the United States) conquered most of the rest of the world. One of the original reasons for this conquest was the European nations' wish to protect their investments in foreign lands and to control these lands as markets for themselves. European countries divided up the rest of the world in order to restrain their own rivalry. Their ascendancy over the indigenous populations was guaranteed by their superior military power.

The United States and most of Latin America won independence between 1775 and 1825, but between 1875 and 1915 about one quarter of Earth's land surface was distributed or redistributed as colonies among a half-dozen imperialist states. Of all the countries in the world today, the only ones never ruled by Europeans or

European Partitioning of Africa

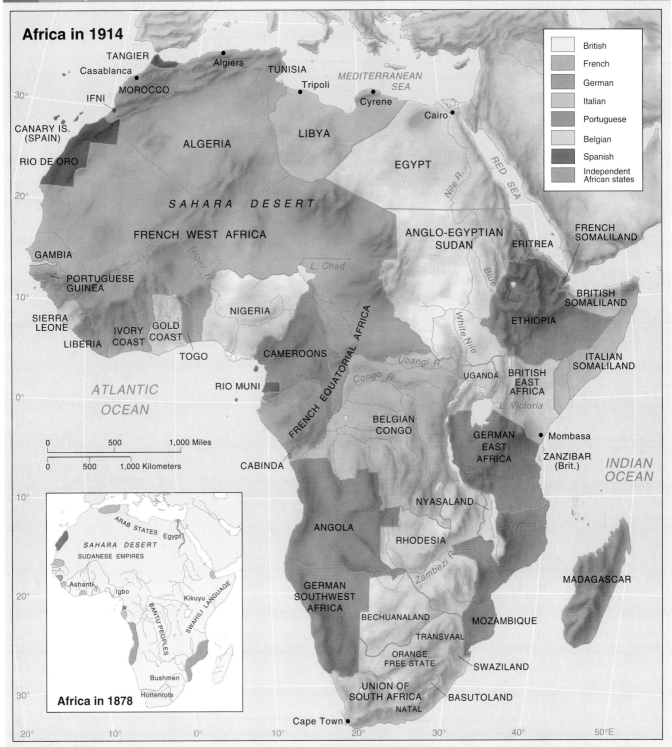

Africa in 1914

	British
	French
	German
	Italian
	Portuguese
	Belgian
	Spanish
	Independent African states

TANGIER
Casablanca
Algiers
TUNISIA
MEDITERRANEAN SEA
IFNI
MOROCCO
Tripoli
Cyrene
Cairo
CANARY IS. (SPAIN)
ALGERIA
LIBYA
RIO DE ORO
EGYPT
Nile R.
RED SEA
SAHARA DESERT
FRENCH WEST AFRICA
ANGLO-EGYPTIAN SUDAN
ERITREA
FRENCH SOMALILAND
GAMBIA
L. Chad
Niger R.
Blue Nile
PORTUGUESE GUINEA
BRITISH SOMALILAND
SIERRA LEONE
NIGERIA
White Nile
ETHIOPIA
LIBERIA
IVORY COAST
GOLD COAST
TOGO
CAMEROONS
FRENCH EQUATORIAL AFRICA
ITALIAN SOMALILAND
ATLANTIC OCEAN
RIO MUNI
Ubangi R.
Congo R.
UGANDA
BRITISH EAST AFRICA
L. Victoria
BELGIAN CONGO
GERMAN EAST AFRICA
Mombasa
CABINDA
ZANZIBAR (Brit.)
INDIAN OCEAN
Lake Tanganyika
NYASALAND
0   500   1,000 Miles
0   500   1,000 Kilometers
ANGOLA
RHODESIA
L. Nyasa
Zambezi R.
MADAGASCAR
GERMAN SOUTHWEST AFRICA
BECHUANALAND
MOZAMBIQUE
TRANSVAAL
ORANGE FREE STATE
SWAZILAND
UNION OF SOUTH AFRICA
BASUTOLAND
NATAL
Cape Town

Africa in 1878
ARAB STATES
Egypt
SAHARA DESERT
SUDANESE EMPIRES
Ashanti
Igbo
Kikuyu LANGUAGE
BANTU PEOPLES
SWAHILI LANGUAGE
Bushmen
Hottentots

FIGURE 6-29 Europeans divided the African continent in order to prevent competitive war among themselves. Treaty ports and trading stations along the African coast expanded into vast inland empires. Industrial Europe demanded African raw materials, and railroads allowed the Europeans to draw them out of the African interior. The native African peoples were not consulted in the political reapportioning, which was completed at a conference held in Berlin in 1884–85.

European Intervention in Asia

FIGURE 6-30 Many of Asia's port cities were at first spigots that Europeans tapped into the continent to draw off Asia's wealth. As happened in Africa, however, the ports expanded into plantations for the production of goods valuable in international trade, and then the plantation regions grew into political colonies.

by the United States are Turkey, Japan (although it was occupied by the United States from 1945–1952 and its constitution was imposed on it), Korea (which was ruled by Japan from 1910 to 1945 and remains split today), Thailand (left as a *buffer state* between the French and English empires), Afghanistan (a buffer between the English and Russian empires), China (which was nevertheless divided into foreign "spheres of influence"), and Mongolia (ruled by China).

Therefore, European cultural attributes linger as a legacy of European rule in most countries. They still predominate or overlay native pre-European traditions.

European concepts of law, for example, drastically changed native societies, especially European ideas of property rights and land ownership. Before the Europeans came, land was generally considered a good that was held in common for all members of the community. Local political leaders apportioned land use and occupation by customs that brought the community together. Neither the leaders nor anyone else *owned* land. The idea that any individual could own land and determine alone how to use it was largely unknown. Europeans introduced their idea of "ownership" of an "estate" that could be bought, sold, or mortgaged by individual contract. Social cohesion was dissolved when land was no longer a common good, and the regulation of its use was no longer a shared community affair.

Native American chiefs, for instance, did not by their own customs have the right to transfer land out of tribal control, and they frequently did not actually understand what Europeans meant when they "bought" Native American lands. In the history of the United States, innumerable wars were sparked when Native Americans returned to hunt or harvest unoccupied land that the Europeans insisted the Natives had sold. Throughout Africa and Asia, the Europeans often simply assigned ownership to the local political leaders. This ended traditional egalitarian systems and created new classes of rich and poor. The descendants of many of these newly enriched leaders remain great landholders today throughout the Near East, for example. Many traditional societies crumbled under the transformation of land from a public asset into a private commodity.

European law tended to transform labor into a commodity, too. Traditional communities were not idyllic; they restricted individual liberties, and slavery and serfdom were not unknown. These constraints, however, were often balanced by strong webs of responsibilities and rights that usually kept anyone from being entirely outcast and starving. The European idea of a self-regulating market for individual labor is more abstract and impersonal. It cut traditional ties of both rights and responsibilities. In addition, Europeans required the use of money as a universal measure of value. Natives were forced to work for wages or to sell goods for money in order to pay taxes.

Europeans also brought their forms of administration, government, centralized state authority, written arrangements, uniformity, secularization, economic planning, public accounting and treasury control, central administration, and decision making. In many cases the civil services that the Europeans left behind remain the pride of new nations, as in India.

Cultural Imperialism

European rule was marked by **cultural imperialism,** which is the substitution of one set of cultural traditions for another, either by force or by degrading those who fail to acculturate and rewarding those who do. Europeans seldom doubted that native cultures were inferior and that native peoples needed "enlightenment." Therefore, the Europeans destroyed other ways of life, including religious and political traditions, physical artifacts such as art and architecture, and even records of history and science (Figure 6-31).

One reason for this is the nature of Christianity. It is a proselytizing religion, which means that its adherents try to convert others to their faith. The natives in many areas accepted that there is truth in all religions, and their toleration opened them to acculturation.

Additionally, Europeans believed that their military and technological superiority presumed European superiority in all other aspects of life. Europeans did learn some things from those they conquered—farming techniques from Native Americans, for example—but they did not learn as much as they could have learned, because they often failed to appreciate the values, science, and technology of the civilizations that they conquered.

European cultural imperialism began with the systematic training of local elites. The missionary schools produced converts who proselytized among their own people, helping to eradicate the local culture. Later the government schools turned out bureaucrats and military officers who helped govern their own people. A second channel of transmission was *reference group behavior.* People who wish to belong to, or be identified with, a dominant group often abandon their traditions in order to adopt those of the dominant group. The returning slaves who carried the first wave of westernization to West Africa wore black woolen suits and starched collars in the tropical heat. In India the native officer corps imitated English officers, complete with waxed mustaches.

FIGURE 6-31 A Mayan book. This page is from a book called the *Grolier Codex,* one of only four books known to survive from the considerable libraries of Central America's Mayan civilization. The rest were burned by the Spanish. This book contains astronomical records regarding the recurrence of the planet Venus. The vertical columns of markings on the left are countings of days. The warrior, perhaps mythological, stands before an incense burner. (Courtesy of Justin Kerr/Barbara and Justin Kerr Studio)

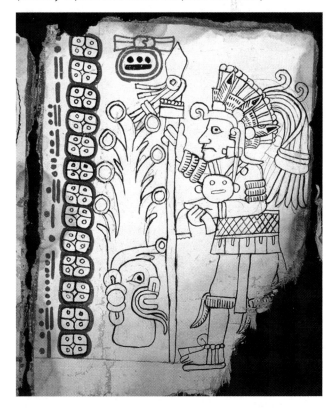

Local elites also adopted Western ways because they were made to feel ashamed of their color and their culture. The rulers' racism and cultural imperialism meant that natives could succeed only by adopting the whites' ways. The later autobiographies of the new national leaders of every African and Asian country recount racial humiliation (Figure 6-32). The colonial school systems implanted in children's minds an image of the power and beneficence of the "Mother Country." Children in the Congo, for instance, knew more about Belgium than about their own land and peoples. Cristopher Monsod, the chairman of the Philippines Election Commission, remembers similarly having been taught from U.S. textbooks and says, joking, "Thanks to my American education, I know the capitals of North and South Dakota." History textbooks used in French African colonies opened with the words, "Our ancestors the Gauls. . . ." Schooling focused on each colony's

ruler, so the native peoples scarcely knew that other countries existed. Residents of the Congo, for example, referred to all whites as Belgians.

Colonial intellectual institutions usurped the power to define local tradition. For example, after Europeans founded the Asiatic Society of Bengal in 1784, European study of Indian art and literature defined Indian culture. Europeans certified what were to be considered "classics" of Indian culture. European archaeological surveys determined what monuments were fit for description and preservation as part of "the Indian heritage." Educated Indians learned about their own culture through the mediation of European ideas and scholarship.

Self-westernization   By the end of the nineteenth century, the elites of the entire non-Western world were taking the Europeans as their reference group. Even the three major nations that were never colonized—the Turks, the Chinese, and the Japanese—were all militarily humiliated, and this experience was traumatic for them. All three had had empires of their own, and their defeats forced them to reconsider all the assumptions on which their institutions and daily lives were based.

Turkey, China, and Japan all preempted or co-opted Western civilization, and to some degree this saved them from Western political rule. In Turkey, Mustafa Kemal (1881–1938) seized power and forced the country to undergo self-westernization (Figure 6-33). Many historians believe that the challenges he faced, the solutions he devised, and the degree of success he achieved

FIGURE 6-32 Imperialism in the movies. This is a scene from the 1935 British movie *Saunders of the River*. In this Eurocentric and racist movie, Commissioner Saunders, a British imperialist officer, on the left, supervises and keeps peace among tribes in a British African territory. The tribes are depicted as dangerous children. In this key scene Saunders prods a submissive native in the chest with his cane while lecturing on the superiority of British civilization. The native character was portrayed by Jomo Kenyatta, who was then a young actor, but who later served as president of Kenya. What could Kenyatta have been thinking when this scene was filmed? What could he have thought years later while viewing the film in the presidential palace of independent Kenya? (Picture Desk, Inc./Kobal Collection)

FIGURE 6-33 Self-westernization. In this historic photograph Turkish president Mustafa Kemal Ataturk, whose surname means "father of the Turks," is teaching the Roman alphabet in an Istanbul public park. Ataturk recognized the technological superiority of the West, and he consciously turned his people toward westernization. One of his changes was the transformation of the writing of Turkish from the Arabic into the Roman alphabet. (Courtesy of the Turkish Tourist and Information Office)

FIGURE 6-34 A Japanese copy of Western technology. This is a Japanese woodblock cutaway view of a German battleship that visited Yokohama in 1873. This woodcut would later provide a virtual blueprint for the technologically backward Japanese Navy to copy. (Arthur M. Sackler Gallery, Smithsonian Institution, Washington, D.C.: Gift of Ambassador and Mrs. William Leonhart, S1998.65 a-c.)

defined a prototype for the leaders of all nations that emerged from colonialism after World War II. In China, the Republican Revolution of 1911 attempted to modernize the country, but China's subjection to the West continued. Later another leader, Mao Zedong, applied an alternative brand of westernization—communism—although in a uniquely Chinese form. The Japanese were forced by the United States to open their society to Western trade in 1854, after centuries of near-total isolation. The Meiji Restoration in 1867 launched the self-westernization of Japan; the Japanese decided to become thoroughly Western but to retain control of the process. Japan sent students to the United States and Europe and adopted Western science, technology, and even many cultural traits so rapidly and so successfully that by 1904 the country was able to defeat a Western power—Russia—in war (Figure 6-34).

Japan was later defeated in its attempt to extend its empire in World War II, but its early military successes discredited the Western powers and encouraged non-Western colonial subjects to dream of freedom. Jawaharlal Nehru (1889–1964), the first Prime Minister of independent India, wrote in his memoirs of how, in 1904, when he was 15 years old, "Japanese victories stirred up my enthusiasm . . . Nationalistic ideas filled my mind. I mused of Indian freedom. . . ." Chinese revolutionary Sun Yat-sen recorded, "We regarded the Russian defeat by Japan as the defeat of the West by the East. We regarded the Japanese victory as our own victory." Despite Japan's military defeat in 1945, it later developed into one of the world's dominant economic powers, and it continues to provide a

non-Western model for economic development without complete surrender to Western culture.

The period of European relations with Africa and Asia that began 500 years ago is ending. Only two tiny Spanish holdings survive on the African continent—Ceuta and Melilla. The Western empires in Asia have been surrendered except for a few islands still held by France and by the United States. Great Britain yielded Hong Kong to China in 1997, and in 1999 Portugal returned the Chinese territory of Macau, which it had held since 1557.

The West and non-West since independence   The fixation with the West among the elite in the non-European world did not end when these countries won political independence. Those who assumed power in Africa and Asia were mostly educated in the West, and they demanded independence quoting Western political writers whom they had read at Western universities. They often expressed anger that the West did not achieve the values it espoused and had taught them—racial equality, for example, and the equality of all people before the law. Their reference groups were entirely European, not their own people. Nehru himself was eventually to say to U.S. Ambassador J. K. Galbraith, "You realize that I am the last Englishman to rule in India."

Few of the Western-educated elites developed indigenous models of development. They continued the diffusion of Western models. They started building in the middles of their commercial cities and went on building outward. The technical term for the favored cities is *growth poles*. The leaders hoped to convert their whole national territories to modern Western societies, but they did not realize how long it would take the majority of their people to benefit.

By the time the colonial rulers left, the traditional sources of social status in the societies, such as religion, family, and customs, had faded. The new politicians, bureaucrats, and business executives defined their status in the only way they knew, which was to exhibit Western material goods. Thus their homes, clothes, and cars mimicked Western status symbols. In 2003 the parliament of Kenya still forbade its members to wear African clothing. European-style suits and ties were required. This mimicry may inspire corruption in some of the world's poorest countries; their elite pauperize their own countries and funnel money to banks or to buy prestigious properties in the rich countries. They flaunt their wealth in Paris and New York, and they vacation in villas along the French Riviera and in great mansions in London.

In many countries the new rulers have practiced a sort of internal colonialism on their own people. For example, the rulers who assumed power in Latin America following the independence movements of the early nineteenth century were of European stock, and their descendants still dominate this region. Peoples of Native American or African backgrounds have been treated as subject groups, forced to adopt European culture, religion, and language, and subjected to discrimination. The situation is not very different in Africa and Asia. The "colonizers" are generally the westernized elite; the "colonized" are all those who do not belong to this group, which is often the majority of the population.

Westernization Today

The diffusion of Western culture continues today. Western culture diffuses from the top of societies, from the examples and activities of the local elites. Young people also diffuse it by adopting Western dress and lifestyles as status symbols. The rich and the young are everywhere the most cosmopolitan consumers, and most consumer items in international trade are artifacts of Western popular culture.

Even the schools have become instruments of westernization. Their syllabuses emphasize modern, urban activities and values. The young sometimes emerge oblivious to their traditional culture or even despising it in favor of Western popular culture.

The media reinforce the message. Western television programs, movies, advertisements, and videos penetrate millions of homes and implant Western values. Night after night, on television screens around the world, the images of the good life are images of the life among the wealthy in the United States. This imagery has dramatically changed behavior. Chapter 5 described how popular media affect even national birth rates. People want what they see, so new felt needs are created, regardless of explicit advertising.

Western media have been accused of entirely supplanting traditional culture, at least among urban middle-class adolescents. Sumner Redstone, the head of VIACOM, the parent company of MTV, has said, "Kids on the street in Tokyo have more in common with kids on the street in London than they do with their parents." Historian William McNeill has seconded Redstone, writing that the diffusion of television to virtually every place on Earth "is a very deep transformation of human life. I would rank what is happening now with man's transition from a hunter and gatherer into a settled farmer. Television has replaced inherited culture." (See Figure 6-35.)

Tourism provides still another channel of westernization. Westerners are attracted to "different" and "unspoiled" places, but the presence of tourists changes the places they visit. National cultures can degenerate into commercialized spectacles and shoddy souvenirs. Many local people abandon their own material culture and adopt that of the visitors.

Global flows of professionals and of professional education are another powerful force for the diffusion of westernization. The rich Western countries export professional services, and people from around the world attend Western schools for professional education. The elite and professionals of most countries have been educated in Europe or the United States. These graduates, who hold influential roles in their societies, have been acculturated to Western ways.

Western architects, civil engineers, and urban planners—or non-Western individuals educated in Western schools—are transforming built environments.

FIGURE 6-35 Television brings mixed messages. Television, particularly Western television, penetrates practically every region. How do these people interpret what they see? How does it transform their cultural inheritance? The truth is that we do not know the answers to these questions. (Courtesy of Raghu Rai/Magnum Photos, Inc.)

Critical Thinking

THE DIFFUSION OF "NEWS"

Western media dominate the gathering and dissemination of news to such an extent that most people in non-Western lands learn about the affairs of other non-Western lands and even about their own national affairs through Western media. Nevertheless, Western countries veto efforts to achieve international information gathering and reporting as "censorship."

When the United States attacked Afghanistan in the fall of 2001, U.S. citizens learned what it is like to lack one's own reporting from an international hot spot. The only network with coverage from inside Afghanistan was al-Jazeera, a Qatar-based Arab-language television news service that has satellite uplink stations throughout the Islamic world. The U.S. government asked U.S. media to censor their rebroadcasts of al-Jazeera's original broadcasts, but anyone in America with an inexpensive dish antenna could have received al-Jazeera broadcasts directly. U.S. officials continue to criticize al-Jazeera for broadcasting interviews with America's enemies or even with dissident Americans, and al-Jazeera's independence continues to startle many Muslim governments, too. It criticizes them, for example, for their lack of democracy and subjection of women. These topics are avoided in the state-owned media monopolies in most Muslim states.

America's government-subsidized Voice of America (VOA) has won tens of millions of listeners in 53 languages. Its Charter states that the service should be "a reliable and authoritative source of news" and that it should be "accurate, objective and comprehensive," but some American critics demand that it present only the American point of view.

The U.S. press often feels constrained to question or criticize the government during times of lofty patriotism or national challenge. Several years after the 1991 war in Iraq, however, the American people learned that the U.S. government had lied about the accuracy of U.S. bombs and other matters during that war. In October 2001 Defense Secretary Rumsfeld quoted Winston Churchill: "In wartime, truth is so precious that she should always be attended by a bodyguard of lies." We may say that the American people have been warned that the American government might lie to them.

Questions

1. Do you have or could you obtain direct access to sources of international news other than U.S. sources?

2. If you have such access, have you ever been surprised by the difference between what you learned from them and the news as reported in U.S. sources?

3. Do you think that the U.S. government should be allowed to censor your access to such sources?

Non-Western countries' cultural landscapes are increasingly "modern," and there is no prototype other than the Western. For example, the largest homebuilders in both Thailand (Anant Asavabhokhin) and in the Philippines (Manuel Villar), two of the richest men in the world, worked for Los Angeles homebuilder Kaufman & Broad, Inc., before returning to their own countries with blueprints to build whole subdivisions for the rising middle classes. Some countries save "traditional" landmarks only as tourist attractions. For example, Singapore bulldozed historic Bugis Street, which offered a mix of old shops and small bars, and then the government re-created it for tourists.

Countries are being transformed by world flows of capital investment, and capital is invested according to the standards of the societies that export the capital. This activity will be examined in Chapter 12.

Under this barrage of westernization, many traditional cultures and social structures are being radically transformed, and these transformations are not always in accord with the people's conscious wishes. In some cases whole cultures disappear, and their disappearance reduces cultural diversity and impoverishes all of us.

America's Role

The terrible events of 9/11 and the political realignments and wars since then have jolted Americans and many other world citizens into reflecting on the unique role of the United States. Since the collapse of Communism in the early 1990s, economic, political, and cultural dominance within the West has moved decisively to the United States. In all human history, there is no example of world dominance comparable to that of the United States today.

In the dozen years between the fall of the Berlin Wall in 1989 (see p. 556) and September 11, 2001, Americans tended toward self-congratulation that was grating to other peoples. Americans even stopped paying attention to the rest of the world. Whereas in the year 1989 the ABC, NBC, and CBS evening news programs gave a total of 4,032 minutes of coverage to foreign news, in the year 2000 they gave only 1,382 minutes to foreign coverage.

On September 20, 2001, President Bush said, "Americans are asking, 'Why do they hate us?'" If some others do hate America, then Americans must investigate both "what's wrong with them," but also what America may be doing wrong, or what America could do better to communicate America's goals and hopes and to help others achieve theirs. Some haters may hate America for what it is, but others may hate America for what it does, and the problem is not just that people hate Americans, but that it *matters more* now.

The Scottish poet Robert Burns asked (freely translated), "Oh! Would some power give us the gift to see ourselves as others see us! It would free us from many a blunder and foolish notion." The U.S. government does not always interpret its own actions the same way those actions are viewed by many others. For example, the U.S. government sees its "war on drugs" as a crusade against a global scourge. Around the world, however, many view the effort first as a war against the American people themselves, imprisoning an unacceptable number of them (roughly 150 Americans per 100,000 adults, which is double the incarceration rate for *all* crimes in European countries) for what is best viewed and treated either as a sickness or as a petty crime. (U.S. state governments are coming to agree, but the federal "war" continues.) In addition, many see the "war" as an attempt to export an American problem to other countries. The combination of Americans' insatiable appetite for drugs ("Why can't Americans cure their craving?") and the U.S. government's outlawing of drugs has raised drugs' value to astronomical heights, encouraging farmers in many countries to grow drugs instead of nourishing foods and sabotaging several countries' agricultural economies. Furthermore, the U.S. government has sent troops that have intervened in civil wars and destabilized governments and societies in Panama, Colombia, Peru, Ecuador, Afghanistan, Pakistan, and still other countries. U.S. Coast Guard ships patrol the rivers of Colombia today. Cynics suggest that U.S. agencies are actually involved in the international drug trade to finance covert operations, or that the U.S. government uses the "war on drugs" simply as an excuse to occupy other countries. Whoever is correct in these descriptions of U.S. government motives and actions, the American people could at least better understand other countries' actions if Americans understood other peoples' views.

U.S. leaders have spoken of the need to study and understand other cultures, although "understanding" does not necessarily mean agreeing with them. It is, Americans have learned, *dangerous* not to understand other peoples. U.S. Secretary of Defense Donald Rumsfeld has recommended the ideas of writer Thomas Schelling. Schelling has written that America's ability to be surprised by others' actions is the result of "poverty of expectations. . . . There is a tendency in our planning to confuse the unfamiliar with the improbable. The contingency we have not considered seriously looks strange; what looks strange is thought improbable; what is improbable need not be considered seriously." Rumsfeld has mused on this, "The poverty of expectations—the failure of imagination—I found this just so interesting."

U.S. military power   Militarily, there is simply no other superpower. The United States spends more money each year on defense that the next ten leading countries combined, assuring American armed forces' technological superiority. The United States has assumed the role of global peacekeeper, and it stations troops in a growing number of foreign countries: 95 in 1950, but 148 in 2001. Many governments invited U.S. troops to protect them against either foreign enemies or their own people. America's intentions may be peaceful, but many peoples feel intimidated by such a widspread military presence. America's participation in political and military organizations and alliances will be detailed in Chapter 13.

U.S. economic power   The U.S. economy represents about one quarter of the world economy, and most Americans enjoy a high standard of material comfort. Therefore, some poor and oppressed people throughout the world may inevitably resent the United States, whether or not American wealth is the cause of their own poverty. U.S. investment and U.S. products are found virtually everywhere, and conversely, the United States is the greatest market for many other countries' goods.

The diffusion of U.S. popular culture   Most of the world's most recognized brand names are American, and American corporations pursue sales abroad aggressively. Many people view the marketing of U.S. popular culture—that is, American salesmanship—as the deliberate destruction of their own traditional folk culture and its replacement by American popular culture. For example, from a distribution center in Singapore, the Disney Company floods the countries of Southeast Asia with over 16,000 Disney products. Can the children of Southeast Asia resist Donald Duck T-shirts and lunch boxes? If not, will they also remember their traditional folk fables and mythical characters? (See Figure 6-36.)

U.S. popular culture incorporates American cultural and political values, so it can challenge traditions and initiate cultural change even unintentionally. For example, the American women's magazine *Cosmopolitan* seeks profit, not revolution, yet its championing of women's rights and sexual freedom is revolutionary in many of the 60 countries in which it is distributed. U.S. dominance of the world's television market has shrunk as the market has grown, but cultural observers have noted that the very format of television shows—dramatic or comedic shows, for example—are an American cultural product. U.S. movies still capture about 80 percent of total global box office receipts.

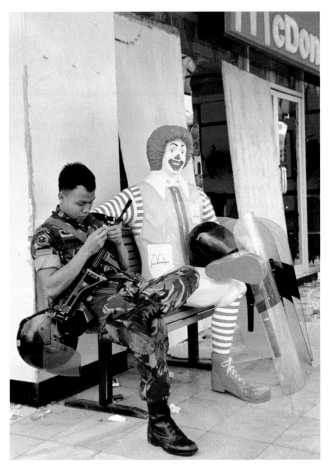

FIGURE 6-36 Who's your friend? This Indonesian soldier is guarding an American hamburger restaurant. Ubiquitous and conspicuous symbols of American private enterprises are often targeted during anti-American riots, even if the local business is a locally owned franchise, as it is here. (AP/Wide World Photos)

Many peoples see this onslaught of U.S. popular culture as cultural imperialism. Americans, however, generally accept the idea of *the democracy of the marketplace*. According to this theory, consumers' action in selecting and purchasing goods is comparable to voters exercising their right to vote. People "vote" by buying products or tickets to performances. Thus, the market democratically reveals what people want. People watch Hollywood's latest movie because they want to, not because they are forced to. Americans interpret actions limiting consumer choice as comparable to restricting citizens' civil rights.

No one accepts this analogy completely—Americans debate it among themselves—but Americans do generally accept it more completely than many other peoples do. Many other peoples argue more strongly that the marketplace does not always protect values that must be preserved for the good of a society. The marketplace may not protect tradition; it may widen disparities in income or opportunity, and it may fail to protect the general welfare above economic results. Indeed, no matter how strongly we may believe in democracy as the best system, we may well ask: Is the politician who wins the most votes, *really,* always the best statesman, and, similarly, is the hamburger that sells the most, *really,* the best meal? Both voting and buying involve complicated compromises and trade-offs.

Accepting the theory of the democracy of the marketplace, American trade representatives argue that when a government limits cultural imports, its purpose is not to protect culture, but to protect markets for local producers. Other peoples say that Americans "just don't get it." Some governments insist that they have the right to speak for their people, to ban films or products in the name of their people. The Chinese government, for example, reserves the right to determine whether U.S. cultural products are "catering to the tastes of the Chinese people," and it bans many American films. To Americans, this is simply preemptive censorship.

America's insistence on consumer selection and freedom even offends many cultures' religious teachings. For example, many Americans still resent Iran's Ayatollah Khomeini having called the United States "The Great Satan" in 1979, yet few Americans understand the original interpretation of this label. In the *Koran,* the sacred text of Islam, Satan is not portrayed as the great conqueror or exploiter familiar to most Americans, but as "the insidious tempter who whispers in the hearts of men." Many fundamentalist Muslims find offensive the degree to which the U.S. economy and culture thrive on the creation of ever-new felt needs and the way in which the United States spreads that consumer culture around the world. Once this definition of Satan is understood, even American religious individuals of all faiths may admit to feeling ill-at-ease with the energy that goes into the pursuit and accumulation of material possessions in the United States. Americans themselves debate the proper role of consumerism, advertising, and even consumption-driven debt in American life.

The global dominance of U.S. popular culture has triggered a backlash. In Iran, for example, men known as *Bassijis,* "those who are mobilized," patrol the streets battling prostitution, drugs, alcohol, and atheism, as well as objects and values they interpret as imported from the West, including stereos, pop music, videos, lipstick, and indecorous dress among women. Their commander, Ali Reza Afshar, has said, "This war goes to the root of our existence. While physically there is no loss of life, our young people are being felled by cultural bullets, and this cultural corruption makes our young impotent to rebuild the nation." Many of those "cultural bullets" say "Made in the U.S.A." on them. The case of Iran, however, illustrates that those who battle American cultural influence may have an agenda they themselves wish to impose upon their people, rather than to allow true freedom.

Even the other rich countries resent U.S. cultural diffusion. France requires theaters to reserve 20 weeks of screen time for French feature films. Australia demands that 55 percent of a television broadcaster's schedule be filled with domestic programs. Canada insists 60 percent of television programming be Canadian. In July 1998 representatives of 19 nations gathered in Ottawa, Canada, to discuss the threat to their nations' cultures from free trade in cultural products—specifically, those from the United States. The United States was pointedly not invited, nor has it been invited to the follow-up meetings held annually. U.S. cultural predominance, however, may even increase if, as predicted (see Chapter 5), it is virtually the only rich country whose population will increase in coming years, and America's evolving demographic mix may bring surprises for the evolution of America culture itself.

U.S. cultural dominance is not just a matter of popular culture. U.S. educational institutions attract the world's leaders and future leaders. American universities educate the elite from around the world. In 2000, 11 ministers in the Cabinet of the government of Taiwan held U.S. degrees. U.S. intellectual, academic, and scientific journals set world standards.

It must be reemphasized that not every American cultural export is an original U.S. cultural product. The United States is the apex of global information diffusion, so U.S. global corporations find product ideas around the globe and then introduce them into other places, as demonstrated in the example of Coca-Cola above. Endless additional examples could be cited: In June 1997 the U.S. corporation Häagen-Dazs developed a dulce de leche ice cream in Argentina. Within one year dulce de leche was the second most popular flavor in the United States and Europe. The U.S. corporation Nike has introduced around the world a shoe designed by Kenyans: It has a separate big toe, like a mitten, to simulate running barefoot. East German coaches developed in-line skates for competitive ice skaters to practice without ice, but the skates' popularity around the world today is a product of U.S. marketing. Even MTV is less distinctively American than when it began in 1981. It is increasing the local-content percentage of its programming everywhere. In late 2000 it had 22 versions around the world, being watched by about two million people every second, some 1.2 million of them outside the United States. MTV's chairman Tom Freston said that in opening new markets the company "starts with expatriates [from America] to transfer company culture and operating principles," but then the company surrenders control of programming to local executives. "We're always trying to fight the stereotype that MTV is importing American culture," says company president Sumner Redstone. "We aren't. To do so would be cultural imperialism." MTV is "cultivating and nurturing local" artists and shows.

Perhaps the United States has not been sufficiently sensitive to other peoples' fears for their cultures, and to others' shock at the inequities that come from capitalism and globalization. These processes

have triggered resentment and protest even within the United States (p. 540).

The U.S. political example and influence

Beyond these military, economic, and cultural considerations, U.S. political institutions and forms also sweep the Earth. The United States was born in revolution, and the authors of the U.S. Constitution produced the modern world's first republic—that is, a government without a king. They trusted in the people to be able to rule themselves without one, and they saw themselves as setting an example that would sooner or later be followed around the world. In 1836, poet Ralph Waldo Emerson was asked to commemorate the first battle of America's War of Independence, the Battle of Concord Bridge, where patriot farmers had fired upon British soldiers. Emerson wrote, "By the rude bridge that arched the flood/Their flag to April's breeze unfurled/Here once the embattled farmers stood/And fired the shot heard round the world." Emerson was already voicing the popular belief that the American War of Independence launched the idea of republican constitutional democracy—a "shot" that would eventually diffuse around the world, challenging and eventually bringing down all other forms of government (Figure 6-37).

The freedom offered by U.S. political institutions, however imperfect, remains the envy of most people on Earth. The American example of democracy, the rule of law, women's rights, and other aspects of American life is still profoundly destabilizing in many places. By embodying, demonstrating, defending, and even

FIGURE 6-37 The Statue of Liberty in Tiananmen Square. This papier mâché copy of the Statue of Liberty (although with her arms incorrectly placed) was erected by Chinese students demanding democracy in Tiananmen Square in front of the imperial palace in Beijing in June 1989. Thus "the shot" fired at Concord Bridge in 1775 had diffused even to the site traditionally respected by Chinese as the very center of the world. (Alan Reininger/Contract Press Images, Inc.)

promulgating these values, the United States, as a lightning rod, attracts the hatred of people who feel threatened by these values or who hate them.

In global political affairs, however, the role of the United States has been to a degree self-contradictory. On the one hand, the U.S. sets a revolutionary example, and the United States often justifies its actions with regard to other countries by citing the lofty principles of democracy and the right of nations to self-determination (p. 451). At the same time, however, the United States has supported and continues to support many nondemocratic governments. During the Cold War, the United States felt it necessary to accept as allies many governments that suppressed the rights of their own citizens (see p. 556).

The United States has also repeatedly intervened militarily in what would normally be regarded as the internal affairs of other countries (such as Iran, Grenada, and Panama) when its own interests called for such actions. United States foreign policy has often switched sides on international disputes, favoring, for example, Iran or Iraq in turn in disputes between those two nations. Some people call such behavior by the German word *Realpolitik* (realistic politics), but others call it hypocrisy, and it fuels resentment of U.S. power.

In the late 1980s and early 1990s the United States promoted democratization, most notably across Asia and Europe. In 1986, for example, the United States stopped supporting dictator Ferdinand Marcos in the Philippines when it became clear that tolerating his dictatorship was too high a price to pay for military bases there and for "stability" that was in fact toppling into chaos. The United States supported democratic revolutions (or evolution) in South Korea, Taiwan, China, and eventually, in 1989, the collapse of communist government across Eastern Europe. In the Western hemisphere, Africa, and the Near East, by contrast, the United States has less forcefully promoted democracy.

Since 9/11, U.S. officials have been willing to admit this. In December 2003, President George W. Bush admitted that we "have in the past been willing to make a bargain: to tolerate oppression for the sake of stability. . . . Now we're pursuing a different course, a forward strategy of freedom in the Middle East. We will consistently challenge the enemies of reform. . . ." In 2003 U.S. Secretary of State Colin L. Powell admitted that America's role in the coup that overthrew Chile's elected President Salvador Allende in 1973 and brought military dictator General Augusto Pinochet to power for 17 years was "not a part of American history we are proud of."

Overall, the wealth, power, and example of the United States dictate that it is the only country that is involved in the affairs of virtually every other country on Earth. People around the world consume American products and debate America's government policies, and their homelands are often host to American troops. Many people even feel that the United States is more important to them than their own national government is. Sometimes, for example, American corporations can be sued in American courts by foreigners for the corporations' acts in foreign countries, such as pollution or violation of labor laws, more easily than the corporations can be sued in foreign courts. American ubiquity in the world places unique responsibility on Americans to understand other cultures and America's impact on them.

CONCLUSION: CRITICAL ISSUES FOR THE FUTURE

Today many non-Western peoples are trying to defend or to revive their own traditional cultures and values. In societies that were colonized, however, European intervention was so profound that it is virtually impossible to reconstruct what was there before, and today change occurs so fast that it seems unstoppable. Much of what is today called "traditional" is in fact the result of European codification or impressions of rules. When broadcast television was first introduced in Malawi, many people feared an influx of Western culture, but Steven Mijiga, the postmaster general, said, "If you go into any bookstore, the books are all brought here from the West. Look at our religion, our legal system. They are all brought from the West." Today European museums hold African ritual objects for which Africans have forgotten the rituals. Many non-Western people study their history as a fierce reclaiming of their history, of their independent identity, of their self-respect (Figure 6-38). Nigerian writer Chinua Achebe has quoted the African proverb, "Until the lions produce historians, the stories of the hunt will glorify the hunters." In other words, people who were colonized must produce historians and writers to tell their side of history.

All non-Western national leaders today face a problem. Many might like to re-create some idealized model of what their society was like in the past, but that is impossible. They also reject total conversion to Western ideas and standards. Therefore, they must produce some synthesis of civilizations. This requires difficult decisions and compromises. The Japanese, for example, adopted much of Western science and technology, but they have protected and maintained other aspects of traditional Japanese culture, such as an especially high regard for group effort and teamwork.

Retreat into a local cultural past can sustain local morale and cohesion in times of troubles—even when the past is a reconstructed past that never really existed.

> Overall, the wealth, power, and example of the United States dictate that it is the only country that is involved in the affairs of virtually every other country on Earth.

FIGURE 6-38 Telling national history. The extraordinary 1981 international hit movie *Lion of the Desert* was financed by the government of Libya to teach Libya's history from Libya's point of view rather than Hollywood's. Anthony Quinn starred as Omar Mukhtar, who defeated Italy's attempted conquest of Libya from 1911 to 1931. Not all countries have been able to tell their own versions of history this way, either to international audiences or even to their own people. (Museum of Modern Art/Film Stills Archive)

If, however, the resulting defensive attitude disregards the ideas and skills of alien peoples and cultures, the result is to be disastrously left behind by the rest of the world. Even a civilization as vast and successful as the Chinese had to face up to this hard fact beginning in the early nineteenth century. The Chinese people have yet to recover from the shock to their self-esteem. The 2003 SARS epidemic (discussed in Chapter 5) eventually brought home to China the real meaning of globalization so forcefully that the Chinese government publicly referred to it as "our 9/11." Many people in the Islamic world, too, are rankled by the knowledge that from the ninth through the thirteenth centuries, their realm was the world's leading economic power and achieved great heights of artistic and scientific achievement; beginning in the fifteenth century, however, residents of Islamic lands found themselves unable to keep pace with the West, and eventually falling under its domination.

If a nation is to have long-term success, cultural continuity must somehow be combined with attention to useful new imported ideas, practices, and technologies. Many of the problems that countries face today—problems such as urbanization, pollution, and cultural confusion—are problems that Europe faced first but

that European culture has been unable to solve. Some observers have spoken of the "exhaustion" of Western modernity, and they divide world history into three periods: (1) the *premodern world* of relative cultural isolation before 1500; (2) the *modern world* of Western expansion and cultural diffusion from 1500–1950; and (3) the *postmodern world,* in which more world societies are no longer passive receptacles of Western influence, but active shoppers in a global cultural bazaar, picking and choosing what they want and then turning it into something of their own (Figure 6-39).

As we noted earlier, new technology, such as the Internet, lowers the cost of producing and disseminating information, so virtually anyone can be an information producer. Movies made for just a few thousand dollars win international acclaim. The biannual Panafrican

FIGURE 6-39 Global sources of artistic inspiration. The strong, bold forms of this sculpture by Jacques Lipchitz (1891–1973) demonstrate that African art was one of the main influences on the twentieth-century Western art movement called cubism. Lipchitz was one of the founders of cubism, along with Pablo Picasso and Georges Braques, all of whom acknowledged their artistic debt. (Jacques Lipchitz "Figure" 1926–1930 (cast 1958–1961). Bronze, 85⅜ × 38½ × 29¼ inches. Hirshhorn Museum and Sculpture Garden, Smithsonian Institution, gift of Joseph H. Hirshhorn, 1966. Photography: Lee Stalsworth. © Estate of Jacques Lipchitz/Licensed by VAGA, New York, NY/Courtesy of Marlborough Gallery, NY)

Film and Television Festival in Burkina Faso attracts entries from many countries, and many of the entries find global distributors. If one of the world's poorest countries can become a hearth for the production of global culture, then almost anything is possible. The worldwide revival of non-Western cultures and the new cultural permutations coming into being every day allow everyone to contribute to the new world we are creating.

Chapter Review

SUMMARY

An isolated society depends entirely on its local environment for all of its needs, but the evolution of individual human cultures in isolation has, through history, yielded to the increasing interconnectedness of human societies.

Cultures diffuse, cultures change, and people can transform themselves and their behavior, but cultures and culture realms also have elements of stability. Any cultural pattern or distribution maps a current balance between the force of change and of stability. Cultures are constantly changing and evolving through time, and these changes may result either from local initiatives and developments or else as the result of influences from other places. Each local culture is a mix between what originated locally—through cultural evolution—and what has been imported through cultural diffusion. As global communication and transportation have increased, the balance of factors that explain the local activities and culture at any place has tipped steadily away from factors of evolution and toward factors of diffusion. What happens at places depends more and more on what happens among places.

Today's rapid pace of cultural innovation and diffusion requires us to make a distinction between folk culture and popular culture. Folk culture preserves traditions. Most folk culture groups are rural, and relative isolation helps these groups maintain their integrity, but folk culture groups also include urban neighborhoods of immigrants struggling to preserve their native cultures in their new homes. Popular culture, by contrast, is the culture of people who embrace innovation and conform to changing norms. Popular culture may originate anywhere, and it tends to diffuse rapidly, especially wherever people have time, money, and the inclination to indulge in it. The number of aspects of this culture that geographers can study is virtually limitless.

In studying economies and trade, geographers note how the people at any place make their living, where specific economic activities locate, and how economic activities and trade affect other aspects of cultural geography.

All human activities find territorial expression, and maps of culture realms, regions of economic specialization, or political jurisdictions reveal current distributions of human activities. These activities, however, are dynamic, continuously organizing and reorganizing, forming and reforming. Therefore, their distributions and patterns are disrupted and reshaped repeatedly. Geographers need to understand what forces make things move and redistribute themselves.

Despite the rich variety of indigenous local cultures around the globe, the European cultural model is widespread. Europeans came to play a central role in world history and geography because they paved the way for the modern system of global interconnectedness. Global diffusion was fixed hierarchically, with Europe as the apex. Europe conquered most of the rest of the world, and European political domination imposed European concepts of government, law, property, and other aspects of culture. The spread of Western culture continues today, and many traditional cultures and social structures are being radically transformed. In some cases whole cultures disappear.

Many non-European peoples are attempting to revive their own cultural history and values. Perhaps world cultural diffusion will be less hierarchical in the future.

KEY TERMS

acculturation p. 221
Agricultural Revolution p. 250
behavioral geography p. 233
cognitive behavioralism p. 233
Commercial Revolution p. 249
cultural diffusion p. 221
cultural ecology p. 223
cultural imperialism p. 253
culture realm p. 235
culture region p. 235
cyberspace p. 245
diffusionism p. 221

economic geography p. 241
electronic highway p. 245
environmental determinism p. 223
ethnic group p. 231
ethnocentrism p. 232
evolutionism p. 221
felt needs p. 243
folk culture p. 225
footloose activities p. 245
historical consciousness p. 241
historical geography p. 241
historical materialism p. 222

hunter-gatherers p. 221
Industrial Revolution p. 249
inertia p. 240
infrastructure p. 241
mental maps p. 233
pastoral nomadism p. 221
popular culture p. 227
possibilism p. 223
proxemics p. 235
race p. 231
racism p. 231
territoriality p. 235

QUESTIONS FOR REVIEW AND DISCUSSION

1. How did Alexander von Humboldt disprove Varro's theory of unilinear social evolution?

2. Many people believe that we have reached the end of a 500-year period of human history. What has characterized this period? What might happen next?

3. Define historical consciousness and give an example of how it can act as a force of inertia fixing cultural geography.

4. What four things usually happen when you build a road to connect three formerly isolated villages?

5. Define cultural imperialism and give specific examples of how much of the world's cultural geography is the legacy of European cultural imperialism.

6. According to the theory of historical materialism, why are political institutions almost always "behind the times"?

7. Differentiate diffusionist influences on cultural development from evolutionist influences on cultural development and give specific examples of each. What forces or factors at work in the world today favor either diffusionist influences or evolutionist influences?

8. Differentiate folk culture from popular culture. Give a few examples of each that you know about personally.

THINKING GEOGRAPHICALLY

1. What were a few distinctive products of your region 50 years ago? Religious observances? Food products? Clothing or costumes? Architectural styles? Games? Have they been exported from the region? Are they still typically produced? Attend a local street fair or celebration. What aspects of the festival are different from such a festival 100 miles away?

2. Could there be any isolated people left on Earth? Where would they most likely be found?

3. Identify developments that are contributing to the formation of one global culture.

4. Clip a few photos from magazines and challenge your fellow students to identify the location on the basis of attributes of the cultural landscape.

5. How many cities are served by nonstop flights from your town?

6. Henry Louis Gates, Jr., has written, ". . . for those . . . who value the survival of diverse human cultures in just the way naturalists value the diversity of flora and fauna—the march of international corporate capitalism is not without its mournful aspect. In Tibet, an intricate culture of Buddhist worship has been tragically disrupted and weakened by a brutal, deliberate state policy of cultural extirpation. In neighboring Nepal, a highly evolved practice of Hindu worship has been similarly disrupted and weakened, not by a hostile state regime but by the BBC World Service, Coca-Cola, Michael Jackson—in short, the encroachments of Western consumer culture." Do you agree that the processes he speaks of are similar or equivalent? Are they equally malevolent?

7. Explain the saying that what happens *at* places is increasingly dependent on what happens *among* places.

8. Find in today's newspaper an example of civil strife somewhere in the world. Can the strife be understood in terms of the identities or self-perceptions of the parties involved?

SUGGESTIONS FOR FURTHER LEARNING

Ayers, Edward L. *All Over the Map: Rethinking American Regions.* Baltimore: Johns Hopkins University Press, 1996.

Baranowski, Shelley, and Ellen Furlough. *Being Elsewhere: Tourism, Consumer Culture, and Identity in Modern Europe and North America.* Ann Arbor: University of Michigan Press, November 2001.

Bell, Morag, Robin Butlin, and Michael Heffernan, eds. *Geography and Imperialism: 1820–1940.* Manchester, UK: Manchester University Press, 1997.

Berry, Kate A., et al. *Geographical Identities of Ethnic America: Race, Space, and Place.* Las Vegas: University of Nevada Press, 2001.

Black, Jeremy. *DK Atlas of World History.* DK Publishing, 2000.

Black, Jeremy. *Maps and History.* New Haven: Yale University Press, 1997.

Blaut, J. M. *The Colonizer's Model of the World.* New York: The Guilford Press, 1993.

Braudel, Fernand. *Civilization and Capitalism: Fifteenth–Eighteenth Century* (English translations). vol. 1, *The Structures of Everyday Life,* 1981; vol. 2, *The Wheels of Commerce,* 1982; vol. 3, *The Perspective of the World.* 1984. New York: Harper & Row.

Burenhult, Göran, ed. *Traditional Peoples Today: Continuity and Change in the Modern World 5, The Illustrated History of Humankind.* San Francisco: Harper, 1995.

Carney, George O., ed. *The Sounds of People and Places: Readings in the Geography of American Folk and Popular Music.* Lanham, MD: University Press of America, 1993.

Carney, George O. *Baseball, Barns, and Bluegrass: A Geography of American Folklife.* Lanham, MD: Rowman & Littlefield, 1998.

Curtin, Philip. *The World and the West: The European Challenge and the Overseas Response in the Age of Empire.* New York: Cambridge University Press, 2000.

Everson, Paul, and Tom Williamson, eds. *The Archaeology of Landscape.* Manchester, UK: Manchester University Press, 1998.

Fernandez-Armesto, Felipe. *Civilizations.* New York: Macmillan/ Free Press, 2000.

Foote, Kenneth, et al., eds. *Re-Reading Cultural Geography.* Austin, TX: University of Texas Press, 1996.

Franklin, Wayne, and Michael Steiner, eds. *Mapping American Culture.* Iowa City: University of Iowa Press, 1995.

The Futures of Cultures. New York: UNESCO Publishing (Future-Oriented Studies Series), 1994.

Geertz, Clifford. *After the Fact: Two Countries, Four Decades, One Anthropologist.* Cambridge, MA: Harvard University Press, 1995.

Green, D. Brooks, ed. *Historical Geography: A Methodological Portrayal.* Lanham, MD: University Press of America, 1991.

Groth, Paul, and Ted Bressi, eds. *Understanding Ordinary Landscapes.* New Haven: Yale University Press, 1997.

Herman, Andrew, and Thomas Swiss, eds. *The World Wide Web and Contemporary Cultural Theory: Magic, Metaphor, Power.* New York: Routledge, 2000.

Hirsch, Eric, and Michael O'Hanlon, eds. *The Anthropology of Landscape: Perspectives on Place and Space.* New York: Oxford (Oxford Studies in Social and Cultural Anthropology), 1995.

Hodgen, Margaret T. *Early Anthropology in the Sixteenth and Seventeenth Centuries.* Philadelphia: University of Pennsylvania Press, 1964.

Low, Setha M., and Denise Lawrence-Zuniga, eds. *The Anthropology of Space and Place: Locating Culture.* New York: Blackwell, 2003.

MacDonald, Mary N. ed. *Experiences of Place.* Cambridge: Harvard University Press, 2003.

Nisbett, Richard E. *The Geography of Thought: How Asians and Westerners Think Differently . . . and Why.* New York: Free Press, 2003.

Noble, Allen G., ed. *To Build in a New Land: Ethnic Landscapes in North America.* Baltimore, MD: Johns Hopkins University Press, 1992.

Pile, Steve, and Nigel Thrift, eds. *Mapping the Subject: Geographies of Cultural Transformation.* New York: Routledge, 1995.

Mitchell, Donald. *Cultural Geography: A Critical Introduction.* New York: Blackwell, 1999.

Parker, Philip M. *National Cultures of the World: A Statistical Reference.* New York: Greenwood Publishing Group, 1997.

Rooney, John F., Jr., and Richard Pillsbury. *Atlas of American Sport.* New York: Macmillan, 1992.

Silberfein, Marilyn, ed. *Rural Settlement Structure and African Development.* Boulder, CO: Westview, 1995.

Stea, David, and Mete Turan. *Placemaking: Production of Built Environment in Two Cultures.* Brookfield, VT: Avebury Press, 1996.

Upton, Dell, ed. *America's Architectural Roots: Ethnic Groups that Built America.* Washington, D.C.: National Trust for Historic Preservation Press, 1986.

Wallach, Bret. *Losing Asia: Modernization and the Culture of Development.* Baltimore, MD: Johns Hopkins University Press, 1996.

Weiss, Michael J. *The Clustered World: How We Live, What We Buy, and What It All Means about Who We Are.* Boston: Little Brown and Company, 2000.

The World Atlas of Architecture. London: Mitchell Beazley Publishers, 1984. (This is the English edition of *Le Grand Atlas de l'Architecture Mondiale,* published by Encyclopedia Universalis, Paris, 1981. An American edition was published in 1988 by Portland House in New York, NY.)

Yamin, Rebecca, and Karen Metheny. *Landscape Archaeology: Reading and Interpreting the American Historical Landscape.* Knoxville, TN: University of Tennessee Press, 1996.

Zelinsky, Wilbur. *A Cultural Geography of the United States: A Revised Edition.* Upper Saddle River, NJ: Prentice Hall, 1996.

WEB WORK

Many Web sites relate to cultural studies. You might wish to begin at

http://www.popcultures.com/

The American Studies Web is a rich source of wide-ranging resources on the entire range of cultural topics. At this site you can find resources on many aspects of art and material culture, as well as good coverage of race and ethnicity, including African American, Asian American, Native American, and Latino and Chicano studies:

http://www.georgetown.edu/crossroads/asw/

Another interesting site is the Virtual Library and Museums of the World Homepage sponsored by the International Council of Museums:

http://www.icom.org/vlmp/

The European Research Centre on Migration and Ethnic Relations is a university-based research institute that focuses on movements of people around the world, interrelationships among ethnic groups, and problems of social integration and social exclusion. It can be found at

http://www.ercomer.org

The Research Centre maintains the World Wide Web Virtual Library of Migration and Ethnic Relations:

http://www.ercomer.org/wwwvl/

Culture on the Edge is a web-based magazine devoted to the study of vanishing cultures

www.culturesontheedge.com

Practically every large corporate producer of popular culture has a home page. Among the more interesting are

http://www.warnerbros.com/pages/main/index.jsp
http://www.sony.com/
http://www.viacom.com/online.tin

Culture is a broad topic that touches on every aspect of human existence throughout the world. A visit to the Web sites mentioned here will lead you to thousands of other interesting and rewarding sites.

A Look Ahead

DEFINING LANGUAGES AND LANGUAGE REGIONS

The great variety of languages spoken today testifies to the relative isolation of groups in the past, and the distribution of any language illustrates the pattern of dispersal of its original speakers or their cultural impact on others.

THE DEVELOPMENT AND DIFFUSION OF LANGUAGES

If groups of people who speak a common language disperse, then each breakaway group will develop its own language but retain elements of the common ancestral language. Languages that are thus related make up a language family. The geography of written language also records the origins and dispersal of peoples and cultures. Toponymy, the study of place names, records natural features or the origins or values of a place's inhabitants.

LINGUISTIC DIFFERENTIATION IN THE MODERN WORLD

The use of some languages is expanding because their speakers are diffusing around the world, gaining

7

THE GEOGRAPHY OF LANGUAGES AND RELIGIONS

◄ This new mosque in Rome, Italy, stands just a few miles from Vatican City, the seat of the Roman Catholic Church. Immigrants are bringing Islam to Europe, but some Europeans are converting to it, too. No Christian churches are allowed anywhere in Saudi Arabia, the seat of Islam, but Christianity allows the separation of Church and State. Religious teachings carry ramifications across many aspects of human geography. (Origlia Franco/ Corbis/Sygma)

greater power and influence in world affairs or winning new adherents to their ideas. Other languages are disappearing.

THE TEACHINGS, ORIGIN, AND DIFFUSION OF THE WORLD'S MAJOR RELIGIONS

Each religion originated in one place and spread out from there, but it is difficult to identify precise reasons why any religion diffuses. Some religious groups try to convert others to their religious beliefs, but others do not.

THE POLITICAL AND SOCIAL IMPACT OF RELIGION

Religion is a major determinant of human behavior, so the geography of people holding and following a set of teachings affects politics, economics, agriculture and diet, and many other aspects of human life.

The Soviet Union was officially atheistic, so since its 15 constituent republics broke into 15 independent states, each has had to define a new relationship between religion and government. The interplay between religion and nations' politics is often one of the most perplexing aspects of national cultures.

The Russian Constitution prohibits the establishment of an official religion, but many Russians view Orthodox Christianity as an essential element of Russian identity. The Orthodox Church has been in Russia for over 1,000 years. The Russian language is written in the Cyrillic alphabet, devised by Saint Cyril in order to convert the Russians to Christianity. After Rome fell to non-Christians in 476 and Constantinople in 1453, Moscow for 500 years called itself "The Third Rome" and center of the Christian world. President Vladimir Putin is the first Russian leader since Czar Nicholas II to have been baptized and remained faithful to the Church. Czar Nicholas and his family were canonized as saints by the Church in August 2000. The neighboring Republic of Belarus passed a law in 2002 according the Orthodox Church "a determining role in the culture and government" of that nation. The law recognizes other faiths, but their practice is circumscribed by regulations, and their followers are denied oportunities in society.

Several of the Central Asian republics, where the population is Muslim, have had difficulty defining church-state relations. In Uzbekistan, for example, the government has persecuted members of the Islamic Party of Liberation, which promotes Islam as a religious and social force. President Karimov, the former Communist Party leader who has held office since first elected in 1991, insists that religious fundamentalism is a "plague and sickness capable of infecting the minds of young people." Thousands of Muslim men have been imprisoned, and many more have disappeared. The President insists that his actions are necessary to preserve democracy in Uzbekistan. Liberation Party members, however, claim that while they criticize corruption, they do not preach violent overthrow of the government.

Neighboring Tajikistan demonstrates another face of political Islam. The government fought a war against Muslim insurgents from 1992 to 1997 that left at least 50,000 dead, but today the Islamic Revival Party holds seats in Parliament and controls several local governments. Party members insist that anyone who prevents Muslims from teaching about Islam invites war, but there is no reason for rebellion if Muslims are allowed to preach.

Language and religion are two of the most important forces that define and bond human cultures. Their influences are so pervasive that many people take them for granted and cannot objectively observe their influences in the lives of others—or even in their own lives. Nevertheless, peoples who share either of these two cultural attributes often demonstrate consistencies in other aspects of behavior, and often they can more easily cooperate with one another in other ways, such as in international affairs. Sharing these attributes may help them understand one another in ways that are unique and profound.

Each language and each religion originated in a distinct hearth, and although the carriers of each of these cultural attributes have diffused around the world, each language and each religion still predominates within a definable realm. These realms of language and of religion are two of the most important of all types of culture regions.

DEFINING LANGUAGES AND LANGUAGE REGIONS

Many social scientists believe that language is the single most important cultural index. A **language** is a set of words, plus their pronunciation and methods of combining them, that is used and understood to communicate within a group of people. Each language has a unique way of dealing with facts, ideas, and concepts; and variations in languages result in variations in how people think about time and space, about things and processes. Exact translation from one language to another is virtually impossible. The language an individual speaks has an influence in structuring his or her perception and logic, and the comparative study of languages is one of the richest ways of understanding human psychology.

Any group of people who communicate exclusively with one another will soon develop "their own language," whether they are nuclear physicists or a social clique at a high school. If human beings were to appear suddenly in today's highly interconnected world, one worldwide language might develop. The great variety of languages spoken today testifies to the relative isolation of groups in the past. The distribution of any language illustrates the pattern of dispersal of its original speakers or their cultural impact on others.

The number of different languages recognized varies with the accepted definition of language. The term language is usually reserved for major patterns of difference in communication. Minor variations within languages are called **dialects,** but scholars do not agree on the amount of distinctiveness necessary for a pattern to be considered a language. Some scholars, for instance, accept Danish, Swedish, and Norwegian as distinct languages, but a speaker of any one of them can understand the others. Therefore, other scholars insist that these are three dialects of one language. A **standard language** is the way any language is spoken and written according to formal rules of diction and grammar, although many regular speakers and writers of any language may not always follow all those rules. A country's **official language** is the one in which official records are kept and government business is normally conducted.

A pidgin language is a system of communication that has grown up among people who do not share a common language, but who want to talk with each other. Pidgins are marginal or mixed languages, and they usually disappear after a few years or else they evolve into a **creole.** A creole is a pidgin language that has survived long enough to become a mother tongue. That usually takes a generation or two. Examples of creoles include Gullah, the English-based language used among some African Americans living along the U.S. Southeast coast; Haitian French Creole; Guyanese Creole; and Seychellois, a French-based creole spoken on the Seychelles Islands.

A **lingua franca** is a second language held in common for international discourse. Today English is the world's leading lingua franca. Air controllers and pilots in international aviation, for example, all speak English. Other languages have served as lingua francas in the past. Latin long served Western civilization, and Swahili served throughout East Africa. Swahili developed among black peoples and in communication with Arab traders, so it has many Arab words. Today it is an official language in Tanzania and Kenya.

Individual languages change through time, but religious classics or classics of literature can exert a powerful force for stabilization. In English, for example, the works of William Shakespeare and the 1612 King James translation of the Bible have molded the language, yet parts of even these works may be difficult for many English speakers to read today.

Linguistic Geography

The study of different dialects across space is called *dialect geography,* or *linguistic geography.* Dialects usually diverge more in the way they are spoken than in the way they are written. This is because writing is often widely dispersed, but sounds are localized only among a group of people who speak together, called a **speech community.** Probably no student of dialects has ever developed the ability of the fictional Henry Higgins in G. B. Shaw's play *Pygmalion*: "I can place any man within six miles. I can place him within two miles in London. Sometimes within two streets." To this day, however, there are distinct dialect regions across the United States.

Sometimes researchers survey speech and draw lines around places where speakers use a linguistic feature in the same way. These boundary lines are called **isoglosses.** Figure 7-1 reproduces a map from the early twentieth-century *Atlas linguistique de la France.* The atlas's makers identified a bundle of isoglosses running across France from east to west, dividing the country into two major dialect areas. These two areas are known as *langue d'oil* (the North) and *langue d'oc* (the South) based on the words used for "yes" in these areas during the thirteenth century, when the division was first recognized.

FIGURE 7-1 These six isoglosses represent the dividing lines between regions to the north and to the south of which in France the people pronounce six individual words differently: *essayer,* to try; *vingt,* twenty; *sève,* vigor; *nous étions,* we were; *perdue,* lost; *raison,* reason. Clearly this bundle of isoglosses represents a significant border between two spoken dialects. (Recreated from the *Atlas linguistique de la France*)

Isoglosses frequently parallel physical landscape features, because physical features often act as barriers to human migration and diffusion. For example, the Pyrenees mountains divide Spain from France, and the Pripet marshes separate Belarus from Ukraine. By contrast, languages often diffuse quickly across broad lowlands, and languages have historically diffused along river valleys or other routes of trade and transportation.

Researchers often map what is called a *geographical dialect continuum*. This is a chain of dialects or languages spoken across an area. Speakers at any point in the chain can understand people who live in adjacent areas, but they find it difficult to understand people who live farther along the chain. An example of such a chain is the North Slavic continuum, which links Czech, Slovak, Ukrainian, Polish, and Russian.

The Official Languages of the Countries of the World

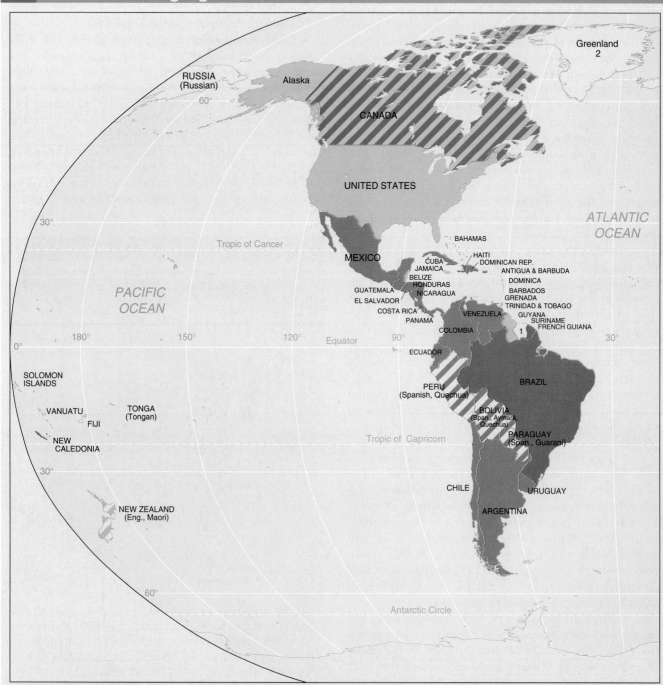

FIGURE 7-2 This map reveals the worldwide distribution of some European languages. This distribution is one legacy of European imperialism.

The World's Major Languages

Most scholars agree that there are more than 3,000 distinct languages, at least 30 of which are spoken by 20 million or more people each as a first language (Table 7-1). About 50 percent of all the world's people, however, speak one of the 12 major languages listed in the table. The language with the most speakers is Mandarin Chinese (Guoyo), with about 885 million native speakers. English is the primary language of about 350 million people worldwide, and it is either the only official language or one of several official languages in about 50 countries (Figure 7-2). Arabic derives special transnational importance as the language of the Koran, the sacred scriptures of Islam. The Koran has been translated, but Muslims are still encouraged to study

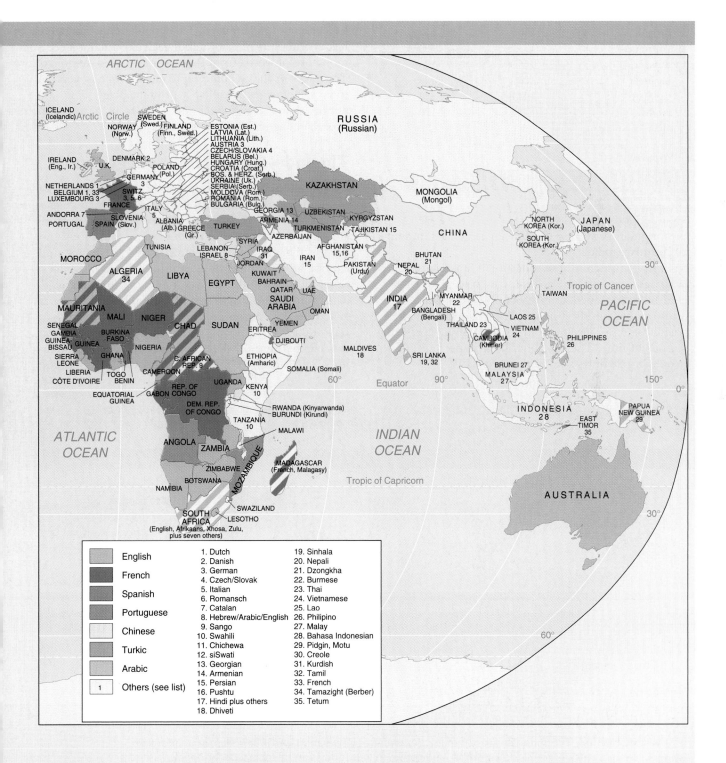

TABLE 7-1 The World's Leading Languages and the Number of Speakers of Each (in millions)

Language	Native speakers	Total speakers
Mandarin	885	1,075
Hindi	375	496
Spanish	358	425
English	350	514
Arabic	211	256
Bengali	210	215
Portuguese	178	194
Russian	165	275
Japanese	125	126
German	100	128
French	77	129
Malay-Indonesian	58	176

Note: A native speaker is one for whom the language is his or her first language.

Source: U.S. Department of State

THE DEVELOPMENT AND DIFFUSION OF LANGUAGES

Any isolated group of people develops a language of its own. This language describes everything that those people see or experience together. If groups of these people break away and disperse, then each group discovers new objects and ideas, and the people have to make up new words for them. After thousands of years, the descendants of each of these breakaway groups have their own language. Each descendant language has a vocabulary of its own, but each also retains a common core of words from that earliest shared language. The ancestor that is common to any group of several of today's languages is called a **protolanguage.** The languages that are related by descent from a common protolanguage make up a **language family.**

The Indo-European Language Family

In 1786 the English philosopher Sir William Jones first pronounced his theory that a great variety of languages spoken across a tremendous expanse of Earth demonstrate similarities among themselves so numerous and precise that they cannot be attributed to chance and cannot be explained by borrowing. These languages, then, must descend from a common original language. The group of languages first identified by Sir William is called the *Indo-European family* of languages, and about half of the world's peoples today speak a language from this family (Figure 7-3). Sifting through the vocabularies of all Indo-European languages yields a common

the original. Arabic is the official language in some 20 countries today.

In contrast to these widespread languages, some languages are extremely local. Linguists have discovered fully developed languages in New Guinea spoken by only a few hundred people in certain valleys. These languages have developed in total isolation over long periods of time. They are utterly incomprehensible to people just 20 miles away in the next valley.

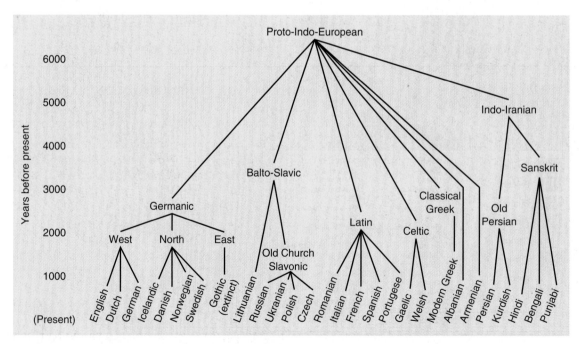

FIGURE 7-3 The Indo-European family of languages. The languages in this "family tree" are all descendants of proto-Indo-European, and common word roots can be found among them.

core vocabulary, which is the common ancestor of these languages, *proto-Indo-European*.

Jacob Grimm (1785–1863), one of the brothers who collected children's fairy tales, formulated rules to describe the regular shifts in sounds that occurred when the various Indo-European languages diverged from one another. There is, for example, a regular sound shift between words beginning with "p" in Latin and "f" in Germanic languages (as in pater and father). These rules are known as *Grimm's law*.

The vocabulary of proto-Indo-European tells us a surprising amount about how proto-Indo-European society was organized and how the people lived. It also hints at the language's hearth. Reconstructed proto-Indo-European has words for distinct seasons (one with snow), woody trees (including the beech and the birch), bears, wolves, beavers, mice, salmon, eels, sparrows, and wasps. These things can be found together around the Black Sea, but proto-Indo-European also includes words borrowed from the languages of the Near East. The word for wine, for instance, seems to descend from the non-Indo-European Semitic word *wanju*. Thus, the hearth area for Indo-European languages was probably in today's Turkey, some 8,000 years ago. Archaeologists disagree as to whether proto-Indo-European diffused quickly, carried by warriors, or slowly, carried by individual farmers.

Hunting common Indo-European roots of words provides a fascinating study. The proto-Indo-European root *aiw*, for example, which means "life" or "the vital force," descended into Hindi as *ayua*, "life," but it also shows up as *aetas* in Latin, *aion* in Greek, *ewig* in German, and the words *ever* and *age* in English. Words that are related appear in surprising places. *Maharajah*, for example, a Hindi word for a great ruler, may seem exotic, yet *maha* is a distant cousin of the English words "major" and "magnitude." *Rajah* is a distant cousin of the English "reign" and "royal." In this case the cousins clearly look or sound alike, so they are called **cognates.** One must never assume any connection between two words from different languages, however, until it can be proven that they share a common root. The study of word origins and history is called **etymology.**

Proto-Indo-European provided the basic stock for all Indo-European languages, but that does not mean that one ethnic or racial group spread out to live where all these peoples are today. Sometimes a few Indo-Europeans conquered and imposed their language on a much larger group of people that had previously developed a language of their own—as occurred throughout Latin America and Africa. Sometimes peoples adopted Indo-European just to be able to communicate with Indo-Europeans. Not everyone who speaks English in the world today necessarily has an ancestor from England.

Also, cultures borrow things and the words to name them from one another. Western culture gave

Japan the word *erebata* along with the elevator itself. Japanese culture gave the West edible raw fish and a name for it: *sushi*.

Other Language Families

More people speak Indo-European languages than languages of any other family, and the Indo-European family has been studied more than any other. Researchers have benefited from written sources dating back 3,700 years—tablets in the extinct Indo-European language of Hittite. These sources give researchers considerable confidence in including individual languages in the family and in reconstructing proto-Indo-European.

The classification of all Earth's languages into families is an enormous and difficult task. Comparative linguists do not agree whether it has yet been satisfactorily completed. Figure 7-4 reflects the most generally agreed-upon state of understanding.

The same principles apply to each of these families as to the study of Indo-European. Scholars sift vocabularies of today's individual languages to compile vocabularies of a protolanguage. The protolanguage tells us about the origins and culture of the people who originated that protolanguage, and the sound shifts help us trace how and when various individual languages and peoples broke off linguistically and geographically from the main branch of the group. We know, for example, that the Finns, Estonians, and Hungarians are distant cousins; their languages are among the Uralic languages, and their common ancestors long ago emigrated from the region of the Ural mountains in today's Russia. Most peoples of sub-Saharan Africa speak Bantu languages, whose origin can be traced to today's Nigeria about 500 B.C. The ancestors of the Turks of today's Turkey left their homeland near Asia's Altai Mountains 1,000 years ago, while their cousins—the Uzbeks, Kazaks, Kyrgyz, Azerbaijani, and Turkmen—stayed in Central Asia. Scholars believe that early migrants to Madagascar came from the South Pacific rather than from nearby Africa, because the language of Madagascar is related to those of the South Pacific.

Some comparative linguists believe that after all languages have been classified into families, protolanguages can be constructed for each language family and that study of those protolanguages will eventually yield superprotolanguages. They argue that the Indo-European family, for instance, is only one of six branches of a larger group, which they call *Nostratic*. Reaching even further back into prehistory, Nostratic itself is only one descendant of a protohuman language, or Mother Tongue, that was spoken in Africa 100,000 years ago and then diffused around the globe.

In 2001 a study of the genomes of humans and chimpanzees revealed the first human gene specifically involved in language, the FOXP2 gene. The gene had remained largely unaltered during the evolution

The World's Language Families

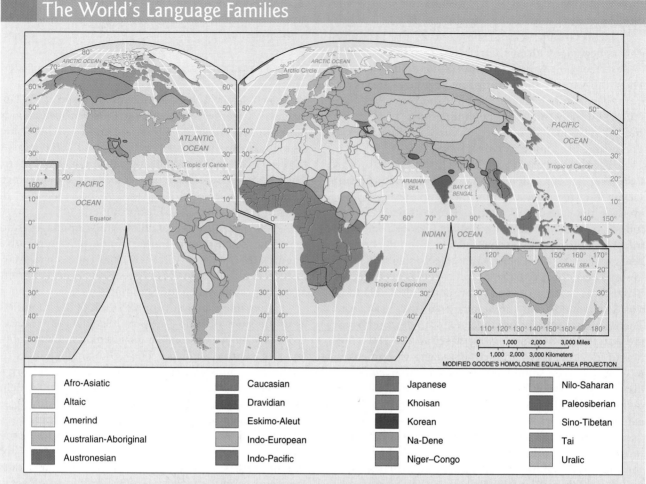

Afro-Asiatic	Caucasian	Japanese	Nilo-Saharan
Altaic	Dravidian	Khoisan	Paleosiberian
Amerind	Eskimo-Aleut	Korean	Sino-Tibetan
Australian-Aboriginal	Indo-European	Na-Dene	Tai
Austronesian	Indo-Pacific	Niger–Congo	Uralic

FIGURE 7-4 Language families are groups of languages that have common linguistic ancestors. In some areas, languages other than those shown on this map have been adopted as official languages, even though they may be spoken by minorities.

of mammals, but it suddenly changed in humans after the hominid line split off from the chimpanzee line of descent about 100,000 years ago. These discoveries suggest that we now have both linguistic and biological evidence that all humans can be traced back to one stock that evolved in Africa. The further back into prehistory that scholars attempt to reconstruct individual languages, however, the more controversial the work becomes. Some linguists believe that the time depth of such studies is too great to reconstruct more protolanguages, let alone one prehistoric mother tongue. Geography, linguistics, anthropology, and biology each contribute to this fascinating but controversial research into prehistory.

> ... we now have both linguistic and biological evidence that all humans can be traced back to one stock that evolved in Africa.

There have been two indisputably independent inventions of writing—by the Sumerians in Mesopotamia before 3000 B.C. and by the Central American Olmec peoples about 650 B.C. Egyptian writing of 3000 B.C. and Chinese writing (by 1300 B.C.) may also have arisen independently, but probably all other peoples who have developed writing have borrowed, adapted, or at least been inspired by existing systems (Figure 7-5).

Most languages are written in alphabets, which are systems in which letters represent sounds. There are several alphabets in use today, and the alphabet in which a language is written can reflect a historic diffusion. Modern Western European languages are written in the Roman alphabet because Western Europeans were converted to Christianity from Rome. In Eastern Europe, in contrast, Russian, Belarussian, Ukrainian, Serbian, Bulgarian, and a few other languages used in Russia are written in the Cyrillic alphabet. This is the Greek alphabet as it was augmented and taught by the

The Geography of Writing

The geography of languages is complicated by the geography of **orthography,** which is a system of writing.

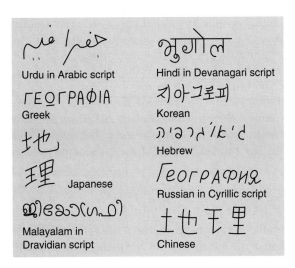

FIGURE 7-5 Examples of orthography. This figure shows the word "geography" written in several different orthographies.

missionary Saint Cyril (d. 869), who converted many of these peoples to Orthodox Christianity. Serbian and Croatian are one spoken language, but the Serbs write it in Cyrillic, and Croats write it in Roman letters. Thus, the line between Serbs and Croats demarcates a cultural borderline across Europe (Figure 7-6).

A similar orthographic-religious divide separates Pakistan from India. Spoken Urdu, the language of Pakistan, is the same as spoken Hindi, the language of northern India. The Muslim Pakistanis write it in Arabic script, but the Hindus of India write it in Devanagari script. Several other Indian languages are also written in Devanagari.

Sometimes a people can change its orthography as a self-conscious political act. Romania, for example, switched from the Cyrillic to the Roman alphabet early in the twentieth century. This change reflected a choice to be a "Western European" nation. In the same way, the Turkish languages of Central Asia came to be written in Arabic when these people converted to Islam, as discussed later in this chapter. Early in the twentieth century, the Turks of the country of Turkey

The Distribution of Alphabetic Scripts in Eurasia

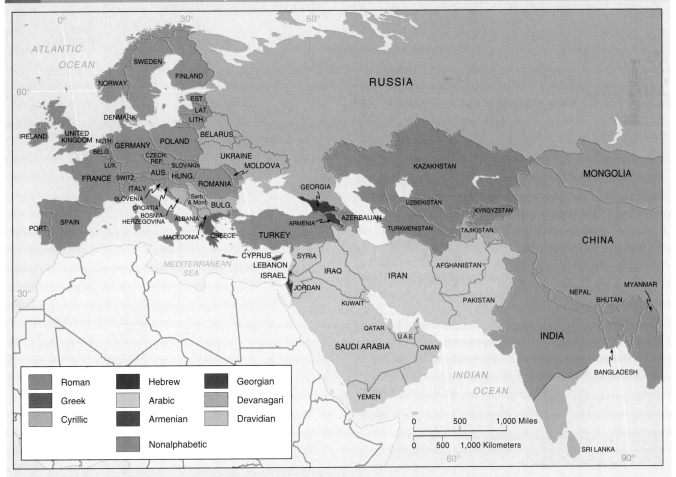

FIGURE 7-6 The distribution of scripts reveals the diffusion of religions and other cultural attributes.

chose to replace Arabic with the Roman alphabet to signal a deliberate Western-style modernization (look back at Figure 6-33). In 1939 Soviet dictator Joseph Stalin forced the Turkish peoples in Central Asia to replace Arabic script with Cyrillic script. His objective was to cut off these peoples from their cultural heritage and to intensify relatively minor linguistic differences among them. In 1992, however, with the collapse of the U.S.S.R., the newly independent republics of Kazakhstan, Kyrgyzstan, Uzbekistan, Turkmenistan, and Azerbaijan all chose to abandon Cyrillic script. They did not choose to revert to Arabic, but to adopt Roman. This reflected a repudiation of their previously colonial status and a commitment to join the cultural community of the West. A Russian law of 2002 prescribes the Cyrillic alphabet as mandatory for Russian and for the languages of all ethnic republics within Russia; the restless region of Chechnya, however, defiantly uses the Roman alphabet as a symbol of cultural independence.

Not all languages are written in alphabets. The major non-alphabetic forms of writing are Chinese, in which each character represents a word or concept, and Japanese, in which each character represents a syllable. Korean writing was modeled on Chinese, and the National Academy of the Korean Language is currently revising the proper English spellings of Korean words so that the sounds are more accurate when pronounced in English. Words that have long started with the letters k, t, p, and ch, for example, should now start, respectively, with g, d, b, and j. Thus the name of Korean President Kim Dae Jung, should now be Gim Daejung, and the city of Pusan is now Busan. The new spellings were given the force of law in 2000, and the change is to be completed by 2005. Computer software has eased communication in ideographic languages, but the demands of contemporary international communication are nevertheless pressuring these countries to adopt the Roman alphabet. The Vietnamese changed from Chinese writing to Roman orthography in the eighteenth century when they were Christianized by missionaries.

Many traditional Native American, African, and Asian languages never had written forms until Christian or Islamic missionaries set out to translate the Bible or the Koran into these languages, using the missionaries' own orthographies. Somaliland, for example, received a written language in 1972.

These peoples may have benefited from having their languages written, but imposing any orthography on a people whose culture had been oral is an act of cultural imperialism. The transition from an oral to a literate culture is arguably the greatest transition in human cultural history.

Toponymy: Language on the Landscape

Toponymy is the study of place names. Place names record natural features in the present or as they were in the past, as well as something about the origins or values of a place's present or past inhabitants.

Some names describe environments as they are today (Oak Bay, British Columbia), whereas others describe environments as they were in the past. Many treeless towns in the North China plain, for example, are named Wang, which means oak tree. We assume that oaks grew there in the past, but either climate change or human impacts have eliminated oaks from the landscape.

Names often reveal what the people in a particular location do or believe. If cities have names like St. Paul, we are in the Christian realm; conversely, Islamabad ("the place of Islam," the capital of Pakistan) indicates the Islamic realm. City names in Russia honored Communist heroes during the period of Communist government, but today many of those places have reassumed the names they had before the Communists came to power (Figure 7-7). St. Petersburg, for example, was renamed Leningrad through the Communist period, 1917–1991. Newly independent countries often replace colonial names. In Africa, for example, Rhodesia, named for Cecil Rhodes (1853–1902), an English administrator and financier in South Africa, became Zimbabwe (the name of an ancient native city there). South Africa has recently Africanized many formerly European-language place names. The city of Bloemfontein, for example, has become Mangaung, "the place of leopards," in the Sesotho language.

Communist era name	Previous and now restored name
Andropov	Rybinsk
Brezhnev	Naberezhnye Chelny
Chernenko	Sharypovo
Frunze	Bishkek
Georgiu-Dezh	Lisky
Gorky	Nizhny Novgorod
Gotvald	Zmiev
Kalinin	Tver
Kuibyshev	Samara
Kirovbad	Gyanja
Leninabad	Khodjent
Leningrad	St. Petersburg
Mayakovsky	Bagdati
Ordzhonikidze	Vladikavkaz
Sverdlovsk	Yekaterinburg
Voroshilovgrad	Lugansk
Zhdanov	Mariupol

FIGURE 7-7 Politics changes place names. During the period of Communist government in the Soviet Union, all these historic cities' names were changed to honor Communist leaders. The cities have reassumed their earlier names.

Toponymy can also reveal history when other evidence has been erased. Spain, for example, was long occupied by Arabic peoples, and today "guada," as in Guadalquivir or Guadaloupe, lingers as a corruption of the Arabic *wadi*, which means river. Guadalquivir is Wadi al Kabir, the great river.

The majority of place names in North America today are possessives and personal names, such as Jones Creek, but descriptive names are also common (the Red River, for example). Many Native American place names survive (Winnipeg is Cree for dirty water), but areas explored or settled by the Spanish, French, or English can be traced by trails of place names. The arc of French settlement from Quebec to New Orleans can be traced through Montreal (Mount Royal); Detroit (the "straits" between Lakes Erie and St. Clair); Fond du Lac, La Crosse, and Prairie du Chien (on routes across Wisconsin); down the Mississippi past Dubuque (named for settler Julien Dubuque); St. Louis; and Baton Rouge. Spanish-named settlements stretch across the South from St. Augustine (originally San Augustino), Florida; to San Antonio, Texas; to Sante Fe, New Mexico; to San Diego, California. Other place names recall early settler groups (Figure 7-8). Commemorative names honor historic personages. Alexander von Humboldt, the geographer introduced in Chapter 1, visited North America, and he is commemorated by counties in three states; towns in six states and in Saskatchewan; and several features in the U.S. West, although he did not visit all of those places.

Today all place names in the United States are recorded by the Board on Geographic Names, created in 1890. The board has a committee to certify foreign place names for American usage, such as all names used in this book.

LINGUISTIC DIFFERENTIATION IN THE MODERN WORLD

Diffusion and differentiation of the major languages continues. Communication within individual speech communities (usually countries) tends to individualize each dialect from the common mother language, but at the same time international communication among these groups tends to homogenize the dialects. For example, English has differentiated into many dialects. Today an English-speaking American, Nigerian, and Indian will not be able to understand one another completely without effort. Sometimes even the same words will convey different meanings when they are used by people from such diverse cultures as the United States, Nigeria, and India. In 1877 the linguist Henry Sweet (the model for Henry Higgins in *Pygmalion* and later the musical version, *My Fair Lady*) predicted that by 1977 even the English, Americans, and Australians would speak mutually incomprehensible languages, because of their isolation from one another. These dialects have not, however, differentiated to the degree Professor Sweet predicted. This is partly because English, American, and Australian cultures have not greatly differentiated from each other in their other aspects. Also, Sir Henry never could have foreseen the great increases in world communication, nor the fact that rising percentages of national populations would receive formal education, and formal education tends to stabilize languages.

Other major languages continue to diffuse and differentiate. Both French and Spanish are widely used. Both the French and the Spaniards try to "purify" their languages of linguistic borrowing but also to add new words so that their languages stay useful in the modern world. The French Academy, a scholarly institution dating back over 300 years, devises new French words for new concepts and items in international trade and discourse, and these terms become official for French. Despite the efforts of the Academy, however, the French spoken in the 35 Francophone countries and Quebec are different dialects. Spanish also experiences differentiation. In 1997 Mexico, the world's most populous Spanish-speaking nation, sponsored the first International Congress on Spanish. Scholars and writers from around the world discussed the national variations of Spanish. International conferences bring together Germans, Austrians, and Swiss to harmonize the spelling and grammar of German.

National Languages

There is no exact correspondence between languages and the countries of the world. A few languages, such

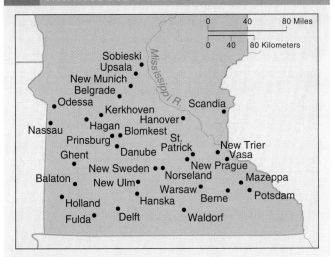

FIGURE 7-8 Some North American cities retain the names of explorers who just passed by, but these Minnesota place names supply a good clue to the origins of the immigrant settlers in the area. Each is a city in Europe or a hero to some national group. How many origins can you identify?

Critical Thinking

COMPETITIVE EXPANSION AND SHRINKAGE

The map of the geography of languages is not static. The use of some languages is expanding because the speakers of those languages are diffusing around the world, are gaining greater power and influence in world affairs, or are winning new adherents to their ideas. English is the world lingua franca partly because much of the world's science and business uses it. In the 1970's Malaysia's then-Minister of Education Mahathir Mohammad expelled English from all national schools and required the use of Malay, Chinese, or Tamil. In January 2003, however, English was reintroduced for teaching all math and science classes in Malaysia. Mahathir Mohammad, Prime Minister in January 2003, said, "We have to accept English whether we like it or not." In 2002 over 92 percent of secondary school students in the European Union's non-English-speaking countries were studying English, whereas only 33 percent were studying French and 13 percent German.

Many multinational corporations have designated English their corporate language, whatever the languages of their home countries might be. In computer terminology, there is virtually no language but English. Another reason for the spread of English is the global appeal of U.S. popular culture, especially movies and music. The genuine merits of English provide still other reasons for its success. It has a huge vocabulary but a simple grammar. Furthermore, it is open to change and absorption of new words: foreign words, coinages, and grammatical shifts. Nevertheless, as English grows as a second language, it is losing ground as a first language. The percentages of Earth's population speaking Hindi, Urdu, Arabic, Spanish, and Chinese as first languages are all growing faster than the percentage speaking English first.

Anybody anywhere can communicate via the Internet with people who speak his or her own language, but the Internet was born in the United States, so about 80 percent of home pages (Web sites) are in English. The second greatest number of sites are in Japanese (only 2.5 percent), followed by German, French, and Spanish. English accounts for about 97 percent of all pages linked to secure servers, however—that is, pages used in electronic business transactions. The English-language advantage in access will probably decrease as more international users log on and as computer translation improves. A unicode devised in 1995 allows computers to represent the letters and characters of virtually all the world's languages, and commerical services now offer translations on the Web. Ministers from French-speaking countries meet regularly to promote French-language use of the Web, and the number of Chinese Internet users and Web pages is increasing rapidly.

Japan has recently won such wealth and economic power that Japanese might be expected to become an international language. Nevertheless, despite the multiplying number of Japanese Web sites, this has not happened. The reason is that the Japanese language developed over centuries in isolation, and it is so complicated and difficult that the Japanese actually suffer handicaps in international communication. More computer capacity is required to define non-alphabetic languages than alphabetic languages, and special computer software is required to visit Japanese-language Web sites. Japanese have been quick to adopt the languages of others, most notably English. Today Japanese power and influence actually contribute to the diffusion of English.

While some languages are growing and spreading, devising new ways of expressing new concepts and winning new speakers, others are dying out. For example, among the Celtic languages of Western Europe, Manx, the original language of the Isle of Man, and Cornish, the original language of Cornwall in western Great Britain, both have been overwhelmed by English and have disappeared. Welsh is confined to a small area and population. Countless languages or dialects throughout Africa, South America, and Asia are disappearing before they are even recorded. In 2003, researchers estimated there were 46 different people who were the last remaining speakers of their native languages. Each language constructs perception and reality differently, so the loss of a language forecloses an opportunity to explore the workings of the human mind.

Questions

1. What government policies might preserve the national language of a small country? Investigate the policies of the Netherlands or Iceland.

2. A shrinking percentage of Americans is studying foreign languages. Can you hypothesize why? Is this good, bad, or of no significance?

as Icelandic and Japanese, are associated almost exclusively with one country, but several languages are shared by many countries. The relationship between languages and nationalism is very complicated.

In European history, language has been interpreted as the basis of nationalism. As early as 1601 Henri IV of France seized French-speaking territories from the Duke of Savoy and declared to his new subjects, "It stands to reason that since your native tongue is French, you should be subjects of the King of France." This logic, however, never stopped the French from seizing the territory of non-French speakers whenever they could.

In 1873 the Third International Statistical Congress recommended that all censuses henceforth include a question of language. The language that a person speaks at home was thought to be the only aspect of nationality that could objectively be counted and tabulated. Therefore, a person's language was "definitive" of that person's nationality. That decision played a key role in augmenting feelings of competitive, self-conscious nationalism that eventually helped cause World War I.

Language and nation building   *Philological nationalism* is the idea that "mother tongues" have given birth to nations (Figure 7-9). This idea persists despite the fact that standard languages usually were, and still are, actually the product of self-conscious efforts at nation building by centralizing governments. Standardized languages cannot emerge before mass schooling and mass literacy or, alternatively, universal service in a national army. Usually these centralizing pressures transform the language of a small percentage of the population, the political or cultural elite, into the national language. In 1789, 50 percent of the population of France did not speak French at all, and only 12 to 13 percent spoke the standard language.

Several national languages have developed as the result of deliberate political pressure. In 1919 Hebrew was actually spoken by about 20,000 people, but the Israeli state has nurtured it for nationalistic reasons. Today Israel's Academy of Hebrew monitors the language. Some national languages were virtually invented, such as Romanian in the nineteenth century. Other new states have cultivated new national languages to unite diverse populations. The Indonesian government, for example, has succeeded in codifying and establishing one common language: Bahasa Indonesian. Since the breakup of Yugoslavia into several countries, the new governments are encouraging the differentiation of Bosnian, Serbian, and Croat, which were formerly virtually one language. The governments are reviving the use of old words in official documents, and education ministries are printing checklists of forbidden words and fining people who use them. In 1996 President Tudjman of Croatia himself made a mistake. He greeted U.S. President

Clinton to Zagreb using the Serbian version of happy, *srecan*, rather than the Croat word *sretan*. This mistake was broadcast live, but the government edited it out of later rebroadcasts of the event. Since the breakup of Czechoslovakia into the Czech Republic and Slovakia in 1992, the Czechs have stopped studying Slovak, but many Slovaks (who still receive their television programming from Czech regions) have continued to study Czech. Not all countries succeed in reviving or creating national languages. When Ireland achieved independence in 1922, the government tried to enforce the use of Irish Gaelic rather than English, but the people themselves did not accept it. They continued to use English.

A minority people in a country often clings to its language as a gesture of cultural independence. For example, from 1937 until the mid-1950s, the Spanish

FIGURE 7-9 **Luther's German Bible.** This is the title page of the last edition of Martin Luther's German translation of the full Bible printed before his death in 1546. This book launched the Protestant Reformation and fixed the German language. It can still be read by German speakers today. The page's rich decoration includes references to Christian symbols, including the rewards of heaven and the terror of hell. (Courtesy of the Pierpont Morgan Library, New York. PML19474/Art Resource, NY.)

Regional Focus on

LANGUAGE IN THE NEW STATE OF EAST TIMOR

East Timor faced a language problem when it won independence in 2002 after 4 centuries of Portuguese rule, 24 years of Indonesian occupation, and 2 years of U.N. administration. Only 57 percent of its population were literate in any language. The country's new Constitution designates Tetum (an indigenous lingua franca) and Portuguese as official languages, plus English and Bahasa Indonesian as "working languages." The long-term survival of Tetum is questionable: Its written form is incomplete, and there are few written materials in it. English and Bahasa are used in business. About 25 percent of the population speak Portuguese, and strengthening it could help promote tourism, but the Roman Catholic Church adopted Tetum for its services.

government tried to stamp out the Basque language by forbidding its use in all public places. Through the 1960s and 1970s, however, Basques won new rights, and in 1980 the first Basque Parliament was elected, with Basque as a second official language in Spain's Basque provinces.

Language in postcolonial societies   European colonial powers imposed their languages on their colonies, regardless of how many native languages were already spoken and whether many Europeans actually settled in any given colony. The conqueror's language became the language of government, administration and law, economic development, and usually education. When the colonies gained their political independence and had to choose official languages, many kept their former ruler's language. Figure 7-2 partly reflects the map of former European empires. Some former colonial peoples find the acceptance of a major international language useful. It facilitates ties of trade, cultural exchange, travel, and even diplomacy with its former ruler and with other countries that are former colonies of that ruler. The English-speaking countries of the Caribbean, for example, find English a useful tie among themselves.

Across much of the world, however, the official language is not actually spoken by most of the population. Figure 7-2 shows a few international languages covering most of the globe, and yet in some of these areas the majority of the people do not actually speak that language. The difference between official languages and the languages actually spoken is particularly great in Africa. The principal official languages of Africa are those of the continent's former European rulers, but there is an extraordinary variety of native languages spoken, by some counts more than 1,000 (Figure 7-10). The great number reflects the minimal interaction of its many native peoples before European conquest. Many of these languages do not have a written form, and only 40 or so have as many as 1 million speakers. Some African governments are reviving their use or study to rekindle appreciation of their peoples' pre-European heritage.

In some cases the newly independent countries retained their former ruler's language to prevent arguments about which of the native languages should become the new national language. Many African countries have retained English or French for this reason. The Berber minority in Algeria speaks French and Tamazight, and Berbers protested violently after 1998, when Algeria accepted only Arabic as official, until 2002, when Tamazight was accepted as official. Over 250 languages are spoken in Nigeria, and advocates of a national language could not decide among Hausa, Yoruba, or Igbo, the three major native tongues. Therefore, English was chosen, although only about 20 percent of the population speaks it well and still fewer use it as their first language. Another reason for the continuing preferences for European languages is the high cost of schooling in the many indigenous languages. For example, Tanzania's choice of Swahili for primary education allowed the government to bridge linguistic divisions among the people, but it restricted the books available to Tanzanian schools.

Linguistic rivalries maintain English as an official language of India, although it is spoken by a small fraction of the total population. The leaders of India's independence movement intended Hindi one day to be India's official language, and Hindi is spoken by over 300 million people. Hindi speakers, however, are concentrated in northern India, and other Indians are unwilling to accept the domination of Hindi. Whenever Hindi has been pressed on India's southern states, they have threatened to secede. India recognizes English as 1 of 16 official languages, and English remains the preeminent language of the upper classes and the upwardly mobile.

A former colony can rise to greater wealth and influence than its former ruler, in which case its dialect sets the international standard. This has happened with

The Languages of Africa

FIGURE 7-10 Figure 7-2 shows that most of the official languages in Africa are the languages of the former European rulers, but this map reveals major language groups and a sampling of the extraordinary number of native languages, many of which are actually spoken by a majority of the local population.

the U.S. version of English. Noah Webster published his *Spelling Book*—a self-conscious American cultural declaration of independence—in 1783, the same year that the United States won its political independence. Today American English competes with British English

for precedence worldwide. Philologists in Canada, Australia, South Africa, and other English-speaking countries have published dictionaries of each national variant. The French Academy rigorously defends its privileges, but in 2003 the French government accepted the Quebecois term "courriel" for e-mail, rather than the French "courrier électronique."

Portuguese boasts 178 million native speakers, most of whom live in Brazil, a country with 17 times Portugal's population. In 1990 the governments of Brazil, Angola, Mozambique, São Tomé and Príncipé, Guinea-Bissau, and Cape Verde—all former Portuguese colonies—accepted the Brazilian version as the standard. Brazilian is much simpler in spelling than is the Portuguese of Portugal, and it incorporates many Native American, African, and even American words.

The fact that languages are often a cultural legacy of imperialism explains why the breadth of distribution of a language is not always directly related to the number of people who speak it. Table 7-1 lists among the leading languages in terms of speakers several Asian languages that are geographically restricted. Hindi and Bengali are examples. These languages are not widespread across the world map, because the speakers of these languages did not create vast empires.

Polyglot states   India is not the only country that grants legal equality to two or even more languages. The several states that do so are called **polyglot states.** If various languages predominate in distinct regions of a country, the country might accept each language as official in its region. Belgium, and even its capital city of Brussels, is legally divided into French and Flemish zones. In the center of Brussels, all signs are in both languages (Figure 7-11). The precedence of languages

is hotly contested, and several Belgian governments have fallen if they have been seen as giving preference to one language or the other. Canada is also officially polyglot, and language is a source of friction between the Anglophonic (English-speaking) majority and the Francophonic (French-speaking) minority centered in Quebec. South Africa's Constitution names 11 official languages.

Polyglot states usually select one language as official for the federal center and for communications among the states, and whatever the country's constitution or laws might say, one language will generally be the preferred language of the country. Those who do not use it may find their opportunity restricted or their upward mobility blocked. In Tanzania, for example, Swahili is recognized, but knowledge of English is necessary to rise in the national power structure.

Different language communities within one country may be defined not only geographically but also socially, by class. Throughout Central and South America the Native American populations and their cultures were smothered under European rule. Vestiges of this cultural imperialism remained even after independence. For example, the original Constitution for Bolivia, written by Simón Bolívar, accepted as citizens only those who could speak Spanish. Today some countries are granting official recognition to indigenous languages as a way of enhancing national identity and community. Therefore, more and more countries today recognize two or more official languages—one international language plus one or more local indigenous languages. Today Bolivia, for example, recognizes three official languages: Spanish, plus the Native American languages Aymará (spoken by 25 percent of the people) and Quechua (34 percent). Peru accepts both Spanish (spoken by 68 percent of the population) and Quechua (27 percent).

Languages in the United States   The population of the United States has always been composed of a great variety of peoples speaking a great variety of languages. English was the language of the principal colonial ruler and of the greatest number of European settlers, and it has always served as a lingua franca. A distinct American English nevertheless evolved. From the days of the earliest settlers, the American language adopted terms from Native American languages. These included the names of native animals and plants unknown in Europe or Africa, products derived from them, and also place names. Each of the many immigrant groups has in turn learned American English, but each has also contributed vocabulary, grammar, and diction to American English.

Three major dialects had developed in the 13 English colonies along the Eastern Seaboard by the time of the American Revolutionary War: northern, midland, and southern American English. Toponymy reveals the paths that settlers from each seaboard

FIGURE 7-11 Bilingual signs in Brussels. All street signs in Brussels, the capital of polyglot Belgium, must be in both official languages: French and Flemish. The name Häagen-Dazs, which is advertised in the window, is a linguistic curiosity. The name means nothing. It was concocted by a New York businessman to sound vaguely Scandinavian to profit from associations of Scandinavia with cleanliness and health. The shop sign is in Flemish, reflecting the personal preference of its owner. (Courtesy of Robert Mordant)

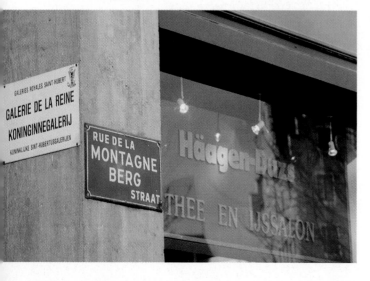

U.S. Dialect Regions

FIGURE 7-12 The three regional dialects of northern, midland, and southern American English developed already in the eighteenth century, and as pioneers headed westward from each dialect region, they took place names with them. (Adapted with permission from E. Bagby Atwood, *The Regional Vocabulary of Texas*. Austin: University of Texas Press, 1969.)

some people fear that it may also reflect a resentment of the changing immigration trends discussed in Chapter 5. Immigration has increased the number of people who speak a language other than English. In 2000, 47 million people claimed to speak at home a language other than English (Table 7-2). Fully 28 million of them (60 percent) spoke Spanish. Chinese ranked a distant second (2 million speakers) followed by French, German, Tagalog (the native language of the Philippines), and Vietnamese—further evidence of the newest immigrant streams into the United States.

Language has become a civil rights issue in education. If, as is generally agreed, the inability to communicate in English is a handicap in the United States, then the teaching of English becomes a key route to equal opportunity for all children. In a case involving Chinese American children in San Francisco, the U.S. Supreme Court ruled that "students who do not understand English are effectively foreclosed from any meaningful education" (*Lau* v. *Nichols*, 1974). This ruling triggered a national concern to identify local school districts in which English was the students' second language and to sponsor bilingual educational programs in those districts. Arguments persist, however, over whether the only purpose of bilingual programs is to ease the students' transition to English or whether they also should preserve the languages and cultures that immigrant children bring to school. Some people argue that using the education system to acculturate all children to the English language denigrates the

region took toward the West (Figure 7-12). Users of the northern dialect frequently named places Brook, Notch, and Corners. People moving westward from the midland areas commonly named places Gap, Cove, Hollow, Knob, and Burgh. Southern settlers favored names using Bayou, Gully, and Store.

Regional variations in American English persist, both in grammar and in pronunciation, despite the nationwide reach of popular media. Television homogenizes national speech patterns much less than scholars originally thought it would.

Should American English be the Official National Language? The U.S. Constitution did not specify English as the official language, and many local governments through history and still today have found it useful to provide services and even keep official records in other languages (Figure 7-13).

Toward the end of the twentieth century, however, a movement grew to declare English the country's official language. In 1996 the U.S. House of Representatives approved such a bill, but the Senate did not act. Sixteen states already had statutes or constitutional clauses declaring English their official language. This movement may reflect a dedication to a national language as a bonding force of a diverse population, but

FIGURE 7-13 **A bilingual city.** The Florida State Constitution insists that "English is the official language of the State of Florida." In 1980 Dade County (metropolitan Miami) passed an ordinance prohibiting the use of any language other than English for government business. By 1990, however, more than half (50.1 percent) of county residents spoke Spanish at home, so in 1993 the county ordinance was repealed. (Donald Dietz/Stock Boston)

TABLE 7-2 The Leading Non-English Languages in Use in the United States, 2000

Language	Number of speakers
Spanish or Spanish Creole	28,101,052
Chinese	2,022,143
French (incl. Patois, Cajun)	1,643,838
German	1,383,442
Tagalog	1,224,241
Vietnamese	1,009,627
Italian	1,008,370
Korean	894,063
Russian	706,242
Polish	667,414
Arabic	614,582
Portuguese or Portuguese Creole	564,630
Japanese	477,997
French Creole	453,368
African languages	418,505
Hindi	317,057
Persian	312,085
Urdu	262,900
Gujarathi	235,988
Serbo-Croatian	233,865
Armenian	202,708
Mon-Khmer, Cambodian	181,889
Navajo	178,014
Miao, Hmong	168,063
Laotian	149,303
Thai	120,464
Hungarian	117,973

Source: U.S. Census Bureau

richness of their native inheritances, while others argue that preservation of other languages threatens to lock the youngsters into second-class citizenship.

This debate intensified in 1996, when the school board in Oakland, CA, declared that many of its black students spoke a distinct language, *Ebonics* (ebony phonics). Scholars have recently begun to study how some aspects of Ebonics, sometimes called Black English Vernacular (BEV), do demonstrate that the speech of parts of the African American population preserved elements of West African languages. In Ebonics, for example, "he busy" translates into standard English as "he is busy right now," whereas "he be busy" means "he's always busy." Such grammatical structures may have diffused from West Africa to the southern United States and from there around the country. Most Americans speak at least a couple of words of Wolof, the

language of Senegal: *degan,* "to understand" (as in, "Can you dig it?"), and *hippie,* "to open one's eyes" (a "hippie" is someone with his eyes open, that is, aware). Political battles continue to be fought over bilingual education in school districts across America, and referenda generally favor intensive English lessons to guarantee all students English proficiency as fast as possible.

Equal linguistic access to the political process is another volatile issue. A 1975 Amendment to the Voting Rights Act of 1965 mandated bilingual ballots in voting districts where voters of selected language groups reached 5 percent or more. Today those electoral districts include vast areas of the country, including both inner-city neighborhoods and rural areas. Elections in them are observed by the Department of Justice.

From the 1880s until the 1950s, federal policy tried to discourage or eliminate Native American languages. When the Europeans first arrived, more than 500 Native American languages were used throughout the territory of today's United States, but only about 200 survived to 2000. In 1990, however, federal legislation had been passed to "encourage and support the use of Native American languages as languages of instruction." Language is such an important cultural index that many Native Americans view linguistic survival as cultural survival. Many social scientists would agree.

THE TEACHINGS, ORIGIN, AND DIFFUSION OF THE WORLD'S MAJOR RELIGIONS

A religion is a system of beliefs regarding conduct in accordance with teachings found in sacred writings or declared by authoritative teachers. Most religions involve personal commitment to worship a god or gods, but the teachings of several Asian systems of belief, including Confucianism, Taoism, and Shintoism, are basically ethical and psychological. That is, they focus on appropriate behavior (*orthopraxy*) rather than belief in a set of philosophical or theological arguments (*orthodoxy*). Several do not even address theological questions such as the nature of God or the gods, or life after death. The strictest adherence to traditional beliefs is called **fundamentalism. Secularism,** by contrast, is a lifestyle or policy that purposely ignores or excludes religious considerations.

> Religion is a major determinant of human behavior, so the geography of people holding and following a set of teachings affects politics, economics, agriculture and diet, and many other aspects of human life.

Religion is a major determinant of human behavior, so the geography of people holding and following a set of teachings affects politics, economics, agriculture and diet, and many other aspects of human life. Most religions involve joining an organized body

The World's Major Religions

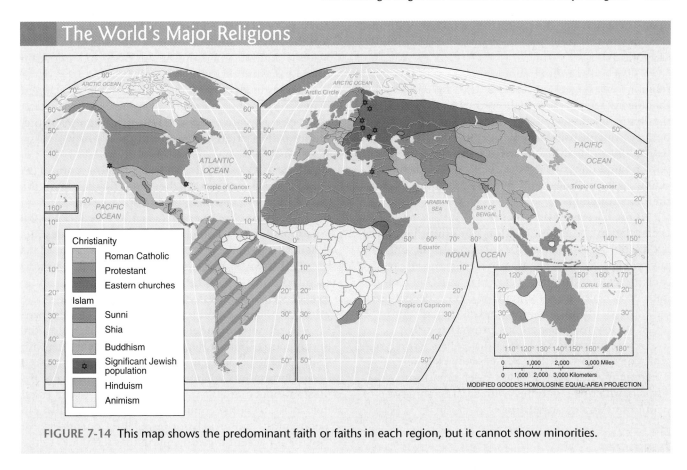

FIGURE 7-14 This map shows the predominant faith or faiths in each region, but it cannot show minorities.

of fellow believers, and local administrative patterns, such as Roman Catholic dioceses, often organize and focus socializing, community spirit, and in many cases education.

Each religion originated in one place and spread out from there, but today communicants of various religions mingle around the globe, and religious affiliations cut across lines of politics, race, language, and economic status. Figure 7-14 shows the predominant faith or faiths in each region, but it cannot show minorities. At the most local scale, individual religious communities can be found grouped in identifiable neighborhoods within most of the world's large cities.

No worldwide religious census is taken, but Table 7-3 is one estimate of the adherents of each of the world's principal religions. These figures cannot indicate the depth of anyone's belief nor the degree to which a person's religion actually affects his or her behavior. Also, several systems of belief listed are not exclusive; that is, a person may adhere to more than

TABLE 7-3 Adherents of the World's Major Religions (mid-2001 estimates in thousands)

Religion	Total	Africa	Asia	Latin America	Northern America	Europe	Oceania
Christians	2,019,052	368,244	317,759	486,591	261,752	559,359	25,343
Roman Catholics	1,067,053	123,467	112,086	466,226	71,391	285,554	8,327
Protestants	345,855	90,989	50,718	45,295	70,164	77,497	7,478
Orthodox	216,314	36,038	14,219	564	6,400	158,375	718
Anglicans	80,644	43,524	735	1,098	3,231	26,628	5,428
Buddhists	387,167	139	356,533	660	2,777	1,570	307
Confucians	6,313	250	6,277	–	–	11	24
Hindus	819,689	2,384	813,396	775	1,350	1,425	359
Jews	14,111	215	4,476	1,145	6,045	2,506	97
Muslims	1,207,148	323,556	845,341	1,702	4,518	31,724	307

Source: Adapted from the 2003 Encyclopaedia Britannica Book of the Year one. The following discussion will explain additional reasons why this table must be viewed with considerable caution.

This book cannot examine the fine points of each religion's message. For that, a text in comparative theology is needed. We will, however, briefly summarize the basic teachings of the world's major religions, because the distribution of people who hold those beliefs affects the distributions of other things: forms of government, dietary habits, women's rights, the organization of economies, and peoples' relationship with their environments.

Our coverage will identify the hearth region of each religion and review the story of its diffusion, but it is difficult for scholars to identify precise reasons why any religion diffuses. Believers in a religion will say that their religion diffused because that was God's or the gods' will. This kind of argument cannot be criticized or analyzed, for it is grounded in a faith that cannot be tested. A few religious groups **proselytize**—that is, try to convert others to their religious beliefs. Other religions do not proselytize. Some scholars differentiate *universalizing religions,* which proselytize, from *ethnic religions,* which are identified with a particular group of people and do not proselytize.

Judaism

Judaism has only about 14 million adherents, but it was the first of the great **monotheisms**—religions that preach the existence of only one God—to emerge in history. Many scholars suggest that **polytheism,** the worship of many gods, is a primitive form of religion and that among some peoples polytheism slowly developed into *henotheism*—the worship of one god without denying the existence of others. Henotheism then evolved into monotheism.

Judaism rests on a belief in a pact between God and the Jewish people that they would follow God's law as revealed in the *Pentateuch*—the first five books of the Old Testament section of the Bible. This covenant was granted to Abraham. Both Jews and Arabs claim descent from Abraham—Jews through his son Isaac, and Arabs through another son, Ishmael. Judaism is divided among a great variety of Orthodox (fundamentalist), Conservative, and Reform *sects,* or subdivisions of a religion.

Judaism developed historically in the Near East over many centuries. Under the Roman Empire, Jewish communities were established outside the Near East, but in A.D. 70 the Romans destroyed the temple in Jerusalem, and the Jews scattered in the Diaspora. During the Middle Ages, many European Christian rulers persecuted and expelled Jews, but Jews returned to Western Europe during the Enlightenment of the seventeenth and eighteenth centuries. On their return, they were required to live in segregated communities called *ghettoes.* Legal emancipation came only in the nineteenth century. Millions of Jews came to the United States from Eastern Europe at that time, and migration from Russia and surrounding lands continues.

An estimated 5.2 million Jews lived in the United States in 2002. The majority live in New York state, other eastern states, and in California; but Jewish communities are scattered throughout the country.

Israel as a Jewish state During the Diaspora many Jews visited Jerusalem if they could, but when the nation-state idea developed in Europe (see Chapter 11), many Jews came to accept **Zionism,** the belief that the Jews should have a homeland of their own in Palestine. When Israel came into existence in 1948, many Jews who had survived the Holocaust in Europe and many Mediterranean Jews migrated there. Israel has continued to welcome Jewish immigrants from around the world, although several Jewish sects repudiate Zionism.

Israel was intended to be a homeland for the Jewish people, but the degree to which Israel is "a Jewish state" is still being defined. In 1997, Israel declared Arabs born in Jerusalem to be aliens and subject to expulsion, but in 2000 the Israeli Supreme Court ruled that an Arab couple could not be barred from living in a community built solely for Jews. The Court declared: "We do not accept the conception that the values of the state of Israel as a Jewish state justify discrimination by the state between citizens on the basis of religion or nationality." The balance of Jews and non-Jews in Israel has continuously shifted as a result of Israeli territorial expansion, immigration to Israel, and the birth rates of the various groups in Israel. In 1994 Israel surrendered some territory inhabited by Palestinians to the Palestine Liberation Organization (Figure 7-15). That action increased the "Jewishness" of Israel's remaining territory to about 82 percent, but Israel cannot remain distinctly Jewish unless it surrenders more territory occupied by non-Jews or expels non-Jews.

Since 1967, Israel has held on to East Jerusalem, the Gaza Strip, and the West Bank of the Jordan River (a region previously part of Jordan). Israeli governments, spurred by Zionist extremists, have continued to occupy and to build new settlements in these areas in violation of international law and despite admonitions against the occupation by the United States, the United Nations, and virtually all other countries and international agencies. By mid-2003 about 200,000 Israelis lived in East Jerusalem and another 200,000 on the West Bank and in Gaza. In each case, Israeli settlers occupy 20 percent of the land but are allocated, by Israeli authorities, over half of the water, and in each case over one million Palestinians crowd the rest of the land. Israel allowed self-rule to the Palestinians in Gaza in 1994, but Israel retains authority for security. The Palestinian people have fought this occupation—often violently and even with terrorism—but the military

Focus on

FUNDAMENTALISM AND TERRORISM

The people responsible for the attacks on the United States of September 11, 2001, are often referred to as Islamic fundamentalists. Their actions were, in fact, repudiated by Muslim religious leaders throughout the world, but we must still ask what force of furious hatred this concept can carry that it can trigger slaughtering thousands of innocent people, whatever one's religious beliefs. The 9/11 attacks were defined as acts of **terrorism**, that is, violent acts intended to frighten and to intimidate for political ends a civilian population beyond their immediate victims.

Religious fundamentalism of any kind sometimes provides meaning and direction to people who feel lost in a confusing world. Many people find comfort, or at least some kind of psychological "anchor" in strictly adhering to religious doctrines that are accepted as literal truth. These texts are followed absolutely, and sometimes reason, judgment, and even conscience can be subjugated to dogma. Fundamentalism has led people to extraordinary acts of both good and evil. A "true believer" must encourage others to obey, too, or even coerce others, because other people's sins can corrupt you, too. Sin must be punished and purged, even by force. When people feel that an issue is beyond matters of life and death, that the issue affects immortal souls, then they may commit acts other people would call "extremism."

The use of religion as an excuse for repression and terror is not unique to Islam. The Christian Gospels teach peace, love, and forgiveness; but through history, Christianity arguably has a worse record of persecution than Islam does. Christian history records centuries of persecution of Jews, mutual slaughter by Protestants and Roman Catholics (continuing in Northern Ireland, p. 547), hunting for religious deviation by the Inquisition and other bodies, the murder of "witches," Crusades, forced conversions, and other similar actions.

Today Christianity is seldom associated with terrorism, but the actions taken by some Christians could be interpreted as such. The bombings of abortion clinics and murder of abortion providers by people who call themselves Christians, for example, demonstrate that some Christians continue to practice evil deeds to enforce what they see as the greater good. These Christian extremists truly believe that abortion is mass murder; therefore, the murder of abortion providers is logical to them. Such violence is seen as justified because it is viewed as the result of a choice between taking action and eternal damnation.

If religious texts written in primitive societies a thousand years or thousands of years ago are interpreted literally, the modern world can be terrifying. In recent decades, some Islamic fundamentalists have begun to argue that all of modern Western culture and society is sinful and evil. Believers of this fundamentalist perspective may argue that this world must be destroyed, or at least that its destruction is the consequence of its sin. Even some Christian fundamentalists hold this view. After the attacks on September 11, the American Baptist minister Jerry Falwell argued, "God continues to lift the curtain [of His protection] and allow the enemies of America to give us probably what we deserve. I really believe that the pagans and the abortionists, and the feminists, and the gays and lesbians who are actively trying to make that an alternative lifestyle, the ACLU [American Civil Liberties Union—a civil rights organization], People for the American Way—all of them who have tried to secularize America—I point the finger in their face and say, 'You helped this happen.'" In other words, for Reverend Falwell, as for many American Christian fundamentalists, the September 11 massacre was divine retribution for American behavior.

Some people have been confused by hearing that the hijackers lived in America for years, went bowling with their neighbors, and even drinking. Weren't they, then, "won over" to an accommodating American way of life? Obviously not. There is no room in the fundamentalist psyche for accommodation. The very dynamics that lead repressed homosexuals to murder homosexuals or that drive sexually promiscuous preachers to preach chastity are the same that lead vodka-drinking fundamentalists to destroy symbols of modern life. The act is not designed to achieve anything, but blindly to destroy the thing they feel threatened by. It is a violent acting-out of an internal conflict.

force of Israel is so great that this contest of peoples has been an uneven one in which the "score" of dead and wounded, of destruction and persecution, enormously "favors" the Israelis.

The Palestine Liberation Organization (PLO), with both Muslims and Christians among its leaders (about 20 percent of Palestinians are Christian), has sought to establish an independent Palestinian state since its founding in 1964. In 1999, the Internet Corpo- ration for Assigned Names and Numbers assigned the Palestinian areas a domain code (.ps). This action seems to suggest some unofficial international recogni- tion of Palestine as a state. Thus, Palestine exists in cyberspace, but the creation of an independent Pales- tinian territorial entity is the object of ongoing political confrontation and debate. The process of achieving lasting peace between Israel and its Arab residents and neighbors has been frustratingly slow and is marked by periodic retreats into violence, but a just settlement might encourage rising standards of living and security for everyone in the region.

Christianity

Christianity, the belief that God lived on Earth as Jesus Christ, emerged from Judaism, but part of the wide- spread appeal of Jesus' teachings rests on his emphasis on the nearness of the Kingdom of God and God's love for all. Jesus lived his entire short life in Judea; but after the death of Jesus, Saint Paul preached in many major cities in the eastern Mediterranean (Figure 7-16). Paul's letters (*epistles*) to early Christian groups make up much of the New Testament. Many Christians believe that proselytizing is a duty (Matthew 28:19).

The Bible records that the first non-Jew to convert to Christianity was an Ethiopian (Acts 8). Today many scholars believe that man was probably from Makurra, in what is today the Sudan. The Christian church of Egypt, called the Coptic Church, was founded in Alexandria in A.D. 41, and Christianity did diffuse south from Egypt to today's Ethiopia by the fourth century. It has survived there despite having been cut off from the Mediterranean world by the later conversion of Egypt and the Sudan to Islam (to be discussed shortly). Jesus' disciple Thomas may have traveled to India, and al- though that is not certain, we do know that followers of

Israel and Palestinian Territories

As of January 1, 2001.

☐ Occupied by Israel

▨ Area surrendered to PLO Administration, 1994

▨ Area under full or partial Palestinian control

● City surrendered to PLO Administration

0   25   50 Miles
0   25   50 Kilometers

FIGURE 7-15 Israel was carved out of the Mideast in 1948, and it occupied larger areas during a war in 1967. Jewish settlers have occupied these territories, but their international legal status remains unclear. Israel may return the Golan Heights to Syria. In May 2000, Israeli troops pulled out of southern Lebanon, which they had occupied for 22 years. Israel has surrendered about 20 percent of the West Bank to Palestinian control and another 22 percent to joint control, but the "Palestinian" areas are criss-crossed with Israeli-held roads and checkpoints.

Zionists insist that Jerusalem is the capital of Israel, and in 2002 the U.S. Congress insisted that President George W. Bush recognize it as such. The President, however, has refused to do so, arguing that Congress cannot breach his ability to make foreign policy. No other nation recognizes Jerusalem as Israel's capital, nor does the United Nations, and the Congressional action triggered a Palestinian declaration of Jerusalem as capital of Palestine. Prime Minister Mahathir Mohamad of Malaysia called the Congressional action "pouring oil on a fire."

The Early Diffusion of Christianity

FIGURE 7-16 Saint Paul visited the Christian communities of major cities from Rome to Jerusalem. By the year 350 five patriarchs—co-equal leaders of the Church—had been recognized, with their seats at Alexandria, Jerusalem, Constantinople (today's Istanbul), Antioch (today's Antakya), and Rome. To the south, a Cathedral of Saint Mary was built at Axum in 330, where the Christian emperors of Ethiopia were crowned until 1931.

Thomas had established Christian churches in India as early as the second century. Christianity also spread to the northeast, and very early in the fourth century Armenia became the first Christian nation. The Roman emperor Constantine (288?–337) converted to Christianity and favored the religion; and by the end of the fourth century, Christianity was the Roman Empire's official religion.

Under the Roman Empire, the Christian Church adopted a geographical organization that paralleled that of the empire's secular administration. Bishops ruled over dioceses. When the western half of the Empire fell to invaders in 476, the church organization survived as one of the few stabilizing and civilizing forces in Western Europe. Bishops, including the pope as bishop of Rome, assumed civil authority as well as

Focus on

ISRAEL AND U.S. FOREIGN POLICY

Since 1967, the United States has stood behind the formula stated in Security Council Resolution 242, a formula that would guarantee peace in exchange for Israeli surrender of land, but many people around the world feel that the United States, as Israel's principal ally, has not brought sufficient pressure to bear against Israel to force it to stop its settlement activities. The United States has consistently supported Israel with both military and economic aid. This aid has been essential to Israel's ability to win wars and maintain control of its territory. Thus, Palestinians in particular and Arabs generally associate the United States with Israeli actions, and they blame both countries.

On October 5, 2001, the terrorist Osama bin Laden condemned America's support of Israeli actions. "Anyone who lives in the United States," he said, "will not feel safe before the people of Palestine feel safe." Even many of America's friends who were prompt to denounce

the September 11 attacks coupled their condemnation with condemnation of Israel. In Kuwait, for example, a country that owes its continuing independence to the United States, the Director of Kuwait University's Center for Strategic and Future Studies, Ghanim Alnajjar, said, "The story is not bin Laden. The story is the injustice to the Palestinian people." Iran's supreme leader Ayatollah Ali Khamenei insisted, "America's hands are soiled with all the crimes committed by the Zionist regime. . . ."

In 2003 President George W. Bush announced U.S. support for the establishment of an independent Palestinian state and hosted the opening of another round of peace talks between the PLO and the Israeli government. Points of contention remain the amount of land that Israel might have to surrender (and whether the surrender of *any* land would be acceptable to Zionist extremists), explicit acceptance by Arab nations of Israel's right to exist, how Israeli security could be guaranteed, Israeli acceptance of or reparations to millions of Palestinians refugees, the full sovereignty of a new Palestinian state, and water rights.

---

ecclesiastical authority in areas where there was no other effective government. The first Western monastery was founded by Saint Benedict at Monte Cassino about 529, and as monasteries were established across Europe, the pope brought them under his protection (Figure 7-17). Throughout the West during

FIGURE 7-17 Monte Cassino. Monte Cassino, near Naples, was the first Western Christian monastery. Founded by Saint Benedict in the sixth century, it has played an immeasurable role in the history of European civilization. Although destroyed repeatedly in wars (most recently in 1944), it has been rebuilt. (Robert W. Madden/National Geographic Image Collection)

the Dark Ages, the Church was the focus of life, and monasteries and cathedrals rose as the monuments of the civilization.

In the eastern half of the Roman Empire, secular imperial control survived, with Constantinople (also called Byzantium, today's Istanbul) as its capital. The Patriarch of Constantinople headed the Eastern Christian Church, and missionaries from Constantinople (including Saint Cyril, mentioned earlier) converted the peoples of the Balkan Peninsula, Eastern Europe, Ukraine, Russia, and the Near East to Christianity. The Western and Eastern churches split finally when the pope and patriarch excommunicated each other (officially declared each other to be outside the community of Christians) in 1054. Thus, Roman Catholic Western Europe came to be culturally divided from Orthodox Eastern Europe, although the mutual excommunications were revoked by the patriarch and pope in 1965. The Uniate Church of Ukraine bridges the gap by using the Cyrillic alphabet and following the Eastern rite, but recognizing some spiritual authority of the pope. The fall of Constantinople to the Muslim Turks in 1453 opened eastern Europe to Islam, but the office of the Patriarch of Constantinople survived, and all Orthodox Christians still recognize its spiritual leadership. Orthodox Christians are subdivided into national churches, such as the Serbian, Greek, and Russian, although the Orthodox Christians in the

United States and in Canada cling to their historic national affiliations.

The Protestant Reformation of the fifteenth and sixteenth centuries split the Christian Church in the West. The Protestant denominations (Lutherans, Anglicans or Episcopalians, Baptists, and others) were named for their protest against the Church of Rome, and a variety of Protestant denominations and national denominations were established in Western and Northern Europe.

When Europeans explored and conquered other lands, they often cited religious conversion as their purpose. Christian missionaries were the chief agents of the partial Europeanization—of religion, language, social mores, and the acceptance of secular authority—that made many natives more tractable to European rule. Christianity diffused rapidly and almost completely throughout the Western Hemisphere. Certainly its message of all-embracing love carried strong appeal in lands in which many of the gods had been portrayed as terrifying and cruel. Furthermore, the Christian heaven after death is democratic and accessible to all classes of people. Christians also flexibly adapted some native practices. For example, in the fourth century, Western European Christian churches timed the celebration of Christmas to replace the celebration of the winter solstice among non-Christians. Christians still accept indigenous practices in order to bring peoples to Christianity. The 1992 Conference of Latin American Roman Catholic bishops pledged to incorporate the traditional religious symbols, rituals, and cosmologies of Latin America's Native Americans and blacks whenever "compatible with the clear sense of the faith" and "the general discipline of the church" (Figure 7-18).

In Latin America, some natives might have adopted Christianity because they were astonished and demoralized by the Spaniards' invulnerability to the diseases that were wiping out the Native Americans. Neither the Spaniards nor the Native Americans understood that those diseases had been brought by the Spaniards themselves. The Spaniards convinced the natives to acquiesce to Spanish secular authority and to seek both physical and spiritual salvation in the Spaniards' church. Similar beliefs prevailed among both the Native Americans and the whites in North America.

Christian missions to Asia were less successful. In 1542, exactly 50 years after Christians conquered Granada, the last Arab Muslim outpost in Europe, and Columbus's first voyage to the West, Saint Francis Xavier arrived in Goa in India to proselytize Asia. By 1549 Saint Francis was in Japan, which nearly converted to Christianity in the early seventeenth century but then expelled or persecuted Christians when it closed its doors to the world in the mid-seventeenth century (Figure 7-19). Christians had some initial success in China, but in 1723 Christianity was banned there as well.

Conversion to Christianity, and as will be seen later, to Islam, usually weakens the continuity of a people's cultural inheritance. This is because both

FIGURE 7-18 Jesus as a Native American holy man. This image of Jesus as a Mescalero holy man dominates the altar at the St. Joseph Mission church on the Mescalero Apache Reservation in New Mexico. His left palm holds the symbol of the Sun; his right hand a deer hoof rattle; at his feet an eagle feather, a grass brush, and bags of tobacco and cattail pollen used in ceremonies. He stands on Sierra Blanca, the Mescalero sacred mountain. The inscription, in Apache at the bottom and Greek at the top, reads, "Giver of life." (© 1990 Robert Lentz. Used with permission of Bridge Building Images, P.O. Box 1048, Burlington VT 05402.)

Christians and Muslims believe that conversion is a spiritual rebirth. It is a deliberate repudiation of much of what came before, so conversion can destroy a people's earlier culture more completely than political conquest.

FIGURE 7-19 The Basilica of Goa. The Basilica of Bom Jesus in Goa, India, is the final resting place of the missionary Saint Francis Xavier (1506–1552). Saint Francis is today the patron saint of Roman Catholic missionaries. (Courtesy of Father Vasco do Rego/EFB)

Christian sects and their distributions

Roman Catholicism is the largest single denomination among Christians. It is headed by the pope in Vatican City, a tiny independent city-state within Rome (Figure 7-20). Roman Catholicism has historically dominated in the Mediterranean basin, throughout Latin America, and wherever else Mediterranean peoples colonized or converted, as in former French or Portuguese African colonies and in the Philippines and Vietnam.

Protestant denominations dominate in Northern Europe and wherever Northern Europeans have settled or converted: North America, Australia, and New Zealand, and the parts of Africa that were either formerly English colonies (today usually Anglican) or German colonies (Lutheran).

Several theological points differentiate Protestantism from Roman Catholicism, but among the most important is that Roman Catholics believe that Jesus gave Saint Peter unique responsibility for founding the Christian Church, that Saint Peter became the first bishop of Rome, and that the popes retain this special responsibility. The crossed keys on the Vatican flag represent Jesus's having given Saint Peter the keys to heaven. This belief that a church or priests intercede between God and humankind is called *sacerdotalism*.

Protestants deny sacerdotalism. Evangelical Protestants (from the Latin for "bringing good news") emphasize salvation by faith through personal conversion, the authority of Scripture and each individual's responsibility to read the Scriptures, and the importance of preaching as contrasted with ritual. The evangelicals emphasize the ability of individuals to change their lives (with God's help), and this message offers to many people a new sense of personal empowerment. Thus, Evangelical Protestantism frequently brings a revolutionary force into traditionally rigid or stratified societies.

Today Evangelical Protestantism is growing and spreading throughout the world. In some places it is replacing non-Christian religions, but in many places it is replacing Roman Catholicism. The population counted as Roman Catholic in Table 7-3 is the population baptized Roman Catholic. The Church insists that once a person has been baptized as a Roman Catholic, that person is always Roman Catholic, even if the person converts to another denomination or is excommunicated. Nevertheless, Evangelical Protestantism has replaced Roman Catholicism in several countries as the most widely practiced faith, and as a rule the ardor of converts is strong. Evangelical Protestantism is also spreading in Pacific Asia (especially into South Korea and the Philippines), across Africa, and throughout Latin America. About one quarter of the total Latin American population is Protestant today. This percentage is even higher in Brazil, which counts from baptism the world's largest Roman Catholic population, but where today full-time Protestant pastors outnumber Roman Catholic priests, and as many as 600,000 people convert to Evangelical Protestantism each year.

In July 2000 more than 10,000 Protestant Evangelists from more than 200 countries and territories assembled in Amsterdam, the Netherlands. The "Amsterdam Declaration," which they adopted, urges evangelists to be sensitive to the societies in which they work and to avoid equating Christianity with any particular culture. It obliges evangelists to present the Christian gospel as authoritative, even while respecting people of other religions. A clause that encourages women to be gospel teachers could trigger change

FIGURE 7-20 Vatican City. Vatican City is one of the world's smallest independent states. It is entirely within the city of Rome, but it is ruled by the pope according to a treaty signed between the papacy and Italy in 1929. The official language here is Latin, and in 2003 the Vatican produced a new Latin dictionary to keep up with the times (a motorcyle is a *birota automataria levis*). (Dan Budnik/ Woodfin Camp & Associates)

in many traditional societies that do not accept women in leadership roles and hasten conversions from Roman Catholicism.

The spread of Evangelical Protestantism affects many other aspects of human geography. In most Latin American states the Roman Catholic Church has traditionally enjoyed special privileges and a role in education, so its teachings have been enacted into law. Many Latin American countries, for example, prohibit divorce. Today, however, a rising share of elected officials throughout Latin America is Protestant, and elections are explicitly referred to as "Holy Wars."

The diffusion of Evangelical Protestantism is also at least partly responsible for slowing population growth. Evangelical Protestants are often as opposed to abortion as Roman Catholics are, but they are not always so adamantly opposed to other forms of birth control. As Roman Catholicism has lost political power in Latin America, Quebec, Spain, Italy, the Philippines, and elsewhere, governments have begun sponsoring family planning, distributing birth-control devices, and introducing sex education into school curricula. The Philippines, for example, elected its first Protestant president, Fidel Ramos, in 1992, and his government introduced birth control and sex education. Jorge S. Elias, elected president of Guatemala in 1991, was the first Protestant ever elected president of a Latin American country. Chile, which is historically Roman Catholic, elected a socialist agnostic president in 2000, Ricardo Lagos. His platform included a promise to legalize divorce.

It should be noted that not all Roman Catholics obey church teachings. The president of Peru through the 1990s, for example, Alberto Fujimori, a baptized Roman Catholic, nevertheless sharply criticized Roman Catholic teachings and introduced sex education into Peruvian schools. In North America, democratic culture confronts an authoritarian church structure: Canadians warmly welcomed the pope in 2002, but a national opinion poll of Catholics revealed that 82 percent of them felt priests should be allowed to marry; 80 percent wanted female priests; 73 percent wanted divorced people to be able to remarry inside the church; and 68 percent favored birth control. All of these beliefs are strongly forbidden by the church. In 1995, 78 percent of U.S. Catholics said that one can "make up one's own mind on moral issues" and "disagree with church teachings," but still be "a good Catholic."

Other possible results from the spread of Evangelical Protestantism have been hypothesized, and they bear watching through coming years. One is the spread of literacy. This follows from Protestants' individual responsibility to read and study Scripture. Literacy has risen in areas where Evangelical Protestantism has spread, but the exact cause-and-effect relationship is difficult to measure. If Evangelical Protestantism does raise literacy rates, this trend may carry further

ramifications—greater political participation, for example. Still other possible consequences of the spread of Protestantism in the areas of women's rights and even economics will be discussed below.

In addition to Roman Catholic, Protestant, and Orthodox Christians, many smaller sects include the Georgian and Armenian churches, the Maronite churches in the Near East, the Copts of Egypt, the Christians of India, Ethiopia, China, and a few other groups.

The future of Christianity   The geography of Christianity is undergoing significant changes. The observance of Christianity has increased throughout Eastern Europe with political liberalization since the fall of Communism. Nevertheless, within the lifetime of most living Christians, Christianity has shifted from being a religion of the Western industrialized nations to being a religion of Asia, Africa, and the Pacific. In Africa, for example, the number of Christians has grown, chiefly by conversion, from fewer than 10 million in 1900 to 360 million today (Figure 7-21). In Latin America, Christians rose from 62 million in 1900 to 463 million today; in Asia from 19 million to 284 million. Today there are more Lutherans in South Africa than in North America, and at the 1998 world conference of Anglican bishops, bishops from Great Britain,

FIGURE 7-21 The world's largest Christian church. The Basilica of Our Lady of Peace in Yamoussoukro, the Ivory Coast, is the world's largest Christian church. The country's first president, Félix Houphouët-Boigny (1905?–1993), converted to Roman Catholicism as a teenager (although he always insisted on his right to have several wives), and he built the basilica as a pilgrimage center for Africa's Catholics and as a bulwark against Islam and native religions in his country. Ivory Coast is about 12 percent Christian, 60 percent Muslim, and 28 percent animist. The air-conditioned basilica can hold 18,000 people. (Yann Arthus-Bertrand/Getty Images, Inc.—Liaison)

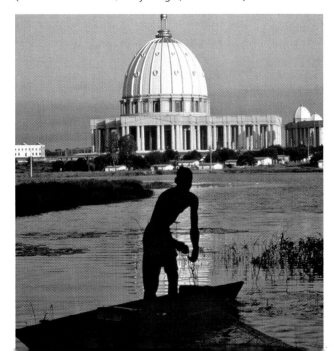

Canada, the United States, and Australia were for the first time outnumbered. South Korea is about 30 percent Christian—80 percent of them Protestant—and it may become the second Asian nation to have a Christian majority, after the Philippines. In 2003 the World Council of Churches elected its first African Secretary General, Rev. Samuel Kobia, a Methodist from Kenya.

Christian missionary activity is increasing: There are today about 400,000 Protestant and Roman Catholic missionaries in the world, six times the number in 1900, and about 100,000 of these missionaries are from Protestant churches in non-Western countries. Korea, for example, has sent almost 10,000 Protestant missionaries outside Korea. Many missionaries witness their faith only through good works, such as building educational and charitable institutions and providing humanitarian assistance. More aggressive efforts toward conversion, however, are meeting stiffening resistance from other religions, and the divisions within Christianity aggravate the tensions. The spread of evangelical Christian missionaries throughout the Muslim world, for example, has provoked anger not only among Islamic clerics, but also among the native Christian groups. Trying to convert Muslims is often interpreted as a provocation that invites violence, and conversion from Islam is in many places punishable by death. Also, missionary activity by Americans has acted as a lightning rod for anti-American political sentiments. American missionaries have been killed in the Muslim Near East and in Africa, and in India by Hindus. In 2003 the U.S. National Association of Evangelicals, representing 43,000 congregations, reaffirmed its commitment to proselytizing, but issued a "loving rebuke" to some evangelical leaders, such as the Rev. Franklin Graham (who said Islam is "a very evil and wicked religion"), Rev. Jerry Falwell ("Muhammad was a terrorist"), and Rev. Jerry Vines (who described Muhammad as "demon-possessed") for remarks that "tarnish Christianity" and jeopardize the safety of missionaries and the indigenous Christians in some countries.

Pope John Paul II has doubled the number of Roman Catholic saints, and virtually all of the new saints have been African and Asian. In 2002, he canonized Juan Diego, a Native American who supposedly sighted the Virgin (Mary) of Guadalupe in today's Mexico in 1531. Historians insist that Juan Diego never existed, but the pope hailed Diego as "the first indigenous saint of the American continent."

The Vatican maintains that the only purpose of any interfaith dialogue is to convert others to Roman Catholicism, and the pope's conviction of the unique correctness of his religion has triggered resentment and even political and social repercussions. For example, on a trip to India he declared that "the peoples of Asia need Jesus Christ and his gospel." Prominent Hindu, Sikh, Muslim, and Buddhist priests walked out of the room. Similar statements made in the Near East

during visits have triggered strong objections and even street riots. Roman Catholic proselytizing in Russia—trying to convert Russians from their traditional Orthodox Church to that of Rome—has triggered political restrictions on Roman Catholicism, such as the denial of permission to build new churches.

Chapter 6 noted that any cultural system or artifact may adapt and permutate as it diffuses across local cultures, and the shift in Christianity's population center of gravity has inspired changes in Christian theology and practice. As Roman Catholic bishop Bonifatius Hauxiku of Namibia has said, "Our African people have accepted Christ, but this Christ walks too much among them in a European garment." Local churches are adopting local foods for the sacred rite of communion, although, historically, Christians' need for wine and bread for this rite triggered the global diffusion of grape and wheat crops. Roman Catholic bishops around the world have recommended ordaining married men, a practive forbidden by the Vatican. The bishop of Nassau, the Bahamas, emphasized that on such questions "the church should not be tied to cultural vestiges typical of the European experience." Some liberalizing trends in Western countries (women's and gay rights, for example) have met resistance in more traditional societies, threatening new splits within Christianity.

In many areas Christianity is challenged by **syncretic religions,** which combine practices and beliefs from two or more religions. The African and Caribbean religions of voodoo and Santería, for example, identify African deities and spirits with Christian saints. When Mathieu Kerekou, a Christian, was sworn in as president of Benin in April, 1996, he left out the italicized words in the oath: "Before God, *the spirits of the ancestors,* the nation and before the people ... I ... swear to respect the constitution." The public of this largely voodoo nation insisted he retake the oath with the italicized words. Chapter 5 noted the Native American ceremony at which Alejandro Toledo, a Christian, was sworn in as president of Peru in 2001. Many Brazilians practice a syncretic religion called Candomblé, and in 1999 Brazil's Minister of Culture declared a Candomblé temple a national monument. Other syncretic religions include Cao-Dai from Vietnam, and Chondo-Kyo from Korea. Christian leaders struggle to express Christian belief in terms meaningful to a great diversity of cultures without abandoning essential Christian distinctiveness.

Islam

Through the centuries after Abraham and Ishmael, the Arabs fell away from monotheism to polytheism, but they were brought back to monotheism by Muhammad (c. 570–632). He founded the religion called Islam, which means "submission [to God's will]." "One who

Focus on
HOLIDAYS

One aspect of American commercial culture that has roused worldwide objections is the commercialization of the celebration of holidays, particularly the celebration of holidays that began as religious observances. Throughout Europe, for example, Roman Catholic and Protestant leaders have denigrated Halloween, an increasingly popular American import, condemning it as a pagan ritual and a celebration of Satan. Their efforts, however, have been futile. Shops decked out in orange and black now offer fancy dress competitions and pumpkin decorating displays, while nightclubs plan parties. In 2000, France's Roman Catholic Bishops' Conference derided "skeletons and witches, pumpkins and ghosts," asking, "How long

can this marketing operation called Halloween continue to distort our sense of life and death?" The celebration of Valentine's Day now circles the globe, with fanciful greeting cards and boxes of sweets, although Hindu protesters in India, for example, have burned down shops offering such novelties. In Muslim countries, the end of the holy month of Ramadan is now routinely celebrated with strings of lights, parties, and gifts.

Questions

1. What do you say to the French bishops' question?
2. What holidays do people you know celebrate in ways that overlook the holidays' original religious nature?

---

submits" is a Muslim. The Arabic word for the one God is *al-elah*, or Allah, a cognate of the Hebrew *eloh*, "god." The five essential duties of a Muslim, called the *Five Pillars*, are belief in the one God, five daily prayers, generous giving of alms, fasting during one month (called *Ramadan*), and, if possible, a pilgrimage (*hajj*) to Mecca at least once in one's lifetime.

Despite Muhammad's flight from his native city of Mecca in 622 (the *Hegira*), a temporary setback and the year from which Muslims date their calendar, he had converted and united most Arabs by his death. Just as Christians see their religion as building on Judaism, adding the New Testament to the Old, so Muhammad envisioned his teachings as a continued evolution of monotheism. Muslims believe that Muhammad was the last of God's prophets, who also include Adam, Noah, Abraham, Moses, and Jesus. Muhammad expected Christians and Jews to be among the first to embrace Islam. He first directed Muslims to face Jerusalem in prayer, but he later changed the direction toward Mecca (Figures 7-22 and 7-23).

As Islam diffused, the Arabic of the Koran became the language of peoples throughout the Near East and across North Africa who were ethnically quite different. They had not previously been Arabic speakers, but as they converted, all came to be known as Arabs. To the East, Persia (today's Iran) had its own ancient culture, and although the Persians converted to Islam, they retained their own language, Farsi, an Indo-European language. Therefore, today's Iranians are not Arabs, nor are any of the peoples to the east of Iran.

Within slightly more than a century after Muhammad's death, Islam stretched from the Atlantic coast

FIGURE 7-22 **Mecca.** This black shrouded building is the Kaaba in Mecca, where, Muslims believe, the biblical patriarch Abraham and his son Ishmael built the first sanctuary dedicated to the one God. Through millennia the temple was given over to the worship of idols, but Muhammad threw out the idols, rededicated the Kaaba, and had its walls covered with paintings of Abraham, Mary, and Jesus. That building has since been replaced, but this is probably the longest continuously revered spot on Earth. To the left of the Kaaba a small stone canopy covers what believers accept as the well of Zamzam (Genesis 21: 19), where Abraham's concubine Hagar and their son Ishmael quenched their thirst. Ishmael is thought to be buried just in front of the Kaaba. (Mehmet Biben/Photo Researchers, Inc.)

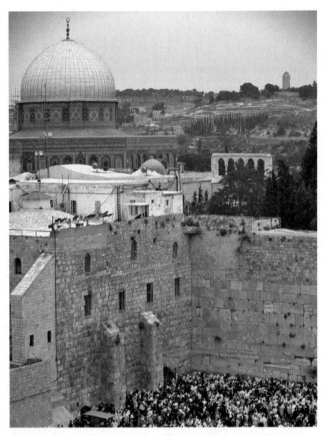

FIGURE 7-23 Dome of the Rock, Jerusalem. Many sites in Jerusalem are sacred to Jews, Christians, and Muslims. All believe God prevented Abraham from sacrificing his son on this rock (Mount Moriah), thus ending human sacrifice. King David later built the Jews' Temple here. Muslims believe that Muhammad ascended to heaven from this spot. The beautiful building with the golden dome is the Dome of the Rock (begun in 643), the oldest existing monument of Muslim architecture. Its builders adopted the Byzantine style when they conquered this region. Contrast it with the Dyula-style mosque in Figure 6-16. Christians believe that Jesus ascended to heaven from the Mount of Olives in the background. In the foreground is the Western Wall, the only wall remaining of the Jewish Temple, which was destroyed by the Romans in A.D. 70, triggering the Jewish Diaspora. (A. Ramey/Woodfin Camp & Associates)

of Spain across North Africa and through Southwest and Central Asia to the borders of China. An Arab army met a Chinese army in the Talas River Valley in today's Kyrgyzstan in 751, just 19 years after another Arab army had faced a Frankish army in Tours in today's France thousands of miles to the West. Within this vast realm, dominance eventually passed from the conquering Arabs to the non-Arab majority, including the Persians and the several Turkish peoples. Islam later expanded down into South and Southeast Asia (Figure 7-24). This culture realm greatly exceeded the Christian culture realm in extent, power, and riches for hundreds of years. Learned travelers crossed its length and breadth, and their descriptive writings constitute

some of the greatest works of historical geography (Figure 7-25).

Islam denigrates the earlier cultures of its converts, just as it was noted that Christianity can. Everything before Islam was, in Arabic, *jahiliya*, "from the age of ignorance." This leaves little room in these peoples' historical consciousness for their pre-Islamic past, so they often lack interest in it. For example, despite Persia's brilliant antique history, for contemporary Iranians the glory began with the coming of Islam. Pakistan is a new Muslim state, but the land contains ruins of civilizations thousands of years old, yet contemporary Pakistanis disdain them. Many people in Muslim countries view their own ancient cultural landscapes without interest. They may even discourage tourists from viewing pre-Islamic ruins. Similarly, the study of the history and art of pharaonic Egypt is the result of Christian European historical interest. The fact that Egyptians are interested in it today demonstrates that Egyptian nationalism, a newer cultural force, competes with Islam for the primary loyalty of the Egyptian people.

As Islam filtered south across the Sahara desert from North Africa, its message reached the black peoples of the sahel and savanna. There it competes still today with Christianity, which entered sub-Saharan Africa from the Europeans' coastal incursions. From Senegal to the Congo the coastal areas are Christian, and inland areas are Muslim. When several of these countries first won independence, the westernized political leaders were Christian (for example, President Léopold Senghor in Senegal and President Félix Houphouët-Boigny in the Ivory Coast), but the majority of the population was Muslim. Christian politicians continue to play prominent roles, but as democracy matures in these countries, Muslims often come to power.

Within Europe, only Albania, long ruled by Turks, is predominantly Muslim. Bosnia and Herzegovina is about 40 percent Muslim.

Islamic sects   The two principal sects of Islam date back to a struggle over rule of the Islamic world that occurred shortly after Muhammad's death. Sunni Muslims accept the tradition (*Sunna*) of Muhammad as authoritative and approve the historic order of Muhammad's first four successors, or *caliphs*. About 85 percent of Muslims worldwide today are Sunni. Most of the other 15 percent are Shia Muslims, or Shiites. They believe that Muhammad's son-in-law Ali was the rightful successor to the Prophet, and they commemorate the martyrdom in Muslim civil war of Muhammad's grandsons in a battle at Karbala in today's southern Iraq in 680. Through the centuries, differences in ceremony and in law have further differentiated the Sunnis from the Shiites. The ruling family of Saudi Arabia is of the Wahabi sect, a puritanical movement that arose in the early eighteenth century and is the only modern separatist Sunni sect.

The Diffusion of Islam

FIGURE 7-24 The first nine centuries of Islam were a period of almost continual expansion, and by 1800, Arab and Indian traders had spread the faith through many of the islands of Southeast Asia.

The Travels of Ibn Battuta

FIGURE 7-25 In the fourteenth century the scholar Ibn Battuta traveled from Morocco to China, and everywhere rulers happily received him. Several even appointed him as a judge thousands of miles from his own home. His geographic writings are a superlative description of life across the vast Islamic realm of his day.

Shiites form the majority in Iran, Azerbaijan, and in Iraq and important minorities in Kuwait, Lebanon, Bahrain, Syria, Saudi Arabia, and Pakistan. Animosity between Sunni and Shia can be fierce. Countries that contain significant shares of both sects are split by the enactment into national laws of either Sunni or Shiite interpretations of Muslim religious teachings, and several, including Lebanon and Pakistan, have suffered civil disturbances between the two groups. Even the hajj to Mecca has regularly been disrupted by violent clashes between Sunnis and Shiites.

The Muslim world today Today Muslims are distributed from Morocco to Indonesia, north to the frontiers of Siberia and south to Zanzibar, with outposts throughout the world. Even Rome boasts a splendid mosque. The leading Muslim states, in numbers, are Indonesia, Bangladesh, India, and Pakistan.

Muslims believe in proselytizing, as Christians do, and this necessarily injects an element of competition into the relationship between the two religious communities. Many Muslims in the Islamic realm are fundamentalist, but this fundamentalism may evolve or modify as more Muslims move outside the Islamic realm, just as Christianity has evolved as it has diffused. For example, speaking in 1994 at the dedication of a new mosque in Lyons, France, the mosque's imam (Islamic prayer leader), Algerian-born Chellali Benchellali, proclaimed that he stood "between two cultures. . . . I think this is the beginning of a French Islam." Unfortunately, two of Benchellali's sons were in 2003 being held in terrorist investigations. One was captured by American forces in Afghanistan in 2001, and the other was arrested in Paris in December 2002.

In other non-Arab regions, local practices differ in Islam. Many Indonesians, for example, retain ancient local gods and goddesses and folk practices in a form of syncretism. These may be declining as a result of urbanization and the preaching of a more fundamentalist Islam by Arab-financed foundations.

Although many Americans tend to equate Islam with Arabs, Arabs actually constitute only a fraction of Muslims. There are more Muslims in Indonesia and Pakistan than in all the Arab countries combined. Americans' association of Islam with Arabs is a political preconception, caused by the distinctive role of Islam in Arab countries and by U.S. news media focus on Arab lands as sources of oil and antagonists of Israel.

Hinduism and Sikhism

Hinduism is the most ancient religious tradition in Asia. The oldest Hindu sacred texts (the *Vedas*) date to about 1800 B.C., but the religion originated somewhere in Central Asia long before that. It entered the Indian subcontinent with the coming of Central Asian peoples about the time of the writing of the *Vedas*. Today it is

Religions in South Asia

FIGURE 7-26 Hinduism is today restricted almost exclusively to India, Nepal, and to places where Indians have migrated, including Jamaica, Trinidad, Guyana, and Fiji in the Pacific Ocean. Sikhs form a majority in the Indian state of Punjab, but smaller Sikh communities can be found throughout India. The regions labeled Christian converted from animism in the nineteenth century.

confined almost exclusively to India and Nepal; in Nepal it is the official state religion (Figure 7-26).

Hindus believe in one Supreme Consciousness, Brahman, whose aspects are realized in three deities: Brahma, the creator; Vishnu, the preserver; and Siva, the destroyer. These are coequal, and their functions are interchangeable. All other Hindu "gods," saints, or spirits are emanations of Brahman.

Hinduism classifies people in a hierarchy of classes called **castes.** The four main castes are (1) the Brahman, or priestly caste; (2) the Kshatriya, or warrior caste; (3) the Vaisya, or tradesman and farmer caste; (4) the Sudra, or servant and laborer caste. Each of these is split into hundreds of subcastes. Many castes are defined by occupation. People are expected to mix socially, marry, and stay in the caste into which they were born. A group of people called **untouchables** (*Dalit*) is considered so low that their status is below the formal structure of the four castes. To some degree caste discrimination still structures Indian life, but it was legally abolished in the Indian Constitution of 1950, and an untouchable, K. R. Narayanan, was even sworn in as president of India in 1997. (India's president is elected by parliament, not by popular vote.)

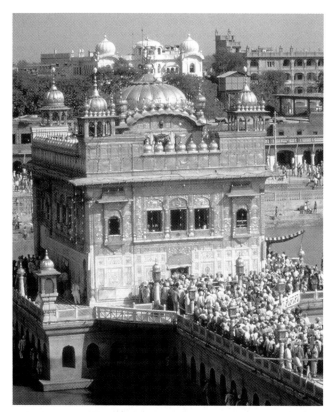

FIGURE 7-27 The Golden Temple in Amritsar. The Sikhs' 400-year-old Golden Temple in Amritsar, India, stands in the middle of a sacred lake (in Sanskrit *amrita saras*, "pool of immortality"). The building was occupied by armed Sikh extremists in 1983, and the following year Indian troops stormed it and drove out the occupants in a bloody encounter. Sikh fanatics retaliated by assassinating India's Prime Minister Indira Gandhi. The world's second largest Sikh temple, seating 3,000 worshippers, opened in London in 2003. (Jenhangir Gazdar/Woodfin Camp & Associates)

Hindus believe in *reincarnation*—that is, individual rebirth after death. The caste into which you are born is not haphazard, but depends upon your behavior in an earlier life. This teaching, called *karma*, discourages ambition, because only by docilely keeping to your place in this life can you hope to enjoy a better position in your next life. The goal of Hindus is liberation from the cycle of death and rebirth.

Sikhism is an offshoot of Hinduism based on the teachings of Guru (teacher) Nanak (c. 1469–1539). Nanak tried to reconcile Hinduism and Islam, teaching monotheism and the realization of God through religious exercises and meditation. Nanak opposed the maintenance of a priesthood and the caste system. Under a series of gurus, the Sikhs had their own state in northern India, but they were eventually conquered by the British. Many Sikhs dream of the restoration of an independent Sikh state. The Sikh's holy temple is in the city of Amritsar in the Indian state of Punjab (Figure 7-27). This state is largely Sikh, and it might provide a territorial base for independence.

Buddhism

Siddhartha Gautama (c. 563–483 B.C.) was a Hindu prince born in today's Nepal who, through meditation, achieved the status and title of Buddha, or Enlightened One. He taught Four Noble Truths: (1) life involves suffering, (2) the cause of suffering is desire, (3) elimination of desire ends suffering, and (4) desire can be eliminated by right thinking and behavior. This cessation of suffering is called *nirvana*, or total transcendence.

As Buddhism diffused out of India, sects and schools arose. The Theravada school ("doctrine of the elders") diffused to the South (Figure 7-28). This school centers on the idea of a monk striving for his own deliverance. Mahayana Buddhism (the "great vehicle" because it carries more people to nirvana) diffused northward. Mahayana idealizes the concept of the *bodhisattva*, someone who merits nirvana but postpones it until all others have achieved enlightenment. In Bhutan, Tibet, and Mongolia, Buddhism evolved a special form called *Lamaism*, which is known for its elaborate rituals and complex priestly hierarchy (Figure 7-29). Chinese Buddhists produced a new theory of spontaneous enlightenment, called *Ch'an*. This diffused into Japan as *zen*.

Buddhism has several hundred million followers, but its adherents are hard to count because it is not an

FIGURE 7-28 Buddhism was localized near its hearth for hundreds of years, but it spread widely through the support of the third-century Indian emperor Asoka. It has largely been abandoned in India.

FIGURE 7-29 The Dalai Lama. The Dalai Lama, the Buddhist spiritual leader of Tibet, was forced into exile by the government of China. He has won worldwide acclaim—including the Nobel Peace Prize in 1989—for his efforts to promote religious freedom. Hollywood stars, including Penelope Cruz, have become devoted to him. (AP/Wide World Photos)

exclusive system of belief. Its practice has declined in India, but today it is the state religion in Thailand and Sri Lanka, and it may achieve that status in Mongolia. Buddhist philosophy has also won considerable influence in the modern Western world. Much popular "New Age" philosophy derives from Buddhism.

Other Eastern Religions

Confucianism is a philosophical system based on the teachings of K'ung Fu-tzu (c. 551–479 B.C.). He taught a system of "right living" preserved in a collection of sayings, *The Analects,* which governed much of China's political and moral culture for 2,000 years.

Confucianists hold that people may attain heavenly harmony by cultivating knowledge, patience, sincerity, obedience, and the fulfillment of obligations between parents and children, subject and ruler. These moral precepts permeate life in many Eastern societies, and today the word Confucian is often popularly used as a synonym for these qualities. Confucianism influenced Western philosophy and political theory at the time of the Enlightenment, when it appeared as the realization of Plato's utopian dream of a state ruled by philosophers.

Taoism is derived from the book *Tao-te Ching* (third century B.C.) attributed to Lao-tze. It advocates a contemplative life in accord with nature (Figure 7-30).

Shinto is the ancient religion native to Japan. It includes a set of rituals and customs involving reverence for ancestors, celebration of festivals, and pilgrimage to shrines, but there is no dogma or formulated code of morals. Shinto is nationalistic, and it traditionally

FIGURE 7-30 Lao-tze and Confucius with the infant Buddha. This fourteenth-century Chinese painting shows Lao-tze (on the left) and Confucius protecting the infant Sakayumi, the future Buddha. Lao-tze's appearance reflects his legendary life spent wandering and communing with nature, in contrast to the courtly scholar Confucius. Images such as this, emphasizing the compatibility of Confucianism, Taoism, and Buddhism, became popular as Buddhism spread across East Asia. (© The British Museum)

recognized the emperor as divine. The Emperor Hirohito renounced this divinity in 1946, following Japan's defeat in World War II. Nevertheless, signs at the Yakasuni Shinto Shrine still proclaim, "Some 1,068 people, who were wrongly accused as war criminals by the Allied court were enshrined here. . . . War is a really tragic thing to happen, but it was necessary in order for us to protect the independence of Japan and to prosper together with Asian neighbors." Periodic visits to Yakasuni by Japan's Prime Ministers roil antagonism between Japan and Japan's World War II enemies, especially China and the Koreas.

Animism and Shamanism

Animism is a belief in the ubiquity of spirits or spiritual forces (Figure 7-31). Animistic religions may be basically monotheistic, but they recognize hierarchies of divinities who assist God and personify natural forces. Millions of Africans believe in animism.

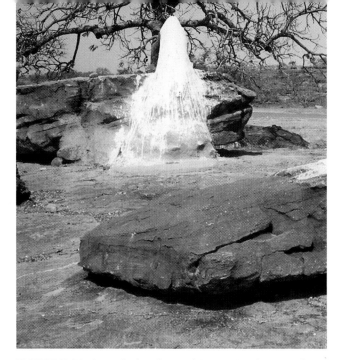

FIGURE 7-31 An animist altar. This animist altar stands in a Dogon village in the West African country of Mali. It is covered with a patina of millet gruel laid on as an offering to the spirits of the harvest. (Photograph by Eliot Elisofon 1972. Eliot Elisofon Photographic Archives, National Museum of African Art, Smithsonian Institution.)

Animism is frequently accompanied by **shamanism**. A shaman is a medium who characteristically goes into auto-hypnotic trances, during which he or she is thought to be in mystical communion with the spirit world. Shamanism is noted among the peoples of Siberia, the Inuit, some Native American tribes, in Southeast Asia, and in Oceania. Almost everywhere both animism and shamanism are yielding to Muslim and Christian proselytizing. Meanwhile, the world's pharmaceutical companies race to learn shamans' ethnobotanical knowledge.

THE POLITICAL AND SOCIAL IMPACT OF THE GEOGRAPHY OF RELIGION

The distribution of religions affects the distributions of many other things. Religion influences what people eat; their tolerance for others; what men think of women, and vice versa; when and how they work, and when and how they celebrate; which types of behavior they encourage and discourage; and many other aspects of life. The influence of religion is so pervasive that it is difficult to isolate and identify precisely. Clearly, however, many people choose to obey their religion's teachings. The following pages catalog just a few of the many ways that the geography of religion influences the geography of other aspects of human lives and societies.

Religion and Politics

Almost all countries guarantee freedom of religion, and most governments observe a form of secularism. Many governments nevertheless favor one religion over others implicitly or explicitly. Therefore, the world political map partly fixes or stabilizes the map of world religions, just as it fixes the map of world languages.

Theocracy is a form of government where a church rules directly. The Vatican is a theocracy. Morocco may be a modified theocracy, because the king's legitimacy derives partly from his descent from Muhammad (Figure 7-32). Many theocracies have existed in history: Tibet, for example, and Massachusetts Bay Colony in British North America. Utah was once a Mormon theocracy, and although it is today increasingly cosmopolitan, no Utah politician can ignore Mormon Church leaders (Figure 7-33).

In Israel Israel has steadily enacted more religious strictures into law since its founding in 1948. Recent laws, for example, ban the production or sale of pork and prohibit activities on the Sabbath. Also, the government has become more Orthodox, recognizing, for example, conversions to Judaism performed only by Orthodox rabbis.

FIGURE 7-32 King Muhammad VI of Morocco. King Mohammed VI alternates between modern westernized clothes and the robes traditional to his role as hereditary religious leader. (Benito-Buu-Benyatouille Hires/Getty Images, Inc.—Liaison)

FIGURE 7-33 The citadel of the Church of Jesus Christ of Latter-Day Saints (often called the Mormon church) in Salt Lake City. In 1847 this site was surrounded by hundreds of miles of wilderness, but Mormon leader Brigham Young recognized it as an oasis that Mormon industriousness could make bloom and where Mormons might escape the persecution they had suffered in the East. The building at left in the form of an overturned hull of a ship is the Tabernacle, and the neo-Gothic building in the center is the Temple. Mormonism is one of the fastest-growing religions in the world. In 1950 there were 1 million Mormons, nearly half of whom lived in Utah. Today the religion counts 12 million members around the world, fewer than 15 percent of whom live in Utah. (© The Church of Jesus Christ of Latter-Day Saints. Used by permission.)

This has distressed some U.S. Jews, 85 percent of whom are Conservative or Reform. Many of the Russian Jews who migrated to Israel in the 1990s are not Orthodox, and they press for secularization, including civil marriage, eliminating the "nationality" designation on identity cards, and the operation of public transit on the Sabbath—all policies opposed by the Orthodox.

In Christianity The Christian Bible records Jesus' words, "Render therefore unto Caesar the things which be Caesar's, and unto God the things which be God's" (Luke 20:25). For 2,000 years, governments among Christian peoples have argued about exactly where to divide responsibility between religion and government, but they accept the idea that there is a dividing line, and that church and state, religion and politics, must be separated.

The separation between religion and the political order allowed Europe to develop secular government, secular knowledge, and a secular culture. Nevertheless, several countries are today explicitly Christian or at least support various Christian sects (called *established* churches) with public funds. These include Argentina, Peru, Ireland, Norway, Denmark, Iceland, the United Kingdom, Finland, and Germany (although in 1998 Gerhard Schroeder became the first-ever German

Chancellor to refuse to end his oath of office with the words "so help me God."). At the same time, other countries are disestablishing official churches. For example, Italy disestablished Roman Catholicism in 1984, Spain in 1988, and Colombia in 1991. Sweden disestablished Lutheranism in 2000, and the government of Greece stopped listing religious affiliation on state identity cards in 2000. Many countries in Eastern Europe have revived their national churches as a means of defining and emphasizing their national identities since the fall of communist governments. The Uniate Church of Ukraine, for example, survived persecution by the communists, and it revived in the late 1980s.

Liberal democracy, as it is known in the West, was shaped by a thousand years of European history and, beyond that, by Europe's double heritage of Judeo-Christian religion and ethics and Greco-Roman statecraft and law. No such system ever originated in any other cultural tradition, and it remains to be seen whether such a system can survive when transplanted and adapted in another culture.

In Islam Theoretically, no distinction can be made between Church and state in Islam, which teaches that the only purpose of government is to ensure that each person can lead a good Muslim life. Church and state should be the same, and thus Islam is inherently political. (The centrality of religion in Islamic cultures is illustrated even in city planning on p. 418.) Therefore, many countries with largely Muslim populations are today officially Islamic, and fundamentalist groups in several others are struggling to achieve such recognition. Countries that are officially Islamic include Mauritania, Afghanistan, Libya, Saudi Arabia, Yemen, Oman, Qatar, Bahrain, Iraq, Iran, Comoros, Maldives, Pakistan, Bangladesh, and Malaysia. These countries may be considered modified theocracies. Islamic states enact Islamic teachings into law, called **Sharia** law, and establish Sharia courts to rule whether secular law conforms to Islamic teaching. If the Sharia court rules against a secular law, that law is usually repealed. All legislation in Iran, for example, must be approved by a body of 12 religious judges, who enforce fundamentalist beliefs. Indonesia, which has the world's largest Muslim population but which is officially a secular country, has recognized Sharia rulings on family law.

Fundamentalist Islamic political parties are contesting elections in several countries that are not now officially Islamic. Many observers fear that these parties see democracy as a means to an end—the creation of an Islamic state—rather than as a system to be valued for itself. Turkey is officially secular, but the government pays for mosques, supervises the (compulsory) religious education, specifies qualifications for imams, appoints all imams to mosques, and pays their salaries. The Turkish Constitution declares the army to be the official guardian of the country's secular status, and Turkey's army has repeatedly intervened in democratic processes to frustrate the rise of fundamentalists. Elections in 2002 brought to power the Justice and Development Party, which has a background in Islamic activities, and its government has negotiated policies with the Army. The new government has maintained Turkey's traditional strong support of Israel. It refused to cooperate with the U.S. war against Iraq in 2003, but it has cooperated in the reconstruction effort.

The Algerian Army canceled 1995 elections for fear that fundamentalists would win, and it was accused of meddling in subsequent elections. Through the 1990s, more than 100,000 people died in fighting in Algeria, and some observers fear that Algerian terrorists are allying with other international terrorist groups. Some Algerians were found stockpiling poison in London in 2003. Egypt has suffered a civil war between the government and fundamentalists that took the life of the country's president, Anwar Sadat, in 1981. Fundamentalists massacred 58 foreign tourists and four Egyptian guides at the ancient site of Luxor in November 1997.

The role of education   Some observers insist that the rise of fundamentalist Islamic parties is due to the failure of the secular governments in many Muslim countries to serve the people—to provide basic services, such as health care and education. Many see education as secularists' strongest weapon against fundamentalism. Mohammed Charfi, who served as Education Minister in Tunisia from 1989 to 1994, presided over hundreds of Arabists, historians, philosophers, and other specialists to remove aspects of political Islam from Tunisian schoolbooks. "I left not a single schoolbook untouched," he wrote. "The reform is not against Islam, but rather designed as a modern version of Islam consonant with democracy. School is the best place to fight fundamentalism, and school is where it is born and dies. No one could be a fundamentalist if they read Spinoza, Voltaire, and Freud."

In many Muslim lands, by contrast, the only education available (and only to males) is in Islamic schools, called *madrassas,* where modern mathematics and science, as well as humanistic concerns, are repudiated. Students spend hours poring over the Koran, and the "A" students memorize the most passages. Fundamentalist governments, individuals, and Islamic foundations have financed madrassas in many countries in Africa, Southeast Asia, and other places where secular public education is weak, rousing fears of networks of anti-Western indoctrination. In 2002, Pakistan's President Pervez Musharraf acknowledged that much of the backwardness of the Muslim world was rooted in the education systems. "Do we believe," he asked, "that religious education alone is enough for governance, or

do we want Pakistan to emerge as a dynamic Islamic state? The verdict of the masses is in favor of a progressive Islamic state." President Musharraf promised educational reform.

The Muslim Umma   To a Muslim, the international community of all Muslims is called the *Umma.* To a fundamentalist, the division of this community into many political states is blasphemy. In 1998, the terrorist Osama bin Laden issued a *fatwa* (holy directive) saying:

> The call to wage war against America was called because America has spearheaded the crusade against the Islamic community, sending tens of thousands of troops to the land of the two holy mosques over and above its meddling in its affairs and its politics, and [because of American] support of the oppressive, corrupt and tyrannical regime that is in control. . . . To kill Americans and their allies, both civil and military, is an individual duty of every Muslim who is able, in any country, where this is possible, until the Aqsa Mosque [in Jerusalem] and the Haram Mosque [in Mecca] are freed from their grip and until their armies, shattered and broken-winged, depart from all the lands of Islam.

Bin Laden referred to an "Islamic community," that spans many ethnicities and states, and to the mosques in Jerusalem and Mecca as being in the same "land." Bin Laden thus saw all the governments across the entire Muslim world (including Israel) as illegitimate, and he denied them recognition. This belief made bin Laden, ultimately, a threat to all individual Muslim rulers and states (compare Figure 7-25, p. 297). By his definition, the establishment of Israel and the existence of U.S. troops in the Near East are blasphemy, even if those troops—first based on Saudi territory in 1991—were sent to protect Saudi Arabia and to free Kuwait, which are both Islamic states, from Iraq. Any government is not truly Islamic if it allows these troops, even if it calls itself Islamic. Bahrain and Oman have also provided bases for U.S. troops, and the 2003 conquest and occupation of Iraq has fired many Muslims' indignation. U.S. troops did leave Saudi Arabia in 2003.

Bin Laden saw his own homeland as the most egregiously impious state in the Muslim world because Muslims have the right to expect the most of it. Since 1932, when Standard Oil of California won the first oil concession, Saudi Arabia has been closely tied to the United States. Nevertheless, the royal house of Saud rests its legitimacy on its strict Wahabist principles, and the government has funded thousands of fundamentalist mosques and madrassas throughout the Muslim world. Many of the thousands of members of the royal

Some observers insist that the rise of fundamentalist Islamic parties is due to the failure of the secular governments in many Muslim countries to serve the people.

Focus on

ARE WE SEEING AN ISLAMIC REFORMATION?

The early twenty-first century is witnessing a war of ideas within the Islamic world comparable to Western Europe's Protestant Reformation. One leader is Hashem Aghajari, a former Islamic revolutionary and college professor in Iran. Professor Aghajari is not the only significant theologian in the Islamic world, nor are his ideas necessarily the best, nor will they necessarily triumph, but his ideas exemplify the birth of a new conversation among Muslims. Developments in Iran are particularly important not only because Iran is a populous and powerful country, but, because of the population policies of the 1980s (see Chapter 5), the median age of its population is only 20. This is too young to have known any government before the current regime, but today's repressive government often views rebellion as reactionary.

In November, 2002, Professor Aghajari gave a speech on the 25th anniversary of the death of Ali Shariati, a leader of the Iranian revolution. Professor Aghajari began by noting that "the Protestant movement wanted to rescue Christianity from the clergy and the church hierarchy," so Muslims must do the same. The Muslim clergymen who dominate their faith today

were never, he argued, meant to have a monopoly on religious thinking or interpetation. "Just as people at the dawn of Islam conversed with the Prophet, we have the right to do this today. Just as they interpreted what was conveyed [to them] at historical junctures, we must do the same. We cannot say: 'Because this is the past we must accept it without question.' This is not logical. For years, young people were afraid to open a Koran. They said, 'We must go ask the mullahs [religious leaders] what the Koran says.' Then came Shariati, and he told the young people that those ideas were bankrupt. [He said] you could understand the Koran using your own methods. The religious leaders taught that if you understand the Koran on your own, you have committed a crime. They feared that their racket would cease to exist if young people learned the Koran on their own. . . . We need a religion that respects the rights of all—a progressive religion, rather than a traditional religion that tramples the people. . . . Today, more than ever, we need the 'Islamic humanism' and 'Islamic Protestantism' that Shariati advocated."

Professor Aghajari was in effect accusing the mullahs of introducing sacerdotalism into Islam. He was arrested and sentenced to death for *apostasy,* renunciation of his faith, but in February 2003 the Iranian Supreme Court lifted the death sentence.

---

house, however, have become notorious playboys and worse. In October 2001, Sheik Hamoud bin Oqla al-Shuaibi, the leader of the Wahabis, virtually excommunicated the Saudi royal family. Only a small minority of Saudi Arabia's people, however, are Wahabi, and the majority of Saudis might actually want to repudiate Wahabi influence. Saudi Arabia, despite its immense oil revenues, has run a national budget deficit for over a decade and now carries the burden of a large national debt.

Osama bin Laden never represented the Islamic community in general. The Organization of the Islamic Conference, a coalition of 56 Islamic nations representing 1.2 billion Muslims, did meet in Doha, Qatar, on October 10, 2001, in a special emergency session. Representatives insisted that the "terrorist acts [of 9/11] contradict the teaching of all religions and human and moral values." Later, in July 2003, at a conference of Muslim scholars in Malaysia, Grand Sheik of Al-Azhar Mohammed Sayed Tantawi, considered by many to be the Sunni Muslim world's highest religious authority, said Muslims should "wholeheartedly open our arms to

the people who want peace with us. . . . I do not subscribe to the idea of a clash of civilizations," he insisted, revealing his familiarity with popular Western thought (see Chapter 6).

The strife within the Islamic Umma between nationalism and internationalism, and its efforts to define the role of religion in modern life, will occupy headlines for the foreseeable future.

The geography of religion in the United States
Several of the American colonies were founded as theocracies, and most retained established churches until the War for Independence. In forging the new United States, however, James Madison included in the Bill of Rights a prohibition against the federal government's establishing any religion—the world's first such clause: "Congress shall make no law respecting an establishment of religion, or prohibiting the free exercise thereof. . . ." The 1797 treaty between the U.S. and Tripoli—ratified by Congress—declares that "The government of the United States is not, in any sense, based on the Christian religion." Congress added the words

Church Membership in the United States

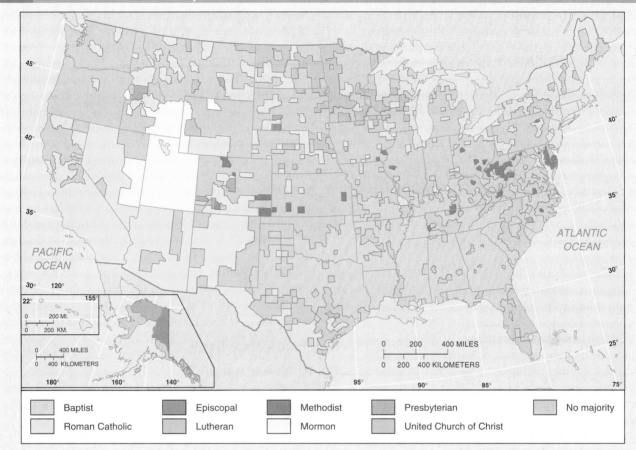

FIGURE 7-34 This map of Americans' church affiliations reveals the predominance of Baptist membership in the Southeast and of Mormons in Utah and adjacent regions. The predominance of Lutherans in the northern Midwest reflects the Scandinavian and North German backgrounds of that region's pioneer settlers, and the distribution of Roman Catholics reflects, among other factors, the migration to the United States of Hispanics and Southern Europeans. (Modified from Martin B. Bradley, Norman M. Green, Jr., Dale E. Jones, Mac Lynn, and Lou McNeil, *Churches and Church Membership in the United States 1990*. Nashville, TN: Glenmarry Research Center, 1992.)

"under God" to the pledge of allegiance in 1954, but in 2002 the Federal Appeals Court's Ninth Circuit ruled that action unconstitutional, and litigation continues. American coins do insist "In God We Trust."

People of every faith can worship in the United States as they wish, and they can even proselytize. The consequence of U.S. separation of church and state has not been the death of religion, but, rather, the birth of a vibrant religious society, arguably the most religious nation in the world. In a 2001 Gallup Poll, 82 percent of Americans identified themselves as Christian; 10 percent of other faiths, and only 8 percent as agnostic or atheist. This is a degree of religious commitment, and of religious homogeneity, almost unique in the world. Surveys in different countries do not always ask the same questions, but for comparison about 2 percent of people in England regularly attend church, and fewer than one French person in 10 goes to church as often as once a year. In 2002, 49 percent of Danes, 52 percent

of Norwegians, 55 percent of Swedes, and 64 percent of Czechs said God "does not matter" to them, whereas 82 percent of North Americans said God was "very important." The U.S. Census Bureau does not study religions, but Figure 7-34 reproduces one map of the variations of church membership across the country. Americans' religiosity, despite the Constitutional separation of Church and State, is, perhaps, particularly baffling to Islamic fundamentalists, for whom the unity of Church and state is important.

Immigration is changing the religious composition of the U.S. population. In one survey of immigrants in the late 1990s, 41 percent described themselves as Roman Catholic (compared to about 25 percent of the current national population), and 9 percent were Protestant (compared to about 50 percent of the current population). Nearly 30 percent describe themselves as something other than Protestant, Catholic, or Jewish; 7 percent claimed to be Muslim. In

1995, Paterson, New Jersey, with a total city population about 15 percent Muslim and a school-age population almost 25 percent Muslim, became the first U.S. city ever to close its public schools for a Muslim holiday.

Churches in the United States, particularly in central-city African-American neighborhoods, often provide an array of social services, including counseling, meals, education, and even health care. Thus, churches are potential building blocks of government community-assistance programs, and bills funding "faith-based" programs first passed Congress in 1996. Such programs have increased, and today the federal government even pays to maintain houses of worship if social services are provided in them. The constitutionality of these programs is under legal review.

Indirect Religious Influences on Government

In all countries the prevailing religion can dictate national ethics or morals. Religion so deeply affects what people assume to be natural or desirable human behavior that religious prejudices are taken for granted and unconsciously translated into laws. Religion should never be underestimated as an institutional, economic, and political force. Any organized religion has its own bureaucracy, sources of income, and, in the pulpit, its own channel of communication that often reaches more of the population than does government information (Figure 7-35).

> Religion so deeply affects what people assume to be natural or desirable human behavior that religious prejudices are taken for granted and unconsciously translated into laws.

In several countries churches operate a school system parallel to that of the government. In the United States, religious schools are accredited, but not funded, by the government. The provinces of Canada vary, but in Ontario, for example, the provincial government finances both Roman Catholic schools and secular public schools. In virtually all countries but the most fundamentalist Muslim, the governments insist that religious schools cover such modern subjects as science and math along with religious studies.

Relations between the government and the leading religious organizations constitute a major political issue in most countries. A national church may sustain a suppressed nationalism, as in Ukraine or in Ireland. Religious leaders may lend legitimacy to rulers or, conversely, provide a rallying point for political opposition. The Polish Roman Catholic Church stood against the Communist Party and state as an alternative repository of Polish identity between 1945 and 1989. Attending mass was an act of political defiance. From 1959 until 1992 the constitution of communist Cuba insisted that Cuba was officially atheist, and the Roman Catholic Church has remained the only force outside the state-party machine with any significant structure or loyalty. The government has granted more freedom to the church recently, however, and Pope John Paul II visited Cuba in 1998. In Hong Kong, Roman Catholic Bishop Joseph Zen emerged at the forefront of opposition to the diminution of civil liberties there after Hong Kong merged with China in 1997. In 2003, U.S. Secretary of State Colin Powell singled out Archbishop Pius Ncube of Zimbabwe for his bravery in the face of government oppression there.

In many countries a church or religion may form a political party, as the Christian Democrats have in several European countries, or as various Islamic groups have. The rise of a Hindu nationalist party in India, Bharatiya Janata, threatens India's status as a secular country, although the party has headed coalition governments.

In many countries a religion can become a big landowner or financier, and religious organizations do not always maximize the productive capacity of their property. Real property that is restricted in ownership or in the purposes for which it can be used is called *mortmain*, literally "dead hand." Church property may also enjoy preferential tax rates or escape taxation entirely. Throughout Latin America the Catholic Church is a principal landowner. In the United States, church-owned properties are usually concentrated in the cities. Because they are free from taxation, their concentration reduces any city's ability to raise property-tax income. In turn, however, the churches themselves may provide a wide range of services to a public beyond their own communicants.

FIGURE 7-35 The main square of Mexico City. Here in Mexico City, as in many other Latin American capitals, the cathedral (to the left) and the presidential palace (to the right) confront one another across the main square. (Courtesy of the Mexico Government Tourism Office)

Critical Thinking
LIBERATION THEOLOGY

A philosophical dispute within the Roman Catholic Church has brought about revolutionary changes in politics throughout Latin America. **Liberation theology**—named after a 1968 book by the Peruvian Father Gustavo Gutiérrez—puts the problems of overcoming poverty at the heart of Christian theology. It recommends political activism, using the Church institutional framework to organize the population in the struggle for social justice and equality. Following this idea, many Latin American bishops have proposed national redistributions of land and income, provided legal services for the poor, and attacked elite power structures from their pulpits. Cuban leader Fidel Castro has praised liberation theology and supported liberation theologians throughout Latin America.

Liberation theology provoked political repercussions throughout Latin America. In some countries it triggered militant opposition among conservatives threatened by changes on either religious or political grounds. For example, in El Salvador during the 1980s, nuns, priests, and even Archbishop Oscar Romero were assassinated by right-wing elements of the military regime. The most conservative Latin American governments have favored

Evangelicals as a counterforce to liberation theology. In Guatemala, for example, government death squads murdered politically active priests during the 1980s, and one Roman Catholic bishop was forced into exile. Meanwhile, Protestant sects, with government encouragement, have converted more than 30 percent of the population.

Pope John Paul II has tried to arrest liberation theology. He forbade Father Gutiérrez to write or teach, and he has appointed conservative bishops who have closed church legal-services offices, land-rights offices, and liberal seminaries.

The arguments over liberation theology are ostensibly religious arguments restricted to church affairs, but in fact this theological dispute may activate politics in Latin America and other areas for years to come.

Questions

1. Investigate the official position papers of the U.S. Council of Roman Catholic Bishops. Do you think they tend toward liberation theology?

2. How has the Catholic Church interacted with Cuban leader Fidel Castro? Has it supported him?

In some countries a community of citizens might be alienated from their national government if co-religionists in an adjacent country enjoy more complete religious freedom. For example, the Islamic vigor of Iran provided religious inspiration to the Muslim populations of the Central Asian Republics when they were under the rule of the Soviet Union.

France has long enforced a rigid separation of church and state, but the growth of Islam in France (today about 5 million adherents) has forced the government to rethink its role. In 2003 a government-sponsored elected Consultative Council of Islam was welcomed to the presidential palace, and, although the law currently forbids the government from building houses of worship, the influx of foreign subsidies to mosques and madrassas has been so great (mostly from Saudi Arabia), that the government may change its position. In some areas local authorities are already skirting the law by giving land for mosques or renting it cheaply.

States split by two or more religions   When two religious communities compete to write a nation's laws, the country must devise compromises or suffer internal

conflict. Attempts to enforce Sharia in the Sudan, for example, have intensified conflict between the Arab Muslim-dominated north and the black Christian and animist south. Similarly, in Nigeria, several northern states have adopted Sharia, triggering rioting between Christians and Muslims in which an estimated 10,000 people died from 1999–2003. Christians have migrated out of Sharia states, and Muslims into them. Religion-based civil war broke out in the Ivory Coast in 2002. Muslims are the majority, but Christians have monopolized political power since independence in 1960. In the Philippines, tensions between Christians and Muslims have fostered revolts on several islands against the central government, and since the 9/11 attacks, the United States has increased aid to the Philippine Army to battle insurgents.

Sometimes states will enforce two separate systems of law, depending on the religion of the people involved in a case. Senegal and several other African states will allow Muslim men—but only Muslim men—more than one wife. In India, Muslims are allowed to follow Muslim law in many matters of education, marriage, divorce, and property. The government of Malaysia enforces mandatory tithing (giving 10 percent

Regional Focus on

RELIGIOUS TENSIONS ON THE INDIAN SUBCONTINENT

The Indian subcontinent illustrates how religious divisiveness can cause problems both within countries and among countries. The British ruled the subcontinent (today's India, Pakistan, and Bangladesh) as one colony, "India," and they intended to grant it independence as one secular state. Many of the Muslims within this immense territory, however, demanded that the colony be divided so that they could create a Muslim state. When

independence came in 1947, the Muslim population established the country of Pakistan. It originally included a western territory (today's Pakistan) and a separate eastern territory. In the months before the partition was realized, millions of Muslims fled toward areas that were designated to become Pakistan, and millions of Hindus to regions scheduled to become parts of India. Tens of thousands died in terrible acts of violence. Muslim-Hindu tensions were exacerbated by the fact that for hundreds of years Muslim minorities had in many areas ruled Hindu majorities.

Border Disputes in South Asia

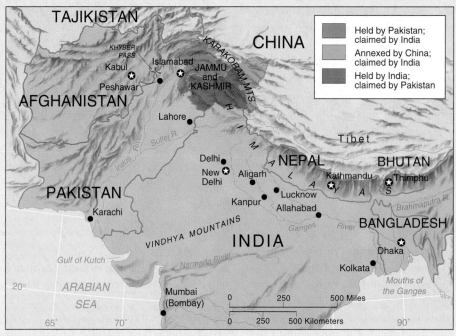

Antagonism and border disputes persist between India and Pakistan and between India and China. These three nuclear powers contain about 40 percent of the human population.

of one's income to charity) among Muslims, about half of the country's total population.

In some countries, struggles that are ostensibly religious actually conceal other social divisions. Tension in Northern Ireland is, on the surface, a conflict between a Protestant majority and a Roman Catholic minority. That split, however, involves much more than theology. It is a competition for jobs and opportunity, and it is complicated by deep historic roots and the fact that both sides draw on outside support.

China recognizes five "approved" religions: Protestant Christianity, Catholic Christianity, Buddhism, Islam,

and Taoism. The government oversees religious activity carefully, and the Chinese Catholic Church does not recognize the authority of the Vatican.

Religion and Women's Rights

Religions differ in the attitudes that their teachings advocate toward women, and national laws often reflect religious teachings. In many cases, however, the actual practice of any religion varies from place to place. These variations reflect variations in other aspects of the cultures among places. Therefore, it is difficult to

In 1971 eastern Pakistan broke off from Pakistan to form a separate Islamic country, today's Bangladesh, but religio-political turmoil continues in what is left of Pakistan. Sunnis (about 70 percent) scorn Shiites (about 20 percent), and tribal loyalties and identifications subdivide the population still further. Non-Muslims (about 10 percent of the population) are guaranteed 10 of the 307 seats in Parliament, but they do not represent geographical districts, so they have to campaign across the entire country.

India itself emerged as a secular state with a majority Hindu population. Between 10 and 13 percent of India's population, however, is Muslim—a greater number of Muslims than in Pakistan. Muslims and Hindus regularly battle on the streets of India's cities, and India's Muslims were persecuted during India's wars with Pakistan in 1965 and again in 1971. The rise of the Hindu nationalist party Bharatiya Janata has coincided with a growing climate of intolerance of minorities and the introduction of new Hindu-oriented syllabi in schools. Avil J.P. Abdul Kalam, a Muslim nuclear scientist, was inaugurated as India's 12th president in July 2002 (his family converted to Islam, but Hindus still consider him a Hindu with, in fact, Brahmin status), but growing hostility between Muslims and Hindus in India threatens domestic peace. Furthermore, the conversion of many untouchables to Islam or Buddhism further upsets India's internal religious balance.

Conflict continues on the Indian-Pakistani border. In 1947 India and Pakistan fought over the border state of Jammu and Kashmir. Its population was mostly Muslim, but the state was eventually split, with India taking about two-thirds. The population of Indian Kashmir today is about 60 percent Muslim, but India refuses to allow the people of Kashmir to vote on whether they want to be a part of India or Pakistan. It is important to India's image as a secular country to have a state with a Muslim majority.

Water is another source of contention between India and Pakistan. A 1960 treaty regulates sharing of the water of the Indus River and its eastern tributaries which rise in China or in the Indian-controlled part of Kashmir. India, however, has threatened to revoke the treaty and cut off the supply of water to Pakistan. Throughout much of the Indus watershed region, local aquifers are being depleted, water tables are falling, waterways are polluted, and soils are becoming saline from overuse of underground water supplies.

The Taj Mahal. A Muslim ruler built the Taj Mahal, a mausoleum in Agra, India, for his beloved wife, in 1648. India's Muslim rulers were so rich and powerful that their name, Mughals, has come down to us in English as a word for a business executive: a mogul. (Courtesy of U.S. Agency for International Development)

determine exactly how much difference in women's rights can be attributed to the predominant local religion and, in turn, how much of that difference has been translated into local law.

For example, the Catholic Church is centralized, and it imposes a worldwide ban on the ordination of women. Women ordained by bishops in Argentina and Austria have been excommunicated by the Vatican. The worldwide Anglican Church, however, is less centralized. Its official head, the Archbishop of Canterbury, has approved the ordination of women, but local communities of Anglicans have different attitudes, and these reflect local variations in general social attitudes toward women's rights. In U.S. culture, for example, women's rights are protected in many areas, and the American Anglican Church (the Episcopal Church) ordains women. Anglican communicants from some other world regions, however—most notably Africa, where women's rights are not always assured—have refused to ordain women or to recognize ordained women.

Variations in practice can also be found across the Islamic realm. Islam originated in an Arabic culture that granted women few rights, but the list of women's rights in the Koran was liberal for the time and place.

Focus on
TURMOIL IN PAKISTAN

Pakistan cooperated with the United States in 2001 when the United States overthrew the Taliban in Afghanistan and attempted to root out Al Qaeda, but Pakistan is a troubled state. Pakistani-Indian animosity periodically threatens to explode in full-scale war, and some observers doubt the amount of control that the government of Pakistan has over its territory, its people, or even its own army.

During the long anti-Soviet war in Afghanistan (see p. 576), refugees from Afghanistan overloaded Pakistan's ability to absorb and help them. The Afghan fighters, in an effort to raise funds, turned Afghanistan into a virtual heroin farm. Pakistan became the transshipment point, leaving an estimated 600,000 heroin addicts in the Pakistani city of Karachi alone. The violence also engendered what is called a "Kalashnikov culture," from the name of a popular automatic weapon left behind in the hands of thousands of poor, unemployed, hardened young men. Still today Pakistan seems incapable of controlling or sealing its borders with Afghanistan, Iran, or India; drug runners, guerilla fighters, and religious extremist agitators seem to circulate freely. Within Pakistan, the traditional animosities between Sunnis and Shiites also continue to disturb the country.

Pakistan's government supported radical madrassas in both Pakistan and Afghanistan in order to create a cadre of devout Muslims who would strengthen Pakistan's hand against India. This plan succeeded, perhaps, too well. Extremist graduates of madrassas have embroiled Pakistan not only in repeated attacks on India, but, it is believed, in the Near East. Graduates of madrassas even make up a substantial percentage of Pakistan's own armed forces, and they may secretly back the attacks on India. Some observers fear that Islamic fundamentalists might spark a coup d'etat in Pakistan, placing nuclear weapons in the hands of extremists.

U.S. aid to Pakistan dried up during the 1990s, falling from several hundred million dollars per year in the late 1980s to just $6 million in 2000. In October 2001, however, in order to win Pakistan's cooperation in Afghanistan, the United States granted special trade rights, and the United States has continued to extend friendship to Pakistan in the early twenty-first century. Nevertheless, anti-American fundamentalist groups have consistently won local and regional elections, threatening the stability of Pakistan and continued cooperation with the United States.

Islam outlawed female infanticide, made the education of girls a sacred duty, and established a woman's right to own and inherit property. At the same time, Islam also fixed certain discriminatory practices. It teaches that a woman's testimony in court is half that of a man's and that men are entitled to four spouses, whereas women may have only one. In practice, women's rights vary throughout the countries where Islam predominates. Different countries interpret Sharia differently: In Dubai, for example, Sharia courts have ruled that a man may divorce a wife by sending her a text message over a mobile phone; Sharia courts in Singapore have ruled that he may not. Women are frequently secluded, or veiled in public, in Islamic countries, but this is compulsory only in Saudi Arabia and Iran. Also, only in the most fundamentalist Arab societies are women generally forbidden to work outside the home.

Future struggle is nevertheless almost inevitable between the expansion of women's rights in Islamic societies and an Islamic fundamentalist backlash that demands the repeal of laws that had banned polygyny (having more than one wife), permitted birth control, and given women the right to divorce in some countries. Almost everywhere, women struggle for equal access to education.

The Old Testament of the Bible contains restrictions on women's rights that are comparable to those found in the Koran, but few Jews or Christians today observe them. In Israel, however, the law denies women equality with men. For example, Israel has no civil marriage or divorce, and in Jewish law only a husband can initiate a divorce.

It has been hypothesized that religious teachings about women's rights may affect women's role in politics, but this hypothesis is difficult to prove. Women have served as head of government in Islamic countries, as well as in countries that are officially Christian (Catholic and Protestant), Jewish, and Buddhist. What sort of evidence for such a hypothesis could be sought? Reading scriptures alone does not prove whether those teachings are obeyed. Could we examine the percentage of national legislatures that are female? Geographers have noted that within the United States, the populations of Utah and of the states of the Southeast are, by some measures, the most committed to organized religion. The legislatures of these states have the smallest percentages of women legislators. Perhaps there is a cause-and-effect relationship between these facts, but clearly more research is needed on this topic.

Religion and Dietary Habits

Different religions teach different beliefs concerning the sacredness of plant and animal life, and this affects the farming and dietary practices of their adherents. Buddhists are generally vegetarians. Hindus believe that cattle are sacred; therefore, they will not eat beef, and they use cattle only as draft animals and as a source of milk. McDonald's restaurants in India feature lamb-burgers and vegetarian sandwiches. Outside observers have tried to measure in economic terms the advantages and disadvantages of preserving India's cattle. Aging and unproductive cattle may overgraze the land or may compete with the human population for limited food supplies. Sacred cows roam the streets of Indian cities unmolested, befouling them and congesting them. On the other hand, the cattle provide dung as a fuel and ma-nure, and those that die of natural causes are eaten by non-Hindus anyway. Whatever the balance of these economic arguments might be, to a religious Hindu this is a religious matter not open to economic debate.

Other religions prohibit certain foods. Jews and Muslims both refuse pork, so there are few hogs in Israel or anywhere across the Islamic realm (Figure 7-36). Muslims are prohibited from drinking alcohol. Alto-gether, a significant share of humankind restricts its diet because of religious beliefs.

This affects even world trade in food. Several of the greatest importers of plants and seeds that are rich in protein (peas, beans, and lentils, for example) are countries that limit their consumption of meat for religious reasons. Annual per capita consumption of meat in India, for example, is only 2 kilograms (4.4 pounds), and India is a major importer of protein-rich vegetable products.

Religion and Economics

The religion that predominates in any region may have a significant impact on that society's economy. Elaborate burial practices, for example, can drain a society of capital. Many peoples have buried their dead along with their possessions. Some societies lavish expendi-ture on houses of worship and other religious institu-tions. In some Buddhist societies high percentages of the male labor force spend several years, or even their whole lives, as monks.

Furthermore, religious teachings may affect the way people view the accumulation of money. For exam-ple, the scriptures of most religions bar the charging of

FIGURE 7-36 The virtual absence of hogs across North Africa, the Near East, and all the way to the borders of China is not caused by any attribute of the physical environment, but by the predominance of Islam and Judaism across this region. Both religions prohibit the consumption of pork. An increase in protein supply would be beneficial to the local populations, but the people demand other meats or protein-rich plants, such as lentils and peas. (Data from U.N. Food and Agriculture Organization)

interest on a loan as taking unfair advantage of another person. Christian teachings, however, have evolved to differentiate between two reasons for borrowing money: for needs and for investment. It may be sinful to charge interest on money borrowed to buy food, but it may not be sinful to charge interest if the borrower invests the money for his or her own profit—as when the borrower buys a truck to increase the profit in his or her own business. The religious leaders in some societies do not accept such a distinction. Islamic financial systems forbid interest, but encourage instead risk sharing. Eight Muslim nations support the Islamic Financial Services Board in Kuala Lumpur, Malaysia, which brings together scholars to debate the fine points of stockbroking and money management. In practice, financial structures and compromises have to be defined in each Islamic country.

Weber and the Protestant ethic   German sociologist Max Weber linked Protestantism and capitalism in *The Protestant Ethic and the Spirit of Capitalism* (1904). According to Weber, Protestantism encourages individualism, and with the rise of Protestantism in Western Europe, acquisitiveness, as the result of exercising individual ability in the marketplace, became a recognized virtue. Today this ethic seems to characterize adherents of almost any religion, so it is usually referred to as the *work ethic*.

If, however, there is any true advantage in the Protestant Christian attitude toward moneymaking, then it might be hypothesized that the worldwide expansion of Protestantism might encourage individualistic capitalism. This explains why the spread of Protestantism is widely reported and happily received in America's business press, which clearly accepts the hypothesis of a link. *The Wall Street Journal* and other publications all devote extensive coverage to the spread of Protestantism. They report it as economic news, not just religious news.

The Catholic Church and capitalism   In 1991 Pope John Paul II wrote, "On the level of individual nations and of international relations, the free market is the most efficient instrument for utilizing resources and effectively responding to needs." The Catholic Church's 1992 *Universal Catechism* notes, however, that markets do not always meet human needs, and it insists that governments regulate markets "according to a just hierarchy of values." It also demands that every worker receive a "just salary." The Church's attitudes toward markets, speculation, and worker-management relations may influence the behavior of individual Roman Catholics and of governments in Roman Catholic lands.

Religion and economics in Asia   The Confucian tradition in East Asian societies may exemplify another religious influence on economic development. Confucianism recommends societal leadership by an intelligent elite with the moral obligation to guide the peo-

ple. Several East Asian leaders have praised their own societies' traditional communitarianism and decried Western Christian individualism. Confucianism enhances the status of jobs in government bureaucracies, so ambitious young graduates compete for positions with the Japanese Ministry of Finance or the Korean Economic Planning Board the way ambitious young Americans compete for jobs at investment banks. The ability to attract talent gives these Asian governments more legitimacy and competence in dealing with businesses than the U.S. government can bring to bear, which may have helped to build successful economies.

Also, diligence, obedience, and high savings rates characterize the peoples of neo-Confucian Japan, South Korea, Taiwan, and Singapore, and of every Chinese settlement in the world. It is risky to hypothesize a cause-and-effect relationship, but the coincidence must be noted.

Religions and the Environment

Different religions teach different attitudes toward environmental issues, and cultural landscapes often reflect the religions of their creators. Most religions address the questions of how the world came into existence, what power created it, and what humankind's relationship should be to it and to other living things. Many major faiths not normally classified as animist nevertheless designate specific features of the environment as holy. Some religions revere rivers (Hindus revere the Ganges, Christians the Jordan); some revere high places (the Shinto revere Mount Fuji, Jews and Christians Mount Ararat, Hindus and Buddhists Mount Kailash) (Figure 7-37).

Early religious beliefs often formed as attempts to explain natural phenomena, such as the changes in the seasons or floods. Many religions celebrate natural events, time their annual ceremonies in accordance with astronomical events, or celebrate human affairs connected to the environment. Most religions celebrate harvest festivals of thanksgiving, for example.

Different religions recognize different holidays and pace human activities through the year differently, and this is often reflected in national laws across religious realms. Most Christian countries use the solar calendar. Jews use a calendar of 12 months, with an additional month added seven times in 19 years. Muslims use the lunar calendar. Muslims celebrate the Sabbath on Friday, Jews on Saturday, and most Christians on Sunday. Many governments prohibit work or other activities on that day. Muslims recognize an entire holy month and accordingly curtail their activities.

Most religious traditions award humankind a special role in the world, but, it has been argued, some religions suggest adaptive approaches to the physical world, while others suggest more exploitative approaches. Today, however, environmental degradation afflicts areas of all religious persuasions.

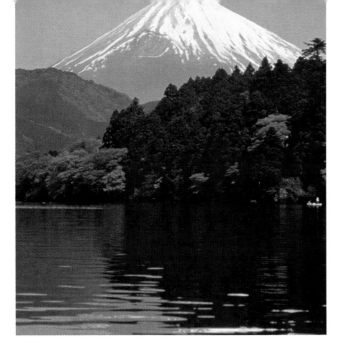

FIGURE 7-37 Mount Fuji. Volcanic Mount Fuji, Japan's highest peak at 3,776 meters (12,389 feet), is sacred to many Japanese. It has been celebrated for centuries in Japanese paintings and verse. Today the mountain and Lake Ashi ("foot"), shown in the foreground, are within the Fuji-Hakone-Izu National Park. (Courtesy of Japan National Tourist Organization)

Religious landscapes Most landscapes reveal the predominant local religion. The architectural styles of houses of worship, monasteries, and similar institutions are often revealing—muslim mosques, for example, usually have free-standing towers, called minarets.

Any religion's places of worship may become goals of pilgrimages, and these attract visitors and play a major role in cultural diffusion (Figure 7-38). Fairs and markets grow up at these sites. Muslims' hajj pilgrimages bring together people from around the world. At the local level, religious administrative patterns may organize and focus community spirit. More Roman Catholic New Yorkers can name the parish they live in than can name their city council district, and it is through the parish that they make friends and meet potential spouses. As noted earlier, the diffusion of religions can affect the diffusion of languages. Muslims everywhere are encouraged to read the Koran in its original Arabic, and the translation of the Bible into German by Martin Luther defined the German language and, arguably, the German nation.

Different religions' burial practices create different mortuary landscapes. Hindus and Buddhists cremate their dead, whereas Christians and Muslims usually bury their dead and erect monuments to them. Buddhist landscapes are marked by *stupas* (called pagodas east of India). These serve as temples today, but the form was originally devised as a funeral mound to contain the remains of a revered bodhisattva (Figure 7-39).

FIGURE 7-38 The Cathedral of Santiago, in the northwest corner of Spain, is believed to hold the relics of Saint James. Pilgrims from throughout Europe and the world have sought it out. In medieval times these traditional routes and staging points played important roles in trade and in the diffusion of ideas. The routes are marked by a chain of historic churches, inns, universities, and trade fairs. Still today a Pilgrim's Certificate is given to anyone who walks at least 100 kilometers (62.5 miles) along the medieval pilgrim's trail, and it must be stamped by parishes along the way.

FIGURE 7-39 Great stupa in Nepal. The eyes on this great stupa in Nepal emphasize that God sees everything. The symbolism is virtually universal. An eye can even be found on the back of the U.S. dollar bill. (Courtesy of Julia Nicole Paley)

Religion means "rebinding" of people with their gods and of people with one another. The geography of such binding will always be a key consideration of human geography.

CONCLUSION: CRITICAL ISSUES FOR THE FUTURE

This chapter opened by noting that peoples who share either language or religion can often more easily cooperate with one another in other ways, such as in international affairs. Sharing these attributes may help them understand one another. Much of the detail in this chapter, however, cataloged not cooperation and sharing but divisiveness. The same strong ties that bond groups of people together often separate them from other groups and cause conflict.

Martin Luther, who triggered the Protestant Reformation and split the Christian Church, noted the divisiveness of language: "The Frenchman," he wrote, "has only hatred and scorn for the German. The Italians have only hatred and scorn for all others. So we can see that this division of languages has led to division of habits, ways of thought and priorities that have put barriers between the very essence of peoples. It can justly be called the source of all misfortunes."

Religious contention also disturbs the world's peace. Groups' efforts to impose the teachings of their religion on other people often provoke violence. The Sudan, it was noted earlier, is rent by war between the Islamic government that wants to impose Sharia on the entire country, and the Christians and animists of the South. Pope John Paul II, speaking of Sharia law while visiting the Sudanese capital Khartoum in February 1993, said, "You absolutely cannot impose this law on those of other faiths. . . . Islamic law can be applied only to the Muslim faithful." The Church of Rome nevertheless assumes that its teachings should be written into law wherever it can accomplish that purpose.

What of the future? We can only hope that people can live together in peace.

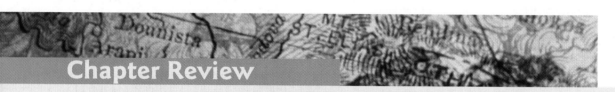

Chapter Review

SUMMARY

Language and religion define and bond human cultures. Peoples who share either of these two cultural attributes often demonstrate consistencies in other aspects of behavior and often cooperate with one another in other ways.

The distribution of any language illustrates the pattern of dispersal of its original speakers or their cultural impact on others. If human beings were to appear suddenly in today's highly interconnected world, one worldwide language might develop. The great variety of languages spoken today testifies to the relative isolation of groups in the past. There are more than 3,000 distinct languages, but about 60 percent of all people speak 1 of just 12 languages. The language with the most speakers is Mandarin Chinese. English is the world's lingua franca. In contrast to these widespread languages, some languages are extremely local.

A few languages are associated almost exclusively with one country, but several languages are shared by many countries, and some polyglot countries officially recognize several languages. The idea that languages gave birth to nations persists despite the fact that standard languages usually were the product of self-conscious efforts at nation building by centralizing governments.

Today's world distribution of European languages as official languages partly reflects patterns of colonization. Europeans forced their languages on their colonies, and when the colonies gained their independence, many kept their former ruler's language. Across much of the world, the official language is not actually spoken by most of the population.

Orthography can also reflect a historic diffusion process. Toponymy records information about the environment or the beliefs of a region's inhabitants.

Languages can be categorized into families on the basis of common ancestry, and they continue to diffuse and differentiate today. The use of some languages is expanding because their speakers are diffusing around the world, gaining greater power and influence in world affairs, or winning new adherents to their ideas. Other languages are dying out.

The population of the United States has always been composed of a great variety of peoples speaking a great variety of languages. Regional differences persist, and some people think English should be the official national language.

Religion is a vital force in human affairs, and understanding the messages of different religions contributes substantially to understanding many other aspects of human life.

Judaism, Christianity, and Islam form one ancient religious tradition, originating in the Near East and diffusing from there. Christianity came to predominate in Europe, and it spread worldwide with European influence. Today most Christians are outside Europe and North America. Christianity continues to win new converts, but practices and doctrines that derive from European culture are under pressure for change. New syncretic religions are thriving. Islam is newly influencing the politics of the countries where it predominates.

East Asia is dominated by Hinduism and its offshoots Buddhism and Sikhism. Other Eastern religions include Confucianism, Taoism, and in Japan, Shinto. The teachings of several of these systems of belief are more exclusively ethical and psychological, and they do not address theological questions. Many people are adherents to more than one of these. Animism and shamanism have many followers in the less developed parts of the world.

Religious teachings bear great influence on political organization. Theocracy is a form of government where the church rules directly, but religious teachings will often be translated into law in societies that claim to be secular. Religions' teachings differ regarding politics, women's rights, government, diet, economics, and environmental attitudes, so the geography of religious communities influences the geography of these other aspects of human culture. Christianity recognizes a distinction between church and state, but Islam teaches that the only purpose of government is to allow people to be good Muslims, and fundamentalist Islamic societies enforce sharia law. The actual practice of any religion varies from place to place, often reflecting variations in other aspects of the cultures among places. Each society defines a balance among the teachings of its religion, other aspects of its cultural tradition, and the demands of modern life, but that balance may shift through time.

KEY TERMS

caste p. 298
cognate p. 273
creole p. 269
dialect p. 269
etymology p. 273
fundamentalism p. 284
isogloss p. 269
language p. 268
language family p. 272
liberation theology p. 307

lingua franca p. 269
monotheism p. 286
official language p. 269
orthography p. 274
polyglot state p. 282
polytheism p. 286
proselytize p. 286
protolanguage p. 272
secularism p. 284
shamanism p. 301

Sharia p. 302
speech community p. 269
standard language p. 269
syncretic religions p. 294
terrorism p. 287
theocracy p. 301
toponymy p. 276
untouchable p. 298
Zionism p. 286

QUESTIONS FOR REVIEW AND DISCUSSION

1. Why is it that several former colonies retained their former rulers' language? What advantages and disadvantages did this offer?

2. Why are more countries today recognizing more official languages? What does this tell us about the internal politics of those countries?

3. Name a language that is expanding in use today and one that is shrinking. In each case, explain why.

4. What does it mean to say that Judaism, Christianity, and Islam are three religions, but form one religious tradition?

5. What are the observed and hypothesized consequences of the conversion of much of Latin America from Roman Catholicism to Protestant Christianity?

6. How do religious teachings affect economic behavior? Systems of national laws?

7. What is a lingua franca? What languages have served as lingua francas?

8. What is the caste system?

9. What is the Indo-European family of languages? Who first identified it? What is the geographic distribution of these languages? What is the history of the family?

10. What is Liberation Theology, and why does it have important political ramifications today?

11. Describe the diffusion of Islam from 622 to 751 A.D.

12. From what religious tradition did the Sikh religion spring? Where? When?

13. Briefly describe a map of the geography of religion in the United States.

14. Where did Christianity diffuse before it diffused throughout Western Europe?

THINKING GEOGRAPHICALLY

1. How many languages are spoken in your community? Does the local school system have bilingual or English as a second-language programs? Where do your community's non-English speakers come from?

2. The Dutch have considered making English the official language for college-level education in the Netherlands. Can you consider why?

3. If you speak two or more languages, can you think of specific ideas or images that may be difficult to translate from one to another? Try to explain this difficulty to your fellow students.

4. How many religions have temples, churches, or synagogues in your community? What, roughly, is the statistical breakdown of religious affiliations of the people of

your community? Does this reflect the source areas of settlers to your region? Does each place of worship anchor a residential neighborhood of that religion's communicants? How many religiously oriented parochial schools exist in your community? With what religions are they associated?

5. Almost the whole world uses Arabic numerals, but the use of the Arabic script is largely restricted to Islamic lands. Why?

6. Investigate the competition between Christianity and Islam for converts in an African country. Try Nigeria, Senegal, or the Ivory Coast.

7. The fourteenth-century Islamic scholar Ibn Battuta could travel from Morocco to China unmolested—even being welcome across most of the region. What political, religious, and other criteria of territorial organization have fragmented that region today so that a contemporary traveler might have trouble retracing Ibn Battuta's travels?

8. What do the place names in your community reveal about its environment or history?

9. If you are studying a foreign language or if you speak two languages, think of examples of cognate words.

SUGGESTIONS FOR FURTHER LEARNING

Abate, Frank. *American Places Dictionary.* 4 vols. Detroit: Omnigraphics, 1994.

Allsopp, Richard, ed. *Dictionary of Caribbean English Usage.* New York: Oxford University Press, 1996.

Bhardwaj, Surinder Mohan. *Hindu Places of Pilgrimage in India: A Study in Cultural Geography.* Berkeley, CA: University of California Press, 1992.

Clark, Victoria. *Why Angels Fall: A Journey Through Orthodox Europe from Byzantium to Kosovo.* New York: St. Martin's Press, 2000.

Crawford, James. *Hold Your Tongue: Bilingualism and the Politics of "English Only."* New York: Addison-Wesley, 1992.

Crystal, David. *The Cambridge Encyclopedia of Language.* Cambridge, England, and New York: Cambridge University Press, 1997.

Crystal, David. *Language Death.* New York: Cambridge University Press, 2002.

Diamond, J., and P. Bellwood. "Farmers and their languages: the first expansions." *Science* 300 (2003): 597–603.

Dictionary of American Regional English. vol. 1, A-C. Edited by Frederic G. Cassidy and Joan Houston Hall, 2003; vol. 2, D-H. Edited by Frederic G. Cassidy and Joan Houston Hall, 1999; vol. 3, I-O. Edited by Frederic G. Cassidy and Joan Houston Hall, 1996; vol. 4, P-Sk. Edited by Joan Houston Hall, 2002. Cambridge, MA: Belknap Press (Harvard University).

Fardon, Richard, and Graham Furniss, eds. *African Languages: Development and the State.* New York: Routledge, 1994.

Gaustad, Edwin, and Philip Barlow. *New Historical Atlas of Religion in America.* New York: Oxford University Press, 2001.

Haynes, Jeff. *Religion in Global Politics.* New York: Addison-Wesley, 1998.

Haynes, Jeff. *Religion, Globalization and Political Culture in the Third World.* New York: St. Martins Press, 1999.

Holloway, Joseph, and Winifred K. Vass. *The African Heritage of American English.* Bloomington, IN: Indiana University Press, 1993.

Kachru, Braj B. *The Other Tongue: English Across Cultures.* Champaign: The University of Illinois Press, 1992.

Krantz, Grover. *The Geographical Development of European Languages.* New York: Peter Lang, 1988.

Labov, William, ed. *Atlas of North American English: Phonetics, Phonology and Sound Change.* Berlin: Mouton/de Gruyter, 2004.

Lewis, Bernard. *What Went Wrong: Western Impact and Middle Eastern Response.* New York: Oxford University Press, 2001.

Lewis, Bernard. *The Political Language of Islam.* Chicago: University of Chicago Press, 1988.

Marty, Martin E., and R. Scott Appleby, eds. vol. 1, *Fundamentalisms Observed,* 1991; vol. 2, *Fundamentalisms and Society: Reclaiming the Sciences, the Family, and Education,* 1992; vol. 3, *Fundamentalisms and the State: Remaking Polities, Economies, and Militance,* 1993; vol. 4, *Accounting for Fundamentalisms: The Dynamic Character of Movements,* 1994; vol. 5, *Fundamentalisms Comprehended,* 1995. Chicago: University of Chicago Press.

Moseley, Christopher, ed. *Atlas of the World's Languages.* New York: Routledge, 1994.

Murphy, Joseph M. *Working the Spirit: Ceremonies of the African Diaspora.* Boston: Beacon Press, 1994.

Naipaul, V. S. *Beyond Belief: Islamic Excursions Among the Converted Peoples.* Boston: Little, Brown, 1998.

Nettle, Daniel, and Suzanne Romaine. *Vanishing Voices: The Extinction of the World's Languages.* New York: Oxford University Press, 2000.

Park, Chris. *Sacred Worlds: An Introduction to Geography and Religion.* Routledge, 1994.

Perica, Vjekoslav. *Balkan Idols: Religion and Nationalism in Yugoslav States.* New York: Oxford University Press (Religion and Global Politics Series), 2002.

Regenstein, Lewis. *Replenish the Earth: A History of Organized Religion's Treatment of Animals and Nature.* New York: Crossroad, 1991.

Robinson, Francis. *Atlas of the Islamic World Since 1500.* New York: Facts on File, 1982.

Ruhlen, Merritt. *The Origin of Language: Tracing the Evolution of the Mother Tongue.* New York: John Wiley & Sons, 1994.

Scott, Jamie, and Paul Simpson-Housley, eds. *Sacred Places and Profane Spaces: Essays in the Geographics of Judaism, Christianity, and Islam.* New York: Greenwood Press, 1991.

Smart, Ninian. *Atlas of the World's Religions*. New York: Oxford University Press, 1999.

World Wide Fund for Nature. A series of volumes "World Religions and Ecology," including volumes on Islam, Christianity, Judaism, Buddhism, and Hinduism. New York: Cassell, 1992.

Yavuz, M. Hakan. *Islamic Political Identity in Turkey*. New York: Oxford University Press (Religion and Global Politics Series), 2003.

WEB WORK

The World Wide Web is rich in resources for the study of material about languages and religions. The University of Wisconsin maintains a site for people learning or teaching languages at

http://polyglot.lss.wisc.edu/lss/lang/langlink.html

The latest redesign of The Human-Languages Page and the Languages catalog of the WWW Virtual Library is called

http://www.ilovelanguages.com

Another privately-maintained guide to web resources about languages and learning languages is

www.languages-on-the-web.com

For research on religion, you might start at the Academic Info: Religion, Mainpage at

http://www.academicinfo.net/religindex.html

At the Yahoo home page you can search for sites dealing with religions at

http://dir.yahoo.com/Society_and_Culture/Religion_and_ Spirituality/

The award-winning Facets of Religion, a Virtual Library, can be found at

http://www.edunet.ie/resources/religioninfo/religion.html

The Virtual Religion Index of Rutgers University is at

http://religion.rutgers.edu/vri/

A guide to Jewish teachings and traditions, with links to other related sites, may be found at

http://www.torah.org/

The Vatican's home page offers information in several languages—but not Latin:

http://www.vatican.va/

A good source of information on Hinduism is the home page of Hinduism Today:

http://www.himalayanacademy.com/

Islami-City is a Web site presented by the Islamic Information Network. It contains a wide variety of news and information and has an internal search engine for researching specific topics:

http://www.islamicity.org/

One attempt to count followers of world religions is at

http://www.adherents.com

The Center for the Study of Islam & Democracy (CSID) in Washington, D.C. maintains a page:

http://www.islam-democracy.org

A Look Ahead

FOOD SUPPLIES OVER THE PAST 200 YEARS

Humans have managed to avoid the global starvation predicted by Thomas Malthus 200 years ago by opening new lands to agriculture, redistributing and improving crops, and applying technology to agriculture.

AGRICULTURE TODAY

Ten principal types of agriculture may be identified today. Four of these (nomadic herding, low-technology subsistence farming, intensive rice farming, and Asian mixed cereal and pulse farming) are basically subsistence farming—that is, just to feed the farmers. Five other types (mixed farming with livestock, prairie cereals, ranching,

Mediterranean, and plantation agriculture) are predominantly commercial. Irrigated lands may be either subsistence or commercial.

LIVESTOCK AROUND THE WORLD

Humans consume most grain directly, but much grain is also consumed indirectly when people eat grain-fed livestock. Livestock present many environmental problems. Dairying is a special kind of livestock raising, and value can be added to the raw milk by processing it into milk products.

FOOD SUPPLIES IN THE FUTURE

We can be cautiously optimistic about food supplies for the future, but some people fear the newest

8

THE HUMAN FOOD SUPPLY

◄ The graceful curves of hillside contour plowing exemplify prudent farming techniques that prevent soil erosion and sustain productive agriculture. Modern scientific methods of farming—including the promise of biotechnology—have probably secured our ability to feed Earth's population, but raising food in environmentally sustainable ways and assuring all humans a healthy diet remain challenges.
(© Jim Richardson/CORBIS)

biotechnologies. Increasingly intensive agriculture also carries environmental hazards. Challenges remain in spreading technology around the world and designing political and economic systems that will allocate food to all.

THE WORLD DISTRIBUTION OF FOOD SUPPLIES AND PRODUCTION

As many as one billion people are undernourished today, with sub-Saharan Africa as the most troubled region. Problems in increasing food production include inefficient application of fertilizers, lack of incentive for many farmers, and problems of land ownership.

THE HARVEST OF FISH

We may have reached the limit of fish we can draw from the sea, triggering angry confrontations among fishing nations, but aquaculture offers new possibilities for fish supplies.

I n the summer of 2003, the 15 nations of the European Union enacted new rules requiring that all food products that result from genetic modification (called GM food) be labeled. These crops have been biologically altered to build in desirable characteristics from insect resistance to faster growth to greater vitamin content. Despite scientific studies that have found GM food to be safe, many European consumers fear their possible long-term effects on human health and on the natural products of fields and farms. In several countries pollen from GM crops has already, in fact, spread to other crops.

American trade representatives were angry. Americans—especially Argentines, Canadians, and farmers in the United States—have been quicker to adopt GM foods, and a high percentage of American agriculture is already modified. In the United States in 2003, for example, about 34 percent of the corn crop and 75 percent of soybeans were grown from GM seeds. Americans accused the Europeans of requiring labeling only to scare customers away from American products. Furthermore, argued the Americans, the European requirements would worsen Africa's hunger crisis. Many African countries, fearing the loss of markets in Europe, are reluctant to plant GM crops that could otherwise ease famine. In 2002, Zimbabwe even refused U.S. food aid that included GM corn.

The United States, Canada, and Argentina filed a protest against the European action with the World Trade Organization, which adjudicates world fair trade rules. These countries were originally joined by Egypt, but Egypt withdrew, fearing European retaliation, and the United States abruptly "punished" Egypt by terminating negotiations for a free trade treaty.

Science is achieving unprecedented strides in its ability to understand and manipulate life. The application of this new knowledge to food production is realizing astonishing possibilities, including increasing productivity, plant and animal resistance to diseases and pests, and even food's inherent nutritional content. Some people, however, fear the scientific manipulation of something once thought so "natural" as food.

Furthermore, the political and economic context for the development of these possibilities will determine when they might be applied, where, and for whose benefit. Even today, producing enough food for everyone on Earth is not as much of a problem as distributing food is. There are great disparities in the map of food production and supplies. Some countries produce substantial surpluses of food, while others lack supplies sufficient to feed their entire populations a nutritious diet. Even in many of the countries with surpluses, some people go hungry. Problems of food production and distribution can be understood only as the result of interaction among environmental, technological, political, and economic factors.

FOOD SUPPLIES OVER THE PAST 200 YEARS

Chapter 5 examined a pessimistic prediction that Thomas Malthus made 200 years ago. Malthus argued that the population would always tend to increase faster than the food supply, so cycles of mass starvation would limit the population increase.

Since Malthus published his theory, the human population has increased from one billion to over six billion. The mass starvation he predicted, however, has not occurred. Chapter 5 noted that today there are as many overweight people as undernourished people. Humankind has avoided starvation not by lowering the rate of population increase, but by increasing the food supply faster than the population, so per capita supplies have actually increased. How has this been done?

New Crops and Cropland

Extensive areas that were scarcely utilized during Malthus's lifetime have been opened to productive agriculture (Figure 8-1). The vast prairies of both North and South America, as well as food surplus regions in Australia and South Africa, have been developed since Malthus published his theory. Since the middle of the last century alone, over 900 million hectares (2.2 billion acres) of Earth's surface have been converted to croplands. Most of these lands were opened by irrigation.

A second factor is that many food crops have been transplanted to new areas where they have thrived, in some cases better than in their areas of origin (Figure 8-2, p. 324). The world's first farmers raised barley and wheat in the Middle East about 10,000 years ago. Later, Mexicans began cultivating corn and beans

(9,000 years ago), Chinese began cultivating rice (8,000 years), and South Americans potatoes (7,000 years). Natives in what is today the Eastern United States raised sunflowers 5,000 years ago. These five areas have traditionally been listed as the hearth areas of agriculture, but recent discoveries suggest a sixth: We now believe that peoples in the highlands of New Guinea cultivated taro and banana between 7,000 and 10,000 years ago. It is unknown whether the first Sub-Saharan African farmers began to cultivate millet, sorghum, and rice independently about 4,000 years ago, or whether they learned from Near Eastern teachers.

Even before Malthus wrote, the Western Hemisphere had contributed important food crops to the Eastern Hemisphere. For example, the potato is native to the Andes region of South America. It yields the second highest number of calories per acre of any crop, although its protein content is only 6 percent. By Malthus's day it had already become a major food in Northern Europe (Figure 8-3, p. 324). Today it is the world's fourth most important food crop, measured by total tonnage harvested, after corn, wheat, and rice. China has recently recognized the potato's versatility and is today, after Russia, the world's second-largest producer. India is not far behind, and the crop is becoming a mainstay throughout Africa and Asia. Potato yields have increased due to **genetic engineering,** which is the manipulation of species' genetic material. Two main techniques of genetic engineering are selective breeding, which has been carried on for millennia, and recombinant DNA, a new form of genetic modification discussed below. New genetically modified potatoes are being introduced to the tropics, where they are ready to harvest only 40 to 90 days after planting. Potato planting continues to spread, and the value of the world's potato crop increases each year. The true treasure of the Andes was not the gold that the Spanish conquerors sought, but the potatoes they trampled.

Maize is another Western Hemisphere native. Corn is the name English-speaking people have traditionally given to the major cereal crop of any given region. Therefore, in England "corn" means wheat; in Scotland and Ireland, oats; but in North America it means maize, or Indian corn. Maize was cultivated in Central and South America long before the arrival of Europeans, but Europeans introduced it into northeastern North America, where over one-third of the world's maize is grown today (Figure 8-4, p. 325). Maize is susceptible to frost, so the length of the frost-free period limits its distribution. Maize is a staple for people in South America and Africa, but about 70 percent of the U.S. crop is fed to livestock. Maize's average protein content is 10 percent.

Still another transplant is manioc, or cassava. Portuguese traders introduced manioc into Africa in the sixteenth century. It has a very low protein content and should be supplemented with other high-protein foods, but it remains a staple across lowland Africa.

Transportation and Storage

Improvements in worldwide transportation have allowed regional specialization in food production, and specialization can multiply productivity. Railroads, trucks, and cargo ships—many of them refrigerated—move quantities of food with a speed and efficiency that could not have been imagined in Malthus's day (Figure 8-5, p. 325).

Transportation also allows the shifting of food from surplus to deficit areas, so fewer people need die from local famines. In the past it was not uncommon for surplus food to rot in the sun 100 miles from a starving population because the food could not be transported. Improved transportation also allows us to maintain smaller inventories of food at any specific time and place, and yet to be secure.

An increasing share of the world's food production is entering world trade—$547 billion worth in 2001. Table 8-1 (p. 326) lists the largest importers and exporters of some of the major foods traded. Some of the major exporters of certain crops are not among that crop's major producers. This indicates that in those countries, food production for export is a highly commercialized business. For example, the United States produced only about 2 percent of the total world rice crop in 2001, but the United States accounted for 10 percent of all world rice exports because most rice harvested in Asia is consumed where it is produced. As much as one-sixth of world wheat production enters into world trade.

The technology of food storage has also continually improved, decreasing both spoilage and loss to pests (Figure 8-6, p. 325). Improvements have been continuous from the introduction of the silo in the nineteenth century to freeze drying—actually, the Inka did this 500 years ago, but the rest of the world learned how only recently—and antiseptic packaging.

The Green Revolution

Starting about 1950, an intensive effort was launched to develop new grain varieties and associated agronomic systems and to establish them in developing countries. It focused on certain crops (rice, wheat) and certain techniques (breeding for response to fertilizer inputs). It was driven largely by private foundations such as the Ford and Rockefeller Foundations. This focused effort was known as the **green revolution.** Many projects continue. For example, specialists at Texas A&M University and at the International Rice Research Institute in the Philippines have produced higher-yielding strains of rice that are more resistant to disease and pests and that mature in only 110 days. This allows for two or even three crops per year. New potato strains are developed at the International Potato Center in Lima, Peru. Wheat yields have multiplied from new hybrids. India, where 1.5 million people died in a 1943 famine, became a net grain exporter in 1977, even though its population doubled between those years.

Productivity continues to rise: World yield of cereal crops per hectare rose a third just from the period 1979–1981 to 1999–2001, from an average of 1,608 kilograms per hectare to 2,143 kilograms per hectare.

Other Technological Advances

The green revolution is just one aspect of the continuing application of science to agriculture that began in the late eighteenth century known as the **scientific revolution in agriculture.** New pesticides save crops from insects that once wiped out entire harvests, and new fertilizers multiply yields per acre. Parallel scientific research has been directed to livestock. As with crops, the developments include world redistribution, increases in total numbers, improved breeding, and even "engineering" for greater hardiness and higher yield from each animal. In 1950 the average U.S. dairy cow gave 2,339 liters (618 gallons) of milk per year; by 2000 that number more than tripled to almost 8,000 liters (2,114 gallons). In 1969 the average pig used for breeding in the United States produced 6.7 piglets per year. By 2000 that had risen to 16. So far, however, only a fraction of the world's livestock herds is this productive.

Farm machinery invented since Malthus's day reduces the number of fieldworkers and at the same time increases yields by improving the regularity of plant spacing and the efficiency of harvesting. Heaters rescue many crops threatened by freezes. During the past 200 years, humans have literally re-formed Earth with drainage projects where there was too much water and with irrigation projects where there was not enough. Projects such as these are not new in theory, but they are achieved in a scale beyond anything that Malthus could have foreseen.

AGRICULTURE TODAY

Humankind has thus managed to prevent the widespread starvation Malthus predicted. Not all of these developments, however, have taken hold equally everywhere. The following pages will first catalog regions of different types of agriculture today, and then we will examine the distribution of food supplies. What regions produce surpluses, and where are people hungry? Finally, we will address the question of whether and how we will be able to produce enough food for an increasing human population in the future. The pessimistic neo-Malthusian point of view introduced in Chapter 5 will again be set against the optimistic Cornucopian point of view.

Chapter 6 introduced Varro's theory of the evolution of human societies. The theory was based, principally, on the evolution of agriculture, for humankind's

The World of Agriculture

FIGURE 8-1 These are regions of different types of agricultural land use. The success of agriculture depends largely upon the possibilities of the physical environment, so we should not be surprised that

first concern is to get enough to eat. According to Varro, humans were at first hunter-gatherers before they domesticated animals and developed pastoral nomadism. The later domestication of plants encouraged settled agriculture, which is more productive. That is, settled agriculture yields a greater and more secure quantity of food. **Subsistence agriculture**—agriculture to feed oneself and family—eventually yielded to **commercial agriculture**—raising food as a product for sale.

A few hunter-gatherer bands still do exist and provide evidence of how all humans lived in prehistoric times, before the invention of agriculture. Isolated

Non-agricultural regions	Mixed farming with livestock
Irrigated areas in dry lands	Prairie cereals
Nomadic herding	Ranching
Low-technology subsistence	Mediterranean mixed grains, fruits, and vegetables
Intensive rice	Plantation agriculture
Asian mixed cereals and pulses	

the categories mapped here largely coincide with the climates mapped in Figure 2-31 and the biomes mapped in Figure 4-23. "Nomadic herding" and "low-technology subsistence farming" cover impressive areas on this map, but their productivity is low. These areas are generally sparsely populated, as you can see by comparing this figure with the map on the rear endpaper.

groups inhabit the periphery of world settlement—the Arctic, the Amazon basin in South America, central and southern Africa, and in Papua New Guinea. Examples include the Bushmen of Namibia and the Aborigines of Australia. Anthropologists believe, however, that these few remaining bands will probably abandon their ways or die out within decades. During 2002 the government of Botswana forcibly rounded up and settled the last of the Basarwa, or San Bushmen—about 2,000 people. "Hunting and gathering," decreed President Festus Mogae, "is no longer a viable way of life."

Subsistence Farming Contrasts with Commercial Farming

Chapter 6 discussed how societies evolve from self-sufficiency to participation in trade in general terms by examining three villages connected by a road. Once peoples begin to exchange products, they have an incentive to specialize and increase the production of a good that can be exchanged. Farmers can specialize in one crop and import other foods from elsewhere. Thus, agricultural **polyculture,** the raising of a variety of

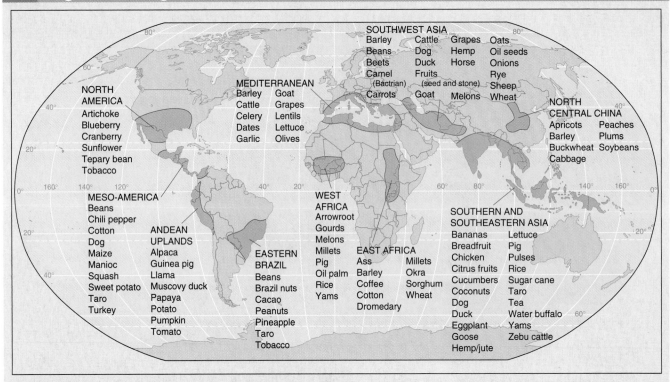

FIGURE 8-2 This map identifies ten regions where significant numbers of plants or animals were domesticated. A few examples, such as barley, were domesticated in two places. The fact that the people in so many different regions and biomes learned to exploit the different plants and animals they found locally is a tribute to human ingenuity. Many of these plants and animals, however, have been widely redistributed. Therefore, comparing this map with the following maps of crop and livestock distribution today reveals that the area that produces the most of any crop or animal today is not necessarily a place where that plant or animal was originally domesticated.

Current Production of Potatoes and Rice

FIGURE 8-3 The potato is native to South America, but its greatest production today is in Europe. Rice remains the great staple food of Asia, where it was first domesticated. (Data from United Nations Food and Agriculture Organization)

World Maize Production

FIGURE 8-4 Maize was first domesticated in Central America, but world production today is concentrated in North America. (Data from FAO)

crops, may yield to **monoculture,** specialized production of one crop.

There is no clear dichotomy between subsistence agriculture and commercial agriculture. They are the endpoints of a spectrum. The most self-sufficient hunter-gatherer may trade extra berries for fish, and the most specialized wheat farmer in Kansas may raise his or her own tomatoes. Nevertheless, many farmers today are called subsistence farmers because they do

consume most of what they produce, and they produce most of what they consume. Subsistence farmers may sell some of their output, but surplus production is not their primary purpose, and in years of poor harvests they may not even have a surplus.

FIGURE 8-5 Fresh food year-round. Modern techniques of food transportation and storage allow Alaskans to enjoy fresh produce even when outside temperatures read −40° C (−40° F). (Courtesy of Carr Gottstein Foods Company, Anchorage, Alaska)

FIGURE 8-6 New food-storage facilities. This officer of the U.S. Agency for International Development (AID) is overseeing the construction of a silo in Peru. Many countries produce enough food to nourish their people, yet much of the crop is wasted or lost to pests or spoilage because of inadequate storage and distribution facilities. (Courtesy of U.S. Agency for International Development)

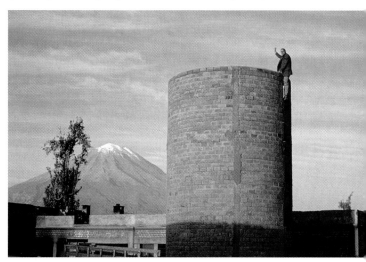

TABLE 8-1 Top Ten Importers and Exporters of Wheat, Rice, and Corn, 2001 (in thousands of metric tons; data from FAO)

Leading Exporters	Exports	Leading Importers	Imports
Wheat (total traded 113,705)			
U.S.	25,782	Italy	7,526
Canada	17,658	Brazil	7,016
France	15,621	Iran	6,438
Australia	15,542	Japan	5,521
Argentina	10,789	Egypt	4,412
Rice (total traded 26,761)			
Thailand	7,685	Ivory Coast	1,948
Vietnam	3,729	Nigeria	1,508
U.S.	2,622	Iraq	950
Pakistan	2,423	Philippines	810
India	2,193	Iran	778
Maize (total traded 78,909)			
U.S.	47,943	Japan	16,221
Argentina	10,909	Korea (South)	8,481
France	7,046	Mexico	6,174
China	5,997	China	5,234
Hungary	1,568	Egypt	4,797

The world is too interconnected, however, for there to be many completely self-sufficient subsistence farmers left anywhere. Farmers generally respond to economic incentives, so if a farmer has a market opportunity, he or she will try to produce something for it. In virtually all countries, urban populations offer markets for farmers to satisfy. The day after a dirt road or a railroad reaches the smallest village, many peasants are on the way to market something in the nearest city. The 800 million peasants of China are often referred to as subsistence farmers because most of what they raise is for their own consumption. They do, however, manage to raise and market enough of a surplus to feed China's 500 million urbanites and still have enough left over for China to achieve net exports of food.

Several characteristics differentiate subsistence agriculture from commercial agriculture. Subsistence farming usually relies on enormous investments of human labor, but little animal or machine power. It is usually low in technology and, therefore, in productivity, and most of the food raised is consumed by the farmers themselves.

Modern commercial farming, by contrast, relies on substantial capital investment in machinery, chemicals, improved seeds, and livestock. The high price of farm machinery necessitates that it be used maximally,

and this encourages ever-increasing average farm size. Commercial farm products are seldom sold directly to consumers, but to large companies that process, package, store, distribute, and retail food. Individual companies are called agribusinesses—agricultural businesses—and many of them, such as Ralston Purina and General Mills, are household names in North America. Most farms may be owned by individual families, but a high and rising percentage of food is produced on gigantic farms owned by the agribusinesses themselves.

Types of Agriculture

According to the United Nations Food and Agriculture Organization (FAO), in 2000 about 1.5 billion hectares (3.75 billion acres) of Earth's land area were devoted to cropland (11 percent of the total, excluding Antarctica); and an additional 3.4 billion hectares (8.4 billion acres, 26 percent) were devoted to permanent pasture. Cereals (grains) and potatoes are the world's basic foods. Cereals have been grown since the dawn of history, and every major civilization has been founded on them as the principal source of food. Wheat, rice, maize, barley, sorghum, oats, rye, and millet are all members of the grass family of plants.

The preeminence of cereals as a food is also partly explained by the ease with which they can be produced, stored, and transported.

FIGURE 8-7 **The shipment of grain.** In the mid-nineteenth century, grain was still stored and transported in individual bags. This required enormous amounts of labor, and a lot of grain was lost to spillage. Late in the nineteenth century, however, methods were devised to store and ship grain as if it were a liquid. These techniques greatly reduced the cost and improved the efficiency of transportation. (Courtesy of the Kansas Wheat Commission)

They yield more food, both in bulk and in nutritive value, than most other crops. The preeminence of cereals as a food is also partly explained by the ease with which they can be produced, stored, and transported (Figure 8-7).

Most of Earth's "nonagricultural land" is hot, cold, or dry (compare Figure 8-1 with Figure 2-31). Agriculture has not been productive in these areas, nor have humans settled in them in large numbers (compare Figure 8-1 with the map on the rear endpaper). In Figure 8-1, the categories of nomadic herding and low-technology subsistence agriculture cover enormous areas, but these activities support very few people. The category irrigated agriculture covers a wide range of farming styles, from the intensely worked and productive fields of California to subsistence oases in the Sahara (Figure 8-8).

The nine other categories of agriculture in Figure 8-1 are defined by the way that five variables interact in each place: (1) the natural environment, (2) the crops that are most productive in that environment, (3) the degree of technology used by local farmers, (4) the degree to which local farming is market-oriented, and (5) the degree to which crops are raised either for human consumption or to feed animals.

Four of the nine categories are basically subsistence: nomadic herding, low-technology subsistence farming, intensive rice farming, and Asian mixed cereal and pulse farming. The five other categories are predominantly commercial: mixed farming with livestock, prairie cereals, ranching, Mediterranean, and plantation agriculture.

Nomadic herding Pastoral nomads depend primarily on animals rather than crops, and pastoral nomadism may still be a successful low-technology adaptation to arid or semiarid environments. It may be found across North Africa, the Middle East, and into parts of Central Asia. The Bedouins of North Africa and Saudi Arabia and the Masai of East Africa are examples of nomadic groups.

The animals upon which nomads depend may include horses, camels, goats, sheep, or cattle. The nomads will consume them or else sell them to sedentary farmers in exchange for grains. Sometimes, however, a nomadic group will plant crops at a fixed location and later return to harvest them. Pastoral nomads do not wander aimlessly; they have a strong sense of territoriality.

There are probably no more than 12 to 15 million pastoral nomads today, and many are settling down. Many governments are forcing them to settle in order to control them. The government of Kenya, for example, is forcibly settling the Masai people. Some nomads are today allowed to move about only within limits fixed by governments. In the future, pastoral nomadism will probably be confined to areas that cannot be irrigated or that do not contain valuable mineral resources.

Low-technology subsistence farming The least intensive type of farming is shifting cultivation, called *slash-and-burn* cultivation or **swidden** cultivation, from a word meaning "to singe." It is usually called "cultivation" rather than "agriculture," because the word agriculture implies greater use of tools and animals

FIGURE 8-8 **Oasis in Algeria.** The Grand Erg Occidental Taghit oasis in the Sahara Desert in Algeria supports a small population and feeds an occasional caravan passing through the surrounding sands. (RAGA/Corbis/Stock Market)

FIGURE 8-9 Swidden burning. This fire was started to clear forest land for agriculture in Brazil, but when the soil is exhausted, the site will be abandoned and reclaimed by forest. (James P. Blair/National Geographic Image Collection)

and more extensive modifications of the landscape (Figure 8-9). Nevertheless, swidden cultivation is quite sufficient in tropical areas: the Amazon region of South America, Central and West Africa, and Southeast Asia, including Indochina, Indonesia, and New Guinea.

Small patches of land are cleared with machetes and other bladed tools, trees are stripped and killed, and fires are set to clear the land for planting. The ashes contain a simple fertilizer, potash (potassium). Various subsistence crops are planted with the use of a simple digging stick or hoe. Polyculture is practiced even in a single swidden, which may contain a large variety of intermingled crops. Soil nutrients are rapidly depleted, so the cleared land can be used to grow crops for only a short time, usually three years or less. Then it must be left for many years, although the villagers may return someday.

Shifting cultivation can support low levels of population without causing environmental damage, but it is being replaced by commercial logging, cattle ranching, and cultivation of cash crops. These activities, however, require cutting down much greater expanses of forest. Wholesale destruction of the rain forests reduces global biological diversity and may be contributing to global warming, as discussed in Chapter 4.

Intensive rice farming   The greatest number of farmers living in the large population concentrations of East, South, and Southeast Asia practice intensive subsistence agriculture. Even tiny, irregular plots are planted, and individual farmers' holdings are often fragmented as a result of subdividing holdings among children through generations. Most of the work is done by hand or with

animals, largely because human labor is abundant, but capital is scarce. Livestock are rarely permitted to graze on land that could be used to plant crops, and little grain is grown to feed the animals. The intensive agriculture region of Asia can be divided between areas where wet rice dominates and areas where other crops predominate. Wet rice is characteristic of Asia; dry rice is a grassy staple largely confined to West Africa.

About half the world's population subsists wholly or partially on wet rice. Wet rice occupies a relatively small percentage of the land, but it is the most important source of food in Southeast China, East India, and much of Southeast Asia. It has a protein content of 8 to 9 percent. Successful production of large yields of rice is an elaborate, intensive process. Wet rice should be grown on flat land, because the plants are submerged in water much of the time. Therefore, most wet rice cultivation is located in river valleys and deltas, although additional suitable land may be created by terracing the hillsides of river valleys (Figure 8-10). A farmer prepares the field using a plow drawn by water buffalo or oxen. The plowed land is flooded, and dikes and canals must be maintained to control the quantity of water in the field. Flooded fields are called **sawah** in Indonesian. Rice is either broadcast through the field, or else seedlings that have been nursed elsewhere are transplanted into the field. Rice plants grow submerged in water for approximately three-fourths of the growing period, and they are harvested by hand.

In places with relatively warm winters, such as South China and Taiwan, two rice crops can be harvested per year from one field, a process known as **double cropping.** Most parts of India have dry winters, so double cropping there, where possible, usually involves alternating between rice in the wet summer and another cereal such as wheat in the drier winter.

FIGURE 8-10 Rice terraces. These rice terraces in China increase the amount of land available for rice cultivation by turning steep slopes into flat steps. (Keren Su/CORBIS)

Production of Barley, Oats, and Rye

FIGURE 8-11 Barley, oats, and rye can serve either as human food or as feed for livestock. (Data from FAO)

Asian mixed cereals and pulses   Wet rice cannot be grown where summer precipitation levels are too low and winters too harsh. Therefore, other crops predominate in the interior of India and in northeast China. In milder parts of the mixed cereals and pulses region, crop rotation may allow double cropping. Wheat is the most important crop, followed by barley. Barley is the world's fourth most important cereal crop (Figure 8-11). It has the shortest growing season of all cereals and can be grown farther north, at higher altitudes, and in more arid regions than any other cereal. Barley is an important food crop in parts of Asia and in Ethiopia.

Pulses is the botanical name for a large family of herbs, shrubs, and trees, also called the pea or legume family. The seeds, which include beans, peas, lentils, peanuts, soybeans, and carob, are rich in protein and may form the principal source of dietary protein in poor regions or in regions where religion prohibits the consumption of meat. In China and Japan, tofu, made from soybeans, is a major food source. The root nodules of pulses have bacteria that absorb nitrogen out of the air and release it into the soil (a process called *fixing* nitrogen). This ability makes pulses valuable as crops grown temporarily to reduce erosion and build up a soil's nitrogen content.

Mixed farming with livestock   Mixed farming with livestock is usually a commercial undertaking, for most of the crops are fed to animals rather than consumed

directly by humans. Mixed farming allows the farmer to distribute the workload throughout the year (fields require less attention in the winter) and also to maintain his or her income by selling animals or animal products throughout the year. Crop rotation allows the farmers to maintain the soil fertility because different crops deplete the soil of different nutrients and restore others.

Mixed farming is the dominant form of agriculture in much of the world today, and, although it is still widespread in North America, North American farming is evolving toward more specialized production. Farms that are truly mixed, including both feed production and animal production on the same farm, remain common; but today corn and beans are increasingly grown on farms that no longer keep more than a few animals, and the feedstock is sold to feeding operations that produce the meat. Specialization has separated the two components of the system in both ownership and location. In the United States "corn belt," which stretches from Indiana across Iowa to the Dakotas, approximately half of the cropland is planted in corn. Some of this corn is consumed by people directly as oil and other processed foods, but most of it is fed to livestock (either on the farm or concentrated in feed-lots), including pigs, cattle, and poultry. The second most important crop in this region, soybeans, is grown almost exclusively as animal feed. The third most important crop in the United States by area harvested (although not by value) is hay, which is also an important component of mixed farming systems.

This farming is enormously productive. Corn production in the United States represented 38 percent of world total production in 2002; soybeans grown in the United States accounted for 40 percent of the world total.

Prairie cereal farming   Crops on commercial grain farms are grown primarily for consumption by humans rather than by livestock. Large-scale commercial grain production is found in only a few countries, including the United States, Canada, Argentina, Australia, Russia, Ukraine, and France. This production is mostly, but not exclusively, in regions that are too dry for mixed crops and livestock. Wheat is today the most important cereal in world food production. It supplies about 20 percent of the total calorie consumption of the human species (Figure 8-12) and provides a staple for over one-third of the world's population. The protein content varies between 8 and 15 percent.

Wheat generally has more uses as human food than other grains, so it is more valuable. It can be stored relatively easily without spoiling and transported long distances both easily and economically. Hard-kerneled varieties yield flour with a high content of *gluten* (a protein substance) and are used to make breads; soft-kerneled ones are starchier and are used to make cakes and biscuits. *Durum* wheat is an especially hard-kerneled wheat used in pasta products. Wheat grain, *bran* (the fibrous outer coat of the wheat kernel that remains as a residue after milling), and the rest of the plant are all valuable livestock feed.

In North America, large-scale grain production is concentrated in three areas. In the winter wheat belt that extends through Kansas, Colorado, and Oklahoma, the crop is planted in the autumn and develops a strong root system before growth stops for the winter. The wheat survives the winter and is ripe by the beginning of summer. In the spring wheat belt of the Dakotas and Montana in the United States, and in Southern Alberta, Saskatchewan, and Manitoba in Canada, winters are usually too severe for winter wheat, so spring wheat is planted in the spring and harvested in the late summer. A third important grain-growing region is the Palouse region of eastern Washington State. Large-scale grain production is highly mechanized and conducted on large farms, but the effort required is not uniform throughout the year. Therefore, some individuals or firms may own fields in each belt in order to keep the expensive machinery in use longer. Combine companies start working in Texas in late May; they move to Oklahoma in early summer and farther north through the season.

Ranching   Ranching is the commercial grazing of livestock over extensive areas. It is best suited to arid or semiarid land, where the vegetation is too sparse or soil too poor to support crops. Cattle raised on ranches are frequently sent for fattening to farms or to local feedlots before going to meat processors (Figure 8-13).

World Wheat Production

FIGURE 8-12 Today some wheat is grown almost everywhere except in the tropics, and it is a major commodity in international trade. (Data from FAO)

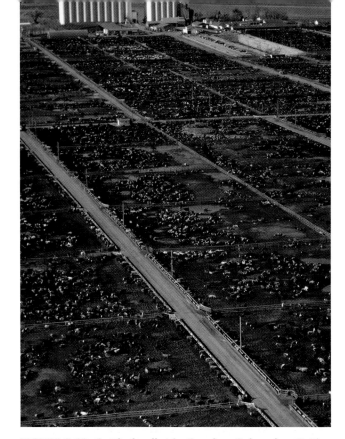

FIGURE 8-13 Cattle feedlot in Greeley, Colorado. Cattle that have been grazing over extensive areas are usually fattened at feedlots such as this before slaughter. (Lowell J. Georgia/Photo Researchers, Inc.)

Spaniards introduced cattle to the Western Hemisphere in the sixteenth century, and the cattle thrived on the grazing lands in both North and South America. In the late nineteenth century, British capital developed the commercial cattle industry in Argentina in areas convenient for shipping to overseas markets. Today a large portion of the prairies of Argentina, southern Brazil, and Uruguay are devoted to grazing cattle and sheep (compare Figure 8-1 with Figures 2-31 and 4-23).

The interior of Australia was opened for grazing at the same time, although sheep are found along with cattle (see Figure 5-28). Ranches in the Middle East, New Zealand, and South Africa are also more likely to raise sheep. Commercial ranching is found in other relatively developed regions of the world, but it is rare in Europe, except in Spain and Portugal.

Mediterranean agriculture Mediterranean agriculture is a type found in the distinctive Mediterranean climate regions—usually on the west coasts of continents at about 30–40° north and south latitude around the Mediterranean Sea, in southern California, central Chile, the southwestern tip of South Africa, and southwest Australia. The Mediterranean climate has hot, dry summers and cool, rainy winters. The lack of summer water and good grazing land hinders livestock raising, so Mediterranean-style farmers traditionally derive only a small percentage of income from animal products. Many of these areas are mountainous, or at least hilly,

so some farmers do practice transhumance—keeping sheep and goats on the coastal plains in the winter and transferring them to pastures in the hills in the summer.

Most crops in Mediterranean lands are grown for human consumption, including most of the world's olives, grapes, and other fruits and vegetables. Hilly landscapes allow farmers to practice polyculture, with some crops facing the Sun, others away.

Two-thirds of the world's wine is produced in countries that border the Mediterranean Sea, especially Italy, France, and Spain, while other Mediterranean regions elsewhere produce most of the rest (Figure 8-14). About half of the land may be planted in cereals, especially wheat for pasta and bread. Seeds are sown in the fall, and the crops harvested in early summer. Land is periodically left fallow to conserve moisture in the soil.

California has less land in cereals than is traditional in these regions, because California growers concentrate on profitable citrus fruits, tree nuts, and fruits produced by deciduous trees (apples, peaches, plums, and similar fruit). Rapid urbanization along the coast has driven agriculture inland, where irrigation is required. Competition for water among farm needs, urban users, recreational uses, and environmental demands constitutes one of the most important political issues in the West.

Plantation farming A plantation is a large farm that specializes in the commercial production of one or two crops. Plantations found in tropical Latin America, Africa, and Asia are often owned or operated by Europeans or North Americans, and they are worked by corporate employees. The crops are grown primarily for sale in the rich countries. Latin American plantations grow coffee, sugarcane, and bananas, while Asian plantations may provide rubber and palm oil (Figure 8-15).

FIGURE 8-14 Grapes in Chile. Chile exports substantial quantities of grapes, but it also adds value to this rich agricultural harvest by converting the grapes to wine and exporting wine. (Courtesy of the Chilean Fresh Fruit Association)

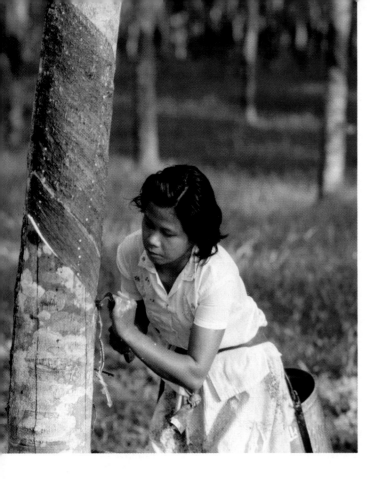

What Determines Agricultural Productivity?

The various types of agriculture described above all concentrate on different sorts of crops, but we can learn something about the great range of productivity in agriculture around the world by looking at yields of grain (Figure 8-16). This variation is partly attributable to variations in the physical environment. For example, Western European countries enjoy mild climates, and their agriculture is highly productive, whereas the Sahara Desert region of northern Africa is generally unproductive. Surprising exceptions, however, can also be found. For example, several Near Eastern Arab countries record some of the highest grain yields anywhere on Earth.

When we compare Figure 8-16 with Figures 8-17 and 8-18, however, we see considerable coincidence between the variations in productivity and the application

FIGURE 8-15 Tapping a rubber tree in Indonesia. This woman is tapping a rubber tree on a plantation on the Indonesian island of Sumatra. (© Vince Streano/ CORBIS)

Agricultural Productivity

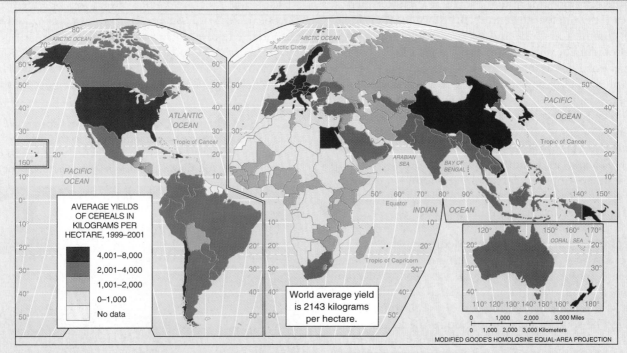

AVERAGE YIELDS OF CEREALS IN KILOGRAMS PER HECTARE, 1999–2001

- 4,001–8,000
- 2,001–4,000
- 1,001–2,000
- 0–1,000
- No data

World average yield is 2143 kilograms per hectare.

MODIFIED GOODE'S HOMOLOSINE EQUAL-AREA PROJECTION

FIGURE 8-16 Yields per hectare vary greatly around the world, and these variations are not always clearly related to natural environmental conditions. This map does not reproduce the pattern of precipitation (Figure 2-30), climates (Figure 2-31), or biomes (Figure 4-23). Factors other than natural environmental conditions must be found to explain agricultural productivity. (Data from FAO)

The Use of Tractors

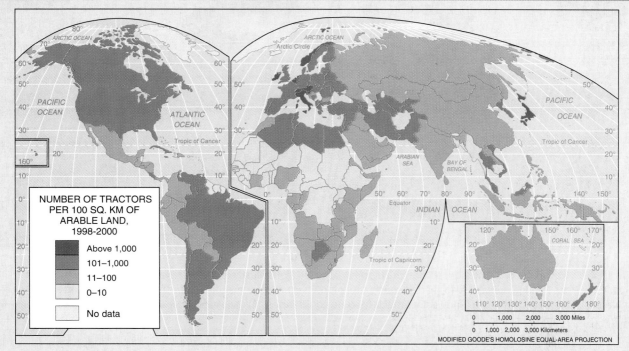

FIGURE 8-17 Capital inputs in agriculture vary tremendously. The use of tractors is a good measure of modern capital-intensive farming. Does this map reproduce the pattern of Figure 8-16 and, therefore, help explain variations in yield? (Data from FAO)

The Application of Fertilizer

FIGURE 8-18 Fertilizer is another important capital input in agriculture. Its application can override deficiencies in the local soil. (Data from FAO)

(a)   (b)

FIGURE 8-19 Extremes of capital investment. These two photographs illustrate the extremes of capital investment in agriculture. If the Senegalese farmer (a) had as much capital to invest in farm machinery, fertilizer, and improved seeds as the Nebraska farmers (b) do, who knows how much food he could raise? ([a] Carl Purcell/U.S. Agency for International Development and [b] U.S. Department of Agriculture)

of fertilizers and the use of tractors. The countries that enjoy the highest yields rank high in at least one of these two investments. Investment in agriculture seems to be a principal factor determining productivity. Many of the rich places that are able to invest in tractors and fertilizer can also be assumed to be the places where farmers have the best hybrid seeds. These are the product of careful interbreeding of the best strains of a crop through generations. These countries also have the best irrigation, pesticides, and all the other modern technological apparatus of agriculture. Therefore, we can conclude that capital input is a principal determinant of agricultural productivity today. Capital investment may be as important as the natural environment, or even more important. If investment in agriculture could be increased in the regions where there is little investment today, total world agricultural output would undoubtedly soar (Figure 8-19).

> ... capital input is a principal determinant of agricultural productivity today.

LIVESTOCK AROUND THE WORLD

Humans began to domesticate animals about the same time as they did plants, some 8,000 to 10,000 years ago. Dogs were probably domesticated first, independently at a number of places around the world. Dogs helped in hunting, and they were themselves eaten for food. The domestication of a great variety of other animals followed through centuries at various locations (look back at Figure 8-2 again). As humans learned to domesticate plants, they found it convenient to herd

plant-eating animals and to pen them close to settlements. From then on, growing crops and raising livestock advanced side by side.

A wide range of animals have been domesticated as livestock, and their numbers totaled almost 20 billion in 2002—over three times the human population. These numbers included about 16 billion chickens, 3.5 billion ruminants (cud-chewing animals, mainly cattle, sheep, goats, buffalo, and camels), and about 950 million hogs (Figure 8-20 and also look back at Figure 7-36). Domesticated livestock provide high-quality protein, either as meat or milk. They also supply hides, wool, and other raw materials, and they serve as draft animals used to pull plows and wagons; large numbers of people in poor regions still depend on oxen and buffalo. In India alone, there are over 80 million draft animals. Among pastoral peoples, livestock represents wealth on the hoof (Figure 8-21).

The Direct and Indirect Consumption of Grain

Many domesticated animals eat grass, foliage, and other plant material that humans cannot digest. Some grains, however, called *feed grains,* are fed to livestock. Eventually either the livestock or their products (milk, butter, cheese, and eggs, for example) are consumed by humans, so we say that feed grains are consumed by humans indirectly. Over 40 percent of the world's total grain harvest is fed to livestock. This percentage is higher in the richer countries. In the United States, for

World Distribution of Cattle

FIGURE 8-20 Cattle are widespread, but these 11 countries counted almost two-thirds of the world's herds in 2002. (Data from FAO)

example, the figure is approaching 80 percent, whereas in India the figure is about 4 percent.

As a country develops economically, the population consumes a greater amount of grain, but an increasing percentage of that grain is consumed indirectly as meat and dairy products. The people of the United States and Canada, for example, consume as

FIGURE 8-21 Masai cattle. Cattle represent wealth among the Masai people of Kenya. As the government requires these traditionally nomadic people to settle, multiplying cattle overburden the environment. (Photograph by Eliot Elisofon, 1972. Eliot Elisofon Photographic Archives, National Museum of African Art, Smithsonian Institution)

much as 2,000 pounds of grain per person per year, but only about 150 pounds of this is consumed directly as bread or cereal. The rest is consumed indirectly. In poor countries, on the other hand, only about 400 pounds of grain are available for each person per year, and the people consume most of it directly. Meat is the center of the diet only in the richest countries, and some nutritionists argue that the consumption of so much meat and dairy products leads to increases in debilitating diseases such as heart disease. Nevertheless, most people want to enjoy meat and dairy products. Therefore, rising incomes in some countries multiply their appetite for the indirect consumption of grain as meat.

Global per capita consumption of meat rose from 26 kilograms (57.2 pounds) in 1970 to 34 kilograms (74.8 pounds) in 2000. Rapidly increasing quantities of meat—mainly pork and chicken—consumed in China account for much of this. Neo-Malthusians believe that this increase in meat consumption threatens world food supplies because it multiplies the demand for grain. The threat is greatest to the world's poor people. Rising demand for grain-fed meat drives up grain prices. Rich people can afford to pay much more for food than poor people can, so they can outbid poor people for grain on the world market. As long as rich people are willing to pay high prices for meat, grain will be fed to animals instead of being available to poor people. Cornucopians, however, doubt whether the trend of increasing meat production threatens food supplies, and they predict a different future. They argue that if grain prices rise, the price of meat will also rise. Rising meat prices will reduce the demand for meat and increase grain production. If this occurs, then more inexpensive grain will be available for human consumption. Improving cereal yields might cause prices to continue to fall in any case. In fact, we do not know which of these economic events will occur, but such future price variations will affect the diet of billions of people.

Some kinds of animals transform grain into meat more efficiently than others do, and humankind overall could greatly improve its food supply by concentrating on raising more efficient animals. Chickens are the most efficient. They yield 1 pound (45 kilograms) of edible meat for every 4 pounds (1.8 kilograms) of grain they consume. Pigs produce 1 pound for every 7 pounds (3.2 kilograms) of grain, and beef cattle 1 pound for every 15 pounds (6.8 kilograms) of grain. In addition, chickens reach maturity and can be consumed in just seven weeks—much faster than pigs or cattle (Figure 8-22). There are already more than twice as many chickens on Earth as people, and today many countries are concentrating on increasing their chicken populations. China, for example, more than tripled its chicken population between 1984 and 2002. In many societies, however, cattle are a status symbol, and this preference delays the switch into more productive livestock.

FIGURE 8-22 Raising roasting chickens in Florida.
Chicken production in advanced countries has moved from the barnyard into virtual factories in which chickens are mass-produced. Hog production is undergoing a similar transformation. Factory-like conditions maintain control over feed, meat characteristics, breeding and reproduction, size, and other characteristics. (Norm Thomas/Photo Researchers, Inc.)

In some places the people avoid meat and dairy products because of religious prohibitions. Figure 7-36 noted the absence of pigs, for example, in Jewish and Muslim regions. These people meet their protein requirements from beef, lamb (mutton), fish, grains, or pulses (Figure 8-23).

Problems Associated with Animal Production

There are about 1.4 billion cattle on Earth today, and some scientists criticize these herds as inefficient and even harmful. The indictment against them is long. In the United States, runoff from feedlots is a major water quality problem. In sub-Saharan Africa, cattle contribute to desertification by denuding arid lands of fragile vegetation. In Central and South America, ranchers are felling rain forests for pasturage for cattle. Some agronomists argue that the millions of acres of land dedicated to feedstocks should be rededicated to crops for direct human consumption.

In addition, cattle contribute to greenhouse gases. The bacteria that live in every cow's gut enable the animal to break down cellulose, a tough fiber found in grass and other plants that humans cannot digest. As a byproduct, however, these bacteria produce methane, which is one of the most important greenhouse gases. The amount of methane produced by livestock is annually estimated to be almost double the amount produced by all termites (which have the same bacteria) or from landfills (which give off gas from decomposition).

Other livestock contribute to environmental devastation. Sheep and goats, for example, have overgrazed substantial areas in the Mediterranean basin, as well as parts of Africa and India. The global populations of hogs and chickens are rising faster than the cattle population, and their proportionate contributions to environmental problems, especially water pollution associated with feedlots, are increasing.

The pollution associated with livestock farming on industrial scales is increasingly a problem in the United States. Industrial farms housing as many as tens of thousands of hogs generate millions of gallons of waste each year. Hog waste is collected in cesspools because it must be treated before it can be used as fertilizer—sprayed on fields or into the air in liquid form. Seepage from unlined waste ponds pollutes groundwater, and the runoff causes water pollution. Cesspools the size of football fields also emit the toxic gases hydrogen sulfide and ammonia. Furthermore, concentration of livestock increases their vulnerability to disease, as well as the chances of any disease jumping to humans. Enormous dairy farms present the same problems. These huge facilities are truly factories, but they are not treated as such in the law and subject to the same environmental regulations.

World Distribution of Diets

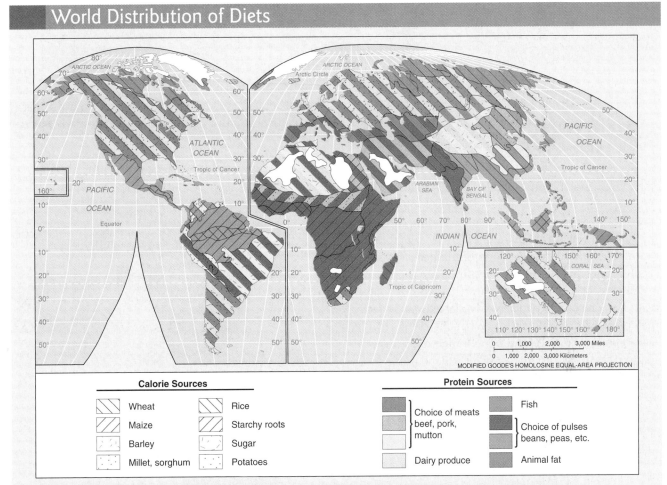

Calorie Sources

Wheat		Rice	
Maize		Starchy roots	
Barley		Sugar	
Millet, sorghum		Potatoes	

Protein Sources

Choice of meats beef, pork, mutton

Dairy produce

Fish

Choice of pulses beans, peas, etc.

Animal fat

MODIFIED GOODE'S HOMOLOSINE EQUAL-AREA PROJECTION

FIGURE 8-23 Everybody needs both calories and protein. The stripes and patterns on this map convey the predominant local source of calories, and the colors convey the predominant source of protein. The meat and dairy diet of Europe and of Europeans' settlements around the world stands out sharply. East and Southeast Asians rely more on fish, while pulses are important in South America, Africa, and in Southwest and South Asia. (Re-created from *The Times Atlas of the World,* Times Book Division at The New York Times Co., 1983, by John Bartholomew & Sons Limited and Times Books Limited, London)

Dairy Farming and the Principle of Value Added

Nearly 90 percent of the world's supply of cow's milk is produced in developed countries. In the past, fresh milk was available only on farms and in nearby towns, but in rich countries urbanization created a demand for large-scale commercial milk production. Milk is heavy and spoils quickly, so the cost of transporting it is high. Therefore, most cities receive their supplies from dairy farms in surrounding regions referred to as the cities' *milksheds.*

Some areas that are remote from urban centers but that have cool, damp climates unsuited to grain farming may specialize in dairy farming. Wisconsin in the United States and Switzerland in Europe provide examples. Much of the milk from these regions, however, is first transformed into butter, cheese, or dried, evaporated, or condensed milk. These products are not only lighter and less perishable than milk; they are

more valuable. The difference between the value of a raw material and the value of a product manufactured from that raw material is called the **value added by manufacturing.** Adding value to raw materials produces wealth. Wisconsin cheese and Swiss cheese and milk chocolate are exported worldwide, providing incomes to those two regions.

There are many steps and activities between the milk from a cow and a milk chocolate bar in a child's hand. In economic discussions, it is common to use the image that all products "flow" from their sources as raw materials to their ultimate consumers; the economic activities closer to the consumers are called **downstream activities.** Downstream activities add more value, and therefore they are profitable. Figure 8-14 notes how Chileans add value to their grapes by converting them to wine. We shall see in Chapter 9 and again in Chapter 12 that the possession of either agricultural or mineral raw materials may not

Focus on

HOW FARMERS DECIDE THE WAYS IN WHICH TO USE THEIR LAND: VON THÜNEN'S "ISOLATED CITY" MODEL

Johann Heinrich von Thünen (1783–1850) was a Prussian aristocrat who wondered what product he could grow on his suburban estates and market most profitably. Thinking about this problem led him to question how various crops became distributed across any countryside. To answer his question, von Thünen devised one of geography's earliest models. He began with the idea of an imaginary city market in a perfectly flat plain with absolutely no variations on it. This is called an **isotropic plain.** Von Thünen drew a model pattern for the distribution of different land uses.

Von Thünen noted that the different uses to which parcels of land are put result from the different values placed upon the land, called its rent value. On an isotropic plain, transport costs are the only variable, so differences in rent reflect the transport costs from each farm by straight-line distance to the market—as if everything moved by helicopter. The greater the transport cost, the lower the rent that can be paid if the crop produced is to be competitive in the market. In addition, perishable products such as milk and fresh vegetables need to be produced near the market, whereas less perishable crops such as grain can be produced farther away.

Von Thünen deduced that a pattern of concentric zones of land use will form around a city market. The intensity of cultivation—that is, the amount of labor and capital applied, will decline with distance from the market. Perishable crops that have the highest market price

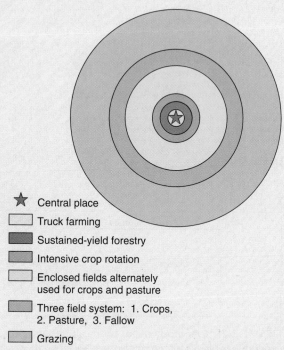

★ Central place
▢ Truck farming
▢ Sustained-yield forestry
▢ Intensive crop rotation
▢ Enclosed fields alternately used for crops and pasture
▢ Three field system: 1. Crops, 2. Pasture, 3. Fallow
▢ Grazing

Land use around an isolated city. Von Thünen's model shows rings of decreasingly intensive land use concentrically outward from the city. The ring of sustained yield forestry may be surprising, but forestry was then an intensive land use. The forest provided grazing for livestock, fuel, and raw materials for building and many other purposes.

necessarily enrich their possessor significantly. The key to wealth is adding value to raw materials and taking a share of that value added as profit. The geography of wealth is largely the geography of adding value to resources.

FOOD SUPPLIES IN THE FUTURE

All the factors just discussed have held off the specter of worldwide starvation. Is it possible that humankind is now, at last, at the end of its ability to increase food supplies? The answer to this question is a cautious "probably not." If demographers are correct in their projections of Earth's future population, the population can be fed. Humankind has scarcely begun to maximize productivity with the best contemporary

technology, and that leading technology has been applied to only a small portion of Earth.

Several of the other factors listed above that have increased total production since Malthus's time still offer potential for advance. Spreading urbanization is replacing agriculture in many places, but more lands can still be farmed. In 1995 the FAO estimated that developing countries of Latin America and sub-Saharan Africa could more than quintuple their cropland; the countries of East Asia could double their cropland. Irrigation can open new acreage (Figure 8-24), but in most dry areas where irrigation is most effective, water is already in short supply. Greater opportunities lie in increasing the efficiency of irrigation. Only about 37 percent of all irrigation water in the world, for instance, is actually taken up by crops; the rest is lost in runoff, evaporation, or leakage. The percentage lost could be and the highest transport costs per unit of distance, such as vegetables, will be grown closest to the market. Today a hint of this zone around a city survives in New Jersey's nickname, The Garden State. New Jersey farms historically supplied the cities of Philadelphia and New York. Today New Jersey is heavily industrialized, yet it also still specializes in greenhouse products, dairy products, eggs, and tomatoes, which are typical market garden commodities. In von Thünen's model, fields farther away from the market will be dedicated to less perishable crops with lower transport costs per unit of distance. In the rings farthest outward from the city, livestock grazing and similar extensive land uses predominate.

Von Thünen elaborated his model by adding a navigable river flowing through the town. He noted that the areas along the river's banks would enjoy greater accessibility to the city market. In other words, the cost distance to the city market was lower along the riverbank, so each of the zones would extend out along the river.

Some scholars have tried to apply von Thünen's model to entire countries. They have plotted the use of cropland in the United States, for instance, as if the Eastern Seaboard cities or Chicago were a central market. Trying to apply von Thünen's model to vast territories under contemporary conditions, however, strains the model's usefulness. The model focuses on distance from city markets, but in the real world proximity to the market is only one consideration. Physical environmental conditions, governmental regulations, the economic system, the pattern and the regulation of the transport system, and still other factors must be considered. Also,

transport costs have fallen dramatically relative to other costs of providing food. Most of the fruits and vegetables consumed in New York and Philadelphia, for example, come not from New Jersey but from California, Florida, Chile, and other distant places where production costs are lower and growing seasons longer.

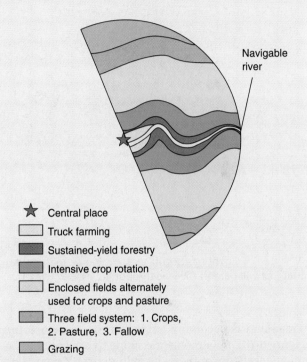

☆ Central place
☐ Truck farming
■ Sustained-yield forestry
■ Intensive crop rotation
☐ Enclosed fields alternately used for crops and pasture
■ Three field system: 1. Crops, 2. Pasture, 3. Fallow
■ Grazing

The influence of a river. If a navigable river flows through the town, the activity characteristic of each zone extends outward along the river.

reduced. Drip irrigation conserves water by piping only as much as crops need and delivering it directly to their roots. This method reduces the average amount of water needed per irrigated acre, boosts crop yields, removes the threat of parasitic disease spread by irrigation canals, and reduces soil salination.

Research is continuing on *halophytes*, plants that thrive in saltwater. Samphire (Salicornia), for example, is a tasty vegetable that not only contains more and healthier oil than soybeans, but also produces a valuable animal feed as a byproduct. It thrives in saltwater, and it is now cultivated commercially in Mexico, Arizona, Egypt, Iran, Syria, and around the Arab Gulf (Figure 8-25). Interbreeding halophytes with conventional crops has made these crops more salt resistant, which means that they can grow in more diverse environments. Farmers are today harvesting lands once

thought too salt-soaked to support crops in Egypt, Israel, India, and Pakistan. Conventional crops may someday be grown in saltwater.

Techniques of storage and transport could still be improved in many areas. China loses at least one quarter of its wheat harvest, for example, due to inadequate drying and storage, rats and mice, and poor transport. Improvements have been achieved, but in 1998 the U.N. World Food Program estimated that 40 percent of the world's crops were destroyed as they grew or before they left the field.

New Crops Offer New Potential

The green revolution focused on the improvement of just a few of humankind's most important crops, and the success in improving these has raised concern that

Egypt's "New Nile"

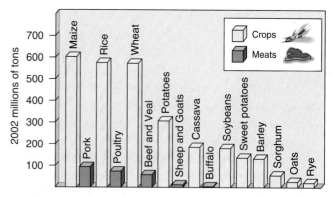

FIGURE 8-26 **World food production.** The great bulk of the human diet is today based on just a few crops and animal meats, but many others offer food potential. (Data from FAO)

FIGURE 8-24 Egypt has begun construction of an irrigation canal that will parallel the Nile through its Western Desert.

humankind is becoming too reliant on too few crops (Figure 8-26). If a new disease suddenly appeared today and attacked any of these, it could destroy a significant percentage of humankind's total food supply. A model of such a disaster occurred in the U.S. corn crop in

FIGURE 8-25 **Halophyte cultivation.** Samphire is today cultivated in India. (Courtesy of Planetary Design Corporation)

1970. A new fungus suddenly appeared that was well matched to the T-cytoplasm that had been incorporated into 80 percent of the country's seed corn, and U.S. corn production fell 15 percent.

A diversity of crops offers protection against catastrophe in case any one crop should fail, but modern commercial agriculture is increasingly specialized. Only "primitive" agriculture, which is in fact very sophisticated, has preserved diversity. Ethnobotanists have reported more than 50 kinds of potato grown in a single village in the Andes. At least 75,000 plant species have edible parts, and many of them are superior to the plants cultivated today. The New Guinea winged bean, for instance, is entirely edible—roots, seeds, leaves, stems, and flowers. It grows rapidly, up to 4.5 meters (15 feet) in a few weeks, and it offers a nutritional value as high as that of any known crop. Several Brazilian rain forest fruits—including the acerola, camucamu, and açaí—are both tasty and extraordinarily rich in vitamins and minerals. The National Academy of Sciences has recommended the cultivation of 36 other crops, including amaranthus, buffalo gourd, tamarugo, guar, mangosteen, and soursop. Most of us have undoubtedly never heard of these, yet each offers enormous potential for food. The potential of raising food from more and different plants is a reason why the extinction of plant species threatens humankind's welfare. The steady loss of genetic diversity before botanists can study each species sacrifices potential new crops or genes for interbreeding.

Agricultural technology offers many alternative foods, but what people eat, or refuse to eat, is to a great degree cultural (Figure 8-27). Culinary imagination could probably make a wide variety of alternative crops or animals appealing and palatable. For example, the menu for the New York Entomological Society's 100th anniversary banquet in 1992 featured live honey-pot ants as an appetizer, cricket and vegetable tempura, mealworm balls in zesty tomato sauce, roasted Australian kurrajong grubs, waxworm fritters in plum

FIGURE 8-27 Would you eat this? Peasants bring live iguanas into downtown Hanoi, Vietnam, for sale for Saturday night dinner. Many people around the world find nutrition in and enjoy eating things most Americans do not find tempting. (Courtesy of Julia Nicole Paley)

sauce, giant Thai waterbugs sautéed in olive oil, and for dessert, chocolate cricket torte. All these insects are very nutritious, and they are common fare in many countries. South Africa and Botswana do in fact annually export hundreds of tons of mopanie worms. A dried worm is 56.8 percent protein, and 15 worms provide a full daily supply of calcium, iron, and riboflavin for an adult. The worms taste like peanuts.

> Some scientists argue that [the scientific revolution in agriculture] has just begun . . .

The Scientific Revolution in Agriculture Continues

The scientific revolution in agriculture is not over. Some scientists argue that it has just begun and that new scientific advances will multiply future food yields. The economist Henry George (1839–1897) succinctly contrasted the rules of nature with the multiplication of resources through the application of human ingenuity. "Both the jayhawk and the man," he said, "eat chickens, but the more jayhawks, the fewer chickens, while the more men, the more chickens." This principle is key to understanding and counting all resources.

Biotechnology is the term given to a variety of new techniques for modifying organisms and their physiological processes for applied purposes. One aspect of biotechnology is **recombinant DNA,** or **gene splicing.** Scientists can today join the DNA of two organisms to produce a recombinant (or "recombined") DNA. Scientists can then introduce this recombinant DNA into another organism, permanently changing the genetic makeup of that organism and all its descendants. Such products are said to be **genetically modified (GM).**

Another technique of biotechnology is **cloning,** the production of identical organisms by asexual reproduction from a single cell of a preexisting organism. Still other techniques of biotechnology develop new microorganisms for diverse purposes ranging from pharmaceuticals to cleaning up oil spills. Biotechnology can be applied to both plants and to animals.

Biotechnology offers genetically altered crops that can be custom-designed to fit the environment and produce bountiful harvests. Biotechnology can even replace chemical pesticides (Figure 8-28). Breeding natural herbicides and pesticides into the plants themselves reduces the enormous quantities of chemicals and energy consumed in the manufacture, packaging, distribution, and application of chemical herbicides and pesticides. For example, a new type of Russet Burbank potato contains its own protection against the Colorado potato beetle. In 1995 an estimated 5,000 barrels of oil went into the manufacture and distribution of insecticides to fight this one insect pest in the United States, and less than 5 percent of the insecticide reached the target insect. In the single year 2001, the use of GM crops reduced U.S. use of pesticides by an estimated 21 million kilograms (23,100 tons). A GM corn introduced in the United States in 2003 significantly reduces the need for pesticides on that crop.

The first GM food to reach U.S. groceries was the Flavr Savr tomato, in 1994, engineered to stay firm longer than regular tomatoes, but GM soybeans, cotton, potatoes, corn, and rape soon followed. Most historic advances in the green revolution addressed improving yields of wheat, potatoes, and rice, but today many

FIGURE 8-28 A natural insecticide. The rows of potato plants on the left and right were devoured by the Colorado potato beetle. The plants in the center row, however, had a gene introduced into them from a common soil bacteria. Those plants produce a protein that acts as a natural insecticide. Biotechnology offers the promise of protecting many crops in this way. (Courtesy of Pharmacia Corporation)

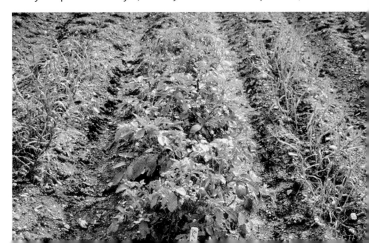

Critical Thinking

CONTINUING CROP REDISTRIBUTION

Different peoples still harvest different indigenous crops and enjoy different diets, and food crops continue to

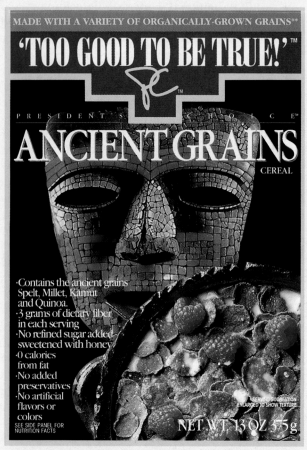

MADE WITH A VARIETY OF ORGANICALLY-GROWN GRAINS**

'TOO GOOD TO BE TRUE!'™

PRESIDENT'S CHOICE™

ANCIENT GRAINS
CEREAL

·Contains the ancient grains Spelt, Millet, Kamut and Quinoa.
·3 grams of dietary fiber in each serving
·No refined sugar added— sweetened with honey
·0 calories from fat
·No added preservatives
·No artificial flavors or colors
SEE SIDE PANEL FOR NUTRITION FACTS

SERVING SUGGESTION ENLARGED TO SHOW TEXTURE

NET WT. 13 OZ 375g

Quinoa is gradually being introduced into the North American diet. (Courtesy of Loblaw Brands Limited)

migrate around the world. *Chenopodium quinoa,* for example, a member of the genus goosefoot, is a native of South America. It is an annual, broadleaved herb usually standing about 3 to 6 feet high. Leafy flower clusters rise from the top of the plant. The dry, seedlike fruit is about 2 millimeters in diameter and enclosed in a hard, shiny, four-layered fruit wall. Quinoa seeds are flat and pale yellow, and they can be steamed, ground into flour, or fermented to make a mildly intoxicating beverage. The leaves are highly edible and often used for livestock feed. Quinoa is sometimes called Inka rice.

Of all the world's grains, quinoa is the highest in protein content, and it has an amino-acid profile that parallels the ideal standard set by the FAO. New varieties of this plant are being bred that are better suited to North American conditions, and quinoa has recently been appearing in North American food products.

Questions

1. Can you think of the first time you tried some new food? How about a kiwi or a Japanese buckwheat noodle?

2. Find exotic foods in local markets. Why were these first imported to your local store? Were they brought for a local immigrant population?

3. New foods from around the world are almost daily introduced into heterogeneous urban centers. How has the change in the source areas of immigrants to North America discussed in Chapter 5 caused the introduction of new foods?

other traditional crops, including the tropical crops such as sweet potatoes and cassava, are winning new attention (Figure 8-29). Today lentils provide one of the world's leading sources of protein, and the new GM Mason lentil is up to 18 percent more productive, and it is hardier.

Today's food has been improved not only in quantity but in quality. For example, scientists have doubled the vitamin content of most vegetables and produced soybeans that produce heart-healthy oils and are easier to digest. This will be increasingly important as the human population ages.

GM crops have even been bred to produce pharaceuticals, in a process called biofarming or farma-

ceuticals. DNA science has spawned a generation of drugs made from human antibodies, but the antibodies can be produced more cheaply by splicing antibodies into the genetic fabric of plants, growing them in fields, and then extracting and purifying them. Hemoglobin is being bred into corn and soybeans. Corn-bred enzymes stimulate insulin production in diabetics. Bananas and potatoes can now deliver a dose of vaccine. Scientists have increased tomatoes' content of lycopene, which seems to prevent some types of cancer, and have introduced genes that produce beta-carotene, the precursor of vitamin A, into rice grains. The rice contains sufficient beta-carotene to meet human vitamin A requirements. Other GM crops are

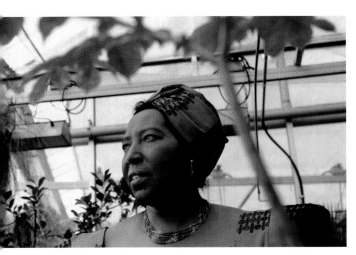

FIGURE 8-29 A leader in research on GM tropical crops. Dr. Florence Wambugu of Kenya has won world renown for developing a virus-resistant sweet potato that more than doubles yields for subsistence farmers. She herself was raised on sweet potatoes on a subsistence farm. "I wasn't even supposed to be educated," she has said with a laugh. "My mother had to go to a tribal tribunal to get permission. I was a girl who was supposed to be married, and that was it." Today she heads a Nairobi-based scientific foundation and is an advisor to many scientific boards around the world. (Louie Palu/The *Globe* and *Mail*)

FIGURE 8-30 Dolly, the cloned sheep. On February 22, 1997, Scottish embryologist Ian Wilmut stunned the world by announcing that he had created an exact copy—a clone—of an adult Dorset sheep. The historic lamb, created from DNA extracted from a sheep's mammary gland, was named in honor of Dolly Parton. Dolly died February 14, 2003, of a lung infection. (Remi Benali/Stephen Ferry/Getty Images, Inc.—Liaison)

drought-resistant. Scientists from the United States and South Korea have together borrowed a gene from *E. coli* bacteria and engineered rice to withstand drought, salt water, and cold. Even a decaffeinated coffee bean was introduced in 2003.

The application of biotechnology to livestock itself is yielding hardier and more productive livestock. Biotechnologists are even improving animal feed. One product helps animals absorb phosphorous better, thus reducing the ecologically-damaging phosphorous in animal waste. Cloning first achieved publicity with the birth of Dolly, a sheep, in 1997 (Figure 8-30). Dolly was the first mammal cloned from an adult cell, but since then mice, goats, monkeys, cats, and rabbits have followed.

Resistance to Biotechnology

In April 2000 the U.S. National Academy of Sciences announced that there is no evidence GM foods are unsafe to eat. It did, however, insist that rigorous monitoring of future research must be maintained.

Focus on
GOOD-BYE TO THE BANANA?

Without a miracle of genetic engineering, bananas may be extinct within ten years. The banana is one of our oldest crops, the first edible variety having been propagated about 10,000 years ago from a rare mutant of the wild banana, which, with a mass of hard seeds, is virtually inedible. All edible bananas are genetically decrepit sterile mutants—effectively clones of that first plant. Bananas are, therefore, unable to evolve to fight off new diseases.

Today we are seeing the global spread of black sigatoka, a fungal disease that cuts yields up to 75 percent and reduces the reproductive lives of banana plants from 30 years to only two or three. The fungus reduced yields in Uganda 40 percent in one year, and by 2003 it was spreading across Brazil.

Genetic engineering may be only way to save this popular fruit. Scientists hope to sequence the genetic blueprint of the banana by 2008, focusing on the inedible wild banana, which remains resistant to black sigatoka. Some producers fear that consumers will not accept a GM banana, but some 500 million people in Asia and Africa depend on bananas for up to one-half of their daily calories.

History does demonstrate that technological solutions to problems can trigger unexpected new problems, so some people fear that GM products may have unforeseen effects in nature (outcompeting and destroying natural species, for example) or on our own bodies. In 2000 a GM corn that had been approved for cattle but not for human consumption did find its way into commercially available taco shells. The corn contains a gene that causes the plant to produce a substance that is toxic to the corn borer insect pest. This gene—found in over one-quarter of the corn grown in the United States today—reduces the need for pesticides, but fear of unknown and unintended side effects has triggered widespread controversy. The cloning of Dolly reminded some people of the novel *Frankenstein,* by Mary Shelley (1797–1851). In this classic, Dr. Frankenstein's confidence in science leads him to create a living creature that he hopes he can control but that turns on him and kills him. Some critics refer to GM products as *Frankenfood.* The idea survives in our imagination that science can create "monster problems" more serious than the problems that science solves. In 2001 scientists in Australia did, inadvertently, create a modified mousepox virus that destroys the animal's immune system. This scientific breakthrough may lead to better understanding of AIDS, but it may also inspire new developments in biowarfare. Since Dolly, cloned animals have often exhibited severe and unpredictable health problems, including lung, heart, and immune system defects.

Opponents of GM foods have orchestrated demonstrations and even destroyed fields of experimental crops. Major food manufacturers, supermarkets, and fast-food chains have already stopped stocking food with GM ingredients. As told in the anecdote opening this chapter, Europeans insist that GM foods be labeled, but Americans oppose labeling as unnecessarily frightening. There are, however, no clear definitions of what GM means. For example, are hogs that have been fed GM corn necessarily GM hogs? Lack of agreement means that contradictions in distribution abound: Frito-Lay will not use GM corn in its corn chips, yet its parent corporation, Pepsi Cola, uses syrup from GM corn in its soda. McDonald's bans GM potatoes, but the company cooks potato fries in vegetable oil made from GM corn and soybeans. Some American opponents of GM foods have begun labeling things that are not GM, but "natural" and "organic" are vague terms and in dispute. The U.S. Department of Agriculture has been working on definitions of "organic" labels usable as a sales pitch to U.S. consumers. Meanwhile, Monsanto chemical company, however, has sued a Maine dairy for advertising that its products contain no "growth-inducing hormones." Monsanto is the maker of the only major artificial growth hormone, Posilac, which is used in about one-third of U.S. dairy cows.

Delegates from 130 nations met in Montreal, Canada, in 2000 and adopted a treaty regulating trade in GM products. The new treaty, known as the Cartagena Protocol (after the city in Colombia where talks started) is an outgrowth of the Convention on Biological Diversity forged in Rio de Janeiro, Brazil, in 1992, and the new treaty went into effect September 1, 2003, having been ratified by the requisite 50 signers. The treaty allows countries to bar imports of genetically altered seeds, microbes, animals, and crops if they deem the product a threat to their environment. Labeling is required on products only in international shipments, not when the product is on store shelves. Exporters must obtain permission in advance from an importing country before the first shipment of a particular "living modified organism" is meant for release into the environment—like seeds, microbes, or fish which are to be put into a river. Advance notice and permission are not required for exports of commodities meant for eating or processing. The United States has never ratified the 1992 Biodiversity Convention, so it cannot be a party to this treaty. It will, nevertheless, have to abide by the terms of this treaty when exporting products.

To some degree, popular resistance to biotechnology is not entirely a technological question but a religious one. Many people feel that nature is immutable or that tampering with it is sacrilegious. Prince Charles of the United Kingdom, for example, has said he will never eat any GM food. "That takes mankind into realms that belong to God, and to God alone," he insists. Monsignor Elio Sgreccia, however, president of the Vatican Bioethics Institute, has said, "We are open to the use of genetic technology in agriculture and with animals, as long as we don't do it with man. We believe that man has a primacy on this planet. . . . Nature is here for him."

Whatever happens in the rich countries, many of the developing countries are betting on GM foods to increase their agricultural yields and food supplies. China, for example, is second only to the United States in biotechnology research, and it is forging ahead with productive and pest and disease-resistant strains of rice, cotton, tomatoes, tobacco, sweet peppers, green peppers, and still additional crops. China, India, and Indonesia are among the leaders in increasing plantings. Some observers fear that in GM foods and in other aspects of biotechnology, religious scruples may cause the West to fall behind in twenty-first century science.

Concerns about the application of biotechnology in agriculture are greatly magnified when biotechnology is applied to humans themselves, as in the controversy surrounding research with DNA, human embryos, and stem cells. Therapeutic applications offer great promise (such as the elimination of hereditary diseases), but biotechnology will undoubtedly soon be used for physical enhancement. Augmented athletic prowess, for example, will almost certainly be demonstrated in sports contests in the near future. Some people may find these possibilities literally monstrous.

Other social obstacles remain to technological advances in agriculture. For example, biotechnology is expensive. Will all farmers have access to this technology, or only those with sufficient capital? Some observers worry that the relatively few large corporations that own the rights to seeds and other advances will exercise increasing economic control over agriculture.

Another threat seems like an ironic paradox: If biotechnology truly unleashes the productivity the optimists foresee, then just a fraction of the farmers on Earth today will be able to feed everyone. Millions of farmers—hundreds of millions—will be unemployed. What will they do for a living? How will humanity be kept busy? Multiplying our ability to produce food threatens to impoverish billions of people, to launch even greater migration to cities that are under-equipped to absorb the influx, and thus to trigger civil unrest on scales never before contemplated in history—as well as, perhaps, to worsen the health problems associated with obesity outlined in Chapter 5. Will our political and economic systems be able to transform human societies fast enough to avoid chaos? Chapter 10 examines the pressures on cities already caused by the migration of farmers from the countryside, and Chapter 12 examines economic development possibilities.

Global Warming

Forecasts of changes resulting from global warming might modify our optimistic scenarios of continuing increases in food production. In 2001 the FAO estimated that by the year 2080, the world's poorest 40 countries (in 2000) could lose 20 percent of their ability to raise food because of global warming. Food-growing capability would also fall in the United States, France, Romania, Hungary, Belgium, the United Kingdom, the Netherlands, the Czech Republic, the Ukraine, all of Central America and the Caribbean region, and India. In Canada, by contrast, agricultural production could double, and other "winners" would be Finland, Norway, New Zealand, Russia, and China.

WORLD DISTRIBUTION OF FOOD SUPPLIES AND PRODUCTION

In 1968 Professor Paul Ehrlich of Stanford University wrote in his bestselling book *The Population Bomb:* "The battle to feed humanity is already lost. . . . We will not be able to prevent large-scale famines in the next decade." Local famines did occur, but they were not caused by a lack of food on Earth. World excess food stocks relative to consumption have steadily outpaced population growth. Nevertheless, the debate between the pessimists and the optimists, the neo-Malthusians and the Cornucopians, continues.

Figure 8-31 maps the prevalence of undernourishment around the world. A balanced diet includes carbohydrates (derived from staples such as rice, corn,

wheat, and potatoes), proteins (from meat, fish, poultry, eggs, dairy products, or pulses), vitamins (from fruits and vegetables, as well as other sources), fats, and minerals. The map suggests that, overall, the food supplies in some countries fall short of per capita needs. Many people in those countries are well fed, but many others are hungry. Even in the countries in which overall supplies are adequate, however, some people are hungry.

No country is completely self-sufficient in food. Most countries both import and export food, and a few countries are net exporters of food despite the fact that portions of their own populations are undernourished. This may be due to injustice or civil strife. Political instability contributes to hunger. Several African countries, for example, are environmentally richly endowed, yet a great many of their people go hungry. Food-supply problems are often problems of economics or politics, not problems of geography or technology. In poor countries, rich people have plenty to eat, while poor people are starving. An increase in wealth that allows rich people to eat more meat without improving the standard of living of the poor only increases political instability. For everyone on Earth to be adequately fed today and in the future as Earth's population continues to rise, we need both an increase in production and improvements in distribution, and these must occur both within countries and among countries.

Peter Rosset, director of the Institute for Food and Development Policy, wrote in 1999, "There is no relationship between the prevalence of hunger in a given country and its population. . . . The world today produces more food per inhabitant than ever before. Enough is available to provide 4.3 pounds to every person every day: two and a half pounds of grain, beans and nuts, about a pound of meat, milk and eggs, and another of fruits and vegetables—more than anyone could ever eat. The real problems are poverty and inequality. Too many people are too poor to buy the food that is available or lack land on which to grow it themselves." These are the problems we must examine in the following pages.

Problems in Increasing Food Production

Technology creates the potential for increasing food production, but it cannot guarantee that this will occur or that the food will be produced or distributed where it is needed. Other potential problems in the effort to increase food supplies include diminishing returns to fertilizer, lack of financial incentives for farmers, and inequitable land ownership or unsuccessful land redistribution.

Fertilizers: overuse and diminishing returns   Fertilizers can improve crop yields significantly when used in appropriate amounts. For many years, however, the

Prevalence of Undernourishment

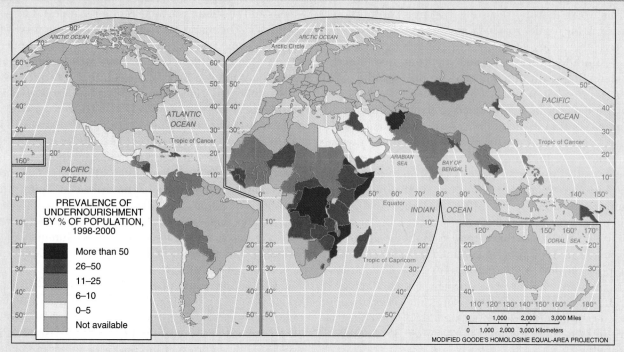

FIGURE 8-31 The amount of food needed per capita varies from country to country, depending on the age distribution of the population, activity levels, health, climate, and other factors. Nevertheless, this map reveals that in many countries, substantial percentages of the populations are undernourished. In virtually all places where data are unavailable, undernourishment afflicts only a very small percentage of the population.

quantities of fertilizer applied in the advanced countries have far exceeded the point of **diminishing returns.** Diminishing returns exist when, in adding equal amounts of one factor of production, such as fertilizer or labor, each successive application yields a smaller increase in production than the application just preceding. Table 8-2 applies this concept to a hypothetical example of fertilizer. In this table the point of diminishing returns is 4 pounds of fertilizer. This is true because adding the second and third pounds of fertilizer increased the yield by 11 bushels, but adding the fourth increases the yield by only 10 bushels. At some point, then, the diminishing increases in crop yields no longer justify the application of greater amounts of fertilizer at one place. That fertilizer could more effi-

ciently be applied in another place that has not reached the point of diminishing returns.

The U.S. uses about 110 kilograms per hectare (20 pounds per acre) of cropland. If each U.S. farmer's last 5-kilogram bag of fertilizer were given to a farmer in a poor country, the world food supply would increase dramatically. In other words, the same fertilizer that yields only diminishing returns in the United States could be used much more productively in a poor country.

Improvements being made in the techniques of fertilization, however, offer the promise of improving the efficiency of its application. Global positioning systems (GPS) installed in the cabs of farmers' tractors allow the farmers to target the amount of fertilizer appropriate to each small subsection of a field (Figure 8-32). This precision greatly reduces the common error of applying more fertilizer than the optimal amount, thus causing pollution.

Many farmers lack financial incentives Chapter 6 referred to the agronomists' dictum: "The market produces the surplus." In other words, if a market exists for any surplus that farmers can produce, and if the political and economic system sufficiently rewards farmers for their effort, production of a surplus is all but assured (assuming adequate technology is available).

In many countries, however, farmers are so heavily taxed that they have little incentive to increase food

TABLE 8-2 Diminishing Returns in the Addition of Fertilizer

Fertilizer (pounds)	Total yield (bushels)	Production increase (bushels)
1	10	10
2	21	11
3	32	11
4	42	10
5	51	9
6	59	8

FIGURE 8-32 A GPS monitor in a tractor cab. GPS satellite navigation plus GIS field mapping and data analysis allow farmers to distribute seeds and apply fertilizers mixed specifically for each square meter of a large field as the tractor drives across it, according to the soil characteristics of that small area. (Courtesy of Ag-Chem Equipment Co., Inc.)

production, so they remain at the subsistence level. Many governments force the farmers to sell their crops to the government at fixed low prices, charge high taxes on food exports, support high exchange rates for their national currencies, and impose high import duties on the tools and agricultural chemicals the farmers need. These are all ways of taxing farming. The wealth that is squeezed from the farmers in these ways is often used to subsidize urban food supplies or to support urban civil service bureaucracies, military spending, or unprofitable state-run industries. In some countries, deteriorating conditions in the countryside trigger migration to the cities (a topic we shall examine in Chapter 10), and swelling urban populations increase the demand for subsidized food. In some countries in which recent political changes have given farmers incentives to produce more food, the increases in production have been astonishing. India lifted some agricultural price controls in 1991, and since that year grain production has risen about 4 percent annually. This rate is more than twice the rate of population increase, so per capita food supplies have improved markedly. When farmers enjoy higher profits, more rural jobs are created both in agriculture and in rural craft industries.

Systems of land ownership can discourage production   In many countries, systems of land ownership reflect broad political and social inequalities. In Brazil in 2002, for example, about half of the country's arable territory was held by only three percent of all landowners. This concentration of land ownership spurs the landless to attempt to farm new claims in the rain forest regions, degrading the environment but producing only disappointing yields. The government has announced land redistribution schemes, but it has

been slow to act. Furthermore, to be successful, newly landed peasants require infrastructure (such as roads), technical support, and even assistance marketing and financing their production. Hundreds of landless peasants have died in clashes with armed landowners and police in Brazil in recent years, but, at the same time, the modern, highly capitalized sector of Brazilian agriculture achieved great success increasing exports of sugar, soybeans, and beef.

In many countries the landholding systems are holdovers from colonial periods. Colonial governments replaced communal systems with systems of private property, and the distribution of the property was highly inequitable. This created one class of large estate holders and another, much larger, class of landless poor. The large landowners often moved to the cities, becoming absentee landlords unwilling to make the investments necessary to maximize the return from their own agricultural holdings.

In the Philippines, for example, large blocks of land were granted to the Spanish colonial elite, and the natives on these land grants became tenant farmers. After a few generations the landlords were mostly mestizos (persons of mixed native and colonial ancestry), and their descendants owned the best agricultural land. This land-grant system prevailed in all areas colonized by Spain, and it is at least partly responsible for political unrest in Central and South America today.

Indonesia presents a contrast to the former Spanish colonies. For more than 300 years under Dutch rule, only native Indonesians could own land, except for city lots. Dutch, Arabs, Chinese, and English could rent agricultural land, but they could not acquire title to it. Thus, land ownership among Indonesian peasants remained much more widespread than it was in the nearby Spanish Philippines. Indonesian peasants who have inherited their land and expect to leave it to their children are much less likely to join peasant guerrilla movements than are landless Filipinos. In addition, individual farmers of small personal holdings are likely to maximize their productivity.

When land is leased to those who actually work it, the conditions of tenancy and of payment determine whether the farmers are encouraged to produce significant surpluses. If the landlord takes too large a share, surpluses will be small.

Communist regimes in Eastern Europe collectivized agriculture in large state-owned farms and in cooperative enterprises. Neither provided sufficient incentive to encourage farmers to maximize productivity. Since the fall of Communism, however, private farming is being encouraged. Poland ended food subsidies and restrictions on direct marketing in 1989; farmers' markets sprang up overnight, food shortages eased, and prices fell. The other formerly communist Eastern European countries are privatizing the large communal landholdings. Vietnam ended collectivized agriculture in 1990, and the country swiftly rose to become a significant rice exporter. Giving or selling government assets to private individuals or investors is called **privatizing** an activity. Small private market gardens were always allowed in the former Soviet Union, and although these accounted for less than 2 percent of total Soviet farmland, they yielded 60 percent of the country's potato crop and around 30 percent of its vegetables, meat, eggs, and milk. Today the government of Russia is slowly but steadily privatizing agriculture. In 1978 China relaxed its state-directed agricultural system and allowed peasant farmers decision-making power and profit incentives. Within 10 years farm output rose 138 percent, and China turned from a net importer of food products into a net exporter. Privatizing of industry and of other aspects of national economies is proceeding around the world; we will examine this important development in more detail in Chapter 12.

Agricultural productivity can also be hampered when title to land—legal ownership—is unclear, as is the case in many countries. Generations of peasants may have tilled small plots, but families have never registered title, and bureaucratic complications make it difficult to establish title now. It has been estimated that in poor countries only about 20 percent of all land has clear title. Lack of clear title discourages the occupants from making improvements on the land. Furthermore, as we have seen, successful farming today requires capital investment in irrigation pumps, chemicals, improved seeds, and machinery. Farmers can borrow money to make these investments only if the land can serve as *collateral* (a thing of value that can be seized by a creditor in case of nonpayment of a debt). Farmers without clear title to the land lack collateral. The satellite-based GPS and GIS have greatly reduced the cost of land registration, and new programs have been launched around the world. Establishing farmers' title to land triggers investment and increases productivity.

Unsuccessful land redistribution: Mexico

Mexico demonstrates how communal land ownership can restrict productivity because communal holdings cannot serve as collateral. More than one half of the country's arable land is held in **ejidos,** a form of land tenure in which a peasant community collectively owns a piece of land and the natural resources and houses on it (Figure 8-33). Mexican law states that ejidos are "inalienable, nontransferable and nonattachable." They cannot be used as collateral, so banks will not extend loans. This system, combined with a government tradition of paying farmers low prices for their crops while at the same time subsidizing food for urban consumers, has kept farmers poor and productivity low. These conditions in turn intensify the migration of peasants to the cities. Mexico imports 35 percent of its total food supplies, including nearly half of its staples of corn, wheat, and beans, from the United States.

The government is struggling to overhaul the ejido system, but official recognition of ejido ownership was a principal issue of the Mexican Revolution

FIGURE 8-33 A Mexican ejido. Ejidos represent the traditional Mexican form of communal landholding. The land is inalienable, so the ejido cannot borrow money for investment, and thus productivity is low. (Carl Frank/Photo Researchers)

(1910–1920), in which more than 1 million Mexicans died. Many Mexicans remain attached to the idea. In 1992 the government introduced a program of certifying individual land rights, but by 2000 scarcely 1 percent of ejido land had been sold. Since the NAFTA trade pact between the United States and Mexico was launched in 1994 (see Chapter 13), Mexican farmers have found it difficult to compete with U.S. farmers. The Mexican government has ended price guarantees without helping farmers market their produce, and it costs about twice as much to get a ton of wheat to Mexico City from Mexicali, near Mexico's northwest corner, as it does to get a ton of wheat to Mexico City from Kansas. Some Mexican farmers are switching to new commercial crops, or they are contracting to agribusinesses. Agribusinesses provide the needed capital investment, as well as an assured market. The Mexican government has also experimented with guaranteeing loans made to the ejidos, thus encouraging capital investment and improving agricultural productivity.

Barriers to increasing production in Africa

Many of the problems found in Mexico's agricultural economy can also be found in Africa. Landholding is often communal, so successful farmers cannot expand their holdings or borrow money to invest in greater productivity. Many African governments themselves hold ownership of agricultural land and lease it to farmers. In Zimbabwe, for example, the government nationalized numerous large white-owned private farms that were exporting food. The government relocated black settlers onto the properties but retained ownership, so the farmers cannot borrow to invest in increasing productivity. Productivity has fallen, and Zimbabwe has lost export income.

In addition, farmers in many countries are required to sell their crops to state marketing boards.

These boards often pay the farmers very little. In Nigeria, for example, the government owns almost all land, leases it to farmers, and then pays the farmers fixed low prices for their crops. Many African state marketing boards have proved so corrupt that the farmers' returns for their work are virtually stolen from them. Many governments have failed to invest in rural infrastructure or in agricultural research to develop technologies or native crops suited to local conditions.

These conditions discourage farmers and reduce agricultural output. Many farmers migrate to the cities, but because land tenure is based on occupancy, they must leave their families in the villages. In some parts of rural Zimbabwe, 40 percent of the households have lost the father to the town. The overworked women, left to tend children as well as crops, can scarcely rise above subsistence. The United Nations Children's Fund (UNICEF) estimates that women grow 80 percent of Africa's food. AIDS has also taken a terrible toll on African farming, killing over 7 million farmers by mid-2003, thus greatly reducing food output.

Civil wars are still another factor that has reduced food production in some areas of the world, again particularly in Africa. The continent was feeding itself in 1960, but today Africa imports about 40 percent of its food supply.

Summing up, the reasons for inadequate food production in many countries are not necessarily environmental or even technological. Food production could be greatly increased if many governments could end their civil wars, change national economic policies that economically discriminate against farmers while subsidizing urban food supplies, stop discriminating against women, reduce import taxes or quotas on imported tools and chemicals, improve infrastructure (particularly roads and water supplies), and reform landholding systems. All of these factors influence food production.

Commerical crops in the poor countries Today many farmers in poor countries do not concentrate on growing staple crops for local consumption. Instead, they raise cash crops for sale. Senegal, for example, exports peanuts, the Ivory Coast cocoa, Angola coffee, Zimbabwe tobacco, and Kenya tea and flowers, while food availability in those countries is low (Figure 8-34). We must investigate whether these countries would not be better off growing food for their own consumption.

If any country chooses to dedicate its efforts to specialty crops, it can buy basic foodstuffs on world markets. In that way, the country can achieve what is called *economic self-sufficiency* in food. Farmers who choose to depend on cash crops, however, face two risks: the price for their product could collapse because of overproduction elsewhere; prices for the foods they will have to buy could rise faster than their own incomes.

Raising cash crops may increase farmers' disposable incomes rapidly, while maintaining food produc-

tion. Guatemalan farmers raising snow peas for New Yorkers, for example, can make 14 times as much per acre as they would raising corn. Even after taking into account the increase in their costs, their return for each day of work more than doubles. A great share of the increased costs are for hired labor, so more people find work.

The rise in income may not, however, improve the lives and health of farm families. Studies have revealed that in many poor countries, women control the production of food crops, but men control the production of cash crops. The men often spent "their" money on things such as alcohol and tobacco, which did not increase family welfare.

Several Caribbean countries have found it profitable to surrender self-sufficiency in food and to dedicate their cropland to grass—that is, to fairways for golf courses that attract tourists. High earnings from tourism allow the countries to buy food on world markets. The role of tourism in national economies will be examined in Chapter 12. Unfortunately, the world's most successful cash crop remains the one most governments try hardest to discourage—illegal drugs (Figure 8-35).

The Rich Countries Subsidize Production and Export of Food

Many rich countries have erected walls of high *tariffs* (taxes on imported goods) to protect their markets against food imports from the poor countries. This requires the urban consumers in the rich countries to pay inflated prices for food. These inflated prices are indirect subsidies to the rich countries' farmers. Many rich countries also give their farmers direct subsidy payments. In 2002 the 15 countries of the European Union spent $112.6 billion dollars in agricultural price supports, the United States $90.3 billion, and Japan

FIGURE 8-34 Flowers as a cash crop. More than 330 tons of flowers and fresh vegetables are air-freighted out of Kenya to Europe every day. The industry employs some 90,000 Kenyans and earns Kenya about $500 million per year, an amount that exceeds that earned today from tourism and coffee, which were previously principal exports. (Francesco Broli/The *New York Times*)

International Trade in Illegal Drugs

FIGURE 8-35 The value of illegal drugs is so high and the market demand so strong that international trade flourishes despite international police efforts to stop it. The flow of opium from Afghanistan has increased since the U.S.-led war ousted the fundamentalist Taliban from rule there (see Chapter 13). (Information from the U.S. Drug Enforcement Agency)

$55.7 billion (Figure 8-36). As a result, farmers in many rich countries produce food surpluses, which the countries then export or sell on world markets at prices below the costs of production (a practice called *dumping*).

These subsidies distort the world geography of agriculture. They encourage production of agricultural surpluses in many rich countries and discourage increasing production in the poor countries. As long as the rich countries sell their surpluses on world markets at low prices, the governments of many poor countries neglect to invest in their own rural areas. Farmers in the poor countries cannot export to the rich countries or even compete in their own national markets; they remain at the subsistence level, or they are driven off their farms entirely.

Why do some rich countries subsidize agriculture?

Many rich countries subsidize their farmers and erect import barriers for several reasons:

1. Some countries pursue self-sufficiency in food production as a national security measure. The governments want some defense against grain embargoes and crop failures in those countries that normally have surpluses to market. Subsidizing a

> [Rich countries'] subsidies distort the world geography of agriculture. They encourage production of agricultural surpluses in many rich countries and discourage increasing production in the poor countries.

country's own farmers offers some protection against these threats. Following U.S. threats to halt grain shipments to the Mideast in the 1970s, for example, Saudi Arabia spent heavily to improve its agriculture. Saudis give farmers subsidies to raise grain, import irrigation equipment, and even raise dairy cattle in air-conditioned sheds where outside temperatures exceed 48° C (120° F) (Figure 8-37).

2. Many rich countries, particularly in Europe, subsidize agriculture to preserve traditional agricultural communities and keep their farmers from migrating to the cities in search of work. Rural-urban migration can overwhelm the ability of cities to absorb the new labor force.

3. Some rich countries subsidize farming as part of national land-use plans. They enjoy agricultural landscapes and want to preserve green areas around cities, called *greenbelts*. Their urban citizens are willing to subsidize farms in order to preserve them.

4. Many rich countries' political systems favor farmers. The farmers in rich countries constitute a decreasing percentage of the national populations, yet, particularly in Europe and Japan, the electoral

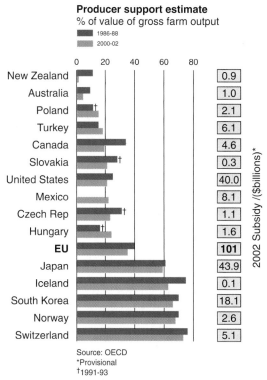

Producer support estimate
% of value of gross farm output
- ■ 1986-88
- ▨ 2000-02

Country	2002 Subsidy /($billions)*
New Zealand	0.9
Australia	1.0
Poland †	2.1
Turkey	6.1
Canada	4.6
Slovakia †	0.3
United States	40.0
Mexico	8.1
Czech Rep †	1.1
Hungary †	1.6
EU	**101**
Japan	43.9
Iceland	0.1
South Korea	18.1
Norway	2.6
Switzerland	5.1

Source: OECD
*Provisional
†1991-93

FIGURE 8-36 Farm subsidies. Among the rich countries, farmers earned about two-thirds more in 1999 than they would have earned if they had sold their produce at world market prices. The OECD's method of measuring agricultural subsidies may not agree with individual countries' methods. (Data from Organization for Economic Cooperation and Development)

districts have never been redrawn to reflect the relative depopulation of the countryside and the urbanization of the populations. The U.S. Senate disproportionately favors farm-state populations, as we shall see in Chapter 11. Thus, farmers enjoy disproportionate political power over urbanites. The contrast is sharp between the rich countries, in which the national governments generally drain wealth from the cities to support farming, and the poor countries, in which governments often heavily tax farmers and subsidize the urban populations.

5. Furthermore, in the rich countries only a small percentage of what urban consumers pay for food actually goes to farmers; most represents value added by processing and packaging. In 1980 the cost of farm commodities in the United States—the raw material—represented 37 percent of the price of food; by 2000 that had fallen to 20 percent. Food processors and middlemen, including marketing and advertising, take an ever-larger share. For example, the cost of the corn in a box of corn flakes is only about 6 percent of the retail cost of the box. The rest of the retail price represents processing, packaging, distributing, and advertising the cereal. Advertising rates affect the retail price of cornflakes more than corn prices do. Therefore, most of the increase in urban food costs is attributable to these activities, not to farmers' profits. Most urban consumers remain unaware of how much the prices they pay for food are inflated to subsidize their nation's farmers. They may also be largely indifferent because the cost of food as a percentage of urban incomes in most rich countries is falling anyway, due to increases in agricultural productivity.

The American diet has actually become so rich and so processed that the country has experienced periodic trade deficits in food products since 1998. That is, the United States has been importing more food—by value—than it has been than exporting. This is astounding for a country so rich in agricultural resources. The reason for this is that America exports basic foodstuffs (wheat, for example), but Americans are importing food products of ever-higher value added (notably wines and out-of-season fruits and vegetables).

The rich countries periodically negotiate to end their subsidies, but they cannot even agree on what constitutes a subsidy. The United States heavily subsidizes water for California farmers but denies that that is a "farm subsidy." Throughout California's Central Valley, sprinklers irrigate fields of alfalfa, cotton, and rice—crops more suited to wet lands than to a desert—and more than 50 percent of federally irrigated land in California is devoted to crops that are already in surplus. The issue of farm subsidies is complicated, and

FIGURE 8-37 Food for national security. The desert in the background is a typical landscape of Saudi Arabia, but Saudi Arabia has invested some of its oil income in irrigating the desert to achieve agricultural self-sufficiency. High capital investment in agriculture allows Saudi Arabia to achieve such high productivity that it occasionally dumps surplus food on world markets. This bankrupts farmers in other countries with environments more naturally suited to agriculture. Thus, political and economic considerations can outweigh the role of physical conditions for agriculture. (Courtesy of Aramco World Magazine)

Focus on

THE U.S. FEDERAL AGRICULTURAL IMPROVEMENT AND REFORM ACT OF 1996

In 1996 the U.S. federal government attempted to reform and reduce U.S. agricultural subsidies, but the generally admitted failure of the effort demonstrates how political and economic considerations can overwhelm intentions.

The 1996 Act ended a farm policy in which farmers voluntarily accepted certain government restrictions on farm practices in exchange for subsidies for their crops. These programs began as efforts to ensure farmers a decent living during the Depression in the 1930s. If prices for certain crops dropped below fixed levels, the government made up the difference through deficiency payments. Farmers who agreed to keep land idle and to limit the production of crops could receive government payments. The programs stayed in place through the years, and government payouts for crop subsidies averaged $5.5 billion per year between 1990 and 1996.

The subsidies even redistributed the geography of U.S. farming. Subsidies to the dairy industry, for example, were higher the farther the dairy farm was from Wisconsin, which is the environment most appropriate for dairy farming and the traditional "Dairy State." The intent was to ensure a supply of fresh milk for children everywhere. The subsidies rewarded dairy farming in warm Florida and California, and since the invention of air conditioning for cowsheds, California has become America's leading dairy producer.

Many of the rules and regulations had desirable results, such as encouraging soil conservation and discouraging water pollution. For example, in order to qualify for subsidies, farmers with land classed as "highly erodible" were required to manage that land in ways that would reduce erosion. Highly erodible lands were also targeted in programs that left some cropland idle. Other provisions preserved wildlife habitats. The 1996 legislation continued some of these programs.

By the 1990s, however, critics argued that some government regulations had become rigid and arbitrary. By guaranteeing generous prices for certain crops while at the same time removing cropland from production, the supports may have encouraged ever-increasing uses of fertilizers and pesticides. Furthermore, by making it difficult to switch from one crop subsidy program to another, the policies discouraged experimentation with new crops or new methods and even crop rotation, a good farming practice. Thus, the 1996 program intended to increase productivity, flexibility, and rewards. Skeptics argued that it endangered valuable environmental safeguards and that the elimination of subsidies would accelerate the trend toward capital-intensive industrial forms, eliminating small family-based farming.

The 1996 law covered most major grains but left in place rules to force American consumers to pay prices for tobacco, peanuts, and sugar above international market prices. These three farming interest groups have enormous political power.

The 1996 legislation did not, however, accomplish what its designers had hoped. Prices fell, and Congress provided payments to U.S. farmers that soared to a record $23 billion in 1999, a full 40 percent of farmers' net cash income. Some of the payments provided disaster relief, but huge harvests were a larger factor in the payments. Farmers still get paid for every bushel produced.

In 2002 Congress passed a new bill that actually guaranteed agricultural subsidies of more than $200 billion over a ten-year period, retaining generous subsidies for cotton, rice, wheat, corn, and soybeans, and adding price guarantees for lentils, chickpeas, ginseng, onions, apples, and even catfish. Other provisions seek to preserve farmlands, save wetlands, and improve water quality and soil conservation on working farms. Thus, the new legislation retained farmers' freedom to make planting decisions, but efforts to wean American farmers from subsidies were abandoned; the "revolution" of 1996 was defeated. In an additional effort to protect American producers from rising imports, the law ordered all meat, fish, and produce to be labeled with its country of origin by 2004.

in each country it is a volatile domestic political issue. The overall result, however, is to increase food surplus in the rich countries and to discourage food production in the poor countries. If the rich countries stopped subsidizing their farmers and opened their markets to crops from developing countries, the farmers of the poor countries might be able to sell more food to the rich countries. This action would benefit the poor countries more than direct foreign aid does.

On July 11, 2003, when U.S. President George W. Bush was touring Africa, Amadou Toumani Touré and Blaise Compaoré, the presidents, respectively, of the African countries of Mali and Burkino Faso, published a joint plea to the American people. "Your farm subsidies," they wrote, "are strangling us." Rich countries' subsides to their own cotton producers totaled almost $6 billion in the production year 2002, "distorting cotton prices and depriving poor African countries" of their own ability to market this crop. During that year, they noted, "America's 25,000 cotton farmers received more in subsidies—some $3 billion—than the entire economic output of Burkino Faso. . . . Further, United States subsidies are concentrated on just 10 percent of its cotton farmers. Thus, the payments to about 2,500 relatively well-off farmers has the unintended but nevertheless real effect of impoverishing some 10 million rural poor people in West and Central Africa." The powerful chairman of the U.S. Senate Agriculture Committee, Senator Chad Cochran of Mississippi, however, favors support for U.S. cotton growers. Leaders of many of the world's poor countries could cite the ramifications of rich countries' protectionist policies.

Our study of world food production and availability illustrates many ways that the world's agricultural economy is manipulated and distorted. Thus, we cannot know what quantities of food the world's farmers would be capable of producing if these conditions were eliminated, or what the geography of food production would be. In 1992 George Demko, a former Geographer of the United States, wrote, "All of us could be fed adequately if the world's supply of food were equitably distributed irrespective of cost, politics, hatreds, and corruption." He concluded, however, that "straightening out the system and ending political infighting, disparities in income, and other problems will not occur in our lifetime." Let us hope that this pessimistic conclusion is not accurate.

THE HARVEST OF FISH

Seafood represents about 6 percent of the total human protein intake. Humans harvest four main groups of marine species. *Demersal fish* (bottom dwellers), including cod, haddock, sole, and plaice, tend to concentrate on broad continental shelves, especially of the North Atlantic. *Pelagic fish* (surface dwellers) include herring, mackerel, anchovy, tuna, and salmon. These two groups together make up about 72 million tons of catch per year. *Crustaceans*, such as lobsters, shrimps, and other shellfish, provide more than 11 million tons. And *cephalopods*, including octopus and squid, yield about 2 to 3 million tons. The Japanese harvest more cephalopods than any other nation, but cephalopods are important in the diets of Mediterranean peoples and many poor countries. In 2001, world *capture fishery production*, that is, basically, hunting and gathering seafood, totaled 92 million tons, of which 84 million was from marine fishing and 8 million was from inland fishing.

Traditional Fishing

Traditional, or artisanal fishing, may be distinguished from modern fishing. Traditional fishing is concentrated mainly in the developing countries, particularly in Asia and Oceania. Little capital is invested, and incomes are low, but these activities employ about 80 percent of the world's fishermen. Traditional fishing is important in the lives of the majority of the populations of islands and coastal areas, and it directly supports an estimated 30 million people.

Fishing is frequently a dangerous occupation, and it demands skills and local knowledge. Physical risks are great, and so are the risks in terms of income, which can fluctuate greatly. Many traditional fishing areas are characterized by close-knit communities with distinct customs and rituals. Traditional fishermen from Canada to Madagascar to Indonesia share certain hardships and experiences. Yields of fish from these traditional ways feed substantial numbers of people, but they are only a small fraction of the global catch. This way of life is rapidly being undermined and replaced by modern, highly capitalized fishing.

Modern Fishing

Concentrations of aquatic species suitable for commercial harvesting are called **fisheries.** The world's major commercial marine fisheries are shown in Figure 8-38. Between 1950 and 1989 the annual harvest from ocean fishing rose from 22 million tons per year to 100 million but then it slowly slipped down through the 1990s. In 2001 China, Peru, the United States, Japan, Indonesia, Chile, and Russia accounted for more than half of the total marine catch.

Demand for fish and fish products continues to rise. Even as people eat more meat, they create a demand for animal feed supplements, including fishmeal. Almost one-third of the global fish catch now goes into meal and oil, mostly to feed livestock and pets in the rich countries of Western Europe and North America. The Japanese, by contrast, want to consume more fish directly.

Overfishing and depletion of the seas   It is difficult to regulate the exploitation of a resource that no individual owns because each country or individual fishing vessel harvests as much as it can. This often leads to overexploitation and rapid depletion of the resource, a process called *depleting the common*. This has been the case with fish. The United Nations has sponsored periodic conferences on the world's fisheries, and most observers agree that several major fisheries are already fished to their sustainable limits or beyond, and virtually every fishery is at risk. The major fishing nations have enlarged their fleets and developed new technologies to multiply their harvest: New electronic devices, for example, discover and track wholly new groups of fish. So-called factory ships are equipped with huge drag nets that catch vast numbers of fish and other sea life indiscriminately. In 1999 the world's fishing nations agreed to reduce the size of their fishing fleets, and the World Summit on Sustainable Development held in South Africa in 2002 called for restoring stocks by 2015, but governments have been slow to act.

The World's Major Fisheries

FIGURE 8-38 This map shows each major fisheries' contribution to the total world catch. Many of these are currently fished to their maximum sustainable yield. (Data from the FAO)

We do not fully understand fish populations and their migrations. Systematic fisheries research began only in 1902, with the formation of the International Council for Exploration of the Sea, which studies North Atlantic species. Today a great many international fishery commissions and advisory bodies exist. Some only consult, but others set quotas as well as conduct research. Some factors in changing fish populations are caused by human interference, but some are natural. Changes in water temperature and salinity can wipe out certain stocks, especially the small pelagic species such as sardines, anchovies, pilchard, and capelin.

Overfishing is not the only problem. Marine animals rely on coastal wetlands, mangrove swamps, or rivers for spawning grounds, and the world's wetlands and coasts are being destroyed by pollution and overdevelopment. About a third of the world's urban population lives within 60 kilometers (100 miles) of the sea, contributing to the pollution that reaches the seas. Heavy metals have contaminated fish and damaged the health of people who eat them. Sewage, fertilizers, and runoff from agriculture have overfed algae (tiny marine plants), causing them to grow so rapidly that they use up the oxygen that fish need to breathe.

The depletion of atmospheric ozone lets in ultraviolet radiation that harms life in the sea. Ultraviolet radiation may be responsible for the observation that the amount of plankton (microscopic animal and plant life

that form the lowest link of the food chain) has decreased by as much as 20 percent near the surface of Antarctic waters, and the surface is where most marine growth and reproduction take place. Global warming might also alter ocean currents.

International law and regulation of fishing   International talks focus on two categories of fish: species like pollock in the Bering Sea and cod off Canada's eastern coast, whose wanderings cause them to straddle territorial and international waters, and fish like tuna, swordfish, and billfish, whose seasonal migrations cover thousands of miles. Coastal states dispute rights to these species with countries that operate long-range distant-waters fleets. The coastal states claim that factory ships in international waters are destroying the stocks in their territorial waters. Others argue that the crisis stems from coastal countries' mismanagement. Coastal countries want binding rules to regulate catches in international waters; distant-water countries want nonbinding guidelines drawn up regionally.

About 90 percent of the world's marine fish harvest is caught within 370 kilometers (200 nautical miles) of the coasts, and therefore many coastal countries have recently extended their claims out into the sea. The 1982 U.N. Convention on the Law of the Sea authorizes each coastal state to claim a 200-nautical-mile **exclusive economic zone (EEZ),** in which it controls both mining

FIGURE 8-39 Aquaculture of salmon. These are coastal salmon pens near Chile's salmon-farming center, Puerto Montt. Chile is the second-largest producer of farmed salmon in the world, after Norway. Chile's favorable growing conditions—clean waters, mild climate, long daylight hours, abundant feed, and absence of native salmon species—make it an ideal area to grow salmon. Most of the salmon grown in Chile—primarily Atlantic and coho—are exported to the United States and Japan. (Courtesy of the Association of Chilean National Tourist Board)

and fishing rights (see Chapter 13). Many countries have already restricted foreign fishing vessels in their waters.

Aquaculture Global seafood consumption is on the rise, while the catch from the oceans has peaked. Therefore, humankind has begun to shift from fish capture to **aquaculture,** which involves herding or domesticating aquatic animals (Figure 8-39). By 2001 the world yield from aquaculture was 38 million tons, about half as much as from fish capture, but this percentage is rising rapidly. A full two-thirds of the aquaculture yield is in China. Fish farmers can monitor both the purity of the

aquatic environment and what the fish are fed. Fish waste is a valuable fertilizer. Aquaculture technology is advancing (computer-controlled feeding, for example), and experiments with different species are revealing which are most productive in aquaculture. Mahimahi, for example, can be grown from fingerlings to 1 to 2 kilograms (2 to 5 pounds) in just five months.

Large-scale experiments have also begun with mariculture, which is fish farming in huge cages in the seas. Aquaculture and mariculture offer promising frontiers for the production of high-protein food.

CONCLUSION: CRITICAL ISSUES FOR THE FUTURE

Technology creates the potential for increasing food production, but it cannot guarantee that this will occur or that the food will be produced or distributed where it is needed. Technology is not spread around the whole world, nor is it distributed in the most economic way. Furthermore, agricultural research and development are occurring overwhelmingly in the private sector—for profit. The fruits of the green revolution emerged almost entirely from public-sector laboratories and national breeding programs, allowing easy access for users. The development of new biotechnologies, however, has come by private companies that have shielded their investments with patents. This may enhance the productivity of rich farmers, but not that of the farmers in the poor countries. How can they get this expensive technology? The private corporations that invest in the development of technology cannot give it away. The governments of the rich countries have not indicated willingness to buy the technology in order to provide it to the poor as foreign aid, and the governments of the poor countries are not making significant investments, either. If, then, we rely on technology to boost food supplies, who will pay to transfer the technology to the world's poor farmers?

Chapter Review

SUMMARY

In the 200 years since Thomas Malthus wrote, global starvation has not occurred. This has been because of the farming of more areas of the planet, the transplantation of many food crops, improvements in worldwide transportation and the technology of food storage, the introduction of higher-yielding and hardier strains of crops, new pesticides, improvements in livestock, and farm machinery. These developments are called the green revolution and the scientific revolution in agriculture.

Today ten principal types of agriculture can be identified: four of these are basically subsistence (nomadic herding, low-technology subsistence farming, intensive rice farming, and Asian mixed cereal and pulse farming), and five are basically commercial (mixed farming with livestock, prairie cereals, ranching, Mediterranean, and plantation agriculture). Irrigated farming can be either subsistence or commercial. The richer a nation becomes, the more total grains it consumes, but the higher the percentage of its consumption that is usually in the indirect form of meat and dairy products. Humankind overall could greatly increase food supplies by concentrating on raising the most efficient animals. Some areas that are remote from urban centers may specialize in dairy farming and add value to milk by producing dairy products.

Variation in agricultural yields worldwide are partly attributable to variations in the physical environment, but today capital input is also important. Biotechnology is opening new possibilities, but some observers fear possible injurious ramifications of tampering with plant and animal biology.

Technology creates the potential for increasing food production, but technology is not spread around the whole world, nor is it distributed in the most economic way. Prob-

lems in the effort to increase food supplies include lack of financial incentives for farmers, unsuccessful land redistribution, and inequitable land ownership.

Many rich countries have erected tariff walls against food imports, and they subsidize their own farmers. These subsidies contribute to production of agricultural surpluses in many rich countries and discourage production in the poor countries. We cannot know what quantities of food the world's farmers would be capable of producing or what the geography of food production would be if these conditions were eliminated.

The world's harvest of fish has significantly increased Earth's per capita food supplies, but future increases from fish capture are dubious. Aquaculture offers new possibilities.

KEY TERMS

aquaculture p. 355
biotechnology p. 341
cloning p. 341
commercial agriculture p. 322
diminishing returns p. 346
double cropping p. 328
downstream activities p. 337
ejido p. 348
exclusive economic zone (EEZ) p. 354

fishery p. 353
gene splicing (recombinant DNA) p. 341
genetic engineering p. 321
genetically modified (GM) p. 341
green revolution p. 321
isotropic plain p. 338
monoculture p. 325
polyculture p. 323

privatize p. 348
sawah p. 328
scientific revolution in agriculture p. 322
subsistence agriculture p. 322
swidden p. 327
value added by manufacturing p. 337

QUESTIONS FOR REVIEW AND DISCUSSION

1. What are the world's principal types of low-technology subsistence agriculture, and where can each be found?

2. What current conditions in many sub-Saharan African countries frustrate attempts to raise more food?

3. How have humans forestalled a "Malthusian" catastrophe for the past 200 years?

4. Why do rich countries subsidize agriculture? What effect does this have on the urban populations of the rich countries, and what effect does it have on the farmers in the poor countries?

5. Why are countries increasingly clashing over the rights to fishing at sea?

THINKING GEOGRAPHICALLY

1. If fewer and fewer farmers can produce all the food the world needs, what will the rest of the people do for a living?

2. Why is humankind just now domesticating fish thousands of years after we managed to think of domesticating animals?

3. What percentages of the grain you consume do you consume directly and indirectly?

4. How does your diet differ from that of your parents or grandparents? Compare the amount of meat or fish you eat. How much more processed is the food you buy? Compare the price of the ingredients of a cake with the cost of a cake.

SUGGESTIONS FOR FURTHER LEARNING

Agricultural Statistics. Washington, D.C.: U.S. Department of Agriculture, annual.

Agricultural Yearbook. Washington, D.C.: U.S. Department of Agriculture, annual.

Altieri, Miguel, and Clara Nicholls. Biodiversity and Pest Management in Agroecosystems. New York: Food Products Press, 2nd ed., 2003.

Altieri, Miguel. Genetic Engineering in Agriculture. New York: Food First Books, 2001.

Bains, William. Biotechnology from A to Z. New York: Oxford University Press, 1998.

Barraclough, Solon, and Krishna Ghimire. Agricultural Expansion and Tropical Deforestation: International Trade, Poverty, and Land Use. New York: Earthscan Publishers, 2000.

Conway, G. "Food for all in the twenty-first century." *Environment* 42, no. 1 (2000): 8–18.

Crosby, Alfred W. *Germs, Seeds and Animals: Studies in Ecological History.* Armonk, NY: M.E. Sharpe, 1994.

Cramer, Gail, Clarence Jensen, and Douglas Southgate. *Agricultural Economics and Agribusiness,* 8th ed. New York: John Wiley & Sons, 2000.

Evenson, R. E., and D. Gollin. "Assessing the impact of the Green Revolution, 1960 to 2000." *Science* 300 (2003): 758–762.

Grant, Wyn, and John Keeler, eds., *Agricultural Policy* (The International Library of Comparative Public Policy, 14). New York: Edward Elgar Publishers, 2000.

Grigg, David. *An Introduction to Agricultural Geography,* 2nd ed. New York: Routledge, 1995.

Hoag, Dana. *Agricultural Crisis in America: A Reference Handbook.* New York: ABC-Clio, 2000.

Kiple, Kenneth F., and Krienhild C. Ornelas, eds. *The Cambridge World History of Food.* 2 vols. New York: Cambridge University Press, 2000.

Knutson, Ronald, and J. B. Penn. *Agricultural and Food Policy,* 5th ed. Upper Saddle River, NJ: Prentice Hall, 2003.

Krueger, Anne, Maurice Schiff, and Alberto Valdés, eds. *The Political Economy of Agricultural Pricing Policies.* vol. I, *Latin America* (1991); vol. 2, *Asia,* 1991; vol. 3, *Africa and the Mediterranean,* 1992; vol. 4, *Synthesis of the Economics in Developing Countries,* 1992; vol. 5, *A Synthesis of the Political Economy in Developing Countries,* 1992. Baltimore, MD: Johns Hopkins University Press (for the World Bank), 1991–1992.

Laegreid, M., O. C. Kaarstad, and O. Bockman. *Agriculture, Fertilizers and the Environment.* New York: CABI Publishers, 2000.

Lobell, D. B., and G. P. Asner. "Climate and management contributions to recent trends in U.S. agricultural yields." *Science* 299 (2003): 1032.

Lurquin, Paul. *High Tech Harvest: Understanding Genetically Modified Food Plants.* New York: Westview Books, 2002.

Macrae, R., R. K. Robinson, and M. J. Sadler, eds. *Encyclopedia of Food Science, Food Technology and Nutrition.* 8 vols. San Diego: Academic Press, 1993.

National Research Council. *Lost Crops of the Incas: Little-known Plants of the Andes with Promise for Worldwide Cultivation.* Washington, D.C.: National Academy Press, 1989.

National Research Council. *Lost Crops of Africa.* Washington, D.C.: National Academy Press, 1996.

Paarlberg, R. "Genetically modified crops in developing countries: promise or peril?" *Environment* 42, no. 1 (2000): 19–29.

Pringle, Peter. *Food, Inc: Mendel to Monsanto.* New York: Simon & Schuster, 2003.

Qaim, M., and D. Zilberman. "Yield effects of genetically modified crops in developing countries." *Science* 299 (2003): 900–902.

Robbins-Roth, Cynthia. *From Alchemy to IPO: The Business of Biotechnology.* New York: Perseus Press, 2000.

Sanchez, Pedro. "Soil fertility and hunger in Africa." *Science* 295 (2002): 2019–2020.

Santaniello, W., V. Santaniello, and R. E. Evenson, eds. *Agriculture and Intellectual Property Rights: Economic, Institutional, and Implementation Issues in Biotechnology.* New York: CABI Publishers, 2000.

Sokolov, Raymond. *Why We Eat What We Eat: How the Encounter Between the New World and the Old Changed the Way Everyone on the Planet Eats.* New York: Summit Books, 1992.

Valdes, Alberto, ed. *Agricultural Support Policies in Transition.* Washington, D.C.: World Bank, 2000.

Winston, Mark L. *Travels in the Genetically Modified Zone.* Cambridge, MA: Harvard University Press, 2003.

World Agriculture: Situation and Outlook Report. Washington, D.C.: U.S. Department of Agriculture, Economic Research Service, quarterly.

Zohary, Danile, and Maria Hopf. *Domestication of Plants in the Old World: The Origin and Spread of Cultivated Plants in West Asia, Europe, and the Nile Valley.* New York: Oxford University Press, 1988.

WEB WORK

For information on the Web regarding food and agriculture, you might start with the World Agricultural Information Center of the United Nations Food and Agriculture Organization

http://www.fao.org/

and its statistical index, called faostat, is at

http://apps.fao.org/

the home page of the Cartagena Protocol on Biosafety is at

http://www.biodiv.org/biosafety/

The United States Department of Agriculture has a home page

http://www.usda.gov/

Information about the Department's National Organic Program is at

http://www.ams.usda.gov/nop/indexIE.htm

The Iowa State University's College of Agriculture maintains a directory of sites related to agriculture

http://www.ag.iastate.edu/other.html

For information from an environmental point of view, try the home page of Resources for the Future

http://www.rff.org/

and The World Resources Institute

http://www.wri.org/

For general libraries of information about economics, where you can focus your research about agriculture, look at the World Wide Web Virtual Library in Economics

http://netec.wustl.edu/WebEc/WebEc.html

and Econolink

http://www.progress.org/econolink/

A Look Ahead

WHAT IS A NATURAL RESOURCE?

A natural resource is something that is *useful* to people. Usefulness is determined by a mix of cultural, technological, and economic factors in addition to the properties of a given resource. We often substitute one resource for another as our needs change.

MINERAL AND ENERGY RESOURCES

Mineral resources include metals and nonmetals. As mineral resources are depleted, they are usually replaced with substitute materials. Landfills accumulate used materials and may one day be a source of materials for reuse. Our principal

9

EARTH'S RESOURCES AND ENVIRONMENTAL PROTECTION

◄ The Three Gorges Dam in China under construction. When completed, it will be the largest dam in the world. (AP/Wide World Photos)

energy resources come from three fossil fuels—oil, coal, and natural gas. As fossil fuels are depleted, they are likely to be replaced with nuclear and renewable energy.

AIR AND WATER RESOURCES

Pollution results when a substance is discharged into the air or water faster than it can be dispersed or removed by natural processes. Recycling and pollution prevention are growing in importance as methods for solving pollution problems.

FORESTS

Forests are multiple-use resources that, in addition to providing lumber and paper, are important for recreation, biodiversity conservation, and carbon storage. As world population grows, conflicts among competing uses of forests are increasing.

In 1954, one of the most devastating floods to strike China occurred on the Yangtze River. Thirty thousand people were killed, and 19 million were displaced. Following that flood, Chinese leader Mao Zedong vowed to proceed with plans to build one of the largest dams ever constructed: the Three Gorges Dam. The dam is now under construction. The sluicegates were closed in June 2003, and completion is scheduled for 2009. The dam is designed to control floods, and it will also generate 18,000 megawatts of clean hydroelectricity, boosting China's electric production by 10 percent. This energy source will substitute for pollution-creating coal.

The construction costs are estimated to be several tens of billions of dollars, a hefty burden for a country struggling to meet enormous needs for investment in infrastructure: the roads, railways, and energy and water distribution systems needed in a growing industrial economy. Normally a project of this type would be funded primarily by grants or loans from wealthier countries, either directly or through agencies like the World Bank. In this case, however, the usual funding agencies are steering clear of the Three Gorges project. They have been persuaded by natural scientists and engineers that the environmental impacts of big dams are often severe, and the bigger the dam, the greater the unintended damages. National economic and political considerations loom larger than environmental ones, however, and the Chinese government is pressing ahead with the project. But as with any major resource-use decision, there will be winners and losers.

Everything we consume is extracted from our planet and returned to it in one form or another. As we use Earth's resources of air, water, minerals, energy, plants, and animals, we simultaneously discharge our waste into the environment. As Earth's human population of 6.3 billion nears 10 billion in the twenty-first century, consumption of resources will increase. This expanded population will place tremendous stress on Earth's remaining resources and the ability of the planet's air, water, and land to accommodate human waste.

Consumption and waste vary among cultures and over time. For example, different nations at different times have obtained energy from wood, coal, petroleum, natural gas, running water, nuclear energy, the wind, and the Sun. Issues of natural resource use and environmental quality must be understood in both their physical and human dimensions.

Resource management is exceedingly complex, because each resource varies geographically and physically. Each resource also varies in value, depending on the human factors: culture, technology, beliefs, politics, economics, and style of government. Many resources are publicly controlled, so resource management is a political process.

The issue of the Three Gorges Dam illustrates this complexity. While the immediate justifications for the project are flood control and energy production, the controversy touches on water-quality impacts, untested engineering designs, inundation of agricultural lands and ancient cultural and architectural treasures, displacement of about 1.2 million people, and the benefits of generating pollution-free hydroelectric power in a country that consumes more coal than any other country in the world.

In this chapter, we explore the factors that affect the value of resources. These factors include the physical characteristics of resources and the natural systems in which they exist, the changing technology of resource use, and human value systems. We then consider how changing resource values affect what and how much resources we use. Finally, we look at environmental pollution and resource conflict and management.

WHAT IS A NATURAL RESOURCE?

A **natural resource** is anything created through natural processes that people use and value. Examples include plants, animals, coal, water, air, land, metals, sunlight, and wilderness. Natural resources are especially important to geographers, because they are the specific elements of the atmosphere, biosphere, hydrosphere, and lithosphere with which people interact. Natural resources can be distinguished from human-made resources, which are human creations or inventions, such as money, factories, computers, information, and labor.

We often use natural resources without considering the broader consequences. For example, the burning of oil to generate heat or to power an automobile engine pollutes the atmosphere with exhaust gases. Streams are then polluted with acid rain, which is caused partly by increased nitrogen oxides in the atmosphere. Oil consumption also weighs heavily on international economic and political relations.

Characteristics of Resources

A substance is merely part of nature until a society has a use for it. Consequently, a natural resource is defined by the three elements of society:

- A society's *cultural values* influence people's decision that a commodity is desirable and acceptable to use.
- A society's level of *technology* must be high enough to use the resource.
- A society's *economic system* affects whether a resource is affordable and accessible.

Consider petroleum as an example of a natural resource in North America:

- *Cultural Values.* North Americans want to drive private automobiles rather than use public transport such as trains.
- *Technology.* Petroleum is the preferred fuel in private automobiles because autos are easily powered by gasoline engines.
- *Economic System.* North Americans are willing to pay high enough prices for gasoline to justify removing petroleum from beneath the seafloor and importing it from distant places (Figure 9-1).

The same elements of society apply to the study of any example of a natural resource: rice to the Japanese, diamonds to South Africans, forests to Brazilians, and air quality to residents of Los Angeles. In every case, a combination of the three factors is necessary for a substance to be valued as a natural resource. Differences among societies in cultural values, level of technology, and economic systems help geographers to understand why a resource may be valuable in one place and ignored elsewhere.

Cultural values and natural resources   To survive, humans need shelter, food, and clothing, and we make use of a variety of resources to meet these needs. We can build homes of grass, wood, mud, stone, or brick. We can eat the flesh of fish, cattle, pigs, fish, or mice—or we can consume grains, fruit, and vegetables. We can make clothing from animal skins, cotton, silk, or polyester. Cultural values guide a process of identifying substances as resources to sustain life.

A swamp is a good example of how shifting cultural values can turn an unused feature into a resource. A century ago swamps were seen in the United States as noxious, humid, buggy places where diseases thrived instead of agriculture. Swamps were valued only as places to dump waste or to convert into agricultural land. Eliminating swamps was good, because it removed the breeding ground for mosquitoes while simultaneously creating productive and valuable land.

Net Oil Imports and Exports

NET OIL IMPORTS AND EXPORTS, 2000 (thousands of barrels per day)

Imports
- 1,000 and above
- 100–999
- 0–99

Exports
- 0–499
- 500–999
- 1,000 and above

No Data

MODIFIED GOODE'S HOMOLOSINE EQUAL-AREA PROJECTION

FIGURE 9-1 The largest importers are the wealthy nations of North America and Europe. Many developing nations are also net importers, and although the quantities are much smaller, the cost may be a higher percentage of the total national income for poor countries. (Data from U.S. Energy Information Administration)

During the twentieth century, cultural values changed in the United States. Scientists and environmentalists praised the value of swamps and documented their importance in controlling floods, providing habitat for wildlife, and reducing water pollution. Philosophers increasingly regarded nature as beautiful and praiseworthy. Changing public attitude toward swamps is reflected in our vocabulary: instead of *swamps,* now we use a more positive term, *wetlands.*

As a consequence of cultural change, wetlands now are a valued land resource, protected by law (Figure 9-2). We restore damaged wetlands, create new ones, and restrict activities that might harm them. Even in the Netherlands, where deep-rooted cultural and landscape traditions are based on draining wetlands to convert them into farmland, some of these same farmlands are now being converted back to wetlands to improve water quality and enhance species diversity.

Technology and natural resources   The utility of a natural substance depends on the technological ability of a society to obtain it and to adapt it to its purposes. Metals are elements that are usually heavy, reflect light, can be hammered and drawn, and are good conductors of heat and electricity. A metal ore is not a resource if the society lacks knowledge of how to recover its metal content and how to shape the metal into a useful object, such as a tool, structural beam, coin, or automobile fender.

Earth has many substances that we do not use today, because we lack the means to extract them or the knowledge of how to use them. Things that might become resources in the near future are **potential resources.** As a result of its high level of biodiversity, the tropical rain forest is brimming with plants and animals that North Americans regard as potential resources. New medicines, pesticides, and foods might be developed from these substances. To the indigenous peoples of the Amazon rain forest, some of these plants and

animals are already resources. But deforestation threatens the availability of these resources for indigenous peoples and the potential use of these resources by others even while it creates economic opportunities for Brazilians. By destroying the rain forest, we are diminishing Earth's pool of both current and potential resources.

Human needs can drive technological advances. People living in cold climates invented insulated homes and heating technology. The need to increase the supply of food drove people to develop new agricultural technology.

Because human need drives many technological advances, new technologies may emerge when a resource becomes scarce. New technology for reusing materials is being developed in part because space in landfills has become a scarce resource, especially in large urban areas of relatively developed wealthy countries, where consumption is highest. This scarcity is stimulating development of new methods for reusing and recycling materials. Most waste currently is not reusable. But as we deplete the resource of landfill space, we will make more things recyclable and manufacture new products from waste, and waste materials themselves will become resources.

Economics and natural resources   Natural resources acquire a monetary value through exchange in a marketplace. The price of a substance in the marketplace, as well as the quantity that is bought and sold, is determined by **supply and demand.** Common sense tells us some principles of supply and demand.

- A commodity that requires less labor, machinery, and raw material to produce (bicycle) will sell for less than a commodity that is harder to produce (automobile).
- The greater the supply, the lower the price (corn). The greater the demand, the higher the price (World Series tickets).
- Consumers will pay more for a commodity if they strongly desire it (computer) than if they have only a moderate desire (textbook).
- If a product's price is low (beer and hamburgers), consumers will demand more than if the cost is high (champagne and prime rib).

In general, natural resources are produced, allocated, and consumed according to rules of supply and demand. Water is a good example. In areas where water is plentiful because of high rainfall and low demand—such as northern Minnesota—consumers pay a low price, little more than the cost of pumping it from the nearest well, river, or lake. But in the arid southwestern United States, water rights must be purchased, and scarce water must be carried hundreds of kilometers through pipelines and aqueducts. Thus, prices are generally much higher. Because of these high costs, the

FIGURE 9-2 Wetlands. With the skyline of Sacramento, California, in the background, the wetlands of the Yolo Basin Wildlife Area and other wetlands in the Sacramento Valley provide wildlife habitat, remove pollutants from water, and help control floods. (Bob Galbraith/AP/Wide World Photos)

government may subsidize water provision, and prices paid by consumers may not reflect the true cost.

Many important natural resource problems result from the inability of markets to account for pollution. **Externalities** are exchanges of commodities that take place outside the marketplace and thus have no price attached at the time of the exchange. For example, a coal-fired power plant generates air pollution in the process of producing electricity. The people who buy the electricity pay for it directly. The price they pay reflects their desire to have the electricity and willingness to pay for it. But people downwind from the power plant also receive the pollution, whether they like it or not. The power plant is not charged directly for the privilege of discharging pollution to the atmosphere. This pollution thus constitutes an externality— a hidden cost. If the power plant were charged for its pollution and the people who receive it were compensated, perhaps the power plant would emit less pollution and people would be less offended by it.

> Many important natural resource problems result from the inability of markets to account for pollution.

Example: Uranium In every society, *cultural values, level of technology,* and *economic system* interact to determine which elements of the physical environment are resources and which are not. Uranium is a good example. Until the 1930s, uranium was a resource only because its salts made a pretty yellow glaze for pottery. After German physicists realized that great energy stored in uranium atoms might be released by "splitting" their nuclei (nuclear fission), uranium gained value as a power of weaponry during World War II.

- *Technology.* The United States and its allies, fearing that Germany might develop a nuclear bomb, began what was known as the Manhattan Project, a crash program to develop the bomb technology first. Germany surrendered before the Manhattan Project achieved its goal, but the technology was eventually developed and used to win the war against Japan.
- *Economic System.* After World War II, nuclear technology was applied to generating electrical power. Nuclear-generated electricity was slow to gain acceptance because it was more expensive than alternative sources, but after Middle East petroleum supplies were threatened during the 1970s, more nuclear power plants were built. The cost of nuclear power had little to do with the cost of uranium, which is cheap and abundant. Increases in the price of electricity in 2000 and 2001 and continued tension in the Middle East since then have renewed interest in nuclear power.
- *Cultural Values.* With the construction of nuclear power plants, the public became increasingly concerned about the risks. Following power-plant

accidents at Three Mile Island in Pennsylvania (1979) and Chernobyl in Ukraine (1986), government agencies regulated nuclear plant safety much more closely. Higher safety standards increased the cost of nuclear power at a time when conservation efforts had succeeded in reducing demand for electricity. Fear of the hazards of nuclear power has prevented growth of the industry, even though the hazards are probably less than those routinely accepted in other aspects of life. Orders for new power plants ceased in the United States during the 1980s.

Substitutability

Many natural resources are valued for *specific properties*— coal for the heat it releases when burned, wood for its strength and beauty as a building material, fish as a source of protein, clean water for its healthiness. In most cases, several substances may serve the same purpose, so if one is scarce or expensive, another can be substituted (Figure 9-3). Copper is an excellent conductor of

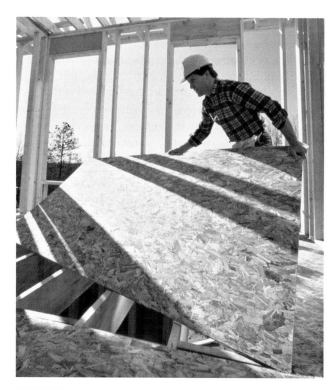

FIGURE 9-3 House construction in the United States. Wood is used in house construction in the United States mostly because it is relatively inexpensive and easy to work with. Here oriented strand board, a manufactured wood product, is being used in new construction. This board is much cheaper than sawn lumber, but has ample strength for covering exterior walls, floors, and roofs. This wood product costs about the same as a foam board made from petroleum that is an alternative to wood-based board in exterior walls. (John Blaustein/Woodfin Camp & Associates)

electricity, but it may be expensive relative to substitute materials. Wire made of other metals can be substituted. For information transmission, such as in computer networks, using light in a fiber-optic cable is much more efficient than using electrons in a copper wire.

The **substitutability** of one substance for another is important in stabilizing resource prices and limiting problems caused by resource scarcity. If one commodity becomes scarce and expensive, cheaper alternatives usually are found. Such substitution is central to our ability to use resources over extended periods without exhausting them and without declines in our standard of living.

However, many resources have no substitutes. There is only one Old Faithful Geyser and only one species of sperm whale, so if we destroy Old Faithful or force extinction of the sperm whale, we have no substitutes waiting in the wings. Other geysers and whales exist, but they are not the same as those we now know. The uniqueness of these resources is the essence of their value.

Renewable and Nonrenewable Resources

In thinking about Earth's resources, we distinguish between those that are renewable and those that are not:

- **Nonrenewable resources** form so slowly that for practical purposes they cannot be replaced when used. Examples include coal, oil, gas, and ores of uranium, aluminum, lead, copper, and iron.
- **Renewable resources** are replaced continually, at least within a human lifespan. Examples include solar energy, air, wind, water, trees, grain, livestock, and medicines made from plants.

Even a renewable resource can be *depleted*, or used to a point where it can no longer be economically used. The only ones we cannot deplete are solar energy and its derivatives: wind and precipitation.

MINERAL AND ENERGY RESOURCES

Mineral resources are substances that we derive from the lithosphere. They are basic materials that we use to construct roads and buildings, manufacture goods, and power transportation. Without minerals, modern industrial societies could not function. Our use of mineral resources in industry and commerce is governed primarily by technology and economics.

We value most minerals for their properties of strength, malleability (ability to be shaped), weight, and chemical reactivity rather than for their aesthetic characteristics. Few car owners care if the engine is made from aluminum or iron—what matters is that it is

powerful, durable, and efficient. Few people care very much whether the roof of their house is made from slate or asphalt shingles, as long as the roof keeps out the rain and doesn't cost too much. Gold has been a rare exception of a mineral valued mostly for its beauty in jewelry, although even gold is increasingly demanded for industrial uses, especially electronics.

Because we value a mineral primarily for its mechanical or chemical properties, our use of mineral resources is continually changing as our technology and economy change. As new technological processes and products are invented, demand can suddenly increase for materials that had little use in the past. When these new processes and products replace older ones, demand may be reduced for minerals important in the past. As a result of changes in consumer demand, remaining supplies, and prices, one mineral may become more favored while another is less desired.

This section discusses metal and nonmetal minerals and energy resources.

Mineral Resources

The terms *Stone Age, Iron Age,* and *Bronze Age* capture the importance of particular minerals at various times in the past. Minerals are as essential to civilization as plant, animal, water, and energy resources. They are present in virtually every product we manufacture, and though their value may be a small part of the total value of a finished good, those products could not be made without minerals (Figure 9-4).

Earth has 92 naturally occurring chemical elements, but most of the crust is made up of 8 elements: oxygen, silicon, aluminum, iron, calcium, sodium, magnesium, and potassium. These elements, as well as rarer ones, combine to form thousands of minerals, each with its own properties and distribution pattern around the world. Each mineral potentially is a resource if people find a use for it.

Metallic minerals, such as copper, lead, silicon, tin, aluminum, and iron, usually occur in ores. Nonmetallic minerals include building stone, graphite, rubies, sulfur, slate, and quartz. Both metallic and non-metallic minerals must be discovered, mined, transported, refined, and manufactured into useful goods.

Variations in Mineral Use

Historically, the use of particular metals and nonmetals has fluctuated between periods of high demand and price and periods of low demand and price. Discovery of a new resource could create a "rush" of people to the area of discovery. The "Gold Rushes" to California, Colorado, and Alaska are nineteenth-century examples. The period between 1970 and 1985 featured especially volatile mineral prices, as a result of rises and declines of industrial output and inflation. For example, the price

U.S. per capita consumption of selected minerals, 1999 (kg)

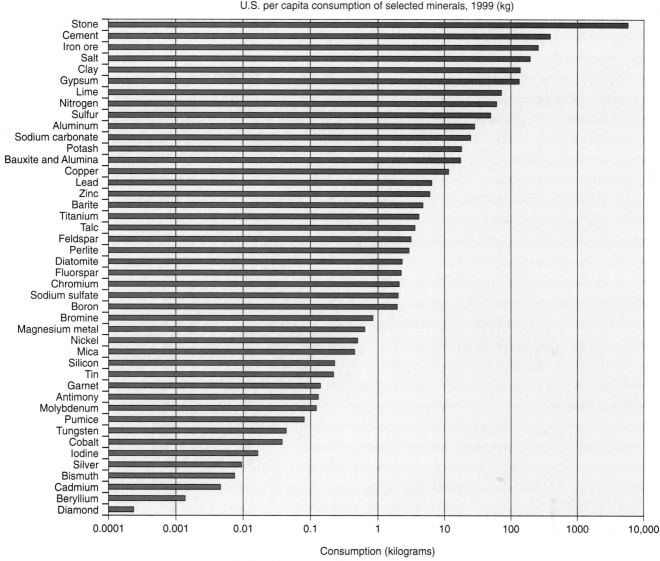

FIGURE 9-4 Annual per capita consumption of nonenergy minerals in the United States, 1999. (U.S. Census Bureau)

of copper doubled between 1973 and 1980, then fell nearly 40 percent between 1980 and 1985. Prices of most minerals have been relatively stable since the mid-1980s.

Mineral deposits are not uniformly distributed around the world. Most of the world's supply of particular minerals is concentrated in a handful of countries. For example, four countries—Australia, Guinea, Jamaica, and Brazil—produce about three-fourths of the world's total aluminum ore (bauxite). Russia, Canada, Indonesia, and New Caledonia together produce about three-fourths of the world's nickel, while China, Indonesia, Peru, and Brazil together produce two-thirds of the world's tin (Table 9-1). The United States, Russia, Canada, China, and Australia are especially rich in metal and nonmetal mineral resources.

The concentration of mineral resources and production in a few countries favors the establishment of **cartels.** A cartel is a group of countries that agree to control the market by limiting production in order to drive up prices. During the 1970s, when world demand for minerals was strong, a few cartels were able to control world markets for brief periods. But weak demand, falling prices, and political instability have limited the strength of cartels in recent years. In addition, the United States—the largest consumer of most minerals—has accumulated a stockpile of important minerals to protect against short-term reductions in supply caused by high prices, political instability, or hostile foreign governments.

> The concentration of mineral resources and production in a few countries favors the establishment of cartels.

TABLE 9-1 Major producers of selected minerals. Countries that produce at least 5 percent of world total production are listed. Note that some countries are listed for several minerals (Australia, Canada, China, Indonesia, Peru, Russia, USA) indicating that they are rich in several minerals.

Mineral	Country	Percent of total world production	Mineral	Country	Percent of total world production
Bauxite and Alumina	Australia	39.0	Lead	Australia	24.7
	Guinea	11.3		China	22.4
	Brazil	9.4		United States	15.5
	Jamaica	9.2		Peru	10.2
	China	7.1		All others	27.2
	India	6.4		**Total**	**100.0**
	All others	17.6	Mercury	Spain	35.7
	Total	**100.0**		Kyrgyzstan	21.4
Cadmium	Japan	13.4		Algeria	15.7
	China	13.1		All others	27.1
	Korea, Republic of	9.9		**Total**	**100.0**
	Mexico	8.0	Nickel	Russia	24.8
	Canada	6.7		Canada	14.2
	Belgium	5.9		Australia	14.1
	Germany	5.3		Indonesia	8.0
	All others	37.7		New Caledonia	7.4
	Total	**100.0**		Cuba	5.5
Copper	Chile	33.2		All others	25.9
	United States	8.4		**Total**	**100.0**
	Indonesia	8.2	Tin	China	34.2
	Australia	6.3		Peru	26.9
	Peru	6.3		Indonesia	18.9
	Russia	5.1		All others	20.0
	All others	32.4		**Total**	**100.0**
	Total	**100.0**	Zinc	Australia	17.1
Gold	South Africa	15.6		China	16.9
	United States	11.9		Peru	12.9
	Australia	11.1		Canada	11.2
	China	6.9		United States	8.3
	Indonesia	6.7		Mexico	5.3
	Russia	6.7		All others	28.3
	Canada	6.3		**Total**	**100.0**
	Peru	5.5			
	All others	44.9			
	Total	**100.0**			
Iron ore	China	20.9			
	Brazil	20.0			
	Australia	17.3			
	Russia	8.0			
	India	7.3			
	Ukraine	5.5			
	All others	21.1			
	Total	**100.0**			

Source: U.S. Geological Survey

Depletion and Substitution

Fluctuations in the price of a mineral as a result of actions by a cartel, a political dispute, or a limited supply rarely continue for a long period of time. If high prices persist for several decades, technological innovations usually result in substitution of cheaper minerals for more expensive ones. For example, when the price of copper rose rapidly in the 1970s, plumbers began to substitute polyvinyl chloride (PVC). Today PVC (a plastic made from petrochemicals) has largely replaced copper pipe for plumbing in new buildings.

The substitution of one mineral for another has an important consequence: Even though world supply of a mineral resource may be limited, we will never run out of it. The reason is that if the supply of a resource dwindles relative to demand, its price will rise. The increase in price has four important consequences:

1. Demand for the mineral will decrease, slowing its rate of depletion.
2. Mining companies will have added incentive to locate and extract new deposits of the mineral, especially deposits that might have been neglected when prices were lower.
3. Recycling will become more feasible.
4. Research to find substitute materials will intensify, and as use of the substitute increases, demand for the scarce mineral will cease before the supply is exhausted.

Although at current rates of use the world would exhaust its remaining supply of lead in 23 years, zinc in 20 years, and copper in 33 years, you should not be worried about running out of these materials in your lifetime. However, if you know that a company has just invented a product that will replace one of these minerals cheaply, consider investing in it!

Disposal and Recycling of Solid Waste

The average American throws away 2.2 kilograms (4.8 pounds) of solid waste per day, twice as much as his/her parents did in 1960. Paper accounts for one-third of all solid waste in the United States. Discarded food and yard waste accounts for another one-third (Figure 9-5). Relatively developed societies generate large quantities of packaging and containers made of paper, plastic, glass, and metal. We dispose of this solid waste in three ways: landfills, incineration, and recycling. Each of these methods poses significant problems, either in environmental degradation or in costs of disposal. Normally the choice of one disposal method over another means that costs are shifted from one group to another, and conflicts are inevitable.

Landfill disposal Over 70 percent of solid waste generated in the United States is trucked to landfills and buried under soil in **sanitary landfills** in which a layer of earth is bulldozed over the garbage each day. They are considered *sanitary*, because burying the garbage reduces emissions of gases and odors from the decaying trash, prevents fires, and discourages vermin. Unlike air and water pollution, which are reduced by dispersal into the atmosphere and rivers, solid-waste pollution is minimized by concentrating the waste in thousands of landfills. However, landfills have been closed in many communities because they can contaminate groundwater, devalue property, or have run out of space. Opening new landfills is difficult because environmental regulations are more stringent today, and local opposition to new landfills is usually overwhelming. The result in many areas has been a solid-waste crisis. Disposal sites are few and costly. Some communities pay to use landfills elsewhere. San Francisco trucks solid waste to

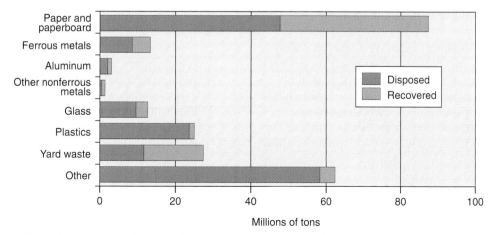

FIGURE 9-5 **Composition of solid waste, 2000.** Paper and paperboard form the largest part of solid waste generated in the United States. About one-third of all paper and paperboard waste is recovered and reused; much of this is cardboard boxes used for shipping. Recovery rates for aluminum are also about one-third. (U.S. Census Bureau)

Altamont, California, 100 kilometers (60 miles) away. Passaic County, New Jersey, hauls waste 400 kilometers (250 miles) west to Johnstown, Pennsylvania.

Incineration   One alternative to landfills is incineration. Incineration reduces the bulk of trash by about three-fourths, and the remaining ash requires far less landfill space. Incineration also provides energy. The incinerator's heat boils water, producing steam that can be used to heat homes or to generate electricity by operating a turbine. More than 100 incinerators now burn about 17 percent of the trash generated in the United States. However, because solid waste is a mixture of many materials, it does not burn efficiently. Burning releases some toxic substances into the air, and some remain in the ash.

Recycling   Recycling solid waste reduces the need for landfills and incinerators and reuses natural resources that already have been extracted. Recycling simultaneously addresses both pollution and resource depletion. Most U.S. communities have instituted some form of mandatory or voluntary recycling.

Several barriers to recycling must be surmounted: separating waste, consumer resistance, market, cost, and indirect loss.

- *Waste Separation.* Solid waste comprises a variety of materials that must be separated to recycle (Figure 9-6). Metals containing iron can be pulled out of the pile magnetically; but paper, yard waste, food waste, plastics, and glass cannot be separated easily from each other. Consequently, many communities require consumers to practice source separation of solid waste. Typically, consumers must place newspaper, glass, plastic, and aluminum in separate containers for pickup. Each type is collected on a different day or by a different truck, and it is shipped to a specialized processor. The procedure is generally more expensive for the community than picking up all the trash together.

- *Consumer Resistance.* Separation for recycling is a nuisance for people in relatively rich developed countries used to throwing things away. To encourage recycling, some communities charge high fees to pick up nonrecyclables, but take recyclables at little or no charge. Bottle and can laws requiring a deposit on beverage containers have been enacted in many states to encourage recycling and reduce roadside litter.

- *Lack of Market.* To succeed, recycled products must have a market. The lack of an assured market for many recycled products is perhaps the most difficult obstacle to increased recycling. Demand for recycled goods among industries and consumers is uncertain. For example, mixed plastics can be used as a substitute for wood in

FIGURE 9-6 Resource recovery. Resource recovery from solid waste requires sorting and separating different types of materials. One approach is to collect them together and separate them at a centralized facility like this one. Alternatively they can be separated at the source (home, school, work) and transported separately to reprocessing facilities. (Gabe Palmer/Kane/Corbis/Stock Market)

products such as picnic tables and playground equipment. The price of wood is not very different from that of recycled plastic, however, and for aesthetic reasons most consumers prefer wood. This helps keep the market for recycled plastic small. To increase consumer acceptance, the quality of recycled material must also improve. Poor quality has resulted from the fact that recycling became mandatory before proper manufacturing methods had been developed.

- *Hidden Costs.* Recycling involves far more than just "melting down and reshaping." For example, to recycle paper, ink must be removed, an additional step compared to conventional papermaking, and therefore an added cost. It is difficult to remove all of the ink, so recycled paper is unacceptable for some uses because it is too gray or speckled. Removing ink may create some pollution, though probably less than would be produced in making virgin paper. The relatively high cost of processing and the lower value of the products results in limited markets for some recycled materials.

- *Indirect Losses.* Trash burns only if it contains enough combustibles. If paper is recycled and yard waste is composted instead of being thrown in the trash, the trash may be difficult to burn. As recycling has increased during the 1990s, some communities have not had enough combustible waste to operate their incinerators.

Many companies are developing manufacturing methods and packaging that facilitate recycling. To reduce packaging volume, detergent is being sold in concentrated form. Refillable containers are available for more products. The 3M Corporation is producing a desk organizer designed to be recycled; when thrown away, its snap-together parts can be easily separated for recycling. In Germany, manufacturers of some goods—including motor vehicles—are required to accept their products back when their useful lives are over. Such regulations are costly, but they encourage manufacturers to design recyclability into their products.

One alternative is to require consumers to bear a greater share of the cost of waste generation. If those costs are clearly associated with the products that generate the wastes, then consumers have an incentive to reduce waste. For example, in 2002 the Irish government imposed a tax of about 17 cents on the use of disposable plastic shopping bags. Within a year use of such bags dropped 95 percent, as consumers either reused bags or carried their purchases in other, reusable containers. The portion of plastic bags in litter dropped from 5 percent to 0.3 percent.

Landfills: An example of changing resource values

Sanitary landfills are building mountains of trash. The Roosevelt Regional Landfill in south central Washington is an example (Figure 9-7). It is the fourth largest landfill in the United States and receives about 35 percent of the State of Washington's solid waste.

FIGURE 9-7 The Roosevelt Regional Landfill, Washington. This landfill is the site of an innovative program to convert waste to electricity, reducing the need for combustion of fossil fuels for that purpose. In addition, the carbon dioxide produced from the landfill will be delivered to greenhouses which otherwise would burn propane to stimulate plant growth by increasing the CO_2 concentrations in the greenhouses. (AP/Wide World Photos)

Garbage dumps are nothing new, as excavations of ancient cities reveal, but they are growing much more rapidly than in the past. Traditionally, food waste was fed to livestock, iron was used in durable goods, packaging was simple, and plastic products were unknown. Far fewer goods were manufactured. Our "throwaway culture" is a modern invention.

Garbage dumps have been sited in swamps or other low-value land. For years garbage simply was dumped and left uncovered. Rubber tires, deliberately burned in landfills to keep fires going, emitted offensive black smoke. Fires smoldered, rats thrived, flies buzzed, and homes situated downwind rarely enjoyed cookouts. Many coastal cities dumped their garbage at sea.

To reduce fire, vermin, and odor, cities have converted open dumps to sanitary landfills, which reduced some environmental problems but aggravated others. Landfilled toxics are leached by groundwater, polluting it. This consequence, plus much publicized problems such as the chemical landfill at Love Canal near Buffalo, New York, heightened public concern. In response to these concerns, the U.S. government has passed several laws and regulations that control handling of toxic substances.

As landfills run out of space, new ones are difficult to site because of the *NIMBY*—Not in My Back Yard—phenomenon. Most people agree that landfills are needed, but few want them near their homes. Stricter regulations have forced some landfills to close and have prevented others from opening. The number of operating landfills in the United States declined from about 30,000 in 1976 to just over 6,000 in 1987. Since that time the few remaining landfills have grown higher and higher as they accept the waste that formerly went to other sites.

At the Roosevelt Regional Landfill an innovative project is underway in which garbage is burned to produce electricity. The CO_2 produced in the process is sold to nearby greenhouses that formerly burned propane to increase the CO_2 levels in the greenhouses to accelerate plant growth. This is encouraging: An industry that disposes of unwanted things also sells a commodity that people need. In this case, not only are we producing electricity but doing it in a way that reduces greenhouse gas emissions. Today landfills are a scarce resource that must be used carefully. To slow the rate of filling, many landfills no longer accept grass clippings and leaves, which can be composted (decomposed harmlessly) at home, or any materials that can be recycled.

Energy Resources

Earth has bountiful and varied sources of renewable energy that humans have been able to harness:

- Solar energy comes from the Sun.
- Hydroelectric power and wind power come from natural movements of water and air caused by solar energy and gravity.

- Geothermal energy comes from Earth's internal heat in volcanic areas, including California, Iceland, and New Zealand.

However, most of the world's energy comes from chemical energy stored in such substances as wood, coal, oil, natural gas, alcohol, and manure. Energy is released by burning these substances. People burn these substances to heat homes, run factories, generate electricity, and operate motor vehicles. Most of our energy comes from fossil fuels, and burning them reduces their supply—they are nonrenewable resources. We can continue to burn them for another few decades, but eventually we must switch to new energy resources if we are to preserve our current standard of living.

Energy from Fossil Fuels

Oil, natural gas, and coal, known as **fossil fuels,** come from the residue of plants and animals buried millions of years ago. Through photosynthesis, plants convert solar energy to the chemical energy stored in their tissues. When plants die, this energy remains in their tissues. The energy may be released promptly if animals or decomposers consume it or if a fire burns the field or forest. Or the plants may become fossilized as coal, and the stored energy may wait millions of years to be released when the coal is mined and burned.

Oil and gas are also stored sunlight. When we burn these fossil fuels today, we are releasing the energy originally stored in microscopic plants (phytoplankton) millions of years ago. When animals eat plants, they store the energy incorporated into the plant tissues. As the sea's countless creatures died over thousands of years, their bodies sank to the bottom, creating an organic sediment. Over time, this was converted to oil, accompanied by natural gas. Coal, oil, and natural gas still are being created, but the processes are so slow that from a human perspective, fossil fuels are nonrenewable resources: Once burned, they are gone forever as useful sources of energy.

From wood to coal, oil, and gas From the time that humans first lived in North America—probably at the end of the last Ice Age about 18,000 years ago—until the mid-1800s, wood was the most important source of energy. Prior to the arrival of European colonists, North American residents used wood almost exclusively, but because total population was small, they did not significantly deplete the resource. When they arrived in North America beginning in the seventeenth century, Europeans harvested the forests for fuel and lumber and cleared the land for agriculture. By the end of the nineteenth century, most of the forests near populated areas of the eastern United States had been cut down, and fuel wood became very expensive. It was also inadequate for providing the large amounts of energy demanded by a growing industrial economy. Wood

still provides the largest portion of energy in developing countries, though supplies are dwindling in areas of dense, rural populations, such as in East Africa and southern Asia.

Coal served as a substitute for fuel wood during the nineteenth century. Although large amounts of coal have been consumed, abundant supplies still remain in the United States and several other countries. Another fossil fuel, oil, was a minor resource until the diffusion of motor vehicles early in the 1900s. Today it is the world's most important energy resource. A third fossil fuel, natural gas, once was burned off during oil drilling as a waste product because it was too difficult to handle, and markets for it were not established. In recent years, however, it has become an important energy source. Today these three fossil fuels provide more than 80 percent of the world's energy and more than 90 percent in relatively developed countries (Figure 9-8).

For U.S. and Canadian industry today, the main energy resource is natural gas, followed by oil and coal. Some businesses directly burn coal in their operations, while others rely on electricity generated primarily at coal-burning power plants. At home, energy is used to generate heat and hot water and to operate diverse electrical devices. Natural gas is the most common source for home heating, followed by petroleum and electricity. Nearly all transportation systems operate on petroleum products, including automobiles, trucks, buses, airplanes, and most railroads. Only subways, streetcars, and some trains run on electricity, much of which is generated from burning fossil fuels.

Distribution of fossil fuels Fossil fuels are not uniformly distributed beneath Earth's surface (see

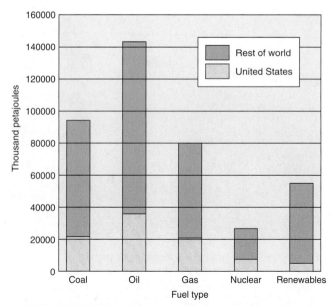

FIGURE 9-8 World energy sources. Fossil fuels make up the major proportion of global energy production; renewable and nuclear energy are relatively minor. (World Resources Institute)

TABLE 9-2 Reserves of Oil, Natural Gas, and Coal, 2000. For each resource only the top 15 countries are listed. The countries listed account for 92 percent of all oil reserves, 84 percent of all gas reserves, and 96 percent of all coal reserves.

CRUDE OIL			NATURAL GAS			COAL		
Country	Billion barrels	Percent of world total	Country	Trillion cubic feet	Percent of world total	Country	Million short tons	Percent of world total
Saudi Arabia	262	25	Russia	1680	31	United States	273656	25
Iraq	113	11	Iran	812	15	Russia	173074	16
United Arab Emirates	98	9	Qatar	509	9	China	126215	12
Kuwait	97	9	Saudi Arabia	220	4	India	93031	9
Iran	90	9	United Arab Emirates	212	4	Australia	90489	8
Venezuela	78	8	United States	183	3	Germany	72753	7
Russia	49	5	Algeria	160	3	South Africa	54586	5
Libya	30	3	Venezuela	148	3	Ukraine	37647	3
Mexico	27	3	Nigeria	124	2	Kazakhstan	37479	3
China	24	2	Iraq	110	2	Poland	24427	2
Nigeria	24	2	Turkmenistan	101	2	Yugoslavia	17919	2
United States	22	2	Indonesia	93	2	Brazil	13149	1
Qatar	15	1	Australia	90	2	Colombia	7328	1
Norway	9	1	Malaysia	75	1	Canada	7251	1
Algeria	9	1	Uzbekistan	66	1	Czech Republic	6259	1
All others	87	8	All others	875	16	All others	47996	4
World Total	**1032**		**World Total**	**5457**		**World Total**	**1083259**	

Source: Energy Information Administration, *Oil and Gas Journal*

Table 9-2). Mineral *reserves* are known deposits that can be extracted profitably using current technology. Some regions have abundant reserves, whereas others have none. This distribution reflects how fossil fuels are formed. Coal forms in swampy areas, rich in plants. The lush tropical wetlands of 250 million years ago are today the coal beds of the world, relocated to midlatitudes by the ponderous movement of Earth's tectonic plates (see Chapter 3). Oil and natural gas form in seafloor sediment, but Earth's tectonic movements eventually elevate some seafloor above sea level to become land. Today we drill for petroleum on both land and the seafloor.

Because minerals are only available in specific geologic environments their distribution is very uneven. Table 9-2 reveals that, like the minerals listed in Table 9-1, fossil fuel resources are highly concentrated in a few places. For each of the fossil fuels listed, the top five countries hold more than 60 percent of world reserves.

Relatively wealthy developed countries—which comprise about one-fourth of the world's population—possess more than 50 percent of the world's coal and more than 40 percent of natural gas. China is the leading producer of coal, but the United States is also a

major coal producer and user. Russia is the leading producer of natural gas. Other than China and India, which have significant coal deposits, most countries in Africa, Asia, and Latin America have few reserves of coal or natural gas.

The distribution of oil is somewhat different. Industrialized countries possess only 12 percent of the world's oil reserves. Two-thirds of the world's oil reserves are in the Middle East, including one-fourth in Saudi Arabia. Mexico and Venezuela also have extensive oil fields.

North America and Europe have much higher per capita oil consumption rates, so they account for nearly two-thirds of the world's energy consumption. The United States, with less than 5 percent of the world's population, consumes nearly 20 percent of the world's commercial energy. In North America and Europe, the high level of energy consumption supports a lifestyle rich in food, goods, services, comfort, education, and travel.

North America and Europe account for nearly two-thirds of the world's energy consumption.

Because relatively rich developed countries consume more energy than they produce, they must import energy, especially oil, from developing countries. The United States imports roughly half of its needs, Western European countries more than half, and Japan

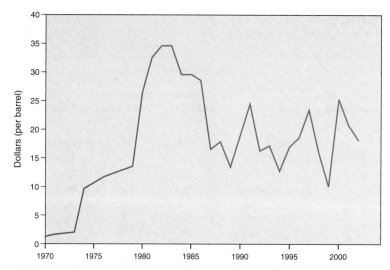

FIGURE 9-9 Crude oil prices, 1968–2002. Crude oil prices vary depending on the characteristics of the oil. The price of Saudi Light Crude, a benchmark commodity, jumped sharply in 1973–74 and 1979–81, but dropped dramatically in the mid-1980s. Price fluctuations since the late 1980s have mostly been the result of changing economic conditions around the world. The effects of the 1998 financial collapse in southeast Asia are particularly evident. (Data from U.S. Energy Information Administration)

more than 90 percent. U.S. dependency on foreign oil began in the 1950s, when oil companies determined that the cost of extracting domestic oil had become higher than foreign sources. U.S. oil imports have increased from 14 percent of total consumption in 1954 to about 50 percent in 2002. European countries and Japan increasingly depend on foreign oil because of limited domestic supplies.

Oil production and prices   Early in the twentieth century, the United States was an oil exporter, but its needs soon exceeded domestic supplies. Europe has always been a net importer of oil. In their search for cheaper oil, U.S. and Western European companies drilled for oil in the Middle East and sold it inexpensively to consumers in relatively developed countries. Western companies set oil prices and paid the Middle Eastern governments only a small percentage of their oil profits. To reduce dependency on Western companies, the countries possessing oil created the Organization of Petroleum Exporting Countries (OPEC) in 1960. OPEC members include eight countries in the Middle East (Algeria, Iran, Iraq, Kuwait, Libya, Qatar, Saudi Arabia, and United Arab Emirates), plus Venezuela in South America, Gabon and Nigeria in Africa, and Indonesia in Asia.

Angry at the United States and Western Europe for supporting Israel during its 1973 war with the Arab states of Egypt, Jordan, and Syria, OPEC's Arab members organized an oil embargo during the winter of 1973–74. OPEC states refused to sell oil to countries

that had supported Israel. Soon gasoline supplies dwindled in relatively developed countries, and prices at U.S. gas pumps soared. Another oil crisis followed in 1979 as a result of a revolution in Iran. Oil prices went from $3 per barrel (42 gallons, or 160 liters) in 1974 to more than $35 in 1981 (Figure 9-9). To import oil, U.S. consumers spent $3 billion in 1970 but $80 billion in 1980, nearly 3 percent of Gross National Income (GNI). This rapid price increase abruptly halted the rise in energy use in the United States and elsewhere (Figure 9-10) and caused severe economic problems in relatively developed countries during the 1970s and 1980s.

In each of these oil crises, long lines formed at gas stations in the United States, and some motorists waited all night for fuel (Figure 9-11). In some cases gasoline was rationed by license plate number; cars with licenses ending in an odd number could buy only on odd-numbered days. Some countries took more drastic action; the Netherlands banned all but emergency motor vehicle travel on Sundays.

Developing countries were especially hurt by the price rises. They depended on low-cost oil imports to spur industrial growth and could not afford higher oil prices. Relatively wealthy developed countries lessened somewhat the impact of higher oil prices on their economies by encouraging OPEC countries to return some of their money by investing it in real estate, banks, and other assets in North America and Western Europe. Poorer nations could not offer this opportunity for reinvesting oil wealth.

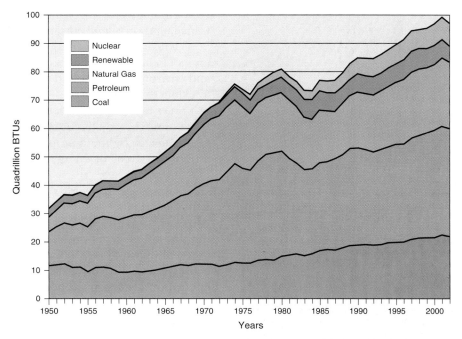

FIGURE 9-10 Trends in U.S. energy use, 1950–2002. Fossil fuels make up the major proportion of energy consumed in the United States. Use of oil and natural gas increased rapidly from the 1950s to the 1970s, but growth has been much slower since the energy crises of 1974 and 1979. (Data from U.S. Energy Information Administration)

But no sooner had the world begun to adjust to high energy prices than the trend was reversed. Energy conservation measures in consuming countries reduced demand. Internal conflicts weakened OPEC's influence: Iraq and Iran fought a war that lasted for eight years, and in the early 1980s some countries broke ranks with other OPEC countries and reduced oil prices. The price of oil plummeted from more than $30 to less than $10 per barrel. Exporting countries flooded the world

FIGURE 9-11 Gas line in 1970s. Cars lined up for gas in Los Angeles during the oil crisis in the 1970s. (Craig Aurness/Woodfin Camp & Associates)

market with more oil in an unsuccessful attempt to maintain the same level of revenues as they had received during the 1970s. While supplies increased, demand in relatively developed countries remained lower than before the boycott. Oil prices briefly rose to about $40 during the 1991 Persian Gulf War but then quickly dropped back to about $20. In early 1998, following a significant drop in oil prices, OPEC members agreed to reduce production in an attempt to drive prices back up, and by late 2000 prices were again above $30 per barrel. They fell back to about $20 by late 2001, falling particularly sharply after September 11, 2001, then rose again as tensions mounted in Iraq, peaking at about $35 at the height of the 2003 Iraq war. Political instability in the Middle East is leading the United States to look elsewhere for oil, perhaps hoping to diversify its sources. There is considerable potential for new oil discoveries in many areas, including Africa, central Asia, and southeast Asia (Figure 9-12). Even if the oil-producing countries of the Middle East remain politically stable in the future, world oil prices are not likely to fall significantly because of increasing demand, driven especially by rising oil consumption in the United States.

Future of fossil fuels How much of the fossil fuels remain? Oil, natural gas, and coal occur beneath Earth's surface, so we cannot see them. Geologists can estimate fairly accurately the *proven reserves* of fossil fuels available in fields that have been explored. Geologists disagree sharply on potential reserves of the fossil fuels—the amount in fields not yet discovered and explored.

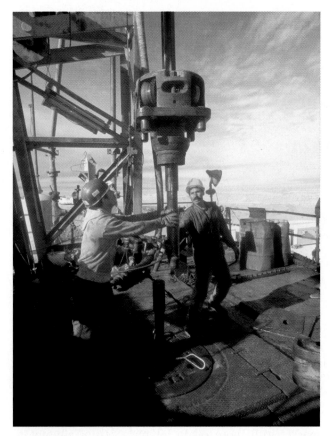

FIGURE 9-12 Oil wells near Hassi-Messaoud, Algeria. Oil exploration and production in Africa and elsewhere outside the Middle East is likely to grow in many parts of the world in coming years. (Parrot Pascal/Corbis/Sygma)

If we divide Earth's current proven oil reserves (about 1000 billion barrels) by current annual consumption (about 24 billion barrels), we get about 40 years of oil supply remaining. Rates of consumption will change and new reserves will be discovered, but proven oil reserves will probably last only a few decades. Thus, unless large potential reserves are discovered, Earth's oil reserves will be significantly depleted during the twenty-first century, possibly in your lifetime.

Every discovery of new deposits extends the life of the resource. But extracting oil is becoming harder—and therefore more expensive. When geologists seek oil, they look first to accessible areas where geologic conditions favor accumulation. The largest, most accessible deposits already have been exploited. Newly discovered reserves generally are smaller and more remote, such as beneath the seafloor, where extraction is costly. Exploration cost also has increased because methods are more elaborate and the probability of finding new reserves is less.

Unconventional sources of oil are being studied and developed, such as *oil shale* and *tar sandstones*. Oil shale is a "rock that burns" because of its tarlike content. Tar sandstones are saturated with a thick petroleum. The states of Utah, Wyoming, and Colorado contain large amounts of oil in shale, more than ten times the

conventional oil reserves found in Saudi Arabia. Canada has very large tar sandstone deposits.

Unconventional energy sources such as oil shale are not being used today, because economically feasible and environmentally sound technology is not currently available to extract the oil. The cost of conventional oil resources must increase dramatically before unconventional sources could contribute significant portions of our needs. By the time prices have risen that high, different sources of energy may be less expensive to obtain than oil.

How long the world's supply of oil will last depends heavily on how we use it today. We can either continue to use it rapidly, as we are now doing, knowing that oil must get more expensive and that we will eventually have to switch to some other fuel. Or we can conserve the oil that we have and switch to substitute energy sources as soon as possible.

If we increase our rate of consumption without finding more fossil-fuel reserves, we will deplete oil resources sooner. In general, since the 1970s, we have made more efficient use of energy. For example, motor vehicles have lighter engines, fewer metal parts, and more efficient transmissions than before the 1970s oil crises. The average fuel efficiency of automobiles in the mid-1990s was about 50 percent greater than it was in the mid-1970s. In the 1990s, however, low gasoline prices helped to increase the popularity of inefficient sport utility vehicles to the extent that the average fuel economy of new vehicles declined in the U.S. in the late 1990s. The technology to improve fuel efficiency is readily available, however, as evidenced by the introduction in 2000 of hybrid vehicles that combine gasoline engines and electric power (Figure 9-13). Such vehicles are capable of 60 or more miles per gallon.

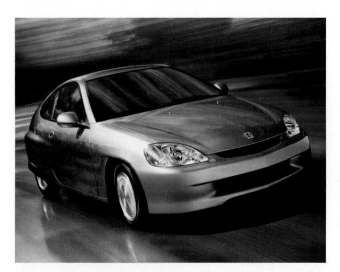

FIGURE 9-13 A hybrid vehicle. The Honda Insight, introduced in the United States in 2000, combines electric power with a small gasoline engine to achieve fuel efficiency of about 65 miles per gallon and very low emissons. Toyota's Prius has been available in the U.S. since 2001, and other manufacturers have plans to produce hybrid vehicles by 2004 (American Honda Motor Co. Inc.)

Critical Thinking
SHOULD WE CONSERVE ENERGY RESOURCES?

The history of energy policy in the United States presents a curious paradox of a country that appears to want to do one thing but behaves in a totally contradictory fashion. While most Americans would agree that energy conservation, like Mom and apple pie, is basically a good thing, our actions tell a different story.

Ever since the energy crisis of the 1970s, energy conservation has been a valued goal both as measured by public opinion and in government policy. The initial reaction to the energy crisis was that the crisis demonstrated that oil supplies were perilously short and that immediate and significant conservation measures were needed. We have since learned that the crisis was a short-term phenomenon, that we have substantial flexibility in our energy-use system, and that oil will probably be around for another half century or so. Nonetheless, the negative consequences of our energy use are well known. It increases pollution, strains our foreign-exchange accounts by requiring massive oil imports, and necessitates increased military expenditures, including the possibility of waging an occasional war to protect our access to foreign oil reserves.

In the 1970s and early 1980s, the U.S. government established a wide range of voluntary and mandatory measures to encourage energy conservation and reduce dependence on foreign oil. These ranged from tax incentives for conservation-related investments to a nationwide 55 m.p.h. speed limit. Most of these programs have since been either abandoned or sharply curtailed in favor of free-market energy choice. Consumers, in an era of cheap gasoline, have acquired a fleet of gas-guzzling sport utility vehicles equipped with four-wheel drive, which for most of us has true utility at most a few hours of the year when plows may have difficulty keeping up with snowfall.

The evidence from past energy prices and consumption patterns demonstrates that a small increase in energy prices, as occurred during the energy crisis, can bring major improvements in energy conservation. A modest energy tax could generate a new wave of energy conservation while still keeping energy prices well below those in most industrial nations. We refuse, however, to give up our cheap energy, and for the U.S. that means substantial dependence on imported oil and the necessity of ensuring access to the oil resources of the Persian Gulf and other areas through diplomatic and in some cases military means—and these are externalities.

Questions

1. Implementation of the Kyoto Protocol (Chapter 13) would probably require an energy tax and other measures to encourage conservation by raising energy prices. Currently this is not politically feasible, as most voters strongly object to tax increases, especially on gasoline. What do you think is the best way to reduce carbon dioxide emissions?

2. Would you support a modest (say, 30 percent) increase in the price of gasoline if it would bring about a 20 percent decrease in fuel consumption over a period of a few years? Would you support a $30-per-month increase in your electricity bill if it allowed us to live without nuclear power, or if it produced a 10 percent decrease in acid rain?

If progress in efficiency continues and alternative energy sources are developed, we probably will not completely exhaust the world's oil reserves. We may be able to continue using oil for those few tasks for which it is best suited. Earth is not in immediate danger of "running out" of oil, but our dependency on this nonrenewable fossil fuel will prove to be a remarkably short period of human history.

> Earth is not in immediate danger of "running out" of oil, but our dependency on this nonrenewable fossil fuel will prove to be a remarkably short period of human history.

Natural gas and coal: Short-term oil substitutes In searching for alternatives to oil, we look first to the other two major fossil fuels—natural gas and coal. Natural gas is important in the United States as the clean-burning fuel of choice to heat more than half of the country's homes. A 1.6-million-kilometer (1-million-mile) pipeline network efficiently distributes gas from the production areas in the Gulf Coast, Oklahoma, and Appalachia to the rest of the country. Demand for natural gas is increasing, in large part due to air pollution considerations.

Natural gas does not offer much of an alternative to oil, because the global distribution of the two resources is similar. At current rates of consumption, natural gas reserves would last for 60 years, although potential reserves may be greater than for oil. Russia and the Middle East have about two-thirds of the world's natural gas reserves; the United States has only a few years of proven natural gas reserves. Continued intensive use of gas will depend on increased imports.

Coal reserves are more abundant than oil or natural gas; at current rates of use, the world has nearly 400 years of proven reserves. Coal can play an especially important role in providing the United States with energy, because the country has a large percentage of the world's proven reserves of coal. Coal is used mainly to produce electricity, and our use of electricity is expanding much faster than our use of other forms of energy. But several problems hinder expanded use of coal:

- *Air Pollution.* Uncontrolled burning of coal releases sulfur oxides, nitrogen oxides, hydrocarbons, carbon dioxide, and particulates into the atmosphere. The sulfur and nitrogen oxides are a major component of *acid deposition.* Acid deposition occurs when sulfur and nitrogen oxides combine with water to form acidic precipitation or when these acids are formed in the soil from pollutants that settle from the atmosphere. Many communities suffered from coal-polluted air earlier in this century and encouraged their industries to switch to cleaner-burning natural gas and oil. The Clean Air Act now requires utilities to use low-sulfur coals or to install emission-control devices on smokestacks. Pittsburgh, Pennsylvania, once noted for severe air pollution from burning coal for steel mills and glass factories, today has air relatively free of particulates and sulfur oxides. But coal-fired power plants still pump large amounts of carbon dioxide and sulfur oxides into the atmosphere.

- *Land and Water Impacts.* Both surface mining and underground mining cause environmental damage, an externality caused by burning coal. The damage from surface mining is visible: Vegetation, soil, and rock are stripped away to expose the coal. Today surface-mined land must be restored after mining, but restoration may not leave the land as productive as it was before. Underground mining causes surface subsidence, and acidic groundwater may also be released.

- *Limited Uses.* Coal is a bulky solid, incapable of powering the internal combustion engines used in cars, trucks, and buses. It is not well-suited to shipment through established pipelines, though it can be mixed with water and sent through pipelines as a muddy *slurry.* Coal-generated power plants can provide the power to run electric vehicles, which are expected to become more common in the next few years.

Nuclear and Renewable Energy Resources

Where can we turn for energy that is safe, economical, nonpolluting, widely available, and not controlled by a handful of countries? In the long run, we must look to energy sources that are renewable—or at least to resources like nuclear energy that are so abundant they are not likely to be depleted. The two most promising energy sources are nuclear and solar. Other alternatives at present include biomass and hydroelectric power. We will now look at each of these sources.

Nuclear energy   Nuclear power can be generated either by fission (splitting an atom into two or more parts) or by fusion (joining two atoms together). Most peaceful generation of nuclear power today relies on fission of uranium, although plutonium also can be used. Someday we may use fusion to generate nuclear power, but as yet such technology is unavailable. The big advantage of nuclear power is the tremendous energy that is available from a small amount of material. A kilogram of nuclear fuel contains more than two million times the energy of a kilogram of coal.

The peaceful use of nuclear reactors to generate electricity began in the 1950s, and today about 400 reactors are operated around the world. Nuclear power is used exclusively to generate electricity, a growing part of our energy needs. Nuclear power supplies about one-third of all electricity in Europe. Japan (which has virtually no fossil fuels), South Korea, and Taiwan also rely on nuclear-generated electricity (Figure 9-14). Nuclear power generates approximately 20 percent of U.S. electricity (Figure 9-15). The United States derives a smaller portion of its energy from nuclear power than other wealthy developed nations, in part because of more abundant coal reserves.

Like coal, nuclear power presents serious problems. These include potential accidents, generation of radioactive waste, public opposition, and high cost.

- *Potential Accidents.* A nuclear power plant cannot explode like a nuclear bomb, although it is possible to have a runaway chain reaction. The reactor can overheat, causing a meltdown, possible steam explosions, and scattering of radioactive material into the atmosphere. An explosion did occur in 1986 in a nuclear power plant at Chernobyl, Ukraine, near the border with Belarus (both countries at the time were part of the Soviet Union). The accident has caused over 3,500 deaths, including about 800 among emergency clean-up workers. Cancer rates remain elevated in southern Belarus. For example, childhood thyroid cancer rates in Belarus increased from fewer than 5 per year before the accident to more than 80 by 1994. The impact of this accident extended through Europe: Most European governments

FIGURE 9-14 Nuclear power plants are clustered in more developed wealthy regions, including Western Europe, North America, and Japan. Nuclear power has been especially attractive in the United States (which produces about 30 percent of the world's nuclear power) and in some European nations that lack abundant reserves of petroleum and coal. (Energy Information Administration)

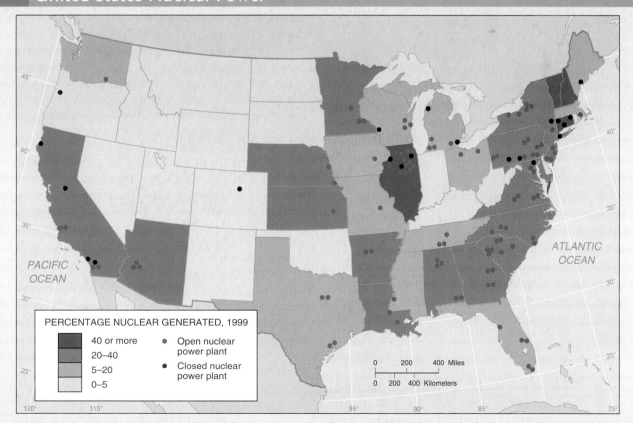

FIGURE 9-15 Nuclear power is an important source of electricity in several northeastern and midwestern states. Some locations have more than one nuclear reactor. (Data from Nuclear Regulatory Commission and Energy Information Administration)

temporarily banned the sale of milk and fresh vegetables because of possible contamination by radioactive fallout.

- *Radioactive Waste.* The uranium-based fuel used in nuclear reactors eventually is broken down to the point that it is no longer usable. Waste materials in this spent fuel remain highly radioactive for thousands of years. Spent fuel can also be reprocessed to extract plutonium, which can be used in nuclear bombs. Pipes, concrete, and water near the fuel also become "hot" with radioactivity. This waste cannot be burned or chemically neutralized. It must be isolated for several thousand years until it loses its radioactivity. Currently, spent fuel generated in the United States is stored in cooling tanks at nuclear power plants, but these are nearly full. Work is under way to develop a long-term storage facility in Nevada, but many problems remain to be solved before waste can be moved from temporary storage sites to a permanent facility. Kazakhstan, which already has radioactive waste storage and disposal facilities dating from the Soviet era, is building a facility to receive nuclear waste from the European Union.

- *Public Opposition.* Public concern about safety has been an obstacle to the diffusion of nuclear power since the technology first emerged. The accident at Chernobyl, as well as less damaging incidents in the United States and other countries, dramatically increased public concern about nuclear power. At Shoreham, Long Island, near New York City, a nuclear power plant was ready to operate and begin generating electricity when, under pressure from worried citizens, the government decided not to allow the plant to open and compensated the electric company for the loss of the plant. In 1992, Italian voters rejected future nuclear power development in that country, and public opposition is similarly strong in Germany and Scandinavia. Even in France, where over 70 percent of electricity is generated from nuclear power, public opposition is a major barrier to new development.

- *High Cost.* Nuclear power plants cost several billion dollars to build, primarily because of elaborate safety measures. Without double and triple backup systems, nuclear energy would be too dangerous to use. As a result, the cost of generating electricity is much higher from nuclear plants than from coal plants.

Biomass Biomass fuel derives mostly from burning wood but includes other plant material and animal waste. Energy is generated by either burning directly or converting substances to charcoal, alcohol, or methane gas. Biomass provides most of the world's energy consumption for home heating and cooking, especially in the developing world. In China, some individual homes have fermentation tanks that convert waste to methane, which is used for cooking and heating.

Forms of biomass, such as sugarcane, corn, and soybeans, can be processed into motor vehicle fuels; Brazil in particular makes extensive use of biomass to fuel cars and trucks. Biomass-to-energy conversion is especially appealing when it also serves to dispose of wastes. In Bangladesh, for example, in 2003 a new pilot plant began generating electricity using the waste from 5,000 poultry farms. Potential for increasing use of biomass for fuel is limited for several reasons. Burning biomass may be inefficient, because the energy used to produce the crops may be more than the energy supplied by the crops. When wood is burned for fuel instead of being left in the forest, the fertility of the forest soil may be reduced. But the most important limitations on using biomass for energy are that it already serves other important purposes—providing much of Earth's food, clothing, and shelter—and that burning wood contributes to deforestation.

Hydroelectric power Flowing water has been a source of mechanical power since before recorded history. In the past, water was used to turn a wheel which in turn could operate machines capable of grinding grain, sawing timber, and pumping water. Since the early 1900s the energy of moving water has been used primarily to generate electricity, called **hydroelectric power.** It supplies about one-fourth of the world's electricity, more than any other source except for coal.

To generate hydroelectric power, water must abruptly change height, as at a dam. The falling water turns turbines that power electrical generators. A hydroelectric plant produces clean, inexpensive electricity, and a reservoir behind the dam can be used for flood control, drinking water, irrigation, and recreation.

The United States and Canada are the largest producers of hydroelectric power in the world, followed by China, Brazil, and Russia. Together, North America and Europe generate about half the world's hydroelectric power. Worldwide, hydropower supplies about 2.5 percent of total commercial energy production. In the United States it supplies about 1.5 percent and in Canada 9 percent of commercial energy production. Most of the best sites for hydroelectric generation are already in use in the United States and Europe, but in many areas there is considerable undeveloped potential. China, Brazil, Indonesia, Canada, and the Congo have especially large hydroelectric potential.

Opposition to construction of big dams and reservoirs is strong among environmentalists who fear the environmental damage they cause, such as loss of farmland or habitat. Hydroelectric dams may flood formerly usable land and displace the people who lived on it. The dam converts a free-flowing stream to a lake, altering aquatic life. Many good sites for generating hydroelectric power remain in the world, but political considerations restrict their use, especially if the river flows through more than one country. For example, Turkey's recently built dam on the Euphrates River was strongly opposed by Syria and Iraq, through which the river also passes. They argue that the dam diverted too much water from the river and increased its salinity.

Despite these problems, hydroelectric power remains an attractive alternative to fossil- or nuclear-fueled power plants, largely because it generates no pollution. In rapidly industrializing countries such as China, increased electric generation can mean an increase in the standard of living not only in industrial cities but in small towns where electricity provides clean energy for indoor cooking and lighting and operates pumps that help provide clean drinking water. In the case of the Three Gorges Dam in China, the choice between clean hydroelectric power and coal-generated electricity, with its accompanying air pollution, is particularly stark.

Solar energy Solar energy—energy derived directly from the Sun—offers the best potential for providing the world's energy needs in future centuries. The Sun is a non-polluting and perpetual source of energy.

At present, solar energy is used in two principal ways: thermal energy and photovoltaic electricity production. Solar thermal energy is heat collected from sunshine. Collection may be achieved by designing buildings to capture the maximum amount of solar energy. Alternatively, special collectors may be placed near a building or on the roof to gather sunlight. The heat absorbed by these collectors is then carried in water or other liquids to the places where it is needed.

Photovoltaic electric production is a direct conversion of solar energy to electricity in **photovoltaic cells.** Each cell generates a small electric current, but banks of them wired together can produce a large amount of electricity. Solar-generated electricity is now used in pocket calculators and where conventional power is unavailable, such as spacecraft and remote places on Earth. As more photovoltaic cells are produced, and as the technology and efficiency are improved, the cost of solar power will decline. Photovoltaic cells are already competitive with conventional energy sources in many new residential and commercial installations (Figure 9-16).

Solar energy can be generated either at a central power station or in individual homes. Many countries are wired for central distribution, so central generation by utilities makes sense. But solar power now makes feasible individual home systems. An installation costing several thousand dollars provides a solar energy system that provides virtually all heat and electricity for a single home. Because the high installation price is offset by low monthly operating costs, home-based solar en-

FIGURE 9-16 Photovoltaic cells supplying electricity to Malibu High School in California. While this technology has been used for remote applications for some years, it is now becoming economically feasible for homes and other facilities. (Felicia Martinez/PhotoEdit)

ergy is economical for consumers who remain in the same house for many years. Individual solar energy users do not face rising electric bills from utilities that pass on their cost of purchasing fossil fuels and constructing facilities. The United States, Israel, and Japan lead in solar use at home, mostly for heating water. Solar energy is likely to become more attractive as other energy sources become more expensive.

Wind generation of electricity is one of the fastest growing solar technologies today. Significant improvements in turbine efficiency in recent years, combined with growing demand for clean electricity, are making commercial-scale wind projects much more attractive. A proposed array of wind generators off the coast of Cape Cod would produce as much electricity as a small conventional power plant, and would increase total U.S. wind electric generation by nearly 20 percent. Though still very small in comparison to other energy sources, this signals a significant trend in the shift toward renewable energy.

> Wind generation of electricity is one of the fastest growing solar technologies today.

Transition to new energy sources The world contains a variety of energy sources. In addition to fossil fuels, people make use of hydroelectric, biomass, solar, and nuclear power. Other technologies are also in commercial use today, including geothermal, wind, and tidal power.

The emergence of new energy technologies suggests that we are beginning the transition from a fossil-based energy system to a new mix of energy supplies. Oil and gas are likely to remain important for several decades, but they will likely diminish in importance relative to other resources, especially renewable alternatives.

At present, the emerging energy sources are less versatile than oil and gas and are likely to find only specialized uses. Solar thermal energy might be used to heat buildings, whereas photovoltaic cells can power small electrical appliances. Centralized electric power plants—coal, nuclear, or hydroelectric—will probably remain the major source of energy for heavy users, such as large factories and shopping malls.

If history is a reliable guide, the mechanism that will drive this transition to new energy supplies is the market. If the price of oil rises significantly, alternative technologies that are currently uncompetitive would become attractive. As these new technologies are more extensively used, their prices will decrease, further encouraging a shift away from oil.

Conservation is equivalent to discovering a new energy source, and in all likelihood the greatest source of new energy resources will be conservation. More efficient use of energy means that we can produce more goods, operate more motor vehicles, and heat more buildings with the same amount of energy. Conservation was one factor in the slow growth of energy consumption since the 1970s. Sluggish demand for petroleum contributed to the decline in oil prices in the 1980s; similarly, slow growth in electricity demand contributed to the demise of the nuclear power industry.

The slow growth of electricity consumption in 1980s and resulting lack of investment in new generation capacity contributed to new problems at the end of the 1990s. In 2000 a crisis of electricity supply in California appeared. Electricity demand grew steadily in California in the 1990s, increasing by 11 percent between 1990 and 1999. In that same time period California's generation capacity decreased by about 2 percent, largely as a result of closure of aging generating facilities. The state deregulated the electric energy system in the hope of stimulating production, but it maintained controls on prices paid by consumers. Utilities were unable to meet rising demand, and they were forced to buy electricity from other states at prices that were higher than they were permitted to charge consumers. When these utilities were unable to afford to pay for added power, they cut supply to consumers, causing blackouts.

The electric energy crisis in California and the 2003 blackout in the northeastern United States and Canada are believed by many to be signs of problems that may be experienced with increasing frequency in the coming years, as electric energy demand grows while production capacity and distribution infrastructure have remained relatively unchanged. One important factor contributing to the growth of electricity demand is the rapid expansion of the Internet. The computers that host Web sites and transmit data across the internet consume large amounts of electricity. In addition, more and more homes and businesses have computers and other electric devices. While individually these consume relatively little power, collectively their expansion increases demand. It is likely that this, along with increasing population, will create the need for many new power plants and transmission lines in the coming years.

The attacks on the United States on September 11, 2001, and the events that followed are prompting the United States to seek new sources of energy. Energy independence is virtually impossible, but we may find more secure sources, albeit at higher prices. Are we willing to abandon the unstable Middle Eastern Islamic regimes and pay this price? Our energy dependence has previously affected our foreign policy. In 1991, then-Secretary of State James K. Baker said that the war against Iraq was really about oil and American jobs. As long as the United States imports oil from unstable Near Eastern countries, or as long as the United States feels responsible for protecting these sources for our Japanese and European allies, the American economy and American foreign policy will remain hostage to these governments. In addition to urging exploration for new energy sources within American territory in places such as the Alaska National Wildlife Reserve, instability in the Middle East has increased U.S. interest in Africa and central Asia—areas in which there is potential for new oil development.

Other world regions could produce much greater quantities of oil, including the Central Asian countries, Russia, and West Africa, but developing these sources would require enormous investment, and several of these areas suffer from their own political instability. The United States has several options for handling a potential energy crisis. The United States might burn more coal, perhaps installing improved devices to protect against air pollution. Nuclear energy might win new proponents. Or the United States might research renewable sources (although from 1980 to 2001 the federal budget for research on such sources fell 83 percent). Improving energy efficiency would be the equivalent of finding new sources. In March 2002, however, Congress rejected a measure to increase fuel efficiency in U.S. vehicles.

AIR AND WATER RESOURCES

Consumption of resources is half of the resource equation—waste disposal is the other half. Burning fossil fuels pumps carbon dioxide into the atmosphere. The sulfur from burning oil and coal also enters the atmosphere and returns as acid deposition. Thus, the geography of resource consumption is also the geography of environmental pollution.

Air and water share two important properties: They are critical to human and other life on Earth, and they are useful places to deposit waste. An extreme environmentalist may argue that we should not discharge *any* waste into air and water, but in practice we need to rely on air and water to remove and disperse some waste. Not all human actions harm the environment, for the air and water can accept some waste. When we wash chemicals into a river, the river may dilute them until their concentration is insignificant. **Pollution** (elevated levels of impurities in the environment caused by human activity) results when more waste is added than a resource can accommodate.

Pollution levels generally are greater where people are concentrated, especially in urban areas. When many people are clustered in a small area, the amount of waste they generate is more likely to exceed the capacity of air and water to accommodate it.

Air Pollution

The purity of air is paramount to life on Earth. Some air pollutants come from natural processes unrelated to human actions, such as dust, forest fire smoke, and volcanic discharges. Humans add to this by discharging into the atmosphere smoke and gas from burning fossil fuels, incinerators, evaporating solvents, and industrial processes.

The atmosphere is constantly stirred by temperature and pressure differences, mixing vertically and horizontally. As it moves from one place to another, air carries with it various wastes. The more waste we discharge to the atmosphere, or the less the air circulates, the greater the concentration of pollution.

Average air at the surface contains about 78 percent nitrogen, 21 percent oxygen, and less than 1 percent argon. The remaining 0.04 percent of air's com-position includes several *trace gases*. *Air pollution* is a human-caused concentration of trace substances at a greater level than occurs in average air. In addition to the carbon dioxide (CO_2) emitted by "clean" combustion of fossil fuels, the most common air pollutants include **carbon monoxide, sulfur oxides** (SO_x, where the subscript x stands for the number of oxygen atoms), **nitrogen oxides** (NO_x), **hydrocarbons,** and **particulates** (very small particles of dust, ash, and other materials). Concentrations of these pollutants in the air can damage property and affect the health of people, other animals, and plants (Figure 9-17).

Each pollutant entering the atmosphere behaves differently. For example,

- SO_x and NO_x combine with water and fall to Earth as acid precipitation.
- *Particulates* in smoke are cleansed from the atmosphere quickly by gravity and precipitation.
- *Photochemical smog*—the product of *hydrocarbons,* as well as NO_x and sunlight—is created by chemical reactions that occur in the atmosphere itself.
- *Chlorofluorocarbons* (CFCs), chemicals used as refrigerants and in a variety of industrial applications, remain in the air long enough to be widely dispersed and carried into the upper atmosphere, where they damage Earth's protective ozone layer.

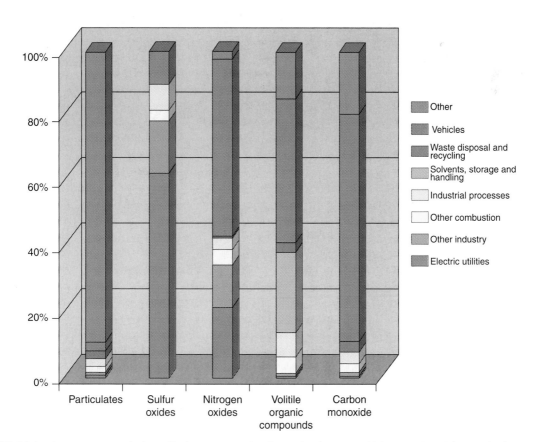

FIGURE 9-17 Major human-caused air pollution sources in the United States, 1997. Nationwide, particulates come mostly from soil erosion, but in urban areas they are derived from vehicles and industrial sources. Sulfur oxides are mostly derived from industrial sources. Transport, mostly cars and trucks, is the most important source of carbon monoxide as well as a major source of nitrogen oxides and hydrocarbons. (U.S. Census Bureau)

We are polluting our air in many ways today. We will focus on two important air pollution issues: acid deposition and urban air pollution.

Acid deposition   **Acid deposition,** commonly known as acid rain, occurs when sulfur oxides and nitrogen oxides, produced mainly from burning fossil fuels, are discharged into the atmosphere. The sulfur and nitrogen oxides combine with oxygen and water in the atmosphere to produce sulfuric acid and nitric acid. When dissolved in water, the acids may fall as *acid precipitation,* or they may be deposited in dust. We use the term acid deposition to include both types of pollution (Figure 9-18).

Acid deposition seriously damages lakes and kills fish and plants. But the most severe damage of acid deposition is to the soil. The acid deposits harm the soil in which plants grow in a number of ways. Some of the acid deposited in soil is neutralized by calcium, magnesium, and other naturally present chemicals, but the amount of acid deposited can exceed the capacity of the soil's chemicals to neutralize the acid. If soil water grows too acidic, plant nutrients are leached away and are unavailable to plants. Acid deposition can dissolve aluminum in the soil, which can be toxic to plants and interfere with their nutrient uptake. Acids may also

harm the soil worms and insects that decompose organic matter.

Acid deposition is a regional problem, most severe in the densely populated industrial regions in Europe, eastern North America, and eastern Asia. It is a major source of damage to forests, especially in the highly industrialized and densely populated regions of eastern United States, central Europe, and eastern China, where high SO_x discharges combine with precipitation to form acids (Figure 9-19). Much of the acid deposition in the northeastern United States is derived from emissions in the Ohio Valley carried eastward on the prevailing west-to-east winds. Damage to individual forests varies widely, depending on its age, tree species, the capacity of the soil to neutralize acids, and interactions between trees and other organisms in the forest. Because the relationship between tree damage and high discharges of SO_x and NO_x has not been documented precisely, some governments are unwilling to impose the cost of controlling emissions on their industries and consumers.

Nonetheless, progress has been made. During the past quarter century the United States has reduced SO_x emissions about 25 percent. Over this same time period, emissions have been cut by larger percentages in other relatively wealthy developed countries, including

Acid Deposition in North America

FIGURE 9-18 Lower values of pH correspond to greater deposition of acids caused by air pollution. The areas of North America most affected by acid deposition are in the Ohio Valley, eastern Great Lakes, and populated areas of the eastern United States. (U.S. Environmental Protection Agency)

FIGURE 9-19 Trees in the Black Forest, Germany, suffering from forest decline probably caused by acid deposition. The Black Forest, in the heart of Europe's industrial region, has been a focus of concern about the effects of acid deposition. (Gernot Huber/Woodfin Camp & Associates)

50 percent in Canada, 60 percent in France, 75 percent in Germany, and 86 percent in Sweden. NO_x emissions, which are more difficult than SO_x to control, have remained at about the same level in the United States during the past quarter century. Although precise figures are not available, SO_x emissions have probably increased in developing countries, especially China, which is responsible for nearly one-fourth of the world's coal combustion, the major source of SO_x emissions (Figure 9-20). China's sulfur emissions are a major source of acid deposition in Japan, again borne by prevailing westerly winds. In an attempt to reduce this problem, Japan is helping pay for the installation of pollution-control equipment in China.

Urban air pollution Urban air pollution results when a large volume of emissions is discharged into a small area. The problem is aggravated in cities when the wind cannot disperse these pollutants. Urban air pollution has three basic components:

1. Proper burning in power plants and vehicles produces carbon dioxide (CO_2), but incomplete combustion produces carbon monoxide (CO). Breathing CO reduces the oxygen level in blood, impairing vision and alertness and threatening persons who have chronic respiratory problems.
2. Hydrocarbons also result from improper fuel combustion, as well as from evaporation of solvents as in paint. Hydrocarbons and NO_x in the presence of sunlight form **photochemical smog,** which causes respiratory problems and stinging in the eyes.

3. Particulates include dust and smoke particles. You can see particulates as a dark plume of smoke emitted from a smoke stack or a diesel truck—not a white plume, which is condensed water vapor. Many particles are too small to see, however.

Three weather factors are critical to urban air pollution: wind, temperature, and sunlight.

1. When the *wind* blows, it disperses pollutants. When the air is calm, pollutant concentrations build up.
2. Air *temperature* normally drops rapidly with increasing altitude. But over cities, conditions sometimes cause **temperature inversions,** in which warmer air lies above cooler air. This limits vertical circulation, trapping pollutants near the surface (Figure 9-21).
3. *Sunlight* is the catalyst for smog formation.

As a result of these three factors, the worst urban air pollution occurs under a stationary high-pressure cell, where a combination of slight winds, temperature inversions formed by descending air, and clear skies allow pollutants to accumulate. A city that experiences frequent stationary highs, such as Denver, Colorado, has frequent pollution problems.

Mexico City is notorious for severe air pollution, especially in winter, when high pressure often dominates, and the surrounding mountains discourage dispersal of pollutants by wind (Figure 9-22). In the eastern United States, pollution problems are worst in summer and autumn, because stationary highs are most common then. In West Coast cities such as Los Angeles and San Francisco, the pollution "season" is also summer

FIGURE 9-20 Air pollution in China. Industrial growth in China is fueled by coal and, more recently, oil. Until recently there has been little concern about the environmental consequences of that growth, and pollution problems are widespread. (F. Hoffmann/The Image Works)

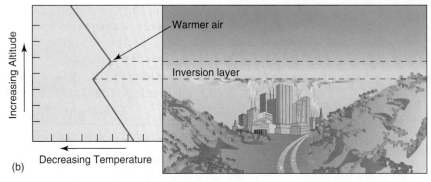

FIGURE 9-21 A normal temperature profile (a) and a temperature inversion (b) above a city. The temperature inversion traps pollutants. (From R. W. Christopherson, *Geosystems,* 3rd ed. Upper Saddle River, NJ: Prentice Hall, Inc., 1997.)

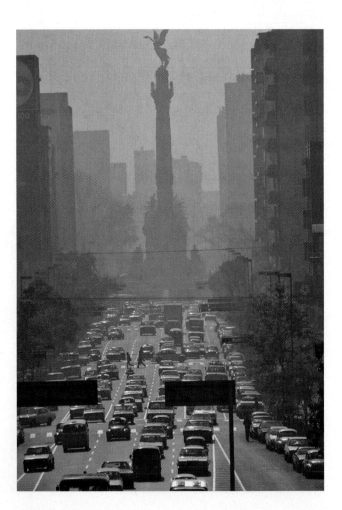

and autumn because inversions and bright sunshine are more persistent then.

Progress in controlling urban air pollution is mixed. In relatively wealthy developed countries with strict regulations, air quality has improved. Controls on use of coal and improvements in automobile engines, manufacturing processes, and generation of electricity have all contributed to higher-quality urban air.

To reduce auto emissions in the United States, for example, catalytic converters that oxidize unburned hydrocarbons have been attached to exhaust systems since the 1970s. As a result, carbon monoxide emissions have declined by more than three-fourths and nitrogen oxide and hydrocarbon emissions by more than 95 percent. Gains in relatively developed countries have been offset somewhat, though, by increased use of cars and trucks in recent years.

In many developing countries, urban air pollution is getting worse (Figure 9-23). Although ownership of motor vehicles is less common than in wealthy nations, the cars and trucks are older and lack pollution controls

FIGURE 9-22 Mexico City suffers from a combination of circumstances that all lead to significant air pollution problems. The city lies in a mountain basin that limits dispersion of pollutants. Automobiles are numerous and traffic is heavy, but there are few emission controls. In the early 1990s the problem reached crisis proportions, and a pollution-control effort is now under way. (Wesley Bocxe/Photo Researchers, Inc.)

FIGURE 9-23 Satellite image of air pollution in northeastern India and Bangladesh. The grey-colored haze seen in this image is common in winter, often remaining for weeks at a time. (Jacques Descloitres, MODIS Rapid Response Team/NASA Headquarters)

found on newer vehicles in relatively developed countries. Instead of relying on gas or electricity, many urban residents in developing countries burn wood, coal, and dung for cooking and heating. These smoky fires can create acute air pollution problems in poorly ventilated areas, even in the countryside. In developing countries an estimated 4 million children die each year from

acute respiratory problems, for the most part caused or aggravated by air pollution.

Water Pollution

Water is our most immediate resource other than air—it is consumed daily and comprises about 70 percent of our bodies. Oceans occupy 71 percent of Earth's surface. From the sea we obtain fish, shellfish, oil, gas, sand, salt, and sulfur. The seafloor may someday yield manganese and cobalt and may become a burial site for nuclear waste. Countries with inadequate freshwater supplies, such as Saudi Arabia, desalinate seawater.

In the United States, about 1.3 billion cubic meters (350 billion gallons) of fresh water are pumped from the ground or from rivers and lakes each year. About 13 percent of the water is used in homes and businesses, 45 percent in industry, and 42 percent in agriculture. About two-thirds of the water we use is returned to rivers and lakes in liquid form, while the remaining one-third is evaporated, mostly from irrigated fields.

Water availability varies greatly around the world. South America has vast water resources (about 17 percent of all the freshwater runoff in the world flows to the sea via the Amazon River) and a relatively sparse population. Asia, on the other hand, includes three-fifths of the world's people, and large parts of this vast continent are relatively dry, so per capita water availability is limited (Figure 9-24). Use of water for irrigation is

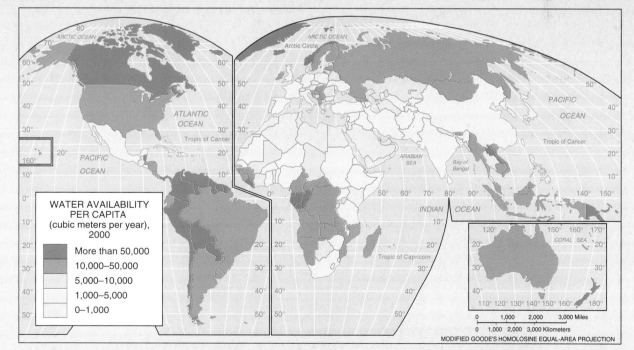

FIGURE 9-24 Water is most available in South America, Central Africa, Russia, and Canada. These places have high amounts of runoff and relatively low populations. Major problems of inadequate water supplies occur in India, China, and other densely populated countries, as well as semiarid and arid regions. (World Resources Institute)

FIGURE 9-25 **The bed of the Yellow River (Huanghe), exposed at low flow.** Diversion of water from the river for irrigation has severely depleted its flow, and today the river barely reaches the sea for much of the year. Water is in critically short supply in much of northern China. (Panorama Images/The Image Works)

especially significant in dry regions because the water is evaporated and not available for reuse (Figure 9-25).

Water is a "universal solvent." It can dissolve a wide range of substances, and it can transport bacteria, plants, fish, sediment, toxic chemicals, and trash of all kinds. As with air pollution, water pollution results when substances enter the water faster than they can be carried off, diluted, or decomposed. Water pollution is measured as the amount of waste being discharged in relation to a body of water's ability to handle the waste.

Pollutants have diverse sources. Some come from a *point source*—they enter a stream at a specific location, such as a wastewater discharge pipe. Others may come from a *nonpoint source*—they come from a large diffuse area, as happens when organic matter or fertilizer washes from a field during a storm. Point-source pollutants are usually smaller in quantity and much easier to control. Nonpoint sources usually pollute in greater quantities and are much harder to control.

The *concentration* of pollution, such as sewage effluent, usually declines downstream from where the waste is discharged (Figure 9-26). This reduction occurs because the waste is diluted, and natural processes decompose pollutants and remove them from the water.

Reduction of oxygen in water Because aquatic plants and animals need oxygen, the **dissolved oxygen** concentration in water indicates the health of a stream and lake. The oxygen consumed in decomposing organic waste constitutes the **biochemical oxygen demand.** If a lake or stream contains an excessive amount of decomposing waste, the oxygen demand is too great and the water becomes oxygen starved—and fish and other animals living in the water may die.

This condition often occurs when water becomes loaded with municipal sewage or industrial waste. The sewage and industrial pollutants consume so much

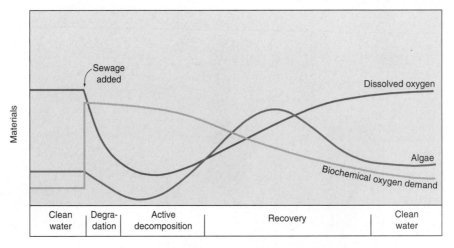

FIGURE 9-26 **Water quality varies downstream from a point where sewage is added.** Dissolved oxygen is depleted as organic matter is consumed but recovers further downstream. Nutrients are released to the water, and algae increase downstream as a result. (Adapted from H. B. N. Hynes, *The Biology of Polluted Waters.* Liverpool: University of Liverpool Press, 1978.)

oxygen that the water can become unlivable for normal plants and animals, and fish die. Similarly, when runoff carries fertilizer from a farm field into a stream or lake, the fertilizer nourishes excessive production of algae, which consume oxygen when they decompose.

Agriculture is the leading consumer of water, primarily for irrigation, and it is also the leading water polluter. Runoff from agricultural land discharges sediment, fertilizers, animal waste, and small quantities of chemicals into nearby streams. The fertilizer and animal waste, which are rich in nitrogen compounds, can nourish aquatic plants to the point of overproduction. Decomposition of these plants consumes so much oxygen from the water that fish may be unable to survive (Figure 9-27).

Wastewater and disease
Wastewater comprises about 13 percent of the flow of U.S. rivers. Improved wastewater treatment is critical, especially in a world with a rapidly growing human population. Rich,

> Wastewater comprises about 13 percent of the flow of U.S. rivers. Improved wastewater treatment is critical, especially in a world with a rapidly growing human population.

industrial countries like the United States generate more wastewater than do poorer nations, but they also have greater capacity to treat this wastewater. In the rich countries, strict legislation required treatment facilities to be upgraded, making their rivers cleaner than a few decades ago. However, many of the ambitious water-quality goals of this country's Clean Water Act, which was first passed in 1972, have not yet been met.

In the developing world, untreated sewage often goes directly into rivers that also supply drinking water. A combination of poor general sanitation, nutrition, and medical care can make the drinking of water deadly. Waterborne diseases, such as cholera, typhoid, and dysentery, are major causes of death in developing countries. Because of improper sanitation, millions of people in Asia, Africa, and South America die each year from diarrhea. As people in these rapidly growing regions crowd into urban areas, drinking water becomes less safe, and waterborne pathogens flourish (Figure 9-28).

Chemical and toxic pollutants   Any waste discharged onto or into the ground may pollute streams or groundwater. Pesticides applied to lawns and golf courses find their way into streams and groundwater. Landfills and underground tanks at gasoline stations can leak

FIGURE 9-27 **Water quality variations in lakes in the Minneapolis area revealed by remote sensing.** This image was created using the relation between light reflectance from water and concentrations of nutrients and sediments. Lakes that are blue have generally good water quality, green and yellow lakes are intermediate in quality, and those that are red have poor water quality. This map was created by a combination of GIS and remote sensing technologies. (Image courtesy Upper Great Lakes Regional Earth Science Applications Center/NASA)

FIGURE 9-28 **Women washing and collecting water at a river in India.** Less than 25 percent of the wastewater generated in India's 12 largest cities is collected by sewers. Most smaller cities have no wastewater treatment. It is estimated that between 5 and 10 million persons die each year from diarrheas caused by waterborne organisms in Asia, Africa, and Latin America. (Macduff Everton/The Image Works)

pollutants into groundwater, contaminating nearby wells, soil, and streams. Petroleum spilled from ocean tankers contaminates seawater. During times of flooding, normally secure tanks of chemicals may be dislodged and broken open and their contents mixed with the floodwaters.

Toxic substances are chemicals that are harmful even in very low concentrations. Contact with them may cause mutations, cancer, chronic ailments, and even immediate death. Major toxic substances include PCB oils from electrical equipment, cyanides, strong solvents, acids, caustics, and heavy metals such as mercury, cadmium, and zinc.

During the 1950s and 1960s, toxic wastes were often buried, but by the 1970s this method of isolating toxic wastes had proved inadequate because many waste sites were leaking. One of the most notorious was Love Canal, near Niagara Falls, New York, where several hundred families were exposed to chemicals released from a waste disposal site used by a chemical company. The chemicals had been left by the Hooker Chemicals and Plastic Company, which had buried toxic wastes in metal drums in the 1930s. In 1953 a school and several hundred homes were built on the site. Eventually the metal drums were exposed, and beginning in 1976, residents noticed a strong stench and

Focus on
MEAT PRODUCTION AND WATER POLLUTION

Farm animals such as cattle, pigs, or fowl produce considerable amounts of waste that can constitute a significant source of water pollution. Many rural streams carry substantial loads of nitrogen, phosphorus, and partially decomposed wastes. In rural areas agriculture is usually the dominant source of water pollution, and feedlots (where animals are kept and fed) are the most obvious and concentrated pollution source in the rural landscape. As just about anyone who lives in a farming region in eastern North America can confirm, pigs are especially noticeable in this regard.

In the late 1980s a radical transformation began in the U.S. pork production industry. Prior to that time, the usual pig production system was a form of mixed agriculture combining grain and meat production. A typical farm may have included perhaps 100–200 hectares (250–500 acres) of land on which grain—primarily corn—was grown. Most of this was fed to a herd of perhaps 100–200 pigs kept in a feedlot, preferably downwind from the farmstead. But a new form of pig production is growing rapidly and replacing the traditional system. Huge industrial-scale feeding operations are springing up across the central and eastern United States. The new farms, if they can be called that, have pig populations in the thousands, fed from grain purchased on the grain market and not necessarily locally produced. These farms tend to be concentrated in specific areas, often supplying a single centralized slaughterhouse. For example, a slaughterhouse in Guymon, Oklahoma, processes 2 million pigs per year,

mostly drawn from a few surrounding counties. Large pig-feeding operations enjoy economies of scale that allow them to fatten animals much more cheaply than can be done on a traditional farm.

How does this concentration of pig-raising affect water quality? In many areas where new industrial-scale operations have sprung up, the impacts have been locally severe. An adult pig produces significantly more waste than an adult human. A facility feeding 5,000 pigs produces more sewage than a small city. And while facilities for collecting and treating the wastes are required, standards for design and operation are usually not as stringent as they are for municipal sewage-treatment plants. Wastes are often stored in lagoons, and in a few instances the dams holding these wastes have failed, turning the rivers downstream into open sewers. Pollution from commercial-scale pig and chicken farms has been blamed for outbreaks of a mysterious organism called pfisteria that is killing fish in many streams in Maryland, Virginia, and North Carolina.

Concentration of animal production, however, does not necessarily reduce water quality. If wastes are produced at a few centralized locations instead of many scattered small farms, then it becomes easier to collect and treat the wastes. It is also easier—legally and politically as well as technically—to regulate and inspect waste-treatment systems. If industrial-scale feedlots are carefully regulated, they might provide an opportunity to manage a form of pollution that otherwise has been largely unmanageable. Unfortunately, the regulations in place have not been strong enough to prevent significant water-quality degradation in the immediate vicinity of these facilities, and locally water quality is suffering.

slime oozing from the drums. They began to suffer from high incidences of health problems, including liver ailments and nervous disorders. After four babies on the same block were born with birth defects, New York state officials relocated most of the families and began an expensive clean-up effort.

Love Canal is not unique—toxic wastes have been improperly dumped at thousands of sites. It was, however, very significant in raising public concern about groundwater pollution and stimulating action to combat it. In the U.S. such polluted sites are slowly being cleaned up, but the cost is tremendous. As safe and legally approved toxic-waste disposal sites become increasingly hard to find, some European and North American firms have tried to transport their waste to West Africa and other areas where pollution regulations are less strict.

The U.S. government has spent well over $100 billion in construction of new sewage-treatment plants since the mid-1960s. Industry has also spent heavily to meet water-quality standards. Many streams and lakes that had become open sewers now are suitable for recreation. But the job is unfinished. Serious efforts are made to control chemical waste in relatively developed nations, but developing nations often give little attention to this problem, and wastes are freely released into streams.

Reducing Air and Water Pollution

The most often used pollution-control strategy is to remove pollutants from water or exhaust gases before they are discharged to the environment. Control of air pollution in the United States and other industrialized nations usually means installing equipment that either reduces the amount of waste gases generated or removes wastes from the exhaust. For example, particulate emissions from coal combustion are often controlled with devices called electrostatic precipitators, which use the principle of electric charge to attract particles or other systems that remove ashes from the smoke. SO_x emissions are controlled both by burning fuels that contain less sulfur and by using devices that remove sulfur from exhaust gases after combustion. Similarly, automobiles are fitted with catalytic converters that reduce harmful emissions in the exhaust and with computer-controlled fuel-injection and ignition systems that generate less pollution by increasing the efficiency of combustion.

The largest source of water pollution—sewage—is also controlled by cleaning water before it is discharged. In a typical sewage-treatment plant, large solid particles are screened from the water or allowed to settle, the remaining water is oxygenated to allow bacteria to break down organic matter, and the water is then discharged to the environment. In advanced systems, some of the products of that breakdown, such as nitrogen and phosphorus, are removed. Most of the world's population living in developed countries is served by some kind of sewage collection and treatment system. In developing countries, however, sewage systems are less common. In Latin America, for example, only about 80 percent of the urban population is served with sewage collection systems, in Asia about 60 percent have such service, and in Africa the figure is only 53 percent. In most poor countries the sewage collected in urban areas is not treated, but simply piped to rivers or the sea. In rural areas of developing countries, sanitation systems are uncommon.

Instead of using these so-called "end-of-pipe" approaches that remove pollutants from air or water *after* they are produced, some pollution can be controlled by simply not producing it in the first place. This is known as **pollution prevention,** as opposed to pollution *control.* Many opportunities exist for reducing or eliminating industrial pollution without sacrificing quality of products or increasing manufacturers' costs. For example, toxic cleaning products are used in manufacturing to remove oil, grease, soldering residues, dust, coatings, and metal fragments. Trichloroethylene (TCE) is an excellent solvent, but it is also very toxic. Less toxic cleaning agents can be substituted, such as alcohol or detergents.

Another approach is to recycle polluting substances instead of discharging them into the environment. Some industrial processes, such as molding plastics and stamping sheet metal, generate recyclable scraps. Normally, larger scraps are recycled, but small fragments, like dust from polishing or sanding, may be placed in landfills as solid waste or discharged in water. Capturing and recycling these fragments has dual benefits: It reduces pollution while increasing the supply of scrap for making new materials. Health benefits also may accrue if the dust presents a health hazard. In this way, environmental externalities are eliminated without increasing producers' costs.

In the early 1990s a fundamental shift in pollution management began, from pollution control to pollution prevention. This change was spurred by several factors:

1. Despite pollution controls, significant water and air-quality problems remained.
2. Polluters faced legal liability if they discharged substances that were *later* found to cause harm, so reducing discharges to the environment reduces potential later liability.
3. Pollution prevention often is cheaper than pollution control.
4. Publicizing its efforts to reduce pollution can improve a company's image and help it market its products, recruit employees, and negotiate with government regulators.

Pollution prevention has quickly been adopted by many industries, especially in the relatively rich developed countries of Europe and North America. Large corporations—which are especially visible to governments, consumers, and the general public—have led the way in reducing discharges into the environment. Companies increasingly recognize that the best way to prevent pollution is to design a product that can be manufactured with a minimum of toxic chemicals, used without generating pollution, and recycled easily when it is no longer wanted.

While industrialized countries have been successful in reducing pollution during the past few decades, pollution problems are increasing in many poor countries. Most of the differences in pollution between rich and poor countries result from differences in wealth and level of industrial development. Where electricity, clean-burning fuels, and pollution-control devices are widely available—as in Europe, North America, and Japan—urban air pollution has largely been controlled. But problems are much more acute in places where people cook and heat with wood and drive old vehicles.

As manufacturing expands in developing countries, new facilities could be built with pollution control in mind. But pollution controls can be costly. Reducing the threat of large-scale pollution in developing nations requires development of pollution-prevention methods that reduce—not increase—costs.

FORESTS

As world population grows and resources are used for more purposes, conflicts over resource use are inevitable. In some cases the conflict concerns *who* has access to a resource in short supply, such as water in a desert region. But increasingly, conflicts concern *how* a resource should be used. Should a valley be dammed to generate hydroelectricity, or should people continue to live there and practice their traditional way of life? Should a wilderness area be protected as habitat for endangered species, or should it be opened for development?

One resource may have competing uses, some of which may be incompatible with others. Balancing competing and incompatible uses requires careful management, and difficult choices must be made. In this section we will review some of the important uses of Earth's forest lands and the reasons why these lands have long been, and will continue to be, controversial.

About 4.1 billion hectares (10.1 billion acres) or roughly one-third of Earth's land surface (excluding Antarctica) is covered with forest and woodland. The amount of forest land has decreased substantially as a result of clearing land for agriculture, and experts estimate that forests may originally have covered as much as half of Earth's land surface.

A forest is part of an ecosystem in which vegetation plays a major role in biogeochemical cycles. By absorbing carbon dioxide and releasing oxygen, a forest is an essential part of a local climate; a forest also stores carbon that otherwise would be in the atmosphere as carbon dioxide. A forest reduces erosion, aids in flood prevention, and provides a place for recreation. A forest is drained by rivers that provide drinking water, irrigation, habitat, and waste removal. People may use a forest as a habitat and a place to obtain food, fuel, shelter, and medicine.

Because forests cover such a large portion of Earth's surface, often bordering populated areas, they have become centers of natural resource conflicts. While agricultural land is typically under the control of individuals, corporations, or local political authorities, forest land is more often communally owned. And because of this communal or semicommunal ownership, many different people may have some legitimate claim to forest land. This further intensifies resource-use conflicts. So as you think about forests, remember that conflicts occur over other multiple-use resources, especially ones like water and oceans that are commonly owned.

Forests as Fiber Resources

Forests, the wildlife they support, and the water that flows through them are all renewable resources. Conflicts occur between those who wish to harvest timber and those who wish to leave the trees standing (Figure 9-29).

The United States, Russia, China, India, and Brazil are major timber-harvesting regions because their citizens demand wood products (Figure 9-30). Only about 4 percent of forest area worldwide, and 14 percent in the United States, is protected from cutting. Nearly all

FIGURE 9-29 Environmentalists and forests. These huge karri logs have been cut in Western Australia to be chipped for paper. Environmentalists decry this destruction of old-growth forest, particularly when paper can easily be made from much smaller logs. (Cary Wolinsky/PN/Aurora & Quanta Productions, Inc.)

World Timber Harvesting

FIGURE 9-30 Most timber harvesting takes place in Asia, North America, and Africa, with lesser amounts harvested in South America and Europe. Data from the former Soviet Union are lumped, with virtually all this harvesting occurring in Russia. (World Resources Institute)

forests in the United States (except those in Alaska) have been cut at least once during the past 300 years, so very little original (virgin) forest remains. In Canada much more original forest remains, especially in the north and west. The timber industry wants to cut remaining original forests because their large, straight trees yield high-quality lumber (Figure 9-31). Harvesting provides jobs and benefits lumber companies by keeping the price of timber low. Restrictions on cutting would threaten to increase unemployment in regions where extensive original-growth forests are being harvested.

Environmentalists argue that when trees are cut, the forest no longer supports the same wildlife or maintains clean water as effectively as it did before. Cutting down a forest places its inhabitants at risk. During the late 1980s, the spotted owl was endangered in the original-growth forests of Oregon and Washington when timber interests wanted to cut the habitat. Environmentalists point out that U.S. laws prohibit the government from actions that could harm an endangered species. U.S. government attempts to find a compromise between the timber industry and environmentalists has proved elusive, because the two uses of the original-growth forest resource—commercial timbering and as habitat for the spotted owl—ultimately are incompatible.

Sustained yield management is a strategy that can maintain the productivity of a resource even as it is being used. In forestry, sustained yield management

FIGURE 9-31 Timber harvesting in British Columbia, Canada. While most of the original forest area of North America has already been cleared, large areas remain in Canada and Alaska. Lumber from these forests is shipped to Asia and Europe, as well as delivered to the vast North American market. Canada exports about 15 billion board feet of lumber per year to the United States, or about one-fourth of U.S. lumber consumption. (Chuck O'Rear/Woodfin Camp & Associates)

means that the number of trees harvested should not exceed the number replaced by new growth. The strategy also emphasizes harvesting in a manner that minimizes soil erosion, and therefore enhances the ability of the forest to regenerate. Sustained yield management is the official policy of the U.S. Forest Service, which manages the national forests. In practice, however, the Forest Service has allowed timber to be cut faster than mature trees are produced in National Forests. As a result, we have a shortage of harvestable, mature timber in key areas such as the Pacific Northwest.

Similar conflicts regarding forest use are taking place in developing countries. For example, during the 1980s tropical deforestation became a global issue, primarily because the rate of deforestation in the Amazon basin was increasing rapidly. Environmentalists argued that once the forest was cut, it could never be restored to its original condition. Species would become extinct, the soil would be ruined, and indigenous cultures would be wiped out. But those involved in clearing the forest responded that most of the forest land in the United States and Western Europe had already been cleared, some of it many centuries ago. By what right can relatively wealthy developed countries tell Brazilians to save their trees?

Ideally, the solution to conflicting uses of forests is sustainable management. Trees can be cut, but not to the extent that the future productivity of the forest is reduced. Some original forest should be protected to provide habitat for native species and prevent extinctions. Resource managers in northwestern North America and in the Amazon are working toward such compromise solutions.

Other Important Forest Uses

Recreation in forest areas   In many countries, especially wealthier ones, forests are important recreational resources. They are relatively undeveloped and thus have open space available for hiking, camping, and other outdoor activities. They offer a strong contrast to the noise, commotion, crowding, and pollution of cities. Their shade and water availability provide relief from summer heat. Many forest regions are located in mountainous areas that offer other amenities such as skiing, climbing, and the cooler weather of high elevations. And because they serve as habitats for game animals, they can be preferred areas for hunting and fishing.

In the United States and Canada many important recreational centers are located in forested regions. Consider, for example, the coasts of Maine and Nova Scotia, the shores of the northern Great Lakes, or the mountains of the West. The most heavily used national parks, such as Acadia, Great Smoky, Yellowstone, Banff, and Yosemite, are all located in forest regions. Many of these are bordered by national forest lands, and

tourists drawn to the parks also enjoy the forests, both en route and when the parks are too crowded. In fact, in the United States, recreational use of the national forests (measured in number of visitor hours) is more than double that of the national parks.

The tourists who use these forest areas appreciate them for their natural (or natural-like) landscapes: tall trees, clean rivers, wildlife. They are not interested in seeing swaths of land with only stumps, sharing the road with lumber trucks, or hearing the buzz of chain saws. Recreational use clearly conflicts with timber harvesting.

Forests and biodiversity   As we learned in Chapter 4, human occupation of Earth has resulted in significant threats to global biodiversity. While damage to biodiversity is probably greater in agricultural lands than in forests, the remaining forests of the world contain some of the most important reserves of species diversity.

The tropical forests are especially important centers of diversity, with both larger numbers of tree species and complex vertical habitats. Foremost among these is the Amazon, a vast region of more than 4 million square kilometers (1.5 million square miles) containing literally millions of species. The Amazon has been a focus of attention in part because of the relatively high rates of deforestation taking place there. Since the 1980s, for example, the Brazilian Amazon has been cleared at rates of slightly less than 1 percent per year, an area roughly the size of the state of Connecticut. Worldwide, perhaps 40,000 square kilometers (15,000 square miles) of tropical rain forest have been cleared annually in recent years.

Concern over this loss is based partly on practical considerations, especially the potential medicinal uses of natural substances. Many important medicines are derived from forest species, and as mentioned in Chapter 6, indigenous peoples occupying these lands often have very good knowledge of such medicinal values. When the tropical forest is destroyed, these cultures may also be lost in the process, so both the species that provide medicines and a valuable source of information about them are at risk.

But a more powerful argument in favor of biodiversity preservation may be a moral one: that humans, possessing unprecedented powers over nature and Earth, have a special responsibility to protect the natural environments and the species they contain from destruction by our own hands. Our great numbers and advanced technology do not bestow on us the right to eliminate other species from the planet.

Forests and carbon storage   Another long-term impact from cutting down forests may be the release of stored carbon into the atmosphere, through decay or burning of unusable parts of trees. Deforestation in the tropics may be responsible for as much as 10 percent of the increase in CO_2 in the atmosphere discussed in Chapter 2. Left alone, forest growth removes carbon from the atmosphere and stores it in the biomass of living trees and soil organic matter. We must therefore consider both deforestation and reforestation.

In much of the world, the rate of deforestation exceeds the rate of reforestation. Net deforestation is occurring, especially in the tropical rain forest, because people are cutting trees faster than they can be replanted or regrow naturally. In some areas cleared forests do revert to second-growth woodlands, while in others the land is converted to other uses more or less permanently. The timber industry does replant forests, especially in the midlatitudes. And in some areas, notably the eastern United States, forests are growing in areas that were farmed or grazed until the early twentieth century, when these activities became no longer profitable. Forests can require more than a century to regrow, and so these second-growth forests lack some important habitat properties of the old growth they replace. Nevertheless, as these forests grow, carbon is gradually removed from the atmosphere. The potential for removing carbon dioxide from the atmosphere in this way is considerable if forest growth is promoted worldwide. Some estimates indicate that as much as 15 percent of fossil-fuel emissions in the next half century could be trapped in biomass. Of course, this would preclude using these forests in some other ways.

Balancing Competing Interests

Political and economic relations are the key to any situation where competing interests battle for control of a scarce resource. Decisions on allocation of forests among various uses will be made both by governmental regulation and by the marketplace. In the case of government-owned lands in the United States and Canada, the political arena is where most decisions are made. The important interest groups include the lumber industry, environmentalists, recreational interests, and occasionally ranchers and mining interests. Each of these has its own power base. Some may have strong allies in government, while others have based more mass support among voters. The fate of the remaining old-growth forest in the northwestern United States and southwestern Canada has been a particularly controversial issue because both of the principal competing interests—the lumber industry and environmentalists—are powerful groups with wide influence.

In market economies the relative values that society places on different forest uses may be expressed through prices of resource commodities. If the price of lumber rises, there is more pressure to harvest timber and sell lumber. If cheaper substitutes are available for lumber, then the forest is more likely to be

available for other uses. Similarly, if lumber from one region becomes expensive because of, say, government restrictions on harvesting, then lumber from other regions will be demanded instead. Japanese demand for lumber leads to forest clearing in Southeast Asia and Alaska, while U.S. lumber interests argue that protecting old-growth forest there will only lead to more imports of foreign lumber from Canada and elsewhere.

Government regulators sometimes make use of market-based principles, as well as political considerations, in deciding how to manage resources. Those who argue for restricting timber harvesting in favor of recreation and forest preservation, for example, note that the value of timber harvests in U.S. national forests in the early 1990s averaged about $1 billion per year. The national forests also provided about 300 million visitor days of recreation. If each of these visitor days were valued at $3.00, then the recreational value of the forests would be about the same as the value of timber harvested. Of course, both of these activities took place simultaneously in different parts of the forests. But as forest resources become increasingly scarce, such comparisons will likely lead managers to restrict logging further.

Similarly, other values may compete with lumber in economic terms. Since the Earth Summit in Rio de Janeiro in 1992, international negotiations aimed at reducing global carbon dioxide emissions have been under way. Industrialized nations consuming large amounts of fossil fuels are arguing that it would be cheaper to limit CO_2 releases by curtailing deforestation than by reducing fossil-fuel use. They are proposing to establish forest reserves in tropical countries as an alternative to reducing fossil-fuel use at home. In theory, this increases the value of those forests for carbon storage and reduces the supply of timber available for harvesting. We will discuss this further in Chapter 13.

CONCLUSION: CRITICAL ISSUES FOR THE FUTURE

Whether we consider forests, water and air resources, energy, minerals, farmland, or any other natural resource, it is clear that these resources are coming under increasing pressure worldwide, and the environmental impacts of resource use are increasing. As long as population grows and/or per capita resource use increases (and in general these trends have been ongoing for centuries), competition for increasingly scarce natural resources will increase.

In the early 1970s an influential book called *The Limits to Growth*, by D. H. Meadows and others, argued that depletion of Earth's natural resources had placed us on a disastrous course. The report predicted that natural resource depletion, combined with population growth, would disrupt the world's ecosystems and economies and lead to mass starvation. If natural resource protection were not in place within 20 years, the report claimed, environmental systems would be permanently damaged, and everyone's standard of living would decline.

Few contemporary geographers accept the pessimistic predictions made by *The Limits to Growth* a quarter century ago. Use of many resources has declined significantly since then. We may still deplete some resources, but substitutes and other strategies are available. Although pollution continues to degrade natural resources, industrial development has been made compatible with environmental protection in some locations. Nevertheless, many natural resource problems are likely to intensify considerably in the coming decades because of continued population growth and economic development.

The developing countries face the greatest challenges. In places suffering from extreme poverty, sound management of resources for the future is difficult. Careful management of Earth's natural resources is even more difficult where population is increasing rapidly. The desire for economic growth, like that occurring in China, is a powerful incentive for increasing resource use and for valuing industrial uses of the environment above environmental protection. Generating hydroelectric power at the Three Gorges Dam is an example. We can only hope that the Chinese are able to reap the benefits of this relatively clean energy and that they will thus be spared the damages caused by the coal combustion it can replace.

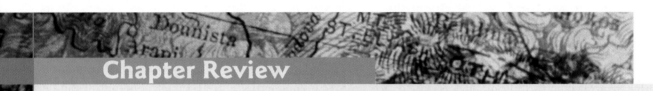

Chapter Review

SUMMARY

A natural resource is an element of the physical environment that is useful to people. Cultural values determine how resources are used. Technological factors limit our use of some resources, whereas economic factors influence whether a resource is used, and how much. Natural resources include air, water, soil, plants, animals, fuels, and non-energy minerals.

Our actions deplete mineral resources. Mineral wastes and other materials accumulate in landfills. If resources are depleted or if our needs change, we usually find or develop substitute resources. At present the world is dependent on fossil fuels for energy. As fossil fuels are depleted, we will need to use energy more efficiently and develop other sources of energy.

Air pollution is a concentration of trace substances at a greater level than occurs in average air. Acid deposition and pollution of urban areas are particularly harmful forms of contemporary air pollution. Water pollution results from both point and nonpoint sources. Pollution prevention is the most promising approach to reducing water pollution.

Forests are an example of a resource with many different uses, and conflicts over which uses are most important often arise. Among the important uses of forests are timber products such as lumber and paper, recreation, biodiversity preservation, and carbon storage. Sustained yield management is a strategy that attempts to balance the productive use of a resource while not depleting its supply.

KEY TERMS

acid deposition p. 382
biochemical oxygen demand p. 386
cartel p. 365
carbon monoxide p. 381
dissolved oxygen p. 386
externality p. 363
fossil fuel p. 370
hydrocarbon p. 381
hydroelectric power p. 378

natural resource p. 360
nitrogen oxide (NO_x) p. 381
nonrenewable resource p. 364
particulate p. 381
photochemical smog p. 383
photovoltaic cell p. 379
pollution p. 380
pollution prevention p. 389
potential resource p. 362

renewable resource p. 364
sanitary landfill p. 367
substitutability p. 364
sulfur oxide (SO_x) p. 381
supply and demand p. 362
sustained yield p. 391
temperature inversion p. 383
toxic substance p. 388

QUESTIONS FOR REVIEW AND DISCUSSION

1. What is a resource? How do political-cultural, techno-logic, and economic factors determine whether things in the environment become valuable resources or not?

2. What is a renewable resource? What is a nonrenewable resource? Give examples of each. Can any resources be considered either renewable or nonrenewable, depending on their use?

3. How has human use of coal, oil, and natural gas changed over the past 300 years? What changes are likely in use of these fuels over the next few decades?

What are some energy resources that are likely to be substituted for fossil fuels?

4. How do emission rates and rates of pollution dispersal interact to determine the severity of air and water pollution? What weather conditions aggravate air pollution?

5. How has water quality in U.S. rivers changed in the past few decades? What have been the major factors in this change? How does water pollution in wealthy nations differ from that in poor ones?

THINKING GEOGRAPHICALLY

1. Where does the drinking water come from where you live or attend school? What are the most significant potential pollution sources affecting your water supply? Your air supply?

2. If you drive, estimate the number of gallons of gasoline your automobile consumes in a year by dividing the number of miles driven by your fuel efficiency in miles per gallon. If each gallon of gasoline contains about 2 kilograms of carbon, how many kilograms of carbon were emitted to the atmosphere by your automobile? Compare this with the number of hectares of growing forest it would take to store this carbon in biomass if a hectare of forest accumulates carbon at the rate of 750 kilograms per year.

3. Of the approximately 90 million new people added to the world's population each year, over 80 million are in developing countries. But the per capita resource consumption of people in developing countries is a small fraction of that in rich nations. If population control is important for limiting global use of resources, should population control efforts be focused in developing nations or industrial nations? Why?

4. What are the most important issues of environmental quality where you live? Has environmental quality there improved or deteriorated in the past 30 years? What is your evidence?

SUGGESTIONS FOR FURTHER LEARNING

Abebe, M. "The Nile—source of regional cooperation or conflict?" *Water International* 20, no. 1 (1995): 32–35.

Allee, D. J., and L. B. Dworsky. "Watersheds: A new picture in an old frame." *Canadian Water Resource* 23, no. 1 (1998): 85–92.

Azzaria, L. M., and R. G. Garrett. "Mercury in the Canadian environment: Current research challenges." *Geoscience Canada* 25, no. 1 (1998).

Balirwa, J. S., C. A. Chapman, and L. J. Chapman, et al. "Biodiversity and fishery sustainability in the Lake Victoria Basin: An unexpected marriage?" *Bioscience* 53 no. 8 (2003): 703–715.

Bedford, D. P. "International water management in the Aral Sea Basin." *Water International* 21 (1996): 63–69.

Binnie, C. J. A. "Climate change and its potential effect on water resources in Asia." *Aqua: Journal of Water Supply Research & Technology* 46, no. 5 (1997): 274–282.

Brown, Lester R., and others. *State of the World 2000.* New York: W. W. Norton (see most current edition of this annual book).

Brown, Lester R., and others. *Vital Signs 2000.* New York: W. W. Norton (see most current edition of this annual book).

Buller, H. "Towards sustainable water management: Catchment planning in France and Britain." *Land Use Policy* 13 (1996): 289–302.

Cannell, M. G. R. "Growing trees to sequester carbon in the UK: Answers to some common questions." *Forestry* 72, no. 3 (1999): 237–247.

Clark, W. C. "Urban environments: Battlegrounds for global sustainability" *Environment* 45, no. 7 (2003).

Clark, W. C., and N. M. Dickson. "Sustainability science: The emerging research program." *Proceedings of the National Academy of Science, USA* 14 (2003): 8059–8061.

Cleveland, C. J. "Resource degradation, technical change, and the productivity of energy use in U.S. agriculture." *Ecological Economics* 13 (1995): 185–201.

Costanza, R. "The ecological, economic, and social importance of the oceans." *Ecological Economics* 31, no. 2 (1999): 199–213.

Costanza, Robert, and H. E. Daly. "Natural capital and sustainable development." *Conservation Biology* 6 (March 1992): 37–46.

Cutter, Susan L., and W. H. Renwick. *Exploitation, Conservation, Preservation,* 3rd ed. New York: John Wiley & Sons, 1999.

Dagnall, S., J. Hill, and D. Pegg. "Resource mapping and analysis of farm livestock manures—assessing the opportunities for biomass-to-energy schemes." *Bioresource Technology* 71, no. 3 (2000): 225–234.

de Montgolfier, J. "Indicators for sustainable development in forestry. *International Journal of Environment and Pollution* 12, no. 4 (1999): 451–458.

Dincer, I. "Environmental impacts of energy." *Energy Policy* 27, no. 14 (1999): 845–854.

Everard, M. "Towards sustainable development of still water resources." *Hydrobiologia* 395/396 (1999): 29–38.

Farrell, E. P., E. Fuhrer, D. Ryan, F. Andersson, R. Huttl, and P. Piussi. "European forest ecosystems: building the future on the legacy of the past." *Forest Ecology and Management* 132, no. 1 (2000): 5–20.

Franklin, J. F. "Lessons from old growth." *Journal of Forestry* 91 (December 1993): 10–13.

Gleick, P. H. "Water in crisis: Paths to sustainable water use." *Ecological Applications* 8, no. 3 (1998): 571–579.

Gupta, S., and A. Gilman. "Canada's contribution to the international reduction of certain persistent organic pollutants." *Environmental Practice* 2, no. 1 (2000): 71–78.

Harrington, W., V. McConnell, and A. Ando. "Are vehicle emission inspection programs living up to expectations?" *Transportation Research* (Part D: Transport and Environment) 5, no. 3 (2000): 153–172.

Hinrichsen, Don. *Our Common Seas: Coasts in Crisis.* London: Earthscan, 1990.

Hirsh, R. F., and A. H. Serchuk. "Will the restructured electric utility system help the environment?" *Environment* (Washington D.C.) 41, no. 7 (1999): 1–9.

Hollander, J. M., ed. *The Energy-Environment Connection.* Washington, D.C.: Island Press, 1992.

Jim, C. Y. "The forest fires in Indonesia 1997–98: Possible causes and pervasive consequences." *Geography* 84, no. 364 (1999): 251–260.

Kellomaki, S. "Forests of the boreal region: Gaps in knowledge and research needs." *Forest Ecology and Management* 132, no. 1 (2000): 63–71.

Kulshreshtha, S. N., B. Junkins, J. C. Giraldez, and R. L. Desjardins. "A systems approach to estimation of greenhouse gas emissions from the Canadian agriculture and agri-food sector." *World Resource Review* 12, no. 2 (2000): 321–337.

Lewandrowski, J., R. F. Darwin, M. Tsigas, and A. Raneses. "Estimating costs of protecting global ecosystem diversity." *Ecological Economics* 29, no. 1 (1999): 111–125.

Mannion, A. "Biotechnology: Its place in geography." *GeoJournal* 31 (1993): 347–354.

Manzini, F., and M. Martinez. "Choosing an energy future: The environmental impact of end-use technologies." *Energy Policy* 27, no. 7 (1999): 401–414.

Maybeck, M., D. Chapman, and R. Helmer. *Global Freshwater Quality: A First Assessment.* Oxford: Blackwell, 1989.

Meadows, D. H., and others. *The Limits to Growth.* New York: Universe Books, 1972.

Moore, T. "Electricity in the global energy future." *EPRI Journal* 24, no. 3 (1999): 8–17.

Naff, T. "Information, water, and conflict: Exploring the linkages in the Middle East." *Water International* 22, no. 1 (1997): 16–28.

Pimentel, David, J. Allen, A. Beers, L. Guinand, R. Linder, P. McLaughlin, B. Meer, D. Musonda, D. Perdue, S. Poisson, S. Siebert, K. Stoner, R. Salazar, and A. Hawkins. "World agriculture and soil erosion." *BioScience* 37 (April 1987): 277–283.

Redford, Kent H. "The empty forest." *BioScience* 42 (June 1992): 412–422.

Schofield, R., and J. Shaoul. "Regulating the water industry: Swimming against the tide or going through the motions?" *Ecologist* 27 (1997): 6–13.

Sharma, N. P., T. Damhaug, E. Gilgan-Hunt, D. Grey, V. Okaru, and D. Rothberg. "African water resources: Challenges and opportunities for sustainable development." World Bank Technology Paper [S.1.], [s.n.], no. 331, 144 pp.

Smil, V. "China's Greenhouse Gas Emissions." *Global Environmental Change* 4 (1994): 325–332.

Szaro, R. C., D. Langor, and A. M. Yapi. "Sustainable forest management in the developing world: Science challenges and contributions." *Landscape and Urban Planning* 47, nos. 3–4 (2000): 135–141.

Thomas, C. E., B. D. James, F. D. Lomax, Jr., and Ira F. Kuhn, Jr. "Fuel options for the fuel cell vehicle: Hydrogen, methanol or gasoline?" *International Journal of Hydrogen Energy* 25, no. 6 (2000): 551–567.

van Kamp I., Leidelmeijer K., Marsman G., et al. "Urban environmental quality and human well-being: Towards a conceptual framework and demarcation of concepts" (a literature study). *Landscape Urban Plan* 65, nos. 1–2 (2003): 7–20.

Westcoat Jr., J. L. "Resource management: Oil resources and the Gulf conflict." *Progress in Human Geography* 16 (1992): 243–256.

World Resources Institute. *World Resources 2002–2004.* New York: Oxford University Press, 2003.

WEB WORK

The Web is an excellent example of a new technology that can improve our lives and increase wealth while reducing the use of resources. We get information faster, and we don't have to drive our cars to get it. Please resist the temptation to waste paper by printing a copy of every page that contains a few words of potential interest!

Many environmental organizations and academic institutions have lists of links to environmental sites. Envirolink is one good one; it also includes an excellent compilation of environmental news items:

http://www.envirolink.org/

U.S. government agencies responsible for resources management are excellent sources of environmental statistics as well as information on government policies and programs. The U.S. Environmental Protection Agency can be found at

http://www.epa.gov/

The U.S. Energy Information Administration is at

http://www.eia.doe.gov/

The U.S. Forest Service site is

http://www.fs.fed.us/

The U.S. Geological Survey Minerals Information is at

http://minerals.er.usgs.gov/

And the Carbon Dioxide Information Center at Oak Ridge National Laboratory can be found at

http://cdiac.esd.ornl.gov/

Among the international organizations with information about global environmental issues are the World Bank at

http://www.worldbank.org/

The Consortium for International Earth Science Information Network at

http://www.ciesin.org/

The World Resources Institute at

http://www.wri.org/

And the United Nations Food and Agriculture Organization at

http://www.fao.org/

A Look Ahead

URBAN FUNCTIONS

Cities may have first developed for cultural reasons, but economic reasons rose in importance. Today we can define basic and nonbasic sectors of urban economies and analyze their roles in the primary, secondary, and tertiary sectors of national economies.

THE LOCATIONS OF CITIES

Some cities were founded to utilize advantageous sites, such as mines, while others exploit favorable situations, such as a crossroads or a bottleneck on transport routes.

10

CITIES AND URBANIZATION

◄ The gleaming skyscrapers in the distance seem to be a mirage, almost taunting the people in the foreground slums. Lagos, Nigeria, typifies the cities in many developing countries: Migration from the countryside plus natural population increase trigger unprecedented population growth. About half of humankind now lives in cities, so governments everywhere are challenged to provide housing, services, a healthy environment, and employment opportunities. (Edgar Clejine/Peter Arnold, Inc.)

WORLD URBANIZATION

Urbanization is occurring everywhere, both because of natural increases of the urban populations and because of continuing in-migration of rural people. Some governments are trying to regulate the growth of cities.

THE INTERNAL GEOGRAPHY OF CITIES

Any city's internal geography is defined by economic considerations, social considerations, and government actions interacting in the local culture.

CITIES AND SUBURBS IN THE UNITED STATES

U.S. metropolitan areas have exploded across the countryside. Job opportunity and housing continue to grow at the periphery, but some central cities that were once hollowed out started growing again at the end of the twentieth century. The many local governments in metropolitan areas must devise ways of cooperating on common challenges.

In Houston, Texas, the stench of the oil refineries has traditionally been referred to as "the smell of money." It never seemed to concern people as the hot summer breezes carried a brown haze over the city, for the haze represented the city's prosperity. Houston prospered from its situation near great oil resources, and the city today has nearly half of America's petrochemical capacity. The price of Houston's success, however, has been that it is today America's smog capital, with more days on which smog levels exceed federal standards than in any other U.S. city. Residents in some neighborhoods are regularly advised to stay indoors.

Perhaps the diffusion of the smog could have been controlled by land-use planning as the city developed, but Houston is America's only major city without any restrictions on land use throughout its vast 617 square miles—enough space to contain Chicago, Philadelphia, Baltimore, and Detroit. Gas stations stand beside proud cultural institutions, and upscale residential neighborhoods aren't far from industrial wastelands.

Thus, Houston exemplifies two of the greatest challenges of the twenty-first century: In an increasingly urban world, how can we reconcile economic growth with environmental quality, and what planning, if any, is necessary to guarantee the quality of life in humankind's urban habitat?

A city is a concentrated nonagricultural human settlement. Every settled society builds cities, because some essential functions of society are most conveniently performed at a location that is central to the surrounding countryside. Cities provide a variety of services, including government, religious services, education, trade, manufacturing, wholesaling and retailing, transportation and communication, entertainment, business services, and defense. The region to which any city provides services, and upon which it draws for its needs, is called its **hinterland.**

In ancient Egyptian hieroglyphics—the earliest writing we can read—the ideogram meaning "city" consists of a cross enclosed in a circle. The cross represents the convergence of roads that bring in and redistribute people, goods, and ideas. The circle around the hieroglyph denotes a moat or a wall. Few modern cities have walls, but cities do have legal boundaries, and within those boundaries a degree of self-government is usually exercised. The process of defining a city territory and establishing a government is called **incorporation.**

Sometimes several cities grow and merge together into vast urban areas called *conurbations*. In the northeastern United States, for instance, one great conurbation stretches all the way from Boston to Washington, D.C. This conurbation has been called *Megalopolis*, which is Greek for "great city." The world's largest urban areas are generally called *metropolises*, Greek for "mother cities" (Figure 10-1).

In several countries one large city concentrates a high degree of the entire national population or of national political, intellectual, or economic life. These cities are called **primate cities.** Paris, for example, is the

primate city of France, and Bangkok is the primate city of Thailand. Not all countries, however, have a primate city. The United States does not. Whether a country has a primate city depends on national history and social and economic organization.

Today in all countries urban populations are growing faster than rural populations. This process of concentrating populations in cities is called **urbanization.** The United Nations says that about one half of the world's population lived in urban areas by the year 2000, compared with 30 percent in 1950. Virtually all population growth expected in coming decades will be concentrated in the urban areas of the world. The degree of urbanization, however, is not the same in all countries (Figure 10-2). One of the purposes of this chapter is to explain why this is so.

The geographic study of cities, **urban geography,** considers three topics:

1. The study of the functions of cities and their economic role in organizing territory;

2. The comparative study of urbanization as it occurred in the past and as it is continuing in different countries today;

3. The study of the internal geography of cities—that is, the internal distribution of housing, industry, commerce, and other aspects of urban life across different cultures.

This chapter will review each of these topics and then examine the growth and internal geography of U.S. metropolitan areas. This examination will illustrate many of the principles of urbanization in situations that will probably be familiar to you.

The World's Largest Cities

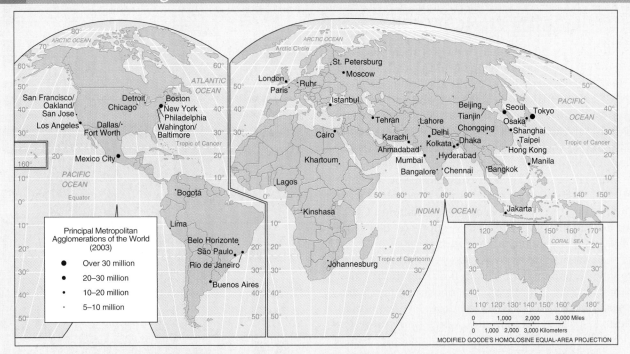

FIGURE 10-1 This map shows the 50 largest metropolitan agglomerations. The percentage of people living in urban areas is higher in the developed countries, but most of the world's large cities are in developing countries. The rapid growth of cities in these countries reflects both increases in overall national populations and migration into these cities from rural areas. (Data from United Nations, 2003)

World Urbanization

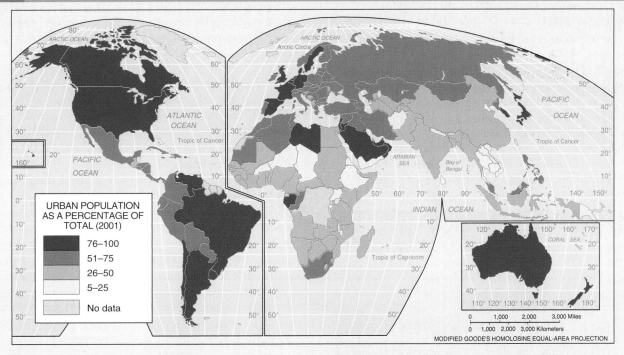

FIGURE 10-2 This map shows the varying degree of urbanization of countries. Different countries use different definitions of "urban area," ranging from settlements as small as 200 persons to as large as 30,000, but most countries use a minimum of 2,000 to 5,000 people. In the United States the minimum threshold population is 2,500; in Canada the urban population is defined as that population living in incorporated places of 1,000 or more and at densities of over 1,000 per square kilometer. (Data from the U.N.)

URBAN FUNCTIONS

The first cities appeared in today's Turkey and Iraq about 4000 B.C. Cities developed in Egypt's Nile Valley about 3000 B.C., in Peru and in the Indus Valley of today's Pakistan by about 2500 B.C., and in the Yellow River Valley of China by 2000 B.C. Mexico supported cities by about A.D. 500. Some early cities reached considerable populations. Ur, which flourished in Mesopotamia by 3500 B.C., may have had 200,000 people, and Thebes, the capital of ancient Egypt, may have had 225,000 in 1600 B.C. Later Rome probably reached 1 million by the second century A.D. Kaifeng, China, housed 1 million by 1150, and London reached 1 million by the early nineteenth century.

Archaeological evidence suggests that settlements probably originated for cultural reasons rather than economic ones. The first permanent settlements may have started as places to bury the dead, or fixed sites for priests to perform ceremonies. Cities came to be embellished as centers of worship or even as the seats of the gods themselves (Figure 10-3). Many religions teach that the largest house of worship should be the tallest building in a city, and when commercial buildings first overtopped church spires in European and American cities in the late nineteenth century, many observers found it symbolic of an unfortunate reversal in society's values (Figure 10-4).

The earliest settlements also may have served as places to house women and children while the men traveled in search of food. Household objects made by

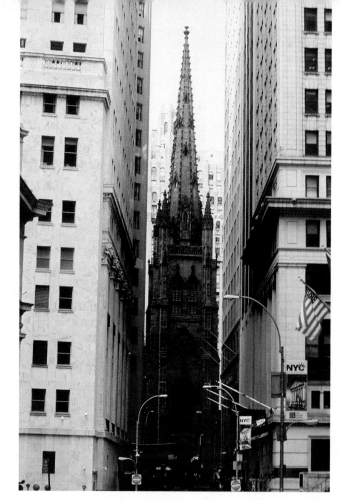

FIGURE 10-4 Trinity Church at the head of Wall Street in New York City. Trinity Church's 281-foot high spire, erected in 1839, was for many years the highest point in New York City. Today the spire is dwarfed by the skyscrapers of New York's financial district. (Rudi Von Briel/PhotoEdit)

FIGURE 10-3 The Todaiji Temple in Nara, Japan. This temple is the world's largest wooden structure. It houses a colossal statue of Buddha 16 meters (53 feet) high. Both temple and statue were originally made in the eighth century, when Nara was Japan's capital city. Today Nara is not a large city or important in Japanese economic life, but much of it has been preserved as a treasure of Japanese history and culture. (Michael S. Yamashita/CORBIS BETTMANN)

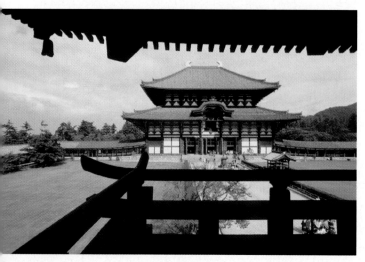

the women, such as pots, tools, and clothing, provided a basis for the creation and transmission of a group's values and heritage. Today settlements contain society's schools, libraries, museums, and archives—the repositories of knowledge and vehicles for passing it along through generations.

Early settlements protected groups' land claims and food sources. Palaces arose to house the group's political leaders, and soldiers were permanently stationed there. Many settlements were surrounded by defensive walls. Long after the introduction of artillery, walls could still hold off an attacker until help arrived or the attacker himself ran out of food, so cities still built them. Paris, for example, surrounded the city with new fortifications as recently as the 1840s and did not completely remove them until 1932 (Figure 10-5). Still today cities are the focus of military and political activities, but few retain walls except for historic interest (Figure 10-6). Most city walls were replaced in the nineteenth century by parks or grand boulevards.

The economic role of settlements may have begun simply as warehousing centers to store food, but as societies develop, production and trade join services as the paramount activities in most cities. Cities

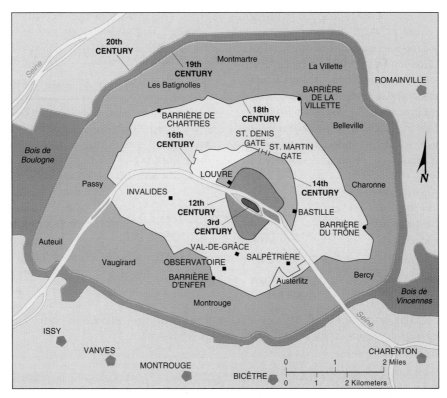

FIGURE 10-5 Concentric rings of growth. Paris periodically built new walls to encompass new neighborhoods as the city expanded, so Paris had concentric rings of protective walls. The old gates of St. Denis and St. Martin have been preserved, but the walls have been torn down.

bring people and activities together in one place for greater convenience. This is called **agglomeration.** Agglomeration promotes the convenient *division of labor,* which is the separation of work into distinct processes and the apportionment of work among

FIGURE 10-6 Quebec, Canada. Two-hundred-year-old walls still surround Quebec, the only walled city in North America, standing above the St. Lawrence River. (Paul von Baich/The Image Works)

different individuals in order to increase productive efficiency. Craft workers flourished within city walls, and specialized occupations emerged, sustained by the peasantry of the hinterland. Settlements serve as convenient sites for trading, and local officials often regulated the terms of transactions, kept records, and created a currency system. Cities thus promote and administer the regional specialization of production throughout their hinterlands. With industrialization, cities become centers of production.

The Three Sectors of an Economy

Economic activities are generally divided into three sectors: primary, secondary, and tertiary. The names used for the three sectors indicate the degree to which each sector is removed from direct involvement with Earth's physical resources.

Workers in the **primary sector** extract resources directly from Earth. Most workers in this sector are usually in agriculture, but the sector also includes fishing, forestry, and mining. Workers in the **secondary sector** transform raw materials produced by the primary sector into manufactured goods. Construction is included in this sector. All other jobs in an economy are within the **tertiary sector,** sometimes called the **service**

sector. The tertiary sector includes a great range of occupations, from a store clerk to a surgeon, from a movie ticket seller to a nuclear physicist, from a dancer to a political leader. This sector includes so many occupations that some scholars have tried to split it into a tertiary services sector and a *quaternary sector*, called an *information sector*. There is no consensus, however, as to what this sector includes, and no agencies of the U.S. government or of the United Nations recognize this sector in statistics. Therefore, it is not used in this book.

As economies grow, the balance of employment and output shifts from the primary sector toward the secondary and tertiary sectors. Fewer people work on farms, and more people work in factories and offices. These activities normally require urban settings, and therefore, as an economy's secondary and tertiary sectors grow, its cities grow. This growth of the secondary and tertiary sectors is called *sectoral evolution*, and it will be examined in detail in Chapter 12.

> Some of the workers in any city produce the city's exports, but others serve the needs of the city's own residents.

The Economic Bases of Cities

Cities depend on their hinterlands for, at the very least, food. The cities must in turn provide services or "export" something to the outside. Many cities produce and export manufactured goods, but the exports of a city are not necessarily things that leave it. They may be things or services that people come to the city to buy. If people go to Houston for heart surgery, for example, then heart surgery is counted as an export of Houston. Vacations are an export of Miami Beach; gambling is an export of Las Vegas. In this economic sense, capital cities export government.

Specialized Economies of Some U.S. Cities

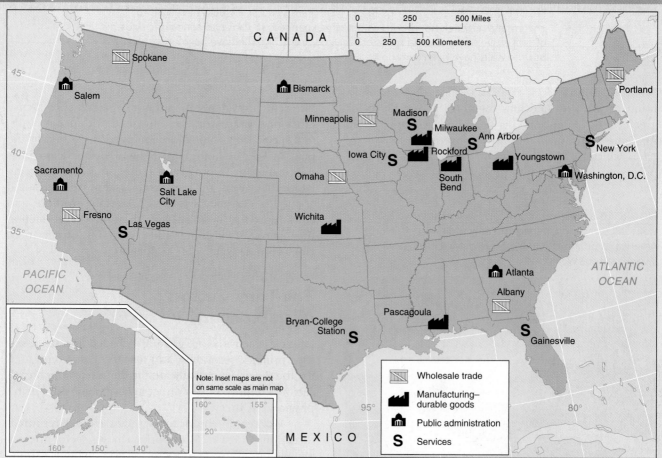

FIGURE 10-7 This map shows some cities that have specialized economies; it does not indicate the cities' individual ranking in specific activities. Los Angeles, for instance, is America's greatest manufacturing center, but it does not appear on this map because its economy is diverse. (Adapted with updating from J. Clark Archer and Ellen R. White, "A Service Classification of American Metropolitan Areas." *Urban Geography* 6 (1985): 122–51.)

Some of the workers in any city produce the city's exports, but others serve the needs of the city's own residents. The part of a city's economy that is producing exports is called the **basic sector,** and that part of its economy serving the needs of the city itself is called the **nonbasic sector.** A city's basic and nonbasic sectors may be in the primary, secondary, or tertiary sectors of the economy as a whole.

Jobs in a city's basic sector create jobs in the city's nonbasic sector. For example, if a local factory makes a product that is sold around the world, the factory workers will spend their earnings by shopping locally, getting their hair cut locally, and purchasing other local goods and services. Each job in the basic sector actually supports several nonbasic sector jobs, because earnings from exports circulate and recirculate through the local economy. When a factory worker buys a shirt, the store clerk can get a haircut; the barber in turn might eat at a local restaurant, and so on. Thus, jobs in the basic sector have a **multiplier effect** on jobs in the nonbasic sector.

Cities can be classified economically by examining each city's basic and nonbasic sectors and by comparing these sectors among different cities. New York, Seattle, and Hollywood, for instance, each contains a number of dry cleaners and doctors. These people, for the most part, work within the nonbasic sectors. In terms of exports, however, workers in New York provide specialized financial services, workers in Seattle write computer software, and workers in Hollywood make movies. A city's employment structure reveals its economic specialization. Any city that has an unusual concentration of workers in a specific job category must be exporting that product or service. A city that has a concentration of autoworkers, for example, presumably exports automobiles (Figure 10-7).

THE LOCATIONS OF CITIES

Today, as in the past, the location of any city depends on a balance of site factors (characteristics of the place itself) and situation factors (its location relative to other places). Choice sites include defensive hilltops, oases, and the locations of mineral resources. Some mining towns virtually sit on top of valuable ore deposits but are otherwise isolated. At the time of the Industrial Revolution, cities often developed at waterfalls to exploit hydropower (Figure 10-8).

The locations of other cities more clearly result from advantageous geographic situations. Cities with the most convenient situations grow (look back at Figure 6-23). Cities frequently grow up at places where two different physical areas meet or along the border between two cultures. In Chapter 6 it was noted that Tombouctou was located on a key trade route, on the border between two physical environments (where the Niger River bends farthest north into the Sahara

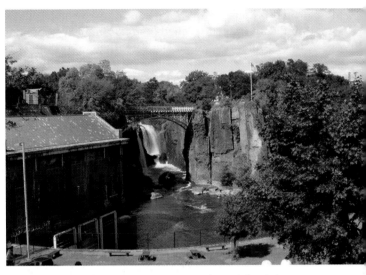

FIGURE 10-8 Paterson, New Jersey. These falls of the Passaic River powered the world's first planned industrial city, Paterson, New Jersey, founded in 1791 by a group of investors lead by Alexander Hamilton. (Peter Bryon/PhotoEdit)

Desert) and also two cultures (nomadic Arabs to the North and settled black peoples to the South). Many cities spring up as transportation hubs or at bottlenecks, such as at a bridge across a river or where two political jurisdictions funnel trade through border checkpoints. Some cities grow at sites where the method of transportation necessarily changes. These are called *break-of-bulk* points. A seaport is an example. Another is the head of navigation of a river. If there is a waterfall on a navigable river, cargo has to be unloaded from ships and then reloaded beyond the waterfall up or downriver, or else the cargo might be shifted to rail or truck.

Louisville, Kentucky, for example, was laid out in 1773 at the falls of the Ohio River. The river is navigable both up and downriver from Louisville. The opening of the Portland Canal in 1830 allowed ships to pass around the falls, but by then the city was well developed. It provided many services to its rich, developing hinterland and to westward-moving pioneers. The bridge across the Ohio River focused north-south traffic, and later the railroad lines also focused on the city.

If a situation is favorable, a great city may arise on an unfavorable site. Many of Asia's coastal cities were built by European merchants or conquerors at sites that provided access to the sea and that may have been defensible, but they sit on deltas or the swampy foreshores of tidal rivers. To this day, Karachi, Pakistan; Madras and Calcutta, India; Colombo, Sri Lanka; Yangon, Myanmar; Bangkok, Thailand; Ho Chi Minh City, Vietnam; and Guangzhou, Shanghai, and Tianjin, China, all face formidable problems of drainage, water supply, construction, and health. New Orleans in the United States is a North American example.

One of human geography's great paradoxes is that one of the world's largest cities, Mexico City, is located on one of the world's most unfavorable places to build: a drained lakebed in an earthquake zone in a basin of interior drainage at a high elevation in a dry climate. These conditions combine to cause physical instability, alternating flooding and lack of water, and air pollution. Both human lungs and internal combustion engines are inefficient when high altitude reduces oxygen levels by 23 percent, and air pollution is aggravated by local windstorms. Today's Mexico City, however, was the site of the Aztec capital Tenochtitlán at the time of the arrival of the Spaniards. The Spaniards maintained the site as their capital, and so have modern Mexicans. Thus, this great city testifies to the power of history and geographical inertia.

Central Place Theory

The relationships between cities and their hinterlands have inspired a model of how cities are distributed across territory. Walter Christaller (1873–1969) began with the simplest imaginary landscape—an isotropic plain on which transportation cost is determined according to straight-line distance. Christaller then asked, "If cities are to serve as convenient centers for exchange and other services across an isotropic plain, how will cities be distributed? What will be the pattern of towns and their hinterlands?"

To answer these questions, Christaller developed his **central place theory.** This model has three requirements: (1) the hinterlands must divide the space completely, so that every point is inside the hinterland of some market; (2) all markets' hinterlands must be of uniform shape and size; and (3) within each market region, the distance from the central place to the farthest peripheral location must be minimal. The only pattern that fills these three requirements is a pattern of hexagons—six-sided figures (Figure 10-9). Therefore, market towns will be distributed across an isotropic landscape as foci of a grid of hexagons.

Hexagons appear repeatedly in both nature and in industrial design. Honeycombs are hexagonal, and so probably is the pencil in your drawer. A hexagonal shape allows the greatest possible number of pencils to be cut out of a block of wood.

Urban Hierarchies

Chapter 1 discussed diffusion up or down a hierarchy of cities. There are many small towns, fewer and more widely spaced medium-sized cities, and still fewer big cities. A hierarchy of cities exists because the more specialized a service or product is, the larger the number of potential customers that is needed for that product or service to be offered. No product or service can be offered without a minimum number of customers. This minimum demand is called the *threshold* for that product or service. A video rental store owner, for example, needs a minimum number of customers to earn a living. Most people periodically rent videos, so the

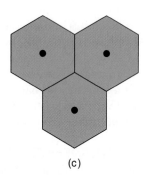

FIGURE 10-9 Hexagons fill space. Circles cannot fill space without overlapping one another, but circles collapse into hexagons that neatly fill space.

threshold of demand for a rental store is low. Each small town or city neighborhood can support one. Similarly, many people frequently buy fresh bread and milk. Therefore, small groceries can be found in each neighborhood.

Few people, however, take tuba lessons or buy diamond bracelets. The threshold of demand sufficient to support tuba teachers or expensive jewelers is high, so only larger towns will be able to support these enterprises. On an isotropic plain on which people and buying power are equally spaced, video rental stores and grocers will be closely spaced; tuba teachers and jewelers will be widely spaced.

In real landscapes, other factors in addition to distance must be considered. If one area has a concentration of rich people, for example, then the density of jewelers will be greater there. Quantity of purchasing power substitutes for numbers of people. If another area has a high population density, then grocery stores would be denser there.

A provider of a service can do one of two things to reach his or her necessary threshold of customers. First, the provider can be *itinerant* (travel from place to place). Alternatively, the provider can set up shop at one convenient place and wait for people to come. Convenience and accessibility are the principal purposes of cities. Agglomeration of services in cities saves travel costs, and it allows the thresholds to be reached for more specialized goods and services.

The range of goods and services in a city offers small businesses the opportunity to rent pieces of equipment and to hire services or temporary employees only when they need them. This saves the businesses the cost of investing in equipment or hiring people full time. These available services or goods are called **external economies,** and their availability lowers initial costs for new business ventures. Thus, cities are incubators of new businesses. If a company grows large enough to justify buying its own equipment or hiring full-time employees, then the company has achieved an **internal economy.**

The Patterns of Urban Hierarchies

Christaller's model of evenly spaced market towns with hexagonal hinterlands can be elaborated to represent a hierarchy of cities. In the isotropic plain the larger cities that are "above" the small market towns in the urban hierarchy must be more widely spaced than the small market towns, but they too must be evenly spaced, and their hinterlands must also be hexagonal. Therefore, the distribution of the larger cities is represented by a grid of larger hexagons superimposed on the grid of small hexagons (Figure 10-10).

Many factors in the real world disrupt Christaller's model; the world is not an isotropic plain. The model has proved useful, however, in planning new cities in countries with unsettled lands. The land that the Dutch have claimed from the sea behind new dikes is similar to an isotropic plain. There, Dutch geographers have planned new market towns on hexagonal grids. Brazil has adapted the model to settle territories in Amazonia. The government built what it calls an "Agrovila" every

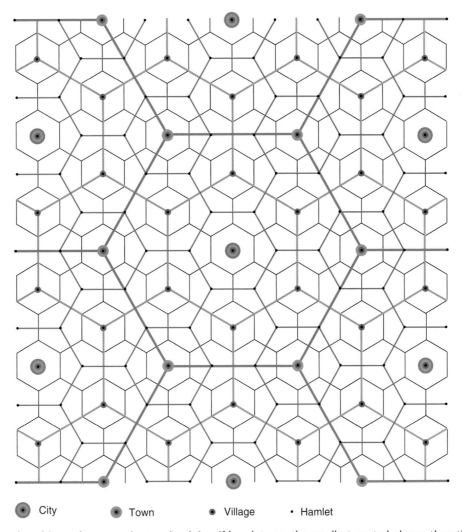

City   Town   Village   Hamlet

FIGURE 10-10 An urban hierarchy on an isotropic plain. If hamlets are the smallest central places, then the distribution of the larger villages could be represented by a grid of larger hexagons (red lines) superimposed upon the grid of the small hexagons that represent the hinterlands of the hamlets (black lines). Geometry defines at least three ways of doing this. The pattern illustrated here is called the K3 model because in it the hinterland of each village includes a hinterland that the village serves as a hamlet plus one-third of the hinterlands of each of the six surrounding contiguous hamlets. When a resident of a hamlet needs a higher-order service available only in a village, he or she finds each of two villages and a town equally convenient. Therefore, that person might visit each of these three settlements occasionally, so news travels rapidly across a landscape organized in this way. Continuing up the hierarchy of central places, town hinterlands (blue lines) are superimposed upon village hinterlands, and city hinterlands (green lines) upon town hinterlands. (From Walter Christaller, *Central Places in Southern Germany,* trans. Carlisle Baskin. Englewood Cliffs, NJ: Prentice Hall, 1966, p. 61.)

10 kilometers (6 miles), with a school, a health-care center, and a post office. Every 40 kilometers (25 miles) the government placed an "Agropolis"; these offer the services of an Agrovila, plus sawmills, stores, warehouses, banks, and other commercial services. Each 136 kilometers (85 miles) in Brazil a "Ruropolis," a city, was established for light industry. The ratio of 10, 40, and 136 kilometers cannot be reconciled with the proportions of a "pure" geometric grid, but they have been found useful by the Brazilian planners. The towns serve as foci for economic development. Venezuela has adopted a program similar to Brazil's.

> When improvements in transportation allow people to travel farther to obtain services or goods, the smallest central places may lose their reason for existence and disappear.

When improvements in transportation allow people to travel farther to obtain services or goods, the smallest central places may lose their reason for existence and disappear. This is happening in the North American farming region from Kansas north into the Canadian prairies, where most small towns developed as commercial centers for farmers. They offered grocery stores, banks, hardware stores, farm implement dealers, automobile dealers, and feed stores. Technological advances, however, starting with the tractor, as well as advances in transportation and communication and other economic forces, have increased the size of farms and cut the number of farmers. As a consequence, there is less need for the small-scale central places. A University of Saskatchewan study revealed that between 1960 and 1990, over 70 percent of the province's 600 towns, villages, and hamlets lost the basic commercial functions necessary to sustain the communities. Farm areas are today punctuated by wholly abandoned towns, and the number of rural towns is expected to continue to decline. Only small towns near metropolitan areas can survive or even prosper. These become **exurbs,** the name given to settlements that make up the outermost ring of expanding metropolitan areas.

WORLD URBANIZATION

Urban geographers compare urbanization as it occurred in the past and as it is continuing today. The rapidity of worldwide urbanization today presents many nations with both challenges and opportunities for the welfare of their populations.

Early Urban Societies

In the seventeenth century Holland became the first modern urban country, with more than half its population living in towns and cities. The principal economic functions of Dutch cities were tertiary-sector functions.

The cities arose as administrative and commercial foci of Holland's global shipping, banking, and trading activity, and the urban populations could be supported by Holland's highly productive agriculture. This urbanization occurred before the Industrial Revolution.

In Britain, a larger country with a more varied economy, urbanization occurred with industrialization. More than half of Britain's people lived in cities and towns by about 1900. Several developments over the previous 200 years had resulted in the concentration of Britain's population. These included the following:

1. Improvements in agricultural technology—part of the Agricultural Revolution—reduced the number of agricultural workers. Landowners found it profitable to release employees and to evict tenants. The resulting rural depopulation is described in Oliver Goldsmith's poem "The Deserted Village" (1770), which laments that "rural mirth and manners are no more."

2. Displaced workers migrated to the cities. There, many were absorbed by the concurrent **labor-intensive** stage of the Industrial Revolution. An activity is labor-intensive if it employs a high ratio of workers to the amount of capital invested in machinery. Other newcomers to the city found work in the tertiary sector. The largest class of urban workers was actually domestic servants—a tertiary-sector job. In 1910 they formed more than one-third of the total British labor force. A similar pattern could be found in the United States and other rich Western countries.

   Rural-urban dislocation caused appalling hardship. The history of urban life in nineteenth-century England records overcrowding in dreadful slums, malnutrition, starvation, crime, and early death. The descriptions of the miseries of England's slum populations in Charles Dickens's novels still haunt our imaginations. Cities have never easily absorbed all those who have flocked to them.

3. Population pressures were somewhat relieved by emigration, or else by the forcible exportation of criminals and debtors throughout the British Empire. The colonies of Georgia and Australia absorbed many of these deported people.

This British experience provided the world's first model for modern urbanization, and most of today's developed countries experienced similar histories. In 1800 the 21 European cities with populations of 100,000 or more held about 4.5 million people, or just under 3 percent of the total European population. By 1900 there were 147 such places with a total population of 40 million, or about 10 percent of the total population.

The word *model* as used here does not mean that the British experience of urbanization was perfect or that it should be imitated. It imposed terrible hardship on millions of people. Britain provides a model only in the sense that Britain experienced these forces of urbanization first, so that experience can be compared with the current situation. In British history the push of rural displacement and the pull of urban job opportunity were not coordinated. Governments have learned how to manipulate both forces in order to ease the process of urbanization, and yet some aspects of contemporary urbanization are causing as much hardship as the experience did in Britain.

Urbanization Today

Today urbanization is occurring in many places without concomitant economic development, especially in the world's poor countries. Burgeoning populations swamp the cities' ability to absorb the people and to put them to work. The populations overload the cities' infrastructures of housing, education, internal transportation, water supply, and sewerage. By one estimate 90 percent of sewage from urban areas in the developing world pours untreated into streams and oceans today. Reliable water, electricity, and telephone services are becoming increasingly rare. From the tops of new skyscrapers in many modern cities, the view presents vast shantytowns of the desperately poor (Figure 10-11). Living conditions in these teeming cities are no better for the majority than those that existed in Europe in the nineteenth century. The stresses of contemporary urbanization have been blamed for the breakdown of family life, recourse to drugs or religious extremism, and the spread of AIDS. The United Nations predicts that by 2025, two-thirds of the world's population will be in cities, most of them in poor countries where infrastructure is inadequate to the demands.

Rapid urbanization presents challenges, but challenges also bring opportunities. New urban populations present concentrated labor forces that might be put to productive work in ways that could raise the living standards of all people.

Rapid urbanization is often caused by deteriorating conditions in the countryside, and in many poor countries the problem is exacerbated by sharp cultural differences between rural and urban populations. Chapter 6 noted that many cities in today's poor countries evolved as outposts of international commerce grafted onto the local societies. Chapter 5 noted that their populations may even be cosmopolitan mixes of ethnic groups that are not native to the region. Throughout Latin America, for example, the cities are predominantly white or mestizo, and measures intended to restrict urban migration are interpreted, often correctly, as racist discrimination against Native

Americans. In East Africa urban populations were often Asian, and in Southeast Asia urban populations were Chinese and Indian. These cultural and ethnic contrasts make it harder to deal with rapid urbanization.

At least seven other circumstances differentiate urban growth today from the historic British experience.

1. The commercialization and mechanization of agriculture has accelerated, rapidly increasing the displacement of rural workers. For example, the introduction of mechanization to Brazilian sugar plantations reduced the labor force from 1.2 million in 1990 to fewer than 700,000 in 2000. Furthermore, modern agricultural technology requires capital investment, so the income gap between rich and poor widens.

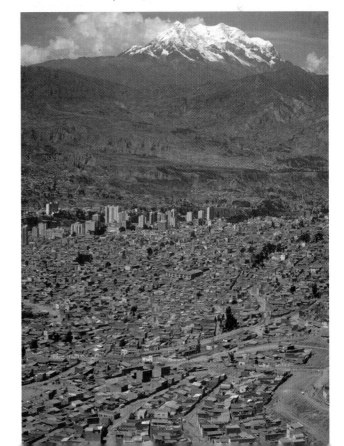

FIGURE 10-11 La Paz, Bolivia. The tall buildings mark the city's downtown core. It occupies a steep canyon in which Spaniards founded this city in 1548 in order to avoid the chill winds of the plateau. The modern city has grown up and down the canyon to the east and west of the historic colonial core. Native American migrants to the city have settled up on the plateau, which is now a distinct city of over 1 million in population, called El Alto. The construction of housing and infrastructure has not been able to keep pace with El Alto's annual 10 percent population increase, so it is a vast slum. Such slums are typical of the growing cities in the poor countries. La Paz is also one of the highest cities of the world, at 3,636 meters (2.3 miles) above sea level, so internal combustion engines are inefficient, and air pollution is a frequent problem. (John Metelsk/U.S. Agency for International Development)

2. National radio and television penetrate the countryside, spreading an image of the city as a place of opportunity. Many rural people feel pulled to the city.

3. In the past, disease and starvation among the poor kept urban death rates above urban birth rates. Without a steady influx of rural people, the town populations would have decreased. Today, however, the introduction of medical care and hygiene lowers urban death rates and triggers natural population increases. These compound the population growth that results from in-migration from the countryside.

4. Urban economies have changed in ways that make cities less able to employ the displaced rural population. Historically, urban economies offered entry-level job opportunities for the unskilled—jobs available for anyone with a strong back and a willingness to work. These jobs concentrated in domestic servitude, manufacturing, and construction. Today, however, domestic servitude has declined, and many of the jobs in manufacturing and construction demand skills. Manufacturing and construction are less labor intensive than they used to be; they are increasingly **capital intensive.** Machinery has replaced workers. Tertiary-sector jobs other than domestic servitude have multiplied, but most of these require job skills.

5. For most societies the safety valve of emigration has been stopped. Migration opportunities for the unskilled and for rural workers are decreasing.

6. The world economy is increasingly interdependent, so local governments have less control over local economies. Factory openings or closings in Bangkok are determined in Tokyo. Worker opportunity in Lima, Peru, is regulated from New York, and even job opportunity in New York is subject to international economic forces, as we will study in Chapter 12.

7. Many governments continue to favor urban projects over rural projects. One reason for this may be status. A shiny new hospital in the capital city, for example, may seem more impressive than a thousand new water pumps in poor villages. Another reason may be that the government fears urban rioting. Urban food riots have occurred in several African countries, including Nigeria, Egypt, and Zambia. Even in relatively well-off Venezuela, urban riots over the price of food left more than 250 dead in March 1989 and over 300 dead in November 1992. As noted in Chapter 8, many governments subsidize urban food supplies while discouraging their own farmers. These actions enhance the perceived opportunity in the cities without providing real opportunity.

Government Policies to Reduce the Pull of Urban Life

Governments might regulate the migration of people from the countrysides to the cities either by reducing the attractiveness of the cities or else by improving the quality of rural life.

Forceful measures to limit urbanization are not new in human history. The Russians have been required to have internal passports since the days of Czar Peter the Great 300 years ago. Recently the world has even witnessed brutal instances of compulsory "ruralization." When North Vietnam absorbed South Vietnam in 1976, the new rulers relocated millions of urbanites to rural areas to raise food using labor-intensive methods. The government argued that the cities had bloated on U.S. financial assistance. The government of Cambodia similarly relocated urban populations in the late 1970s. Some governments use media to discourage migration to the city. Ghana, for instance, broadcasts to the villages films suggesting that life is better there than in the cities.

Many cities try to discourage newcomers by restricting housing and economic opportunity. They pass building codes, for example, that ban substandard housing. These codes are often ineffective in stopping the growth of slums, but they make squatters' settlements illegal. Therefore, their residents do not get city services and will not risk investing to improve the property. The city in turn cannot collect property taxes. These "illegal" communities hold 30 to 60 percent of the total populations—70 to 95 percent of newcomers—in the cities of the poor countries. Cities also try to restrict small businesses in residential areas, but backyard workshops may thrive anyway. Cities may discriminate against new urbanites in education, housing permits, business licenses, or other job opportunities. Hawkers are chased from city centers. In some countries frustrated city authorities have even bulldozed squatters' settlements with only one or two days of warning. This has happened in major cities throughout Latin America, Africa, and Asia, leaving tens of thousands of people homeless. In 1990, for example, bulldozers flanked by army troops, with air-force planes sweeping overhead, leveled the district of Maroko in Lagos, Nigeria, leaving about 300,000 people homeless. One survivor reported that the soldiers brought truckloads of coffins and said that if anyone was ready to die, the army was ready to bury him. In 2001 army bulldozers cleared the Boa Vista slum of Luanda, Angola, which had housed 50,000 people. The residents lost whatever value they may have had in their humble housing, and they could not get jobs and were denied business stalls in the city center. The slum was replaced by expensive homes enjoying the beautiful view ("Boa Vista" in Portuguese.) Other urban governments simply round up squatters and drive them out. Tanzania's Human Resources Deployment Act declares that anyone who cannot show proof of employment in

Dar es Salaam can be deported to a state farm; thousands have been moved.

China has long enforced a household registration system under which people are not supposed to move anywhere without getting their registration changed, but people have moved anyway. As many as 150–200 million Chinese peasants have drifted into the cities, where they are called *floaters*. In major cities they make up more than 10 percent of the population, freely coming and going. Their residence is technically illegal, so they have no right to education, medical care, and other government services. The floaters serve useful functions, such as bringing produce to the city or providing services not provided by central planning, but their continuing influx threatens to overtax China's urban infrastructure. In 1999, in preparation for festivities celebrating 50 years of communism in China, army troops rounded up and imprisoned, or forcibly removed to the countryside, thousands of the floaters in Beijing, where as much as 25 percent of the total population may be floaters. In 2001 the national government announced a relaxation of the registration system in order to create a unified national labor market, but the city governments themselves have already introduced new exclusionary techniques, such as restricting residency permits to those who can afford homes or to skilled workers. From 2004–2009 the Chinese government intends to provide each citizen with a new electronic identification card; it will be difficult to counterfeit, but it will store considerable identification data about each person.

Improving Rural Life

Instead of trying to drive people out of the cities, governments might invest in rural health care and education, housing, roads, and other infrastructure to raise rural standards of living and keep people in the countryside. This might raise national food production. The government can also organize labor-intensive investments in rural infrastructure. Agricultural terraces, roads, and dams can be built by hundreds of thousands of workers equipped with shovels and wicker baskets just as well as by relatively few workers equipped with expensive earth-moving machinery (Figure 10-12). During the Depression in the 1930s, even the U.S. government Civilian Conservation Corps, Civil Works Administration, Public Works Administration, and Works Progress Administration all used labor-intensive methods to alleviate unemployment and at the same time build or repair national infrastructure.

The Economic Vitality of Cities

Most of the previous paragraphs about urbanization in the developing countries may have seemed pessimistic. They presented urban growth as a problem for which solutions were needed. There is another side to the

FIGURE 10-12 Labor-intensive construction. Construction of even massive infrastructure projects such as this dam in India can begin with workers carrying baskets of dirt on their heads. In the rich countries, however, labor-intensive construction methods have yielded to capital-intensive methods. (Courtesy of U.S. Agency for International Development)

coin, however: It is possible to view the backyard shops, the street hawkers, and the Chinese floaters as examples of opportunity and growth. These activities hint that urban migration, balanced between management and liberty, can provide a reservoir of vitality that can be harnessed for national growth.

The book *The Other Path* (1986) was a landmark study of urbanization in Peru by the economist Hernando De Soto. De Soto noted that in many cities the productive activities of a substantial share of the population do not appear in official accounts. The people may not have licenses to do what they are doing, they may be avoiding taxes, or for some other reason their activity escapes official notice. These activities make up the **informal** or **underground sector** of an economy. Every city in the world has such a sector, but the informal sector is particularly important in the cities of the poor countries. One 1999 study estimated that the underground economies total about 15 percent of the measured economies in developed countries but 33 percent for the poor countries—as high as 80 percent in Nigeria, 70 percent in Thailand and Egypt, and 50 percent in the Philippines and Mexico.

De Soto estimated that in Lima, Peru, the informal economy employed fully 60 percent of the population and produced 40 percent of all goods and services. The poor owned and controlled a public transportation network of private taxis and vans, plus land and housing worth billions of dollars. None of this, however, was legal, so it could not be taxed by the government or used as collateral by business owners. If the government simply legalized this activity, these assets would have **liquidity,** which is the quality of being readily convertible into cash. Recognition of ownership would free these assets for use as collateral for investment and business enterprise, and the government could tax the activity. The government's refusal to recognize what was happening handicapped the country's economic growth and vitality.

In Peru from 1995 to 2001 more than 1.2 million households counting 6.3 million people received title to the properties they were inhabiting. They had to prove only that they had lived there for three years. Title reform enabled more people to work outside the home and more family members to join the labor force, because now no one has to stay home to guard the property. The values of the newly registered properties have soared, and mortgage and consumer credit markets have developed. De Soto was invited to the Philippines, where he discovered 60 percent of Filipinos holding real estate assets worth tens of billions of dollars outside the law, and thus illiquid. In Egypt 85 percent of the population lives in homes without property titles. De Soto estimated that in 2000 only 25 countries in the world have genuine contractual urban societies; the rest are informal.

Economic growth will probably occur most rapidly in the developing countries that ensure property rights. A modern market economy cannot develop unless property rights are acknowledged and protected; therefore, De Soto has argued, governments must formalize the spontaneous emergence of informal property. Chapter 8 already noted how productivity in agriculture can be raised by guaranteeing property rights. The same guarantees may develop in the cities: In both instances, GIS and GPS greatly enhance the ability to record and register land holdings, so the spread of these techniques may spark economic growth and vitality around the world.

If governments view urban migration as a problem, they cannot see how urban immigrants' industriousness could be an asset for economic growth. Millions of people continue to choose to migrate to the cities, where they do survive or even thrive. There is something terrifically dynamic going on, and geographers, economists, and government officials at all levels are challenged to understand and measure it, and to harness the energy.

> There is something terrifically dynamic going on, and geographers, economists, and government officials at all levels are challenged to understand and measure it, and to harness the energy.

THE INTERNAL GEOGRAPHY OF CITIES

Urban geographers study not only the distribution of cities in the landscape but also the distribution of activities and of housing within cities. This distribution may be caused by economic forces, social factors, or deliberate actions of the government. In different countries and cultures, these three factors carry different relative weights.

Models of Urban Form

A 1945 book by Chauncey Harris and Edward Ullman, *The Nature of Cities,* proposed three models of the internal patterns of North American cities: the concentric zone model, the sector model, and the multiple-nuclei model. In 1995, the fiftieth year since the publication of *The Nature of Cities,* Professor Harris suggested a fourth model to describe how American metropolitan areas developed in the second half of the twentieth century: the peripheral model.

Figure 10-13 illustrates the *concentric zone model* of urban growth and land use. The core of the city, called the **central business district (CBD),** concentrates office buildings and retail shops. For businesses, accessibility is usually a principal determinant of a location's rent. The success of a department store, for instance, will depend partly on whether customers can reach the store easily. A city's most convenient and busiest intersections are most valuable for commerce. Land owners usually maximize the density of use on this valuable land by building up, so a traditional CBD is identifiable by tall buildings as well as crowded streets (Figure 10-14). Even within the CBD, clusters of functions appear. Lawyers' offices, for instance, cluster near courts or near the offices of their client firms. Retail stores of one type, such as jewelry stores, may cluster so that consumers can comparison shop. The CBD is surrounded by less intensive business uses such as wholesaling, warehousing, and even light industry—that is, nonpolluting industries that require relatively small quantities of raw materials. Residential land use surrounds this urban core.

The concentric zone model can be modified by considering the effect of transportation routes, which affect accessibility. New means of transport—historically canals, then railroads and tramways—spread out radially from the heart of the city, although their paths were modified by topography. Industrial and residential growth took place in ribbons or fingers along these radial routes, and wedges of open land were usually left between these radial routes.

The second model proposed in 1945, the *sector model* (Figure 10-15), assumes that high-rent residential areas expand outward from the city center along new transportation routes such as streetcar and suburban commuter rail lines. Middle-income housing clusters around high-rent housing, and low-income housing lies adjacent to the areas of industry and associated transportation, such as freight railroad lines.

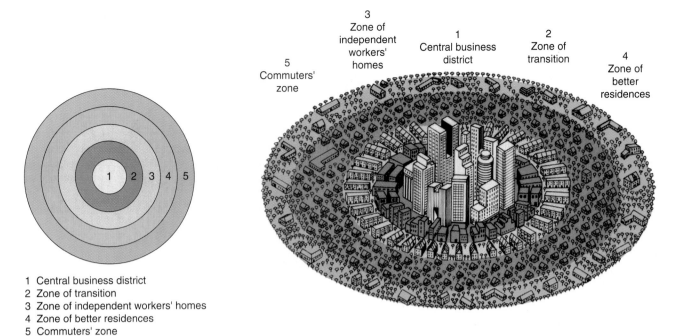

1 Central business district
2 Zone of transition
3 Zone of independent workers' homes
4 Zone of better residences
5 Commuters' zone

FIGURE 10-13 Concentric zone model. According to the concentric zone urban model, a city grows in a series of rings around the central business district.

FIGURE 10-14 Minneapolis, Minnesota. (a) The overhead drawing of the central business district of Minneapolis is dominated by retail and office buildings, but a number of public buildings also cluster downtown, including the city hall, government office buildings, and the central post office. The city's largest stadium, the Hubert H. Humphrey Metrodome, is located just a few blocks to the east. (Adapted from University of Minnesota, Cartography Laboratory, Department of Geography, 1986)

(b) The IDS Center, labeled on the drawing, is the black-topped building in the center of this photograph. (Wes Thompson/Corbis/Stock Market)

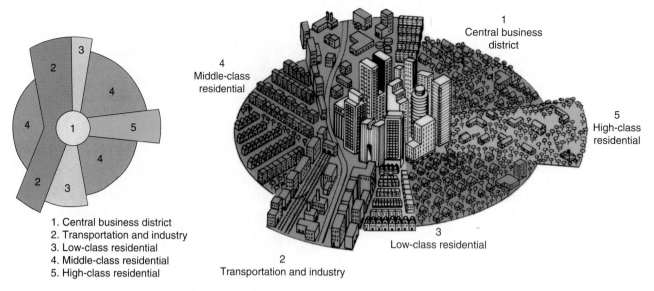

1. Central business district
2. Transportation and industry
3. Low-class residential
4. Middle-class residential
5. High-class residential

FIGURE 10-15 Sector model of urban development. In this model of urban form, a city grows out from the central business district in wedges or corridors of various land uses.

The third model, the *multiple-nuclei model* shown in Figure 10-16, recognized the development of several nodes of growth within an expanding city area. These nuclei may each concentrate on a different special function.

The *peripheral model* defined in 1995 (Figure 10-17), shows how radial and circumferential highways continued to draw activities out of the central city and to disperse them around the region. The story of how this happened will be told below.

Social factors in residential clustering Social considerations play a role in urban residential clustering.

Some people want to live with people like themselves, and this causes a clustering called **congregation.** Ethnic groups or immigrants of a common background, for example, may want certain services, such as grocery stores offering their traditional foods.

In other cases, however, people live together because discrimination forces them to do so. These people suffer **segregation** from others. In any specific case, it may be difficult to determine the degree to which people are congregating or are victims of segregation. Chapter 7 noted that in the past, Jews were legally segregated in ghettoes, but today the word *ghetto* means any residential concentration of any one kind of people. The factors

1 Central business district
2 Wholesale, light manufacturing
3 Low-class residential
4 Medium-class residential
5 High-class residential
6 Heavy manufacturing
7 Outlying business district
8 Residential suburb
9 Industrial suburb

FIGURE 10-16 Multiple-nuclei model of urban development. In this model of urban form, a city consists of a collection of individual nodes or centers around which different types of activities and people cluster.

1. Central City
2. Suburban Residential Area
3. Shopping Mall
4. Industrial District
5. Office Park
6. Service Center
7. Airport Complex
8. Combined Employment &
   Shopping Center

FIGURE 10-17 The peripheral model of urban areas. In this model, describing the growth of U.S. metropolitan areas in the last half of the twentieth century, activities disperse throughout a broad region.

causing residential clustering of any group—of Chinese Americans in "Chinatowns," of Italian Americans in "Little Italys," of Hispanic Americans in "barrios," or of African Americans as discussed later in this chapter—must be evaluated carefully in each case. Sometimes the words placed in quotation marks above are considered insulting, but they usually are not meant to be.

Religion is another social consideration that frequently causes clustering. People who share a religious faith may cluster around their house of worship, and in-migrants of that faith will seek that neighborhood. Language communities frequently form, as do communities of the elderly and communities of gays and lesbians. Almost any factor of social bonding can encourage the creation of an identifiable residential neighborhood.

Urban environments   Chapter 2 listed ways in which urban microclimates may differ from those of the surrounding countrysides: greater heat, lack of natural vegetation, paved surfaces, air pollution, and more. Large cities are also noisy, and noise interferes with communication and sleep, raises blood pressure, triggers muscle tension, and precipitates migraine headaches, elevated cholesterol levels, gastric ulcers, and psychological stress. It diminishes people's ability to concentrate, and if it is loud enough for long enough periods of time, it can damage hearing capability. Cities face enormous challenges in creating livable, healthy environments, as well as in disposing of their garbage and solid waste, problems which were discussed in Chapter 9.

The role of government   Government may determine land use. Restricting or prescribing the use to which parcels of land may be put is called **zoning.** In U.S. history, local government has often been averse to planning, so zoning has more often been *restrictive* (dictating

what cannot be done) than *prescriptive* (dictating what should be done). Industrial and commercial districts, for example, are usually kept away from residential neighborhoods. Each incorporated jurisdiction across vast conurbations exercises independent zoning power. This creates problems that will be discussed shortly.

Today most urban planners believe that the separation of land uses has been overemphasized. It may have been desirable to separate industry from housing when all industry was noisy, polluting, or smelly, but today separating homes from jobs may require excessive commuting. Most new planned communities emphasize the integration of residential and commercial activities, even with some light industry. They offer apartments above downtown shops and offices, for example, as in traditional small towns. The state of New Jersey subsidizes landowners who renovate downtown properties if the renovations create residential space on upper floors. This revitalizes downtowns and encourages the use of public transit—or even walking.

The Western Tradition of Urban and Regional Planning

The process of urban and regional planning applies many of the principles of urban geography to specific situations. The concept of designing an "ideal city" has challenged the best minds for centuries. The founder of Western city planning was probably the Greek thinker Hippodamus of Miletus. He laid out that city in today's Turkey according to a grid plan as early as 450 B.C.

The first modern attempt to formulate the needs of the city as a whole was the work of the British visionary Sir Ebenezer Howard. In *Garden Cities of Tomorrow* (1898), he outlined a plan to stop the unbounded growth of the industrial city and to restore it to a human

FIGURE 10-18 Sir Ebenezer Howard's garden city.
Howard envisioned planned "garden cities" to disperse the
concentrations of population in nineteenth-century cities.
(Reprinted with permission from Ebenezer Howard, *Garden Cities of
Tomorrow.* Cambridge, MA: The MIT Press, 1965.)

scale. He wanted to relocate population into new
medium-sized garden cities in the outlying countryside.
These regional cities would be ringed by greenbelts of
farmland and parks. All land would be municipally
owned, and each town and its surrounding region would
be planned as an interlocking whole (Figure 10-18).

Howard built two "garden cities" just north of
London: Letchworth (1904) and Welwyn Garden City
(1919). These inspired the Regional Planning Association of America, a private nonprofit organization, to
construct two planned communities in the New York
City area: Sunnyside Gardens, Queens (1924); and
Radburn, New Jersey (1928) (Figure 10-19). Neither is
a complete garden city, but both are harmoniously
designed and have greatly influenced urban planning
in the United States and Europe.

Probably the most important city planner of the
twentieth century was the Swiss architect Charles
Édouard Jeanneret-Gris (1887–1965), better known by
his professional name, Le Corbusier. In a celebrated
plan of 1922, he proposed to bulldoze the crowded,
rundown historic core of Paris, preserving only the
central monuments. In its place he wanted to build a
Radiant City of tall glass offices and apartments, spaced
so far apart that each tower would be surrounded by
green space and have a fine, wide view (Figure 10-20).
The concentration of facilities within high-rise slabs
would liberate the city from its environment. It could
be placed anywhere. Le Corbusier brought together
two conceptions: the machine-made environment,
standardized, technically perfect to the last degree and,
to offset this, the natural environment, treated as open
space, providing sunlight, air, greenery, and views.

Paris was never torn down and rebuilt as a Radiant
City, but Le Corbusier planned Chandigarh, a new capital for the Punjab State in India in 1950. The world's
supreme Radiant City, however, is Brasília, the capital of
Brazil, designed by Lucio Costa in 1957 (Figure 10-21).
Unfortunately, the city's gigantic scale demands a completely motorized population. That is the problem with
excessive openness—Corbusier's "city in a park" can
become a city in a parking lot. The Australian capital at
Canberra, planned by Walter Burley Griffin of Chicago,
has less openness. It allows walking, and it is generally
considered superior (Figure 10-22).

In 1930 Le Corbusier planned the town of
Nemours in Algeria. It has a geometric grouping of

FIGURE 10-19 Plan for Radburn, New Jersey. This planned
community, built in 1928, is still a desirable residential settlement.
Key elements of the plan include plenty of park space and the
separation of pedestrian walkways from car traffic and parking.
(Courtesy of the Radburn Association)

FIGURE 10-20 La ville radieuse (Radiant City). Le Corbusier's plan for his Radiant City featured high-density residential areas strictly segregated from other land uses. (Source: Le Corbusier, *La ville radieuse.* Paris: Editions de L'Architecture D'Aujour'hui, 1933, p. 170/French Government Tourist Office)

FIGURE 10-21 Brasília, Brazil. Brasília, Brazil's capital since 1960, arose on a largely unpopulated open plateau 970 kilometers (603 miles) northwest of Rio de Janeiro. It was an effort to shift the nation's political, economic, and psychological focus away from the former colonial cities on the coast toward the interior. As this picture reveals, it sometimes lacks water sufficient to keep the grass green or to operate the elaborate fountains. Slums surround the showplace buildings for 50 kilometers (30 miles) in every direction. Architect Lucio Costa admits, "Of course half the people in Brasília live in [slums]. Brasília was not designed to solve the problems of Brazil. It was bound to reflect them." (Getty Images Inc.—Stone Attstock)

buildings that look like dominoes, and this plan set the international fashion for high-rise slabs for the next 50 years. These ideas were disseminated worldwide by the 1933 Athens Charter of the International Congress of Modern Architecture (CIAM). In the charter's codification, largely by Le Corbusier, the functions of the city—housing, work, recreation, and transport—provided the city's framework. The charter called for separation of high-rise development, industrial zones, parks and sports fields, and streets of different widths spaced for traffic at different speeds. These ideas diffused to dominate urban planning around the world.

Many people believe today that the widespread adoption of Le Corbusier's ideas produced a half century of monotony—not merely of detail and style but of insensitivity to essential differences between one place and another. The high-rise slabs ringing every big city in the world from Mexico City to Singapore look much alike. In 1961 a group of younger architects broke away from CIAM and proclaimed that architecture was more than the art of building. It was the art of transforming people's entire habitat. The School of Architecture at the University of California at Berkeley was reconstituted and renamed the School of Environmental Design.

Many urban planners worldwide have come to criticize the concept of high-rise living. Low-rise dwell-

ings can achieve the same density of habitation as Le Corbusier's "towers in a park" can, and many people feel more content living in low-rise dwellings. High-rise public-housing projects, it turned out, can breed a

FIGURE 10-22 Canberra, Australia. Canberra became home to Australia's Parliament in 1927, but foreign missions and federal government departments moved here only in the 1950s. (Courtesy of Embassy of Australia)

sense of alienation and helplessness, and many have been abandoned and razed across the United States. Today we still work to design more humane urban environments (Figure 10-23).

Other Urban Models in Diverse Cultures

The concentric zone, sector, multiple-nuclei, and peripheral city models discussed earlier in this chapter were all devised to describe North American experience, but other models have been sketched to describe characteristics of other cultures. The three interacting processes identified earlier—economic factors (including transportation facilities), residential clustering or segregation, and government decree—affect all urban settlements, old and new, in all known cultures. These processes sort out the population and the land uses into distinct patterns that can be identified within any city.

The four models just described were first identified in North America, but they do not fully describe even all modern Western cities. The governments of the Western European countries, for example, have always been concerned with preserving the vibrancy and amenity value in the central cities. Therefore, they severely restrict suburbanization, as will be discussed shortly.

Latin American cities offer still another contrast to North American models. There, central business districts thrive (Figure 10-24). This is partly a result of continuing reliance on public transit and partly because high-income populations choose to live close to the central business district. A commercial spine such as a boulevard extends out from the central business district, and amenities such as opera houses, chic stores,

FIGURE 10-24 Latin American model of urban growth. This model contrasts with the North American models. In Latin America an elite residential sector often follows a spine of high-value land use stretching out from the central business district. (Adapted from Larry R. Ford, "A New and Improved Model of Latin American City Structure," *Geographical Review* 86 (1996): 438. Used by permission of the publisher.)

and elegant parks follow this spine. Zones of more modest housing and value surround this elite zone, and the periphery is dominated by squatter slums.

Western forms often overlie indigenous forms Chapter 6 noted that distinct settlement patterns are landscape footprints of distinct cultures (look back at Figure 6-19), but it also noted that many of the world's great cities came into being as a result of trading or political contacts between the native peoples and Europeans. The cities that Europeans initiated and founded were planned and built according to European notions of city planning. Their internal geography still shows port zones; enclaves of former European settlements; barracks; colonial government buildings; and racial, religious, and ethnic ghettoes based on the role each group played in the city's founding and during the colonial period (Figure 10-25 and look back at Figure 6-30).

In other cases, Western or modernizing interests built a new city alongside a preexisting native city. This was particularly common in those parts of Latin America (particularly today's Mexico and Peru), Asia, and North Africa where the local peoples had achieved significant urbanization before the coming of the

FIGURE 10-23 Low-income housing in Newark, New Jersey. The abandoned high-rise public housing in the background (soon demolished) has been replaced by the low-rise development in the foreground. Low-rise housing can achieve equivalent density, and it has been demonstrated to foster more successful communities than high-rise projects. Between 1992 and 2003 the federal government tore down more than 100,000 apartments nationwide in high-rise buildings and replaced them with low-rise housing. (Keith Meyers/*The New York Times*)

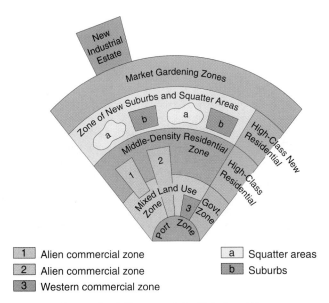

1 Alien commercial zone
2 Alien commercial zone
3 Western commercial zone

a Squatter areas
b Suburbs

**Generalized Diagram of the Main Land Use
Areas of the Large Southeast Asian City**

FIGURE 10-25 Model of a Southeast Asian port city.
Most Southeast Asian port cities are the product of Western
influence, so they reflect Western design forms. (Adapted
from T. G. McGee, *The Southeast Asian City.* Reprinted with
permission of Greenwood Publishing Group, Inc. Westport, CT.
Copyright © 1967 by Greenwood Publishing Group, Inc.)

Europeans. Today the two cities often contrast sharply:
Historic cores in forms traditional to the local culture
stand beside districts of modern commerce, retailing,
industry, and associated residential areas. The modern
or Western cities typically demonstrate grid layouts, just
as the Greek Hippodamus laid out the city of Miletus at
the beginning of Western civilization. In some cases a
third element—vast slums of newcomers—surrounds
both cities. The city of Fez, Morocco, for example, is
clearly two cities (Figure 10-26): A modern Western-style
city of broad, straight avenues lies on the plain below an
older, traditional Islamic city. Western forms continue to
be stamped on the world's cities today. Western architects, engineers, and urban planners, or non-Western
individuals educated in Western schools, continue to
transform even the older traditional built environments.

Individual cities, however, emerged at different
times, for different reasons, and within different cultural contexts. Many great cities still boast historic cores
that illustrate indigenous principles of urban planning,
and they cannot all be squeezed into three or four simplified models.

Islamic urban form The non-Western culture with
the oldest and most articulate urban planning tradition
is the Islamic culture. Traditional Islamic cities well

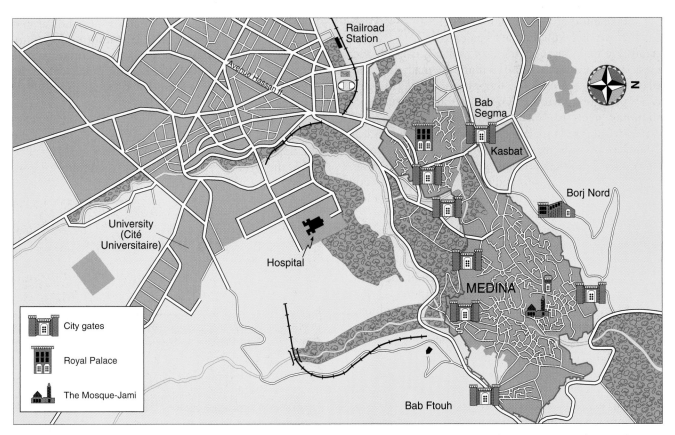

City gates

Royal Palace

The Mosque-Jami

FIGURE 10-26 Fez, Morocco. The Moroccan city of Fez is really two contrasting cities. A modern Western city lies on the
plain to the southwest (left in this figure) of an old Islamic city, the *Medina*. The new city has straight formal avenues, the
railroad station, university, and a modern hospital. The Medina contains old mosques and narrow constricted streets, and it is
surrounded by city walls with great gates.

illustrate the role of culture in urban form. There are regional differences in cities across the Islamic realm, but most nevertheless show surprising similarity. These cities may seem chaotic to Westerners at first glance, especially those of us accustomed to grid patterns, but they present an entirely rational structure. The structure develops from the basic premises that religion is the most important consideration in life and that houses of worship are the most important elements of urban design. The resulting design characteristics can be identified from Seville, Granada, and Córdoba in Spain to Lahore in Pakistan, and elements of these principles can be found from Dar es Salaam in Tanzania to Davao in the Philippines.

The logic of traditional Islamic urban planning is announced in the Koran and has been codified by various schools of Islamic law. Certain basic regulations govern individual rights and the pursuit of the virtuous life in a densely crowded urban environment. For example, Islamic urban planning recognizes the need to maintain personal privacy; it specifies responsibilities in maintaining urban systems on which other people rely, such as keeping thoroughfares or wastewater channels clear; and it emphasizes the inner essence of things rather than their outward appearance. This last principle applies as much to the decoration of houses as to purely spiritual issues.

Take another look at Figure 10-26. At the heart of the traditional Islamic city stands the main mosque, the *jami*, which is typically the city's largest structure. Close to it are the main *suqs*, the street markets and enclosed shopping arcades. These arcades prefigured urban *galleries* in Europe and enclosed shopping malls in North America. Within the suqs, trades are diffused in relation to the mosque. The tradespeople who enjoy the highest prestige, such as booksellers and perfumers, are closest. Farthest away are those who perform the noxious and noisy trades, such as coppersmiths, blacksmiths, and cobblers. The neutral tradespeople, such as clothiers and jewelers, act as buffers.

An immense fortified *kasbat* is attached to the ramparts, on which are located several towers or gates. The *kasbat* was the place of refuge for the governor or sovereign. It had its own small mosques, baths, and shops, in addition to government buildings and barracks.

Everywhere else the city is filled in with cellular courtyard houses tied together by winding lanes. Housing is grouped into quarters, or neighborhoods, that are defined according to occupation, religious sect, or ethnic group. The widest streets usually radiate outward from the core to gates in the city wall. Slightly narrower streets serve the major quarters and define their boundaries, and still narrower third-order streets are used primarily by people who live in the neighborhood (Figure 10-27). Narrow streets provide vital shade, keep down dust and winds, and use little building land.

Interior courtyards, often with trees and fountains, provide shade in hot climates, but, more impor-

FIGURE 10-27 Casablanca, Morocco. In many traditional Islamic cities—as here, in Casablanca, Morocco—occasional straight streets provide views of minarets. The intent is to remind people of the importance of religion. (Courtesy of Moroccan National Tourist Office)

> Traditional Islamic cities well illustrate the role of culture in urban form.

tant, they provide an interior and private focus for life sheltered from public gaze. This is true in Mediterranean architecture in general. The outside of a house may be plain, but the interior and courtyard may display lavish wealth and decoration. The interior vividness parallels the Koranic emphasis on the richness of the inner self compared to a more modest outward appearance.

In the United States, New Orleans is the city closest to exemplifying these values (Figure 10-28). In the old Spanish sections of the city (misleadingly called today the French Quarter), houses generally present

FIGURE 10-28 New Orleans's "French Quarter." Here, as in Islamic countries and Mediterranean architecture in general, houses are not placed in the centers of gardens; courtyard gardens are placed in the centers of houses, not impressing the neighbors and passersby but providing cool private retreats. (© Robert Holmes/CORBIS)

plain fronts to the street, but many enclose beautiful private courtyards. When Anglo-Saxons began to move to the city after its annexation by the United States in 1803, the Anglo-Saxon rich preferred to build grand homes in a new part of town called the Garden District. In the French Quarter the courtyard gardens are inside the houses, but in the Garden District the houses sit in the middle of their publicly displayed gardens. This difference is one of cultural preference.

CITIES AND SUBURBS IN THE UNITED STATES

The dominant feature of metropolitan form in the United States has been the explosive growth of cities out across the countryside. Growing cities have spilled over their legal boundaries into areas called *suburbs*. Some suburbs are entirely residential, but others offer services for the surrounding residential population. Some suburbs are older cities that have been engulfed by the growth of a larger neighbor, but others are newly incorporated settlements. What defines an area as a suburb is its economic and social integration with a larger population nucleus nearby. *Town* and *village* are inexact terms that generally designate settlements smaller than cities, but that may be incorporated.

Many of the developments described in the following discussion are now occurring elsewhere around the world, but they occurred first in the United States—largely because of the nation's prosperity.

The Growth of Suburbs

Early suburbs   Most large U.S. cities included manufacturing districts by the late nineteenth century. These were noisy and dirty, and they often attracted a working class, largely made up of immigrants, whom many long-established residents found to be unpleasantly "different." These biases pushed those who could afford it to move to the suburbs.

At the same time, a cultural preference for rural or small-town life pulled many people out of the city. Many Americans fell in love with the idea of "the country," and they favored a return to nature, to the land or to open spaces—even if only a suburban yard. Therefore, when the railroads brought older rural communities within commuting distance of the city, many people who had the time and the money necessary to live outside the city and to commute to work began to do so. In other cases, the wealthy built new towns (Figure 10-29). "Streetcar suburbs" sprang up when that new means of transportation appeared. Some of these planned suburbs eventually became completely built up, merged into other settlements, and lost their identities.

The automobile ultimately opened the nation's landscape to suburban growth. For those who disliked

FIGURE 10-29 Riverside, Illinois. The designers of New York City's Central Park, Frederick Law Olmsted and Calvert Vaux, planned this real estate subdivision 14 kilometers (9 miles) from the center of Chicago in 1869. The plan included two straight business streets paralleling the railway into the city, but all residential streets were curved to slow traffic (before cars!). Open spaces contribute to the sense of breadth and calm enjoyed by "the more fortunate classes" for whom Riverside was designed. In 1992 Riverside residents refused federal financial assistance for traffic control because federal regulations would have required traffic intersections to be reengineered to 90° angles. (Courtesy of the National Park Service, Frederick Law Olmsted National Historic Site)

urban life, the suburb was the solution. "We shall solve the city problem," wrote Henry Ford in 1922, "by leaving the city."

Government policies and suburban growth   The dispersion of housing in suburbs was slowed by the Great Depression in the 1930s and by World War II in the early 1940s. Following the war, however, government policies established a new balance of push and pull forces that encouraged the movement of investment, residents, and jobs out of the central cities into the suburbs.

The Federal Housing Administration (FHA) guaranteed loans, so down payments shrank to less than 10 percent of the house price. Suddenly thousands of families could afford new houses. FHA benefits did not apply equally everywhere or to everyone. The FHA favored the construction of new single-family houses in the suburbs over the rehabilitation of older houses or apartment buildings in the central cities. Also, the FHA opposed what it termed "inharmonious racial or nationality groups." In some places the presence of one non-white family on a block was enough to cut the entire block off from FHA loans. Thus, government policy helped segregate the suburbs.

The government also granted tax and financial incentives to homeowners. These included the deductibility of both mortgage interest payments and of local property taxes from gross taxable income. These two benefits alone often made buying a new house cheaper than renting. In addition, the government protected homeowners from capital gains taxation—that is, taxes on an increase in the value of the house. By fiscal 2003, the tax loss to the U.S. Treasury of these three benefits totaled well over $100 billion per year. This is, in effect, a subsidy to homeowners that far exceeds government spending on public housing. Furthermore, until 1980 the government allowed savings-and-loan institutions to pay higher interest on savers' money than commercial banks, because money in savings and loans was directed into housing.

The Veterans Administration Housing Program, begun in 1944 and called Homes for Heroes, pumped additional billions of federal dollars into housing programs. By 1947 the Levitt Company was completing 30 new single-family homes in Levittown, formerly a Long Island potato field, each day, and similar developments were springing up on the outskirts of every major U.S. city (Figure 10-30). Nationwide housing starts jumped from 114,000 in 1944 to 1,696,000 by 1950.

The suburbs brought home ownership to an increasing share of U.S. families. The percentage of U.S.

	Cities	Suburbs	Nonmetropolitan
1950	32.9%	23.2%	43.9%
1960	32.3%	30.6%	37.0%
1970	31.4%	37.2%	31.4%
1980	30.0%	44.8%	25.2%
1990	31.3%	46.2%	22.5%
2000	30.0%	51.0%	19.0%

FIGURE 10-31 Distribution of U.S. population. By 1970 a plurality of the U.S. population (37 percent) lived in the suburban portions of the MSAs, and by the century's end, a majority of Americans were suburban. Canada's 2001 census counted two-thirds of Canadians living in 25 "census metropolitan areas," with over 50 percent of the total national population in just four great conurbations (Toronto, Montreal, Vancouver, and Calgary/Edmonton). The Canadian population became predominantly urban in the mid-1920s and predominantly metropolitan in the mid-1960s, about a decade later than the United States. This is explained by Canada's lower population, relatively low density and greater dispersion, and the importance of Canada's resource industries.

FIGURE 10-30 Levittown, Long Island. In building their first Levittown, 40 kilometers (25 miles) east of Manhattan, the Levitt family changed U.S. homebuilding techniques. The land was bulldozed and the trees removed, and then trucks dropped building materials at precise 19-meter (60-foot) intervals. Construction was divided into 27 distinct steps. At the peak of production more than 30 houses were completed each day. Through the years owners have personalized them so much that few visitors today can see that the houses were originally identical. (UPI/CORBIS BETTMAN)

housing that was owner-occupied rose from 44 percent in 1940 to 62 percent in 1960, signaling middle-class status for a rising share of the population. Expanding home ownership increased the number of citizens who profited from the many homeowner subsidies, so it has reduced the political possibility of rescinding them. Some scholars argue that the tax concessions were never necessary. Canada and Australia achieved comparable levels of homeownership without offering such liberal concessions.

As the population spread out, new suburbs incorporated, and the Census Bureau devised a term for these sprawling conurbations: **metropolitan statistical area (MSA).** The Bureau defined a metropolitan area as "an integrated economic and social unit with a recognized large population nucleus." Thus, MSAs are the principal central cities and their suburban counties (except in New England, where the definitions are in terms of cities and towns). When two or more MSAs are contiguous, each is called a primary metropolitan statistical area (PMSA), and the conurbation is called a **consolidated metropolitan statistical area (CMSA).** By 2000 the nation's 274 metropolitan areas contained 80 percent of the total population. These MSAs covered only 20 percent of the country's land surface (Figure 10-31).

The suburban infrastructure   Unplanned sprawl of single-family homes is expensive. It first requires roads. Individualized transportation also demands energy, and so does heating and cooling individual homes. Dispersed housing also requires enormous investment in sewerage, water pipelines, telephone lines, and electrical wiring. The cost of providing electric wiring for 100 people in an apartment building is much less than the cost of wiring for the same 100 people spread out in 20 single-family homes over several hectares.

Infrastructure costs are further inflated by the fact that U.S. suburbs have not expanded contiguously outward from the city, like the waves from a stone tossed into a pond. Each developer wants to buy land as cheaply as possible, so the developer buys land beyond the edge of growth. This is called *leapfrogging*. The infrastructure network cannot be advanced in a regular pattern. This increases initial costs. Later the leapfrogged areas are filled in, and then some infrastructure has to be rebuilt. Many suburbs, for example, originally relied on individual home wells for water, and their sewage was treated in individual home septic tanks. As the suburbs matured and density increased, water supplies became polluted. Homeowners had to pay for wholly new public water mains and sewers. By 1997, 5 percent of septic tanks in Maryland and Virginia suburbs of Washington, D.C., were failing each year, contaminating local groundwater and Chesapeake Bay. Maryland homeowners alone collectively face a $2 billion repair-and-replacement cost. Suburban development also increased stream runoff, so national financial losses from flooding have risen virtually every year.

The high infrastructure costs necessitated high property taxes, and these eventually provoked voter backlashes. The most notable was California's Proposition 13 of 1978, which limited property-tax increases. This has hobbled the government's ability to provide services no matter how much they are needed.

Suburbs are spacious, and since 1945 U.S. urban areas have spread at the rate of about 405,000 hectares (1 million acres) per year. This required that a good share of the country's most productive farmland be paved over, such as the Long Island potato fields covered by Levittown. The agricultural hinterlands of many cities disappeared, so a rising percentage of the nation's food is now grown in conditions requiring expensive fertilizer or irrigation. In addition, the food has to be transported farther, consuming still more fuel, perhaps requiring refrigeration or special handling, and further boosting food prices.

Farmers were virtually forced to sell their land to developers because property taxes are calculated on land's potential value, not its current-use value. A farmer who was making a small profit by farming could not afford to pay property taxes calculated on the land's potential value as housing. Even if a farmer could somehow meet the annual property-tax bill, the farmer's heirs would eventually have to sell the farm to pay inheritance taxes, which are also based on land's potential value. Only recently have these tax laws been changed to preserve greenbelts around some cities.

In some cases, farmers pay use-value taxes, but some farmers have sold development rights to conservation groups or to local governments, thus guaranteeing that the land will remain as farmland.

In the 1950s and 1960s, U.S. citizens did not worry about the costs of creating new suburbs. Between 1950 and 1973 median family income doubled in real terms. Demands for housing continued to mushroom with the baby boom and the splintering of families into separate households, partly as the result of a rising divorce rate. The average number of occupants of a U.S. household shrank from 3.67 in 1940 to 2.6 in 2000.

Today a greater share of U.S. wealth is invested in housing and the necessary infrastructure than in any other nation. A high proportion of Americans enjoy private ownership of spacious, free-standing, well-equipped homes, and this investment has succeeded in bringing a sense of well-being to many Americans. It is becoming clear, however, that this development brought with it some steep social costs.

The Social Costs of Suburbs

The suburban lifestyle has imposed social costs on those who enjoy it and on many others, also. Americans generally sort themselves out residentially so distinctly that sociologists and mass marketers can confidently construct an astonishingly accurate profile of people on the basis of their address alone (Figure 10-32). This residential

FIGURE 10-32 You are where you live. Suburbs may differ greatly from one to the next, but individual suburbs often congregate people of similar income and lifestyle. This clustering allows retailers and advertisers to use demographic data stored in GIS's to locate potential customers. (Henry Martin © 1993 from *The New Yorker Collection.* All rights reserved.)

segregation of the American people—racial, economic, and in many ways cultural—allows advertisers to target potential customers geographically. A great deal of demographic data on peoples' income, education, and other characteristics has been sorted by ZIP code, so the selection of junk mail that arrives in mailboxes can be fine-tuned for each ZIP code. GIS mapping and analysis of neighborhoods and metropolitan areas offers fast-growing career opportunities. Retailers want to know where the most probable potential customers for their product or service reside, what paths they take for work or relaxation and, thus, where the retailer might best locate a new store or service. One marketing consulting firm has categorized the country's ZIP codes and census tracts into 62 "lifestyle types," giving them catchy names such as Gray Power to describe communities of affluent retirees in Sunbelt cities, Towns and Gowns to describe college towns, and Latino America to describe neighborhoods of Latino middle-class families.

In low-density suburbs, it has been argued, local racial and social homogeneity may have caused conservatism and conformity. In addition, property owners in many suburbs established *restrictive covenants*. These were legal agreements that the land would never be sold to people of a designated race or religious group. Such covenants are no longer legal, but they were common as late as the 1970s. Homeowners' associations, however, which regulate some aspects of property ownership, covered in 2003 20 million homes (of the nation's 106 million) housing 50 million people, and those numbers are growing. Most of these associations not only require that homeowners maintain their yards, but some dictate when owners can fly a flag, put up holiday decorations, or park their cars in their own driveways. Major changes in a home's structure or exterior appearance must be approved by the association, which can fine homeowners or even foreclose the offending homes—without the residents' knowing until they are evicted. Courts have accepted the argument that the regulation of "visual pollution" is analogous to the regulation of noise pollution, even though there may be greater disagreement on what constitutes visual pollution. In Portland, Oregon, the front of a house must be at least 15 percent windows and doors, and a garage door may take up no more than half the façade. The purpose of these rules is to maintain neighborhood property values, but some Americans find them stifling. An Arizona homeowner was sued for installing solar panels on his roof; a young suburban couple in Washington State went to jail for painting their house mauve. Homeowners' boards' powers are growing, because financially strained local governments are privatizing responsibility for services.

The suburbs overall are diverse—there are poor suburbs as well as rich ones, and there are suburbs of every ethnic and racial group. Individual suburban communities, however, are usually homogeneous.

The movement of jobs to the suburbs

The suburbs first expanded as bedroom communities for the middle-class workers who left their suburban families each morning to go to work in the city. Soon, however, the interstate highway system (authorized in 1956) and similar limited-access highways not only joined cities but also provided peripheral bypasses around them. These peripheral arteries provided access among suburbs and reduced the geographical advantage of a city's central business district. Retailers soon began to build giant shopping malls at highway crossroads. Developers put up suburban office buildings, and corporations built spacious office parks.

Manufacturing establishments abandoned the central city, too, for several reasons. New light-industrial facilities proliferated in the suburbs because new technology favors horizontal buildings rather than the vertical ones characteristic of the central cities (called *lofts*). In addition, light industries relocated to escape central-city congestion, as well as higher costs for energy, taxes, wages, and rent. Warehousing also relocated out from the inner-city railroad yards to the suburban highway interchanges. Metropolitan airports in the suburbs grew to provide jobs in both freight and passenger services.

In the 1950s, suburban growth was fed by young married couples who wanted to raise children away from the cities. By the early 1970s, however, the proliferation of jobs in the suburbs became the driving force for new housing. The suburbs surpassed the central cities in employment, and job opportunity became a pull factor for continuing suburbanization (Figure 10-33). Employment growth in U.S. suburbs has been greater than in central cities for every year since 1965. As jobs and housing continued to expand outward like waves from a stone dropped into a pool, the extent of any metropolitan area has had to be continually redefined, just as medieval cities periodically built new rings of walls. For a review of all of these developments, look back again at the Regional Focus Box on Utica, New York (p. 10).

Some of the new exurbs, called *satellite cities* or *edge cities*, boast greater volumes of retail sales and contain more office space than the old central cities. They even offer amenities formerly found exclusively in the central cities: art galleries, theaters, sports teams, and fine restaurants.

Changing commuting patterns and problems

When suburban workers commuted in and out of the city, radial mass-transit systems that focused on the central business district could serve transport needs tolerably well. Today, however, suburb-to-suburb commutes account for more than 50 percent of all U.S. metropolitan traffic. Only individualized transportation can serve populations that are spread out at low densities, and as a matter of government policy, gasoline has never been taxed in the United States at rates comparable to those

(a)

Atlanta
Region

CHANGE IN POPULATION
1990–1999

■ Greater than 2,500
■ 501–2,500
■ 0–500
□ Loss

0   5 Miles

0   5 Kilometers

Source: Atlanta Regional Commission

(b)

Atlanta
Region

CHANGE IN EMPLOYMENT
1990–2000

■ Greater than 2,500
■ 501–2,500
■ 0–500
□ Loss

0   5 Miles

0   5 Kilometers

FIGURE 10-33 Population and employment growth in metropolitan Atlanta. These maps show that in metropolitan Atlanta, Georgia, as in most U.S. metropolitan areas, the greatest population growth (a) and job growth (b) are still occurring in the periphery. The city of Atlanta is at the center of the ring highway visible on the map. Central-city neighborhoods began losing population 40 years ago, and inner suburbs began to lose population already 30 years ago.

in other rich countries. Already in 1969, 82.7 percent of workers drove to work, but by 2000 over 90 percent did. More family members have been going to work, so the number of cars has increased faster than the local populations.

The Texas Transportation Institute has been studying traffic in the 75 largest U.S. metropolitan areas since 1982. The Institute has noted that the time penalty for peak travelers in the 75 areas jumped from 16 hours per year in 1982 to 62 hours per year in 2000. The period of time each day during which travelers might experience congestion increased from 4.5 hours to 7 hours, and the percentage of metropolitan roadways on which travel was congested rose from 34 percent to 58 percent (Figure 10-34). The total cost of congestion for the 75 areas in 2000 came to $67.5 billion, which was the estimated value of 3.6 billion hours of

delay plus 5.7 billion gallons of excess fuel consumed. This accounting does not include the costs of air pollution or of highway accidents. Commuting costs now consume more than 20 percent of the average household budget—as high as 23 percent in metropolitan Atlanta and Houston—and the percentage is rising. Older people often have difficulty driving, so as the American population ages, older people may be isolated or traffic accidents may increase.

Potential solutions include more roadways, mass transit, ramp metering, reserved high-occupancy lanes, incentives to make trips at different times of the day, methods of broadcasting information on traffic jams, in-car guidance systems that include live traffic data, better signal timing, and improved clearance of accidents. These will, however, take time and expense to achieve, and meanwhile congestion will continue to

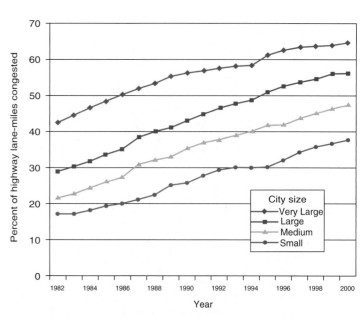

FIGURE 10-34 Increasing U.S. metropolitan highway congestion. The Texas Transportation Institute divides 75 U.S. metropolitan areas into four size ranges, and drivers in all are suffering lengthening delays on increasingly congested roads.

grow (Figure 10-35). Satellite, GPS, and GIS technologies that are currently available only to the military and to security forces would allow tracking of every vehicle and could improve traffic flows, but citizens might be wary of such scrutiny.

Since 1991, in an attempt to stimulate experimentation, the federal government has allowed states to use federal transportation funds for any means of transportation they choose. Salt Lake City, Utah, continues to build highways and to develop 400 hectares (1,000 acres) of land per month. The built-up area is expected to double in size, reaching 800 square miles by 2020, while the population is expected to increase by only 50 percent. Milwaukee, Wisconsin, by contrast, is using federal funds to tear down highways in an effort to revitalize the central city. Other metropolitan areas are experimenting with traffic monitoring and new methods of informing drivers of traffic jams. The technology for electric cars has not proved practicable. Hybrid cars, however, having both electric and internal combustion engines, are already seen on America's highways, achieving significant economies and lowering pollution. Several cities are rediscovering mass transit. Los Angeles built a 60-mile subway and light-rail system in the 1990s, although many observers think that it is ill-suited to such a dispersed, centerless place. In 1999 the state of Georgia created a

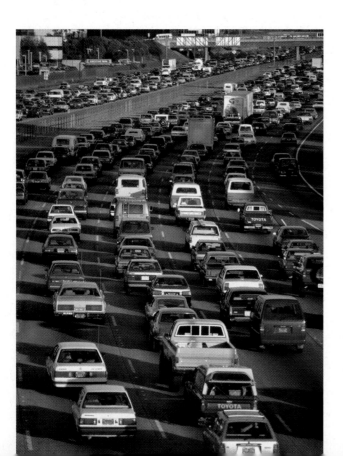

FIGURE 10-35 Road rage. As the frustration of sitting in traffic jams has increased on U.S. highways, the National Highway Traffic Safety Administration estimates that one-third of all fatal car crashes, in which almost 50,000 people die per year, could be attributed to "road rage," which is defined as "aggressive and even violent behavior by drivers caused by frustration or by the actions of other drivers." Road rage may win entry into the American Psychiatric Association's official *Diagnostic and Statistical Manual of Mental Disorders.* (Vince Streano/Corbis/Stock Market)

Critical Thinking

THE QUESTION OF PUBLIC PLACES ON PRIVATE PROPERTY

One significant difference between traditional downtowns and most new villages and suburban malls is that the latter are private property. People cannot be banned from a traditional downtown, and the right to petition on a public sidewalk is constitutionally protected. People may, however, be banned from private property, and constitutional rights, such as political pamphleteering, may be restricted. In many communities today, malls are the only public gathering places, so if the mall owners are allowed to decide who may speak in them, mall owners can determine the public's access to ideas. Candidates for political office have been banned from busy malls owned by their opponents, and so have labor union organizers. In March 2003, during the second U.S. war against Iraq, a shopper was arrested for trespassing at a mall in Albany, New York, when he refused to take off a T-shirt reading "Peace on Earth."

Some communities are even locating schools in malls. The Landmark Shopping Center in Northern Virginia, for example, includes the Fairfax County Mall School. The government insists that the students "feel comfortable" coming to the mall; they can shop during class breaks, and the mall offers opportunity not only for after-school employment but for "career-training programs" in retail sales. Mall restaurants replace costly cafeterias, and heating and air-conditioning expenses are also reduced.

Private environments are patrolled by private security officers, one of the country's fastest-growing occupations. (Security guards constitute almost 2 percent of the nation's total labor force, which is triple the number of public police officers.) The privatization of space is a new form of economic and social segregation that carries profound influence throughout U.S. political and social life.

U.S. law recognizes that private properties can perform functions traditionally associated with government. This is called the *public function doctrine*. The U.S. Supreme Court has ruled that the U.S. Constitution does not protect citizens' access to shopping centers against the wishes of the owners (*Pruneyard* v. *Robins*, 1980). Some states, however, have upheld the right of

A security guard in a shopping mall. Does he enforce the law? The mall owner's rules? Can he arrest people? Evict them? What for? (L. Mulvehill/The Image Works)

public access under their state constitutions. In 2000, for example, New Jersey struck down malls' limits on political leafleting. Each state has balanced public and private rights differently.

Questions

1. Are political candidates allowed to campaign at any malls where you shop? Would you be allowed to set up a table and gather signatures to run for mayor yourself? Where else could you go in your community?

2. Are beggars or charity fundraisers found in your downtown? At the mall? Might their presence distract you from shopping?

3. Does the public function doctrine cover malls in your state?

new Regional Transportation Authority with power over all transportation projects in the 13-county Metropolitan Atlanta region. It can finance and operate public transport, take state money away from localities that refuse to enact its plans, and veto major projects such as new malls.

The "New Urbanism"   As suburbs continue to expand, several new developments have tried to re-create small-town community life, with walkable streets, compactness, downtown shopping and services, and even front porches. This effort has been called *The New Urbanism*. Most such developments try to free residents from dependence upon the automobile (Figure 10-36). Seaside, Florida, a small Victorian town on Florida's Gulf Coast featured in the movie *The Truman Show*, was constructed in the 1980s, but Seaside is just a vacation resort. Celebration, Florida, was developed by the Disney Corporation near Orlando, Florida, in the 1990s. It tries to re-create small-town American life: Oak trees and white picket fences line the town's quiet streets, and all buildings are in traditional styles of architecture. Living in the town means obeying the rules of the Celebration Company, a Disney subsidiary created to run the community instead of elected officials.

Improving Metropolitan Rail Links

FIGURE 10-37 A crucial link between two suburban rail stations in the New York metropolitan region that was opened in 2002 shortened commuting time, increased housing prices, and stimulated commercial development, with new stores and businesses opening all along the line. The convenience of suburban rail service—introduced in the nineteenth century—is enjoying new appreciation.

FIGURE 10-36 A new planned town. Valencia, California, is a new town 48 km (30 miles) north of downtown Los Angeles, where it occupies a 15,000-acre portion of an historic ranch owned by Newhall Land since 1875. The company has been planning and developing Valencia since 1965, and Valencia currently is home to approximately 48,000 residents in 17,000 homes. Its master plan balances residential uses with employment, education, recreation, open space, shopping, and services, so many Valencia residents live, work, play, and shop entirely within the community. There are 48,000 jobs within Valencia. Its residential villages, business parks, neighborhood retail and recreation centers, schools, parks and shopping center are all interconnected by a 45-kilometer (28-mile) system of landscaped pedestrian walkways, bike trails, and bridges called *paseos*. (Carol Maglione, Newland Land)

Some of these suburban centers form self-contained villages but offer mass transit into the central city, thus reproducing the advantages of the original planned suburbs such as Riverside. Orenco Station, Oregon, 30 minutes from downtown Portland, links traditional design with light-rail public transportation, much as the earliest suburban developments did in the nineteenth century. Seattle, Washington, opened a 23-mile light-rail line in 2001, with villages spaced at intervals along it. Minnesota's Twin Cities Metropolitan Council is mapping traditional town centers for suburbs that lie in the path of new transportation projects being built there. Across America, those suburbs that enjoy rail service into the central city are growing, and in many cases rail service is reviving (Figure 10-37). For example, in the 1990s the Massachusetts Bay Transit Authority restored public rail service within 50 miles of Boston that was terminated in the 1970s. The result was a boom in each community affected—both in new developments and in historic towns such as Newburyport, Massachusetts. Newark, New Jersey, is considering restoring a network of trolleys that honeycombs the city and reaches out into the suburbs. The system was abandoned in favor of private automobiles 60 years ago. The 1988 classic film *Who Framed Roger Rabbit?* is a work of comic fiction, but it tells the story of how automobile companies bought and deliberately destroyed Los Angeles's mass-transit network to boost car dependency. Auto interests did the same in many U.S. cities until courts declared such purchases illegal.

Suburbs at Century's End

Advances in information technology, rising incomes, population growth, and infrastructure spending continue to drive residential and business development to the fringes of metropolitan areas. In America's metropolitan areas, suburban population continues to grow faster than the central city population, so that by 2000 central cities housed only 30 percent of the national population. The number of jobs continues to grow faster in suburban areas, too, so about 60 percent of metropolitan jobs are now in suburbs. Furthermore, the rate of development of suburban land actually increased in the 1990s. The national population is growing at 1 percent per year, but land occupied by single-family houses is rising at 2 percent per year. Suburban airports are multiplying and expanding to serve personal and corporate aircraft, but this activity is launching new battles over land use, air, and noise pollution.

The geography of retailing Highway congestion, more family members going to work, fear of crime, toll-free telephone numbers, credit cards, television and Internet shopping, and other factors are changing the geography of U.S. retailing. More Americans are shopping from home by telephone, catalog, or computer. Elaborate catalogs fill mailboxes, and today almost two-thirds of the adult population buys at least occasionally from home. These developments introduce *virtual shopping* (see Chapter 6), which reduces the length of time before orders are filled, the quantity of goods sitting in inventories, and the number of middlemen standing between a product and the consumer.

In 1999 the U.S. government estimated that electronic commerce between businesses would reach $1.3 trillion by 2003; it actually reached $2.4 trillion. The government estimated that U.S. consumer e-commerce would reach $108 billion, and it reached $95 billion. So many music lovers are buying CDs on the web, or simply downloading music, that stand-alone music retailers such as Tower Records, Sam Goody, and the Wherehouse were all virtually bankrupt by 2003. Many travelers now arrange their trips through the Internet, so in 2002 alone 13 percent of traditional U.S. travel agents closed their doors. About five percent of U.S. retail shopping occurred on the Internet in 2002, with eBay ranking as one of U.S.'s largest 15 retailers (it's the nation's leading used car dealer) and Amazon in the top 40.

Many web-based retailing companies went broke early in the twenty-first century, but Amazon succeeded in lowered marketing, inventory, and warehousing costs (Figure 10-38). Its revenues reached $4 billion in 2002 and were growing at 20 percent per year. The company facilities are not footloose, but technology can significantly lower costs. Company founder Jeff Bezos insists, "In the physical world it's the old saw: location, location, location. [But] the three most important things for us are technology, technology, technology."

U.S. retailers doubled shopping space per capita between 1970 and 1996, but as shopping at home increases, stores have begun to suffer. Almost 20 percent of the shopping malls in the United States in 1990 were out of business by 2000. Malls are today being rebuilt as destinations for a wider range of activities than just shopping. Today malls offer movies, skating rinks, games, and a greater variety of restaurants and other

FIGURE 10-38 Amazon's computerized warehouse. Amazon's warehouse in Fernley, Nevada, a quarter-mile long and 200 yards wide, holds over three million different books, CDs, toys, and housewares. Computers send orders to workers' wireless receivers, weigh and sort packages, and apply shipping labels, so the facility can box 11,000 orders per hour. Consumers can order any two items out of the three million and have them delivered quickly in one box—usually at a discount price. (Macduff Everton Studio)

FIGURE 10-39 **New attractions at malls.** A new aquarium and marine science center draw visitors to the Mall of America—the largest in the United States, in Bloomington, Minnesota. (Courtesy of the Mall of America)

attractions (Figure 10-39). Even with such diversions, many of the 16 million sales jobs in the United States may be vulnerable to elimination.

Telecommuting It is easier to move information than to move people, so more people are working at home at computer terminals. This is called **telecommuting.** About 12 million employees in the United States work from home on computers, and several state governments have telecommuting programs. Some government bureaus have distributed telecommuting centers throughout the suburbs, complete with office equipment.

Telecommuting eases the pressure on transport facilities, saves fuel, reduces air pollution, reduces the demand for office space, reduces absenteeism, lowers the fixed costs of a business enterprise, and has been shown to increase workers' productivity. It also allows employers to accommodate employees who want flexible work arrangements, thus opening employment opportunity to more people, such as women who are still homemakers. Telecommuting tends to reduce central-city employment, but we do not yet know what other ramifications increased telecommuting will mean for metropolitan geography. Americans' fear of terrorist acts may also stimulate both virtual shopping and telecommuting, if people fear crowded public spaces and prefer to disperse activities.

Developments in the Central City

America's central cities suffered economic decline through most of the second half of the twentieth century, but several began to show signs of revival at the end. The drain of jobs from the central city and the concomi-

tant development of the suburbs hollowed out many U.S. metropolitan areas. Many central business districts lost their purpose. Commercial, professional, and financial offices relocated to the suburbs, and upscale retailing followed. Many U.S. downtowns came to consist only of a government center, a convention center, and a few hotels; the streets were deserted after 6 P.M. The total populations of many central cities fell. Between 1970 and 1995, population fell by the following amounts in major U.S. cities: Chicago, 19 percent; Philadelphia, 22 percent; Detroit, 34 percent; Baltimore, 22 percent; Washington, 25 percent; Boston, 15 percent; Cleveland, 34 percent; St. Louis, 41 percent; Pittsburgh, 31 percent; and Cincinnati, 21 percent. We shall see below that some revived at the end of the century.

Economic decline Central cities can thrive as long as (1) their economies offer a complete range of job opportunities from entry-level jobs for the unskilled up to specialized jobs for skilled workers, and (2) family stability, education, and other social systems help urbanites ascend the socioeconomic ladder. In other words, cities do not have to retain their middle classes, but they have to offer the lower classes opportunity to become middle class. Unfortunately, the departure of the middle classes for the suburbs occurred at the same time as two other developments.

First, urban economies were transformed by the out-migration of entry-level jobs, particularly in manufacturing, construction, and warehousing. This out-migration broke the rungs of the ladder of upward mobility. In 1968 economist John Kain first suggested that the removal of manufacturing jobs to the suburbs and the concentration of the poor in the central cities created a spatial mismatch between the suburban job opportunity and central-city low-income housing. This spatial mismatch, he argued, could explain the high unemployment found in the central cities. This **spatial mismatch hypothesis** has until today dominated analyses of inner-city unemployment—and therefore the formulation of potential solutions. We will discuss a new alternative hypothesis shortly.

Second, at the very time that many unskilled jobs were leaving the central cities, new waves of unskilled workers were pouring into them. The stream of new migrants to the cities included African Americans from the rural South, Latinos, and other immigrants. This influx continued long after the numbers of entry-level job opportunities began to shrink.

Much of the inner-city housing stock began to deteriorate. This was because the government incentives still made it profitable to give up a house in the city and to move to the suburbs. Policy makers thought that if new suburban houses were available to middle-class people from the cities, the urban poor could move into the older city dwellings. This succession, called *filtering*, would solve the housing problem for lower-income families. Much of the central-city population remaining

(b)

FIGURE 10-40 Pittsburgh has been transformed. No U.S. city illustrates a transformed economy better than downtown Pittsburgh, which was transformed from a dirty and smoky industrial area (a) into a new park and gleaming service center, a "Golden Triangle" (b). This is where the Allegheny River (at left) meets the Monongahela (at right) to form the Ohio. ([a] UPI/CORBIS BETTMAN; [b] Chuck Savage/Corbis/Stock Market)

behind, however, was financially incapable of maintaining the inherited housing stock. Therefore, many central-city neighborhoods deteriorated. The term "inner-city neighborhood" became a euphemism for slum.

The service economies   At the end of the twentieth century, however, some central cities began to enjoy new growth in financial, information, and specialized technical services. In these areas skyscrapers replaced rusty factories (Figure 10-40). New York's leading export, for instance, was for decades garments, but today it is legal and financial services. In 1975 Baltimore's leading employer was the Bethlehem Steel Company; today it is Johns Hopkins University Medical Center. In 2002 almost 10 percent of all jobs in the Megalopolis from Boston to Washington, D.C., were in health care. This shift in job opportunity exemplifies a switch from *blue-collar jobs,* performed in rough clothing and usually involving manual labor, to *white-collar jobs,* salaried or professional jobs that do not involve manual labor.

Analysts of urban economies suggest that cities increasingly compete for jobs and growth on the basis of creativity. Surveys have found, for example, that growth in the biotechnology and computing industries is clustered in and being driven by a few metropolitan areas, including Austin, San Francisco, and Boston. These cities do not enjoy traditional economic advantages such as proximity to raw materials, cheap energy, or low costs of living. They do, however, share two features: a thriving arts scene (reflected statistically by a high number of artists, writers, and other arts workers) and a dense, highly diverse, and tolerant social character—characterized by, among other things, a high number of immigrants and gays and lesbians. Business journals periodically rank cities' relative

attractiveness, and although the mix of criteria chosen is not scientific, a high ranking can boost a city's fortunes (see Table 10-1). Analyst Richard Florida identified "the three T's" of an economically successful city: tolerance, talent, and technology. These features attract the people who are crucial to economic success: creative workers, the engineers and scientists who develop new products and industrial processes, and the creative businesspeople, financiers and other workers who start new businesses and improve the old. Such people have the skills and the means to live wherever they choose, and they are attracted to cities that offer the amenities and broad quality of life they desire.

The holders of the new white-collar jobs— sometimes called *yuppies,* young urban professionals— triggered a rediscovery and revival of urban life. They first occupied and restored select older residential neighborhoods in a process known as **gentrification,** but they soon began to convert even former industrial and warehouse buildings into residences. These conversions were usually in owner-occupied forms (condominiums or cooperatives) so that owners could profit from the tax advantages afforded to suburban homeowners. Some observers argued that the conversion of lofts to residences damaged the cities' chances for industrial resurgence, while others argued that the industries were never coming back, so the conversions were beneficial. Gentrified neighborhoods mix old and new architecture and a vibrant street life with upscale commercial activities such as book shops, gourmet food and wine shops, and art galleries. Yuppies have been joined by *empty nesters*—that is, older people who had raised families in large suburban homes but found urban attractions and activities more appealing in later life. Historic preservation movements have assisted central cities by winning tax advantages for reuse of older

TABLE 10-1 Different Ways of Ranking Cities' Desirability

In 2002, Professor Richard Florida of Carnegie Mellon University ranked the following "city-regions" as the most "creative" in the United States:

San Francisco, CA	Houston, TX
Austin, TX	Washington, DC/Baltimore, MD
San Diego, CA, and Boston, MA (tie)	New York, NY
Seattle, WA	Dallas, TX and Minneapolis-St. Paul, MN (tie)
Raleigh-Durham, NC	

By contrast, in 2003 the American Association of Retired Persons listed the following 15 cities as being America's best places to settle in retirement. The empty nesters referred to in the text choose to move into central cities and enjoy their cultural attractions, but most of these smaller cities offer a relatively low cost of living and a relaxed pace of life. The increasing numbers of retired persons may initiate a revival of many smaller towns. Young upwardly mobile people might find some of these cities relatively boring, but note that two cities are on both lists!

Loveland/Fort Collins, CO	San Diego, CA
Bellingham, WA	San Antonio, TX
Raleigh/Durham/Chapel Hill, NC	Sante Fe, NM
Sarasota, FL	Gainesville, FL
Fayetteville, AR	Iowa City, IA
Charleston, SC	Portsmouth, NH
Asheville, NC	Spokane, WA

buildings. New zoning for mixed use revives the cities day and night. Even many businesses that left the central cities in the period 1950–1980 are moving back downtown, discouraged by suburban traffic tie-ups and attracted downtown by relatively low rents and the high-quality workers. Thus, rising educational levels, shrinking family size, and aging all contributed to central cities' revival.

> Most job and population growth is still in the suburbs, but signs of central-city revival are appearing everywhere.

Most job and population growth is still in the suburbs, but signs of central-city revival are appearing everywhere. The 2000 census revealed that Chicago, New York, Atlanta, and many other central cities enjoyed population growth for the first time in decades. Violent crime dropped 34 percent in America's 10 largest cities in the 1990s, and young people increasingly choose the urban lifestyle. In the newest editions of guidebooks to the best colleges, most of the top-rated schools are urban institutions. Higher education is a mainstay of many central cities' economies, and the cities retain the graduates. Among the largest 100 U.S. cities, the 25 that began the 1990s with the highest percentage of college graduates ended the decade with even greater concentrations. These 25 cities saw the college-educated share of their populations jump 6 percent, twice the average growth of the other 75 cities.

At the same time, many cities have cultivated the tourist sectors of their economies and attract tourists with historic quarters, fine cuisine, shopping opportunities, performing arts, and art exhibitions and festivals (Figure 10-41). Memphis, Tennessee, for example, attracts 8–10 million tourists each year, about 14 per-

cent of whom are foreigners. The city has borrowed exhibitions of artworks from around the world, and the city's role in American musical history draws many tourists to exhibitions and performances. Many cities are building new museums and performing-arts centers just to pull in suburbanites.

The economic downturn that cost the United States millions of jobs early in the twenty-first century afflicted all areas of the country, but, for the first time

FIGURE 10-41 Graceland. The home and grave of musician Elvis Presley remains one of the most popular tourist attractions in Memphis, Tennessee. Tourists coming principally to visit Graceland spend an estimated $40 million per year in Memphis. That is $65 per city resident, and tourists' dollars have a high multiplier effect. (Liz Gilbert/Corbis Sygma)

in 50 years, the central cities suffered no more than the suburbs, but only proportionately.

The role of immigrants As noted in Chapter 5, immigrants to both the United States and Canada concentrate in major cities, giving them a new cosmopolitan sophistication. Immigrants have also played an important role in the cities' economic rejuvenation. Many bring capital or job skills, so for them the city's traditional advantages of agglomeration and external economies confirms the city's function as incubator of new businesses. In Los Angeles County, for example, corporations employing fewer than 100 people offer more than half of all jobs. Virtually all of the firms making clothing and textiles, toys, processed foods, furniture, and biomedical supplies are owned by foreign-born individuals. Furthermore, many immigrants have moved into declining neighborhoods and repaired deteriorating homes themselves—a process called investing *sweat equity* rather than money.

Latinos today outnumber blacks in seven of the ten biggest cities in the United States—New York, Los Angeles, Houston, San Diego, Dallas, Phoenix, and San Antonio. In Los Angeles, Houston, and San Antonio, Latinos outnumber non-Latino whites as well. (Table 10-2). The changing demographic mix in major U.S. cities carries political ramifications, as Latinos, Asians, and other minorities form new political

TABLE 10-2 A Decade of Racial Change in the Cities. A new analysis of 2000 census data shows that non-Hispanic whites are now a minority of the total population of the 100 largest cities, and 18 of those cities became majority nonwhite over the decade.

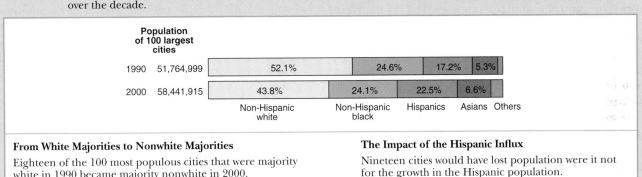

Population of 100 largest cities		Non-Hispanic white	Non-Hispanic black	Hispanics	Asians	Others
1990	51,764,999	52.1%	24.6%	17.2%	5.3%	
2000	58,441,915	43.8%	24.1%	22.5%	6.6%	

From White Majorities to Nonwhite Majorities

Eighteen of the 100 most populous cities that were majority white in 1990 became majority nonwhite in 2000.

City	Percentage Non-Hispanic White		Decline 1990–2000
	1990	2000	
Anaheim, Calif.	56.6%	35.9%	**20.8%**
Riverside, Calif.	61.3	45.6	**15.7**
Milwaukee	60.8	45.4	**15.4**
Rochester	58.3	44.3	**14.0**
Sacramento	53.4	40.5	**12.8**
Fort Worth	56.5	45.8	**10.7**
Augusta-Richmond, Ga.	54.0	43.7	**10.3**
Philadelphia	52.1	42.5	**9.6**
Boston	59.0	49.5	**9.5**
San Diego	58.7	49.4	**9.3**
Mobile, Ala.	58.9	49.8	**9.2**
Montgomery, Ala.	56.1	47.1	**9.0**
Columbus, Ga.	57.3	48.6	**8.7**
Norfolk, Va.	55.6	47.0	**8.5**
Albuquerque	58.3	49.9	**8.4**
Baton Rouge, La.	52.9	44.7	**8.1**
Shreveport, La.	53.6	45.9	**7.7**
St. Louis	50.2	42.9	**7.3**

The Impact of the Hispanic Influx

Nineteen cities would have lost population were it not for the growth in the Hispanic population.

City	Growth in Total Population, 1990–2000	Growth in Hispanic Population, 1990–2000
Anaheim, Calif.	61,608	69,619
Boston	14,858	23,134
Chicago	112,290	207,792
Corpus Christi, Tex.	20,001	21,029
Dallas	181,703	212,347
Des Moines	5,495	8,509
El Paso	48,320	76,206
Grand Rapids, Mich.	8,674	16,424
Hialeah, Fla.	38,415	39,891
Jersey City	11,518	12,557
Kansas City, Mo.	6,399	13,587
Long Beach. Calif.	32,089	63,673
Los Angeles	209,422	327,662
Miami	3,922	14,387
Minneapolis	14,235	21,275
Oakland, Calif.	27,242	35,756
Riverside, Calif.	28,661	38,489
Santa Ana, Calif.	44,235	65,714
Yonkers	8,004	19,376

Source: Analysis of Census Bureau data by the Center on Urban and Metropolitan Policy at the Brookings Institution alliances. The growing diversity of cultures and interest groups tends to submerge America's historic black-white dialogue in a new chorus of voices demanding political expression.

The shortcomings of service economies Upscale urbanites are only a fraction of the total inner-city population. The new white-collar jobs being created do not always equal the number of blue-collar jobs being lost, and while most high-paying jobs are in the service sector (doctors, lawyers, executives, sports stars, etc.), most jobs in the service sector are not high-paying. Creative Web site designers, for example, earn higher incomes than janitors do, but usually janitors outnumber Web site designers. In New York City between 1989 and 1999, a time of national prosperity, the only net growth in numbers of jobs was in jobs paying less than $25,000 per year, which is just about the national median for a full-time worker. The average annual income of New York City securities and commodities brokers in 2001, for example, was $147,867, whereas that of workers in the growing accommodation and food services industry was $13,674. Furthermore, workers who lose their blue-collar jobs often require retraining or education before they can capture one of the new opportunities.

The median household income in U.S. central cities hovers between 70 and 75 percent of the figure for the suburbs, and central-city unemployment rates hover about one-third above those in the suburbs. Furthermore, the percentage of jobs in most central cities held by commuters is rising—especially the percentage of the best jobs. Some 60 percent of the salaries earned in Washington, D.C., for example, go to suburbanites, and federal law protects them from D.C. taxes. In 1999 the New York state legislature, responding to suburban voting power, ended New York City's income tax on commuters. This gesture crippled the city's ability to provide services to them. The labor force participation rates (LFPR) of central-city populations—that is, the percentage of the population that is currently employed or even looking for a job—fall behind those of the nation and of the entire local metropolitan areas. For example, the LFPR for the city of St. Louis in 2001 was 58 percent, whereas for the St. Louis metropolitan area it was 69 percent. This suggests that a smaller percentage of the city's population was working and paying taxes, and also that a higher percentage was receiving public assistance.

Twenty-first century security concerns also distress central cities. As the most likely targets, central city populations suffer increased stress levels, and America's federal system of government functions so that even when the federal government and the state governments are financially healthy, the individual cities must bear the burden of high security costs.

The urbanization of African Americans and segregation by race and income Chapter 5 recounted how some 1.5 million African Americans migrated from the South to the cities of the North between 1910 and 1945, and an additional 6.5 million migrated north and west between 1945 and 1970. That second wave arrived in the central cities just as the central cities were losing their ability to provide entry-level job opportunities, and the migrants often faced discrimination in employment and segregation in housing. Deteriorating conditions in new African-American ghettoes eventually triggered civil unrest. A 1965 riot in the black Watts section of Los Angeles left 34 people dead and more than a thousand injured, and it required military occupation of 119 square kilometers (46 square miles) to halt the violence. The Watts riot was followed by 150 major riots and hundreds of minor ones in cities across America that summer and the next three summers.

Since the 1960s, civil rights have made great advances, and a great many of the black urban in-migrants have achieved success. Others, however, have been left behind. If a family's first urban generation failed to find employment, skills, and upward mobility, the second and third generations may have failed also. Los Angeles erupted again in April 1992. In 2001, black neighborhoods in Cincinnati erupted in three days of rioting after the fatal shooting of the fifteenth unarmed black man by police since 1995, and in 2003 20 houses burned in Benton Harbor, Michigan, during riots which broke out after the death of a black motorist killed during a police chase.

Residential racial segregation continues. The 2000 census revealed that a typical black U.S. city resident lived in a census tract that was at least 60 percent black. Statistics on America's 8.2 million individual blocks disclose that about one-third of blacks and more than half of whites live in blocks that are at least 90 percent of their own race.

During the 1990s, blacks' overall proximity to jobs improved slightly, but the 2000 census revealed that no demographic group remained more physically isolated from jobs than blacks. Some decline in the spatial mismatch occurred between 1990 and 2000, and that was principally due to the residential movement of black households within metropolitan areas. Nevertheless, in nearly all metropolitan areas with significant black populations, the separation between residences and jobs was much higher for blacks than whites.

Another changing pattern appeared in the 2000 census: Whereas one of American's most dramatic trends from 1960 to 1990 was the movement of the American poor into urban neighborhoods of concentrated poverty, that trend reversed itself between 1990 and 2000, and poverty became less concentrated. Concentrations of poverty magnify the problems associated with poverty in general: crime, delinquency, joblessness, drug trafficking, breakdown of the family, and low-performing local schools. Furthermore, when a neighborhood is a known concentration of poverty, middle and working-class people often see it as "dangerous." They avoid the neighborhood, which becomes increasingly isolated, socially and economically.

In 1990, 10.4 million people, which was 15 percent of all poor people, lived in census tracts in which at least 40 percent of the households were in poverty. By 2000, however, the number had declined to 7.9 million people (down 24 percent), about 10 percent of all poor people. Among blacks the concentrated poor population fell from 4.9 million people (30 percent of poor blacks) to 3.1 million people (19 percent of poor blacks). The total number of poverty tracts fell from 3,417 in 1990 to 2,510 in 2000. Observers believe that poverty dispersed because public housing projects were torn down, central cities enjoyed some economic resurgence, immigrants revived some urban neighborhoods, and millions of poor people left urban slums for other neighborhoods.

Poverty had not disappeared between 1990 and 2000; it had in fact increased, but some of it had moved to the inner ring of suburbs. Today fewer central city neighborhoods resemble the nightmarish scenes portrayed in the popular films *Fort Apache, the Bronx* (1981) or *Boyz N The Hood* (1991). Today's metropolitan reality was better captured by rapper Eminem's fictionalized biography *8 Mile* (2002), in which poverty is located in the dreary inner suburbs of Detroit—the American city that in fact saw the most dramatic reduction of concentrated poverty 1990–2000, down by almost 75 percent. Suburban poverty, which rose fastest in metropolitan Detroit, Chicago, Cleveland, and Dallas—is not pleasant, but, as a general rule, a mix of income and ethnic groups in suburbs reduces the isolation of the poor.

Only in pockets of the central cities are conditions at their worst, and even the worst slums of New York City, Chicago, Detroit, and Los Angeles do not compare with the conditions of life for many in Mexico City, Lagos, Nigeria, or Calcutta, India. The inner-city second or third generation living in deprivation, sometimes called the *underclass,* numbers less than 3 million people, which is less than 1 percent of the national population. Still, the conditions of deprivation and the lack of opportunity contrast starkly with the national self-image.

The network hypothesis Scholars are formulating a new alternative to the spatial mismatch hypothesis to explain some urban unemployment. They have learned that the primary qualification employers seek in new unskilled workers is reliability. The best way to find reliable new employees is to ask current employees for recommendations. Therefore, networks of family or friends provide unskilled urbanites seeking entry-level positions with information about jobs, sponsorship for jobs, and role models for work. Hiring is referential, not residential. Many urban black communities, after a generation or two outside the mainstream of labor opportunity, may lack these crucial links. The social networks conducive to upward mobility have deteriorated. Philip Kasinitz, a sociologist, has written, "The

primary reason for ghetto unemployment is not the lack of nearby jobs but the absence of social networks that provide entry into the job market." The theory that it is the lack of these networks that causes unemployment may be called the **network hypothesis.** Scholars investigating this hypothesis note that significant numbers of inner-city jobs, even those in manufacturing, have often been captured by groups whose residences are distant, even though local residents cannot get work. For example, Kasinitz found that jobs in a manufacturing district of Brooklyn, New York, were captured by Hispanics who commuted from New Jersey, even though unemployment stayed high in contiguous black neighborhoods. A 1993 study of Indianapolis by geographer Thomas Cooke of Indiana University similarly concluded that "it is not possible to argue that census tract African American male unemployment rates are related to the number of local job opportunities."

These studies do not conclusively replace the spatial mismatch hypothesis, but they introduce a new partial factor of explanation. Assumptions about what causes a situation will determine approaches to altering that situation. In this case our understanding of the causes of inner-city unemployment will determine what solutions we propose.

Efforts to Redistribute Jobs and Housing

The spatial mismatch hypothesis has dominated thinking about urban unemployment since the 1960s. Therefore, several government programs designed to deal with the problem of inner-city unemployment have addressed the spatial mismatch. Three approaches have been suggested: (1) Bring new blue-collar jobs to the cities, (2) Move inner-city residents to the suburbs, or (3) Transport inner-city workers to suburban jobs.

1. Bringing new blue-collar jobs to the cities involves efforts to reindustrialize the central cities by establishing **urban enterprise zones,** where manufacturers receive government subsidies. The Federal Budget Act of 1993 called for the designation of 65 urban enterprise communities and 30 rural zones as well. Several states and cities had already designated zones and granted manufacturers assistance in their poorest communities. Some of these zones have enjoyed success, exploiting the cities' advantage of existing infrastructure (water, sewerage, and so forth), but the factors that drive industries from the central cities are difficult to overcome.

Many central cities are scarred by abandoned industrial facilities called **brownfields.** There are an estimated 400,000 brownfield sites across the United States, including industrial properties, old gas stations, vacant warehouses, former dry cleaning establishments, and abandoned residential buildings that may contain lead paint or asbestos. They were generally developed in the 1920s and abandoned in the 1960s and 1970s, leaving scarred neighborhoods. Developers shunned these deserted industrial sites because of liability for buried wastes and other pollution-related risks. It was easier to buy undeveloped sites at the edges of cities, where new roads and water and sewage lines then encouraged sprawl. The most difficult brownfields to clean up are often within two miles of the downtown core, thus offering the greatest potential for reclamation and rededication to new land uses. The U.S. Environmental Protection Administration began providing grants for clean-ups in the early 1990s, and a 2002 law expanded the program. A 2003 survey of 244 cities found that federal help to renovate brownfields had cleaned 19,000 acres, boosting local tax revenue by as much as $1.5 billion and adding up to 570,000 jobs (Figure 10-42).

As many central-city districts were virtually abandoned, some scholars recommended that urban density shrink in a planned manner. In New York City's South Bronx and in parts of Detroit, Cleveland, Houston, and other central cities, decrepit apartment buildings have been replaced by new single-family dwellings, surprisingly resembling a typical suburban landscape (Figure 10-43).

2. The second approach is to move poor people from the central city out into new subsidized suburban housing. The federal government actually built suburbs for low-income people in the 1930s: Greendale outside Milwaukee, Greenhills outside Cincinnati, and Greenbelt outside Washington, D.C. Today, however, most suburbs zone to exclude subsidized housing or even private apartments with young families. Suburbs prefer

FIGURE 10-43 Central-city transformations. In December 1997 President Clinton visited a street of new suburban-style single-family homes in a neighborhood of New York City's South Bronx. Here low-income high-rise apartment buildings had been devastated by fires and virtually abandoned for 20 years. (Stephen Crowley/New York Times Pictures)

expensive single-family homes, because only these pay property taxes sufficient to cover the costs of the services that they and their residents require. Owners of undeveloped property, however, usually want to maximize development on their land, so they battle such *exclusionary zoning*.

A clause in the Fifth Amendment to the U.S. Constitution, called the *takings clause*, confirms governments' right to take private land, but requires "just compensation" for that taking. Courts have upheld public takings in order to build infrastructure such as parks and highways. In 1984 the Supreme Court extended the right of eminent domain to take land for any project "rationally related to a conceivable public purpose" (*Hawaii Housing Authority* v. *Midkiff*). Many local governments have condemned land for private ventures that would arguably enhance a community's tax base, although such actions have roused complaints of political favoritism. The city council of Riviera Beach, Florida, for example, approved the condemnation of 1,700 houses and apartments housing 5,100 people for a new development of shops, a hotel, a conference center, and yacht slips.

Today property owners are insisting that even short of complete seizure of land, exclusionary zoning restricts the use to which owners may put their land so severely that the zoning is in effect an unconstitutional taking. Other court battles have sprung up around the country in which property owners protest either environmental or historic preservation legislation as takings. The takings clause is a fine point in the interpretation of law, but in recent years several state supreme courts have banned exclusionary zoning. Several have actually required local communities to

FIGURE 10-42 A brownfield reclaimed. South End, in Charlotte, North Carolina, was formerly a blighted area of abandoned textile mills and warehouses; it now boasts new restaurants, stores, and condominiums. (Courtesy of Meca Properties)

zone to provide all sorts of housing: high-density apartments and low-density single-family homes, as well as housing affordable to people of all income levels.

An alternative is to allow people to subdivide their suburban homes into *accessory apartments* if their homes are bigger than their needs. Most suburbs, however, ban this for fear of lowering overall property values. This political battle is spreading.

Recent federal progrms tried to break up the concentrations of subsidized housing in the central cities and build more in the suburbs. The government encouraged some suburban areas to accept a greater share of public-housing by offering grants for other community facilities. One experimental program provided inner-city public-housing residents with subsidies or vouchers that can be used as money to rent private apartments in the suburbs, and in another, more than 6,000 black families were moved out of Chicago housing projects to comply with a 1976 consent decree after a discrimination suit. The Gautreaux Program, named after plaintiff Dorothy Gautreaux, moved families randomly within a six-county region—about half to the suburbs and half within the city. The relocated families have become better off than the families that stayed behind. Another federal program, called Moving to Opportunity, has given poor familes in five cities housing vouchers to move to middle-income areas. Again, those who moved demonstrated improvements in their employment experiences and in their health.

3. The third approach to solving the spatial mismatch is to provide inner-city poor people with transportation to jobs in the suburbs. The American Automobile Association puts the average cost of car ownership at over $7,000 per year, which is more than many working Americans can afford. Therefore, several city governments, private employment agencies, and even suburban employers have instituted dedicated bus services for this purpose. Wisconsin and California are experimenting with interest-free loans to allow working poor to buy or maintain cars, and a federal level program is called Bridges to Work.

These three efforts to solve the problem of inner-city unemployment follow from the acceptance of spatial mismatch as the cause of the unemployment. The network hypothesis has only recently been offered as an alternative or complementary explanation. Therefore, few government programs have yet been devised to address the lack of social networks. Perhaps we will see proposals in the future.

Still another potential solution to inner-city unemployment among the unskilled is to educate and train the central-city population to capture the skilled tertiary-sector jobs that do open in the central city. The quality of education in the central cities, however, is generally discouraging. Schools in upper-income suburbs usually have more money to spend on modern facilities than inner-city schools do.

Governing Metropolitan Regions

Political geography is the subfield of geography that studies the interaction between political processes and the distributions of all other activities and transformations of the landscape. Political geography can be studied at any scale from local community politics up to international boundary disputes and international law. Conurbations present special problems of interest to political geographers.

The legal boundaries of most U.S. cities were originally drawn to include some surrounding land for future growth, and when the city outgrew its boundaries, it annexed suburban areas. By the 1920s, however, the suburban populations had begun to incorporate themselves to avoid annexation (Figure 10-44). Suburbanites argued that their action upheld the U.S. tradition of local self-government, and many may have felt that by incorporating their own communities, they were escaping the problems (and people) of the old city, including

FIGURE 10-44 **Metropolitan Minneapolis and St. Paul.** Minneapolis and St. Paul, Minnesota, were surrounded by tiers of independently incorporated municipalities, like the rings of growth of a tree. This pattern of suburban incorporation typifies U.S. metropolitan expansion.

high social costs and politics that were often corrupt. Only a few central cities—including Austin, Texas; Charlotte, North Carolina; and Oklahoma City—can still expand by annexing new suburban areas.

Today autonomous municipal units form a legal retaining wall around almost every large city in the United States. Metropolitan areas cover a great number of municipalities and also myriad special district governments, which are incorporated to deal with specific problems. For example, Nassau and Suffolk counties in suburban New York include two cities, 13 towns, 95 villages, 127 school districts, and more than 500 special districts, most of which exercise taxing powers for services from garbage collection to hydrant rental. The five-county Los Angeles metropolitan region contains 160 separate governments. Los Angeles County alone has 82, and even entirely within the boundaries of the city of Los Angeles there are seven independent city governments.

The boundaries among the many jurisdictions are not obvious to everyone, but each government has its own agenda and marks the landscape, and governments jostle for authority and tax dollars. Property taxes soar, but many metropolitan area residents have no idea which government is responsible for which service.

The governing bodies of many of the special districts are not chosen in elections in which each citizen exercises an equal vote. Instead, they are either appointed or else chosen in elections in which votes are weighted in terms of payments for a service, use of a service, or in some other way. As a result, the percentage of public funds spent by officials directly responsible to the voters shrinks. Furthermore, the boundaries of special districts may not conform to those of general-purpose governments but instead may overlap them. Overlapping boundaries multiply the difficulties in coordinating the provision of services. All these factors discourage voter turnout.

Decisions regarding metropolitan land use and the location of industries, recreation facilities, transport facilities, or new housing cannot be made in the best interests of the entire metropolitan population. They are made on the basis of competition among the local governments, each of which wants to enhance its own property-tax base by attracting commercial developments that pay high property taxes but demand little in the way of local services. For example, fast-growing Santa Clara County, California, has zoned for 250,000 new jobs but only 70,000 new homes. Each community hopes to let surrounding towns cope with the additional costs of schooling, pollution, and congestion.

Many metropolitan areas have created councils of governments (COGs). These are committees of officials representing each of the local governments in the region. COGs, however, exercise limited powers. Each local government can veto any proposed area-wide action.

Some U.S. metropolitan regions are creating regional governments to address area-wide problems. State governments in Washington, Minnesota, Connecticut, Virginia, and elsewhere are devising new schemes to redistribute or equalize both the revenues and the costs of housing and schooling throughout metropolitan areas. Dade County, Florida, has assumed responsibility for many public services for Miami and the 29 other cities in the county, and county voters chose to rename the county Miami-Dade. Other state governments are assuming increasing direct responsibility for governing metropolitan areas. Some states are even drawing and applying state-level land-use plans (Figure 10-45).

The governing of metropolitan areas in Canada presents many of the same problems as in the United States, and the province of Ontario created a government for metropolitan Toronto in 1953 that has since served as a model throughout North America.

FIGURE 10-45 New Jersey is among the states that are preparing land-use plans and realizing them through zoning and purchase. (Courtesy of New Jersey Division of Travel & Tourism)

Regional Focus on
METROPOLITAN PORTLAND, OREGON

Oregon has long championed laws that restrict urban sprawl. Metropolitan Portland, Oregon, has a "Metro" government that controls land use, conservation, and transportation for 24 cities and three countries holding 1.3 million people. Metro requires cities and towns of a certain size to bound their urban areas, keeping forests, farmlands, and open space free of development. The law has permitted new housing while protecting a rural character at the edges of cities. As a result, Portland is perhaps America's most "European" city, with a fine mass transit system, compact neighborhoods, and protected forests and farms outside the city limits. Critics argue that the law has pushed up housing prices by reducing the land available for building, but voters have turned down repeated initiatives that would have weakened the government's powers since the law was first passed in 1973.

Today many city and suburban governments—backed by downtown business executives, environmentalists, farmers, and church leaders—are banding together to fight continued sprawl by pressuring state and federal governments to end the many subsidies that encourage suburbanization. The government of Ohio, for example, has cut back on building new roads, begun repairing old ones, and prevented new communities from offering tax breaks that draw businesses from older core metropolises. In 1997 Maryland legislated to confine state spending on infrastructure to existing municipalities. Similar political initiatives—called metropolitics—have been launched in Minnesota, Pennsylvania, Indiana, and Oregon.

The weakness of metropolitan regional governments in North America contrasts sharply with models elsewhere (Figure 10-46). In most European and Latin American countries, either strong metropolitan

FIGURE 10-46 Suburban growth patterns in the United States and Great Britain. This schematic drawing contrasts (a) the leapfrogging typical of U.S. suburban growth with (b) the tighter, more controlled suburban growth typical of Great Britain. The British system saves open space and allows for better and cheaper planning and installation of infrastructure. (Adapted from Marion Clawson and Peter Hall, *Planning and Urban Growth: An Anglo-American Comparison*. Reprinted by permission of Resources for the Future.)

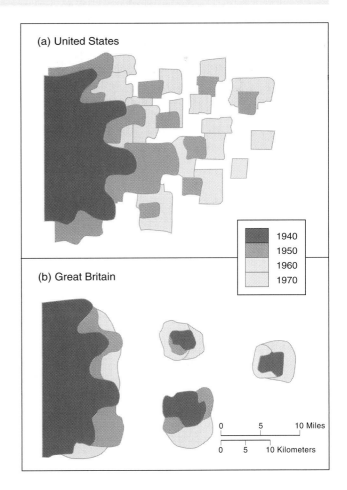

governments exist or else the national governments themselves oversee land use and the growth of metropolitan regions. Those governments generally concentrate activities in city centers and regulate land use very strictly to prevent urban sprawl. Also in contrast to the North American model is the fact that homeownership is relatively low. Europeans seem more content to rent their dwellings.

CONCLUSION: CRITICAL ISSUES FOR THE FUTURE

Today the majority of people on Earth, for the first time in history, lives in cities. Rapid urban growth, driven by demographics and economics, has transformed the face of the planet. Chapter 1 diffentiated a natural landscape, one without evidence of human activity, from a cultural landscape, one that reveals human modification of the local environment. That chapter noted that evidence of human activity is so ubiquitous on Earth that we might be justified in saying that the entire Earth is today a cultural landscape.

Rapid world urbanization has required us to rethink our common definition of "environment." Up to now, in many people's minds the word has meant Earth's natural landscapes. The phrase "environmental preservation" triggered images of struggles to preserve forests and wetlands, beaches and grasslands. Today, however, the transformation of Earth's surface and the actual circumstances in which most people live force us increasingly to direct our attention to urban environments. The environmental problems of the world's settlements are appearing on the political agendas of most countries and even international agencies. Urban environmental problems increasingly impact the health and livelihoods of people in both the rich and the poor countries.

For example, unexpected recurrent epidemics of cholera have focused attention on health threats in human settlements. It had been assumed that the threat of cholera had been eliminated through improvements in water supplies, sanitation, sewage treatment, and food safety, but urban crowding and overloading of the urban infrastructure have brought the problem back. Other urban environmental problems of increasing concern include poor or inadequate housing, safe water supplies and adequate sewage treatment, air quality, exposure to chemical hazards in industrialized areas, physical hazards, including road accidents, the collection and management of solid waste, and noise. Even education becomes an environmental issue because hygiene education is an essential safeguard of public health.

Cities transform natural landscapes not only within the built-up areas but for considerable distances around them. Ecologist William Rees has called the total area of land required to sustain a city its *ecological footprint*, and the United Nations has singled out four regional impacts of particular concern: uncontrolled and unplanned city expansion, disposal of liquid wastes, solid-waste disposal, and acid precipitation. Cities concentrate demands for the products of fertile land, watersheds, and forests. Mining and the extraction of bulky, low-value materials that go into building materials, roads, foundations, and other parts of the built environment also disrupt local ecology, as do the waste materials dumped as a result of excavations. The United Nations notes that there is considerable potential for creating new jobs in reducing resource use and wastes and in recycling or reusing the wastes that are generated.

The challenge is great. We must transform our increasingly "cultural" landscape into a healthy and sustainable physical environment.

Chapter Review

SUMMARY

Every settled society builds cities because some essential functions of society are most conveniently performed at a central location for a surrounding countryside. The region to which any city provides services and upon which the city draws for its needs is its hinterland. Some cities' locations result from site characteristics; others are more the result of convenient situations. Urban environments often suffer from heat, noise, air pollution, and water- and waste-management problems.

In many traditional societies, cities' ritualistic and political importance outweighed their economic importance. As societies develop, cities' economic role usually becomes paramount. Cities offer agglomeration for the division of labor, and they promote and administer the regional specialization of production. Cities can become centers of industrial production and tertiary-sector functions. Some workers in any city produce basic sector exports, whereas workers in the nonbasic sector serve the needs of the city's residents. Jobs in the basic sector multiply jobs in the nonbasic sector. Cities can be classified economically by examining each city's basic and nonbasic sectors and by comparing these sectors among different cities. Central place theory models how cities and hinterlands are distributed across an isotropic plain.

Britain provides the first model for urbanization, but urbanization imposed hardship on millions of people. Today urbanization is occurring in many places without economic development. People migrate from the countryside because of the balance of push and pull factors. Governments can regulate this migration both by reducing the attractiveness of cities and by improving rural life. Urban migrants' industriousness could be an asset for economic growth.

Urban geographers also study the distribution of activities within cities, which may be caused by economic forces, social factors, or government actions.

The concept of designing an "ideal city" has challenged planners for centuries. Sir Ebenezer Howard's garden cities plan tried to stop the growth of the industrial city and to restore it to a human scale. Le Corbusier proposed cities of tall office buildings and apartments spaced far apart and surrounded by green space.

Many of the world's cities were planned and built according to European notions, and this is still reflected in their internal geography. Western forms continue to be stamped on the world's cities. Many great cities' historic cores illustrate indigenous principles of urban planning. Islamic cities present one alternative to Western forms.

In the United States, growing cities have spilled into suburbs. This dispersion has resulted from government policies, cultural choices, and other economic and social forces. Both housing and jobs have dispersed, and commuter traffic has been redirected. The central cities suffered economic decline, but some are reviving with new tertiary-sector activities. Attempts have been made to coordinate governmental activities across conurbations.

KEY TERMS

agglomeration p. 403
basic sector p. 405
brownfields p. 435
capital-intensive activity p. 410
central business district (CBD) p. 412
central place theory p. 406
congregation p. 414
consolidated metropolitan statistical area (CMSA) p. 422
external economies p. 407
exurbs p. 408
gentrification p. 431

hinterland p. 400
incorporation p. 400
informal or underground sector p. 411
internal economies p. 407
labor-intensive activity p. 408
liquidity p. 412
metropolitan statistical area (MSA) p. 422
multiplier effect p. 405
network hypothesis p. 435
nonbasic sector p. 405
political geography p. 437

primary sector p. 403
primate city p. 400
secondary sector p. 403
segregation p. 414
service sector p. 403
spatial mismatch hypothesis p. 430
telecommuting p. 430
tertiary sector p. 403
urban enterprise zones p. 435
urban geography p. 400
urbanization p. 400
zoning p. 415

QUESTIONS FOR REVIEW AND DISCUSSION

1. Define and explain the three sectors of an economy.
2. What are the basic and nonbasic sectors of an urban economy?
3. How are cities distributed across a landscape?
4. In what ways does world urbanization today differ from the model provided by the historical experience of England?

5. How can governments slow urbanization?
6. In what ways does U.S. tax policy subsidize people who buy houses?
7. Contrast modern Western urban-planning ideas with the ideas of traditional Islam.

THINKING GEOGRAPHICALLY

1. What are the site characteristics of your city? Why was the site selected?
2. What are the situational relationships of your city? What principal transport routes converge on your city?
3. What range of external economies is available in your town's yellow pages?
4. What are the biggest employers in your town? Which enterprises do the biggest dollar volume of business? What are your town's basic economic activities?

5. Are any planned suburbs or exurbs located around your town? What forms of transportation do their residents depend on?
6. How many local general-purpose governments and special-purpose governments are in your metropolitan region, or the metropolitan region nearest to you? Study a map of all the forms of independent government in the area. How are the special-district government ruling boards chosen? What kind of regional planning body or government does the region have?

7. Some people enjoy living in cities; some people flee cities. What are the benefits and the drawbacks of living in either cities or suburbs?

8. What would you do to stop rural-urban migration in any given poor country?

9. Quality-of-life measurements can affect a city's prosperity. What are the basic measurements of urban or regional quality of life? How would your community score?

10. Compare the rural movement for the landless in Brazil to seize land discussed in Chapter 8 with urban squatters' struggles for land tenure discussed in this chapter. In what ways are the movements similar? Different?

SUGGESTIONS FOR FURTHER LEARNING

Allen, John, Doreen Massey, and Michael Pryke, eds. *Unsettling Cities: Movement/Settlement (Understanding Cities).* New York: Routledge, 2000.

An Urbanizing World: Global Report on Human Settlements 1996. New York: Oxford University Press for the United Nations Centre for Human Settlements (HABITAT), 1996.

Baxandall, Rosalyn, and Elizabeth Ewen. *Picture Windows: How the Suburbs Happened.* New York: Basic Books, 2000.

Buisseret, David, ed. *Envisioning the City.* Chicago: University of Chicago Press, 1998.

Cronon, William. *Nature's Metropolis: Chicago and the Great West.* New York: W. W. Norton, 1991.

Chapman, Graham P., K. Ashok, and Robert Bradnock, eds. *Urban Growth and Development in Asia.* 2 vols. Brookfield, VT: Ashgate, 1999.

Cities in Transition: World Bank Urban and Local Government Strategy. Washington, D.C.: World Bank, 2000.

De Soto, Hernando. *The Mystery of Capital: Why Capitalism Triumphs in the West and Fails Everywhere Else.* New York: Basic Books, 2000.

Drakakis-Smith, David, and David W. Smith. *Third World Cities,* 2nd ed., New York: Routledge, 2000.

Florida, Richard. *The Rise of the Creative Class.* New York: Basic Books, 2002.

Hodge, David C., ed. A special section of articles on the spatial mismatch theory, in *The Professional Geographer* 48, no. 4 (November 1996): 417–468.

Kaplan, Robert D. *Empire Wilderness: Travels Into America's Future.* New York: Random House, 1998.

Kirdar, Üner, ed. *Cities Fit for People.* New York: United Nations Publications, 1997.

Kotkin, Joel. *The New Geography: How the Digital Revolution is Reshaping the American Landscape.* New York: Random House, 2000.

Lemann, Nicholas. *The Promised Land: The Great Black Migration and How It Changed America.* New York: Alfred A. Knopf, 1990.

McKenzie, Evan. *Privatopia.* New Haven: Yale University Press, 1996.

Main, Hamish, and Stephen W. Williams, eds. *Environment and Housing in Third World Cities.* New York: John Wiley & Sons, 1994.

Mumford, Lewis. *The City in History.* New York: Harcourt, Brace & World, 1961.

Norquist, John. *The Wealth of Cities.* New York: Addison-Wesley, 1999.

Orfield, Myron. *American Metropolitics.* Washington, D.C.: The Brookings Institution, 2002.

Portney, Kent E. *Taking Sustainable Cities Seriously.* Cambridge, MA: MIT Press, 2003.

Shutkin, William, and David Brower. *The Land That Could Be: Environmentalism and Democracy in the Twenty-First Century (Urban and Industrial Environments).* Cambridge, MA: MIT Press, 2000.

Solinger, Dorothy. *Contesting Citizenship in Urban China: Peasant Migrants, the State, and the Logic of the Market.* Berkeley: University of California Press, 1999.

Sorkin, Michael, ed. *Variations on a Theme Park: The New American City and the End of Public Space.* New York: The Noonday Press, 1992.

The State of the Cities 2000: Megaforces Shaping the Future of the Nation's Cities. Washington, D.C.: U.S. Department of Housing and Urban Development, June 2000.

Urban Mobility Report. College Station, TX: Texas Transportation Insititue of Texas A&M University, annual.

Valle, Victor M., Rodolfo D. Torres, and Saskia Sassen. *Latino Metropolis.* Minneapolis: University of Minnesota Press, 2000.

Von Hoffman, Alexander. *House by House, Block by Block: the Rebirth of America's Inner Cities.* New York: Oxford University Press, 2003.

Wiewel, Wim, and Joseph Persky, eds. *Suburban Sprawl: Private Decisions and Public Policy.* New York: M.E. Sharpe, 2002.

Wilson, William Julius. *When Work Disappears: The World of the New Urban Poor.* New York: Alfred A. Knopf, 1996.

Wolf, Peter. *Hot Towns: The Future of the Fastest Growing Communities in America.* New Brunswick: Rutgers University Press, 1999.

World Resources 1996–97: The Urban Environment. New York: Oxford University Press, 1996.

WEB WORK

The United Nations Human Settlements Programme is at

http://www.unchs.org/

A special Web site provides a study of the most successful efforts to solve certain urban problems common around the world:

http://www.bestpractices.org/

The United States Federal Highway Administration's National Traffic and Road Closure Information site can be found at

www.fhwa.dot.gov/trafficinfo/

The United States Department of Housing and Urban Development and its Office of Policy Development and Research maintain home pages that provide data and useful links:

http://www.hud.gov/ and **http://www.huduser.org/**

Several organizations studying urban developments maintain home pages. Among the more interesting is the Center for Urban Policy Research at Rutgers University:

http://policy.rutgers.edu/cupr/

The Planning Commissioners Journal page provides a large and useful array of information and articles about contemporary urban planning problems:

http://www.plannersweb.com/

You may generate maps of urban poverty in the United States at a site operated by The Bruton Center at the University of Texas at Dallas and the Brookings Institution Center on Urban and Metropolitan Policy:

www.Urbanpoverty.net

The Citysearch site provides information about individual cities around the world:

http://www.citysearch.com/

A Look Ahead

THE DEVELOPMENT
OF THE NATION-STATE IDEA

A nation-state is a sovereign territory that includes
a group of people who want to have their own
government. The nation-state idea originated in
Europe and diffused from there—or was applied—
to virtually all the rest of Earth.

EFFORTS TO ACHIEVE A WORLD MAP
OF NATION-STATES

Relatively few states on today's world political map
are nation-states. A map of nation-states could be
achieved only by redrawing the map, by expelling
people from some countries into others, or by forging
national identities in the populations of the existing
states.

11

A WORLD OF STATES

◄ The Canadian flag flies over the Canadian Parliament in Ottawa. Canada and the United States are close allies, sharing the world's longest undefended border and a great deal of their history, culture, democratic beliefs, and economies. Nevertheless, each organizes its territory differently, apportions power differently between the central government and lower levels, distributes activities differently within its territory, and maintains a national culture. The partition of Earth into sovereign countries affects almost all human activities. (Mitchell Bieier/Peter Arnold, Inc.)

HOW STATES DEMARCATE AND ORGANIZE TERRITORY

States maintain their borders and subdivide their territory for governmental purposes. Democratic states have a variety of ways to design representative government, but these techniques can be manipulated for political ends.

MEASURING AND MAPPING INDIVIDUAL RIGHTS

States vary greatly in the degree to which they guarantee all of their citizens equality and rights.

In October 2000 Abdiqasim Salad, the man internationally recognized as the president of Somalia, held a press conference in Somalia's capital, Mogadishu. This may not seem like a big deal, but the truth is that no one was sure where Mr. Salad exercised any real power—or even whether he had any power at all. Rule over the territory labeled "Somalia" and colored one homogeneous hue on world political maps has been unclear for almost 15 years. The territory was ruled ruthlessly by a dictator named Siad Barre from 1969 until 1990, but after his fall, the country dissolved in civil war. The United Nations sent forces to guarantee the distribution of international food-relief supplies and to restore central civil government in 1992, but that mission failed disastrously, and chaos continued to reign in Somalia. There was, quite simply, no central government, although the United Nations continued to recognize one faction and to pretend that faction spoke for Somalia in world affairs. In May 2000 several of the warring factions sent representatives to a conference in the neighboring country of Djibouti. That conference elected a national assembly and a president, Mr. Salad. Mr. Salad was sworn in, while still in Djibouti, but the situation in Somalia was still so unclear and dangerous that Mr. Salad did not travel to Somalia until October. The international community heaved a sigh of relief, but not all of the peoples and regions of Somalia have recognized Mr. Salad. The Northern regions of Somaliland and Puntaland have declared their independence, but they dispute almost half of their territories, and their governments are not internationally recognized. In 2003 neighboring Kenya continued to encourage peace talks among the many factions, but the talks broke into fistfights.

The world political map is probably the most familiar of all maps, because Earth's division into countries, or states, is the most important territorial organizing principle of human activities. **States** are independent political units that claim exclusive jurisdiction over defined territories and over all of the people and activities within them. The governments are not always able to exercise this jurisdiction completely, but states can encourage or even force patterns of human activities to conform to the political map. Chapters 6 and 7 noted how many aspects of culture are affected by the political partitioning of the world. Patriotism, which is a strong emotional attachment to one's country, is itself a powerful cultural attribute. A country can be in many ways a fixed culture realm.

Several of the countries shown on the world political map (Figure 1-6, p. 12) are not really centrally organized. A great many people do not accept that pattern of countries, and much of the territory is not ruled by the central governments. The existing governments try to consolidate their control over all their territory, but not all are able to do so.

The idea that the whole world should be divided up into countries seems natural to us, but it is relatively new in human history, and many of today's countries are young (Figure 11-1). This chapter will explain how the idea originated in Europe and how it was diffused worldwide with European conquest. The neatness of the units on today's world political map nevertheless exaggerates the degree to which all peoples accept the current pattern of countries. The map suggests that all the borders are clearly demarcated and that they divide the activities on their two sides, but in fact some activities overlap the borders. The map also suggests that the areas within those borders are politically homogeneous, and this is also false. Governments are only more or less successful in organizing their territory, and no territory can be sealed off. Border wars among countries continue to afflict the world, and civil wars within countries are even more common.

This chapter also reviews the political geography of states, analyzing how they maintain their borders and subdivide their territories. In addition, it tries to measure and map the degree to which various governments either place restrictions on their peoples' freedom or encourage full development of their peoples' potential.

Countries also try to organize their territories economically—that is, to integrate and build national economies. That will be the subject of Chapter 12.

THE DEVELOPMENT OF THE NATION-STATE IDEA

The idea that a state claims exclusive sovereignty over a demarcated space and all the people and resources within it originated in medieval Europe. Under the Roman Empire, the Roman Catholic Church was geographically organized into dioceses, and when the

Empire fell, the Church and its system survived. As the Church converted people to Christianity, it also converted them to the idea of territorial political organization. Before conversion, the kings of the peoples in Europe had not ruled over fixed territories, but over groups of followers, wherever they wandered. Government over a group of people rather than a defined territory is called *regnum.* The Church taught the principle of rule over a defined territory, which is called *dominium.* The Merovingian kings (fifth through eighth centuries), for example, called themselves Kings of the Franks, but the later Capetians (tenth through fourteenth centuries) had settled down and called themselves Kings of France. Most scholars mark the 1648 Treaty of Westphalia, which ended Europe's Thirty Years' War, as establishing in Europe the general principles of territorially sovereign states and of non-interference by rulers in states other than their own.

In the nineteenth century, anthropologists insisted that conversion from regnum forms of government to dominium is an evolutionary step in human society. This argument, however, might be interpreted as an excuse for nineteenth-century imperialism. The argument that native forms of government everywhere were "backward" compared to European forms offered a justification for European conquest. At least it served as a rationalization after the conquests. British sociologist Herbert Spencer (1820–1903) extended Darwin's theory of evolution to insist that "Nature's law" called for "the survival of the fittest," even among individuals, cultures, and whole peoples. This famous phrase was not Darwin's. Spencer's theory is called **social Darwinism.**

The Idea of the Nation

A state is a territory on a map, but a **nation** is a cultural entity. It is a group of people who want to have their own government and rule themselves. The feeling of nationality may be based on a common religion or language, but it does not have to be, as the Swiss and many other nations demonstrate. A group sharing a sense of nationalism may or may not share any other attributes. Nationalism is a cultural concept in its own right.

Nationalism is one expression of **political community,** which is a willingness to join together and form a government to solve common problems. Today the state is in most places the most powerful level of political community, but Chapter 13 will ask whether worldwide concern over environmental pollution or terrorism may nurture global political community. The evolution of nationalism is part of the story of the evolution of *legitimacy,* the question of who has the right to rule any group. Upon what sort of consent of the governed, if any, is rule based? Even the

most totalitarian states today claim to represent the people.

Chapter 7 noted that Christianity explicitly distinguishes religion from secular life. European emperors and kings signified the state; they ruled over earthly matters with the sanction of the Church—by divine right—but they required only obedience, not loyalty or personal identification with the state. That is why kings could often assign and reassign thrones among themselves and redraw the political map without significant protest by the people. France's King Louis XIV (reigned 1643–1715) insisted flatly, "I am the state." The Church, however, as protector of people's souls, demanded a greater degree of personal commitment than the sovereign did.

The Protestant Reformation challenged this traditional church-state accommodation. Martin Luther preached that every person was his or her own priest and carried individual responsibility for his or her own soul. This belief sabotaged the divine right of consecrated priests and, by extension, that of consecrated kings. In 1581 the Dutch, who were then subjects of the king of Spain, adopted an Act of Abjuration that renounced (abjured) the theory of divine right and argued that a king had an obligation to rule for the welfare of the people. If he did not, then the people could abjure his rule over them. This turned the notion of divine right upside down. It suggested that the king, or any government, served the people.

The U.S. Declaration of Independence of 1776 includes a list of charges against King George III. Prior to that time, even the English themselves had risen up against their king, Charles I, defeated his forces in battle, tried him for crimes against "his" people, and executed him in 1649. The subsequent English Bill of Rights (1689) recognized that sovereignty did not lie in the king but in the people.

The Nation-State

The Swiss philosopher Jean Jacques Rousseau (1712–1778) laid the foundation for allegiance to the state as the people. Rousseau believed that in nature people were merely physical beings, but when they united in a *social contract,* they were capable of perfectibility. For Rousseau, politics was a means to moral redemption. Rousseau's ideas swayed France, and the French Revolution gave birth to the French nation. The 1789 Declaration of the Rights of Man stated: "The principle of all sovereignty resides essentially in the nation; no body nor individual may exercise any authority that does not proceed directly from the nation."

The nation demanded personal dedication and allegiance from its citizens. Therefore, the perfect state

> The theory of the nation-state assumed that nations develop first and that each nation then achieves a territorial state of its own.

The Ages of the World's States

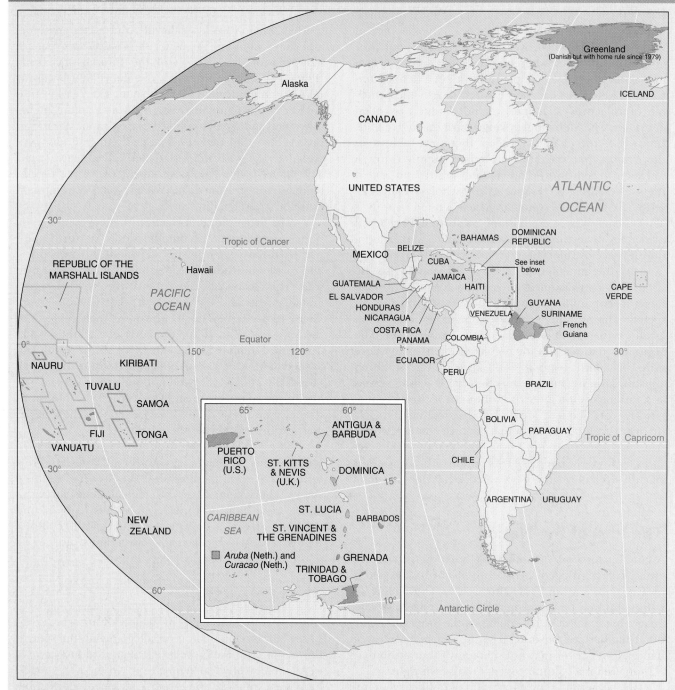

FIGURE 11-1 The current political partitioning of the world is very recent. In many states citizens remember the day their country achieved independence.

was a **nation-state,** a state ruling over a territory containing all the people of a nation. The theory of the nation-state assumed that nations develop first and that each nation then achieves a territorial state of its own.

Some scholars have argued that several of the nation-states had historic **core areas,** or historic homelands. France, for example, has long been focused on the region of Paris, and Ethiopia on the highlands of the horn of East Africa. In many other cases, however, core concentrations of settlement or activity developed only after the state had come into existence.

Constitutions and laws never fully explain how any government works, because each political community has a unique **political culture.** A political culture is the set of unwritten rules—the unwritten ways in which written rules are interpreted and actually enforced.

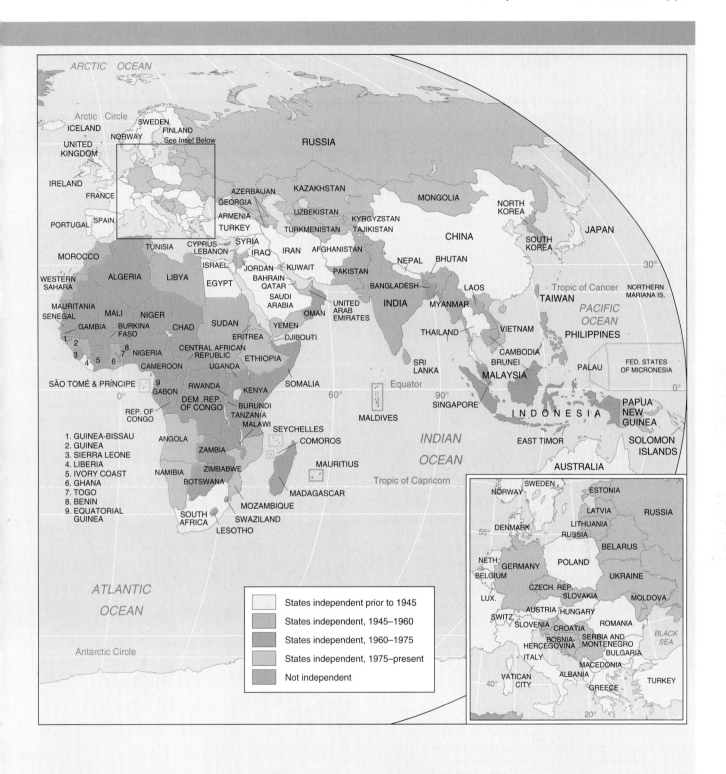

**Map legend:**
- States independent prior to 1945
- States independent, 1945–1960
- States independent, 1960–1975
- States independent, 1975–present
- Not independent

Political communities differ widely, for example, in whether they honor elders, rich people, or religious leaders; in their tolerance of bribery; and in the rigidity or laxity with which they enforce laws. Political culture reflects other aspects of a people's culture, such as their religion. This is true everywhere and at every level of government, from city council district up to the United Nations General Assembly.

For example, one scholar examined political life in Italy and found that there are within that one country distinct regional differences in political culture, which he defined as "civic community"—patterns of social cooperation based on tolerance, trust, and widespread norms of active citizen participation (Figure 11-2). In Italy, the distribution of civic community among the regions was already evident as long ago as the thirteenth

Variations in Civic Community Within Italy

FIGURE 11-2 This map reveals the sharp differences in *civic community* that the scholar Robert Putnam found in each of the regions of Italy. He has hypothesized that economic development and welfare can come about only where local political culture is based on strong civic community. In the regions north of Tuscany, a sense of political community has threatened to split these regions off from Italy into a new state. (From Robert Putnam, Robert Leonard, and Raffaella Nanetti, *Making Democracy Work: Civic Traditions in Modern Italy.* Princeton, NJ: Princeton University Press, 1992.)

century. Economists have translated these findings into economic theory: The more citizens trust one another, the better off their society. These findings suggest that patterns of civic community explain both a region's capacity for economic growth (a topic to be analyzed in the next chapter) and the region's capacity for democratic self-government. Political leaders in regions lacking civic community may lack the fundamental building blocks upon which stable democracy can be built, and civic community may be difficult to create where it does not exist. If this hypothesis is correct, similar studies of whole countries could indicate which will most probably enjoy economic growth and democracy in the future.

The European Nation-States

The Napoleonic Wars that followed the French Revolution carried the idea of nationalism across Europe. Armies had formerly been composed of hired professionals, but now the entire male population could serve, and armies were called "the school of the nation." Napoleon was defeated in 1815, but the code of laws that he had imposed left a widespread legacy. It swept away aristocratic privileges and strengthened the middle class (Figure 11-3).

The idea of nationalism matured in Europe during the nineteenth century, and new nations struggled to emerge from old empires and feudal states. In some cases this produced a competitive nationalism, which has caused many wars through the nineteenth and

FIGURE 11-3 Napoleon Bonaparte. This 1812 painting by Jacques-Louis David was Napoleon's favorite portrait of himself. He is shown neither in imperial robes nor astride a horse in battle, but as a lawgiver. The pen and scattered documents, the hour on the clock, and the dying candles reveal that the emperor has worked all night on the Law Code. The Napoleonic Code remains today the basis of law in 30 countries in Europe and beyond, including Quebec and the U.S. state of Louisiana. (Jacques-Louis David (French, 1748–1825) "Napoleon in His Study," 1812, oil on canvas, 2.039 × 1.51 [80 ¼ × 49 ¼]; framed: 2.439 × 1.651 × 152 [96 × 65 × 6]. National Gallery of Art, Washington: Samuel H. Kress Collection. © Board of Trustees, National Gallery of Art, Washington. Photo by Lyle Peterzell.)

twentieth centuries. For example, several countries claimed the maximum extent of territory over which their people had ever wandered or ruled. This led to overlapping territorial claims.

After World War I, U.S. President Woodrow Wilson advanced the ideal of the nation-state, which he called **national self-determination.** The victors redrew the map of Europe to break up the defeated German, Austro-Hungarian, and Ottoman Turkish empires (but not their own empires) and to grant self-determination to several new European nation-states, including Poland (Figure 11-4). Russia's new communist ruler, Vladimir Lenin, had criticized the czarist Russian empire as "a prison-house of nations." He and his successor, Stalin, reorganized the empire under a new totalitarian government disguised as a union of nations—the Union of Soviet Socialist Republics, or Soviet Union.

It can be argued that several European nations did exist before they achieved their own independent territory and governments—that is, their own states. Even in Europe, however, national governments conscientiously inculcated patriotism in their citizens, and few states have ever achieved a clean match between people and territory. The European map was redrawn again after World War II (Figure 11-5), but dissident groups in several states still claim parts of their

Europe in 1946

FIGURE 11-5 After World War II the Soviet Union expanded considerably. It retook the territory that Poland had won in the earlier war and gave Poland some German territories, thereby effectively shifting Poland about 240 kilometers (150 miles) to the west. It gave the southern half of German East Prussia to Poland, but kept the northern half (with the important city of Königsberg, which the Russians renamed Kaliningrad), and it swallowed up the three Baltic states. They had been part of the old Russian Empire but had enjoyed independence since 1919. The U.S.S.R. also detached Ruthenia from Czechoslovakia and incorporated it into Ukraine, and Bessarabia from defeated Romania. Bessarabia was combined with part of Ukraine to form a new Republic of Moldavia—today's Moldova. The Soviet Union also seized territory from Finland.

Yugoslavia took territory at the head of the Adriatic Sea from Italy, but the port of Trieste, which Yugoslavia wanted, remained Italian. Compare this map with the current political map (Figure 11-1) to see the changes that have occurred since the end of World War II.

Europe in 1920

FIGURE 11-4 After World War I, the defeated German, Austro-Hungarian, Russian, and Ottoman empires were dismembered, and several new states appeared. Some of them, such as Yugoslavia and Czechoslovakia, did not represent nations but were composed of diverse populations. The new Poland was unsatisfied with the eastern border it was originally awarded, and it seized more territory in a war against Russia.

neighbors' territories. Territorial claims on a neighbor are called **irredenta,** from the Italian word for "unredeemed." For example, many Austrians still claim the South Tyrol from Italy, and many Hungarians claim the Romanian province of Transylvania (see Figure 11-4). In both these cases, however, the state governments officially accept existing borders. By contrast, the Bolivian government still claims the coastline that Chile took from it after an attack in 1879. George Bernard Shaw well-described the emotional tug of irredenta in the preface to his 1904 play *John Bull's Other Island:* "A healthy nation is as unconscious of its nationality as a healthy man is of his bones. But if you break a nation's nationality it will think of nothing else but getting it set again. It will listen to no reformer, to no philosopher, to no preacher, until the demand of the Nationalist is granted. It will attend to no business, however vital, except the business of liberation and unification."

The Formation of States Outside Europe

At the same time as the idea of nationalism was maturing in Europe, the Europeans were actually enlarging their empires. They were not willing to recognize that their colonial subjects had national rights. They argued that non-Europeans were inferior or "not yet ready" for political independence. After World War I the Europeans did not offer national self-determination to their subject peoples outside Europe. The winners just took the losers' colonies while retaining their own colonies. In fact, individual colonies seldom represented political or cultural communities. When the European imperialists carved up the world among themselves, they drew borders that ignored any political organization among the native peoples. Some of these new **superimposed boundaries** split native political communities, whereas others combined two or more in one colony (Figure 11-6).

The Europeans often used native rulers as intermediaries between themselves and the people, especially if a colony included several groups. This form of government was called **indirect rule.** In many cases the imperialist powers even amalgamated groups into new tribes and appointed new kings or, in India, maharajahs. These new groupings soon gained the force of "tradition," leaving today's scholars and citizens a bewildering challenge to define what were genuine pre-colonial communities. Indirect rule hindered the native peoples from uniting against the imperialists, and later the only feeling that brought native groups together into one nationalism was often shared indignation against the

Superimposed Borders in Africa

FIGURE 11-6 This map superimposes the borders of today's African states upon the territories of just a few of Africa's many indigenous peoples, just as the European colonial powers superimposed these borders at the Berlin Conference of 1884–85. When the colonies attained independence as new states, the new states seldom contained unified nations.

colonial power. It has been argued that still today the principal basis of some nationalisms is hatred of foreigners, of "traditional enemies," or even of internal minorities, and that in some countries these hatreds are skillfully manipulated by local rulers.

The colonies were administered by bureaucracies made up in some cases of Europeans, in other cases of acculturated natives, and in still others of foreign peoples imported by the Europeans. When the colonies received their political independence, these bureaucracies had a vested interest in maintaining the existing units and borders. Therefore, today's world political map is not a map of nations that have achieved statehood. It is a vestige of colonialism.

The threshold principle As late as the 1920s, most diplomats and statesmen insisted that a nation had to have some minimal population and territory to merit self-determination. This principle, called the **threshold principle,** was responsible for yoking together, for instance, the Czechs and the Slovaks, who had never formed a political community and who eventually split apart again.

The threshold principle was abandoned after World War II, so today the world map reveals many tiny, independent nation-states (Table 11-1). Some, such as Bahrain, thrive economically, but others rely on foreign subsidies. Few are militarily defensible. Small, weak, or poor states often negotiate defense or economic treaties among themselves or with more powerful neighbors, but they usually guard their ultimate sovereignty jealously.

Cultural subnationalism When the entire population of a state is not bound by a shared sense of nationalism but is split among several local primary allegiances, then that state is said to suffer **cultural subnationalism.** In some states many people grant their primary allegiance to traditional groups or nations that are smaller than the population of the whole new state. These traditional identifications may be strong enough to trigger civil war. In other cases, a group's bonds of affinity may extend beyond the state's borders, and this may inspire international disputes. For example, the Kurds were promised a national homeland at the end of World War I, but they were split among several new countries (Figure 11-7). Since the United States defeated Iraq in the first Iraqi War in 1991, it allowed the Kurds in northern Iraq to establish a de facto independent state there. The independence that Kurds enjoyed there has infuriated the Turks, who are trying to squelch demands of the Kurds in Turkey for independence—or even substantial autonomy. After the Second War against Iraq in 2003, the United States refused Turkish demands that Turkish troops be allowed to occupy northern Iraq. Iraq itself had been carved out of the Ottoman Empire by European imperialists, but before 2003 it was never bonded by a nationalism; in the early twenty-first century, Iraqi nationalism may be fostered by the United States.

Subnationalism is among the forces called **centrifugal forces,** which tend to pull states apart. A strong sense of nationalism shared throughout the whole state is among the competing **centripetal forces,** which bind a state together. Multinational empires have been held together by military force throughout history, but states containing several nations—**multinational states**—usually suffer divisive politics.

Civil wars demonstrate that many countries on the world map are not nation-states. At the same time, however, there are many groups that are politically self-conscious and arguably are therefore nations, but these groups are politically submerged and do not show up on the map. Hundreds of African and Native American groups might claim national self-determination. Submerged nationalities often present threats of local violence, as events in Southeast Asia, the Mideast, Africa, the former Yugoslavia, and throughout the former U.S.S.R. have demonstrated. The idea of national self-determination retains its powerful attraction, so such local violence may continue.

TABLE 11-1 A Few of the World's Smallest Independent States

	Area (km²)	Area (mi²)	Population (in thousands)	GDP (in billions of U.S. dollars)
Bahrain	620	239	643	8.2
Barbados	430	166	266	2.8
Grenada	340	131	96	0.3
Maldives	300	116	263	0.5
Monaco	1.9	0.733	32	0.8
Nauru	21	8.1	11	0.1
San Marino	60	23.16	25	0.5
Tuvalu	26	10	11	0.008
Vatican City	0.438	0.169	0.860	NA

NA – Not available.

Source: U.S. Bureau of the Census

Nations Without States

FIGURE 11-7 These are a few of the nations that do not have states of their own but may achieve statehood in the future. The present states, however, may succeed in dissolving these nations. Note that most of these nations overlap two or more existing states, so pressures by these submerged nations to achieve statehood are international issues.

Afghanistan's many ethnic groups, clans, and tribes are quite independent of one another, and few observers would argue that there is an "Afghan nation." The country's total population in 2000 was about 38 percent Pathan (also called Pashtun); 25 percent Tajik, 19 percent Hazara, and 8 percent Uzbek. Some scholars suggest that there is no real reason for Afghanistan to exist as a country, but in October 2001, the United States committed itself to building a nation there after the Taliban had been defeated and Al Qaeda destroyed. The United States sponsored talks among Afghan factions that organized a new government in December 2001, but nation-building is proving more difficult.

There is probably very little of a Pakistani nation either. Pakistan was founded as a faith-based country, Muslim in contrast to secular India, but the civil war that split East Pakistan (today's Bangladesh) from West Pakistan revealed that the religion could not bind the nation. For many Pakistanis—perhaps most—tribal and clan loyalties still today surpass loyalty to the "Pakistani nation."

Rebels in most civil wars claim to be fighting to correct religious, ethnic or political wrongs, but a World Bank report issued in 2000 suggested that those claims are often either coincidental or ex post facto justifications for war. A study of 47 civil wars between 1960 and 1999 concluded that the most probable risk factor for civil war was actually economic dependence on commodities available for plunder. Prime examples of wars of plunder include the wars in Sierra Leone (over diamonds), Nigeria (oil), and Colombia (drugs). Therefore, the way to end civil wars such as these is for countries to develop away from dependence on commodity production, or for international markets to embargo sales of commodities by rebels. The United Nations did embargo the sale of diamonds from Sierra Leone on world markets in July 2000.

Many observers fear a world in which the internationally acknowledged governments cannot claim a monopoly on force within their territories. Colombia, for example, suffered a civil war that took over 35,000 lives and displaced over 1.5 million people in the 1990s. By the end of the decade the government admitted that it did not exercise sovereignty over its total area, and it recognized guerrilla insurgencies' control over two areas. In 1998 it ceded control over 16,000 square miles to *Farc* (Revolutionary Armed Forces of Colombia), and in 2000 it ceded control over a second zone to *ELN* (Army of National Liberation) (Figure 11-8, p. 457).

Regional Focus on
CIVIL WAR IN SRI LANKA

Over 65,000 people have died in a long civil war in Sri Lanka between the 75 percent of the population that is Sinhalese and Buddhist and the 18 percent that is Tamil and Hindu. The 60 million Tamils in India sympathize with Sri Lanka's Tamils, so India has intervened militarily in the civil war. The war has continued so long that the Tamil area has evolved a shadow state, called Eelam, complete with a flag, national anthem, police, army, taxes and courts. In December 2002 Tamil leaders dropped their demand for complete independence, but held out for "substantial autonomy." Talks are complicated by the fact that the U.S. government has labeled the Tamils "terrorists," and not even the government of Sri Lanka can get that designation withdrawn. Sri Lanka may adopt a new federalist form of government.

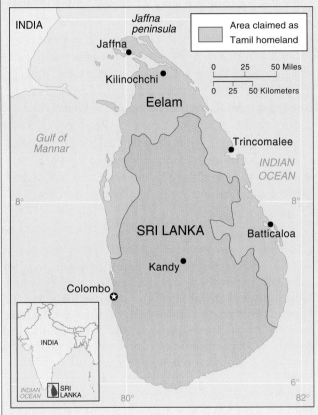

The Tamils call the area they rule Eelam.

This billboard in the Tamil-controlled area of Sri Lanka praises a woman suicide bomber for killing government soldiers. (Photo by Sriyantha Walpola for *The New York Times*)

Subnationalism and civil wars in Africa Subnationalism has particularly plagued African states. Along the southern borders of the Sahara in West Africa, for example, the Taureg people ruled the Mande, Fulani, Songhai, Zarma, and Hausa peoples for centuries before the coming of the European imperialists. Today the Tauregs remain a self-conscious nation, but they are a minority in each of the new countries formed among their former subjects: Algeria, Libya, Mali, and Niger. The Tauregs disrupt all these countries and even encourage fighting among them. Many African countries have suffered almost endless civil strife since receiving independence after World War II. Foreign rule had blanketed and frozen hostilities that were released again with independence. As noted in Chapter 8, civil wars have been partly responsible for recurring famines in several African countries. In some cases, government collapsed in a war of each against all, as in the early 1960s in the Democratic Republic of the Congo. (That country was called Zaire from 1971 until 1997.) In

Focus on
GEOPOLITICS

Chapter 10 defined political geography as the study of the interaction between political processes and the distributions of all other activities and transformations of the landscape. The term was coined by German geographer Friedrich Ratzel as the title of a book he wrote in 1897. Ratzel's studies focused on the political geography of states, and he suggested that each nation needed *Lebensraum,* room to live. As national populations increased, he wrote, nations might need more Lebensraum.

This idea of Lebensraum was adopted by a school of writers who were at that time drawing analogies between states and living things. They argued that states are a form of biological organism. To explain why some states thrive while others perish, these writers combined Ratzel's idea of Lebensraum with Spencer's theory of social Darwinism, and they concluded that competition among states for territory resulted in "the survival of the fittest." In 1916 the Swedish political scientist Rudolf Kjellen published *The State as a Form of Life,* in which he wrote of "the natural and necessary trend towards expansion as a means of self-preservation." Kjellen proposed a new "science" to study states' competitive strategies. He called it **geopolitics.** In 1924 a German professor named Karl Haushofer founded an Institute for Geopolitics in Munich, where he taught that it was natural for strong states to expand at the expense of the weak. These ideas impressed Adolf Hitler, and when Hitler became ruler of Germany, he quoted geopolitical theories to justify Nazi aggression in World War II.

The term geopolitics is still today loosely applied to studies of global military strategies, and it is often confused with political geography. Chapter 12 will demonstrate, however, that any state's possession of extensive territory—or even of natural resources—does not strictly determine the welfare of its people. Already in 1899 the British scientist Sir William Crookes had refuted geopolitical theories. His book *The Wheat Problem* suggested that technological progress can replace territorial aggression in raising a nation's standard of living. Crookes coined the metaphor "scientific frontiers," but even he would probably be amazed to learn of the technological advances in agriculture described in Chapter 8.

---

other cases, one or more regions have attempted to break away from a new state. The Igbo (or Ibo) people of Nigeria, for example, declared their region to be the independent state of Biafra in 1967, and the central government fought for three years to restore its authority over the region. Fatalities were estimated to be more than one million. In Ethiopia a military junta overthrew the emperor in 1974, but it was in turn overthrown in 1991 by separatist armies representing the two provinces of Eritrea and Tigre. Eritrea received its independence in 1993, and Tigre may achieve independence as well. In Rwanda and Burundi the Hutu and the Tutsi peoples have fought periodically. Angola and Mozambique dissolved in civil wars after gaining independence in 1975. They were early examples of what we today call **failed states** or **collapsed states,** that is, countries that have proved incapable of providing their citizens with either economic development or even peace and security.

Because tribal and other allegiances spill across international borders, civil wars often spill over into neighboring countries. This has been demonstrated again early in the twenty-first century, as fighting spread across Sierra Leone, Mali, Guinea, Liberia, and the Ivory Coast.

In some cases outsiders have prolonged wars for economic or ideological reasons or just to keep potentially unfriendly neighbors busy at home. Belgium and France, for example, interfered in Congo and Nigeria. In 1997 troops from Burundi, Rwanda, Uganda, and Angola all played a role in overthrowing a dictator in the Congo, Mobutu Sese Seko, because of complicated loyalties of peoples that overlap all these countries' superimposed borders. After overthrowing Mobutu, these countries' armies turned against one another. The fighting in the Congo, which continued into the twenty-first century, has been called Africa's First World War, involving troops from Congo, Rwanda, Uganda, Burundi, Sudan, Zimbabwe, Zambia, South Africa, Namibia, Angola, and Tanzania. The troops of several of these countries supposedly came as peacekeepers, but they only complicated and intensified the fighting, thus earning the epithet "arsonists masquerading as firefighters."

Subnationalism helps explain the number of authoritarian governments in Africa. Authoritarian rulers argue that iron rule is the only alternative to tribalism tearing their countries apart, but this argument is at least partly a rationalization to enable them to hold on to power. Several have been accused of deliberately

Rebels' Zones in Colombia

FIGURE 11-8 The government of Colombia could not defeat the rebel forces in these areas, so it formally surrendered their control to rebel groups. The United States fears an increase in drug supplies from these zones, so it has encouraged the government of Colombia to reconquer them.

fomenting and manipulating subnational strife. In the 1990s and into the twenty-first century the political scene in Africa demonstrated unending turmoil, partly as a result of the end of the Cold War. Authoritarian governments that had been supported by one side or the other tumbled in Benin, Zambia, Madagascar, Niger, Lesotho, Malawi, and People's Republic of Congo. The future will test the argument that authoritarian rule is necessary to consolidate African countries.

The existence of many failed states in Africa, as well as in Asia and in Latin America, provides breeding grounds for civil wars, guerilla wars, and the multiplication of independent rogue military forces. These have proved capable of projecting violence to the heart of the developed world, as the attacks on the United States on September 11, 2001, demonstrated. (The agents of those attacks were introduced in the Preface and will be examined in greater detail in Chapter 13.)

Therefore, since the attacks, the United States has demonstrated greater involvement in the problems of nation-building and state-building.

EFFORTS TO ACHIEVE A WORLD MAP OF NATION-STATES

The territorial state is the highest level of political sovereignty on Earth. The United Nations organization promotes cooperation among states and in Antarctica, which is the only portion of Earth not divided into an independent state or states (both of these topics are discussed in Chapter 13). The complete division of the world into stable and peaceful territorial nation-states may or may not be an ideal goal, but the ideal holds powerful attraction.

Nations and Ethnic Groups in the Balkans in 1990

FIGURE 11-9(a) This pattern has defied attempts to delineate nation-states since the nineteenth century. World War I began in 1914 when a Serb assassinated an Austrian archduke in Sarajevo, which was then an Austrian possession but was claimed by Serbs. After World War I, Yugoslavia was formed, composed of six constituent republics. That state dissolved in 1991, and the constituent republics of Slovenia, Croatia, Bosnia-Hercegovina, and Macedonia emerged as independent countries. Serbia and Montenegro have agreed to stay bound in one federal republic until at least 2006.

Beginning in 1991, Serbian residents of Bosnia-Hercegovina (pink on this map), backed by troops from Serbia, began corralling the Bosnian-Hercegovinin Muslims (green) into ever-shrinking territories or else driving them out of Bosnia-Hercegovina altogether. Croatia has driven the Serbs out of its territory and resettled the area with Croats. In 1998 Serbs opened an offensive to drive ethnic Albanians out of Kosovo, the province of Serbia that has an Albanian population. During the rule of Slobodan Milosevic (1987–2000), Serbia fought and lost five wars, bringing upon itself terrible destruction, charges of international crimes, and international disgrace. Milosevic was eventually brought down by nationwide rioting and brought to trial for crimes against humanity (see Chapter 13). In 2001 Macedonia, by contrast, peacefully granted Albanians cultural rights within Macedonia.

States use three major strategies to make the map of nations fit the map of states: (1) to redraw the international political map; (2) to expel people from any country in which they are not content, or to exterminate them; and (3) to forge nations in the countries that exist now.

(1) Redrawing the World Political Map

Theoretically, the world political map could be redrawn until everybody is content being in the state in which he or she resides. Unfortunately, this would open endless disputes and provoke new wars. Some countries would split apart, and numerous territories would have to be transferred from one country to another. No satisfactory solution could exist for the states in which different national or ethnic groups intermingle or in which the cities are populated by people of one group, and the countryside by people of another group.

Disputes would also arise over the distribution of natural resources. Each group would claim the most generously endowed territory as its own, or at least a "fair

The Dayton Accords of 1995

FIGURE 11-9(b) An accord concluded at Dayton, Ohio, in 1995 provided that: (1) Bosnia will be a single country divided in two parts, a Muslim-Croat Federation and a Serb Republic with a central government. (2) Each entity will have its own president and legislature; the central government will have a collective presidency; the president and parliament of the central government will be chosen in free elections. The central government in Sarajevo will be responsible for foreign and economic policy, citizenship, immigration, and other issues. (3) Refugees will be allowed to return to their homes, with free movement guaranteed. (4) The peace will be implemented by a NATO force. (5) All parties will cooperate with an international war crimes tribunal.

East and West Germany joined into one country in 1990. Yugoslavia broke up and then fell into war among the remnant republics in 1991 (Figure 11-9), and Czechoslovakia split into two countries—the Czech Republic and Slovakia—on January 1, 1993.

Few borders have been redrawn in Africa or Asia since decolonization (Figure 11-10). African countries could have chosen to redraw their borders at the time they received their independence, but at a meeting of the Organization of African Unity (OAU) held in Cairo, Egypt, in 1964, the African states pledged themselves to respect the existing international borders. They made this pledge even though they resented the borders as a colonial legacy and found it difficult to govern within them. In 1994—after 30 years of civil wars and tragedy—Nigeria's Nobel prizewinning author Wole Soyinka became the first prominent African to insist, "We should sit down with a square rule and compass and redesign the boundaries of African nations. . . . [Continuing wars demonstrate] in a very brutal way that we cannot evade this historical challenge any longer." The Nigerian government responded by exiling Soyinka and threatening his life. In 1994 Rwanda's president, Pasteur Bizimungu, demanded that the neighboring Congo (then called Zaire) surrender territories occupied by the Banyamulenge, a people related to the Tutsi of Rwanda. As previously noted, three years later Rwandan troops helped bring down the Congolese government. President Bizimungu soon openly called for "a Berlin II, so that we can reflect on the disorders created by Berlin I [where colonial powers divided Africa]" (see Figure 6-29). In 2000 Edem Kodjo, former Prime Minister of Togo and Secretary General of the OAU, said, "A certain number of principles that date from the independence era are already being shattered. . . . The intangibility of borders is likely to be next, because many of our states are inviable, either for economic or political reasons, or because of their population or ethnic composition."

Thus, 40 years after the Cairo Conference, African leaders are considering redrawing the map. Some African states may either break up or else recognize subnational communities by adopting federal forms of government (discussed later in this chapter). Some remapping might occur in Asia as well; the most probable candidates would be the Koreas, Iraq, Afghanistan, Pakistan, or even China (as will be noted later in this chapter).

(2) Mass Expulsions or Genocide

A second strategy used to make a nation "fit" a state territory is to expel from the state people who are not accepted as members of the nation, or to exterminate them. These policies are tragic and abominable, but both have actually been implemented when groups have

share" of what had been the entire state's endowment. The Ogoni people of Nigeria, for example, occupy one of the world's richest oil-producing regions, yet the central government claims all mineral rights and distributes profits among all regions of the state. The Ogoni profit little from the oil but live instead in toxic pollution and poverty. Ogoni independence movements have been squashed quickly.

Despite inevitable problems, political maps continue to be redrawn. In Europe in recent years

Despite inevitable problems, political maps continue to be redrawn.

Border Changes Since Independence

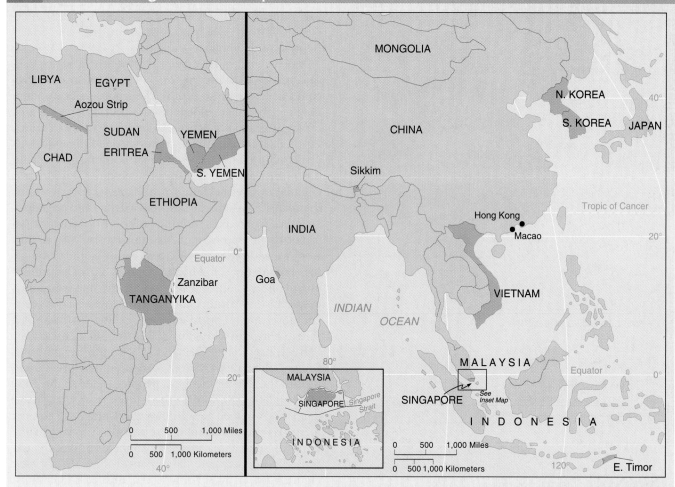

FIGURE 11-10(a) In Africa, Tanganyika and Zanzibar formed a union as Tanzania in 1964, although the terms of that union may be renegotiated. Zanzibar's historic ties link it to the Arab world; Zanzibar was for many years the capital of Oman on the Arabian Peninsula, over 2,200 miles (3,600 kilometers) away. Independent Eritrea appeared in 1993, and from 1998 until 2000 Eritrea and Ethiopia fought over their border. The Aozou Strip between Libya and Chad, plus the borders between Egypt and Sudan, have been disputed, but not redrawn.

A few colonial borders have been redrawn in Asia. In 1965 Singapore separated from Malaysia. In 1976 North and South Vietnam joined into one Socialist Republic of Vietnam. Korea, which had been annexed by Japan in 1910, was divided by victorious Russian and American troops in 1945, and it remains split into South and North Korea. North Korea tried to conquer the South, but South Korea rebuffed it with U.N. assistance (1950–1953). (See Figure 11-10b.) Yemen, which had received independence from Turkey in 1918, and South Yemen, which received independence from the United Kingdom in 1967, merged in 1990. India annexed the Portuguese colony of Goa in 1961 and the independent country of Sikkim in 1975. In 1975 Indonesia seized East Timor from Portugal, but East Timor achieved independence in 2002. China reabsorbed Hong Kong from the United Kingdom in 1997 and Macao from Portugal in 1999.

All in all, these are surprisingly few border changes from what were originally colonial boundaries.

attempted to carry the logic of territorial nationalism to its illogical extreme.

Mass expulsions and genocide occurred in Southern Europe during and after World War I. The Turks massacred over 1 million Armenians in 1915 and later expelled about 1.5 million Greeks from Asia Minor. Greece responded by expelling 400,000 Turks. During the 1930s and 1940s, Germany attempted to bring into the country many Germans who lived in other coun-

tries and to eliminate Jews within Germany. After World War II, Germans were expelled from Poland and Czechoslovakia. Some 12 to 15 million people fled from Pakistan to India or vice versa at the partition in 1947.

Mass expulsions continue. In 1989 Bulgaria expelled about 100,000 ethnic Turks. In 1991 a million Kurds fled or were driven out of Iraq. In 1993 tens of thousands of people of Nepalese origin were driven out

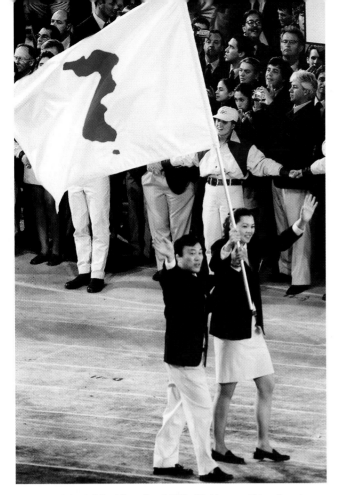

FIGURE 11-10(b) After the 1950–53 Korean War, South Korea developed into a prosperous democratic capitalist state, and North Korea remained a poor communist dictatorship. In 2000, however, the two states opened negotiations to achieve reconciliation. As a symbol of that hope, their teams entered the opening ceremonies of the Olympics in Sydney, Australia, under a single flag—a blue depiction of the Korean peninsula on a white background. Their teams did not compete as one. A few weeks later South Korea's president, Kim Dae Jung, was awarded the Nobel Peace Prize, but it was revealed in 2003 that he had secretly had over $500 million paid to North Korea for consenting to the meeting that boosted Kim's prestige. Relations between the two Koreas are a point of considerable concern for the United States, which in 2003 still had 37,000 troops based in South Korea (see Chapter 13). (Chang Lee/NYT Pictures)

since it won independence from Belgium in 1962. The Tutsi minority (15 percent) traditionally ruled the Hutu majority (85 percent), and in 1993 the first democratically elected Hutu president was assassinated by a Tutsi military uprising. Civil war raged, and by the end of 2000 over 200,000 people were dead and an additional million were internally displaced. The Arab majority of Mauritania continues to expel the country's minority blacks into Senegal. In the Caucasus region through the 1990s, Abkhazia, Ossetia, and Naggorno-Karabakh all suffered ethnic killings and expulsions involving more than a million people combined.

Censuses taken in the Soviet Union in the late 1980s counted some 60 million "displaced Soviets" living outside the regions of their ethnic identity. When the Soviet Union broke apart, not all of the newly independent states guaranteed full civil rights to their minorities. Ethnic Russians, especially, were subject to discrimination in some countries. Millions of people have migrated into the homelands of their own ethnic origins.

Probably the most ethnically homogeneous major states today are Japan and the two Koreas; they are 99 percent homogeneous. China is 92 percent Han. These countries exist more or less within their historical frontiers. Japan may be the closest realization of the concept of a nation-state. The Japanese virtually closed their islands to foreign trade and cultural exchange for hundreds of years and, as noted in Chapter 5, allow very little immigration today.

(3) Forging National Identities

Today many existing states are struggling to weld their populations into nations. This reverses the theoretical order in which the formation of a nation precedes the

FIGURE 11-11 The ruined national library of the Sarajevo, Bosnia, University. "Cultural cleansing" of a landscape includes the deliberate destruction of libraries and other cultural monuments. (Noel Quidu/Getty Images, Inc.—Liaison)

of Bhutan. Serbians referred to the slaughter or exile of Muslims in Bosnia-Hercegovina as *ethnic cleansing;* this policy extended to *cultural cleansing* of the landscape. The Serbs destroyed mosques, libraries, and museums and even bulldozed cemeteries (Figure 11-11). In 1999 Serbs drove about 1 million ethnic Albanian Muslims out of the Serb province of Kosovo. Many may never return. In Rwanda in 1994, Hutus rose up and slaughtered more than 500,000 Tutsis. The next year the Tutsis regained the upper hand, drove several hundred thousand Hutus into Congo, and confined hundreds of thousands more in "resettlement camps." Next door, Burundi has suffered ethnic violence ever achievement of statehood. The significance of this transition for world cultural geography can scarcely be exaggerated. The states are struggling to make the maps of other human activities conform to their pattern. Chapter 7 noted that governments can affect the geography of religion and language. In addition, governments can affect the geography of political community, of law, of land use, and as will be seen in Chapter 12, of economic activity. Countries can encourage the circulation of people, goods, and ideas within their territories, and they can restrict or discourage circulation across their borders. Countries can never control these activities entirely, but modern governments exercise incomparably greater power over human activities than governments did in the past.

We have already emphasized that many units on today's world political map are relatively new. No one can predict how long any world political map will remain fixed, but existing governments are always trying to stabilize it. The history of the United States recounts many struggles to weld one nation out of many diverse groups. In Europe a few nation builders faced their task self-consciously. The Polish hero and later president Joseph Pilsudski (1867–1935) said flatly, "It is the state that makes the nation, not the nation the state." Italy was politically unified in the 1860s, and then the statesman Massimo D'Azeglio observed: "We have made Italy; now we must make Italians." As noted in Figure 11-2, Italian politicians have not completely succeeded.

European experiences were echoed by Julius Nyerere, the first president of Tanzania (served 1961–1985). He commented on the African states: "These new countries are artificial units, geographical expressions carved on a map by European imperialists. These are the units we have tried to turn into nations." Many Asian states face the same difficulty, and even many Latin American countries that have been independent for over 150 years still have not welded their populations into nations.

The destruction of subnational loyalties Before some states can forge their populations into nations, they often try to undermine the cultural subnationalisms within their borders. This destruction of traditional communities can be profoundly destructive psychologically to at least the first generation. Typical is Kenya's campaign to abolish the ancient traditions and distinctiveness of the Masai people, once one of Africa's most powerful nations, and to integrate them into the modern state. The Masai must go to school and conform to a new style of life dictated by the government. They must surrender their age-old lifestyle of nomadic cattle herding, settle down, and take up farming (Figure 11-12). Chapter 8 noted that Botswana has forcibly settled the Basarwa people.

Some countries have absorbed the leaders of the traditional subnational groups into the new political

FIGURE 11-12 Masai people. The westernized Kenyan on the left appear to regard these Masai men visiting downtown Nairobi with bewilderment (and, it seems, hostility). Kenya is forcibly acculturating the Masai to modern norms, and Masai are prohibited from dressing like this on the streets of Nairobi today. (Courtesy of Robert Caputo)

structure (Figure 11-13). Cameroon, for example, still allows the 17 traditional kings within its borders powers of judgment over minor crimes, and it defers to their ceremonial importance. In Zimbabwe local chiefs appoint 10 members to the 150-seat parliament. In the 1990s South Africa's president Nelson Mandela tried to educate and train the country's 300 Zulu chiefs in law and administration. His efforts were criticized as a centralization of power, but the chiefs' powers had in fact been created by the British colonial authority. This struggle provides a fine example of the confusion of pre-colonial tradition with imperialist legacies. Members of many African chiefly families retain positions of power and influence in the modern states. Mandela himself was a traditional prince, and Ghanaian diplomat Kofi Annan, today Secretary General of the United Nations, is the prince of the Fante people, and through his mother heir to the paramount chieftancy of the Akwamu. His father was a thoroughly modern man, working in fact as an executive of a European global corporation. In Malaysia a paramount ruler is elected every five years from among the nine traditional Malay sultans. In 1993 these sultans' powers were for the first time curtailed; this act symbolized Malaysia's emergence from a traditional monarchical country to a modern democratic country.

FIGURE 11-13 The Alake (king) of Abeokuta and his son. These two men represent the transition from tradition to modern statehood in Africa. The son of the Alake (king) of Abeokuta was educated in Britain and returned home to serve as the Chief Justice of the new state of Nigeria. (Photo by Eliot Elisofon, 1972, Eliot Elisofon Photographic Archives, National Museum of African Art/Smithsonian Institution)

India abolished the privileges of its many maharajahs in the 1970s, although many of them subsequently won elective offices. India is today the world's largest experiment in bringing diverse groups of people together under democratic political structures. India is afflicted by centrifugal forces based on religion, ethnicity, caste, and language. Several of these distinctions are regional and might encourage separatism, but the Indian civil service, a legacy of British rule, remains a strong centripetal force.

Urbanization, which is occurring rapidly everywhere, often dissolves traditional rural-based identities. This development may reduce subnational loyalties.

Instruments of nation building Different countries rely on different instruments of nation building, and the choice reflects differences in other aspects of their cultures. Among the instruments most often used are

> Different countries rely on different instruments of nation building, and the choice reflects differences in other aspects of their cultures.

religion, the armed forces, education, symbols, media, political parties, and labor unions.

Religion Chapter 7 noted countries in which a national church was a building block of nationalism: Ireland, Ukraine, and Poland. The Orthodox churches are traditionally national churches, and some Protestant denominations are also rooted in specific nations: Lutheranism among the Germans, Presbyterianism among the Scots, and Anglicanism in England. The Roman Catholic Church is transnational, but the 1965 Vatican Council (called Vatican II) encouraged countries to form national councils of bishops. These councils have begun to reflect cultural differences among countries that are significant enough to surprise and upset the Vatican. The U.S. National Council of Bishops, for example, reflecting U.S. culture, has expressed willingness to allow women a greater role in church services than the Vatican will accept. Chapter 7 noted that the Emperor of Japan renounced his claim to divinity after World War II, but in May 2000 Japanese Prime Minister Yoshiro Mori described Japan as "a divine nation with the emperor at its core." When the remark provoked a storm of criticism, the Prime Minister explained that he had meant only that the nation should be more spiritual. Many Russians view Orthodox Christianity as an essential element of Russian identity, and missionary activity of other faiths is severely restricted.

Armed forces In many countries the armed forces are still "the school of the nation," and a high proportion of the people may be serving (Figure 11-14). The armed forces may consume a high percentage of the national budget, but defense against external enemies is not always their principal function. The army holds the state together, either by force or just by training and socializing young people. When Hutus were slaughtering Tutsis in Rwanda in 1994, U.N. Secretary General Kofi Annan suggested publicly that African governments were not contributing troops to peacekeeping efforts because "they probably need their armies to intimidate their own populations." The remark stirred outrage in many African countries.

In many cases the armed forces provide a disciplined labor force for building infrastructure. The military forces may even exercise an independent role in the government. In Chile, for example, the armed forces are guaranteed their own seats in the national legislature. Chapter 7 noted that the Turkish constitution grants the army a special role in preserving the government. Even the U.S. armed forces have been assigned nation-building tasks. In 1947 President Harry Truman ordered their racial integration. This was brought about successfully decades before the society at large was ready to integrate.

FIGURE 11-14 "The School of the Nation." This young member of the army of the Congo uses a bag printed with cartoon figures to store his grenades. Many national armies—as well as guerilla armies in civil wars—recruit or kidnap child soldiers, despite international convenants against children in warfare. Many countries draft high percentages of their people just to keep them under control. (Michael Kamber/*The New York Times*)

Education   National school systems are another tool with which states create nations. Rousseau wrote: "Education must give souls a national formation, and direct their opinions and tastes in such a way that they will be patriotic by inclination, by passion, by necessity." Today in France, it is said, the Minister of Education in Paris can look at his watch at any time of day and say precisely what all French schoolchildren are studying. In 2003, the United Nations administrator of Bosnia and Hercegovina mandated the integration in local high schools of Serbs, Croats, and Muslims, children of the groups who had fought so hard to live apart. It was hoped that integration in school could dispel age-old antagonisms.

Most countries have national curricula and textbooks. The interpretation of national history in those texts is a political issue, and a change in the government can change that interpretation. Mexico, for example, introduced in 1992 a new edition of the history book for fourth graders. After 20 years of teaching that Porfirio Díaz, the turn-of-the-century dictator, "was very bad for the life of Mexico, because the people were not given the chance to elect their leaders," the new version emphasizes that Díaz achieved stability, tolerated the Catholic Church, built railroads, fostered industrial growth, and attracted foreign investment. These are all programs supported by the current government. The 1992 text also introduces criticism of Emiliano Zapata, a revolutionary hero who defended the ejido agricultural system (see Chapter 8, and the Critical Thinking box in this chapter, "Can a Country Insulate Its Borders and Isolate Its People?"). When Hong Kong was returned to Chinese sovereignty in 1997, its school textbooks were rewritten. They shifted away from British interpretations of history—and even descriptions of

plants and animals typical of the British countryside—to a Chinese interpretation of history and a focus on Chinese geography. In 2003, U.S. agencies began rewriting Iraqi schoolbooks in which maps depicted Kuwait as Iraqi territory and which insisted, "Before the Baath Revolution, landowners were dictators controlling the land and the people, and that is why we produced so little. After the revolution, everything went perfectly."

Schools instill in youngsters the society's values and traditions, its political and social culture. This is called the process of **enculturation,** or **socialization.** National schools propagate national culture, literature, music, and artistic traditions. National history and geography lessons leave the rest of the world outside that focus of concern. Classroom maps often promote hostility to neighbors by claiming territory held by a neighboring country.

The United States has traditionally relied on the schools to forge a nation, yet it allows a diversity of programs and standards across the country. Education is not mentioned in the Constitution, so responsibility for it remains with the individual states, which have generally delegated the task down to local school districts. International comparisons indicate that U.S. elementary and secondary schools are relatively weak, and this has spurred demands for nationwide standardization of curricula and testing. A 1989 conference of all 50 state governors concluded: "We believe that the time has come . . . to establish clear national performance goals, goals that will make us internationally competitive." The U.S. Congress has mandated the National Assessment of Educational Progress to achieve that goal. Under a standardized system, local districts could still experiment in designing school programs, but standardization could end one of the nation's oldest and most revered traditions of local government.

Manipulating symbols   Every country has a set of national symbols, called an **iconography,** and these items are emphasized to schoolchildren. The national flag is one of the most important of these icons (Figure 11-15),

FIGURE 11-15 The flag of East Timor. The field of the flag of newly independent East Timor is red, signifying "the struggle for national liberation." The black triangle represents "the obscurantism that needs to be overcome," the yellow triangle represents "the traces of colonialism in East Timor's history," and the five-pointed white star represents peace, or "the light that guides." The new flag was first raised at midnight on May 19, 2002.

Critical Thinking

CAN A COUNTRY INSULATE ITS BORDERS AND ISOLATE ITS PEOPLE?

Many countries try to limit the information that world citizens—or even their own people—are able to learn about events inside the country. Such censorship, however, is becoming increasingly difficult. New technologies of communication sabotage governments' efforts to control the flow of information across borders.

In the 1970s, for example, the Mexican army launched a campaign against a small guerrilla force in the state of Guerrero. Hundreds of peasants were killed, but the Mexican government suppressed press coverage of the events, and few outsiders ever learned of them. By the 1990s technology had changed. On New Year's Eve 1994 a guerrilla organization calling itself the Zapatista National Liberation Army rose up in Mexico's southern state of Chiapas. Chiapas is one of the poorest states in Mexico, and largely Indian in population. This time, with the world's eyes on Mexico, the government did not crush the rebellion. It negotiated land tenure and other issues with the rebels while nevertheless maintaining strong military pressure on them.

Today hundreds of dissident political movements in countries and regions stretching from Mexico to

Zapatista soldiers. Members of the Zapatista Liberation Army wore masks to disguise their identities but carried banners picturing Emiliano Zapata (1879–1919). Marlon Brando played Zapata in the classic 1952 film *Viva Zapata*. (AP/Wide World Photos)

Cambodia to Chechnya (a dissident region inside Russia) have access to their own satellite transmission facilities or even host their own Internet sites. There they can reach the whole world audience with their interpretation of political developments. Refugees in exile can communicate with resistance movements back home and disseminate information subversive of a repressive government. No government can completely guarantee the security of its borders against the flow of information and ideas. In 1999 the Serbian government seized the Belgrade studio of B92, a dissident radio station, but the station was soon reconfigured as a Web site hosted in Amsterdam, the Netherlands. The English-language Egyptian newspaper *The Cairo Times* is published in Cyprus; in Egypt, government agents censor it, so the publishers put the censored material on a Web site for Egyptian readers. In China only four state-controlled entities can legally connect with the Internet, and commercial Web sites cannot hire their own reporters or publish news not already published by the state news agency.

Even the United States has laws and governmental directives that are intended to keep out of the country the presentation of opinions that the government considers subversive. Many foreign artists, writers, and political figures have been prohibited from entering the United States personally, even though they are interviewed abroad and featured on U.S. television and newsmagazine covers. Legislation requires schools and libraries with Internet access to install software that blocks access to materials that censors consider "pornographic, obscene, or harmful to minors." A 2003 study found that even at their least restrictive settings, however, these filters blocked 24 percent of all health education sites and 50 percent of safe-sex information sites.

Questions

1. What methods do governments use to prevent the influx of information and foreign ideas?

2. Can these methods be entirely successful, given modern technology?

3. Do governments have the right to restrict the flow of information? Why or why not?

and several countries have laws against its desecration. Each nation's national anthem is another icon. In a unique instance of international solidarity, however, the nonpolitical prayer for God's blessing, "God Bless Africa," serves Namibia, Tanzania, Zambia, Zimbabwe, and South Africa. It was written by a South African schoolteacher and taken up by the African National Congress as its anthem.

The map is another icon of great power. Children can be taught to respect—even to cherish—the size, shape, topography, resources, and variety of their land. The U.S. weather map has demonstrated appeal on television, and the editors of *USA Today* were surprised to discover that the weather map is one of the paper's most popular features. Few people actually read the map or even care about the weather at the other end of the country, but the map itself is an icon that wins allegiance.

Iconography also includes all the pomp and circumstance of national ceremony—the celebration of holidays, costumes, the designs on national postage stamps, and international sports competitions. In 2003 the U.S. Treasury introduced a redesigned $20 bill to foil counterfeiters, but also specifically to add "symbols of freedom," meaning more eagles and flags. The words "In God We Trust" were added in 1956, at the height of the Cold War against communism (see Chapter 13).

Many countries ban public displays of the iconography of past regimes, including fascist symbols in Germany, Hungary, and Romania and Communist symbols in several formerly Communist countries.

Media   National media may reach and sway more of a national population than any other shared experience, even the army or the school system. Half of India's people, for example, are illiterate, but 80 to 90 percent of them can be reached by state-controlled radio and television. Government media monopolies may inform the population, educate it, win its allegiance, or command it. The United States is the only country in the world in which commercial television broadcasting preceded public programming, and today only the United States, Sri Lanka, and Norway do not require television stations to donate air time to political candidates.

Many countries protect their national airwaves against foreign programs as a way of safeguarding their national culture. Many countries enforce quotas on the amount of radio and television programming and the percentage of movies shown that can be imported.

Media that spill over international borders can exert a powerful centripetal force. For example, one reason that East Germany never created an East German nation, despite 40 years of trying, was that the government could not prevent East Germans from watching West German television.

Political parties   Political parties can politicize and mobilize populations, recruit people, and give them a sense of participation. They can broaden the govern-

ment's base of support, but they can also be divisive. If people identify themselves more closely with their tribe or faith than with their country, political parties can become means of promoting narrow communal interests, or, worse, of fomenting ethnic grievances. In Uganda, President Yoweri Museveni (serving since 1986), long banned them as divisive, but he first allowed them to form in 2003.

Political parties play different roles in different countries, and these roles reflect local political cultures. For example, most states have adopted one of two forms of government that grew out of European political experience, but the forms do not actually work the same way everywhere. In one model, originated by the United States, the chief executive is a president elected independently of the legislature. The chief executive and the legislature can check one another's power. This model was copied for the most part throughout Latin America and later in France's former African colonies. In the alternative model, called the parliamentary form, the chief executive (usually called prime minister) is elected from among the members of the legislature, and as long as he or she retains the votes of the legislative majority, the prime minister can exercise greater power than an elected president does. This form was left behind in Britain's former colonies, including Canada.

Both forms assume the existence of competing political parties. Multiparty democracy may be appropriate in some cultures, but in others one-party government does not necessarily mean totalitarianism. Traditional African political techniques, for instance, are based on consensus building through dialogue. This technique is called *existential democracy,* as distinct from the *adversarial democracy* that Westerners practice. Existential government conforms with traditional African legal systems and religion, which emphasize the maintenance of social harmony rather than retribution. An example of existential democracy took place in the People's Republic of Congo in 1991, when 1,500 delegates met for three months to revise the country's constitution. Revolutionary changes were introduced, yet few formal votes were taken. The conference reached agreement through debate and consensus, and at its end all delegates dipped their hands into the reflecting pool in the conference hall ceremonially to wash away any ill feelings that might have accumulated.

Asian politics are also notably existential, in keeping with traditional Confucian values. The Japanese constitution, for instance, was imposed during the U.S. occupation of Japan, yet the Japanese govern themselves under that constitution in light of traditional Japanese values. Consensus on most issues is reached before any vote is taken, so formal votes are seldom contests or measures of power. Japan has been ruled by the same party since 1955.

Labor unions   Labor unions may serve as still another building block of nationalism. "Labor" or "workers'" political parties can even govern. In totalitarian states the government may sponsor official labor unions and curtail the formation of independent unions. In Poland, under Communist party rule, the independent Solidarity labor movement nevertheless overwhelmed the Communist party candidates for parliament and deprived the Communists of any pretense of legitimacy. Independent unions also played a key role in bringing down the Communist government in the former Soviet Union.

Each of these six institutions—religious institutions, armed forces, schools, media, political parties, and labor unions—may be either a centrifugal force or a centripetal force in any country. Geographers study their presence or absence, the mix or balance among them that is unique to each country, and their interaction with other aspects of each country's culture. Furthermore, each institution has a geography within each country. One or more institutions may be equally influential across the entire territory, concentrated in only one region, or spill over the international borders. In many countries, for example, including the United States and Canada, the political parties show regional strengths. In several African countries the army draws disproportionately upon one ethnic group or region for troops.

Nationalism can be threatened either by regionalism or by internationalism. Whereas regionalism tends to split a country into many parts, internationalism tends to dissolve a country's external borders. Today many global social and economic developments exert centrifugal force against the centripetal forces of nationalism. Chapters 12 and 13 will examine those global developments.

Democracies and False Democracies

"Democracy is not like instant coffee, where you can just add water and stir," said France's foreign minister, Hubert Védrine, as he declined to sign a U.S.-sponsored international declaration committing the signers to the principle. In 2000, 107 other nations—democratic or not—did sign. Virtually all governments today govern in the name of the people, but some are more genuinely representative than others. The world has many imperfect or just plain fake democracies. Some countries have "elected" presidents, but media (often government-owned) cover only the incumbent; opposition candidates' supporters (or even the candidates) are beaten, arrested on dubious charges, or even killed; and constitutions are routinely rewritten to prolong terms and powers of presidents. In many countries authoritarianism is tempered by populism. Popularly elected presidents of both Peru and Venezuela, for example, actually seemed to rise in popularity when they uncon-

stitutionally closed their countries' parliaments and fired their Supreme Courts in the 1990s.

People living in genuinely democratic countries unconsciously assume that most governments represent their people and work for their welfare. A great many governments, however, do not represent the people. Government by an elite privileged clique is called an **oligarchy.** The governments of many countries have even been called **kleptocracies**—government by theft. The national wealth may be funneled off and even taken out of the country by a privileged few (called "briefcase bandits"), while the mass of the people remain in poverty. In these cases foreign aid, or any system of providing assistance to the poor through their governments, cannot guarantee that the needy will receive the assistance. For example, General Sani Abacha ruled Nigeria from 1993 until 1998, and in 2000 the Swiss government found $670 million in one Swiss bank account in his name, $600 million in another account in Luxembourg, and similar sums in other accounts in France and Germany. In 2002 the U.S. Department of Justice published documents obtained from American oil companies demonstrating that President Nursultan Nazarbayev of Kazakhstan regularly accepted large bribes in connection with dispensing his country's oil concessions. In 2003 the Supreme Court of the Philippines awarded to the government $700 million in Swiss bank accounts of the late president Ferdinand Marcos, whose heirs could not justify ownership of the money. The presidents or rulers of many of the poorest countries in Latin America, Africa, and Asia are among the world's richest individuals.

Some authoritarian rulers insist that their people "are not ready for democracy." Eritrean president Isaias Afewerki said in March 2003, "We have not yet institutionalized social discipline, so the possibility of chaos is still here. Remember, we have nine language groups and two religions. No one in Africa has succeeded in copying a Western political system, which took the West hundreds of years to develop. Throughout Africa you have either political or criminal violence. Therefore, we will have to manage the creation of political parties, so that they don't become the means of religious and ethnic divisions, like in Ivory Coast or Nigeria."

Are there, in fact, preconditions for democracy, and states where the people are not ready for it? Without those conditions, do authoritarian rulers have the right to rule, presumably for the good of the people, to guide the people toward democracy? Lee Kuan Yew, Senior Minister of Singapore, wrote in 2000, "The necessary preconditions [for democracy] are near-universal education and a large middle class that can stabilize voting patterns." Farouk Adam Khan, today the Prosecutor-General of Pakistan, led a military coup that failed in 1973. During the five years he spent in prison, he read the great works

> Are there preconditions for democracy, and states where the people are not ready for it?

of Western political philosophy, including *The Federalist Papers* and John Stuart Mill's *On Liberty*. In 2000 he said, "Every single ingredient that the authors of those books say is required for a civil society—education, a moral code, a sense of nationhood; you name it, we haven't got it! Just look at our history. . . . We need someone who will not compromise in order to build a state."

The French observer Alexis de Tocqueville analyzed the young United States in his classic *Democracy in America* (1835). He noted that one of American democracy's great strengths was the number of nongovernmental organizations among the people: clubs, youth organizations, political parties, sports groups, business executives' leagues, church groups, and so forth. In totalitarian states, by contrast, rulers feel threatened by any organization that brings people together under any auspices but the state's own.

China is an example of a state that severely represses independent citizen groups. Falun Gong is a Buddhist faith-healing sect that has spread across China. Its estimated 70 million members outnumber the 60 million members of the Chinese Communist Party. One morning in April 1999, 10,000 Falun Gong members suddenly appeared in Beijing's Tiananmen Square for a silent demonstration. They sat quietly for a couple of hours, and then the crowd melted away as mysteriously as it had appeared. To have brought together so many people required a tremendous amount of planning and coordination, all of it completely unknown to the government. This demonstration so frightened the government that it has ever since cracked down on the organization, tried to infiltrate it, and arrested its identifiable leaders. Western democracies consider such organizations harmless or even beneficial to a healthy civic life, but totalitarian governments consider a threat any organization that has its own internal information network and the ability to organize people.

In some Western democracies, however, citizen participation rates have fallen so low that observers wonder whether elections are valid. Italian courts invalidated seven national referenda in 2000, as only 32 percent of potential voters had turned out. The Serbian Supreme Court invalidated three successive presidential elections in 2002 because none of them achieved the constitutionally required 50 percent voter turnout. In the United States an average of just 44.1 percent of the voting-age population turned out for national legislative elections through the 1990s, putting the United States in 139th place among 163 countries surveyed by the International Institute for Democracy and Electoral Assistance. In 2002 British Prime Minister Tony Blair said that "failure to vote is the mark of a satisfied citizen," but many observers argue that this remark interprets the death of democracy as a vote for the status quo.

The early twenty-first century has demonstrated that there are areas in which most people simply do not want democracy. Many Arab rulers in Near East, for example, are autocratic, corrupt, and heavy-handed,

but they are nevertheless more liberal, tolerant, and pluralistic than a democracy that would likely replace them. Democracy is a set of rules, not a set of values. Elections in many Mideast countries would produce undemocratic totalitarian or fundamentalist results. Thus, paradoxically, a more democratic country can be a more repressive one. A majority of Iraqis, for example, probably favor restrictions of the rights of women and minorities. For the United States to assume responsibility to build a "democratic" state there is more than just holding elections. Unless majority rule is accompanied by legal protections, tolerance, and respect for minorities, the result can be populist oppression.

Furthermore, in any democracy, the demographic description of those who actually vote may differ from a description of the people at-large. In the United States, for example, the higher one's age and income, the more probably one votes. Popular public opinion polling, by contrast, usually includes all ages and income levels. Therefore governmental activities, which usually respond to the voters' preferences, do not always reflect the population's wishes as counted in opinion polls.

The detailed analysis of governments is the field of political science, but a geographer comparing the world's countries must be careful never to be deceived by appearances and instead look deeper for processes at work.

HOW STATES DEMARCATE AND ORGANIZE TERRITORY

Each state maintains its borders and subdivides its territory for political representation or administration. This section will examine the methods used and some of the results achieved.

The Shapes of States

The shape of a state may affect its ability to consolidate its territory and control circulation across its borders (Figure 11-16). A circle would be the most efficient shape on an isotropic plain because a circular state would have the shortest possible border in relation to its territory, and that shape would allow all places to be reached from the center with the least travel. States with shapes the closest to this model are sometimes called *compact states*. Bulgaria, Poland, and Zimbabwe are examples. *Prorupted states* are nearly compact, but they have at least one narrow extension of territory. Namibia and Thailand are examples. If these extensions reach out to navigable waterways, the extensions are called *corridors* (corridors' special importance to landlocked states is discussed in Chapter 13). *Elongated states* are long and thin, such as Chile or Norway, and *fragmented states* consist of several isolated bits of territory. *Archipelago states,* made up of strings of islands, such as Japan or the Philippines, are fragmented states,

FIGURE 11-16 Shapes of states. Examples are shown of states that are compact, prorupted, elongated, fragmented, and perforated. The five states are drawn to the same scale. In general, compactness is an asset, because it fosters good communication and integration among all regions of the country.

but Azerbaijan is also fragmented, having a territory, Nakhichevan, separated from the rest of it and enclosed by territory of Armenia. Nakhichevan is called an *exclave*. Still other states, called *perforated states,* are interrupted by the territory of another state enclosed entirely within them. South Africa, for example, is perforated by Lesotho, and Italy is perforated by the Vatican and by San Marino.

The shape of a state's territory may influence the government's ability to organize that territory, but this is not always true. A topographic barrier such as a mountain chain may effectively divide even a compact state. Bolivia and Switzerland, for example, are compact in shape, but mountain chains disrupt their interiors. For some of their regions, trade across international borders is easier than trade with other regions of their own country. The people throughout an archipelago state, by contrast, may be successfully linked by shipping. Before drawing any conclusions about political control from the shape of a state alone, one must consider the distribution of topographic features, of the state's population and resources, and whether any centrifugal forces such as economic or cultural ties straddle the state's borders.

International Borders

Many people think of rivers and mountain ranges as "natural" borders, perhaps because on a map these features often resemble lines. This simple assumption, however, is false, and attempts to enforce it have triggered numerous wars through history. Rivers bind the peoples on their two banks as much as they divide them, and mountain borderlines can be drawn either from one peak to the next or else up and down the mountain valleys that reach between the peaks. Mountain valleys can actually provide foci for political organization. The early Swiss united the valleys on

either side of mountain passes. The passes did not divide peoples, but gave them a common interest.

State borders within the United States illustrate various types that can be drawn. Some are rivers (as the Columbia River between Washington and Oregon, and the Mississippi River), and some follow mountains (Idaho-Montana, Virginia-West Virginia, North Carolina-South Carolina-Tennessee). Most, however, are geometric lines imposed across the landscape.

Some international borders are defended with minefields and watchtowers. The United States has built sections of high wall along the border with Mexico to keep undocumented migrants from sneaking across, and Israel is building a 225-mile wall—consisting of concrete, barbed wire, electronic fencing, motion detectors, and trenches—to separate it from the West Bank territories. This fence does not follow Israel's pre-1967 borders, but encloses some West Bank territory, so it excites Palestinian and Jordanian anger.

Some borders are matters of dispute. Guatemala recognized the independence of Belize in 1991 but claimed most of its territory until a treaty was signed in 2002. In 1992 El Salvador and Honduras accepted a World Court decision about their border, over which they had argued since independence in 1861. India and China have fought over their border several times, and they still have not agreed (see the map in the box on page 308). Other international borders, by contrast, are relatively open and free. The U.S.-Canadian border is the world's longest undefended border. The location of an international border can be clear to anyone flying overhead, however, even if there are no border structures. Borders may be evident by contrasting land uses on the two sides, by different types of landholdings, or by discontinuity in the transportation networks (Figures 11-17, 11-18, and 11-19.)

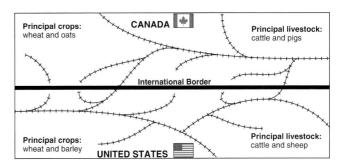

FIGURE 11-17 The U.S.-Canadian border. This schematic drawing illustrates two effects of different policies on the two sides of the U.S.-Canadian border. One feature is that the railroad network frequently approaches the border but seldom crosses it. Countries often try to restrict circulation and trade across national borders. The second feature illustrated here is that different crops and livestock predominate on the different sides of the border. This difference does not result from different environmental or market conditions but from the two countries' agricultural support policies.

Historic Landholding Patterns

FIGURE 11-18 This early map of the region of Green Bay, Wisconsin, demonstrates principles of both political geography and historical geography. By juxtaposing two patterns of landholding, it reveals the history of the region through two different legal systems. The first Europeans who settled here were French. They depended on the river for transportation, so they defined the landholdings that you can see extending inland from the riverbank. This system of landholding is called the *long lot system*. Later this territory came to be part of the United States, so it was surveyed, and landholdings were assigned according to the system defined in the U.S. Ordinance of 1785. This system, called the *township and range system*, is a purely geometric pattern imposed regardless of natural features. U.S. surveyors recognized the preexisting long lot claims, but they subdivided all land not already claimed into a pattern of regular squares. Such regular squares of property holdings are one of the most distinctive features of the U.S. landscape and give a "checkerboard" look to U.S. agricultural regions. (Map by S. Morrison, E. Swelle, and J. Hathaway. Courtesy of the State Historical Society of Wisconsin. Neg. #WHi[X3]39431)

An international border across a sparsely populated area may not be marked or supervised at all, and its exact location may become a matter of dispute only if valuable resources are discovered in the region. This has happened along the inland borders of South

The Border Between Mexico and Guatemala

FIGURE 11-19 The photograph reveals the devastation of the rain forest on the Mexican side of the border. This devastation has been caused by chopping down the forest and then trying to farm the poor soils too intensively and by extensive commercial logging. Food production has not kept up with population growth. (Map courtesy of NYT Graphics/New York Times Pictures; photograph by John C. Stennis Space Center/NASA Headquarters)

American countries. In 1941 Peru attacked Ecuador and annexed 55 percent of Ecuador's territory. The discovery of oil in the former Ecuadorian territory in what is today northern Peru triggered renewed fighting in 1995, but the states agreed on a border demarcation in 1998. Venezuela claims about two-thirds of Guyana, and Suriname claims another 10 percent. Both Suriname and Guyana claim the mouth of the Corentyne River, where new explorations for oil seem promising. The jungle border between Brazil and Venezuela is also disputed. Similar disputes have occurred in Middle Eastern deserts. Under a 1966 agreement, Saudi Arabia and Kuwait share jurisdiction and oil revenues from a 14,763-square-kilometer (5,700-square-mile) "neutral zone" between them. Russia and Japan have not signed a treaty to end World War II, because they dispute possession of four islands. The Russians claim they are part of the Kurile chain occupied by Russian troops during the war and surrendered by Japan; the Japanese claim that the four islands in question are not.

Borders may attract complementary economic activities on the two sides. The zone along the U.S.-Mexican border has developed complementary manufacturing, as we shall see in Chapter 13.

States often want to control the passage of information as well as of people and of goods over their borders. Today, however, direct people-to-people links via telephone lines or satellites have proliferated and destabilized the distinction between what is inside and what is outside. Telephone connections have been used to break into computers in foreign lands and to read, steal, or tamper with privileged information. This action challenges international law. In which country was the crime committed? In May 2000 a computer virus called "I Love You" caused almost $10 billion damage in computers around the world; in the Philippines, however, where the virus originated, its creator was treated as a national hero. Other countries protested, but to no avail. Newspaper and magazine editors and publishers are being held liable to legal action in any country where the Web version of their publication can be accessed. Publishers may obey the laws of their own country, but if "publication" of defamation or of a libel is held to be where the story is read, not where the story was written, then the laws of the most censorious and autocratic governments can be applied to material originating in countries with free speech. Publishers suffer either if the publisher has any assets in the prosecuting country, or if the prosecuting country issues a writ of extradition that the home country recognizes. Furthermore, as geolocation technology improves, censorship and quarantining of space can be achieved that were thought impossible just a few years ago. (Review pp. 32–33.)

Governments can regulate only from within their own borders, and they cannot regulate broadcasts into a country if those broadcasts originate outside its borders. Electronic broadcasts into a country can be blocked, but efforts to do so are expensive and often unsuccessful. Many countries try to ban or regulate the ownership of satellite dish antennas, but these regulations are often ineffective. Globally, radio is still more important than television is, and U.S. broadcasts of Radio Free Europe/Radio Liberty, Voice of America, and Radio Free Asia blanket the world.

Some countries welcome international transmissions. In 1990 Pakistan began to allow transmission of U.S. Cable News Network (CNN) into Pakistani homes. Pakistanis have been astonished to see the range of

debate allowed in the United States on every issue. Programs always contain messages that are not obvious to the programmers. People in some countries may watch a U.S. crime show, for example, and notice that the captured criminals were read their rights. Such direct international dissemination of U.S. political style might represent a triumph for democracy and sow the seeds for democratic awakenings.

U.S. Border Security and Internal Security

Congress responded to concern over American security after the attacks of September 11, 2001, by creating in 2002 a new cabinet-level Department of Homeland Security. It exercises control over 46 federal agencies, including the Customs Bureau and the Coast Guard, but neither the Federal Bureau of Investigation (FBI) nor the Central Intelligence Agency (CIA). The new Department's responsibilities include both border security and internal security.

It is difficult to monitor borders as open and free as those of the United States are. On an average day in 2000, 1.3 million people, 348,000 private vehicles, 38,000 trucks and railcars, 2,600 aircraft, and nearly 16,000 freight containers arrived at U.S. borders, and many Americans are demanding improved inspection and tracking of this massive influx of people and goods (Figure 11-20). The United States has imposed new restrictions on the granting of visas to foreign visitors, tourists, and students, and every one of the world shipping industry's 1.2 million merchant mariners will be required to carry a new biometric identity card. Improved cargo scanners are being installed at many ports, and the United States may even be said to have extended its borders beyond its territory, for it has increased the number of U.S. customs inspectors and migration officers stationed in foreign countries. They inspect ships and planes before they are allowed to depart for the United States.

Within the country, the 2001 U.S.A. Patriot Act (Uniting and Strengthening America by Providing Appropriate Tools Required to Intercept and Obstruct Terrorism Act) and other legislation allow law-enforcement agents to investigate, detain, and deport people suspected of links to terrorist activity or other crimes, as well as to hold suspicious individuals, to obtain wiretaps and search warrants, to monitor computers and bank transactions, and to tap telephone conversations. Authority may be granted to use cameras equipped with facial recognition technology to scan crowds for suspects; to monitor chat rooms, cell phones, and e-mail; and even to establish a national identity card. Most European countries have national identity cards, and the United States has a pilot program—but only for Mexicans. In 2001 the government approved four million special identity cards

FIGURE 11-20 "Invaders" of the United States. On February 7, 2003, when United States was officially on "high risk" for terrorist attacks, this team of armed security agents of a hostile foreign government landed on American soil. Their 30-foot Cuban government patrol boat, flying the Cuban flag, docked at the marina of the Hyatt Hotel in Key West, Florida, a short distance from the U.S. Coast Guard Station. Carrying assault rifles, the men marched into the middle of the resort looking for someone to whom they could present a request for asylum. Their story illustrates the difficulty of policing the United States's almost 161,000 km (100,000 miles) of coastline. (Courtesy of *Miami Herald*)

under a new program for short-term short-range border crossings between Mexico and the United States. Civil libertarians fear that some of these new laws shift the focus of investigations from responding to crimes to trying to prevent them, that is, to preventive measures such as have never before been practiced in America. Western democracy has always been leery of the idea of preventative arrests.

Some civil libertarians fear a repeat of the notorious "Palmer Raids" of 1920. On January 2 of that year, hundreds of agents from the U.S. Bureau of Investigation (it became the FBI in 1935) assisted by local police and volunteers from the American Legion, swarmed across America and arrested nearly 4,000 suspected radicals, Communists, Bolsheviks, anarchists, and aliens. Without arrest or search warrants, the agents invaded political party headquarters, private homes, and even bowling alleys. Many of those arrested were beaten, crowded into filthy detention centers, and not allowed to communicate with their familes or lawyers for weeks. This infamous raid climaxed what was known as the Red Scare, when many Americans believed that a Communist conspiracy was in the works and that a Soviet-style revolution, directed from Moscow, was imminent. The Palmer Raids—named after then-U.S. Attorney

General A. Mitchell Palmer—included mass arrests, deportations, spying, and extreme vigilance. They were on a far larger scale than anything since seen or contemplated in the United States today, but the memory of them inspires caution.

Today's U.S. Attorney General John Ashcroft has tried to silence criticism of government actions by warning: "To those who scare peace-loving people with phantoms of lost liberty, my message is this: Your tactics only aid terrorists, for they erode our national unity and diminish our resolve. They give ammunition to America's enemies, and pause to America's friends." Democracy, however, requires a strong skepticism of letting the government do whatever it thinks is best.

Territorial Subdivision and Systems of Representation

All countries subdivide their territory, and each country has a unique balance of powers between its local governments and its national government. We generally call those countries in which the balance of power lies at the center **unitary governments,** and those in which the balance of power lies with the subunits **federal governments.** Today most national governments are unitary. Few newly independent and poor countries have adopted federalism, perhaps because they fear that if centrifugal forces were officially recognized, they could pull the country apart. Ethiopia, however, adopted a new federal constitution based on regional self-determination for 11 ethnic groups in 1994.

The definitions of unitary states and of federal states represent models, but in fact countries with federal constitutions may be highly centralized through government by a one-party political system or a military regime. India's federal constitution allows the prime minister to impose direct rule from New Delhi when law and order have broken down in a state. Even the U.S. Constitution obligates the federal government to "guarantee to every State . . . a Republican form of government . . ." (Article IV, Section 4). All governmental systems actually leave much open to improvisation and continual redefinition.

Unitary government In unitary states the central government theoretically has the power to redraw the boundaries of the subunits. This offers flexibility to accommodate geographic shifts in national population distribution or economic growth. From time to time it may be useful to redraw internal boundaries in response to changing needs, for example, by extending the boundaries of cities as they expand into the surrounding countryside. Interests become vested in any given pattern, however, so unitary governments seldom redraw internal boundaries. For example, the 50 U.S. states are each internally unitary, but they seldom redraw the boundaries of their counties. This is largely because political parties are organized at the county level, and they resist change. Also, changes may threaten the existing balance of power among groups, so potential losers struggle to maintain existing patterns.

China is constitutionally "a unitary multinational state." Five principal minorities and about 50 smaller national groups exercise some degree of local autonomy, and Hong Kong is guaranteed special political and economic privileges for 50 years after 1997 (Figure 11-21).

The Turkish-speaking Muslim Uighur people of Xinjiang have rebelled periodically; they may be envious of the political autonomy enjoyed by their ethnic relations in Uzbekistan, Kyrgyzstan, Turkmenistan, and Kazakhstan. China has given financial assistance to these neighboring countries in exchange for repressing Uighur independence movements in their own countries. Meanwhile, settlement of ethnic Chinese in Xinjiang has reduced the Uighurs to a minority there.

The Peoples of China

ETHNO-LINGUISTIC GROUPS AND AUTONOMOUS REGIONS IN CHINA

FIGURE 11-21 The traditional limits of Chinese territory were the borders of the Ming Empire—the Jade Gate on the route to Central Asia and the Great Wall against the Mongols. Through the centuries, however, Chinese rule has intermittently expanded out beyond these borders to include non-Chinese peoples. China's contemporary constitution guarantees some political and cultural autonomy to minority peoples in the five peripheral regions that are labeled plus Hong Kong. These regions constitute 42 percent of China's territory but hold only about 8 percent of the Republic's population.

Similarly, the Mongols of Inner Mongolia (a part of China) may envy the Mongols in Mongolia (an independent country) and wish to secede from China, but here too Chinese settlement has reduced the Mongols to a small minority in "their own" territory. Nevertheless, some observers believe that political instability may cause the Chinese multinational state to break apart.

Federal government A federal government assumes that diverse regions ought to retain some local autonomy and speak with separate voices in the central government. Federalism also allows each area to serve as a laboratory for legislation that can be adopted elsewhere if it works. Wisconsin, for instance, might devise a successful school program that other states might copy, New Mexico a highway program, or Maine environmental legislation. The protection of diversity, however, may perpetuate economic or social inequality.

It is dubious whether the states of the United States are today culture regions or even genuine political communities. Each state controls its own education system, which usually inculcates state history and geography, just as education systems do in sovereign nation-states. Each state boasts an extensive iconography: state flags, birds, flowers, and so forth. Texas (where school-children pledge allegiance to the state flag) and Hawaii have histories as independent countries. States have personalities, as revealed by the different types of legislation they pass, and one scholar has even mapped the "dominant political cultures" across the United States (Figure 11-22). If Puerto Rico were to become a state, it would be the most distinct state culture region.

Many observers have suggested redrawing the map of the states. A new pattern could improve the states' conformity with physiographic or economic regions or make each state more compact or equal in population or in economic strength (Figure 11-23). The U.S. Constitution, however, protects the states against restructuring against their will (Article IV, Section 3), although Virginia split into two states during the Civil War in 1863. Strong political currents within California have proposed breaking that state into two or three states.

Despite the United States's federal form, it may be merging into one political arena anyway. Several facts provide evidence. First, voter turnout is much higher for presidential elections than it is for House or Senate contests, or even for elections for state and local officials. This suggests a weak sense of identification with local representatives. Second, in 1994 Republican candidates for the House of Representatives offered

Political Cultures of the 48 Contiguous States

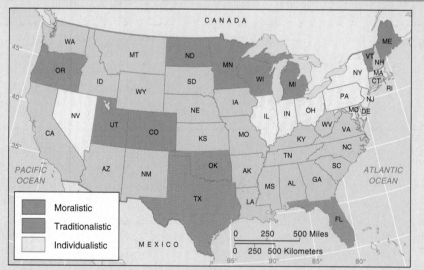

FIGURE 11-22 This map shows the states most strongly identified as exhibiting one of three political cultures defined by Daniel J. Elazar. Individualistic political culture holds that government is instituted for strictly utilitarian reasons; it has no concern with questions of "the good society." Moralistic political culture, by contrast, considers politics a search for the good society—an effort to exercise power for the betterment of society. Traditionalistic political culture accepts a hierarchical society as natural and authorizes and expects those at the top of the social structure to take a special and dominant role in government; politics secures the maintenance of the existing social order. Internal migration and immigration, as well as changes in American society overall, may have changed this map since it was drawn, but traditions persist. Can you see any coincidences between this map and Figure 7-34, the map of religious affiliation? Would you suggest any hypotheses of explanation? (Adapted from Daniel J. Elazar, *American Federalism: A View From the States,* 3rd ed. Copyright 1984 by permission of Harper & Row, Publishers, Inc.)

A 38-State U.S.A.

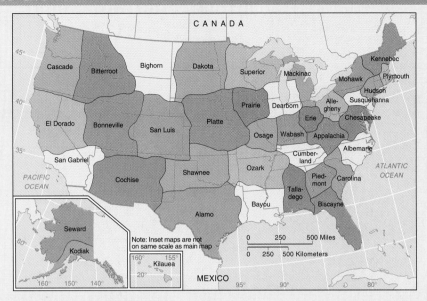

FIGURE 11-23 This redivision of the United States into 38 states was drawn as an intellectual exercise by G. Etzel Pearcy, a distinguished geographer who once served as Geographer to the United States. These territorial units may be more equivalent in population or economic output or may better reflect true economic or cultural regions than do the present states. What state would you be in? Can you think of any advantages from creating such a state? (Adapted with permission from G. Etzel Pearcy, *A Thirty-Eight State U.S.A.*, Monograph No. 2. Redondo Beach, CA: Plycon Press, 1973.)

one platform, a "contract" to the American people. This contract virtually nationalized elections to the House of Representatives. Both parties have since offered such contracts. Third, members of Congress are raising increasing percentages of their campaign funds outside their own constituencies. More than half of the U.S. senators, for example, raise more than 50 percent of their funds outside their home states—several more than 90 percent. This raises the question of whom, or what constituency, these senators represent. Widespread fundraising arguably corrodes the federal character of U.S. government.

Some countries' federal patterns reflect polyethnicity (for example, Russia, Myanmar, and India). Some countries' internal units are older than the federal framework, and they came together only on the condition that they would retain certain powers (the original 13 United States, Germany, and Canada). Federalism may also serve countries that are relaxing central control. Spain's culturally distinctive peripheral provinces, for example, have never been welded into one nation, and in recent years they have demanded and received considerable autonomy. A new Italian Constitution adopted in 2001 devolves considerable power from the central government to the provinces. Belgium is decentralizing into three parts: Flanders, Wallonia, and the city of Brussels. The United Kingdom established new parliaments in Scotland and in Wales in 1999, and France has granted some autonomy to the local government of Corsica. In these European examples, some powers of the national governments are devolving down to the regions at the same time as others are being surrendered up to the multinational European Union (discussed in Chapter 13), so the national governments retain few powers. In states torn between two religions, such as Nigeria, the adoption of Sharia law in some of the states can aggravate antagonisms, and if loosening the federal ties proves insufficient, a federal state may dissolve in civil war, as happened in the former Yugoslavia.

Canada's federal system Canada was formed as much to preserve British institutions in North America as in response to any indigenous nationalism. The British North America Act of 1867 created the Confederation of Canada, and other provinces joined the confederation later (Figure 11-24). The 1931 Statute of Westminster removed all legal limitations on Canadian legislative autonomy, and in 1982 the Canadian Constitution was "patriated," meaning that Canadians obtained the right to amend the constitution in Canada.

In Canada's parliamentary federal government, executive authority is vested in the sovereign (Queen Elizabeth), who appoints a governor general. That office, however, exercises little authority or influence. Legislative authority resides in the Canadian Senate and House of Commons. The 104-member Senate is appointive, and it too exercises little power. It cannot, for example, initiate legislation, although it can delay legislation. Therefore, power is concentrated in the

The Canadian Federation

FIGURE 11-24 The units of federal Canada came together only slowly into the country's present configuration. The 1867 British North America Act united the provinces of Upper Canada (Ontario), Nova Scotia, New Brunswick, and Lower Canada (Quebec). In 1869 the Northwest Territories were purchased from the Hudson's Bay Company; Manitoba was carved from this territory and admitted into the confederation in 1870. British Columbia joined in 1871, and Prince Edward Island in 1873. Alberta and Saskatchewan were formed out of previously provisional districts and admitted in 1905. Newfoundland, previously an independent country, joined only in 1949, and took the name Newfoundland and Labrador in 2001.

The territory of Nunavut formally came into existence April 1, 1999. It covers one-fifth of Canada's land area but contains a population of only 27,000, of whom 85 percent are Inuit. The city that had been Frobisher Bay changed its name to Iqaluit as the new capital. In 2003 the Canadian government granted the Dogrib people local self-government in a territory of 39,000 square kilometers, and three additional self-government treaties are under negotiation.

House of Commons, whose 295 members represent districts based on population.

This concentration of power has allowed the two most populous provinces, Ontario and Quebec (about 62 percent of the total population), to dominate the national agenda. Much of Canadian politics has traditionally been concerned with reassuring the French-speaking (Francophonic) population of Quebec that their culture would be secure in a country with an English-speaking (Anglophonic) majority (see Chapter 7). Many Westerners have felt disadvantaged. In recent elections, however, newly formed political parties or those strongest in the West have risen to power, and the voices of new immigrant groups are making themselves heard.

Canada's provinces have no significant institutionalized recognition at the level of the national government, but they retain more powers than do the individual states of the United States. For example, the Canadian Constitution lacks a "commerce clause" such

as the one in the U.S. Constitution that gives Congress the power "to regulate commerce with foreign nations and among the several states" (Article One, Section 8). Therefore, trade barriers exist among the Canadian provinces. The provinces signed an Agreement on Internal Trade in 1995, but they have failed to meet agreed deadlines to implement it. The people of Quebec have often debated seceding from Canada. In 1998 the Canadian Supreme Court ruled that Quebec cannot secede on its own, but that if a clear majority in Quebec wish to secede, the government of Canada would have to treat that wish as an initiative from Quebec for an amendment to the Constitution. In 2000 the House of Commons provided that a province can secede only after a referendum in which the option of independence is clearly stated and secessionists win a clear majority.

Canada's Native Americans (also called First Nations) are also gaining political significance, in two ways. First, courts are increasingly ruling in favor of their claims to territories and resources. Second, by arguing that they should get any special privileges that Quebec does, they become power brokers in the Anglophone/Francophone dialogue.

Special-purpose territorial subdivisions Several countries divide their territory into economic-planning regions, and these can gather almost as much power as the legal constituent subunits. This has happened in France. In the United States the Federal Power Commission, the National Labor Relations Board, the Bureau of Reclamation, the Federal Trade Commission, the Federal Communications Commission, and the Federal Reserve System have each subdivided the country into a different pattern of districts. Few of these patterns conform to the pattern of the states, so when two different districts enforce different policies in two parts of one state, the state government's own powers can be confused or undermined. When Vladimir Putin became president of Russia in 2000, one of his first official acts was to divide Russia into seven new federal zones, each with an administrator appointed by himself, and to transfer power to these zonal administrators. Thus, Russia's 89 elected regional governors lost much of their power overnight.

The United States is also subdivided by 91 federal district courts and, above them, 11 federal courts of appeal (Figure 11-25). Legal efforts to protect local differences may carry unexpected consequences. Two of today's most controversial topics, abortion and pornography, illustrate this. In 1973 the Supreme Court upheld the legality of abortions, but it allowed the states considerable leeway in defining the conditions under which each state would allow abortions (*Roe* v. *Wade;* upheld in *Planned Parenthood* v. *Casey,* 1992). As a result, wealthy women may travel across state lines to get abortions, but poor women may have to obey the laws of their home state. Dissenters cross state lines to protest abortions in states other than their own. The question remains on the national agenda, and legislation may yet be standardized throughout the country.

Legislative restriction of pornography and obscenity presents another example of differences in local laws. In 1973 the Supreme Court ruled that courts can regulate expression only "when the average person, applying contemporary community standards, would be likely to find that the material appealed mostly to prurient interest, was patently offensive and was entirely lacking in serious literary, artistic, political or scientific value" (*Miller* v. *California*). This ruling is difficult to apply in a country where materials are distributed nationally. The most conservative communities can prosecute what they view as pornography in ways so punitive as to deny the material to other communities. For example, the New York-based Home Dish Satellite

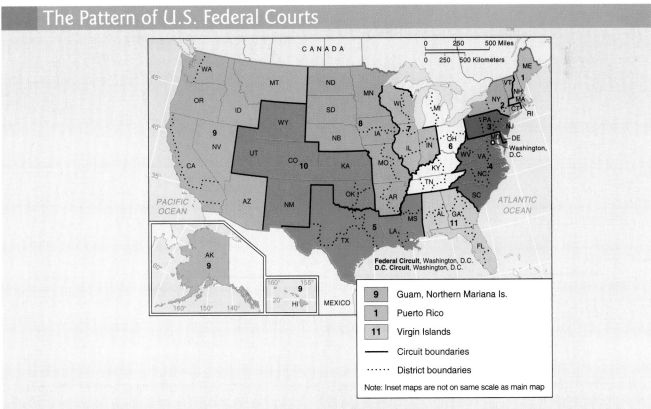

The Pattern of U.S. Federal Courts

FIGURE 11-25 Any two of these courts may interpret federal law differently. Therefore, federal law is not uniform throughout the country until the disagreement is resolved by the Supreme Court.

Network, which beamed X-rated films to 30,000 subscribers around the country, was driven into bankruptcy after a district attorney in Montgomery County, Alabama (where 30 households received the service), brought criminal charges for violating Alabama's anti-obscenity laws. In the same way, the country's most conservative jurisdictions have prosecuted nationally popular recording stars for their lyrics. The Fifth Amendment to the U.S. Constitution prohibits trying the same person twice for the same crime (*double jeopardy*), yet each local jurisdiction can bring charges against nationally distributed materials even if several juries elsewhere have already refused to convict. Today, however, with video rentals, pay-per-view television, and the Internet, the ubiquity of pornography makes it difficult to prove that any community's "standards" are against it. For example, in Provo, Utah, usually thought of as a conservative area, a local video-store owner charged with selling obscene material was able to prove that he had 4,000 regular customers for sex videos, that people in Utah county bought 20,000 sex videos from one satellite programmer alone during the period the store owner was said to have broken the law, and that guests in the local Marriott hotel were paying to watch about 3,000 adult explicit sex videos each year.

Today the federal government has assumed an active prosecutive role. The Anti-Drug Abuse Act of 1988 grants federal prosecutors the right to bring obscenity charges against materials that cross state lines, even if local authorities will not do so. It remains an open question whether the United States is or is becoming one cultural community.

Federal territories   Most federal states contain territories that are not included in any of the subunits but that are administered directly from the federal center with, perhaps, some local government. These territories may constitute a significant share of a state's total area, including capital districts, colonies, strategic frontier areas, and federal territories.

Territories usually have only limited local government until some presumed future time when they will be ready for full statehood. If the territories lack resources, they may remain territories indefinitely. Most of the land area of the United States was once federal territory, but the territories were organized, populated, and admitted into the Union as states. By contrast, today some 40 percent of Canada remains in territorial status. Mexico includes two territories that may become states upon reaching populations of 120,000 and demonstrating "the resources necessary to provide for their political existence."

The U.S. government still owns about one-third of the total national land area (Figure 11-26). The 1976

U.S. Federal Properties

FIGURE 11-26 The U.S. government still owns a substantial share of the total national territory. The largest land trade in U.S. history occurred in 1998. The state of Utah surrendered to the federal government 152,580 hectares (376,739 acres—an area almost half the size of Rhode Island) that had originally been given to the state to finance education; the land was scattered through national forests, parks, and monuments. In return the federal government gave Utah $50 million in cash plus 58,725 hectares (145,000 acres) outside national parks and areas proposed for wilderness.

Federal Land Policy and Management Act requires the federal government to receive full value for any lands traded away, and Congress still wrestles with the question of how the federal government should exploit or preserve its lands.

Capital cities In some countries the federal capital is also the capital of a component state (Bern is capital of both its own canton and of Switzerland). Austria's federal capital, Vienna, is a federal state. Some federal capitals are governed directly by the federal government (Australia's capital territory, Mexico's Distrito Federal). Residents of Washington, D.C., complain that they have no voice in the federal government; if the city were a separate state, it would rank 44th in population.

Several countries have moved their capitals because they believed that their old capitals were international cities grafted onto national life rather than truly representative of national life. A new capital, it is believed, especially if built inland, will symbolize a rebirth and rededication of a national spirit. In 1918 the Russians moved their capital from St. Petersburg inland to Moscow; in 1923 the Turks moved theirs from Istanbul to Ankara; Brazilians moved theirs inland to Brasília in 1960; Tanzanians moved theirs inland to Dodoma in 1975; Nigerians moved theirs to Abuja in 1991. Newly independent Kazakhstan moved its capital from Almaty (which was thought to be too Russian) to Akmola (today called Astana) in 1997. The Japanese, by contrast, moved their capital from inland Kyoto to coastal Tokyo in 1868 to symbolize Japan's opening to the world. Capitals are frequently designed as showplaces of national pride to impress both citizens and foreigners. In 1999 the offices of Malaysia's Prime Minister were moved to the newly built city of Putrajaya, next to the country's new scientific- and computer-research center, Cyberjaya. Kuala Lumpur will, however, remain the official capital until 2012.

How to Design Representative Districts

The subfield of political geography that studies voting districts and voting patterns is called **electoral geography.** In district systems of political representation, the boundaries of the districts can determine the outcomes of the ensuing elections. This can occur in either one of two ways. First, if the electoral districts are unequal in population, then the ballots cast by some voters outweigh those cast by others. Second, district lines can be drawn in ways that include or exclude specific groups of voters so that

FIGURE 11-27 The original gerrymander. The original gerrymander was a district created in Massachusetts in 1812 to concentrate the Federalist Party vote and thus to restrict the number of Federalists elected to the state senate. The district configuration was at first likened to a salamander, but later it picked up the name of Governor Eldridge Gerry, the Anti-Federalist who signed the districting law. ("It looks like a salamander." "No, by golly; it's a Gerrymander!") (Reproduced from James Parton, *Caricature and Other Comic Art.* New York: Harper Brothers, 1877, p. 316.)

one group gains an unfair advantage. This is called **gerrymandering** (Figure 11-27).

The electoral systems in many countries tolerate inequalities in numerical representation. Chapter 8 noted that many countries grant rural populations disproportionate representation in the government. In the United States it was long true at the level of the state and local general-purpose governments, but in 1962 the Supreme Court ruled that for these governments, the number of inhabitants per legislator in each district must be "substantially equal" (*Baker v. Carr*). In 1962 Tennessee was still electing state legislators on the basis of a 1901 apportionment. The population had urbanized, so that by 1962, one rural vote was equal to as many as 19 urban votes. Many other states' legislatures contained similarly disproportionate representation, but nationwide redistricting now follows each decennial census.

After each census, the state legislatures also redistrict for federal representation. The Supreme Court has extended the one-person/one-vote rule to all levels of political representation in the United States except the governing boards of the special-district governments in

> In district systems of political representation, the boundaries of the districts can determine the outcomes of the ensuing elections.

metropolitan areas. Republican legislators in Texas moved to redistrict the state in 2003 shortly after winning the majority in elections in 2002; that attempt could introduce redistricting after each election in every local government all across America.

Representation in the U.S. federal government

The U.S. federal government itself does not perfectly achieve the one-person/one-vote ideal. Neither the House of Representatives, nor the Senate, nor the presidency is based on population count. The Constitution specifies that "each State shall have at least one Representative" (Article I, Section 3). Several states have so few residents that without this constitutional protection, they would have to share a representative. There are 435 seats in the House of Representatives. Therefore, assigning one representative to each state leaves 385 seats to be allocated among the states by population. The Constitution's mandates cannot be reconciled to a one-person/one-vote ideal. In 2000 the 483,000 people of Wyoming had a representative, but each representative from California represented a district of about 651,000 people, a ratio of 1 to 1.3 (Figure 11-28).

Each state is also represented by two senators, no matter how few residents it has. Therefore, each of the senators from Wyoming represented about 246,000 people, whereas each senator from California represented about 17 million people, a ratio of 1 to 69.

The president, in turn, is elected indirectly in the Electoral College, in which each state has as many electors as it has total seats in Congress. The candidate who wins a majority of a state's popular vote wins all that state's electoral vote. In 1888 Grover Cleveland won the popular vote over Benjamin Harrison, but Harrison won the electoral vote, and Harrison became president. At the end of the twentieth century, the populations of some states were decreasing, while the populations of others grew. The national population became increasingly concentrated, and the system of representation thus became increasingly arithmetically undemocratic. By 2000 California's population was greater than that of the 22 least populous states combined, yet those states had twice California's representation in the Electoral College. The presidential election of 2000 repeated the experience of 1888. The Democratic candidate, Al Gore, swept the popular vote by a plurality of over 500,000 votes, but, after the U.S. Supreme Court

Reapportionment Among the States After the 2000 Census

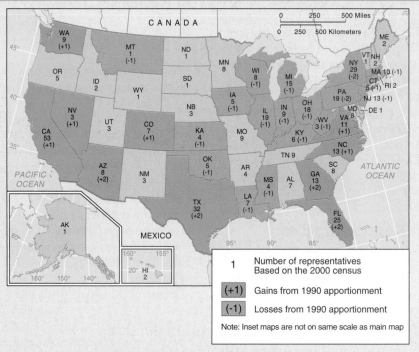

FIGURE 11-28 After the 2000 Census, the seats in the U.S. House of Representatives were reapportioned as shown on this map. Fast-growing states, mainly in the South and West, gained seats, but those with proportionately declining populations, mainly in the North and East, lost representation. Some observers feel that the United States has grown so populous that people are alienated from the Congress. The House of Representatives capped its own number at 435 in 1910, when the U.S. population was 90 million, but if the ratio of members of the House today were the same as it was in 1790, the House would have over 9,000 members.

Critical Thinking

NONTERRITORIAL SYSTEMS OF REPRESENTATION

Not all governments at all levels divide their territories into districts for purposes of political representation. There are at least two nonterritorial systems of representation.

In *at-large systems of representation,* all members of a governing body are chosen by the entire electorate. In such a system, a slight majority of one group can elect a government that is entirely of its group, and a minority may find it impossible to elect any of its candidates. In a *district system,* by contrast, an area with a minority population can elect at least its local representative. Therefore, courts in the United States have often required local governments to switch from at-large systems to district systems of representation.

Another alternative to territorial districts is government by representatives of interest groups. This is called **corporativism.** The representatives may represent labor unions, agricultural cooperatives, chambers of commerce, private or nationalized industries, professionals, and other groups. Corporativism has been tried by Germany, Italy, Ireland, several Latin American countries, France, and several African states that were former French colonies. Candidates for the legislative body of Kuwait may represent interest groups, but no major countries are governed by corporativist systems today. The U.S. Congress, however, does much of its real work in committees that specialize in various topics: agriculture, labor, and so forth. Members of these committees are lobbied by representatives of groups interested in that topic, and the members receive campaign support from those interested groups, whether the groups are based in the representative's district or not. Some observers have suggested that this system in fact imitates corporativism.

Questions

1. Should U.S. representatives and senators be restricted to raising campaign funds within the geographic districts they represent?
2. How would that affect government?
3. Would such a plan hinder free political expression?

---

intervened to stop a recount of votes cast in Florida, the Republican candidate, George W. Bush, won Florida and with it the vote in the Electoral College. Eliminating the Electoral College would require the approval of the low-population states who profit from it, so the possibility is probably slim.

The fact that the government is not truly numerically democratic causes inequitable distribution of federal spending. As a rule, the more-populous states subsidize the less-populous. Even in distributing federal funds to protect against terrorist attacks, for example, the Department of Homeland Security spends more than seven times as much, per capita, in rural Wyoming as it spends in metropolitan New York or Los Angeles, which are certainly more probable targets.

How gerrymandering is done   The following discussion of the techniques of gerrymandering will draw mostly on U.S. examples because they will be most familiar to most readers. These techniques, however, apply wherever government is established with representation by territorial districts. Figure 11-29 illustrates the problem of drawing district boundaries in an area of heterogeneous population. It is easier to define the

(a) Maximum segregation   (b) Maximum integration

FIGURE 11-29 Models of political districts. If we want to draw three districts of roughly equal populations, and if we have three population bases (*x, y, z*) defined by race, income, or any other characteristic, we could draw the boundaries in many different ways: (a) illustrates maximum segregation of the populations among the districts; (b) demonstrates maximum integration of the populations within each district.

(a) Opponent concentration:
A's control one district

(b) Opponent dispersion:
A's control no district

(c) Even division: A's and B's
each control two districts

FIGURE 11-30 Types of gerrymandering. In this figure, the B's have an absolute majority in each case, so they have the power to draw the district lines. Drawing (a) illustrates concentration of the opponents (the A's), so the A candidate wins that one district with an unnecessarily large majority, but A's cannot win anywhere else. Drawing (b) splits up the A's so that they do not form a majority anywhere and cannot elect even one representative. In (c), each group controls two districts.

problems of districting than it is to define equitable solutions. Substantial equality of population count in each district is legally required, and contiguity and compactness of districts is desirable because it eases communication within districts. The question still arises, however, of whether homogeneity or heterogeneity of population within a district is preferable. Some people feel that districts should have a common social or economic characteristic, whereas others feel that balanced or integrated districts are preferable. Gerrymandering to achieve one kind of homogeneity (for example, racial homogeneity) may link people who in fact share no other attributes, problems, or political community.

Computers make it easier to gerrymander, because today GIS makes it easy for any legislator to call up any district on a screen, shift a boundary, and get an instantaneous readout of what the total population, voting behavior, racial composition, and other characteristics would be of the newly drawn district (Figure 11-30). Some observers fear that the easier gerrymandering threatens democracy, as incumbents gerrymander to secure their own reelection. Voters are polarized.

Gerrymandering and civil rights A 1982 amendment to the Federal Voting Rights Act of 1965 banned any redistricting that would have a negative impact on the political representation enjoyed by minority communities. The law insists that when any court is examining a case of redistricting, the court consider "the extent to which members of a protected class have been elected to office." Several states interpreted this to mean that they were mandated to create districts in which the majority of the voters belonged to racial minorities.

These districts are called *majority-minority districts,* and they are drawn by practicing the opponent concentration technique illustrated as (a) in Figure 11-30. Such majority-minority districts have been drawn at every level of government across the nation, from city council seats to congressional districts. In 1986 the Supreme Court defined a three-part test for assessing voter dilution that seemed to encourage the formation of majority-minority districts. The ruling stated: "First, the minority group must be able to demonstrate that it is sufficiently large and geographically compact to constitute a majority of a single-member district. . . . Second, the minority group must be able to show that it is politically cohesive. . . . Third, the minority must be able to demonstrate that the white majority votes sufficiently as a bloc to enable it . . . usually [to] defeat the minority's preferred candidate" (*Thornburg* v. *Gingles*).

Throughout the country, however, these majority-minority districts were challenged as representing racism—favoring a designated group rather than oppressing it, but racism nonetheless. Some people, however, argue that the United States must be color-conscious for a while as a way of making up for past discrimination and segregation. In 1993 the U.S. Supreme Court ruled that creating legislative districts with people of the same race who are otherwise separated "bears an uncomfortable resemblance to political apartheid . . . and reinforces the perception that members of the same racial group—regardless of their age, education, economic status or the community in which they live—think alike, share the same political interests and will prefer the same candidates at the polls" (*Shaw* v. *Reno*). Grouping voters by their ascribed identities and trying to win their votes is commonly called *identity*

Critical Thinking

GARZA V. COUNTY OF LOS ANGELES

In 1990 federal courts ruled in *Garza v. County of Los Angeles* that the Los Angeles County Board of Supervisors had used a combination of unconstitutional gerrymandering methods in drawing their voting districts in order to exclude Hispanic people from representation. Hispanics made up 35 percent of the county's total population, but none had ever sat on the five-member board.

The court recognized that the lack of a Hispanic on the board did not prove that unconstitutional discrimination had occurred. Alternative hypotheses can partly explain the lack of Hispanic representation. Hispanics may be underrepresented because

- a high percentage of them are aliens
- a high percentage fail to register
- a high percentage are children too young to vote
- a high percentage simply are not interested in politics.

Each of these hypotheses has some validity, but the court ruled that their combined explanatory power did not explain Hispanics' total exclusion unless intentional discriminatory gerrymandering had also taken place.

Board district lines were redrawn, and a district was created in which Hispanics made up 71 percent of the population, although only 51 percent of registered voters. Voter turnout in the subsequent 1991 special election was only 21 percent, but a Hispanic won the seat on the board. In the next election for that seat, however, a non-Hispanic individual defeated the Hispanic incumbent and other Hispanic candidates.

Questions

1. Does the fact that a particular group is underrepresented or not represented in a governmental body mean that discrimination has occurred?
2. How can discrimination be proved?
3. How much power should the courts or other institutions have to remedy such situations?
4. What does the defeat of all Hispanic candidates in a Hispanic neighborhood by a non-Hispanic individual suggest about identity politics and voters' interpretation of their own best interests?

---

politics, but racial stereotyping in districting is legally impermissible. In examining one gerrymandered district, North Carolina's Twelfth Congressional District (Figure 11-31), the Supreme Court ruled that a conscious concentration of black voters did not automatically make a district unconstitutional, as long as the state's primary motivation was political (to create a Democratic district) rather than racial (to create a black district). In other words, party-conscious gerrymandering is constitutional, but race-conscious gerrymandering is not (*Easley* v. *Cromartie, 2001*).

When majority-minority Congressional districts have been redrawn so that minorities are no longer in the majority in those districts, minority officeholders running for reelection have successfully defended their seats. Some observers have argued that this proves race is no longer a significant factor in U.S. politics, whereas others have argued it proves only that the incumbent has a great advantage. Arguments over gerrymandering will continue in America's legislatures, and in 2003 the U.S. Supreme Court upheld federal court's right to assume the activist responsibility of stepping in to draw legislative boundaries when state processes fail (*Branch* v. *Smith* and *Smith* v. *Branch*) (Figure 11-32).

Efforts to Gerrymander Districts in North Carolina

FIGURE 11-31 In the early 1990s, North Carolina legislators designed these two congressional districts in order to elect black representatives to Congress. The U.S. Supreme Court upheld the constitutionality of the First District, but, in 1996, it ruled that the Twelfth District was drawn in a manner that was too "race-conscious." In 1997 the North Carolina legislature redrew the Twelfth District in a way that would scarcely be distiguishable on a map at this scale. That new pattern was finally approved by the U.S. Supreme Court in 2001, just as North Carolina set about redrawing all of its congressional districts on the basis of the new data from the 2000 census!

Some U.S. Congressional Districts in 2003

FIGURE 11-32 These are a few "gerrymanders" sending representatives to the U.S. House of Representatives in 2003. In each case, the districts were designed to manipulate the outcomes of elections held in the districts.

MEASURING AND MAPPING INDIVIDUAL RIGHTS

Countries differ greatly in the restrictions they place on their citizens' freedom and in the degree to which they encourage and achieve full development of their citizens' potential. Predominant attitudes toward some issues, such as women's rights, may be determined by the underlying culture, which may cross international boundaries. Arab culture, for example, as noted in Chapter 7, seems relatively insensitive to women's rights. Nevertheless, sovereign states are the units within which laws promote or restrict individual rights, so we can meaningfully analyze and compare the policies of different countries.

A society in which the most capable people can rise to the top on merit alone is called a **meritocracy,**

and most countries claim to be meritocracies. Rigid social stratification can be as unhealthy in a society as hardening of the arteries is in a human body. Hurdles to individual advancement may include stratification of the population by caste, race, or employment restrictions. The modern labor movement has won important victories for working people, but in some countries labor unions may protect their members' jobs at the expense of opportunity for other individuals. In the United States, for example, some labor unions have been bastions of racism.

Many countries of diverse populations have *affirmative action* programs designed to lift to national standards of achievement those segments of the population that have suffered historic lack of opportunity. Laws in both Malaysia and Indonesia, for example, favor ethnic Malays over the more entrepreneurially successful

Chinese. Nigerian law recognizes efforts of school and job-development programs to "reflect the federal character." This is a code for federally mandated ethnic quotas. Some European countries try to achieve diversity in groups in terms of members' age, as well as ethnic composition and sexual orientation. Affirmative action programs have been crafted in the United States in both the public and private sectors. In 2003, in upholding race-consciousness among admissions criteria for U.S. universities, the Supreme Court replaced the traditional appeal to historical justice as a rationale for affirmative action with diversity in itself as a key value in American life (*Grutter* v. *Bollinger* and *Gratz* v. *Bollinger*).

The most pervasive form of discrimination in the world, however, is sexism, and the most powerful boost to individual achievement is education. Individual freedom is a value sought by all individuals. International variations in these three social characteristics—sexism, educational opportunity, and individual freedom—are examined in the paragraphs that follow.

Sexism

Almost everywhere, women are worse off than men. They have less power, less autonomy, less money, more work, and more responsibility. National variations in sexism can be approached through a number of quantitative measures. In Chapter 5 we saw that discrim-

ination against women can begin even against children in the womb, as reflected in higher rates of abortion of female fetuses than male fetuses, and that female-to-male ratios in national populations may reveal discrimination against women. Discrimination against female farmers, discussed in Chapter 8, lowers agricultural productivity and the levels of nutrition in many countries.

U.N. researchers have measured and mapped striking differences in women's welfare around the world. Women in rich countries are generally better off than women in poor countries, but fewer data focus on the contrasts between the welfare of women and that of men in individual countries, and those are the data that must be isolated to compare sexism among countries. One useful statistic might be the measure of female to male enrollments in primary and secondary school (Figure 11-33). A country's failure to educate females matters not just in itself but also because improving female literacy is one of the most effective ways to control population growth and to reduce infant mortality.

Some issues are almost exclusively women's issues, such as the female genital mutilation that is common across Sahelian and East Africa and in the Near East, and maternal death rates. Other issues affect both sexes, but principally women: abortion rights, birth rates, equal pay, rape, domestic violence, maternity care or maternity

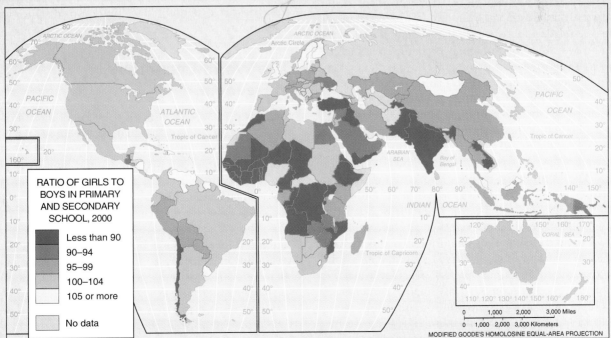

FIGURE 11-33 Where a lower percentage of females than males are in school, we may hypothesize that females do not receive all of the opportunities that males do in that society. (Data from United Nations)

leave, availability of contraception, child welfare and rearing practices, and even marriage and divorce rights. By 2000 some 44 of the United Nations' 189 members had enacted laws against domestic violence, but these laws were not always honored and enforced.

Some researchers have suggested using the percentage of the labor force that is female as one index of women's rights. The problem with this statistic, however, is that statistical measurements of the labor force measure only paid occupations, and in many countries women are not paid for their labor—particularly home-making, farming, or even small-scale domestic manufacturing. Women's participation rate in the labor force is everywhere below that of men. It is about 40 percent in North America; slightly less in Japan, Europe, Australia, and New Zealand; and only 20 percent or even less in South Asia and the Middle East. The figures for Africa are even lower, suggesting that very few African women work. Observation of African life, however, reveals that the women are indeed working; they just are not getting paid. We need better statistical measures of women at work.

Women's participation in politics might serve as another index of sexism. In most democratic countries men got the vote before women, but in 2003 the United Arab Emirates and Kuwait were the only countries where men could vote but women could not. Women have formed political parties in several countries, including Holland, Iceland, Belgium, Spain, and Sweden. Women are proportionately underrepresented in legislatures everywhere, even though several countries reserve legislative seats for women. In Uganda, for example, each of the 34 parliamentary districts reserves an at-large seat for a woman. India adopted a Constitutional Amendment in 1993 that set aside one-third of all seats on village councils and of all village mayoralties for women. A certain percentage of those must go to low-caste women. This has caused revolutionary changes in local government concerns. In France, beginning with municipal elections in March 2001, all political parties must field an equal number of male and female candidates in almost all elections.

The collapse of Communist government in Eastern Europe has paradoxically reduced women's rights. The Communist regimes rigorously upheld job security, maternity benefits, rights of divorce, and other social guarantees as a matter of Communist doctrine. Women lost many of these rights in the transition to democracy and free-market economies. Layoffs in state industries struck women hardest because men's jobs were considered of primary importance. Newly elected parliaments in every country have smaller percentages of women than under communism. Abortion rights have been curtailed, and state-financed child care has been terminated. Party theory, however, never encouraged husbands to do housework, and therefore working women were often still expected to do the shopping, cooking, and cleaning. Women are not necessarily better off if they are "free" to work outside the home.

Regardless of national constitutions and laws, men's and women's relative welfare can be determined by their traditional roles in national culture. Many countries provide formally for sexual equality in the law, but few protect specific job, inheritance, property, or marriage rights or protect women from domestic violence. To learn more about the world geography of sexism, statistical and analytical tools will have to be improved.

The World Geography of Education

The more advanced a society is, the larger the investment it requires in its human resources, specifically education. The U.S. educator Horace Mann wrote in 1846, "Intelligence is a primary ingredient in the wealth of nations."

A country's wealth and the education levels of its citizenry usually rise together. In the rich countries of Western Europe and Asia and in the United States, adult literacy is high. In the rich countries that attract immigrants, significant numbers of people may be literate in a language or languages in addition to, or other than, the official language of the country. The percentages of literacy drop only slightly in Latin America but markedly in Africa. Few African countries boast literacy rates much higher than 50 percent, even for the preferred males (Figure 11-34). School enrollment figures tell the same story. The children in the world's rich countries are in school; those in the world's poor countries are not.

The World Geography of Freedom

Some people in the world live in freedom; others do not. But it is difficult to quantify freedom. Surprises appear when comparing statistics. Many civil libertarians, for example, ask why the United States is second among all countries in the percentage of its population that is in jail (after Russia). In 2002 the number of people incarcerated in the United States reached 2.2 million. This was largely because of the "war on drugs," but states and localities have begun releasing convicts and rescinding harsh drug laws because the cost of jailing so many citizens drains government resources. Does this mean that the United States is a totalitarian country? Or does it mean only that if you break the laws in the United States you are more likely to go to jail than anywhere else? Many observers might respond that just living in many countries is the equivalent of being in jail. How are such statistics to be interpreted?

Literacy Rate of Those 15 and Older

FIGURE 11-34 Literacy is virtually universal in the rich countries, but this map reveals considerable variations in national literacy rates. (Data from United Nations)

The World Geography of Freedom

FIGURE 11-35 This map results from an effort to quantify the degree of freedom exercised by the population in each country. (Rankings courtesy of Freedom House)

Freedom House is a research institute that monitors political and civil freedom around the world. In the measurement of political rights, the institute assigns the highest scores where elections are fair and free and where the people who were elected rule. The institute defines civil liberties as including freedom of expression, assembly, demonstration, religion, and association. It rates highest those states that protect individuals from political violence and from harm inflicted by courts and security forces. These states have free economic activity and strive for equality of opportunity. Freedom House assigns each country a score of 1 to 7 in each of these two categories and then sums up the scores in an overall rating "Free," "Not Free," or "Partly Free." Freedom House findings for 2003 are reproduced in Figure 11-35. These rankings change rapidly as governments rise and fall, but the attempt to rate countries reveals wide disparities in freedom. We shall see in Chapter 13, however, that the very definition of freedom may not be the same around the world.

CONCLUSION: CRITICAL ISSUES FOR THE FUTURE

The complete division of the world into stable and peaceful territorial nation-states may or may not be an ideal goal. At present, however, the state remains the ultimate level of absolute sovereignty on Earth. The difficulties and hardships resulting from trying to make nations fit states, or vice versa, should remind us that the territorial nation-state is only one theory of what might be the best political-geographical division of Earth. Perhaps territorial nationalism is not a positive ideal. A return to regnum forms of government is improbable, but it might be a desirable goal to achieve some universal guarantee of individual rights in any political-territorial framework.

Furthermore, the number of failed states demonstrates that the current political division of Earth is at least partly responsible for continuing war and poverty among hundreds of millions of people. Assuming that we do not want to choose the theoretical option of massacring or expelling the hundreds of millions of people who are unhappy with the current territorial pattern or who are ill-served by it, then should we try to "enforce" the current political boundaries? Who has the authority to do that? (Who is the "we" in the previous sentences?) Who has the power to do that? Alternatively, who has the authority or power to redraw the map? These questions threaten to trouble the world for the forseeable future.

Chapter 13 will investigate whether a genuine world political community exists and whether the world community might assume the responsibility or exercise the power to act in defense of global peace and universal human rights.

Chapter Review

SUMMARY

Earth's division into sovereign states is the most important territorial organizing principle of human activities. The idea that the whole world should be divided into countries originated in Europe and diffused worldwide with European conquest. The neatness of the units on today's world political map nevertheless exaggerates the degree to which all people accept the current pattern.

A nation is a group of people who want to have their own government and rule themselves. The ideal state is theoretically a nation-state. Several European nations existed before they achieved their own states. The political patterns of the colonies, however, were the conquerors' creations. When independence came, only large units were at first granted independence, but later this threshold principle was abandoned.

Many states suffer from cultural subnationalism. States might employ three strategies to fit the nation to the state territory: redraw the political map; expel people from any country in which they are not content, or exterminate them; or forge nations in the existing states. Redrawing the international map presents many difficulties, and expulsion or genocide are abominable, so six national institutions—a country's church or churches, armed forces, schools, media, political parties, and labor unions—become key centrifugal or centripetal forces in any state. Geographers study the presence or absence of these institutions in any country, the unique balance or mix among them in each country, and their interaction with other aspects of each country's culture.

Each state demarcates its territory's borders, and the terrorist attacks on the United States in 2001 brought new concern to matters of both secure borders and of internal security. Countries also subdivide their territory for political representation or administration. Each country has a unique balance of powers between its local governments and its central government. Special districts can undermine the powers of local governments, and most federal states contain territories that are administered directly from the federal center. Capitals are symbolically important. Electoral geographers note that in any system of representative government, voting district boundaries can be gerrymandered to determine the outcomes of the ensuing elections.

Countries vary widely in the degree to which they liberate and train their populations. The major form of discrimination is sexism, and the most powerful boost to individual achievement is education. Individual freedom is a value sought by all individuals. The incidence of each of these three social characteristics varies from country to country.

KEY TERMS

centrifugal forces p. 453
centripetal forces p. 453
collapsed states p. 456
core area p. 448
corporativism p. 481
cultural subnationalism p. 453
electoral geography p. 479
enculturation p. 464
failed states p. 456
federal government p. 472

geopolitics p. 456
gerrymandering p. 479
iconography p. 464
indirect rule p. 452
irredenta p. 451
kleptocracy p. 467
meritocracy p. 484
multinational state p. 453
nation p. 447
national self-determination p. 451

nation-state p. 448
oligarchy p. 467
political community p. 447
political culture p. 448
social Darwinism p. 447
socialization p. 464
state p. 446
superimposed boundaries p. 452
threshold principle p. 453
unitary government p. 472

QUESTIONS FOR REVIEW AND DISCUSSION

1. Differentiate a nation, a state, and a nation-state. Give examples of countries that are not nation-states.

2. What important decision was taken at the 1964 conference of the Organization of African Unity?

3. How do governments try to forge their populations into nations?

4. What are the two principal methods of gerrymandering?

5. What evidence do you have that people enjoy different degrees of rights and liberties in different countries?

THINKING GEOGRAPHICALLY

1. When and how was the pattern of local governments in your state drawn? What responsibilities do the various local governments have? Could you improve the pattern, reassigning responsibilities to new districts or different levels of government?

2. Read the constitutions of a few countries that you believe to be totalitarian and centralized. Do the constitutions reveal that? Does the U.S. Constitution accurately reveal the division of power between state and federal governments today?

3. How much land in your state is owned by the federal government? What is its legal status?

4. Is your state overrepresented or underrepresented in the Electoral College?

5. Do you think regions of a country ought to be able to secede? If you favor a vote, do you think that majority approval should be required throughout the entire country or only in the areas seeking to secede?

6. What would be the advantages and disadvantages if your state withdrew from the Union?

7. Make a list of several mountain chains or rivers around the world that serve as international borders. Which act as effective barriers? Which do not? Why?

8. Investigate the symbolism of the flags of a few countries. Look at Australia, the Comoros, Liberia, Mongolia, Namibia, Venezuela, and South Africa.

9. Why might a unitary government be more successful in Japan than in the United States?

SUGGESTIONS FOR FURTHER LEARNING

Agnew, John and Stuart Corbridge. *Mastering Space: Hegemony, Territory and International Political Economy*. New York: Routledge, 1995.

Anderson, Benedict. *Imagined Communities: Reflections on the Origins and Spread of Nationalism*. New York: Verso Books, 1991.

Barone, Michael, and Grant Ujifusa. *The Almanac of American Politics*. Washington, D.C.: The National Journal, published biannually.

Bell-Fialkoff, Andrew. *Ethnic Cleansing*. New York: Griffin Press, 1999.

Canon, David T. *Race, Redistricting, and Representation*. Chicago: University of Chicago Press, 1999.

Country Reports on Human Rights Practices. Washington, D.C.: U.S. Department of State, annual.

Elazar, Daniel J. *Constitutionalizing Globalization: The Postmodern Revival of Confederal Arrangements*. New York: Rowman & Littlefield, 1998.

Hastings, Donnan and Thomas M. Wilson. *Borders: Frontiers of Identity, Nation and State*. New York: Berg Publishing, 2000.

Herb, Guntrum, and David Kaplan, eds. *Nested Identies: Nationalism, Territory, and Scale*. Lanham, MD: Rowman & Littlefield, 1999.

Historical Atlas of Canada. vol. 1, *From the Beginning to 1800*, edited by R. Cole Harris; vol. 2 (1987), *The Land Transformed 1800–1891*, edited by R. Louis Gentilcore; vol. 3 (1993), *Addressing the Twentieth Century*, edited by Donald Kerr and Deryck W. Holdsworth (1990). Toronto: University of Toronto Press.

Hooson, David, ed. *Geography and National Identity*. London: Blackwell (The Institute of British Geographers Special Publications), 2002.

Marx, Anthony W., *Faith in Nation: Exclusionary Origins of Nationalism*. New York: Oxford University Press, 2003.

Monmonier, Mark. *Bushmanders and Bullwinkles: How Politicians Manipulate Electronic Maps and Census Data to Win Elections*. Chicago: University of Chicago Press, 2001.

Morrill, Richard. "Gerrymandering," an issue of *Focus*. New York: American Geographical Society, 1991.

Mortimer, Edward, and Robert Fine, eds. *People, Nation and State: The Meaning of Ethnicity and Nationalism*. New York: I.B. Tauris, 1999.

Poulsen, Thomas M. *Nations and States: A Geographic Background to World Affairs*. Upper Saddle River, NJ: Prentice Hall, 1995.

Rush, Mark E., Richard L. Engstrom, and Bruce E. Cain, *Fair and Effective Representation?* Boston: Rowman & Littlefield, 2001.

Smith, Anthony D. *Nationalism: Theory, Ideology, History*. Polity Press, 2001.

Shelley, Fred M., J. Clark Archer, Fiona M. Davidson, and Stanley D. Brunn. *The Political Geography of the United States*. New York: Guilford Press, 1996.

Statesman's Year-Book. New York: Macmillan Press and Saint Martin's Press, annual since 1864.

World Boundaries (series coordinated by the International Boundaries Research Unit at the University of Durham). vol. 1, *Global Boundaries*, Clive Schofield, ed.; vol. 2, *Middle East and North Africa*, Clive Schofield and Richard Schofield, eds.; vol. 3, *Eurasia*, Carl Grundy-Warr, ed.; vol. 4, *The Americas*, Pascal O. Girot, ed.; vol. 5, *Maritime Boundaries*, Gerald H. Blake, ed. New York/London: Routledge, 1994.

WEB WORK

Basic information about each country may be found at

http://www.countrywatch.com/

Political news of the world is available at a tremendous number of places on the Web. Most major daily newspapers are posted. Some examples are

http://www.washingtonpost.com
http://www.latimes.com
http://www.nytimes.com
http://globeandmail.com

Some services provide news from specific world regions. Two examples are

http://www.allafrica.com/
http://www.askasia.org/

The Bureau of Consular Affairs of the U.S. Department of State provides travel information (often warnings) about affairs in foreign countries at

http://travel.state.gov/index.html

Getting around in U.S. federal government sources is intimidating, but a couple of libraries and reference points help. A particularly good one is a program of the U.S. Department of Commerce

http://www.fedworld.gov/

The president and each house of Congress have home pages with numerous valuable links:

http://www.whitehouse.gov/
http://www.senate.gov/
http://www.house.gov/

Information about Canada can be found through the University of Waterloo library reference system:

http://library.uwaterloo.ca/

The International Boundaries Research Unit of the University of Durham, England, provides information about borders and border disputes:

http://www-ibru.dur.ac.uk/index.html

The United Nations maintains an Internet Gateway on the Advancement and Empowerment of Women:

http://www.un.org/womenwatch/

Freedom House monitors political developments around the world:

www.FreedomHouse.org

as does the International Institute for Democracy and Electoral Assistance:

http://www.Idea.int

TECHIES AT WORK

A Look Ahead

ANALYZING AND COMPARING COUNTRIES' ECONOMIES

Gross domestic product and gross national income are the principal measures of nations' wealth, but they must be supplemented by non-economic measures to compare the quality of life in different countries. Sectoral analysis reveals how countries' economies may be described as preindustrial, industrial, or postindustrial. National wealth is affected by many things, including natural resources, but natural resources are not the only determinant of national wealth. Adding value to resources or providing valuable services builds wealth. Technology, communications, and government

policies constantly change the locational determinants of economic activity.

THE GEOGRAPHY OF MANUFACTURING

The world geography of manufacturing is the result of identifiable locational determinants, but as these change, some manufacturing is continuously relocating.

NATIONAL ECONOMIC-GEOGRAPHIC POLICIES

Countries have different systems to organize and regulate domestic economic activity in order

12

NATIONAL PATHS TO ECONOMIC GROWTH

◄ This suburban high-technology office park could be almost anywhere in the world. In fact it is just outside Bangalore, India. Here Wipro Technologies, an Indian giant in the field of information technology, devises computer programs for clients around the globe. By exporting this technology, India earns much greater returns than India could by exporting raw materials or even many man-ufactured goods. Providing valuable goods and services to global markets is one economic development strat-egy that some countries em-ploy to raise living standards for their own peoples.
(Mahesh Bhat/India Picture)

to increase wealth and distribute it around the territory.

NATIONAL TRADE POLICIES

Some countries try to build their domestic economies without trade, while others participate in international trade to attract capital and win markets.

THE FORMATION OF THE GLOBAL ECONOMY

International economic links have reduced barriers to exchange and communication, and many activities have expanded their scale of organization to cover the whole globe. Economic globalization has far outpaced cultural or political integration.

In July 2003 the I.B.M. Corporation faced a public relations disaster. A disgruntled employee publicized a recording of a conference call in which executives planned to accelerate the relocation of high-paying white-collar jobs out of the United States. Politicians and the company's own employees condemned the plans, yet the movement of jobs overseas is the inevitable result of corporations' efforts to lower costs and of the technology that knits the world economy—the phenomenon of globalization. The first jobs to leave the United States and other rich countries were the blue-collar factory jobs, which went to new factories in low-wage countries; China has captured the lion's share. The relocation lowered costs to manufacturers and prices to consumers, but at the cost of job opportunity in the rich countries. Many former factory workers retrained and found good new jobs, but many suffered falling incomes—an average 13 percent fall.

Now service sector jobs are following the factory jobs. Everything from relatively low-paying call center jobs to software development, accounting, income tax preparation, and even legal and medical analysis can be performed by lower-paid workers in China, India, the Philippines, Russia or elsewhere and the results communicated to the rich countries instantaneously. Three million service jobs are expected to leave the United States alone by 2015, and an I.B.M. executive said, "Our competitors are doing it, and we have to do it." Will wages in today's rich countries have to collapse to the levels in China and India to compete?

How will economic globalization redistribute good jobs, or wealth, and how will it affect development in the countries where living standards are low today? The answer will determine future job opportunity in today's rich countries.

Most people agree that one of the principal tasks of any government is to promote the welfare of its people, including their economic welfare. Each country devises a program to build its national economy and also to ensure that economic well-being is distributed across the national territory. Therefore, the world political map is also a map of local economies and economic policies. Many states need economic growth not only to raise standards of living, but also because economic growth may convince the population that the state works for them, that it is at least a success economically.

This chapter is divided into five sections. The first section examines several ways of measuring countries'

wealth. Then it maps the rich and the poor countries and examines reasons for wealth or poverty. The second section of the chapter examines the world distribution of manufacturing and reasons for that distribution. The third section of this chapter investigates national economic-geographic policies. Each country has a unique way of organizing its economy and managing the distribution of economic activities within its territory. The fourth section looks at how countries regulate their participation in world trade. We will investigate the meaning of the popular term "the Third World." The closing section examines the formation of a global economy. World trade has accelerated so fast and multiplied to such a degree that international management consultant Peter Drucker has written, "The world economy has become a reality, and one largely separate from national economies. The world economy strongly affects national economies; in extreme circumstances it controls them."

ANALYZING AND COMPARING COUNTRIES' ECONOMIES

Some countries have achieved high incomes and standards of living for their people. Others have not. Throughout this book we have generally referred to the rich countries and the poor countries. The terms "rich" and "poor" are fairly straightforward and descriptive. Other terms often contrast the *industrial* or *developed countries* and the *undeveloped, underdeveloped,* or *developing countries*. This chapter emphasizes that countries' levels of wealth and standards of living are actually difficult to measure and compare. There are many ways to measure economic status and welfare, and these measures do not always rank countries in the same order.

Measures of Gross Product and Their Limitations

The two most commonly used measures of a country's wealth are its gross domestic product and its gross national income. **Gross domestic product (GDP)** is the total value of all goods and services produced within a country. **Gross national income (GNI)** is the GDP plus any income residents receive from foreign investments, minus any money paid out of the country to foreign

FIGURE 12-1 Texaco service station. Texaco is in partnership with Saudi Arabia for refining and marketing in the eastern United States, so profits from this station are part of Saudi Arabia's GNI. They are not, however, part of Saudi Arabia's GDP, because they are not generated inside Saudi Arabia. Many oil-producing states have purchased refineries and distribution systems in order to control supplies from the ground to the ultimate consumer. Thus, they capture profit from value added at each step. (Robert Azzi/Woodfin Camp & Associates)

investors (Figure 12-1). Either GDP or GNI can be measured *per capita*—per person in the country. GDP provides a better guide to analyzing a country's internal economy, because it measures exactly what activities are occurring and what wealth is being created inside the country. GNI is best used when we want to compare and contrast different countries' total incomes as a guide to the total wealth available to the people in that country.

For most countries there is little difference between the two statistics, but differences that exist are important to geographers. GDP describes the economic activity inside a country, whereas GNI reveals the impact that foreign investments have on a country's economy. Some countries either receive large payments from abroad or send money abroad, so these countries will record a significant difference between their GNI and their GDP. For example, as Kuwait sells oil from domestic oil wells, the income is counted as part of Kuwait's GDP. Through the years the Kuwaitis have profited from oil sales and invested profits abroad. Today the profits from Kuwaitis' foreign investments that are returned to Kuwait are part of Kuwait's gross national income, but they are not part of Kuwait's gross domestic product. Kuwait's GNI regularly amounts to 110 to 125 percent of its GDP. This reveals that Kuwaitis have enormous investments abroad. Other countries, by contrast, pay out profits to owners in other countries, so their GNIs are smaller than their GDPs. In 2001 Canada's GNI was 2 percent less than its GDP, Argentina's GNI was about 4 percent smaller than

its GDP, and Ireland's GNI was 16 percent smaller than its GDP. These statistics suggest that significant shares of these countries' economies are owned by foreigners. In the 1960s and 1970s the United States received payments from American investments abroad, so GNI was slightly larger than GDP. In the 1980s, however, the United States began importing more than it was exporting (this is called *running a trade deficit*), and it began to borrow large sums of money around the world. By the mid-1990s foreigners began to draw money out of the United States in the form of corporate dividends, profits, and interest, so GNI slipped below GDP. In 2001 U.S. GNI was 97 percent of GDP.

In 2001, world GDP totaled $31.1 trillion. The United States contributed a full $10 trillion; Japan $4.1 trillion; Germany $1.8 trillion; the United Kingdom $1.4 trillion; France $1.3 trillion; China $1.2 trillion; and Italy $1.1 trillion. Thus, these seven countries accounted for two-thirds of the measured economic activity on Earth.

Limitations of the GNI and the GDP statistics   Both GNI and GDP are deceptive. Both begin with the idea that most goods and services are of little or no value if they cannot be exchanged. They have no value in their simply being used. Therefore, these measuring techniques underestimate the activities of hundreds of millions of people who provide entirely for their own needs or who exchange very little except through barter. For example, many of the world's farmers buy little food but eat food they raise themselves. Therefore, farmers' total output and income are probably both undercounted. A peasant farmer in China whose total annual income is recorded as $250 may eat food that would cost a resident of Chicago well over $1,000 in the local supermarket. Farmers usually make up a greater percentage of the population in the poor countries than in the rich countries, so undercounting farmers' production underestimates the material welfare of many people in the poor countries. By noting this, however, we do not mean to exaggerate the quality of life those people may have or to suggest that they live well. They are still poor by our Western standards.

Because statistics undercount subsistence areas, they exaggerate the degree to which modern areas, especially cities, dominate national economies. In extreme cases, a single city can provide most of a country's measurable output. Abidjan, with 15 percent of the Ivory Coast's population, accounts for 70 percent of all economic and commercial transactions in the country. São Paulo, with 10 percent of Brazil's population, contributes a quarter of that country's measurable economic activity. Statistics particularly overlook the work of women (Figure 12-2).

Government statistics cannot measure activity that is illegal and therefore clandestine. Colombia's profitable drug exports do not seem to exist. Neither do illegal drugs in the United States, yet the federal

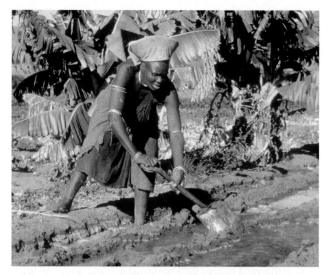

FIGURE 12-2 Subsistence farming. Chapter 11 noted that official statistics in many countries do not record women as working at all, but Chapter 8 noted that women raise most of Africa's food. Only if this woman sold her produce to the neighbors would its value appear in the country's GDP. (Getty Images, Inc.)

government admits that the drug trade is one of the nation's largest industries. Nor does gross product include the activities in a nation's informal sector (discussed in Chapter 10).

Statistics examined per capita through time reveal a relationship between population growth and changes in the statistic measured. A country's GNI can grow, but if the population grows faster than the GNI, GNI per capita will fall. For many countries of tropical Africa, per capita incomes rose in the 1960s, flattened out in the 1970s, and fell from the mid-1980s through the rest of the century. The per capita GNI of Zambia fell at an annual rate of 2.5 percent from 1975 to 1990 and continued to fall at an annual rate of 1.7 percent to 2001. Falls in per capita incomes in many African countries were due in part to the general deterioration of African economies and in part to rapid population growth.

The geography of exchange rates Attempts to compare economies are further complicated by the fact that there are no common measures of value. Gross products are measured in local currencies, and the exchange rates among currencies are changeable and may be manipulated. This is of special concern to geographers not only because it hinders comparison of one place with another but also because variations in exchange rates affect the flows of goods, investment, and people that geographers study. Between 1985 and 1990, for example, the U.S. government deliberately lowered the value of the U.S. dollar by 50 percent relative to the currencies of Japan and the Western European countries. This had extensive geographic results. American goods were cheaper for foreigners to buy, so exports of U.S. manufactured goods rose 80 percent; new exports boosted the output of the nation's manufacturing belt (the

Northeast-Midwest industrial region; look ahead to Figure 12-17). Foreigners invested billions of dollars in the United States, and millions of foreign tourists came to visit. In 1989, for the first time ever, foreign visitors spent more money in the United States than U.S. citizens spent abroad. As the dollar fell, goods flowed out, but tourists and investment flowed in.

These new flows of people and goods diffused culture. Tourists took away impressions of American life, and Americans received new impressions of foreigners. Many more Europeans and Japanese could afford U.S. clothes, cassettes, compact discs, videos, and other cultural exports when those goods' prices fell 50 percent. Millions more saw U.S. movies and television shows.

At the same time, the exports to the United States from foreign trading partners sagged, their economies suffered, and the internal economic geography of their countries changed as their export-dependent regions suffered most. Billions of dollars that U.S. manufacturers had planned to invest in factories abroad were invested in the United States instead, and communities from the Philippines to Brazil went without new vocational training programs, roads, and schools. Places dependent on U.S. tourists suffered recessions. From 1990 to 2001, the exchange value of the dollar generally rose, and all these flows reversed. From 2001 through 2003, the dollar fell again, and the flows reversed once more. Geographers monitor shifting exchange rates to understand changing patterns of travel and trade and of local prosperity.

Furthermore, exchange rates differ from what economists call *purchasing power parity*. Parity attempts to measure what goods or services a certain amount of money will buy at different places. What matters is not how many U.S. dollars a person's income could buy in the foreign exchange market but what standard of living his or her income can support in the place he or she lives. The currency values established in foreign exchange markets often do not accurately reflect purchasing power. In India, for example, the average person's income in rupees in 2000 translated into just $460 per year at the prevailing market exchange rate of 44 rupees per dollar. But because food, clothing, housing, and other necessities were cheaper in India than in the United States, the same amount of rupees was equivalent to an American income of nearly $2,400. Similarly, the average Chinese income in 2000 was $840 at the official yuan/dollar market exchange rate, but more than $3,900 if measured on a purchasing power equivalent base. Therefore, when you read that annual incomes in a certain country are a certain amount of money, do not assume that the people there live at the same level of material welfare that amount of money would buy where you live.

Gross Product and the Environment

Measures of gross product fail to assess environmental damage. In standard techniques of bookkeeping,

FIGURE 12-3 Phosphate mining on Nauru. After years of phosphate mining, the island of Nauru may have to be abandoned. (Michael Friedel/Woodfin Camp & Associates)

If natural resources were treated as assets, then statistics might demonstrate that protecting the environment is sensible economic policy. The United Nations held the world's first conference on sustainable development in 1988, and since then many economists have been working to calculate new statistics that would subtract from gross product the value of natural resources that are destroyed or depleted. The World Bank's 2003 *World Development Report* analyzes several variations. As a result of various statistical manipulations, economists have estimated that many countries' wealth has been growing more slowly than their growing GDPs would suggest. Many of the countries that are exporting their raw materials, such as some oil-exporting states, may not be growing at all, but depleting their national wealth.

The Pacific Ocean island-nation of Nauru illustrates the extreme of unsustainable development (Figure 12-3). The people of this 21-square-kilometer (8-square-mile) island have allowed strip-mining of phosphate for fertilizer exports to ravage their homeland. This has been immensely profitable, and the government trust fund holds over $1 billion for the national population of just 11,000 people. The entire population, however, may have to abandon the island. Nauru would then have a GNI, but absolutely no GDP.

Many economists doubt the value of trying to measure sustainable development. They point out that the countries that are rich today exploited their environments in the past at unsustainable rates, but after these countries had achieved wealth, they were able to repair much environmental damage (Figure 12-4). Perhaps

machines and buildings are counted as capital assets. As they age, their declining value is deducted from income. When a country's natural resources are exploited, however, annual depreciation is not deducted. Many countries sell off their timber and minerals, destroy their fisheries, mine their soils, and deplete their water resources, and national accounting treats the proceeds as current income. The loss of the natural resources is not deducted as a depreciation of national assets.

> If natural resources were treated as assets, then statistics might demonstrate that protecting the environment is sensible economic policy.

FIGURE 12-4 Restoring a forest. It was a natural disaster—the eruption of Mount Saint Helens on May 18, 1980—that felled these trees in Washington State, but human replanting and care is restoring the forest as a productive asset for human use. (Courtesy of the Weyerhaeuser Corporation)

sustainability is irrelevant to a poor population struggling to raise its standard of living to some minimum level, but once the population attains that level of welfare, it can afford to maintain or even repair the environment. Many poor countries resent rich countries' suggestions that they slow their economic activity to a sustainable rate. Imposing conservation measures on poor countries has been derided as *green imperialism.*

Nevertheless, when the world's banks are financing infrastructure projects such as dams, power plants, and pipelines, they are today increasingly accepting the *Equator Principles,* a set of environmental and social-impact standards drawn up by The World Bank. These principles were designed to prevent large construction projects from poisoning air and water, denuding forests, and destroying livelihoods of local residents in

poor countries, where government regulators are often ill-equipped or unwilling to mitigate potential side effects of foreign-financed initiatives. The principles are strict but voluntary, and there is no enforcement mechanism except negative publicity.

The Gross National Product and the Quality of Life

Figure 12-5 is a world cartogram, with each country drawn in proportion to its GDP, and it is strikingly different from a territorial map. Some of the smallest countries—Belgium, for instance—have the largest GDPs. Figure 12-5 also differs from the population cartogram pictured in Figure 5-2. Some countries

The Relative Sizes of Nations' Gross Domestic Products

FIGURE 12-5 The size of the countries on this cartogram corresponds to their total gross domestic product. The United States, Europe, and Japan grow considerably beyond the sizes of their relative territories. Africa, in contrast, shrinks. This cartogram contrasts sharply with the world population cartogram (Figure 5-2). The world's most populous countries are not necessarily those with the greatest economies. (Data from The World Bank)

Per Capita Gross National Income

FIGURE 12-6 Per capita national incomes in many countries in Asia and Africa are a fraction of those in Europe and North America. (Data from The World Bank)

with small populations have enormous economies. Clearly, neither population nor territorial size explains wealth.

Figure 12-6 establishes categories of GNI per capita on a regular world map, but since we are interested in real human beings living at various levels of income, Figure 12-7 shows how much of Earth's population is in each category. Figure 12-7 reveals that 40 percent of the world's population lives in countries in which per capita GNI is less than $745. Many rich people, however, do live in each of these countries, and many poor people live in some of the richest countries.

For most people on Earth, incomes are rising. The proportion of people living on $1 per day has fallen from 20 percent of the world's population in 1975 to just 5 percent in 2001, while the percentage living on only $2 per day has fallen from 44 percent to 19 percent. This is progress—agonizingly slow, but progress.

Noneconomic measures of national welfare   Per capita GNI is only one measure of national welfare. A picture of world standards of living and welfare can be improved by combining it with other statistics. We have already examined infant mortality rates (Figure 5-12), food supplies (Figure 8-31), literacy (Figure 11-34), freedom (Figure 11-35), and the appendix figures on energy consumption. There is a great deal of coincidence among these statistics. The countries with the

lowest infant mortality rates generally have the best food supplies and education, and their citizens consume the most energy. These are all measures of the standard of living, and the highest-ranking countries are much the same as those that have the highest per capita GNIs.

Countries' rankings according to per capita GNI, however, do not always correspond exactly with countries' rankings according to all other measures of welfare. The concentration of wealth, government spending choices, and other factors will affect the quality of life. Therefore, the United Nations has devised a statistical attempt to compare the quality of life among countries. This is called the **human development index (HDI)**. The United Nations argues that living a long and healthy life, being educated, and enjoying a decent standard of living are measures of human development, so the HDI combines statistics of life expectancy, school enrollment, literacy, and income (Figure 12-8, p. 502). The statisticians weigh the relative importance of each of these statistics in order to arrive at one final HDI figure for each country. HDI rankings differ from those based on any one criterion alone. Rankings by HDI and by GNI per capita can differ, showing that high levels of human development can be achieved without high incomes, and that high incomes do not guarantee high levels of human development. For example, Pakistan and Vietnam have similar incomes, but Vietnam has more successfully translated that income

Per Capita Gross National Income Indicated on the World Population Cartogram

FIGURE 12-7 This figure improves on Figure 12-6 by placing the categories of income on the population cartogram that we first saw as Figure 5-2. Therefore, this figure clarifies how many people live at each income level. The colors correspond

into human development. Jamaica has achieved a much better HDI than Morocco has with about the same income. In the 2003 report, HDI values ranged from a high of 0.944 (Norway) to a low of 0.275 (Sierra Leone). The United States, Canada, and Belgium tied for sixth place (0.937). Canada ranked first almost every

year since the United Nations first calculated HDI in 1990, but its ranking has fallen because of the relatively low living standards of some of its Native peoples. The government of Canada gives the Native peoples about $7 billion annually; but, in an effort to preserve Native cultures, the government distributes the money through

KAZAKHSTAN
16.8

MONGOLIA
2.7

UZBEKISTAN
24.4   KYRGYZSTAN 4.6

TAJIKISTAN
6.2

TURKMENISTAN
4.4

IRAN
65.9

AFGHANISTAN
26.7

PAKISTAN
141.1

NEPAL
24.9

BHUTAN
2.0

CHINA
1,256.2

NORTH
KOREA
21.7

SOUTH
KOREA
47.4

JAPAN
126.4

INDIA
1,017.6

BANGLADESH
129.0

TAIWAN
22.3

VIETNAM
78.4

5.6
LAOS

MYANMAR
48.9

THAILAND
61.2

11.9
CAMBODIA

PHILIPPINES
81.0

MALDIVES
0.3

BRUNEI
0.3

MALAYSIA
21.8

SINGAPORE
3.6

SRI LANKA
19.3

INDONESIA
219.3

PAPUA-NEW GUINEA
4.8

FIJI
0.8

AUSTRALIA
19.0

NEW
ZEALAND
3.7

to those in Figure 12-6, so we can see that 40 percent of Earth's population lives in low-income countries, 44 percent in middle-income countries, and 16 percent in high income countries. (Data from The World Bank)

tribal leaders, not all of whom have translated the subsidies into welfare for their peoples.

HDI figures echo world per capita GNI figures in that both are rising for humanity as a whole, but both are nevertheless falling in some individual countries. HDI fell in 21 countries from 1990 to 2003. This fall reflects particularly the impact of AIDS in many African countries, and the result of social and economic turmoil in Russia since the fall of communism (see Chapter 13). In most countries, however, both GNI and HDI are advancing. Both HDI and per capita GNI are advancing in China and in India. These countries hold

The Human Development Index

FIGURE 12-8 This map indicates where life is relatively good, as measured by the U.N.'s Human Development Index, which rates countries on the basis of a statistic that balances life expectancy, school enrollment, literacy, and income. (Data from *Human Development Report 2003*)

between them more than one-third of all of humankind, so improving statistical values for those two countries significantly improves the statistical averages for all humanity.

Any index that weighs and compares different criteria is, admittedly, comparing apples and oranges. Chapter 10 described how popular magazines periodically rank cities based on the quality of life they offer, and we noted then that we may not agree with any magazine's criteria or weighing. Baseball teams, good hospitals, and mild winters may be important to me; golf courses, art museums, and good schools may be more important to you. Therefore, we may not all agree with the ranking. We may not agree that the best criteria were chosen for the Human Development Index, or we may not agree with the weights assigned to each criterion in the final tabulation. Whatever criteria are chosen, however, statistics reveal that conditions of life vary greatly. Furthermore, for most of humanity, conditions seem to be improving, but conditions of life in some countries are poor and even falling.

> . . . for most of humanity, conditions seem to be improving, but conditions of life in some countries are poor and even falling.

Preindustrial, Industrial, and Postindustrial Societies

Geographers want to know what people are doing for a living and which activities are producing the wealth in each country. To do this, they must analyze more than just the total GDP. Chapter 10 noted that over a long period, most countries that are rich today experienced shifts in the distribution of jobs among what are defined as their economies' primary, secondary, and tertiary sectors. This shift was defined as **sectoral evolution.** The best way to examine any country's economy is to study the relative importance of each sector in its GDP.

Sectoral evolution changes national employment

Before industrialization, most of a country's labor force is occupied in the primary sector (Figure 12-9). Societies with the bulk of their employment in the primary sector are called **preindustrial societies.** Many societies still today are preindustrial, and even with high percentages of their workers in agriculture, many can barely feed themselves (Figure 12-10).

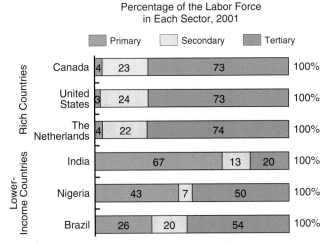

Percentage of the Labor Force
in Each Sector, 2001

FIGURE 12-9 **Sectoral analysis of the labor force.** This figure shows the proportion of the labor force in each of six countries that works in each of the three sectors of that country's economy. Workers in the rich countries (exemplified by the top three countries) work in the secondary and tertiary sectors, whereas higher percentages of workers in the poor countries (exemplified by the bottom three countries) are concentrated in the primary sector. (Data from The World Bank)

As some countries industrialized, many workers found employment in factories, and the proportion of the labor force employed in the primary sector declined. This is true despite the fact that both the total number of people in the primary sector as well as the value of the primary sector's output may have risen. The workers in the sector usually became more productive. The primary, extractive activities remain crucial to many economies, but they provide a diminishing share of jobs.

Many jobs lost in agriculture were initially replaced by new opportunities in industry, and the proportion of workers in the secondary sector increased until, at no precisely defined point, certain societies came to be called **industrial societies.** In the United States, for example, manufacturing employment overtook farm employment during World War I, and by 1925 manufacturing workers had become the largest single occupational group.

Continuing evolution of some countries' economies has drawn a higher percentage of workers into the tertiary sector, producing services instead of goods. Services accounted for almost 25 percent of all jobs in the United States by 1929. Sometime in the 1940s the

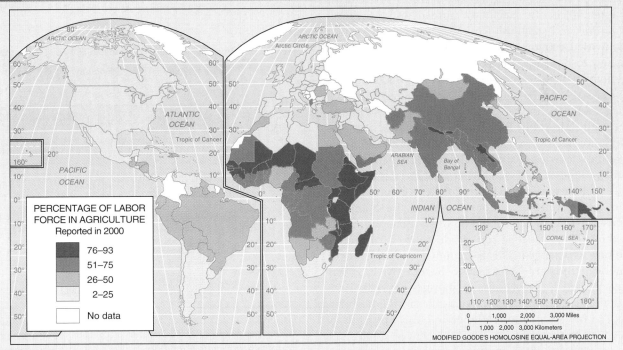

Percentage of the Labor Force Working in Agriculture

FIGURE 12-10 If you compare this map with Figure 8-31, the prevalance of undernourishment, you will see that many countries with high percentages of their labor forces in agriculture are barely able to feed themselves, whereas several countries that have only small percentages of their labor forces working in agriculture enjoy food surpluses. What can explain this paradox? The explanation is that economic and technological development have allowed some countries to invest in agriculture, raise agricultural yields, and at the same time release workers for manufacturing and service activities. Thus, a high percentage of workers in agriculture reveals technological backwardness and poverty. (Data from The World Bank) proportion first exceeded 50 percent, and the United States became the world's first **postindustrial society.** By 2000 tertiary employment represented three quarters of the nation's jobs. The leading job categories in the United States in 2002, by numbers, were retail salespersons, teachers, secretaries and typists, truck drivers, farmers and farm laborers, janitors and cleaners, cooks, nurses, engineers, and freight and stock handlers. All advanced nations have seen the percentage of jobs in the tertiary sector increase. This trend has also been noted in a few poor nations, whose tertiary sectors are well integrated into the international economy—tourist destinations, for example.

At first both the secondary and the tertiary sectors increased their shares of total employment as the primary sector's share dropped, but in recent years in the United States the proportion of workers in the secondary sector has begun to fall as well. The nation began losing manufacturing jobs as a percentage of all employment about 1960, and even the absolute number of manufacturing jobs has been declining since at least 1970.

The "new economy"   The production of computer hardware, software, and services, telecommunications, and of products on and for the Internet are all collectively referred to as the *new economy*. This term, however, like the term *quaternary sector* discussed in Chapter 10, is not specific or agreed upon. Sectoral evolution and technological evolution proceed so rapidly that economic statisticians cannot always agree on how to measure developments. For example, some economists believe that the production of computer software should be counted as a manufacturing activity, and they ranked it as the third-largest manufacturing industry in the United States (after automobiles and electronics). Silicon Valley, the 80-kilometer (50-mile) corridor along Highway 101 from San Francisco to San Jose, California, is home to thousands of computer and software companies (Figure 12-11). The U.S. government, however, using the Standard Industrial Classification System (SIC) that originated in the 1930s, classifies the production of computer software as a service. This is changing, as a new North American Industry Classification System (NAICS) has been phased into government statistics between 1999 and 2004. Professor Rudi Dornbusch of MIT has written, "In the 19th century, the great effort was to draw up a map of each country. The 20th century equivalent is to get a good set of statistics." Even into the twenty-first century, rapid change frustrates achievement of that task.

Measuring how each sector contributes to GDP
The share that each sector of an economy contributes to GDP is not necessarily the same as the share of the national labor force that sector employs (Figures 12-12 and 12-13). Some countries have high proportions of

FIGURE 12-11 Silicon Valley, California. By some measures, California's Silicon Valley is the manufacturing center of the United States, yet no smokestacks are visible. Clearly, technology has progressed from the "dark Satanic mills" of industrialization described by poet William Blake (1757–1827). (George Hall/Woodfin Camp & Associates)

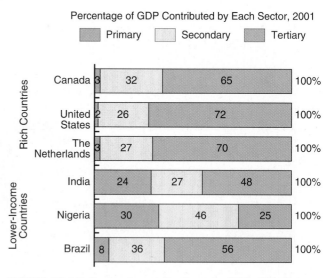

Percentage of GDP Contributed by Each Sector, 2001

	Primary	Secondary	Tertiary

Rich Countries
- Canada: 3 | 32 | 65 | 100%
- United States: 2 | 26 | 72 | 100%
- The Netherlands: 3 | 27 | 70 | 100%

Lower-Income Countries
- India: 24 | 27 | 48 | 100%
- Nigeria: 30 | 46 | 25 | 100%
- Brazil: 8 | 36 | 56 | 100%

FIGURE 12-12 Sectors' contributions to GDP. This figure shows the origin of the GDP, by sector, for the same six countries that were shown in Figure 12-9. The percentage of a country's GDP that is produced in the secondary sector almost always exceeds the percentage of the country's labor force working in that sector. This means that value added per worker is usually highest in the secondary sector, so average workers' incomes are usually highest in that sector. The large percentages of the labor forces that work in the primary sectors of the poor countries produce relatively small shares of GDP. (Data from The World Bank)

Industrial Countries

FIGURE 12-13 The term *industrial society* has no exact formal definition, but this map reveals the countries in which industry—the secondary sector—contributes the highest share of GDP. The industrial sector is relatively small in some of the richest postindustrial countries and high in a few relatively poor countries that do process some mineral ores. (Data from The World Bank)

their workers in the primary sector, but those workers produce relatively little of the measurable national output. In the secondary sector, by contrast, output per worker is usually high, so the secondary sector usually contributes a larger share of GDP than it employs of the labor force.

The continuing evolution of the U.S. economy toward the tertiary sector may explain a widening gap in incomes. Many of the occupations that pay the highest incomes are usually in the tertiary sector (sports star, heart surgeon, lawyer, etc.), but most of the tertiary-sector jobs do not pay as well as factory jobs. Look back a few paragraphs at the list of the most numerous jobs in the United States—retail salesperson, teacher, secretaries and typists, and so forth. The gap has widened between the incomes of America's highest- and lowest-skilled workers. From 1979 to 1999 the average real hourly wage rose 14 percent for those with college degrees and 19 percent for those with advanced degrees, but the average wages *fell* by 4 percent for those with only some college, 10 percent for those with only a high school diploma, and 24 percent for high school dropouts. We do not know whether such a widening wage gap is inevitable in a postindustrial society or whether this phenomenon is the result of uniquely American taxes or other policies. International comparative studies are not conclusive.

Why Some Countries Are Rich and Some Countries Are Poor

If you compare the map of per capita GNI (Figure 12-6) and the map of HDI rankings (Figure 12-8) with the data on location of world natural resources in Chapter 9 (Table 9-1), you can see that the countries most richly endowed with raw materials are not necessarily enjoying the highest per capita GNIs and the highest standards of living. Conversely, some of the richest peoples who do enjoy the highest standards of living live in environments with meager natural resources. The geography of resources alone does not explain the geography of wealth.

Environmental determinists (see Chapter 6) would find this inexplicable, but the study of geography teaches that the key to wealth is not *having* raw materials, but *adding value* to raw materials. The principle of adding value was first introduced in Chapter 8, where we read of dairy farmers converting milk into cheese. Some resource-poor countries are able to import raw materials, process them, manufacture items from them, and enjoy their ultimate use.

The value added to raw materials downstream in the secondary and tertiary sectors surpasses the value of the original raw materials, whether the raw materials are mineral or agricultural. The value added by refining copper ore and manufacturing things out of it, for example, quickly surpasses the value of copper ore. The value added by grinding wheat into flour is greater than the value of the wheat. This principle holds true no matter how valuable the original raw material may be. Diamonds are valuable, but the value added to them by diamond cutters and polishers is greater than the value of the original uncut diamonds. The greater the value added, the greater the potential for profit.

The places that add the greatest value to materials prosper, while the places that export raw materials without adding any value to them do not enjoy the same economic growth. They may even have to buy goods manufactured out of raw materials that they originally exported themselves. As long as Jamaica exports bauxite to the United States, for example, but that bauxite is made into aluminum and then into consumer goods in the United States, and Jamaica buys back manufactured goods, the United States will grow richer than Jamaica. As long as Botswana exports uncut diamonds to Israel and buys back engagement rings, the Israelis will grow richer than the Botswanans.

Even within the secondary sector, there is a hierarchy of value added. The manufacture of textiles, for example, adds less value than does the manufacture of clothing, so places that manufacture textiles try to develop clothing manufacture. Other generally low-value-added manufactured goods include shoes, toys, sporting goods, and today, simple television sets and inexpensive cars. Countries try to progress from production of these items to production of higher-value luxury cars, computers, chemicals, and ever more complex electronic items. An agricultural example is to see how Chile began exporting grapes, then evolved to exporting moderately priced jug wine, and today is exporting high-value fine wines. **Economic development** is a process. It consists of progressively increasing the value of goods and services that a place is able to produce to enjoy or export (Figure 12-14). This process has no foreseeable end, but we do commonly speak of the rich

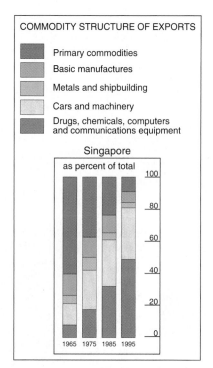

FIGURE 12-14 The developing economy of Singapore. Singapore has steadily reduced the share of its exports that are low-value-added items (primary commodities and basic manufactures) and progressively increased the share of its exports that are higher-value-added items. In 2001, using slightly different definitions, Singapore's exports were a full 60 percent "high-technology goods." (Data from The World Bank)

countries as "developed" and the poor countries as "underdeveloped," or optimistically as "developing." In fact, weekly visits to your local electronic-goods store will demonstrate that the complexity of the products available in even the richest countries continues to increase.

Many of today's poor countries still export only unprocessed raw materials or materials with little value added: agricultural goods, basic metals, or textiles, for example (Table 12-1). Only a small percentage of

TABLE 12-1 A Few Countries' Merchandise Exports 2001 (as percentages; data from The World Bank)

	Food	Agricultural Raw Materials	Fuels	Ores and Metals	Manufactures
Low-income countries					
Bolivia	31	3	24	20	22
Cameroon	17	21	52	5	5
Colombia	18	5	36	1	39
Kazakhstan	7	1	54	18	20
High-income countries					
Austria	5	2	2	3	82
Japan	1	1	0	1	96
The Netherlands	16	3	8	2	70
United States	8	2	2	2	82

Manufactured Exports as a Percent of Total Merchandise Exports

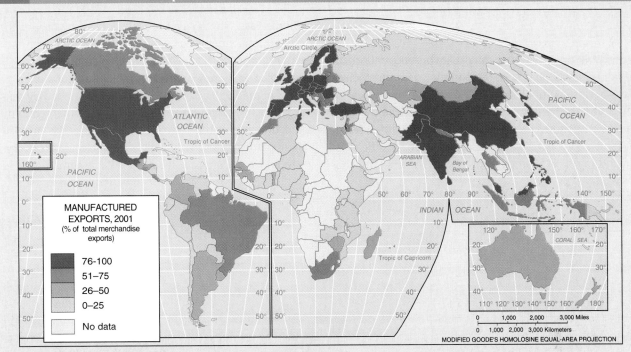

FIGURE 12-15 As a general rule, the higher the percentage of a country's exports that are manufactured goods, the more industrialized that country is. (Data from World Bank)

their exports are manufactured goods (Figure 12-15). The welfare of their people rises and falls with world prices over which they have no control, and world prices of commodities have slowly declined. In 2002 the world market prices of the 33 leading non-fuel commodities in world trade were only 55 percent of what they had been in 1970. Thus, the export earnings of countries exporting these items had dropped by as much as 45 percent.

Exporters of some raw materials have tried to form cartels, such as OPEC (see Chapter 9), but few have succeeded. In 2000 the World Bank launched an International Task Force on Commodity Risk Management in Developing Countries, but it has had limited success. If production of a commodity can easily be increased, then any nation can gain a short-term advantage by dumping stockpiles on world markets, causing prices to collapse. The 14-member Association of Coffee-Producing Countries tries to keep coffee prices high, but they have not been able to cooperate. The price of coffee on world markets periodically collapses, and in the countries dependent on coffee exports, suddenly there is no money for children's schoolbooks; national roads go unpaved; hospitals run out of medicines. Erratic swings in commodity prices can harm the economies of countries dependent on them as much as steadily falling prices do.

In many cases, the production of commodities employs only a small fraction of the national labor force, and this can cause problems of distributing the national wealth. The export of oil, for example, can be very profitable for a country, but the task requires so few workers that having oil to export is often referred to as the "petro curse." Crude petroleum accounts for about half of Venezuelan government revenues, for example, and 80 percent of export earnings. Economists agree that the wisest course is for a government to "sow the oil" (in the words of Venezuelan writer Arturo Pietri) by investing in national education, infrastructure, and developing industries. The temptation is great, however, to spend the income on current consumption.

Can culture cause poverty?   Some peoples are rich and others are poor, and we have seen that this is not necessarily related to the possession of natural resources. If we know the reasons for the distribution of wealth, and if we want to achieve high levels of welfare everywhere, then two paths of action are open to us. Either we could redistribute world wealth (in other words, give foreign aid from the rich to the poor), or else we could teach or encourage the poor places to generate their own wealth locally. For the poor areas to create their own wealth, however, might require them to change local government policies (such as taxation or budgeting) or even to change aspects of their local culture (such as women's rights or education). Culture does have consequences, and some scholars argue that

Regional Focus on
THE NORWEGIAN SOLUTION

Norway is practically the only oil-exporting nation to have achieved balanced economic growth. When Norway first pumped crude oil from the North Sea in 1975, it had the advantages of being a stable democracy with honest civil servants, a well-established legal system, and a large middle class. Median income was behind most of Europe, but above incomes in the Middle East. Norway was at first seduced by oil wealth. The country's generous welfare system was expanded, and the work week declined. Norwegian manufacturing dwindled as money poured into the oil sector alone.

In the 1990s, however, Norway rebounded by segregating the oil business from the rest of the economy and diversifying the rest of the economy. The government made it clear that the oil wealth belonged to everyone and to all regions. It set up a Petroleum Fund and invested in foreign stocks and bonds because investing in domestic companies would have pushed the Norwegian currency too high and distorted the national economy. Norway uses only the returns on the Fund's investments (about $120 billion in mid-2003) to balance the country's non-oil budget. The accumulation in the fund is set aside for Norwegian retirements.

Norway and Its Oil Fields

some cultures hinder economic development. It has even been said that there are "cultures of poverty." In 2002 the United Nations issued a special report on economic development in the 22 member nations of the Arab League, with populations totaling 280 million people. The report, written largely by Arab economists, concluded that the principal barriers to Arab development were shortages of three essentials for development: freedom, knowledge, and womanpower. All of these are matters of domestic politics and culture.

> Culture does have consequences, and some scholars argue that some cultures hinder economic development.

If a people's culture does hinder their economic development, however, the people often nevertheless resent suggestions that they change their culture in order to improve their standard of living. They would rather just be given material assistance. At the first U.N. World Conference on Poverty, held in Copenhagen in 1995, the rich countries argued that poor countries could develop if they would just end corruption, guarantee workers' rights and property rights, and help women and children. Representatives of many poor countries responded that such suggestions were unacceptable intrusions into local customs. They demanded immediate financial aid.

The debate on whether it is necessary to change culture in order to achieve wealth engages not only diplomats but all social scientists—economists, sociologists, anthropologists, as well as geographers. Many of the ideas presented throughout this textbook may help you form your own opinions.

THE GEOGRAPHY OF MANUFACTURING

Most countries strive to industrialize to capture the high value added by manufacturing. One of the earliest scholars to identify the locational determinants of manufacturing was Alfred Weber (1868–1958). Weber's analysis focused on the role of transport costs, and he devised models that differentiate material-oriented manufacturing from market-oriented manufacturing.

Material-oriented manufacturing locates close to the source of the raw material for one of two reasons. The first reason is that the raw material is heavy or bulky, and manufacturing reduces that weight or bulk. The steel industry is an example. The value added in manufacturing steel is low relative to the cost and difficulty of transporting the raw materials (iron ore, coal, and water), so steel mills generally locate where the raw materials can be found or cheaply assembled. The second reason why material-oriented manufacturing locates near the raw material is because some raw materials are perishable and need immediate processing. Examples include canning, freezing, and the manufacture of cheese from milk.

Market-oriented manufacturing, Weber's second category, locates close to the market, either because the processing increases the perishability of the product (baking bread, for instance) or because the processing adds bulk or weight to the product. A can of a soft drink, for example, consists of a tiny amount of syrup, plus water, in a can. The syrup can be transported easily and cheaply, but canned soda is heavy and expensive to move. Therefore, the water should be added and the beverage canned close to the market.

Weber elaborated his models of the location of manufacturing by adding additional considerations, such as the availability of a labor force. In that three-factor model, manufacturers locate factories to minimize the cost distance from three points: the location of the raw materials, the labor force, and the market. The optimum location for any specific manufacturer depends on the balance of these costs in his or her business.

Several examples demonstrate that consideration of Weber's factors is still important in many industrial location decisions. Wood furniture making typifies an industry that may locate near its source of raw materials. In the United States, Grand Rapids, Michigan, became a center for wood furniture manufacture in the nineteenth century, drawing on Michigan's extensive forests. When those forests were depleted, furniture makers relocated to their new sources of wood: Georgia and North Carolina. The copper industry exemplifies a bulk-reducing industry. In the United States most copper ore is smelted (separated into its metallic ingredients) at the mines in New Mexico and Arizona, because smelting removes a high percentage of the original ore as waste. The largest copper refineries and manufacturing facilities, by contrast, are found in Baltimore, Maryland; Norristown, Pennsylvania; and Perth Amboy, New Jersey. These locations are closer to markets and labor forces, and they are also situated to draw on imported copper supplies more conveniently. The automobile industry is an example of a fabricated metal industry, which brings together a number of previously manufactured parts and assembles them into a more complex product. In the United States several new automobile assembly factories have located in the center of the country (Figure 12-16). It should be noted that in addition to geographical convenience, this region is attractive to manufacturers because labor conditions and terms are more favorable to manufacturers in this region than they are in the regions of older, more heavily unionized industry.

Supplying Parts to an Auto Assembly Plant

FIGURE 12-16 Georgetown, Kentucky, was chosen as the location of a Camry assembly plant partly in order to draw on supplies of parts from around the United States conveniently.

Locational Determinants for Manufacturing Today

Weber's studies still help us understand many industrial location decisions, but several things have changed since Weber's time. Transportation costs have steadily fallen (look back to Table 6-1), and at the same time the value added in manufacturing has increased as manufactured products have become more and more complex. The value of the iron, copper, and other raw materials that go into a mobile telephone is only a minuscule percentage of the value of the completed unit. As transport costs fall and the value added in manufacturing rises, high-value-added manufacturing can more easily relocate; it is, as we learned in Chapter 6, increasingly footloose. Furthermore, today trade binds world regions, so locational determinants are no longer tied to local, regional, or even national scales. The entire

globe must be examined as one theater of interconnected operations.

Weber's models considered the locations of the raw materials, the labor force, and the market, and the transportation costs in determining the best site for a manufacturing operation. Since Weber's time the balance among locational determinants for many industrial processes has tipped away from those Weber considered. For example, the value of the labor input is generally a shrinking percentage of the value of manufactured goods. In the consumer electronics industry it is no more than 5 to 10 percent. At least five other considerations, however, must be added to Weber's (Table 12-2). Each of these considerations is an *input*, or ingredient, for manufacturing. One is *capital*, because manufacturing is increasingly capital intensive. A second is *technology*. Third, industries seek places with *hospitable governmental regulations;* that usually means low

TABLE 12-2 Locational Determinants for Manufacturing

Factors Considered by Alfred Weber	Additional Factors that Are Important Today
1. Raw materials	1. Capital
2. Labor force	2. Technology
3. Market	3. Governmental regulations, including environmental legislation, labor rights, and labor union powers
4. Transport costs	4. Political stability
	5. Inertia

World Manufacturing Regions

FIGURE 12-17 This figure shows the greatest concentrations of manufacturing today. (Adapted from Paul L. Knox and Sallie A. Marston, *Human Geography: Places and Regions in Global Context.* Upper Saddle River, NJ: Prentice Hall, 2e, 2001.)

taxes, little environmental regulation, and restraints on labor rights and unionization. *Political stability* is a fourth consideration. Manufacturers hesitate to invest in volatile political environments. The *inertia* of facilities and suppliers already in place is still another consideration in the geography of manufacturing. Factories are major investments, and networks of suppliers and trained labor forces develop around them; therefore, they are not quickly abandoned.

Figure 12-17 is a map showing the world's most important manufacturing regions. A map of manufacturing measured by value added, however, would show that the greatest value added is in those countries that dominate Figure 12-5, the cartogram of the nations' gross domestic products. Much of the world's manufacturing takes place in Europe, and this is for historical reasons. Europeans brought about the Industrial Revolution first, giving Europe a technological lead, and then Europe politically and economically overwhelmed most of the rest of the world. Europeans generally located the manufacturing in their own countries in order to retain the value added. It was not their intention to enrich their colonies. In some cases they actually intended to retard the development of manufacturing in colonies. For example, from 1651 until 1776, at the very beginning of the Industrial Revolution, Britain's Navigation Acts prohibited Americans from manufacturing products from local raw materials. Raw materials had to be shipped to Britain, where value was

added by manufacturing, and then the products were shipped back to the colonies—at higher prices. Even iron from New Jersey had to be shipped to Britain, where it was made into nails, and the nails were then shipped back to New Jersey. The colonists' resentment of this exploitation helped trigger the U.S. War of Independence. Many colonies in Africa and Asia later suffered the same deliberate restraint on their industrialization.

When manufacturers did locate factories in countries other than their own, they generally chose other rich countries where they could find markets and labor forces. Today the rich countries still have the technological lead and the best-trained labor forces. They usually are politically stable, and they provide the eventual market for most goods. This geographical system has prevailed for so long that it has a tremendous inertia. Raw materials continue to flow from the poor countries to the rich countries, where they are transformed into more valuable goods. Still today, for example, Ghanaian cocoa beans may enter Europe tariff-free, but Ghanaian cocoa powder faces a stiff tariff. This pattern has been labeled *economic colonialism* or *economic imperialism*. It is noteworthy, however, that neither Japan nor Switzerland, two countries with wealth but without substantial natural resources, ever had a major political empire.

In their efforts to industrialize, today's poor countries suffer from a number of problems that reinforce one another. Markets in poor countries are not as important as those in the rich countries. In addition, cheap manufactured goods from the rich countries can flood their domestic markets and discourage competitive domestic manufacturing. Their poverty increases their political instability. They need capital, and their people need education.

Poor countries can, however, offer three locational determinants: raw materials, inexpensive labor, and hospitable regulatory environments. Workers in poor countries receive relatively low wages, and they will often work in conditions worse than workers in the rich countries will. Poor countries' governments seldom complain about pollution coming from a new factory that provides jobs. Today's poor countries can exploit these advantages to attract new industry. Many corporations from the rich countries are today investing in the poor countries to take advantage of these factors. Thus, some poor countries are industrializing by attracting foreign investment.

Locational Determinants Migrate

The geography of locational determinants is not fixed. The locational considerations for manufacturing make it possible for manufacturing to migrate around the world as the mixes of attributes among places change and attract it.

There are, in fact, reasons why the geography of locational determinants for industry—and thus the geography of industry—will continually change. Chapter 6 already suggested relevant factors: new products, new technologies, new raw materials, new sources of traditional raw materials, new methods of manufacture, new governments with new policies, growing and shrinking labor supplies, and the opening and development of new markets.

> The locational considerations for manufacturing make it possible for manufacturing to migrate around the world as the mixes of attributes among places change and attract it.

If a place first attracts manufacturing because it offers a low-wage labor force, its local standard of living and local wages will rise. As local wages rise, the industries that developed there to take advantage of the low wages will be driven away. They will migrate to still poorer places, where wages are lower. Thus, ideally, as each location develops, it surrenders low-value-added activities to poorer regions. These regions in turn begin to climb the ladder of economic development. WestPoint Stevens, Inc., for example, America's largest textile maker, reduced its U.S. payroll from 40,000 to 16,000 between 1988 and 2000, while opening or contracting with new mills abroad. It is concentrating on marketing and on adding psychological value (by signing licensing agreements with designers Ralph Lauren, Martha Stewart, and Disney, for example), while leaving the actual production of the textiles to others elsewhere.

The manufacture of men's dress shirts for the U.S. market offers another example. Shirt making is labor intensive, so manufacturers are continuously searching for cheaper labor. In the 1950s, U.S. shirt manufacturers located in Japan to use the low-paid workers there. When the costs of labor and real estate rose in Japan, the companies moved to Hong Kong. As Hong Kong's factories gave way to offices, the shirt makers moved again, first to Taiwan and Korea and then, in the 1970s and 1980s, to China, Thailand, Singapore, Indonesia, Malaysia, and Bangladesh. Toward the end of the 1980s and into the early 1990s, Costa Rica, the Dominican Republic, Guatemala, Honduras, and Puerto Rico became shirt-manufacturing centers. The Nien Hsing Textile Company, for example, founded in Taiwan in the 1970s, moved its jeans sewing factory to Managua, Nicaragua, in the 1990s. Its 1,800 Nicaraguan workers earned (in late 2000) from $65 to $124 per month; the labor input in a pair of jeans that retails in the United States for $22 (under several different brand names) was 50 cents.

The Caribbean basin countries offer cheap labor and U.S. tax advantages that were granted under a U.S. government program, the Caribbean Basin Initiative, that was designed to help these countries industrialize. These countries' proximity to the United States offers several additional advantages. For one, U.S. managers can oversee operations more easily. Proximity also promises speedier delivery, and fashion demands faster introduction of new styles. Also, renting a ship for an additional week's sailing time from the Far East is more expensive than the shorter trip from the Caribbean. Still another advantage of great concern to business executives depends on interest rates, or what we call the *cost of money*. Shirts in the hold of a ship in transit are a form of capital, and that capital is not earning any profit. One shipload of shirts may be worth $15 million. The interest on that capital each day that the ship is on its way to the United States adds a few cents to the retail price of each shirt. "Time is money," and distance is money, too.

Thus manufacturing that is dirty, low in value-added (and, therefore, low wage-paying), and polluting migrates first to the world's poor countries; but as the labor forces of the poor countries are trained, those countries attract the higher-value-added manufacturing. Wages rise, but it takes a long time for wages to reach the levels in the rich countries.

Manufacturing in the United States

No other country yet matches America's total manufacturing output of $1.6 trillion (in 2001). In the United

Focus on
COMPETING FOR A GENERAL MOTORS PLANT

Southeast Asia represents a fast-growing market for goods, and in the early 1990s General Motors decided to build a full-scale auto-assembly plant somewhere in the region. The plant would bring investment of about $1 billion, and it would be expected to provide more than 2,000 jobs and to assemble over 100,000 cars per year. The decision of where to locate the plant would be crucial not only to General Motors but to the economy of the host country. The question was where to build.

At first glance, Thailand seemed the natural choice. Thailand was already a regional auto-assembly center with a growing number of new or planned auto plants, including those of the Ford Motor Company, the Chrysler Corporation, and several Japanese makers. This reveals a broad base of hundreds of automotive component manufacturers—a major attraction. GM would not have to build a network of parts suppliers, but could buy many parts. A growing auto-manufacturing presence, along with a network of roads and port facilities, is turning Thailand's eastern seaboard into a little Detroit. Thailand offered GM a package of attractions, including tax exemptions on export earnings, as well as an exemption from tariffs on machinery imports for the first two years. Thailand also agreed to finance an automotive-training institute that would be open to other manufacturers.

The Philippines competed strongly to attract the plant. Such a facility would be more of a victory for the government of the Philippines, because the auto-assembly industry was negligible in the Philippines, so the plant could generate thousands of additional jobs among subcontractors and suppliers. The government offered GM sufficient land, developed with power, water, and communications needs, rent-free for five years. The government also offered a package of tax breaks, including freedom from all taxes for eight years, duty-free import machinery, a waiver of local content rules for car exports, and a government-financed training program for autoworkers.

On May 30, 1996, GM announced that Thailand won the plant.

Chrysler plant in Thailand. A Thai autoworker assembles a Jeep Cherokee at an assembly plant in suburban Bangkok, Thailand. The Cherokee was the first American vehicle produced in Thailand. (Richard Vogel/AP/Wide World Photos)

Questions

1. Why do you think GM chose Thailand for the location of its new plant?

2. Have any new manufacturing or research facilities opened in your community recently? What attracted them to your community?

3. Many communities in North America compete to attract industrial or research facilities. What does your community offer?

States, however, the share of the labor force in manufacturing and the sector's contribution to GDP are falling—from 31 percent of jobs and 27 percent of GDP in 1960 to 11 percent of jobs and 17 percent of GDP in 2002. American corporations are building new factories abroad, and Americans are increasingly buying imported goods. By 2003, about half of the manufactured goods that Americans bought were made abroad, up from 31 percent in 1987. Economists debate whether this economic evolution away from the secondary sector is good for the United States.

The importation of manufactured goods, notably cars, apparel, electronic goods, computers, and furniture, has lowered their cost to American consumers. Falling prices is called *deflation*. At the same time, however, prices of goods produced locally and of services such as health care, education, and such day-to-day activities as trash collection and auto repair, rise (*inflate*).

In the early years of the twenty-first century, the deflation in the costs of manufactured goods has slowed the rise in the over-all cost of living.

The deflation of prices of manufactured goods comes at the cost of American jobs, and, as we have seen, the substitution of service jobs for manufacturing jobs contributes to the spreading gap in incomes in the United States. Manufacturing jobs pay an average of 20 percent higher wages than tertiary sector jobs. In addition, manufacturing jobs have a higher multiplier effect than service jobs. The average production-sector job produces three times as many additional employment opportunities as does the average service job. Manufacturing also disporportionately employs the non-college-educated, and in 2003 more than 60 percent of U.S. workers lack college degrees. We may applaud that the incomes of female workers, traditionally associated with service occupations, have risen relative to men's wages. Full-time female workers' wages held steady at about 60 percent of men's for decades after World War II, but by 2002 they had risen to 77.5 percent of their male counterparts', but this does not help the nation as long as average wages do not advance. Some observers are disturbed that an increasing share of U.S. military procurement comes from foreign suppliers, even if the suppliers are foreign subsidiaries of U.S. corporations.

Can this evolution away from manufacturing continue? Americans will be able to import manufactured goods only as long as Americans can offer the rest of the world services plus a few sufficiently-high-value goods. In 2003 David Heuther, chief economist at the United States National Association of Manufacturers, stated flatly that "you have to assume that manufacturing will continue to disappear," but he argued that America's high-tech advantage and its ingenuity will sustain a small high-value manufacturing base. Nobel prizewinner George A. Akerlof, by contrast, argued that the value of the U.S. dollar will continue to fall against other currencies. This will make manufacturing operations relatively cheaper in the United States, and that event will entice manufacturing operations back to the United States. How much would the dollar have to fall, and what would be the additional ramifications of that event? We do not know.

The Economy of Japan

Japan is the most astounding and paradoxical success story in economic geography today. The Japanese islands have few natural resources, but on this meager natural geographical endowment, the Japanese have developed a great economy. How?

The answer lies in Japan's cultural resources and in its trading patterns. Japan's economic success demonstrates how a place can prosper by importing raw materials and adding value in manufacturing, or by providing other downstream services. The key cultural resources include the people's education and labor skills, technology, and the country's style and degree of cooperative organization. All these are carefully managed by the government. Japan's participation in international exchange is also carefully regulated. The Japanese import raw materials, transform them into manufactured goods, and export these finished goods, plus an array of valuable services, around the world. In 2001, Japan's merchandise imports were 13 percent food, 3 percent agricultural materials, 20 percent fuels, 5 percent ores and metals, and 57 percent manufactures. Japan's merchandise exports, however, were 93 percent manufactured goods.

In the 1950s Japan was a relatively poor country, but it began to manufacture products of relatively low value added, and whose manufacture does not require the highest technology. These products included cargo ships, toys and games, sporting goods, inexpensive clothing and cars, and simple electronic appliances. Japan built its prosperity by steadily adding value to its exports.

As Japan developed, it surrendered the manufacture of the items that are low in the value-added scale. Japan invests in factories to manufacture these products in less developed countries that are dominated economically by Japan: Vietnam, Thailand, and Indonesia (Figure 12-18). The process continues: Through the 1990s Japan's production of lower-technology consumer electronic equipment began to fall, whereas its production of the highest-technology digital compact cassettes, electronic components and devices, and industrial electronic equipment continued to rise. For example, 18 to 25 percent of the cost of a laptop computer is in its liquid crystal display (LCD)—almost twice the percentage of value that is in its microprocessor chip. By the late 1990s the manufacture of LCDs was essentially a Japanese monopoly.

Entering the twenty-first century, the Japanese economy continued to evolve. By 2002, with about $2 trillion in overseas assets invested around the world and $500 billion in foreign reserves in its vaults, Japan for the first time earned more money from its investments abroad than it did from exporting manufactured goods. Some observers have described this as stagnation. Japan, a nation with a fertility rate below the replacement rate, an aging population, and now beginning to live off its savings, has been called "A nation in retirement."

Technology and the Future Geography of Manufacturing

Because the balance among the inputs to manufacturing is continuously changing, maps of the world geography of wealth, of manufacturing, and of economic development will continue to change. Each nation must tailor its economic policies in order to compete. For example, the importance of technology is increasing. When Japan was industrializing, it bought technology.

Toyota's Widespread Auto Plants

Toyota's average monthly pay for factory workers here is $2,967.

A new pickup truck factory will build 200,000 pickups and SUV's a year. Of those, 100,000 will be exported to 80 countries, including Japan (where those trucks are now built). Monthly manufacturing wage here is $125.

Toyota will build auto parts here for other Asian factories. Monthly manufacturing wage here is $250.

A new SUV assembly plant will build 70,000 SUV's a year, 10,000 of which will be exported outside Asia. Factory will build 180,000 gasoline engines a year. Of those gas engines, 130,000 will be exported to Thailand and around the world. Monthly manufacturing wage here is $40.

FIGURE 12-18 Toyota locates factories throughout Asia, lowering its manufacturing costs by seeking low-wage countries. This map shows Toyota's expansion plans and costs in 2003.

By one estimate the Japanese paid foreign manufacturers a total of only $10 billion for patents and licenses between 1950 and 1980. That must rank as the shrewdest investment any nation ever made. Today Japan holds almost 50 percent of all U.S. patents granted to non-U.S. residents, and it has joined the United States, Germany, Great Britain, France, the Netherlands, Switzerland, and the other most advanced countries among recognized holders of key patents. The world geography of manufacturing tomorrow may be guessed by mapping investment in research today. Some analysts feel that

U.S. investment in research may be falling behind that in Japan and other international economic competitors, measured both as a percentage of GDP and in actual money terms.

Technology and capital contribute the most rapidly increasing shares of the final value of most manufactured goods, and technology and capital are exceptionally footloose. In the manufacture of computer disk drives, for example, the raw materials are an insignificant fraction of the drives' final value, so computer disk drives can be easily manufactured and distributed

Regional Focus on

THE TRANSFORMATION OF A NATIONAL ECONOMY

Costa Rica boasts a new economy. Into the 1970s, the small Central American country's principle exports were coffee and bananas. The country expanded into tourism and the manufacture of textiles in the 1980s, but NAFTA destroyed the textile industry in the 1990s (see Chapter 13).

High-tech workers stretching for exercise contrast with banana pickers stretching for fruits. (Above: AP/Wide World *Photos*. Below: Gilles Mingasson/Getty Images, Inc.—Liaison)

In 1997 Intel, the world's largest maker of microprocessors, chose to locate a new factory in Costa Rica. Costa Rica's attractions for Intel included the highest literacy rate in Latin America (95 percent) and a stable government. Costa Rica did offer concessions to attract the factory, such as reductions in local taxes. One drawback for Intel is that the plant must import 90 percent of its needs, because there is no local electronics industry. An Intel plant in Singapore or Taiwan, by contrast, can buy 80 percent of its supplies locally; this is more convenient for Intel, and it also sends more money into the local economy. Intel's contributions to the Costa Rican economy for now are salaries, electricity bills, and payments to local contractors, but the Intel plant may spur the growth of the necessary suppliers locally. The plant's existence has already helped lure other multinational firms, including Proctor and Gamble and Abbott Pharmaceutical. As Costa Rican president Miguel Angel Rodríguez has said, "We needed to have a presence on the economic stage and to strengthen our ability to attract high-tech foreign investment. Intel did that for us."

Intel employs about 2,000 workers. Technician-level employees make $5,000 per year, which is less than the $20,000-plus such an employee in the United States might earn, but which is well above average Costa Rican incomes. Employees also receive bonuses and stock options.

This one factory has transformed the national economy. In 1986, 88 percent of foreign direct investment in Costa Rica was in agriculture and 10 percent was in industry; by 1996 it was 69 percent in manufacturing and only 7 percent in agriculture. Upon the opening of the Intel plant, Costa Rica's GDP surged by 8.4 percent in one year, and half of that increase was due to Intel, which in its first year accounted for a full 37 percent of national exports, more than coffee and bananas combined.

Questions

1. Do you think Intel chose Costa Rica for its processor plant for the same reasons GM chose Thailand for its auto plant?

2. If you were the president of a small poor country, what would you offer a manufacturer to locate a plant in your country?

3. If you were the president of a company that manufactured computer parts, what locational determinants would you look for?

worldwide from almost anywhere. Why then are more than half manufactured in Singapore? They were not invented there. The decisive locational determinants for this high-value product are the availability of inexpensive skilled labor, technology, capital, and political stability in Singapore. None of these factors, however, is unique to Singapore. How long will Singapore continue to dominate world manufacture of disk drives?

The development of computer software is one of the world's fastest-growing industries, and it is even more footloose than the manufacture of computer hardware. All it requires is a personal computer and a skilled programmer. This multibillion-dollar industry can migrate on an airplane. The United States commands about 50 percent of the global market for software and related services. This value is greater than the total production of almost any of the world's raw materials. India is emerging as a center for software production.

Air-freight costs are falling, and because more manufactured products are becoming smaller and of higher value, they can be moved by air. Electronic components are regularly air-freighted throughout East and Southeast Asia today. Fort Worth, Texas, has opened the world's first airport at which planes can taxi directly from the runway into an industrial park. Sponsors hope that it will attract manufacturers that rely on far-flung foreign and domestic suppliers and that sell their finished goods to buyers around the globe. Could such an airport industrial park just as well be in Alaska or on an island in the South Pacific?

The anecdote about I.B.M. that opened this chapter revealed that today many white-collar tertiary-sector jobs are footloose, and we shall see below how this is remapping the world geography of tertiary service sector industries.

NATIONAL ECONOMIC-GEOGRAPHIC POLICIES

Each country organizes its domestic economy, manages the distribution of economic activities within its territory, and regulates its participation in world trade and investment.

Political Economy

Each country defines a set of principles to organize its economic life. The study of these principles is **political economy.** In all economic and political systems, the government usually provides those services that are unprofitable to the provider but that diffuse benefits throughout the economy. These services include education, transportation, water and sewerage, and other

aspects of the infrastructure. These services benefit many people beyond their immediate providers, so they are called *positive externalities* (see Chapter 9).

Each country also manipulates its national budget and expenditures, interest and exchange rates, and money supply to promote growth and high employment. The tools for manipulating these are imperfect, but national governments' economic planning and management, their revenue and expenditure, and their role in redistributing income have made national economic policies more central to most peoples' lives than ever before.

Direct government participation in the economy ranges theoretically from a *communist system,* in which both the natural resources and the productive enterprises are **nationalized**—that is, owned by the government in the name of the people, to a **capitalist system,** in which the state defers to private enterprise and a stock market serves to raise and allocate investment capital. The economy of each country lies between these two extremes. Most countries have mixed economies.

Even among those countries generally considered capitalist, there is a great range of government involvement in the economy. Each country's political economy

> Each country's political economy usually reflects other characteristics of that country's culture.

usually reflects other characteristics of that country's culture. The United States, for example, generally favors a system that minimizes the government's role. This is called **laissez-faire capitalism,** from the French for "Leave us alone," a phrase supposedly said by the French economist Vincent de Gournay (1712–1759) to a government bureaucrat. The U.S. government has never developed an explicit industrial policy, although its regulations, research funding, and defense budget have greatly influenced the national economy. The United States also designs its tax system to allocate capital. For example, Chapter 10 noted how the tax code favors investment in housing.

The attacks of September 11, 2001, caused some re-evaluation and readjustment of America's laissez-faire stance. Since the attacks, people have come to see the federal government as uniquely capable of performing many urgent tasks: fighting enemies abroad, providing security at home, rescuing bankrupt airlines, protecting Americans from bioterror, and making the skies safe again. These sentiments may cause greater governmental regulation or even governmental assumption of some responsibilities that had been privatized (that is, left to the private sector—see Chapter 8). For example, in November 2001 Congress created a new federal force responsible for airport security, a responsibility that had been surrendered to private security forces. Nevertheless, in 2002 the Bush Administration announced plans to put as many as 850,000 government jobs, nearly half the federal work force, up for competition from private contractors. The president argued that the government could save money by ensuring the lowest cost for many routine duties like mowing lawns, picking up trash, making eyeglasses, and printing paychecks. The pay for all of these jobs, however, is only about 2 percent of the total federal budget.

In several Asian countries, the government plans and regulates the economy, although it does not own many companies outright. This is called *state-directed capitalism,* and it conforms with these countries' Confucian bureaucratic cultural traditions (discussed in Chapter 7). South Korea, for example, is considered a capitalist free-market economy, yet the government proposes national economic plans and subsidizes new industries. Until the late 1990s, industry was dominated by combinations of corporations, called *chaebols,* which were affiliated by interlocking directorships and ownership. Daewoo, Hyundai, and Samsung were examples familiar around the world. The economies of Japan and Taiwan are similarly organized, but in the United States such combinations would be broken up as illegal trusts.

Most Latin American countries, by contrast, have traditions of outright government ownership of the nation's assets and industries. In the 1980s and 1990s, however, these governments began privatizing their assets. Purchasers included both domestic and international investors and corporations. Western European countries have also privatized, although nationalized companies still account for about 25 percent of GDP in France, 11 percent in Germany, and even 3.5 percent in Britain, which began the privatizing trend in the early 1980s. Eastern European countries are rapidly privatizing as they abandon communism, and the special case of China will be discussed shortly. Some African countries long insisted on national ownership of natural resources as a point of national pride, but even they have begun to privatize and invite foreign capital. In 1997 Zambia's president, Frederick Chiluba, said, "We don't care who buys the mines in Zambia so long as the mines make money and contribute to the exchequer [national treasury]." Many countries are today creating new national stock markets (Figure 12-19).

The results of privatization If a country privatizes some assets, it must devise a policy for investing the windfall of money that it receives. This challenge faces many countries that are privatizing assets today. If the country is so desperately poor that it spends the money on immediate needs, such as food, then it will soon be worse off than it was before privatization. The assets will have been sold and the money spent. In other cases, some countries' corrupt rulers have virtually stolen the money from privatization and deposited it in personal accounts in foreign banks. Our present state of economic theory and experience suggests that if a government invests the income from privatization in education, infrastructure, or health care, however, then the country is on the way to economic development.

The privatization of national assets or industries ramifies across a country's economy and politics. This is

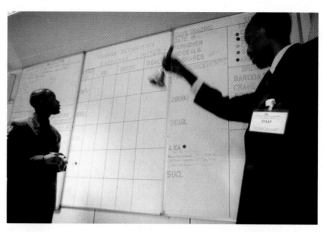

FIGURE 12-19 An emerging stock market. The new Ugandan Stock Market lists stock shares of only five Ugandan corporations, and it is open only a few minutes per week. Nevertheless, the government hopes that its existence will stimulate Ugandans to save and invest in Uganda's growth, and to incorporate new business ventures. (Vanessa Vick)

because nationalized activities fulfill responsibilities beyond the merely economic ones. Nationalized enterprises might be subsidized in order to maintain high employment, to sustain the economies of poor regions, or just to win votes. These political purposes may provide positive externalities across the society, but they are generally difficult to justify economically. The principal goal of a private enterprise, by contrast, is to maximize profits for shareholders. This usually requires the enterprise to be cost efficient in order to compete in world markets. Today, as more countries embrace privatization and capitalism as the surest route to economic development, thousands of workers are being laid off from economically inefficient nationalized industries or from enterprises that are being privatized. In many countries from Argentina to China these workers are heading for the ballot box (or even taking to the streets) in protest. National leaders are gambling that privatization will release economic dynamism and stimulate economic growth fast enough to create new job opportunities for these workers.

For private enterprises to succeed, the government must still perform certain basic functions. In 1996 the World Bank surveyed 3,600 entrepreneurs in 69 countries, asking them about how well such basic governmental functions as securing property rights and creating a reliable judiciary were fulfilled. The results were startling. In Latin America, Eastern Europe, and sub-Saharan Africa, 80 percent of respondents lacked confidence that authorities would protect their property from criminals. Seventy percent reported judicial arbitrariness as a big problem for doing business, and more than 40 percent of entrepreneurs in developing countries said they had to pay bribes as a matter of course, compared with 15 percent in wealthier

Regional Focus on
SOUTH KOREA

South Korea demonstrates that a country can grow under one political economy and then, if continued growth is threatened, it can virtually reverse its course and adopt a new form. South Korea has few natural resources, and in 1962 its economy was still in shambles from the Korean War (see Figure 11-10). Nevertheless, for the next 35 years the South Koreans achieved an annual rate of economic growth of over 9 percent, the highest ever achieved anywhere. South Korea followed

Yoo Jung Hwan, the leader of the labor union at South Korea's Jinro distiller, left, and Kim Seon Joong, the corporation's chief executive, console each other after a South Korean court ordered the company into bankruptcy in 2003. Foreign creditors had demanded the action. Under the directed-capitalism economic system that South Korea had employed until 1997, the government probably would have stepped in and saved the company. Under the laissez-faire system of global capitalism, however, the inefficient distiller went broke, and many workers lost their jobs. (Seokyong Lee/Bloomberg News)

the Japanese model of state-directed capitalism. The Japanese had occupied Korea from 1910–1945, and Japanse-stye institutions were in place. The government owned the banks, which loaned no money to consumers or to small businesses, but only to the chaebols at artificially low interest rates. The chaebols could capture world markets for South Korean goods because neither capital nor labor cost were high. South Korea exported many goods that South Koreans themselves could not afford to enjoy. Workers were satisfied with low wages, but lifetime job security—famously called an "iron rice bowl." The national tax structure encouraged savings and discouraged consumption. The country was governed by virtual dictators.

During two shattering weeks in late 1997, Japan devalued its currency so Japanese goods were cheaper than South Korean goods on world markets. South Korean exports halted, and because there was little domestic demand, the national economy collapsed. The new president, Kim Dae-jung, saw the necessity of reversing economic policies. He privatized the banks and opened the country to foreign investment. The chaebols were broken up, and jobs were no longer secure. By 2001 the country was better off than in 1997. It held over $120 billion in foreign reserves, ranked 54th among countries in GNI per capita, but tied for 23rd rank in the HDI index.

The South Korean experience demonstrates that different types of economic organization may be best for a country at different stages in its own economic development, perhaps at different times in the development of the global economy, or perhaps at different times in each country's own social, cultural, or political development. We really do not know the answers to questions such as these.

countries. These figures suggest how the political culture in many world regions significantly retards economic growth.

Crony capitalism In several countries that practice state-directed capitalism, governments through the 1990s were not fully free and democratic; they were run by oligarchies. The governments favored politically well-connected industries or even friends and family members of rulers in a system called *crony capitalism*. In such circumstances, a political crisis can immediately

become an economic crisis. In the autumn of 1997, for example, an international loss of confidence in the stability of the government of Indonesia triggered a reevaluation of that country's financial strength, and the ensuing turmoil endangered the financial structures of several other East Asian countries—notably South Korea, Malaysia, and Thailand. Political and economic restructuring in each of these countries continued into the twenty-first century, putting in place the regulatory infrastructures and legal safeguards for more laissez-faire capitalist development. These countries experienced a political, social, and economic revolution that forced a greater reliance on markets, democracy, and the rule of law. Meanwhile, however, world financial markets shuddered, and the economic development of these countries slowed.

Events early in the twenty-first century revealed that the United States is not immune to crony capitalism. Corporations were found to have been creating hundreds of millions of dollars in fake profits (Enron), disguising operating expenditures as capital investments (WorldCom), or booking nonexistent revenues to inflate the stock price (Global Crossing, Lucent, and many others). Many large corporations had been virtually pillaged by their boards of directors and executives, while one of Wall Street's most trusted stock analysts had been urging his clients to buy stocks that he had privately been calling "crap." America's most prestigious accounting firms and banks had been colluding. Multi-billion-dollar corporations went bankrupt, hundreds of millions of dollars were levied in fines against corporations, accountants, and banks, the stock market fell, and trillions of dollars disappeared.

New regulatory legislation was passed to rectify these faults in America's economic system. These events, however, significantly damaged the prestige that the American capitalist model had held for 50 years, and the ramifications of this damage will continue in world trade negotiations for decades. The fact that other countries felt freed to develop alternative models may, however, in the long run, produce innovations in economic development theory that will benefit all countries.

Variations in Wealth Within States

Countries try to boost their economic growth by organizing their populations, territory, and resources for production. A strong sense of nationalism can encourage the population to work together, but the population also needs education and training. The territory must be at peace, and the country should constitute a single market for raw materials, goods, and labor.

Internal mobility of the population is another factor that may induce growth. Internal population mobility assists each individual in realizing his or her own potential, and it also allows employers to draw on the potential of the entire national population. Workers would normally be expected to move from areas of high unemployment to areas of low unemployment, so variations in regional unemployment can provide a measure of labor mobility and national cultural homogeneity. Great variations in unemployment suggest that something is discouraging people from moving. Factors preventing mobility may include peoples' attachment to place or family; racial or ethnic animosity among different localized groups; lack of nationwide

availability of certain cultural products and services (religious services, foods, or other consumer goods); regional variations in unionization and difficulty in joining unions; variations in wage scales; variations in workers' benefits and whether workers lose them by relocating; nationwide acceptance of degrees or certification for professionals; lack of knowledge of distant opportunities; and a desire to avoid risks.

Italy exemplifies a country in which unemployment rates in one region, the South, are usually two or more times as high as in the North. This reflects continuing cultural differences between the regions. Unemployment rates vary across the United States, but on the whole, Americans are a mobile population. Each year, almost 20 percent move to a different house; almost 3 percent to a different state. This illustrates that national culture is relatively homogeneous across the country.

States with frontiers   Not all states occupy and utilize their full territory. The world population map (on the rear endpaper) reveals that many countries have core areas of dense population and development but also **frontier** areas, undeveloped regions that may offer potential for settlement. Environmental limitations may hinder exploitation of frontiers. Brazil, for example, contains vast sparsely settled areas in which the government has encouraged settlement, but the results have often been economically disappointing and ecologically disastrous. Similarly, the island of Java is only 7 percent of Indonesia's national territory, but it is home to 60 percent of the population. The Indonesian government has tried to relocate settlers to other islands, but many new settlements have had difficulty supporting themselves.

Some African states seem overpopulated relative to their current ability to feed themselves, yet they are only sparsely populated. With an end to civil wars and with greater investment in agriculture, some of these territories could support greater populations. Canada's northern regions are poorly suited to agriculture (compare Figure 4-23 with Figure 11-24), but they are exploited for mineral, timber, and hydropower resources. Expanses of the Siberian region of Russia have never been densely settled or even fully explored for resources, yet most of Russia's population remains west of the Urals (Figure 12-20). Japan might be a partner with Russia in developing Siberia's potential.

Chinese settlement and development of Xinjiang has been carried out by the Xinjiang Production and Construction Corps, a unique paramilitary organization created in 1949. The Corps' 2.4 million workers and their families have settled hundreds of farms and built factories and entire cities. Local indigenous Uighurs complain that the Corps has been an engine of Chinese colonization of "their" region. Xinjiang's deserts and mountains cover one-sixth of China, but even now the province is home to only 17 million people.

The Resources of Russia

FIGURE 12-20 Siberia remains almost uninhabited, but enormous amounts of capital will be needed to overcome its inhospitable climate and great distances to exploit its raw materials.

How Do Governments Distribute Economic Activities?

Most countries try to maintain a fairly equal standard of living throughout their territory, although many have biases that favor some regions over others. Great disparities of wealth from region to region are a centrifugal force that may pull the state apart (Figure 12-21). Economic competition among regions may dominate national politics. Politicians weigh the domestic regional impact of every program and devise new ones to reduce disruptive imbalances. We have already seen secessionist forces in Italy's rich northern provinces (Figure 11-2). Norway builds bridges and tunnels to reach even the most isolated part of its territory. It has been willing to suffer a huge expense to reach, for example, an island with only 4,000 people. This development strategy is known as District Policy, and its purpose is to keep the population in these farms and villages. Ranghild Queseth Haarstad, Minister of Local Government and Regional Development, said in 2000, "First there is our traditions and our culture. Then there is the fact that our most important export articles—oil, fish, lumber, and tourism—are in those districts. If we're going to exploit them, we have to have people there. . . . We have a broad political agreement in Norway that we should preserve our traditional population patterns." The more isolated areas enjoy relief from taxes to inhabitants and subsidies to local businesses. The apportionment of the Norwegian Parliament even deliberately underrepresents the main cities in favor of rural areas.

The geography of any country's resource endowment may favor some regions over others. The relative fortunes of regions may change, depending on the discovery of new resources, shifting trade patterns, or patterns of innovation. In the area of today's Belgium, for example, the lowland coastal Flemings grew rich

Disparities in Welfare in Brazil

FIGURE 12-21 By applying the HDI formula to the individual states of Brazil, we learn that the populations of the southern states enjoy standards of living as high as those of the populations of Europe or the United States, but other states suffer standards of living among the lowest on Earth. The southernmost seven of Brazil's 33 states account for three quarters of Brazilian GDP, and they support a lively secessionist movement. (Data from the United Nations Development Programme)

from trade and commerce during the medieval and Renaissance periods, and they dominated the highland Walloons. Later the discovery of coal and the industrialization of Wallonia brought Walloons prosperity. Today the decline of coal and steel manufacturing and the development of a trade and service economy has swung the pendulum of prosperity back to the Flemings.

The national distribution of prosperity can also change as the sectors of the national economy evolve.

As employment shifts from sector to sector, job opportunity shifts from place to place. The expansion of the secondary and tertiary sectors, for example, brings urbanization. These shifts might occur faster than workers can be retrained or relocated, so certain regions of a country may suffer while others thrive.

Countries often devise special development programs for poor regions, just as cities or states designate urban enterprise zones (discussed in Chapter 10). For

The Tennessee Valley Authority Region

FIGURE 12-22 The Tennessee Valley Authority was created to bring electric power and jobs to this economically deprived seven-state region in 1933. Despite the Authority's many dams, 60 percent of the electricity it produced in 2002 was generated in its coal-fired plants. The modernization of these plants is an issue of current political debate.

example, the United States created the Tennessee Valley Authority in 1933, an Appalachian Regional Commission in 1964, and a Lower Mississippi Delta Development Commission in 1988 (Figure 12-22). Income and welfare were significantly below national levels in these three regions, and federal programs attempted to boost their economies.

A government may locate factories or other enterprises that it owns in poor regions, or it may distribute offices, research institutes, and military installations. Governments may lure private enterprises by offering subsidies, tax waivers, free development sites, or loans. The Spanish government, for example, subsidizes international businesses that invest in Spain's poorer provinces, and Italy subsidizes investment in its south. India encourages firms to locate in officially designated "backward areas" by offering low-interest loans, tax reductions, and low freight rates on government railroads for products from the designated areas.

No matter what economic policies a government devises, some regions may dominate a national economy while others suffer population losses. As a whole, the central region of the United States has been losing population since World War II. Kansas alone has over 2,000 officially registered ghost towns (Figure 12-23).

Population Change in the United States by County

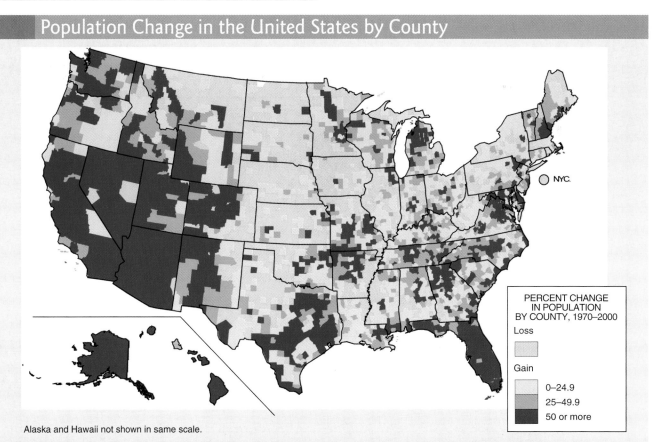

Alaska and Hawaii not shown in same scale.

FIGURE 12-23 Some counties in the central U.S. farming region and some old cotton-farming centers in the South counted their maximum populations before 1900. The populations of many other areas across the Great Plains and in Appalachia peaked by 1940, when the prosperity that had come from farming and mining was past. The national population today seems to be concentrating at the margins of the country. (Reprinted from *Scientific American* August 2000, © Rodger Doyle 2000)

U.S. Plains states have suffered population declines, while the South and Southwest, known as the Sunbelt, have absorbed most U.S. population growth.

National Transportation Infrastructures

Political considerations   Some regions in Southeast Asia and in parts of South America and Africa are in open rebellion against their national governments, and the central governments invest in roads specifically to "occupy" the territory. Colombia, for example, has recently discovered enormous oil deposits beneath its eastern plains, but exploiting these deposits will require the central government to seize control of this region from drug dealers and independence movements.

Transportation and economics   Economic growth, however, is usually the explanation for investment in transportation. Transportation and communication allow different territories to specialize production and to trade. This territorial division of labor is comparable to the division of labor among workers. The design and regulation of a country's transport infrastructure are two important influences on internal geography. In some countries transportation is slow and difficult. Some regions are practically inaccessible (Figure 12-24). In

FIGURE 12-24 Road building in Malaysia. Malaysia's new roads cut through forests and open frontier regions to development. (Permission granted by Caterpillar, Inc.)

other countries virtually everyplace is easily and cheaply accessible from every other place (Figure 12-25).

Whenever anything is in transit, the value it represents is idle. The owner is therefore losing an opportunity to invest that value in some other activity that

FIGURE 12-25 Compare this map with the world population distribution shown in the map on the rear endpaper. Many densely settled regions, such as Eastern China, would benefit from improved road networks. The African continent is three times the size of the United States, yet it contains only one quarter of the mileage of paved roads. (Adapted from *The New Comparative World Atlas*. Maplewood, N.J.: Hammond Inc. 1997.)

might pay a return every minute. A financial return sacrificed by leaving capital invested in one form or activity rather than another is called an **opportunity cost.** Goods in transit cost their owner opportunity costs as well as transportation charges, and that is why people want things to move fast. If a country has a poor transportation infrastructure, much of the country's capital is tied up in raw materials or in materials in transit. This principle applies on a small, local scale: Your local stores do not need to carry large inventories if they can get whatever you want quickly. Furthermore, if factories can receive regular supplies of the materials they need, they can reduce their inventories of parts. Minimizing inventories is the key to the modern manufacturing method called *just-in-time manufacturing*. Fast and efficient transportation releases capital for productive investment, so developing a country's transportation system is key to economic growth (Figure 12-26). In the United States, deregulation lowered both truck and rail freight rates between 1981 and 2002, and new computer systems helped businesses track inventory better. As a result, the share of national GDP accounted for by transportation and logistics fell from 16.2 percent to 8.7 percent—an enormous national saving.

Just-in-time manufacturing is, however, vulnerable to disruption. For example, in 2003 Dell Computer assembled its OptiPlex Business Desktop model in Austin, Texas, but the assembly line for the model's 4,500 parts stretched out the door down Interstate 10 to the monitor plants in Tijuana, Mexico, and back through Dallas-Fort Worth Airport to component manufacturers strung from Malaysia to Korea but clustered in China and Taiwan. On any day, one missing shipment could stop the operation. Disruptions can be natural or caused by humans. For example, on September 21, 1999, an earthquake measuring 7.6 on the Richter scale struck Taiwan, where 90 percent of the world's scanners and over 50 percent of computer motherboards (over 50 percent of those in one industrial park in Hsinchu) were manufactured. Stock prices of global computer manufacturers plummeted. Production was restored at most Taiwanese factories within a week, but there is no guarantee that another disruption might not do more lasting damage to global supply lines. On September 12, 2001, much of U.S. manufacturing activity came to a sudden halt when the United States closed its border and grounded all flights in the wake of the terrorist attacks of September 11. Terrorist threats to

FIGURE 12-26 A communications center for trucking. From this "command center" of Contract Feighters, Inc., in Joplin, Missouri, managers can monitor the locations of all 1,800 CFI trucks in the United States. The map at left shows nationwide weather conditions and can zoom in to show trouble spots in detail. The meters on the center screen show the average monthly mileage, number of trailers awaiting drivers, and quarterly turnover rate of drivers. On the map at right, a red state means that all equipment there is busy; a green state means trucks are available to handle new loads. (Bob Sacha/Bob Sacha Photography)

the international transportation system threaten the development of global just-in-time manufacturing.

India exemplifies a country whose economic growth is held back by an inadequate road system. Traffic in passengers and goods on India's roads has increased thirtyfold since independence, but the total road mileage in India has increased only fivefold. Roads throughout India are overcrowded, and thousands of villages lack all-weather roads. The government hopes to have every village of more than 500 people connected to an all-weather road by 2010. Accessibility would allow regional specialization of production and cash cropping, especially of high-value perishable foods, some 40 percent of which spoils during transit delays

today. Better transportation could also improve the Indian diet.

In many countries the pattern of the transport network is not integrated, but consists of lines penetrating into the country from the ports. These roads or railroads are called **tap routes.** Just as some trees have tap roots that reach deep down into the soil to draw up water, so these tap routes stretch into the interiors from the seaports to bring out the exploitable wealth. Tap routes are often the legacy of colonialism or the product of neocolonialism, and they facilitate getting into or out of a country, but they do not facilitate getting around inside it (Figure 12-27). Australia, too, exhibits a pattern of railroad tap routes, and Australia's routes

Tap Routes in Africa

FIGURE 12-27 Europeans built these railroads to penetrate Africa and to haul out agricultural products and minerals, and still today railway lines provide the only freight service in many roadless areas. Africa's coastline offers few natural harbors, and many of its rivers drop off the continental interior through rapids and waterfalls only a few miles inland from the coast (compare this figure with the map on the front endpaper, world topography). Thus, the continent is relatively impenetrable by water, and this impenetrability presents a major impediment to African economic development. Many lines shown on this map, however, are disrupted by civil strife or else they are inadequately maintained, so they offer only intermittent service. Deterioration of the rail network increases transport costs and lowers income for many African states. Lines in Sudan, for example, and the important line reaching into central Africa from Benguela, Angola, have not operated for over 25 years.

were built at many different *gauges* (widths between the tracks). The Australian government has tried to knit one national rail system, but Sydney and Melbourne, Australia's two largest cities and only 700 kilometers (435 miles) apart, were not linked until 1995, and the first transcontinental north-south line—promised in 1911—is scheduled to open in 2004.

Regulating transportation Mapping a national transport network is only a first step toward understanding what moves, where, and why. Lines on a map are static, but movement on those lines can be manipulated by conditions and charges that are "hidden" behind the map. Most countries blend public and private transportation services, and they manage freight rates on government-owned railroads, for example, as India does, to distribute industry or other activities.

In all countries, trucks, railroads, and airlines compete for traffic. They may also compete for government subsidies, and these subsidies can in turn affect domestic economic geography. In Kenya, for example, the trucking industry has successfully prevented national investment in railroads. In most European countries, in contrast, government-owned railroads receive generous subsidies. In the United States the trucking industry, the oil industry, the highway construction industry, and the car-driving public make up a lobby powerful enough to overwhelm the railroad interests. As a result, trucks pay only a fraction of their real highway-use costs, and they do not pay for the many negative externalities that they impose, such as air pollution, traffic accidents, and costs due to congestion. This subsidy to the trucking industry enables it to capture much long-haul traffic that is actually best suited to railroads. Railroads still handle 38 percent of U.S. bulk traffic, but trucks have captured 28 percent, and railroad trackage is being abandoned. Thus, in the United States as in Kenya, political competition distorts economic competition in the transport system and locks inefficiency into the economy.

Railroads' share of intercity U.S. freight traffic fell from 61 percent in 1940 to under 35 percent by the late 1980s, but freight railroads are enjoying a revival. Computers allow better tracking of shipments and improved timing of schedules, and highway congestion (and the concomitant pollution and accidents) encourage the rail alternative. Furthermore, states and local governments are helping finance improvements in the network. For example, the six major North American railroads announced in 2003 a $1.5 billion project to untangle their lines around Chicago, with the money to come from local, state, and federal governments. A new 20-mile "Alameda Corridor" connects the ports of Los Angeles and Long Beach to the national rail system at a hub east of downtown Los Angeles, and a new overpass in Kansas City speeds transcontinental shipments.

Countries may also manipulate domestic freight rates to direct traffic to favored seaports, and the port facilities themselves may be subsidized to attract business and create jobs. The Netherlands, for example,

Fixed Line and Mobile Phone Subscribers per 1,000 People

FIGURE 12-28 The density of telephone service provides a good measure of the availability of information and communications in a society. (Data from The World Bank)

organizes its national transport network and fixes prices on it to capture an international hinterland for its port of Rotterdam.

These examples provide only a hint of the manipulation of charges built into each country's transportation infrastructure. When geographers investigate economic activities, they often discover these "hidden" policies.

National communications infrastructures The dissemination of information can be as important as the dissemination of goods and materials. The degree to which citizens of a country enjoy access to sources of information can reveal economic development and political liberty, and it can also promote these two values.

The availability of telephone lines is a good measure of access to information (Figure 12-28). Today fax machines, personal computers with Internet access, and other modern communications devices are largely restricted to the rich countries, but we can hardly guess how their widening availability will transform the geography of the twenty-first century.

All aspects of infrastructure—pipelines, electricity generators, highways, telephone facilities, and more—require continuous maintenance and updating. As societies become increasingly dependent on these facilities, and as the facilities become more complex, both maintenance costs and costs in the event of failure rise dramatically. The blackout that darkened much of the Northeast United States and Southern Canada in 2003, leaving over 50 million people without electricity, resulted from the inadequacy of transmission facilities. This event was a warning about infrastructure maintenance.

NATIONAL TRADE POLICIES

Countries differ in the degree to which they participate in world trade and circulation. Each country wants to develop its own economy, to ensure its national security, and to some degree, to protect its national culture or identity. Chapter 8 noted how some countries subsidize domestic food production as a national security measure. Countries also protect national security industries—armaments, for example.

The Import-Substitution Method of Growth

A government may protect a national industry from imported competition as only a temporary measure. This is done in the belief that newly developing industries, called **infant industries,** cannot compete with imports. Infant industries manufacture only small quantities of the product, so manufacturing costs per unit are high. Eventually, however, the industry might build a national market for its product and produce larger quantities of it. Then the cost per unit falls. This is called achieving **economies of scale.** When economies of scale have been achieved in protected national markets, then imports will be allowed to compete. This policy of protecting domestic infant industries is called the **import-substitution method of economic growth.**

This form of economic protectionism helped some countries industrialize in the past, including, to some degree, the United States, but it involves economic risks. If a country raises tariffs to protect a particular industry, all national consumers are subsidizing that industry. Some protected industries may never be able to survive competition with imported goods, so the national public may go on subsidizing them. This may be an inefficient investment of national resources. For example, the United States imposes an average 48 percent tariff on imported apparel; the U.S. Department of Commerce estimates that this protection costs U.S. consumers $145,000 per job saved in the U.S. apparel industry, which is several times what those workers actually earn. The wages apparel workers earn do, however, multiply through the economy, and if the workers had no jobs, some might have to go on welfare. Overall, it is difficult to balance these costs, but the manufacturers and workers lobby for continuing protection from foreign competition. The United States has agreed to drop barriers to trade in textiles by 2005, but production of textiles and clothing in China is so much cheaper than in the United States that Chinese are capturing the U.S. market anyway. China's share in the U.S. market for the 29 leading textile and apparel products jumped from 9 percent in 2000 to 45 percent in 2003.

Export-Led Economic Growth

The alternative to the import-substitution method of economic growth is the **export-led method.** Countries that adopt this program welcome foreign investment to build factories that will manufacture goods for international markets. In that way, they can achieve economies of scale immediately. Export-led policies rely on global capital markets to facilitate international investment, and they rely on global marketing networks to distribute the products. Export-led development is usually accompanied by low tariffs and export duties, market-based domestic economic policies, privatization of most economic activity, and general laissez-faire openness to economic innovation.

The countries that have grown the fastest in recent decades have generally followed export-led programs. The success of these policies demonstrates again how, increasingly, what happens *at* places is the result of what happens *among* places.

For a country's economy to grow by exporting, other countries must be willing to accept the developing nation's products. Through recent decades the United States has been the major importer of products from countries choosing this method of development, and this has been responsible for the United States' substantial trade deficits with several developing countries. At the end of the 1970s, after decades of almost continuous current-accounts surpluses, the United States was a creditor country, with a net stock of foreign assets worth about 10 percent of GDP. Persistent current-accounts deficits turned the country into a net debtor by 1985, and since then it has been getting deeper and deeper into debt. At the end of 2002, net external debt reached 25 percent of GDP. Some economists fear that this debt threatens America's prosperity.

China: How trade policies can affect the national distribution of wealth International trade and investment sometimes multiply regional economic inequalities within a country. If a country adopts protectionism, all regions of the country may not profit equally, and that can intensify regionalism. If a government adopts the alternative export-led growth policies, outside investment might concentrate in favored regions.

China exemplifies on a colossal scale how foreign investment can intensify regional imbalances in economic growth and how the centrifugal force of these imbalances can threaten national stability. China has always had an economic and cultural dichotomy between an outwardly focused southeast coastal region and an inwardly focused central region. Chinese capitals were in the north or center, and the government limited the trading activities of the merchants in the southern coastal cities. In the nineteenth century, however, European powers forcibly "opened" China's southeastern ports to trade, and industrial development, inspired by international trade, clustered along the coast. The Chinese government keenly felt this humiliation.

The Communist government that Mao Zedong brought into power in 1949 set out to boost the inland regions out of their relative economic backwardness by redistributing industry inland. From 1949 to 1979, foreign trade plummeted; China's coastal areas and cities stagnated.

Mao died in 1976, and in 1979 the Chinese government reversed its policies. It designated zones along the coastline where relatively free capitalistic practices were allowed. These areas were soon expanded as international corporations built factories in the zones to exploit the cheap land and labor, and the designated areas leaped to create new wealth (Figure 12-29). China had engaged in very little foreign trade in 1979, but from 1980 to 2000 trade increased by 12–15 percent per year.

As China's seaboard provinces grew in wealth, the inland provinces slipped behind noticeably, and the historic contrast between the interior and the periphery reemerged (Figure 12-30). The disparity of incomes across China was ameliorated only by workers' remittances from the coastal provinces to workers' inland

FIGURE 12-29 The great variations in per capita gross product among China's provinces result largely from foreign investment and astounding economic development in the coastal regions. China has invited foreign investment since 1979, and capitalist-style economic growth has boomed.

home provinces. In several cases the remittances more than doubled inland provinces' incomes. The government has launched programs to stimulate development in the interior: It intends to invest substantially in education, health care, and public works, and to open 20,000 miles of new roads and 2,500 miles of new railroads by 2010.

The rift in the nation's economic geography duplicated a widening rift between the state and private sectors of the economy. China remained dedicated to communism in theory, but in practice it continued to privatize the economy. Chapter 8 noted that agriculture was privatized, and food production soared. Similarly, the industries of the state sector declined as the private industries flourished (Figure 12-31). China continues to privatize state-owned industries, but the government generally offers workers pensions in order to close facilities. China's 1954 Constitution said that business executives must be "used, restricted, and taught." The 1988 Constitution softened those words to "guided, supervised, and managed," but amendments added in 1999 declare private business "an important component of the socialist market economy."

In August 1997, *The People's Daily*, China's principal Communist Party newspaper, announced, "The words 'market economy' have been writ large on the

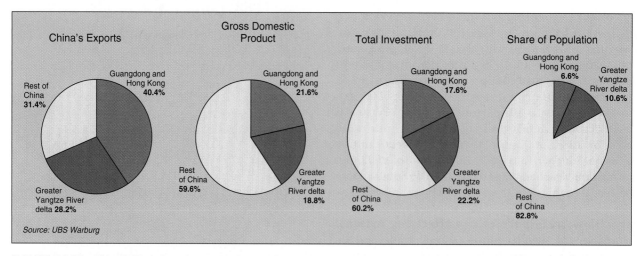

FIGURE 12-30 China's Two Dominant Regions. The Guangdong and Hong Kong Region and the Greater Yangtze River Delta Region (focused on the city of Shanghai) hold only 17 percent of China's population, but they account for about 40 percent of China's GDP.

flag of socialism. . . . We ran a planned economy for more than 20 years and created the foundations of industrialization, but . . . ahead is a new world. There is no way back." Within weeks the Fifteenth Communist Party Congress declared that the government will continue to insist on ownership of only a few key industries, and the definition of ownership will include

holding only a minority stake in enterprises. This is privatization in all but name.

Chinese wages are low, and the country has a seemingly infinite supply of workers, but in 2002 its exports were still 70 percent garments, toys, shoes, and furniture—that is, items of low value-added—but other products of increasing sophistication are expected to follow.

Some observers fear that this avalanche of cheaply made goods might inundate world markets and virtually bankrupt every other country on Earth, but that will not necessarily happen. Chinese exports are already responsible for deflating prices of manufactured goods in the world's rich countries. As China develops, however, the Chinese themselves are absorbing more of their own production and importing more of other countries' products. By 2002, the average annual income of people living in Eastern China had reached about $1,200 per year, creating a lower-middle income market of 470 million people—larger than any other country in the world except India. The domestic Chinese market for cellphones, cars, cameras and film, and DVD players, is already number one on Earth. Toshiba, the Japanese electronics and industrial equipment giant, once used China as an export base, but by 2002 Toshiba was selling over 60 percent of what it made at its 34 Chinese factories to Chinese themselves: televisions, refrigerators, computers, and automation equipment. Thus, Chinese development will lift both their own and the world's standard of living. Measured by purchasing power parity, China already accounts for 11.8 percent of world GDP, a percentage second only to the United States.

Taiwan Taiwan is a principal source of investment funds for China. The 22 million people of this island-nation enjoy a per capita GDP of almost $14,000 per year, contrasted to about $1,000 in China. This contrast underlines a bitter political division.

FIGURE 12-31 Industrial output in China, measured in trillions of yuan. The private sector of the Chinese economy soared after economic liberalization, especially in the early 1990s. The share of the total economy created by state-owned enterprises fell from about 75 percent in 1975 to about 28 percent by 2000. If we accept the economic definition that communism is state ownership of industry, then China ceased being a Communist country about 1993. (Statistical Yearbook of China 1997)

In the 1930s a civil war raged in China between the Communists, led by Mao, and the Kuomintang, a political party led by General Chiang Kai-shek (1887–1975). In 1949 the Communists drove the Kuomintang offshore to Taiwan, where it established a separate government. The ethnic Chinese, although a minority among the Taiwanese, continued to control the government until the elections of 2001. Meanwhile, state-directed capitalism achieved phenomenal economic growth. In the 1990s, Taiwanese began to invest and transfer low-wage, low-technology manufacturing to China, but the sophistication of the technology being transferred is rising. In 2003 Taiwan Semiconductor Manufacturing Company, the world's largest contract chip maker, started to build a huge semiconductor factory in Shanghai. The Chinese government welcomes such investment from its "rebel" province.

Both the government in Beijing and that on Taiwan claim that there is only one China and that they represent it, but at present China and Taiwan are in fact two separate countries. The U.N. General Assembly ousted the Taiwan government and seated the People's Republic in its place in 1971, and the United States recognized the People's Republic as the sole government of China in 1978. China says it will attack Taiwan only if Taiwan explicitly declares independence; Taiwan says it will declare independence only if China attacks. Some political accommodation will probably eventually be reached between the two governments—perhaps following the slogan explaining the merger with Hong Kong: "One country, Two systems." The two might agree on joint economic ventures and joint representation in foreign affairs, for instance, without completely merging their domestic governments. The economic integration of Hong Kong, Taiwan, China, and the considerable economic vitality of the overseas Chinese is proceeding rapidly.

As a result of China's slowing rate of population growth (see Chapter 5), its surge in food production (see Chapter 8), and foreign investment, China has achieved a rate of economic growth estimated at 10 to 12 percent per year. Overall, the Chinese economy quadrupled in size between 1980 and 2000. This has come about with repressive government, and any number of factors might slow or halt this development, including political disruption, environmental disruption, an energy supply crisis, or inadequacies in China's infrastructure. If the growth continues, however, China will overtake the United States as the world's largest economy by 2010. The population of China, a full 22 percent of the human population long living in poverty, might achieve considerable material comfort within 50 years. China's dramatic economic and industrial growth will likely transform many aspects of global political, economic, and environmental patterns in the twenty-first century. The growth is taking place over a period of years rather than suddenly, so it does not make the headlines, but in terms of its impact on the geography of wealth and power in the world, it is one of the most important "news items" of our time.

Where Is the Third World?

The term Third World was introduced by the French demographer Alfred Sauvy in 1952. It reflected attitudes during the Cold War that followed World War II. An ideological line was drawn between the capitalist countries led by the United States and the Communist-ruled countries headed by the Soviet Union. At the height of the diplomatic pulling and tugging, it was expected that all countries should line up on one side or the other. Many did so, but a few, led by presidents Tito of Yugoslavia, Nasser of Egypt, Nehru of India, Nkrumah of Ghana, Makarios of Cyprus, Sukarno of Indonesia, and Emperor Haile Selassie of Ethiopia, clung to a precarious neutrality. These maverick states came to be known collectively as the **Third World,** to distinguish them from the **First World** of the Western bloc and the **Second World** of the Soviet bloc.

Over the years the terms gained economic and sociological connotations, many of which were false and unfair. The First World, for example, was interpreted as a haven of science and rational decision-making: progressive, technological, efficient, democratic, and free. The First World included as allies in the Cold War all reasonably well-to-do non-communist countries even though some of them were politically repressive, such as South Africa, South Korea, and Taiwan. The Second World was also defined as modern, powerful, and technologically sophisticated, but it was dominated by totalitarian Communist governments. In fact, the countries of the Second World never achieved economic growth equivalent to that of the First World, and the collapse of many Communist regimes early in the 1990s exposed their true poverty.

The Cold War rhetoric unfairly characterized the Third World in contrast with the First and Second Worlds, as a world of underdevelopment, overpopulation, irrationality, and political chaos. Much of the Third World had still been organized as colonies in 1939, but it "emerged" after World War II. According to the theory, it was poor but struggling to develop. The rhetoric interpreted the Third World as a zone of competition between the First and Second World models of modernity (Figure 12-32).

Today the terminology of First, Second, and Third Worlds is clearly out of date. A region, by definition, must exhibit meaningful homogeneity. Historically, the unifying element in the Second World was the existence of Communist regimes. When these regimes in Eastern Europe and the former Soviet Union collapsed, however, and the nations individually adopted new political and economic systems (see Chapter 13), the Second World as a political-economic bloc ceased to exist. Furthermore, the term Third World carries a racial, or even racist, aspect. Most Third World

The Third and Fourth Worlds

FIGURE 12-32 The term *Third World* confused Cold War political alliance with economic status, so efforts were made to refine it. Some scholars subdivided the Third World and defined a poorest "Fourth World." Others tried to differentiate a rich "North" of the world from a poor "South."

countries were non-white, and therefore the term suggests, somehow, "all the same," yet the term lumped together peoples with extremely diverse histories and cultures who inhabit a broad range of physical environments that possess a great variety of natural and cultural resources. If we consider just the countries whose leaders made up the original political Third World—Yugoslavia (for its successor states today, see Figure 11-9), Egypt, India, Ghana, Cyprus, Indonesia, and Ethiopia—they cannot be homogenized economically, culturally, or sociologically. It is ultimately insulting to refuse to recognize and to appreciate their differences. The term Third World will eventually be remembered as a condescending relic of a political Cold War mentality. It regarded political alignment in a bygone struggle as the most significant distinction among all the world's nations and cultures.

THE FORMATION OF THE GLOBAL ECONOMY

International trade has been carried on for millennia, and trade always triggers regional specialization and increases in productivity, as well as new cultural possibili-

ties and combinations (see Chapter 6). As new means of transportation and communication have linked world regions at lower cost, more goods are produced for trade, so trade multiplies faster than total production. From 1950–2000 world GDP sextupled, but world exports rose by a multiple of 20, and in 2001 world trade reached a total of over $7.3 trillion. That figure was a 4 percent decline in merchandise trade since 2000, the first annual decline since 1982; the decline from 2000 to 2001 was due to the simultaneous decline in economy activity in almost all of the rich countries, the bursting of the investment bubble in information and communication technology, falling world stock markets, and the tragic events of September 11. By 2003, however, world trade had picked up (Figure 12-33).

International economic and communication links have "shrunk" Earth in terms of time or cost distance, and economic globalization has far outpaced cultural or political integration. Tensions among these economic, cultural, and political organizing activities generate much of today's international news. The globalization of the economy is already redistributing economic activities in every economic sector, and countries are continuously renegotiating conditions of production, trade, and exchange.

World Merchandise Trade by Region

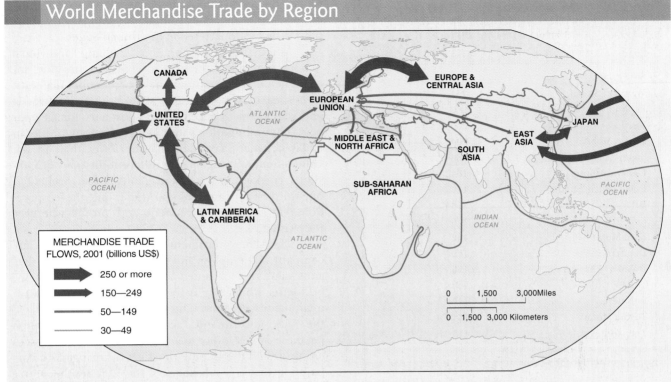

FIGURE 12-33 Most trade is among developed regions, but some developing regions, such as Latin America, are capturing an increasing share. Sub-Saharan Africa's share of world trade remains negligible. Of the $7.3 trillion of world trade in 2001, 61.5 percent was manufactured goods; 20.1 percent commercial services; 10.9 percent mining products; and 7.5 percent agricultural products. (Data from WTO)

Transnational Investment and Production

Two developments in economic globalization accelerated in the late nineteenth century. One was the evolution of a world market for certain primary products, foods, and minerals. Chapters 8 and 9 traced the results of that development. The second development was an increase in international investments. At first these investments were limited to shares of stocks and bonds representing minority holdings in foreign companies, called *portfolio investments*, but foreign direct investment soon followed. **Foreign direct investment (FDI)** means investment in enterprises that are actually operated by a foreign owner.

Modern transnational manufacturing was launched in 1865, when the German chemical magnate Alfred Bayer built a factory in Albany, New York. Today technology allows the integration of manufacturing processes worldwide. For example, in mid-2003 South Korea's Kia Motors Corporation manufactured a popular sport utility vehicle, the Kia Sorrento, named for a town in Italy. The Sorrento was assembled in South Korea, but it was full of components from around the world. The optical pickup units of the vehicle's CD player, for example, which read the CDs, were made in China. Their mechanical structure and electronic components were added in Thailand, and then the units were shipped to Reynosa, Mexico. From there the units were trucked to Matamoros, Mexico, where they were assembled into an audio system. The audio systems got trucked to California and shipped to South Korea, whence the Sorrentos were shipped to the United States.

> The globalization of the economy is already redistributing economic activities in every economic sector, and countries are continuously renegotiating conditions of production, trade, and exchange.

Enormous enterprises have grown up that own and coordinate production and marketing facilities in several countries. These are called **multinational** or **transnational corporations** (Table 12-3). Transnational companies (as defined by the United Nations) with international production facilities now number 60,000 and boast 500,000 foreign affiliates, accounting for an estimated 25 percent of total global production.

TABLE 12-3 Sales of the World's Largest Corporations Compared to a Few Countries' GDPs (in billions of dollars, 2002)

Switzerland	247
Wal-Mart Stores	**244**
Belgium	230
Sweden	210
Exxon Mobil	**204**
Saudi Arabia	186
General Motors	**184**
Royal Dutch/Shell	**179**
BP	**178**
Poland	176
Ford Motor	**162**
Denmark	162
Daimler/Chrysler	**156**
Turkey	148
Indonesia	145
Toyota Motor	**134**
Thailand	115
Egypt	98

Their evolution generally followed four stages. In the first stage, the corporation exported products from its home country to meet demand abroad. In the second stage, the corporation established production facilities abroad to supply specific markets abroad; exports from the home country dropped. In stage 3, the foreign production facilities began to supply foreign markets other than their own local markets. In stage 4, the foreign production facilities began to export back into the home country. These stages appeared first in trade in primary materials such as copper and oil. Later the stages were repeated in the evolution of international manufacturing, and today they are being repeated in the tertiary sector. For example, a U.S. accounting firm—a service industry—will first advise a Belgian company from its New York office. Then the accounting firm might establish an office in Brussels to serve Belgian companies. In stage 3 the firm will serve companies throughout Europe from its Brussels office, and finally, in stage 4, the Brussels office might advise firms back in the United States.

The ownership and range of activities of transnational corporations are so widespread that it is difficult to tell which corporations are corporate citizens of which countries. Nestlé, the world's largest food company, is generally regarded as a Swiss corporation, yet its shares are traded on the London Stock Exchange. The company's board of directors includes citizens of several countries, but no records reveal who Nestlé's actual stockholders are. One economist estimates that U.S. citizens own about 14 percent of Nestlé. Gilbert Williamson, president of NCR Corporation, has said, "I was asked the other day about United States competitiveness, and I replied that I don't think about it at all. We at NCR think of ourselves as a globally competitive company that happens to be headquartered in the United States."

Multinational corporations challenge individual nations to regulate them or to tax them. Imagine, for example, that a multinational corporation manufactures sneakers for $1 per pair at a subsidiary in Malaysia and sells them to a U.S.-based subsidiary for $40 per pair. The multinational has kept enormous profits in Malaysia, where tax rates may be lower than in the United States. Trade among different divisions of individual corporations today accounts for an estimated one-third of all international trade, and the corporations can allocate investment and profits wherever they wish.

The U.S. government takes the position that any company that employs and trains U.S. workers and that adds value in the United States is a U.S. corporation, no matter which flag it flies at its international headquarters or where those headquarters are. Accordingly, the U.S. government, acting on the behalf of a Japanese-owned office machine factory in Tennessee, has protested the trading practices of a Singapore corporation that is a subsidiary of the U.S. Smith-Corona Corporation, which is owned by a British corporation.

The International Tertiary Sector

When most people hear the words "foreign trade," they think of oil and iron, bananas and coffee, cars and clothes being shipped around the world. It is easy to imagine such "things." The international economy, however, has transcended trade in primary products and manufactured goods. Trade in the tertiary sector includes transport, travel, communications, financial services, computer and information services, royalties and license fees, and professional services. Global exports of commercial services totaled $1.5 trillion in 2001, 20 percent of all world trade. Services are also taking a growing share of international investment.

In many cases corporations that provide services to manufacturing corporations in their homelands have followed these out into the world arena. This partly explains the internationalization of legal counsel, business consulting, accounting, and advertising (Figure 12-34). Many professionals, such as architects and physicians, market their skills around the world. Financial services, entertainment, and still other services are produced for global markets. For many nations, welcoming tourists and foreign students have become major elements of their economies. These activities count as exports of services.

The United States is the world's leading exporter of services (17 percent of the total of global service exports in 2002), followed by Britain, Germany, and France. The substantial surplus that the United States earns in this trade offsets part of the country's annual deficit in merchandise trade. In education, for example

The Largest International Law Firm

FIGURE 12-34 The Chicago-based law firm of Baker & McKenzie can provide legal advice on every area of business law in 61 offices in 35 countries. (Courtesy of Baker & McKenzie)

(which students come from abroad to receive, but which is counted as an export), the United States has enjoyed a net profit of almost $5 billion per year. New restrictions on visas to the United States instituted since September 11, 2001, however, have slightly deflected the global stream of students to Canada and to Europe, so U.S. education exports may decrease. Other U.S. tertiary exports include real estate development and management, accounting, medical care, business consulting, computer software development, legal services, advertising, security and commodity brokerage, and architectural design (Figure 12-35). U.S. entertainment productions dominate the world's airwaves and movie screens. Movies, music, television programs, and home videos together account for a significant annual trade surplus.

International trade in services demonstrates the same trends as trade in goods. Countries are losing their autonomy to an increasingly global market. Transnational corporations treat the world as "one place" and exploit resources, locate facilities, and market their products accordingly. For example, computer programmers at Tsinghua University in Beijing can write software for IBM using Java technology. At the end of each day, they send their work over the Internet to an IBM facility in Seattle. There, programmers continue to work on it and then send it 8,400 kilometers (5,220 miles) to the Institute of Computer Science in Minsk, Belarus, and to Latvia. From there, the work is

FIGURE 12-35 Design is a U.S. export. U.S. architect Frank Gehry (born in Canada) designed this spectacular new Guggenheim Museum that opened in Spain in 1997. Architectural design is a significant U.S. export. This building carries a political message, for it was entirely planned, built, and paid for by Basques, a minority group within Spain, in Bilbao, their largest city, to assert their presence in the world (see Chapter 11). On the façade of the building, huge letters proclaim "Museoa," which is not Spanish, but the Basque language. The shimmering metal used to coat the petal-like steel panels is titanium. (Eric Vandeville/Getty Images, Inc.—Liaison)

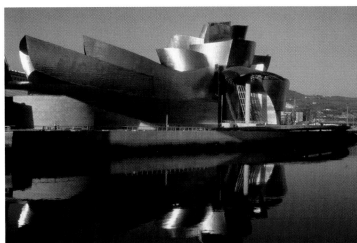

sent east to the Tata Group, a computer firm in India, which passes the software back to Tsinghua by morning in Beijing. This global relay never stops until the work is done.

The global redistribution of white collar work   As noted in the anecdote opening this chapter, many service sector jobs are following the manufacturing jobs out of the rich countries to relocate in countries that are relatively poor. Many developing countries have educated young people willing to work for wages lower than those in the rich countries, but their domestic economies cannot employ them. Multinational corporations need clerical assistants and computer operators, and telecommunications allows the corporations to tap these pools of labor wherever they are. The relocation of these jobs boosts the economies of the developing countries and also spreads cultural values and practices.

U.S. companies have been moving clerical operations out of urban centers for decades. Now they can relocate those jobs around the world. American Airlines set up its first subsidiary to process data from ticket stubs on the Caribbean island of Barbados in 1983. Barbados has since captured data processing for many types of corporations, such as health claims for employers and insurance companies. Today more people work as data processors than on the island's historic sugar plantations. The Boeing and Bechtel multinational corporations have located software-development facilities that provide many jobs in Ireland. In 2003 General Electric Corporation had over 20,000 Indian employees in financial reporting, information technology support, and research and development for medical, lighting, and aircraft production. India's own computer-programing industry is one of the world's largest. AOL's technical staff in the Philippines answers 90 percent of the billing and technical questions that are e-mailed to AOL each month; e-mail customer service responses for many other U.S. companies are in India. Procter & Gamble reports that it has saved over $1 billion between 1999 and 2003 by locating office work in Costa Rica and the Philippines. The relocation has also resulted in better work, because the jobs attract the best local graduates, and worker turnover (and thus training costs) is low. Local accountants in India and the Philippines working for the firm Ernst & Young calculate taxes for the firm's U.S. and global clients. In 2003 several U.S. states's welfare benefits and food stamps "Help lines" were located in Mumbai, India, connected to toll-free help lines (Figure 12-36). Despite the great geographic distances, the time and cost distance to these places is, for electronic transmissions, negligible. Overseas cable costs fell 80 percent just between 1999 and 2003.

In the short run, the transfer of jobs abroad puts deflationary pressure on wages for people doing these jobs in the rich countries; in the long run, these jobs may disappear from the rich countries. For the populations of the rich countries, the crucial question is: Will

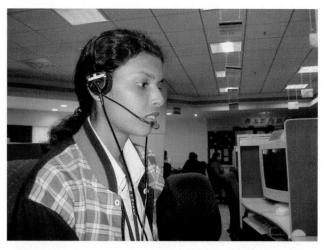

FIGURE 12-36 An Indian Telephone "Help" Line for U.S. Welfare Recipients. Manisha Martin in Mumbai, India, is answering questions telephoned in by welfare recipients in New Jersey. Manisha's monthly $200 salary is five times the per capita income in India, but less than that of the welfare recipients whose questions she answers. (Amy Waldman/*The New York Times*)

the white-collar workers in the rich countries be able to add ever higher value to services, and thus advance into jobs paying ever-higher incomes than they now enjoy, or will they fall into lower-paying service-sector jobs, as most former manufacturing workers have? In other words, how many current or would-be data processors or computer programmers in the rich countries will become movie stars, lawyers, top athletes, or surgeons, and how many, on the other hand, will be flipping fast-food hamburgers or greeting customers at discount stores at miniumum wages? We do not know the answers to these questions. As more and more jobs become footloose and the world becomes one labor market, each worker must compete with every other worker with his or her skills in the world. The only workers protected against the globalization of the labor market are those whose work requires personal contact, and much personal contact is already being replaced in many contexts by videoconferencing and other innovations in communications.

The Geography of Foreign Direct Investment

The amount of FDI is today several times the amount of foreign aid being given around the world, and FDI has been much more successful in triggering economic growth. Developing countries used to seek aid from foreign governments and international lending institutions, but today they are more likely to court foreign companies to invest in their countries. Since the 1980s, global FDI has grown three times faster than world trade and four times faster than total world output, to a total of almost $5 trillion. This flow of FDI greatly increased and significantly redistributed the world's productive capacity. No world agency tracks portfolio investment, but including portfolio investment would multiply the total amount of capital invested internationally.

Companies from an increasing number of countries have become active investors in other countries. Transnational corporations from developing countries, including Brazil, Thailand, South Korea, and Mexico, reached around the world, but three sources—the United States, the combined European Union countries (see Chapter 13), and Japan—still account for about 80 percent of FDI. The United Nations refers to these three sources as the Triad.

U.S. firms account for about 30 percent of global FDI, and the United States is also the world's largest recipient of FDI from other countries. The International Monetary Fund estimated that as of March 2000, foreign investors owned $1.4 trillion worth of U.S. stocks, $900 billion worth of corporate bonds, $1.3 trillion of publicly held federal debt, and $1.2 trillion worth of direct investment. These foreign investments and purchases testified to foreign confidence in the American economy and political stability, and they supported U.S. prosperity. Foreign firms carry on research, introduce new technology and management techniques, and change Americans' daily lives in the workplace as well as Americans' consumption habits.

FDI in the developing countries

Three trends characterize the geography of FDI. First, the greatest share of global FDI—about 75 percent—is investment from one rich country to another. For example, U.S. firms have over $500 billion invested in Europe.

Second, FDI in the developing countries has been geographically selective. The countries that have attracted most investment are countries that have chosen the export-led method of economic growth. Their economies have boomed. Four Asian nations—Singapore, South Korea, Taiwan, and Hong Kong (technically a colony of Great Britain until 1997, but before then considered a country in international trade statistics)—attracted FDI and grew so fast during the 1980s that they earned the nickname The Four Tigers. These areas reached European levels of prosperity within a decade, and they have already become sources of capital and technology for the next tier of Asian developing lands: Thailand, China, Malaysia, and Indonesia. In 2002 China first surpassed the United States as the world's number-one recipient of new FDI.

The areas that have not attracted FDI have been slower to develop. India, for example, was committed to import-substitution growth from independence in 1947 until the early 1990s, and its rate of growth was considerably less than that in countries receptive to foreign investment. India began to invite FDI only in the mid-1990s, but even as late as 2001 its share of total world trade was under 1 percent, the same as it was in the 1970s. Africa has attracted only about 2 percent of total FDI, and almost all of that investment has been in extractive activities—oil and mining. Political

instability was a major factor discouraging international investors there.

A third characteristic of the pattern of investment is that each Triad member has a majority share of the FDI in a cluster of countries that have become its economic satellites. Japanese investment dominates in East and Southeast Asia, U.S. investment in Latin America, and Western European investment in Eastern Europe and Central Asia. These clusters help explain world patterns of trade. For example, in recent years the United States has had a trade deficit in electronic consumer goods with Thailand. Thai electronics factories are mostly owned by Japanese corporations. Thus, the consumer electronics that the United States imported from Thailand profited the Japanese corporations as well as the Thai economy. In a similar fashion, many of Europe's imports from Latin America are the products of U.S. corporations.

Geographers watch capital flows carefully, because the pace of development within individual countries is increasingly determined by their ability to attract foreign capital. The world map of wealth continues to change kaleidoscopically.

Some observers have hypothesized broad political ramifications of the rise in global FDI. To attract private investment, governments must reveal national economic and social conditions they previously may have kept secret. Furthermore, no country can attract investment unless it guarantees consistency in the rule of law. Arbitrary government actions—often typical of totalitarian regimes—will isolate a country from international capital flows. For example, several major international corporations have decided against investing in Malaysia out of concern that Malysia might try to regulate Internet access. Therefore, the substitution of private investment for foreign aid may have a substantial liberalizing and democratizing influence worldwide.

At the end of 1997, for example, the financial setbacks in Thailand, Indonesia, South Korea, the Philippines, and Malaysia mentioned above triggered angry tirades from political leaders in those countries against international bankers. Authoritarian leaders instituted new economic controls and restricted freedom of the press. Each such outburst, however, undermined international investors' confidence in their leadership and spurred investors to get their money out; local currencies fell, and new investment halted. Political leaders soon realized that the authority they exercise over their own people cannot be exercised over international investors. International capital cannot be attracted without acceding to international rules. Tempers cooled, and compromises were reached. This process is often called the *discipline of the market*. This financial crisis slowed but did not stop the economic development of the affected countries.

In 2002 the spread of SARS and global reaction to China's slow reaction to the disease brought home to China the meaning of globalization (see Chapter 5). The disease was less of an embarrassment than was the government's inept and secretive handling of it. The world shunned China, and Chinese government leaders openly referred to their own surprise at this event as "our 9/11." The actions the Chinese government eventually took set a new standard for public disclosure and were, perhaps, a step toward a more open Chinese society.

Political scientists have traditionally contrasted revolutions *from below,* when repressed lower classes rose up in revolt—such as the French Revolution—from revolutions *from above,* such as the reforms that Russia's Czar Peter the Great imposed on a reluctant population. FDI has prompted political scientists to recognize a new form of revolutionary pressure—revolution *from beyond* the national borders.

The Globalization of Finance

The FDI that is redrawing the world economic map is only a fraction of the vast amounts of capital moving around the world. Multinational banking and telecommunications make it possible to monitor and trade in national currencies, stocks, and bonds listed anywhere in the world instantaneously. Today about $2 trillion is traded daily on the world's foreign exchange markets—over 100 times the average daily merchandise trade. An additional $10 trillion worth of stock changes hands on the world's major stock exchanges. This activity is monitored by the International Organization of Securities Commissions (IOSC), but as noted above, no international agency records ownership of global assets, even by national source.

Nations are expected to regulate their own stock exchanges, and the home countries of global banks are responsible for supervising the banks' operations, but countries do not agree on how to do these things. Many giant banks have located their international headquarters in places renowned for lenient banking regulations. Tiny Monaco (population 30,000) counts over 340,000 bank accounts holding almost $50 billion. The 260-square-kilometer (100-square-mile) British Caribbean colony of the Cayman Islands (population 36,000) is one of the world's greatest financial centers, with assets of $700 billion held in over 600 chartered banks. By comparison the state of New York has 154 chartered banks. The Cayman Islands, however, has almost no vaults, tellers, security guards, or even bank buildings. Assets are held electronically in computers. Many banks consist of one-room offices, and local laws protect transactions' confidentiality. For example, international agencies cannot trace the profits of the international drug trade. Shortly after the attacks on the United States on September 11, 2001, the United States and its allies passed laws designed to curtail money laundering as a means of financing international terrorist activities, but tracing the global migration of money remains a difficult challenge.

Today Japan is one of the world's principal sources of capital, so the regulations governing the Tokyo stock market affect investments and jobs throughout the world. The Arab oil states are another source of capital, so a costly Mideast war raises the interest rate on U.S. college tuition loans. Bank policies in Frankfurt, Germany, and Johannesburg, South Africa, affect the checking-account charges of U.S. banks. Financial manipulations in Rio de Janeiro, Brazil; Bangkok, Thailand; and Mumbai, India, affect the security of many U.S. elderly people's pensions, which have been invested abroad. And the interest rates set at the Federal Reserve Bank in Minneapolis affect business taxes in Indonesia. The explanations of the "where" and "why there" of local economic affairs may lie in global finance.

Tourism

Tourism is by some accounts the world's largest industry. Every year a greater proportion of the world population takes part in tourism activity, and in most countries tourism is one of the fastest-growing sectors of the economy. Despite the fears of terrorism and SARS in 2002, international tourist arrivals numbered over 700 million in that year, and international tourist receipts approached $500 billion, excluding air fares. Tourism is the top earner of foreign exchange for many countries, and it is expected to grow everywhere by at least 5 percent per year, with even higher rates in the poor countries. The aftershock of the attacks on the United States in September 2001 and other terrorist attacks around the world were harmful, but the World Tourism Organization believes that world tourism will continue to expand.

Citizens of the rich countries account for the largest tourist expenditures, and the rich countries are also the most popular destinations. Measured by tourist arrivals, the leading destinations in 2002 were France, Spain, the United States, Italy, and China. When tourism income is measured as a share of total national income, however, the importance of tourist income to the poor countries becomes clear. The Czech Republic earned over $300 per capita from international tourism in 2002. Tourism contributed over 10 percent of the total GDP of South Africa.

The tourist potential of any spot depends upon its "three A's":

- *Accessibility,* usually via convenient air routes
- *Accommodations,* at all levels from simple pensiones to grand resort hotels
- *Attractions,* which can include a pleasant climate, as well as beaches, museums, architecture, historic sites, festivals and performances, shopping opportunities, and more (Figure 12-37).

Areas spectacularly endowed with wildlife may feature **ecotourism,** which is travel to see distinctive examples of scenery, unusual natural environments, or wildlife (Figure 12-38). The rich bird life on the islands of Lake

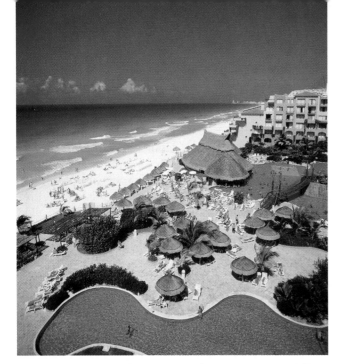

FIGURE 12-37 Cancún, Mexico. This resort was carefully planned. The government of Mexico mapped a great variety of locational determinants, including annual days of sunshine, temperature ranges, attractiveness of local beaches, temperature of the water, water currents, variety of sea life, scenery, and accessibility of archaeological ruins. When these factors all came together at one place, the government invested millions of dollars to develop Cancún as a new vacation destination. (Jose Fuste/RAGA/Corbis/Stock Market)

Nicaragua, the lions of South Africa, and the tropical vegetation in many countries are important tourist destinations. Ecotourism today provides the money needed to save many countries' natural environments as well as many species of rare animals. Overall, the countries best endowed for tourism have both natural and cultural attractions, pleasant climates, good beaches, and reasonably well-educated populations.

FIGURE 12-38 Ecotourism. "Shoot with a camera, not with a gun," insist these intrepid sightseers in Africa. (Martin Harvey/Peter Arnold, Inc.)

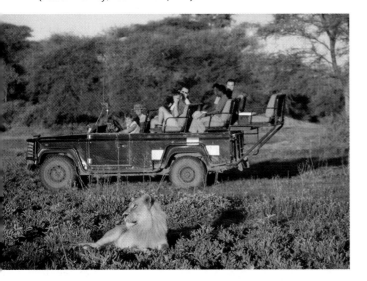

Political stability is a necessity. Despite Africa's wealth of ecological and cultural attractions, political instability and lack of accommodations have restricted its income to only 2.2 percent of global tourist dollars, and those were divided almost exclusively between the countries of Mediterranean North Africa and the country of South Africa.

Tourism creates many jobs and offers wide multiplier effects. For example, a new hotel creates demand for a host of related goods and services. Tourism may also stimulate the local agricultural economy, and it is generally a high-value-added and labor-intensive industry that trains local labor. Tourism may, however, corrupt local culture or provoke inflation, and too many tourists may overwhelm the natural environment.

International Regulation of the Global Economy

The rules regulating business vary so widely from nation to nation that international trade requires cooperation or even political regulation. Under United Nations auspices, a General Agreement on Tariffs and Trade (GATT) was signed in 1947. Under GATT procedures, trade disputes led to the convening of panels of neutral specialists.

A new World Trade Organization (WTO) came into being on January 1, 1995. WTO also establishes panels to investigate trade disputes, but WTO panels have more authority than GATT panels did. Countries found to have broken WTO rules must change their own policies or laws or else offer compensation; if they do not, they can face trade sanctions.

By the beginning of 2003, 144 countries belonged to the WTO. China had just joined (and Taiwan as a "separate customs territory"); but Russia and Iran still stood outside, accused of avoiding the WTO's free-market principles. Membership in WTO brings nations a significant advantage. Each member receives from every other member the terms of trade that that nation grants its most favored partners—called *most favored nation* status.

Critics fear that the WTO will encroach on national sovereignty by defining manufacturing standards, environmental and labor regulations, and other domestic conditions in its efforts to "harmonize" conditions of trade. In the WTO negotiations scheduled to continue until 2005 (called the "Doha Round," after the capital of Qatar, where the talks began in 2001), the rich countries want to guarantee open markets for trade in services and to guarantee intellectual property rights. Many of the poor countries, however, insist that the rich countries open their markets to their potential exports of agricultural products, textiles, and garments.

Other international organizations oversee specific aspects of international commerce, such as the IOSC mentioned earlier. The London-based International Accounting Standards Board works to standardize corporate accounting procedures, and the International Labor Organization (ILO) works to improve international labor conditions.

The Geneva-based International Standards Organization (ISO) carries tremendous responsibility. It determines and publishes standards for manufacture of items in world trade. Virtually all items in international trade must conform to ISO standards, so items manufactured even for nations' domestic markets do, too. For example, the specifications of your tennis racket are covered by eight pages in ISO 11415: 1995. ISO 14000 establishes environmental standards for manufacturing processes and products. It is hoped that this series will reduce global pollution and environmental degradation.

CONCLUSION: CRITICAL ISSUES FOR THE FUTURE

The meeting of the World Trade Organization in Seattle, Washington, in November 1999 was expected to be a festival of celebration for the principles of global trade. Instead, however, the meetings were disrupted by violent demonstrators. Demonstrators have since regularly targeted meetings of international bodies that regulate the global economy. The demonstrators have included a mixture of groups, holding a mixture of opinions, some of them even contradictory. Some demonstrators oppose what they see as the destruction of the world's many cultures by homogenization. Others represent trade unions in the rich countries, objecting to the export of jobs to poor countries. Others object to the "exploitation" of workers in the poor countries. Still others fought to save the global environment from over-exploitation for economic development or from pollution by economic development.

Some observers have asked who elected these street demonstrators to extort agreements from corporations or to change policies of democratically elected governments. In 2000 Mexico's president, Ernesto Zedillo, condemned the "globophobes" for opposing free trade. "Free trade," he said, "gives developing nations a leg up, because with lower wages and less generous social policies, it costs them less to produce products for export. Over the long run, that raises incomes and living standards." In other words, only development allows the poor countries to raise their wages, social policies, and environmental standards. To require the poor countries to observe the same labor and environmental standards that exist in the rich countries would short-circuit this process. Some critics fear that comments such as these constitute a global "race to the bottom," in other words, competition to lower wages, environmental standards, taxes, or any other factor that might affect international locations. Furthermore, as this text has noted, increasing trade or interchange does not guarantee that all countries involved will benefit equally. Some countries whose economies are most completely integrated into the world economy are some of the poorest.

Protestors, or in some cases consumer groups, have succeeded in getting labor or environmental protection standards written into some international contracts. In 1999 Cambodia signed a pact with the United States according to which the United States would import more Cambodian textiles if Cambodia would allow U.S. inspectors to certify that Cambodian factories were meeting labor standards. This pact, the world's first such agreement, set a new standard for investment in poor countries.

For the future, uncertainty threatens the global economy. Political instability frightens capital investment. If terrorists succeed in causing more deaths, damage, or fright, and if civil or international wars are protracted, then corporations will worry about the safety of their staff and facilities, and the free flow of people and goods will slow; this will hurt economic efficiency. The cost of capital will rise because investors will demand higher returns for their risk. Higher capital costs lower the rate of new investment.

Globalization could suffer from insecurity. At the end of the twentieth century, open world markets led to a boom in trade. Globalization helped disseminate capital, technology, and entrepreneurial ideas, as well as products. It lowered prices for consumer goods, enabled manufacturers to boost productivity with just-in-time supply chains, and opened pools of technical talent. The world's markets may be too integrated to break them apart now, but continuing globalization could become slower and costlier. Higher risk means higher cost. Companies will probably have to pay more to insure and provide security for staff and property. Heightened border inspections could slow movements of cargo, forcing companies to stock more inventory. Tighter immigration policies could curtail the liberal flows of labor. Also, companies obsessed with political risk would narrow their horizons for making new investments. The only nations that could attract new FDI would be the most stable and growing. Any place that suffers political instability would be out of consideration for FDI by multinational corporations.

Many Americans remain opposed to globalization. They insist that the decline in many American workers' incomes, the concentration of wealth, the nation's surging current accounts deficit, and the increases in H-1B visas for skilled workers discussed in Chapter 5 are signs that globalization is not really good even for the American people.

One key battle in the struggle over expanding global trade was fought in the U.S. Congress in 2000. The issue was granting permanent trading privileges to China, thus opening the door to China's entry into the WTO. The arguments were as follows: Those against expanding trade with China argued that (1) despite the highly publicized "potential" value of the Chinese market for American goods, China in fact bought less than 2 percent of U.S. exports; (2) trading with China encouraged U.S. firms to relocate factories from the United States to lower-wage China; (3) trade with China helped build China's military power, because most sales to China have been high-tech goods sold to Chinese government agencies or monopolies; (4) China had insufficient environmental protection legislation.

Those in favor of encouraging trade with China argued that (1) China's 1.3 billion people offered a vast potential market for U.S. goods; (2) U.S. jobs in export industries would increase; (3) cheap imports from China lower the cost of living for U.S. consumers; and (4) normalization of relations and trade will ultimately bring about political liberalization in China. In the end, Congress voted to normalize trade relations with China.

Critics of globalization argue that the gap between rich and poor is widening, but this is not clear. Gaps in income between richest and poorest countries may be widening, but rising incomes in China offset declining incomes in many countries with low populations. China and India have done well in recent decades. Income gaps do seem to be widening within countries, but in most cases the rich got richer faster than the poor did, but the poor did not get poorer; they just got rich slower.

Certainly advances in global communication are increasing awareness of gaps in living standards—awareness of gaps within countries and awareness of gaps among countries. This increasing awareness intensifies pressures for migration (discussed in Chapter 5) and may be responsible for political restlessness in many countries.

Poverty among many people is a pressing moral, political, and economic issue, but whether turning back globalization is the answer to this problem is less definite. International economic integration is not inevitable. Governments and large corporations have it in their power to slow, or even reverse, the economic trends of the last 20 years, but economists doubt whether this would be a victory for the poor or for the citizens of the rich countries.

Chapter Review

SUMMARY

Every country devises a program to build and regulate its national economy and to distribute economic well-being across the national territory. Gross national income, gross domestic product, and gross sustainable product are three ways of measuring wealth, although each undervalues subsistence production and unpaid labor. Maps of these values reveal great differences in global incomes. The Human Development Index reveals standards of living. Welfare may vary greatly within countries, and statistics examined per capita and through time may reveal a relationship between population growth and economic growth.

Sectoral evolution reveals that some countries are preindustrial, some industrial, and some postindustrial. Various sectors' contribution to national product is not the same as their share of employment.

The world distribution of wealth is not the same as the distribution of raw material resources. Downstream secondary- and tertiary-sector activities add value and produce wealth. Today's map of global manufacturing is a result of historical developments modified by new balances among relevant locational determinants, including capital and technology, as well as labor and raw materials. Several locational determinants are themselves footloose, so manufacturing will probably be redistributed continually.

Each nation's political economy usually reflects other aspects of local culture. Nations try to knit and develop the national territory and to maintain an equal standard of living throughout it. Transportation networks allow regional specialization and trade, and their improvement releases national resources for production. Communication infrastructures both reflect and boost economic development as well as, potentially, political liberty. The term Third World combines economic measures with Cold War political values.

Import-substitution policies protect national growth behind tariff walls, whereas export-led policies encourage participation in international trade as a source of capital and as markets. Both methods entail both economic risks and political risks.

International trade has multiplied faster than total world production. Capital has been invested internationally, and world markets for primary products, foods, and minerals evolved. Today manufacturing industries are achieving globalization, and flows of FDI are redistributing world productive capacity. China is today the largest recipient of FDI. A global capital market has formed, and international trade in tertiary-sector services is growing. Tourism, including ecotourism, is today arguably the world's largest industry. The World Trade Organization, created in 1995, regulates international commerce.

KEY TERMS

capitalist system p. 517
economic development p. 506
economies of scale p. 528
ecotourism p. 538
export-led economic growth p. 528
First World p. 531
foreign direct investment (FDI) p. 533
frontier p. 520
gross domestic product (GDP) p. 494
gross national income (GNI) p. 494

human development index (HDI) p. 499
import-substitution economic growth p. 528
industrial society p. 503
infant industries p. 528
laissez-faire capitalism p. 517
market-oriented manufacturing p. 509
material-oriented manufacturing p. 509
multinational corporation p. 533

nationalized p. 517
opportunity cost p. 525
political economy p. 517
postindustrial society p. 504
preindustrial society p. 502
Second World p. 531
sectoral evolution p. 502
tap route p. 526
Third World p. 531
transnational corporation p. 533

QUESTIONS FOR REVIEW AND DISCUSSION

1. Define and explain the three sectors of an economy. What is sectoral evolution?

2. What locational determinants for manufacturing did Alfred Weber identify? What new determinants are important today?

3. What is the origin and meaning of the term the Third World?

4. Differentiate laissez-faire capitalism from state-directed capitalism.

5. Name some states with undeveloped frontiers.

6. Why is transportation important to economic development?

7. What is the concept of gross sustainable product? Why do poor nations dispute the idea?

8. Aside from GNI or GDP, what are some alternative noneconomic measures of welfare in different places?

THINKING GEOGRAPHICALLY

1. Americans have difficulty understanding privatization because Americans live in a country in which most activities are and always have been private. Can you, however, imagine selling the national parks to the Disney Corporation? What would be the advantages and disadvantages? Would the corporation run the parks better?

2. What physical and cultural attributes are most conducive to a nation's economic development?

3. Compare Figures 11-34 (literacy) and 11-35 (freedom) with Figure 12-6 (GNI per capita). Can we hypothesize any cause-and-effect relationships?

4. Latin America is rich in natural resources, and yet it has long suffered economic and political problems. Many of these problems stem from class and ethnic structures and from political history. Why?

5. At the opening of international currency markets on August 18, 2003, the following values were quoted as the value of each currency per U.S. dollar:

Australian dollar	0.6583
Brazilian real	0.333
Euro	1.115
Indonesian rupiah	0.0001175
Japanese yen	0.00837
South African rand	.1378

Check these values in today's newspaper. Calculate the percentage each has risen or fallen against the dollar or against one another. What flows of goods, of investment, or of tourists should these fluctuations have triggered?

6. Did the United States develop in a sustainable way, or at a sustainable rate, between 1870 and 1910? Does it today?

7. Could foreign ownership of businesses hurt the economy of a developing country?

8. What countries have frontiers? What barriers exist retarding development of those areas? How might these barriers be overcome? Investigate and compare the real development potential of the frontiers of Canada, Russia, and a South American or African country.

9. How does the quotation from Peter Drucker at the opening of this chapter (p. 494) echo that of Clifford Geertz in Chapter 6 (p. 225)?

10. Find the latest edition of the *Human Development Report* in your library and read the authors' justification for what they have done. If you disagree with their methods, find any one of the other statistical yearbooks published by the United Nations or the World Bank, browse among the quantified descriptions of life in each country, and devise your own overall "ranking of the quality of life" among the countries.

SUGGESTIONS FOR FURTHER LEARNING

Cannon, Terry, ed. *China's Economic Growth: The Impact on Regions, Migration and the Environment.* New York: St. Martin's Press, 2000.

Carley, Michael, and Ian Christie. *Managing Sustainable Development,* 2nd ed. London, U.K.: Earthscan Publications, Ltd., 2000.

Cruickshank, J. "The Rise and Fall of the Third World: A Concept Whose Time Has Passed." *World Review,* February 1991.

Davidson, Eric and George Woodwell. *You Can't Eat GNI: Economics as if Ecology Mattered.* New York: Perseus Press, 2000.

Ginsburg, Norton. "Natural Resources and Economic Development." *Annals of the Association of American Geographers* 47, no. 3 (1957): 197–212.

Grove, Richard. *Green Imperialism: Colonial Expansion, Tropical Island Edens, and the Origins of Environmentalism 1600–1860.* New York: Cambridge University Press, 1996.

Hall, C. M., and S. J. Page. *The Geography of Tourism and Recreation: Environment, Place, and Space.* New York: Routledge, 1999.

Harrison, Lawrence. *Who Prospers? How Cultural Values Shape Economic and Political Success.* New York: Basic Books, 1992.

Holliday, Charles O. *Walking the Talk: The Business Case for Sustainable Development.* New York: Berrett-Koehler, 2002.

The International Bank for Reconstruction and Development. *World Development Report.* New York: Oxford University Press, annually for the World Bank.

Landes, David S. *The Wealth and Poverty of Nations: Why Some Are So Rich and Some So Poor.* New York: W. W. Norton & Company, 1998.

Mabogunje, Akin. *The Development Process: A Spatial Perspective,* 2nd ed. London: Unwin Hyman, 1989.

O'Meara, Patrick, Howard Mehlinger, and Matthew Krain, eds. *Globalization and the Challenges of the New Century: A Reader.* Bloomington, IN: Indiana University Press, 2000.

Reich, Robert. *The Work of Nations.* New York: Alfred A. Knopf, 1991. (The 1992 Vintage Paperback has an addendum.)

United Nations Conference on Trade and Development. *Handbook of International Trade and Development Statistics.* New York: United Nations, annually.

The World Bank. *The World Bank Atlas.* Washington, D.C.: The World Bank, annually.

Yergin, Daniel, and Joseph Stanislow. *The Commanding Heights: The Battle Between Government and The Marketplace That Is Remaking the Modern World.* New York: Simon & Schuster, 1998.

WEB WORK

In Chapter 8 we cited two general libraries of information about economics that are useful for exploring concepts developed in this chapter. You might also want to begin a search for information about economic development at the Yahoo site at

http://www.yahoo.com/Social_Science/Economics/

The United Nations hosts a Center for Sustainable Development at

http://www.un.org/esa/sustdev/

The International Political Economy Network offers links to studies of economic development and economic policies from a political point of view at

http://csf.colorado.edu/ipe/

The World Trade Organization has a home page that can be the source of much useful information on the world economy and world trade:

http://www.wto.org/index.htm

The World Bank Group home page offers innumerable studies, plus links to still others:

http://www.worldbank.org/

The United Nations Conference on Trade and Development offers numerous studies and updates on economic matters:

http://www.unctad.org/

The World Tourism Organization home page may be found at

http://www.world-tourism.org/

and that of the International Labor Organization at

http://www.ilo.org/

A Look Ahead

THE COLLAPSE OF EMPIRES

The collapse of empires through the twentieth century gave birth to many new states. The U.S.S.R. fragmented into 15 countries in 1991.

NEW UNIONS OF STATES

New organizations of states are forming to achieve common ends. The European Union and the North American Free Trade Area are particularly significant, but numerous other international organizations coordinate policies among countries. Cooperation among states can result in redistributions of populations and activities within and among them.

POLITICAL REGIONALIZATION AND GLOBALIZATION

◄ The number of countries represented here in the United Nations General Assembly has risen from 51 in 1945 to 191 in 2003. The United Nations does not have the powers of a real legislature, but the organization coordinates a tremendous variety of international accords; and its peacekeeping body, the Security Council, has often played a key role in defusing world tensions and combating terrorism. All nations face together important political, economic, social, and ecological challenges. (Andrea Brizzi/Corbis/Stock Market)

GLOBAL GOVERNMENT

The United Nations was formed at the end of World War II as an international peacekeeping organization. Its functions have broadened to include economic and social development. Political management of unoccupied portions of Earth's surface, such as Antarctica and the oceans, has been established through international treaties.

PROTECTING THE GLOBAL ENVIRONMENT

Global economic development and population growth have raised concern about global-scale environmental problems. Rapid industrialization in poorer countries is increasing pollution. International negotiations about environmental issues focus on balancing the need for economic development with concerns for environmental protection.

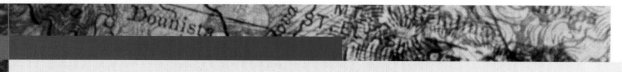

Armaments are one of the principal items in world trade, making up a significant share of the exports of many technologically advanced countries, including the United States, and a significant share of imports of many countries, including many poor countries. North Korea and a few other nations have been accused of supplying arms not only to other states, but to terrorist and guerrilla organizations around the world. Today, when concern is rising about the possible spread of nuclear, chemical, and biological weapons of mass destruction (WMD), the regulation of the arms trade is capturing new attention.

In 2003, under the terms of a new Proliferation Security Initiative, 11 countries—Australia, France, Germany, Italy, Japan, the Netherlands, Poland, Portugal, Spain, the United Kingdom, and the United States—have dedicated effort to preemptive interdiction, seeking to allow ships, aircraft, and vehicles suspected of carrying WMD-related matériel to be detained and searched as soon as they enter member countries' territory, territorial waters, or airspace. The coalition members will also deny overflight rights to suspicious aircraft or ground them when they stop to refuel. The coalition will also try to interdict illegal drugs that finance the arms trade. The effort to interdict arms and drug shipments will severely try international law regarding the law of the sea, national airspace, and territorial waters.

Individual countries remain the highest level of absolute territorial sovereignty, but frameworks do exist to regulate and coordinate activities among countries. Chapter 12 introduced the World Trade Organization, the International Organization of Securities Commissions, and the International Standards Organization. International monitoring of control over WMDs may necessitate greater cooperation or enforcement of agreements. The International Court of Justice, a branch of the United Nations, sits in The Hague, in the Netherlands. It hears cases according to international law, and it issues rulings, but the litigant states must choose to come before it, and the Court has no power to enforce its decisions.

Another topic of increasing concern is potential international clashes over water rights, because the continuing availability of fresh water is a long-range problem in many regions. Many of the world's major river systems are shared by two or more countries, so resolving disputes over international waters may be an important function of international courts in the future.

Organizations that coordinate activities among two or more countries are called **international organizations,** whereas organizations that actually exercise power over countries are called **supranational organizations.** Countries surrender some powers to supranational organizations, but each country can escape that control—by making war, if necessary. Therefore, the allocation of powers in either international or supranational organizations is actually flexible and open to continual renegotiation. A *commonwealth* is another kind of international organization that exercises limited powers over its members, and a **bloc** is a group of countries that presents a common policy when dealing with other countries. Most blocs are regional.

International and supranational organizations are not new in history. Independent governments have always made alliances and coordinated policies. Today some international organizations coordinate military defense, but the most active organizations work for economic or broader cultural goals. They are trying to achieve across several countries many of the same things that sovereign states try to achieve within their individual borders. These goals include creating unified markets for materials, goods, and services; knitting area-wide transport networks; ensuring free travel throughout the territory; and even guaranteeing common standards of political or civil rights. These goals enlarge the territorial scale of organization and thus redistribute people and activities.

This chapter first reviews how the collapse of great empires through the twentieth century gave birth to many new states. More than half of today's countries containing about half of the Earth's population and covering about half of Earth's surface are less than 50 years old (Figure 13-1). The political fragmentation of space triggered economic and even cultural breakup, redistributing and remapping activities.

Even as the number of states has increased, however, the states' genuine independence of action has been undermined by international economic links and other transactions and travels. Therefore, the second part of this chapter examines instances in which nations are joining together to achieve common ends. The European Union and the North American Free Trade Area are particularly significant because the participants have great economic and military power and because in each case the members have surrendered considerable autonomy to a central body or at least to a

The World in 1900

FIGURE 13-1 In 1900 there were far fewer independent actors in international affairs than there are today. Compare this map with today's political map, Figure 1-6, p. 12.

common policy. Numerous other international organizations coordinate policies among countries, although their goals may be limited.

The third section of this chapter takes up the instruments and institutions through which international law may be evolving toward global government. At present no effective global government exists, but at least systems of global agreement are needed to regulate conduct among nations. International agreements also regulate activities in the areas of Earth that all nations hold in common, such as the seas. The fourth section of the chapter notes the ways in which humankind is beginning to agree on rules to protect the global environment.

THE COLLAPSE OF EMPIRES

The twentieth century saw the fragmentation of empires that Europeans built up over hundreds of years. The few remnants of empires can be seen on Figure 13-2. Since the breakup of the U.S.S.R., China is the greatest remaining multinational state.

British Empire to Commonwealth

At its peak in 1900, the British Empire covered over one quarter of Earth's land surface. This empire has gradually been transformed into a loose association of independent states, the Commonwealth of Nations.

Most of the United Kingdom's former colonies have chosen to join, and today the Commonwealth has 54 members. Queen Elizabeth remains head of state in 16 countries in addition to the United Kingdom, including Canada, New Zealand, and Jamaica, but most have become republics. The British legal system, language and education system, the Anglican Church, and other cultural traditions linger in each of the countries. The Commonwealth offers a framework for consultation and cooperation for the achievement of common ends, where they exist; and its achievements have enticed Mozambique, a former Portuguese colony, to join.

Britain retains the Falkland Islands ("Malvinas" in Spanish) against claims by Argentina, which even invaded the islands in 1982; and Britain retains Gibraltar against claims to it by Spain (Figure 13-3). Bermuda is a Crown Colony with its own internal representative government. All three colonies have repeatedly voted to maintain their current status.

Northern Ireland's status is open to question. The United Kingdom gave independence to 26 of Ireland's 32 counties in 1921, but the six counties of Northern Ireland (Ulster) elected to remain within the United Kingdom. Great Britain and Ireland agree that Ulster can remain a province of Britain for as long as most of its people want it to, but it has been wracked by almost nonstop violence since the 1960s. The province's Roman Catholics complain of discrimination by the majority Protestants, and both Catholic and Protestant

Remnants of Empires

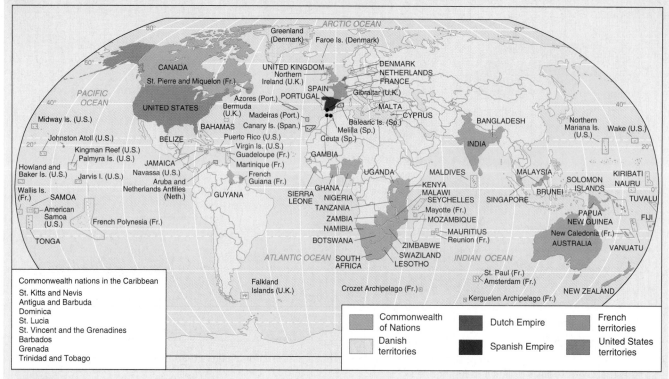

FIGURE 13-2 A few empires linger, whereas the breakup of others left behind cooperative organizations. Zimbabwe, ruled by Robert Mugabe, has been suspended from the Commonwealth for lack of democratic government.

paramilitary groups are guilty of murder and terrorism. Several government schemes have been attempted, but none has yet been able to guarantee peace. In 2000 Ulster was 57 percent Protestant, but birth rates are so much higher among the Roman Catholics that they might soon form the majority.

FIGURE 13-3 Gibraltar. Britain's 6.5-square-km (2.5-square-mile) colony of Gibraltar offers a fine harbor strategically located at the entrance to the Mediterranean Sea. Spain claims the colony, and the United States even fought in alliance with Spain to wrest it away (1779–1783), but Britain will not surrender it unless its 30,000 residents vote to change its status. (Topham/The Image Works)

The French Empire

France built up an empire in the seventeenth and eighteenth centuries but lost most of its colonies during an eighteenth-century rivalry with Britain. During the nineteenth century France built a second empire, beginning with the occupation of Algeria in 1830 and culminating in 1919 with the assumption of rule in Syria, Togoland, and Cameroon. The French believed that their subject peoples would mature not to independence but to full representation in the government in Paris. The four colonies of Martinique, Guadaloupe, Réunion, and Guiana were organized as Overseas Departments of France, and each was granted representation in the French National Assembly in 1946. Algeria won independence in 1962 after a long and bloody war. France granted independence to most of its other colonies in the 1950s, but it continued alliances. France's former African colonies receive aid and even host French military forces.

The present French Republic encompasses, in addition to mainland France and the four Overseas Departments, two Territorial Collectivities and four Overseas Territories. France also possesses a number of islands, including St. Pierre and Miquelon off the south coast of Newfoundland, a remnant of France's once-vast North American holdings. New Caledonia in the southwestern Pacific has major nickel reserves, and in

1998 its people agreed to defer a vote on independence for at least 15 years. France's other territories are generally poor and sparsely populated.

The Successor States of the Ottoman Empire

When the Ottoman Turkish Empire collapsed and a new Turkish nation was born in the 1920s, most of the Turkish Empire in the Middle East was divided between France and Britain. New states were eventually granted independence, but the map of today's states results from the interests of the colonial powers rather than the sentiments of the local people, and several borders have been disputed for more than 80 years. Israel was carved out to provide a homeland for the Jewish people, and bitter confrontation over this land—historic Palestine—continues. Furthermore, the region has remained a cauldron of competing loyalties. Many Arabs bestow their loyalty on units that are much smaller than states (clans, tribes, or families) or on ideas that are much bigger than states. These include *pan-Arabism,* the notion that there exists some bond among all Arabic-speaking peoples that is more legitimate than the modern Arabic states, and a universal bond of Islam. The Middle East is plagued by border disputes, rulers whose power is based on foreign interests rather than popular support, religious animosities, and political and economic intervention by international oil companies and banks. Several wars have troubled the region since World War II, frustrating many hopes for democracy and economic development.

The Russian Empire, Revolution, and Reorganization

At the same time that the Western European countries were building overseas empires, Russia built an empire across Asia (Figure 13-4). During World War I, the empire suffered internal rebellion as well as external

The Historic Expansion of Russia

Principality of Russia, 1300
Territory added, 1300–1462
Territory added, 1462–1505
Territory added, 1505–1584
Territory added, 1613–1800
Territory added, 1800–1855
Territory added, 1855–1955

FIGURE 13-4 Russians' expansion from their homeland around Moscow began in the mid-sixteenth century, and by 1647, Russian cavalrymen had reached the Pacific. During the seventeenth and eighteenth centuries, Russians pushed south to the Black Sea. Russian explorers crossed over the Bering Strait to incorporate Alaska (which Russia sold to the United States in 1867) and to locate settlements as far south as California. In the nineteenth and early twentieth centuries, czarist power advanced still further into Eastern Europe and Central Asia.

The Political and Population Geography of the U.S.S.R. in 1990

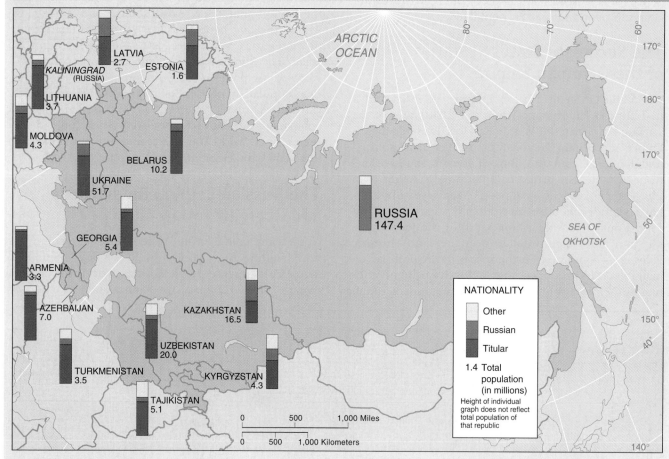

FIGURE 13-5 Most of the states of the former Soviet Union are dominated by nationalities for which the republic was named, here shown as "titular" nationality. It can be seen that Kazakhs were a minority in "their own" republic, and in several republics immigrant peoples—often Russians—formed substantial minorities.

attack. The Communists, led by Vladimir Lenin (1870–1924), seized control in 1917 and offered a new propagandistic ideal of federal union among the former subject nations. This union, the Union of Soviet Socialist Republics (U.S.S.R. or Soviet Union), came into being in 1922.

The configuration and internal organization of the U.S.S.R. changed through the years, but in 1990 the Soviet Union covered 17 percent of Earth's land area (excluding Antarctica), and its population of about 290 million made it Earth's third most populous country, after China and India.

The Soviet Union was subdivided into 15 union republics, each of which was theoretically the homeland of a national group and was named for that group (Figure 13-5). Several Republics contained within them the homelands of less numerous ethnic groups, and these groups enjoyed lesser territorial autonomy. Russia was the largest republic, occupying fully 76 percent of the total territory of the U.S.S.R., and Moscow was the capital of both the Russian Republic and also of the

U.S.S.R. Within Russia itself, 88 constituent units, 20 of which were ethnically based republics, enjoyed varying degrees of autonomy (Figure 13-6). Russia is, therefore, still technically referred to as the Russian Federation.

The individual Soviet republics ostensibly enjoyed considerable autonomy, but in fact all power was centralized in Moscow. Russian Communists dominated in the Union government and even in the non-Russian republics. Russians attempted to acculturate the other peoples. The Russian language, for example, was preferred for central government affairs, and it was one of two official languages in all non-Russian republics. Migration of Russians into the other homelands was still another form of Russian imperialism.

The Soviet Union contained 55 million Muslims in 1990—in Kazakhstan, Turkmenistan, Uzbekistan, Tajikistan, and Kyrgyzstan in Central Asia plus Azerbaijan in the Caucasus. Birth rates were high among these peoples, and many of them were politically dissatisfied. Their dissatisfaction threatened the Union's political stability.

Russia's Autonomous Republics

FIGURE 13-6 These 20 ethnically based "autonomous republics" within Russia are renegotiating their legal powers with the government in Moscow.

The economic geography of the U.S.S.R. The two principal characteristics of Soviet economic development were economic self-sufficiency for the Union as a whole and centralized state control. Natural resources were the property of the state, and no one was allowed to own "means of production" in such a way as to profit from the labor of others. Only a little private property and a few private-sector services were allowed.

Russia's treasury and supplies of raw materials subsidized inefficient industries in other republics. Agriculture was brutally collectivized. New lands were opened to agriculture, notably the steppes of northern Kazakhstan, but conditions across much of the U.S.S.R. territory—particularly northern Russia and the desert regions of central Asia—are unfavorable for agriculture. Agriculture also suffered from mismanagement, lack of worker incentive, lack of investment, poor transport, and government policies that kept farm prices low. Heavy industry, guided by planning bureaucracies, made immense strides, but the Soviet Union lagged behind the West in light industry, in the production of consumer goods, and in high-technology products. Both production and internal trading were highly

centralized, and there was no flexibility. Central planning dictated the geographical concentration of production, so between 30 and 40 percent of the total value of Soviet manufactured goods was produced on single sites. Thus, these three factors—centralization, rigidity, and geographic concentration of production—bound the republics together in trade.

Environmental protection was sacrificed for industrialization. In 1990 Alexei Yablokov, director of the Institute of Biology at the Academy of Sciences, estimated that 20 percent of Soviet citizens lived in "ecological disaster zones" and 35 to 40 percent more in "ecologically unfavorable conditions."

The breakup of the U.S.S.R. Mikhail Gorbachev assumed power as Union president in 1985, and he tried to boost the Union's economy by launching two liberalizing initiatives: *glasnost,* a loosening of restraints against freedom of speech and the press, and *perestroika,* a restructuring of the economy and of politics. Gorbachev gambled that his policies could unleash individual initiative and productivity while retaining the centralized control of the Communist Party and of the central Union government. Gorbachev lost. The tensions unleashed brought down the old governmental system before Gorbachev could bring a new one into place.

In August 1991 the central government proposed a new treaty to grant the constituent republics more autonomy, but reactionary elements of the Communist Party, the army, and the KGB (state police) arrested Gorbachev and attempted to take over the central government. By that time, however, the governments of the individual republics had already gained too much power, and the coup failed. The government of Russia itself, whose president, Boris Yeltsin, had earlier denounced communism and then won reelection, held out against the coup. Within weeks most of the republics had declared their independence from the U.S.S.R. and outlawed the Communist Party. On December 26, 1991, the Soviet Parliament formally voted the U.S.S.R. out of existence. After the dissolution of the Soviet Union, representatives of the three Slavic republics (Belarus, Russia, and Ukraine) created a new Commonwealth of Independent States (CIS), which was eventually joined by nine more former republics, but the Slavic states remain the most closely tied to each other.

Estonia, Latvia, and Lithuania are called the Baltic Republics. They were independent democracies after World War I, but the Soviet Union forcibly reincorporated them in 1939. They achieved relative prosperity under the Union government, but since the collapse of the Soviet Union they have turned their hopes toward the European Union, discussed below.

Developments in Russia   After the fall of communism and the dissolution of the Soviet Union, Russia suffered a virtual implosion of calamitous proportions from which it seems to be emerging only now, early in the twenty-first century. Every measure of income, health, or welfare still reveals a society facing enormous challenges. The economy has shrunk, the birth rate and the total fertility rate have dropped, and the death rate has risen. Life expectancy has shortened, and Russia's population, which was 147 million in 1999, fell to 145 million in 2001 and is projected to fall to 121 million in 2050. Neither attempts to achieve democracy nor to achieve capitalism seem to have succeeded.

Boris Yeltsin resigned as President of Russia in 1999 and appointed Vladimir Putin, then Prime Minister, his successor. Putin was elected independently in March 2000, but the elections would not have been considered open and fair in any Western state. State-owned media, for example, relentlessly attacked Putin's opposition. Most of the basic governmental framework of the Russian Federation has been maintained, but one ethnic unit, Chechnya (formally, Checheno-Ingushetia on Figure 13-6) launched a long and bloody struggle to achieve full independence. The 20 million Muslims in Russia constitute the majority in 7 of the autonomous republics, so in 2003 Russia joined the Organization of the Islamic Conference. Russia's Muslims, said President Putin, "have every right to feel part of the Muslim world."

Meanwhile, the privatization of the state-owned economy meant seizure of many assets by the members of the old communist government bureaucracies, by organized crime, or by a few ruthless oligarchs. *Privatizatsia* (privatization) turned into *prikhvatizatsia* (to grab). An estimated $300 to $500 billion was siphoned out of Russia and deposited in foreign banks during unscrupulous privatizations of state-owned companies between 1900 and 2000. Only history will decide how many leaders or profiteers in Russia in the 1990s were naïve, incompetent, corrupt, or all three.

Russia still has immense natural resources, a large skilled and educated population, and scientific institutes second to none on Earth. Russia is making slow progress, but it may be many years before it fully reassumes a place as a great power. The experience in Russia contrasts to developments in China. China has been criticized in the West for evolving toward a capitalist economy more slowly while holding a tight rein on freedoms, but the Chinese insist that the Russian experience testifies to the correctness of the Chinese path.

The other formerly Soviet republics   Experience in most of the other formerly Soviet republics (aside from the Baltics) has echoed the Russian example. Politically, they have renegotiated their relationships with the former autonomous regions within them; and civil wars broke out in several newly independent republics, including Moldava, Tajikistan, and Georgia. Almost 15 years after winning independence, most continued to be ruled by autocrats or oligarchs, with few real freedoms. The highly centralized command economy of the U.S.S.R. left a legacy that none of the successor states has a balanced economy. Most are dependent on other former republics for essential supplies, and most rely on enterprises in other former republics as their sole markets. Russia continues to supply most of them with natural resources at prices below those on world markets. It is proving very difficult to unravel this economic interdependence, and the process of unraveling has imposed considerable hardship. Most of the new countries are privatizing their economies, but not at the same rate (Figure 13-7). As in Russia itself, the process in several has been tainted by corruption. Several of the states have skidded toward worsening poverty, instability, and international isolation. Russia still maintains military bases in several of them.

Ukraine was the second most powerful of the Union republics, but since the breakup, its industrial might has proved largely illusory. Many factories were obsolete and represented more of a liability than an asset to a new economy. The government has been slow to reform the economy and reluctant to privatize, so inefficient state-owned industries still pile up losses. Income in the Ukraine fell approximately 50 percent between 1990 and 2000. Ukraine remains dependent upon Russia for fuel and raw materials, and the

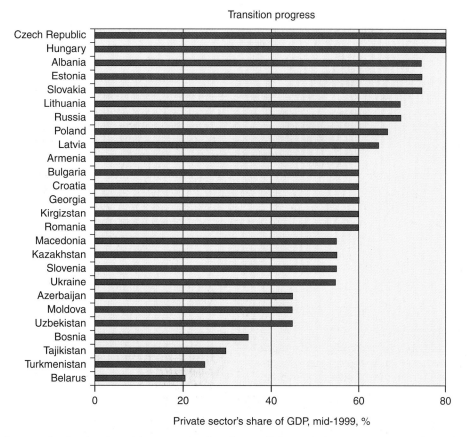

Transition progress

FIGURE 13-7 **The degree of privatization.** The countries that threw off Communist governments in the early 1990s had achieved different degrees of privatization by the end of the century. (Data from the European Bank for Reconstruction and Development)

"breadbasket" of the former Union and potentially one of the richest agricultural regions on Earth has even had to import food.

Many former republics want to diversify away from economic dependence on Russia by giving priority to trade and investment with other partners in the West, Asia, and the Middle East. The newly independent Central Asian Islamic countries retain economic links to Russia, but religion and other cultural ties pull them away. These countries are landlocked, and the several surrounding countries compete for influence. Turkey offers a democratic and secular model. But the Iranians, who are related to the Tajiks in language and culture and who share Shiite Islam with the Azerbaijani, have paid for new mosques and schools in Azerbaijan, and they are initiating joint economic ventures (Figure 13-8). To the east, China has cultivated economic and political contacts, and even Japan and South Korea have invested in Kazakhstan and Kyrgyzstan.

Some of the Central Asian Republics—particularly Kazakhstan—may prove to be rich in oil, so the questions of who will take the lead in developing that resource and by what routes the oil will reach world markets have launched diplomatic initiatives among many concerned countries. Alternative routes for oil from the Caspian Sea and Central Asia to world markets might cross Russia to the Black Sea; Iran to the Persian

Gulf; Iran and Turkey to the Mediterranean; or the republics in the Caucasus Mountains (Azerbaijan, Armenia, Georgia) to the Black Sea. Each possible route carries political and economic complications.

Even the United States has assumed a role in Central Asia. In 1997 and 1998, the United States held joint military exercises with both Kazakhstan and Uzbekistan. In the fighting against the Taliban in Afghanistan in 2001 (see p. 576), the United States was allowed to use military bases in Tajikistan and in Uzbekistan in exchange for U.S. long-term commitment to friendship and assistance. This was partly to protect these governments from their own Islamic fundamentalist organizations. These agreements triggered concern among many regional governments, especially Iran (see p. 576), that the United States intends to establish a permanent military presence in Central Asia—perhaps to obtain the oil.

One of the most volatile issues among the new Central Asian republics is their competitive demands for the limited regional water supplies. Problems associated with the unequal water resources of mountainous Tajikistan and Kyrgyzstan and their low-lying neighbors Uzbekistan, Kazakhstan, and Turkmenistan broke up a U.N.-sponsored conference on regional environmental issues in 2003 and have even raised fears of armed conflict.

Central Asian Transport Routes

FIGURE 13-8 The Iranian government completed a rail link from Mashad to Tedzhen in 1996, so today Iran's port of Bandar Abbas is Central Asia's principal outlet to the sea. Completion of a railroad from Mashad to Bafq will shorten that route by 900 kilometers (560 miles). When the line is completed from Bafq to Zahedan, goods from Central Asia will be able to bypass war-torn Afghanistan to reach Pakistan and India. A pipeline is also under construction from oil-rich Turkmenistan around the southern shore of the Caspian Sea through Iran to Turkey.

The dissolution of the Union of Soviet Socialist Republics created new units on the world political map and released powerful new political, religious, cultural, and economic forces. These forces continue to cause global redistributions of cultural, political, and economic activities.

The Empire of the United States

The United States organized most of its territories and admitted them to the Union fairly quickly. With the granting of statehood to Alaska and Hawaii in 1959, the United States incorporated even overseas territories, as France has.

The United States took Cuba from Spain in 1898 but granted it independence in 1934 and today retains only a naval base in the island's Guantanamo Bay. The United States held the Philippines from 1898 until 1946.

The United States provoked Panama's uprising for independence from Colombia in 1903, and then by prearrangement leased the Canal Zone from the new country (Figure 13-9). The United States surrendered control of the canal to Panama in 1999. The United States invaded Panama in 1989 to interdict drug traffic to the United States and to remove from power and

bring to the United States for trial a military dictator, Manuel Noriega, as an international drug trafficker. U.S. troops occupied Panama until the end of 1999, but U.S. officials report that narcotics traffic through Panama has not been reduced. Panama is one of the ten countries that accept the U.S. dollar as legal tender in an effort to stabilize their own currencies. Some 14 percent of the world's merchant ships are registered in Panama. This is an astonishingly high percentage for such a small, poor country, but many multinational corporations register their ships in Panama because of its low taxes and regulation. The Panamanian flag is known as a *flag of convenience*.

Liberia was founded as a haven for freed U.S. slaves and received independence in 1847. Inequities in wealth and political power fuel antagonism between the dominant Americo-Liberians (about 5 percent of the total population of 2.1 million) and the indigenous Africans, and the indigenous population is itself split

The Panama Canal

FIGURE 13-9 In 1977 the United States agreed to turn the Panama Canal over to Panama by the year 2000. On December 14, 1999, former President Jimmy Carter, who had signed the treaty, presided at the ceremonies. The Panama Canal is too small to accommodate many of today's ships, so an international group representing both governments and private interests in Panama, the United States, and Japan is considering reconstructing the canal.

among ethnic groups. Civil war broke out in the 1980s, and even intervention by forces from a consortium of West African states has been unable to stop the bloodshed. In 2003 U.S. troops operated in Liberia to assist Nigerian forces.

The Liberian flag is another flag of convenience. Liberia registers about 6 percent of the world's merchant marine fleet, representing 13 percent of total world gross tonnage. The threat of global terrorism has stimulated efforts by the London-based International Maritime Organization to make ownership of ships flying flags of convenience more transparent; U.S. intelligence officials insist that Al Qaeda owns a substantial fleet.

The United States still retains a modest empire of islands. Their total population is about 5 million, most of whom are citizens of the United States. American Samoa, Baker Island, Guam, Midway Islands, the U.S. Virgin Islands, Wake Island, and a few other spots are dependent areas. The Northern Mariana Islands is a commonwealth in political union with the United States, while Palau, the Republic of the Marshall Islands, and the Federated States of Micronesia are independent states with Compacts of Free Association with the United States.

Puerto Rico The United States granted Puerto Rico territorial status in 1917 and elevated it in 1952 to the status of a free commonwealth. Puerto Rico enjoys internal autonomy, but Puerto Ricans are U.S. citizens also under the jurisdiction of the federal government. They do not vote in presidential elections, and their resident commissioner in Congress can vote only in committee meetings. The United States and Puerto Rico share a common market and monetary system, but Puerto Ricans pay no federal taxes, and some federal customs and excise duties are paid back into the island's treasury. Puerto Ricans are divided over their political status. In a nonbinding 1993 plebiscite, 48 percent of voters opted to continue the commonwealth status, 46 percent preferred statehood, and 4 percent voted for independence.

Most arguments in the political debate are economic. The densely populated, resource-poor island now enjoys a standard of living higher that of any other Caribbean or Latin American nation, so it may be considered to be either a rich Caribbean country or, conversely, a poor part of the United States. Puerto Rico's 4 million people would rank it twenty-fifth among the states in population, but in 2002 its per capita income of $8,509 was less than half that of Mississippi, which ranked last among the states. The United States granted tax benefits for U.S. corporations that located in Puerto Rico until the late 1990s, so today manufacturing accounts for 42 percent of Puerto Rico's economic output and for about 16 percent of employment. Agriculture accounts for only 1.5 percent of output and 4 percent of employment.

Puerto Ricans are Spanish speaking, and many wish to retain their linguistic and cultural traditions. This might make it difficult for Puerto Rico to assimilate fully into U.S. life. If the Puerto Ricans ever choose independence, however, terms of continuing financial assistance plus leases on U.S. military bases (a full 13 percent of the territory) will have to be renegotiated.

NEW UNIONS OF STATES

The breakup of empires fragmented and reduced the scale of political organization of territory and, often, cultural and economic organization as well. New organizations of states, however, are binding activities across regions. The European Union is the international organization with the most ambitious goals—nothing less than eventual political unification. The North American Free Trade Area has more limited goals.

The European Union: Nations Knitting a Region Together

World War II ended in Europe in 1945 with the surrender of the Axis powers: Germany, Italy, Romania, Bulgaria, and Hungary. Japan had been an Axis Asian ally. The principal victorious Allied powers were the United States, Canada, Great Britain, France, the U.S.S.R., and China. In Europe the Allies had squeezed the Axis powers from West and East, and their victorious armies met in the middle of the continent.

The Allies themselves, however, were already divided by mistrust, and Europe soon split into two competitive blocs. British Prime Minister Winston Churchill said, "From Stettin in the Baltic to Trieste in the Adriatic, an Iron Curtain has descended across the continent." To the east, Soviet armies installed satellite regimes in Poland, Hungary, Romania, Czechoslovakia, and Bulgaria. Local communist leaders seized power in Yugoslavia and Albania without Russian help. The constitutions of all these countries were amended to guarantee Communist Party dominance in national politics, and their economies were largely collectivized.

Germany itself was split into four zones of military occupation. Already by 1949, however, the United States, French, and British zones were united functionally, and a new Federal Republic of Germany (West Germany), with Bonn as its capital, gained sovereignty in 1955. The Soviet Union reacted by granting its occupation zone independence as the German Democratic Republic (East Germany). Germany's capital city of Berlin had also been jointly occupied, so it too was split between West and East Germany, even though it lay about 201 kilometers (125 miles) inside East Germany. Its western sector was recognized as a part of West Germany, and East Berlin became the capital of East Germany. West Berlin showcased Western prosperity in the middle of East Germany. As a result, so many

East Berliners fled into West Berlin that in August 1961, East Germany built a wall across the city to seal in its citizens. Austria, which had been annexed by Germany, was also split into four occupation zones, and they were reunited in a newly independent state in 1955.

The Iron Curtain across Europe followed fairly closely the far older cultural division that Chapter 7 demarcated in orthography and religion: The West was defined by Roman Catholicism or Protestantism and the use of the Roman script, and the East by Orthodox Christianity and the Cyrillic script. The Iron Curtain deviated from this line, however, in that Estonia, Latvia, Lithuania, Poland, Eastern Germany, Czechoslovakia, Hungary, Croatia, and Slovenia had never been "Eastern European" culturally, yet they were east of the Iron Curtain.

The division of Europe into two competing blocs was furthered by Soviet seizure of territory during and after World War II (look back at Figure 11-5). In 1975, 35 countries—mostly European, but including the United States and other World War II participants—met in Helsinki, Finland, and agreed to accept European borders as inviolable. The conference did not, however, rule out revision by peaceful agreement, and European borders have been redrawn since then. In 1990 the Federal Republic of Germany and the German Democratic Republic peacefully united and recognized Berlin as the capital of the united country (Figure 13-10). Eastern Germany still lags behind the Western part, but standards of living are rising. Unification has been judged a political success overall, but Germany's economic slowdown in the early twenty-first century revealed continuing difficulties in absorbing the East. In 1991 Yugoslavia split apart into warring new states. In 1993 Czechoslovakia peacefully split into two countries.

Military pacts   The postwar political division of Europe was accompanied by its division into two military alliances. In the West the **North Atlantic Treaty Organization (NATO),** established in 1949, united Belgium, Denmark, France, Great Britain, Iceland, Italy, Luxembourg, the Netherlands, Norway, and Portugal with Canada and the United States. Greece, Turkey, West Germany, and Spain joined later. In 1955 the *Warsaw Pact* linked the U.S.S.R., Albania (which withdrew in 1968), Bulgaria, Czechoslovakia, East Germany, Hungary, Poland, and Romania. The U.S.S.R. used this treaty to justify the continued occupation of Eastern European countries by Soviet troops, which suppressed occasional anticommunist uprisings. The Western and Eastern blocs faced one another in a protracted *Cold War.* Although the two blocs never engaged directly in combat, they competed for economic growth and for influence among the new countries that gained their independence after World War II.

The pattern of alliances in Europe dissolved in 1989. The U.S.S.R. directed its attention to its own internal problems and released its grip on the countries of Eastern Europe. These countries, in turn, repudiated the privileged role of the Communist Party

FIGURE 13-10 Germans celebrate the opening of the Berlin Wall. At 7 o'clock on the evening of November 9, 1989, an East German official mistakenly handed a television speaker a draft of a new regulation suggesting that East Germans could go to the West without a visa at once. By 10 o'clock, hundreds of thousands stood at the wall, shouting, "Open the Gate!" Confused guards did, and East Germans surged through, to be met by West Germans and doused with beer and champagne. The Berlin Wall had fallen, and East Germany was soon swept away by history. (Anthony Suau/Getty Images, Inc.—Liaison)

in their own countries, and within a few months they scheduled free elections. The Warsaw Pact formally dissolved in 1991. NATO expanded to include Poland, the Czech Republic, and Hungary in 1999, and seven more countries will join in 2004: Latvia, Lithuania, Estonia, Bulgaria, Slovakia, Slovenia, and Romania. In 2002 NATO signed an agreement with Russia on terrorism, arms control, and international crisis management, so today even Russia joins NATO conferences. In March 1999 NATO bombed Yugoslavia; this was the first time NATO ever attacked a sovereign nation.

In September 2001, NATO accepted U.S. evidence that the attacks on the United States had been carried out by Al Qaeda; and it invoked, for the first time, Article 5 of the NATO pact. This Article binds the members to recognize an attack directed from abroad on any member as an attack on all members. This article was designed to protect Europeans from the U.S.S.R.; few would have ever suspected that it first would be invoked to protect the United States from terrorists. Britain and Turkey most immediately provided assistance, but other European allies traced Al Qaeda cells throughout Europe and arrested many collaborators. In 2003, NATO contributed a force to U.N. security operations in Afghanistan. NATO's expansion to include former adversaries and its redirection of effort as far afield as Afghanistan have caused many observers to wonder whether, with the end of the Cold War, NATO has changed its focus, or even its purpose.

FIGURE 13-11 George C. Marshall. George C. Marshall (1880–1959) is shown here (at right) with Mrs. Marshall and President Eisenhower in 1957, at ceremonies commemorating the tenth anniversary of the Marshall Plan. During World War II, Marshall served as Army Chief of Staff and rose to five-star rank, as in the painting behind him. As Secretary of State (1947–1949), he devised the Marshall Plan for the recovery of Europe, for which he won the Nobel Peace Prize in 1953. He also served as Secretary of Defense (1950–1951). During World War II Dwight D. Eisenhower (1890–1969) was Supreme Commander of the Allied Expeditionary Force in Europe, and then Eisenhower served as U.S. president (1953–1961). (George C. Marshall Foundation, Lexington, VA, GCM Photograph Collection #307)

Economic blocs In 1947 U.S. Secretary of State George C. Marshall (Figure 13-11) announced a plan of financial aid to war-shattered Europe. Aid was offered to the Soviet Union and to the countries of Eastern Europe, but Soviet dictator Joseph Stalin (1879–1953) rejected it. The **Marshall Plan** became a program for the economic rehabilitation of Western Europe.

The Soviet Union and its Eastern European satellites formed a competing economic pact, the Council for Mutual Economic Assistance (COMECON), that achieved a partial integration of their economies. The Soviet Union supplied raw materials, including energy supplies, to its satellites. Some of the East European economies grew under their communist regimes, enough to surpass the Soviet Union itself in per capita measures, but they did not keep up with Western Europe. COMECON formally disbanded in 1991, and the East European countries began to privatize their economies and to negotiate new economic links with the more prosperous countries of Western Europe.

The European Union today The Marshall Plan encouraged economic cooperation among the European countries, and in 1952 six of the Western European nations united their industrial economies to form the European Coal and Steel Community: Belgium, the Netherlands, Luxembourg, West Germany, France, and Italy. The success of this union encouraged further cooperation, and the European Economic Community (EEC) and European Atomic Energy Community (Euratom) came into existence in 1957. All three merged into one European Community (EC) in 1967. Denmark, Ireland, the United Kingdom, Greece, Spain and Portugal joined in the 1970s and 1980s. In 1993 these 12 countries signed a new treaty in Maastricht, the Netherlands, and changed the name of their organization to the **European Union (EU).** In 1995 Sweden, Finland, and Austria voted to join, although Norway voted not to join. In 2004, ten more countries will join the Union: the Czech Republic, Estonia, Hungary, Latvia, Lithuania, Poland, Slovakia, Slovenia, Malta, and Cyprus (Figure 13-12).

There are three kinds of economic groupings. A **free trade area** has no internal tariffs, but its members are free to set their own tariffs on trade with the rest of the world. A **customs union** enforces a common external tariff. A **common market** is a still closer association of states; it is a customs union, but in addition, common laws create similar conditions of production within all members. The EU is forming a full common market, and it even declares political merger to be an eventual goal. It has been noted that the United States developed a political community before merging in an economic union, whereas the European states are trying to reverse that process.

The members of the European Union share many broad cultural patterns, and these cultural foundations provide a basis for their successful cooperation. Common traditions include democratic ideals and parliamentary institutions, civil rights and legal codes, Judeo-Christian ethics, respect for scientific inquiry, artistic traditions, and humanism and individualism. It may be difficult to extend Union membership to some states, such as Turkey, that are pressing for entry but do not share these European traditions.

The European Union has several goals. One is to end the wars that have plagued the continent. Another is economic growth, and a third reason for European unification is the urge to play a role in world affairs. Economic strength has largely replaced colonialism or military might as a measure of prestige, and no European nation alone can command prominence on the world stage. United Europe, however, can assume a leading position.

European Union government The Union's Constitution is being rewritten for the enlargement scheduled for 2004, but it will probably continue to be governed by four institutions: the European Commission, the Council of Ministers, the Parliament, and the European Court of Justice. The European Commission proposes legislation and is responsible for administration. It sits in Brussels, Belgium, which has enjoyed considerable

The European Union

FIGURE 13-12 Germany has both the largest population and largest economy in the European Union.

growth and prosperity as de facto capital of the Union. Germany, France, Britain, and Italy, each of which has had two seats on the Commission, will lose their second seat in 2005. Each new entrant to the EU will have one commissioner until there are 27 member states. The final legislative body is the Council of Ministers. Seats in the Council of Ministers are awarded roughly according to the populations of the members, but no decision can be taken by a coalition representing less than 62 percent of the total EU population. This condition in effect recognizes the predominance of Germany. The Council decides many issues by majority vote, thus considerably reducing individual countries' sovereignty. The European Parliament sits in Strasbourg, France. The apportionment of its 567 seats is a compromise between numerical democracy and equality among nations. Its members are elected directly by the citizens of the member countries, and the members who represent affiliated political parties from different countries join together as international voting blocs. Therefore, the Parliament and to a degree the Commission are genuinely supranational.

The European Parliament is not equivalent to a national legislature. It can scrutinize proposed Union laws, offer amendments, vote against directives, veto the actions of any branch of the Union, or dismiss the European Commission, but it cannot initiate legislation, select the executive, or levy taxes. Nevertheless, Parliament is the most democratic of Union institutions, and it has been gaining power.

European law is adjudicated by the 13-judge European Court of Justice. In cases of conflict between Union law and national laws, the supremacy of Union law is being steadily affirmed. Union institutions have become so important to the member states that about 60 percent of each member's domestic legislation is drafted in Brussels and simply translated into national law. Final adoption of the European Convention on Human Rights in 2000 profoundly changed government in several European states that, unlike the United States, had no Constitution or independent judiciary to enforce guarantees against official abuse. After 2000, for the first time, national judges were bound to measure acts of their national legislatures against the EU Convention. This is truly supranational law.

Shortly after the attacks on the United States in September 2001, the EU established one Europe-wide arrest warrant that would enable prosecutors to skip lengthy extradition proceedings. This was seen as a crucial tool in fighting terrorism.

European Union integration   The EU largely achieved the creation of one economic community by the end of 1992. The following six criteria, however, demonstrate that the Union is much more than an economic union. It is uniting its members and binding their territories in many ways.

1. *The Creation of One Market for Goods.* The Union established a common agricultural policy, and Union bureaucracies have standardized business regulations. European manufacturers would like to achieve the economies of scale in one European market that U.S. manufacturers have long enjoyed in the huge U.S. domestic market.

2. *The Creation of One Market for Capital.* In 1999 11 of the EU's 15 members (all but Britain, Sweden, Greece, and Denmark) agreed to adopt one currency, the euro. Greece joined the euro area in 2001, but Denmark and Sweden (most recently in September 2003) voted to stay out. Coordination among the countries that use the euro has been strained, because rules require individual countries to keep down national budget deficits, potentially forcing cuts in social welfare programs and other governmental spending. Nevertheless, the euro has won increasing favor as an international currency.

3. *Transport System.* The members of the European Union are redesigning and rebuilding Europe's transport system from 15 national webs into a single coordinated web (Figure 13-13). All railroads, for example, must use the same gauge and electricity systems, and all highways, bridges,

Europe's Integrated Rail Network

FIGURE 13-13 This integrated network of high-speed trains has been superimposed on the existing national networks. It required laying 7,400 kilometers (4,600 miles) of new track and improving 19,300 kilometers (12,000 miles) more. In an area as compact and densely populated as Europe, city-to-city train travel is a sensible alternative to air travel at little or no additional time cost. When ten more countries join the EU in 2004, this integrated net will be extended eastward. (Data from the European Commission)

and tunnels must have the same construction specifications. One air-traffic control system has evolved.

4. *Regional Policy.* The Single Europe Act commits the countries "to strengthen economic and cultural cohesion." Living standards and quality of life ought to be equalized throughout the Union area, just as nations strive to equalize incomes throughout their territories (Figure 13-14). The Union tailors policies for poor agricultural regions and declining urban regions and furnishes funds for needed infrastructure, such as transport and telephone service.

5. *Social Policy.* The Union members have tried to strengthen cultural ties. International committees are rewriting school history books, for example, to soften nationalistic antagonisms. Union social actions equalize worker health and safety standards, the international transfer of pension rights, the guarantee of rights to migrant workers, and mutual recognition of technical qualifications. The Workers' Charter regulates the workweek, overtime pay, holidays, and other terms of employment. The European Court of Justice has enforced standard civil rights and antidiscrimination legislation. Union citizens may travel, settle, and work anywhere within the Union, whereas non-Union citizens face common immigration and visa requirements. The governments have committed themselves to the creation of a common judicial area. Ever since 1992, EU citizens have been able to vote in any EU country in which they were resident, and today noncitizens can even hold offices throughout the member countries. The concept of "European citizenship" is being made real.

6. *Environmental Policy.* Through the 1970s and early 1980s, much of Europe lagged behind the United States in protecting the environment. Since the late 1980s, however, the EU has become a leader in international efforts to protect the global environment. The Treaty of Maastricht established policies of protecting the quality of the environment, protecting human health, ensuring prudent use of natural resources, and promoting international measures to deal with regional or global environmental problems. These policies have led to new standards for air and water quality, restrictions on pollutant discharges, and procedures for environmental impact assessment.

The Communist governments in most of the Eastern European new members had compelled managers to meet industrial production targets regardless of the cost or hazards to people's health or to the environment. Unbridled industrial spewing and spilling carried disastrous environmental consequences that have not all been stopped, and that will have to be remedied (Figure 13-15).

Through the years, statistical descriptions of life in the EU member countries have converged. These include statistics as diverse as median educational attainment of the population, degree of urbanization, percentage of females in the labor force, crude birth rates, and total fertility rates. This convergence demonstrates that life in the member countries is growing measurably more alike. At the same time, the various European national cultures are growing less distinct than they were just decades ago. Styles of dress, cuisine, architecture, and other artifacts are merging. Already by 1997 over half of the Union's citizens reported speaking at least one European language in addition to their native tongue—one-third of Europeans over 55 years of age did, but 70 percent of those between 15 and 24 did.

> Through the years, statistical descriptions of life in the EU member countries have converged.

Expansion and international relations Bulgaria, Romania, and Turkey remain applicant countries and Switzerland enjoys close links without membership.

The extension of the Union in 2004 will present enormous challenges to the EU. Never before has the Union attempted to "digest" so many new members at once, especially new members with such large populations economically so far behind the current members. Poland, the Czech Republic, the Baltic states, and Hungary seem well on the way to establishing successful capitalist democracies, with Slovenia not far behind (look back at Figure 13-7), but other countries have had only limited success in privatizing, and several have seen their economies crumble. Today the boundary between West and East Europe traces an economic divide almost as deep as that between the United States and Mexico. Eastern European countries generally have lower per capita incomes, shorter life expectancies, higher percentages of their populations in agriculture, and lower measures of HDI. Millions of Eastern Europeans have already migrated to the West illegally, and perhaps this migration stream will increase, legally or illegally. The EU has begun to assist economic restructuring and development in the East. The accession of the ten new members will also present political and even cultural challenges. For example, power has slowly been shifting to the democratic European Parliament, and the accession of the ten new countries means that these poorer countries will have great political power. Will the richer countries yield power to their poorer new partners? Furthermore, most of the new members experienced communism for decades, and lingering political propensities regarding workers' rights, the role of religion in government, or many

Variations in Income Across the European Union

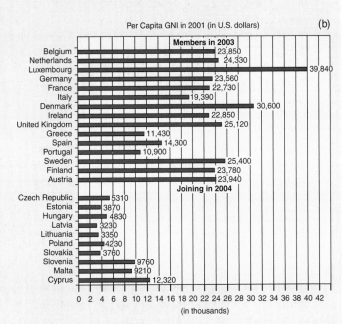

FIGURE 13-14 (a) The map reveals the wide disparities in income that existed as late as 1996 even among the EU members at that time. (Data from the European Union)

(b) This graph shows that the ten countries joining in 2004 are (except for tiny Cyprus) poorer than any current member. They are also relatively poorer than the poorest current members were when they joined. The ten will increase the EU population by 20 percent and its area by 25 percent, but its gross product by only 5 percent. Enormous quantities of economic aid will begin to flow eastward.

Pollution in Eastern Europe

FIGURE 13-15 When Communist governments in Eastern Europe collapsed and Western observers were allowed to examine industrial facilities, they found a degree of pollution they could scarcely believe. The city of Most, in today's Czech Republic (shown in the photograph), was industrialized in the nineteenth century. While the country was under Communist rule in the mid-twentieth century, pollution from the town's vast chemical works and inefficient coal-burning power stations was unchecked. The local population suffers from a high incidence of respiratory problems. (Map: CIA; photo: Simon Fraser/Science Photo Library/Photo Researchers, Inc.)

Critical Thinking
GEOGRAPHICAL INDICATORS

In negotiations among its own members and with other countries, the European Union has become concerned with the integrity of labeling products that take their names from the regions where those products originated. For example, the following place names can now be used only for wines actually produced in those regions: Bordeaux, Chianti, Claret, Madeira, Malaga, Marsala, Medoc, Moselle, Burgundy, Rhine, Sauternes, Chablis, Champagne, Port, and Sherry. Other geographical indicators include Parma ham and Stilton, Parmesan, Asiago, Roquefort, and Gorgonzola cheeses. Many of Europe's trading partners, including Canada, have agreed to stop using these European place names to label their domestic imitations. The United States is negotiating.

Questions

1. How many of the places named in the products above can you find on a map of Europe?

2. How many U.S. products can you find in your local grocery store or wine shop that use the European geographical indicators?

other issues might cause frictions within the wider Union. These are matters to be watched carefully.

The European Union negotiates trade pacts with outside countries and other blocs, but it lacks an overall foreign policy. This reveals how far the countries remain from true union. Nevertheless, the 1997 Amsterdam Treaty created a new post, the Secretary General, who represents the foreign policy decisions of the Council of Ministers on the international stage. The Union has also formed a Defense Force, which took over some NATO peace-keeping responsibilities in Macedonia in 2003. The EU force was more important for its symbolism—320 soldiers serving under the Union flag—than for its military might.

Many U.S. corporations do business in Europe or export to Europe, so European regulations increasingly affect U.S. economic affairs, and European governments are much less laissez-faire than is the U.S. government. The European Commission has intervened in everything from U.S. antitrust policy to food safety. For example, Europeans regulate what Internet companies may do with personal information collected from consumers far more strictly than the U.S. government does. European governmental review of corporate mergers is also much stricter than in the United States, so it was Europeans who in 2001 blocked the proposed merger of the American giant corporations General Electric and Honeywell. American corporations' acceptance of European regulations may make it harder for these corporations to fight the introduction of similar regulation back in America.

The existence of the euro currency also threatens U.S. economic welfare. Many nations have long had such confidence in the United States that they have kept their own national reserves in U.S. dollars—hundreds of billions of them. This has allowed the United States literally to print money to pay its debts and to run high current accounts deficits. If many countries gain sufficient confidence in the EU and the euro that they shift even a fraction of their national reserves into euros, the United States could be challenged to meet its international financial obligations.

The Formation of a North American Trade Bloc

On January 1, 1994, the North American Free Trade Agreement (NAFTA) came into effect, linking the United States, Canada, and Mexico. The agreement does not propose a full customs union or common market, nor are there any plans for a supranational governing body. NAFTA concerns only trade and investment. In 2003 the North American partners outstrip the 15-member European Union in total area (8.2 million square miles to 2.8 million) and in total population (396 million to 375 million), but the two blocs' total gross incomes are about the same (almost $10 trillion each). After 2004, however, the 25-member European Union will outstrip the North American bloc in population (with 450 million citizens) and in total income, though the EU will still be considerably smaller in total area.

The Canada–U.S. free trade agreement Canada and the United States signed a Free Trade Pact in 1988, by the terms of which all barriers to trade in goods and

Focus on
WHAT IS CANADIAN MUSIC?

A Canadian law insists that 35 percent of the daytime playlist on Canadian radio be "Canadian." But what is Canadian music? The nationality of the singer, of the composer, and of the lyricist must all be taken into account. Some results of the government directive may seem bizarre. For example, Lenny Kravitz's 1999 remake of the 1970s song "American Woman" qualifies as Canadian. Kravitz isn't Canadian, and the song isn't about Canada, but the song was written by Guess Who, a Canadian band. Celine Dion, a Canadian, singing "My Heart Will Go On," a song by a U.S. composer and lyricist, does not count as Canadian. Rod Stewart's "Rhythm of My Heart," however, is Canadian because both the music and lyrics are by the Canadian Mark Jordan.

Similar problems of definition plague other cultural products. *Top Cop*, a U.S. television program filmed in Canada, is Canadian, but the Disney Corporation's movie *Never Cry Wolf*, a 1983 film version of a Canadian classic book by (Canadian) Farley Mowat, is not Canadian because it was produced in the United States.

Questions

1. Should Canada feel that it is necessary to protect its culture from the United States?
2. How can the determination be made as to what is really Canadian?
3. Should the United States somehow identify and label which pop stars and film stars are Canadian?

North American Trade Flows

FIGURE 13-16 The United States buys about 85 percent of Canada's and of Mexico's exports each year, whereas Canada takes about 20 percent of U.S. exports and Mexico just over 10 percent. Trade between Canada and Mexico is limited, but increasing rapidly.

services were eliminated between the partners by 1999. The treaty also removed or reduced the hurdles to cross-border investment, government procurement, agricultural sales, and movements of employees.

Canada and the United States have long been by far the world's two leading trading partners (Figure 13-16), as well as major investors in each other's economies. About two-thirds of foreign investment in Canada comes from the United States, representing one-third of all U.S. investment abroad, and Canada is the fourth largest investor in the United States. About 10 percent of the chief executives of the largest U.S. corporations are Canadian-born.

U.S. corporations were generally strong supporters of the pact, because Canadian tariffs were two to three times higher than U.S. tariffs, but Canada's labor unions fought the pact. They pointed out that employment fringe benefits in Canada are generally much more generous than they are in the United States, and the unions feared a drop to U.S. standards. In Canada labor unions represent 40 percent of the labor force, whereas in the United States unions represent only 19 percent. Canada's farmers also fought the pact. Canadian agriculture is at a competitive disadvantage because of the harsh climate, vast distances, and small

scale of regional markets. Canadian small producers suffer competition from big food-processing firms that can buy more cheaply in the United States.

Some Canadians feared that strengthening economic ties with the United States would weaken Canadian nationalism. If you look back on Figure 11-24, you will notice a thin line of Canadian settlement along the border and an almost empty empire stretching away to the north. Some 75 percent of Canadians live within 241 kilometers (150 miles) of the U.S. border, where they absorb U.S. news and culture. Canadian means of transportation and communication have always been designed to foster east-west linkages, as illustrated in Figure 11-17. Nevertheless, the international border to the south is punctured by more than 75 million border crossings per year, as well as by common telephone lines, pipelines, computer links, power lines, contracts, cross-ownership, and even by common professional sports allegiances (Figure 13-17).

The trade pact provides protection for Canadian broadcasting, publishing, and related cultural industries, but it may be difficult even today to distinguish Canadian cultural products from those of the United States. The majority of Canadians do not watch Canadian television, read Canadian books, listen to Canadian music, or go to Canadian movies and plays. Moreover, many U.S. citizens are probably unaware of the Canadian nationality of authors Robertson Davies and Margaret Atwood, reporter Morley Safer, and entertainers Rich Little, k. d. lang, Jason Priestley, Michael J. Fox, Celine Dion, Shania Twain, Alanis Morissette, Sarah McLachlan, the Tragically Hip, Avril Lavigne, Carolyn Dawn Johnson, Remy Shand, Chad Kroeger, Diana Krall, Cathy Fink, Marcy Marxer, Jane Bunnett, Walter Ostanek, the Barenaked Ladies, Mike Myers, David Cronenberg, Atom Egoyan, James Cameron, Norman Jewison, and Jim Carrey. Also

FIGURE 13-17 The blackout of August 14, 2003. The power grids of the United States and Canada are so interconnected that the border itself could not be identified on this satellite image of the international blackout of August 14, 2003. This image was created from the difference between two other images, one collected before the blackout, the other during it. The areas with power outages are either dark or red. (Christopher D. Elvidge, Ph.D./NASA Headquarters)

Focus on

FOOD SAFETY AND INTERNATIONAL FOOD TRADE

Chapter 8 noted the opportunities that the continuing scientific revolution in agriculture plus international trade in food offer. It also noted, however, that biotechnology may carry serious risks. These risks are multiplied by increasing international trade in foods. The explosion in world trade, travel, and tourism is moving diseases around the world, and many of these diseases threaten both humans and other animals. Alien animal and plant pests represent such a broad set of dangers that the U.S. National Intelligence Council has termed foreign animal diseases a threat to national security.

Many of these diseases are transmittable to humans. One of the most worrisome is Mad Cow disease (technically BSE, *bovine spongiform encephalopathy*). In the 1980s and 1990s BSE spread through Great Britain and later much of western Europe via protein feed that contained infected animal parts, requiring the slaughter of almost 4 million cattle. Britain imposed a domestic ban on the tainted feed in 1988, but the feed continued to be sold all over the world. The spread of BSE severely damaged the beef industry in some European countries, and it seems to appear in humans as the related new variant Creutzfeldt-Jakob Disease (nvCJD). By 2003 over 100 people had died of nvCJD, but the rate of occurrence was decelerating. In 2000 three flocks of sheep in Vermont imported from Belgium were killed for fear of BSE. It was later determined that they did not have BSE but did have scrapie, a related disease. In 2003 the diagnosis of BCE in a cow in Alberta, Canada, triggered embargoes against Canadian beef, cattle, and animal feed by the United States and other countries, but the cow found to have BSE in late 2003 had been imported prior to that ban.

BSE is caused by a type of malformed protein called a *prion*. Prions are also believed to cause a contagious, invariably fatal disease called chronic wasting disease (CWD), which occurs in wild and captive deer and elk in the western United States. CWD has not been observed in wild animals in Canada, but it has been diagnosed in Saskatchewan game-farm elk that were imported from the United States. This prompted the Canadian Food Inspection Agency to slaughter thousands of elk on Saskatchewan ranches.

Foot and mouth disease (FMD) is another disease that, while rare in humans, is very dangerous among some meat animals. Japan and Korea have had to slaughter hundreds of thousands of cattle that suffered FMD, which may have come from China. In 1997 Taiwan slaughtered 3.8 million pigs with FMD, and in 2001 an outbreak of FMD in Britain devastated the meat industry. Millions of cattle, hogs, and sheep had to be slaughtered, and tourism was significantly reduced. FMD also appeared in several other European countries.

These events have raised major concerns worldwide about the safety of our food supplies. Such concerns can often rise out of proportion to the actual risks involved, just as many people fear flying on airplanes but are very willing to ride in an automobile, even though cars are much more dangerous than planes. Intense public concern about food safety can lead governments to place restrictions on international trade in food. Cultural and political differences between countries mean that perceptions of food hazards may be very different from one country of another. If a few countries that import significant amounts of food place restrictions on the way such food is produced, then there is pressure on major producers around the world to modify their production techniques. Pressures such as these can lead to disputes between trading nations, and are likely to place major strains on international food trade in the coming years.

Canadian is Alias/Wavefront, which developed the animation software prominent in virtually every special-effects-based feature film in recent years (*Spiderman, Ice Age, Hollow Man, The Matrix, The Perfect Storm*, and the *Lord of the Rings* and *Star Wars* series of films).

The degree of cooperation between the United States and Canada may be represented symbolically by the fact that Canada is the only country that the United States has allowed to build an embassy on Pennsylvania Avenue in Washington, D.C., America's grand ceremo-nial boulevard; and in October 1999 President Clinton traveled to open a new U.S. embassy in Ottawa. This was the first time that a U.S. president has ever traveled abroad for this purpose. As the United States and Canada grow closer together, American security concerns necessarily suggest extending America's security precautions around Canada, too. In 2002, for example, the two countries signed a "Safe Third Country Agreement" in an effort to eliminate differences between the two countries' procedures for granting asylum.

Critical Thinking

INTELLECTUAL PROPERTY AND GLOBALIZATION IN THE DRUG INDUSTRY

As the U.S. population ages, the cost of prescription drugs has become a political issue. Drugs are expensive even for Americans, but in poor countries such as India, where the per capita GNI is just $460, such drugs are simply out of reach of all but the richest citizens. India does, however, possess the ability to manufacture these drugs, and some brand-name products sold by European and U.S. companies are actually wholly or partly manufactured in India under contract to those companies. Generic equivalents of brand-name drugs are also made in India, and one Indian company, Cipla Ltd., makes some 400 important drugs sold elsewhere under patents owned by Western pharmaceutical companies. These are sold in India, under different names, at a fraction of the prices charged by the patent holders. Cipla claims to make an equivalent of every product of America's giant Pfizer pharmaceutical corporation.

The large pharmaceutical companies, based primarily in wealthy countries, insist on the need to protect their investments in research and development to produce new drugs. It costs many millions of dollars and years of research to produce a new drug, and unless those companies can charge high prices for their products, there is no way to recover that investment. Without patents that protect the interests of drug developers, there would be little incentive to develop new drugs. The pharmaceutical companies are concerned about what they see as "pirating" of their property by Cipla and similar companies. That property is indeed valuable; pharmaceutical companies as a group are one of the most profitable sectors of the U.S. economy.

India long pursued an import-substitution development strategy after gaining independence from Britain in 1948. The spinning wheel at the center of the Indian flag symbolizes Indians' determination to manufacture at home products that their former colonial rulers forbade them to make; the British had required Indians to purchase British-produced textiles. Still today, Indian patent and trade law is designed to inhibit Indian companies from purchasing patented technology from abroad; it encourages local production, and drugs produced locally can be sold at prices dramatically lower than imported drugs. If Indians paid U.S. prices for these drugs, the drugs would cost perhaps 20 times the price of the Indian equivalents. This reduction in cost brings the drugs within reach of many people who would otherwise suffer without them. It can also be argued that foreign drug companies would not sell many drugs in India at the high U.S. prices, so the companies' losses may not be as large as they might claim. Improved health care in India has contributed to an increase in life expectancy from only 38 in 1956 to 64 today. Early in 2001 Cipla announced an offer to supply AIDS-fighting drugs to African countries at prices well below those charged in the United States.

When Americans faced the threat of biological warfare with the spread of anthrax spores at the end of 2001 (see Chapter 5), they stampeded to buy Cipro, a patented medicine that provides the best protection. Not enough Cipro was being made to meet the new demand, so the U.S. government found itself questioning the protection of the patent, and it even considered turning to the Indian producer of a generic version for new supplies. Americans were outraged to learn that while Cipro cost $350 per month in the United States, the generic costs only $10 a month in India.

The summit conference of the World Trade Organization held in Doha, Qatar, in November, 2001, eventually agreed that, in cases of national emergency, governments could override the patents on life-saving drugs and produce the drugs themselves or purchase generics on international markets. Within weeks, 10 African countries signed agreements with international pharmaceutical companies that slashed the cost of AIDS drugs in those countries by up to 85 percent. Negotiations continue on providing drugs to poor countries at prices they can afford while preserving drug companies' rights and encouraging research on new drugs.

Questions

1. Should pharmaceutical companies and other technology-developing industries in rich countries allow poor countries to use their intellectual property free of charge, so that those who would otherwise have no access to these goods may enjoy at least some modest improvements in the quality of life?

2. Should we insist that patents and other intellectual property be rigorously protected, so that corporations have an incentive to create new technology?

Mexico A trade agreement among the United States, Canada, and Mexico went into effect on January 1, 1994, under the terms of which tariffs and other barriers to trade will be brought down in four five-year steps.

The integration of Mexico into a North American trade pact might be more difficult than the integration of the U.S. and Canadian economies. The United States and Canada are both rich, postindustrial societies with populations that are growing slowly. Mexico is a less developed nation with a population still growing rapidly—from 25 million in 1950 to 100 million by 2000. To provide work for this rapidly growing labor force, Mexico needs economic growth. Economic growth might reduce the push for Mexicans to migrate, legally or illegally, to the United States. Mexico's Hispanic culture and language contrast with the British traditions and English language held in common in the two countries to the north (except in Quebec), but that cultural contrast may not be as troublesome in building a free trade pact as it would be if the three countries were building a full common market.

Signing the trade pact was part of an evolution of Mexican policies toward export-led economic development. Mexico's economy had been state directed and centralized through most of the twentieth century; but in 1986 Mexico began privatizing its state sector, selling off factories, shopping centers, mines, and public utilities. Only PEMEX, the state oil monopoly created when Mexico's oil industry was nationalized in 1938, was viewed as politically untouchable. Mexico also liberalized rules for foreign investment.

Many U.S. companies were active in Mexico even before the trade pact was signed. They had moved labor-intensive operations to Mexican factories called **maquiladoras** concentrated along the U.S.-Mexican border. Mexico imported components duty free; these were assembled and the products reexported to the United States. The United States charged tariffs only on the value added, which was low because Mexican wages are low (Figure 13-18). The existence of these factories had already by 1994 significantly redistributed Mexican population and urbanization toward the border, but maquiladoras have multiplied since the pact (Figure 13-19). New FDI in Mexico, about $2 billion per year in the 1980s, rose to $12 billion per year in the 1990s. This investment has lessened Mexico's financial dependence on the export of oil and modernized its industrial sector. Exports of manufactured goods rose from 43 percent of Mexico's merchandise exports in 1990 to 85 percent by 2001. In 2000 about 800,000 maquiladora jobs existed in the border region. Mexico has also steadily increased the value added to its exports. For example, Tijuana is the world capital of television manufacture, but many companies that started manufacturing television sets are now assembling computers and other higher-value electronics.

Some of the new FDI in Mexico comes from European and Asian corporations that reexport goods

FIGURE 13-18 A new Mexican industry. These workers are assembling Web televisions in a maquiladora in Guadalajara. (Veronica Garbutt/Panos Pictures)

from Mexico into the United States. This occasionally causes trade disputes. Negotiators of trade pacts must agree what percentage of the total value of a good entering one country must have been added in the second country for that product to qualify as a product of the second country. This is called a **local content requirement.** Arguments over local content requirements typically arise whenever two countries negotiate

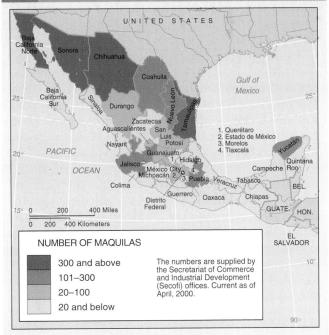

FIGURE 13-19 Maquiladoras (or maquilas) are concentrated along the U.S.-Mexican border, but U.S. corporations have spread facilities throughout Mexico. (Courtesy of Twin Plant News)

free trade pacts that are not customs unions. Disagreements have arisen between the United States and Mexico about whether goods assembled in Mexico out of components made in other countries should enjoy free access to the U.S. market. Such disputes have also arisen with Canada.

Many Americans feared that the pact with Mexico would see a migration of U.S. manufacturing jobs to Mexico, where wages, working conditions, and environmental protection laws are all below U.S. standards. U.S. manufacturers, however, responded that Mexican wages are lower than U.S. wages because Mexican productivity is lower than U.S. productivity; that is, each Mexican worker's output is much lower than the output of each U.S. worker. A worker's productivity is usually determined by the amount of capital investment that has been made per worker, and that investment is much higher in the United States than in Mexico.

Continuing growth of manufacturing in Mexico is threatened by compettiton from China, where wages are so much lower than in Mexico that by 2002 the number of maquiladoras was actually falling. Between 2001 and mid-2003, more than 500 of Mexico's 3,700 maquiladoras shut down at a cost of 218,000 jobs. In the 1990s Mexico rose to the rank of number two among nations exporting to the United States (after Canada), but China may have seized that position already by 2003. Manufacturers insist that Mexico must provide cheaper energy (now a government monopoly), improve the transportation infrastructure, curtail corruption, and improve education.

The maquiladoras are not the only border phenomena that geographers have noticed. Wherever two cities—one Mexican, one U.S.—face each other across the border, these so-called "twin cities" are being compelled to integrate and often internationalize their infrastructures. In 1998, for example, Mexican and U.S. officials opened the International Wastewater Treatment Plant, which treats Tijuana's excess sewage on the San Diego side of the border, the first international facility of its kind in the world. A unified air-quality district covers El Paso, Texas, and Ciudad Juarez.

Results of the NAFTA pact   Since the formation of NAFTA, the individual economies of the three signatories have experienced many changes, so it is difficult to isolate the impact of NAFTA. Trade and investment among the three partners have risen substantially. Imports into the United States from both Mexico and Canada have surged even more strongly than U.S. exports, so that net value of trade between the United States and the two partners has swung from a net surplus to a net deficit.

U.S. labor unions argue that the U.S. has lost jobs and that wages and living standards are down and workers' rights and bargaining power weakened in all three signatory countries. Both U.S. and Canadian labor unions have launched strong recruitment and organizational training drives in Mexico. Some factories are undoubtedly being closed and moved to Mexico. In February 1997, for example, the RCA Corporation announced its intention to close the world's largest TV assembly site, a plant in Bloomington, Indiana, and to move 1,100 jobs to Mexico. The company cited wage costs as the principal reason for the move. RCA is not a U.S. corporation, but is owned by Thomson, a French multinational conglomerate.

NAFTA's overall impact on the United States has been far less important than either its advocates or its opponents predicted. In July 1997 President Clinton reported to Congress that NAFTA so far "has made a modest positive contribution to the U.S. economy in terms of net exports, GDP, employment, and investment." This assessment is probably still true into the twenty-first century.

International manufacturing and marketing have begun to reorganize in scale from national to North American scale planning. The Ford Motor Company, for example, began to make Escorts and Tracers for the eastern U.S. market in Michigan and for the western U.S. market in Hermosillo, Sonora, Mexico.

The pact has not helped Mexico's agricultural economy as much as had been hoped. Agriculture produces only about 5 percent of Mexico's GDP, but some 20 percent of the labor force is directly involved in it. Local production costs of corn are high, and the United States subsidizes U.S. corn farmers by an estimated $10 billion per year, so Mexico has been flooded with imported corn. In 2002 nearly one-third of the corn used in Mexico was imported from the United States, and the price of corn in Mexico had fallen over 70 percent since 1994, reducing the incomes of the 15 million Mexicans who depend on producing corn for their livelihood. Mexico's exports of fruits and vegetables to the United States, however, had increased so much as to double Mexican agricultural exports to the United States overall from 1994–2002. Tariffs were removed on all agricultural products traded between the United States and Mexico on January 1, 2003 (except beans and white corn, which Mexico can protect until 2008), and economists are now analyzing the results of that removal.

The pact's effects on Canada are difficult to ascertain. Total U.S. exports to Canada have risen, and some Canadian industries seem unable to compete, but overall Canada's trade surplus with the United States has doubled. Canada's increase in exports to the United States has created, economists believe, as many as 50,000 new jobs. Nevertheless, Canadian unemployment rose in the first few years; some Canadian jobs may have migrated to Mexico.

Many environmentalists remain critical of NAFTA. NAFTA's Commission on Environmental Conditions has criticized the worst polluters in North America, but it has no powers to enforce cleanups.

The transport infrastructures of the three countries are being knit. In 1998 Canada's largest railroad, the Canadian National, with a main line stretching across Canada from Halifax, Nova Scotia, to British Columbia, bought the Illinois Central Railroad, giving the Canadian line a route south to the Gulf of Mexico ports of New Orleans, Louisiana, and Mobile, Alabama. Canada's deepwater port at Halifax was already winning trade from New York, Boston, other U.S. East coast ports. In 1999 the Canadian National signed a cooperative contract with the Kansas City Southern Railroad, whose line to Mexico City carries 40 percent of Mexico's rail traffic. In 2000 the U.S. government declared a moratorium on a merger between the Canadian National and America's Burlington Northern Sante Fe. That merger, still a possibility, would create a massive web with 50,000 miles of track linking 32 states, eight Canadian provinces, and reaching down to Mexico City. The United States has promised to allow Mexican trucks to haul goods throughout the United States, but the question of harmonizing truck inspections and driver qualifications has slowed implementation of this agreement.

Trade between Canada and Mexico has risen substantially since the pact, but it remained a small fraction of either country's trade with the United States. The pact has had devastating results on the economies of the Caribbean nations. One-third of Caribbean exports to the United States have been replaced by imports from Mexico.

Mexican politics From 1929 until 1997, Mexico was ruled by one political party, the Institutional Revolutionary Party (PRI), and its political apparatus made little distinction between the party and the state. Accusations of corruption and repression abounded. Some Americans hesitated to join in a pact with Mexico because they doubted Mexico's democracy, but others hoped that closer ties to the United States would strengthen democracy in Mexico. In 1997, in largely free and fair elections, the PRI lost its majority in the lower house of Mexico's Congress, two governorships, and the important position of mayor of Mexico City. In July of 2000 the PRI lost the presidency. Many observers attribute this revolutionary political change to the economic changes that have come with NAFTA.

Within Mexico economic integration and investment threaten to intensify the split between the modernizing economic sector and the great share of the population that remains in subsistence poverty. Just as the U.S.-Mexican pact took effect, a native American guerrilla organization calling itself the Zapatista National Liberation Army, in honor of Emiliano Zapata (discussed in Chapter 11), rose up in Mexico's southern state of Chiapas (look back at Figure 11-19). The rebels were motivated in part by fears that imports of U.S. corn and other market pressures would wipe out their traditional agricultural economy. "The free trade agreement," said the leader of the group, "is a death certificate for the Indian peoples of Mexico." Chiapas is

one of the poorest states in Mexico, and largely Indian in population. The insurrection has received support from Roman Catholic priests, followers of liberation theology discussed in Chapter 7. In 2001, the new Mexican government closed military bases in Chiapas and began negotiating land tenure and other issues with the rebels, but the uprising serves as a reminder of how far Mexico has to go to achieve the goals of economic welfare and political equity for all of its peoples.

After the attacks on the United States in September, 2001, Mexico's President Fox offered all possible assistance. Mexico has traditionally remained neutral in international affairs or even expressed distrust of U.S. motives, but it has recently been stepping forward on the international stage: In 2001, Mexico won a two-year seat on the U.N. Security Council (see p. 572). President Fox has suggested that the United States and Mexico seal the two countries' borders against outsiders but open the borders to freer movements between the two countries; some ten percent of Mexico's population already lives in the United States. Mexican refusal to support the U.S. second war against Iraq, however, set back continuing merger of the two countries' economies and activities.

Meanwhile, Mexicans' impatience with the Fox administration, which had, perhaps, promised too much change too soon, had expressed itself in the voters' giving the largest number of seats in the lower chamber of the parliament back to the PRI in July 2003. In his State of the Union address of September 1, 2003, President Fox regretted that he had not been able to achieve all of the "historic transformations our times demand," but he insisted that democracy and the rule of law had made progress under his presidency. Corruption, a feeble system of justice, and the economy presented continuing challenges.

Expanding Western Hemispheric Free Trade

The United States continues to sign free trade pacts with individual countries. By 2003 it had pacts with Mexico, Canada, Israel, Singapore (causing Singapore's Trade Minister George Yeo to boast that "the free trade agreement makes us almost an extension of the United States in Southeast Asia"), Australia, Chile, and Jordan (ratified in September 2001 as a gesture of U.S. solidarity with a friendly Islamic country). The United States has suggested the creation of a Free Trade Area of the Americas, excluding only Cuba, and negotiations continue on that initiative. A number of trade blocs formed in the western hemisphere in the early 1990s. Venezuela, Colombia, Peru, Bolivia, and Ecuador formed the Andean Pact. The English-speaking countries of the Caribbean united in a Caribbean Union and Common Market (CARICOM). Colombia, Mexico, and Venezuela formed a bloc. Argentina, Brazil, Paraguay, and Uruguay united in a customs union called MERCO-SUR; Chile, Bolivia and South Africa are associate members, and Peru and Venezuela may join. In 2002 the four original members plus Bolivia and Chile approved a plan that will allow their 250 million people to live and work in any country and be granted the same rights as citizens of those nations. Ricardo Lagos, the President of Chile, said, "I have always maintained that MERCOSUR is a fundamentally political undertaking and thus cannot be understood simply as a customs union." Perhaps a genuine United States of South America is in the works. Trade within MERCOSUR has soared, and when MERCOSUR's trade with the European Union first equaled its trade with the United States in 1996, President Clinton traveled to South America largely to remind these countries of America's commitment to hemispheric free trade.

The unification of transport infrastructures again reveals the expanding territory being organized (Figure 13-20). For example, a new railroad tunnel and new gas and oil pipelines have pierced the Andes Mountains between Chile and Argentina and link Bolivia and Brazil. New roads join Brazil and Venezuela, and a new 1,400-km (870-mile) power line brings elecricity from Venezuela's Guri dam to Brazil's Amazon cities. Argentina uses Chilean ports for its exports to Asia. Brazil and Argentina have reconciled the gauges of their railroad systems, and a new highway is under construction from São Paulo to Buenos Aires. New roads link Brazil and Peru. A new inland waterway allows barges to travel 3,424 kilometers (2,140 miles) down the Paraguay and Paraná Rivers from eastern Bolivia to the sea.

Only 18 percent of total MERCOSUR trade in 2000 was with other MERCOSUR countries. This is a small amount compared with the level of trade typical between industrialized neighbors, and it suggests that the various Latin American states did not produce items they could exchange to mutual profit. The new trade agreements, however, encourage specialization of production, increases in productivity, and intraregional trade and growth. The United States is the principal supplier of imported goods to most Latin American countries, so growth in Latin America boosts the U.S. economy.

Other Regional International Groups

The European Union and the Western Hemisphere trade areas are the most important regional international organizations today, but many others exist (Figure 13-21). Some of these are primarily military-defensive. Since the end of the Cold War, however, they have been turning into general economic and cultural associations. For example, the Association of Southeast Asian Nations (ASEAN) began as a military alliance, but in 1992 it became a free trade organization. Today China discusses joining. Other organizations, such as

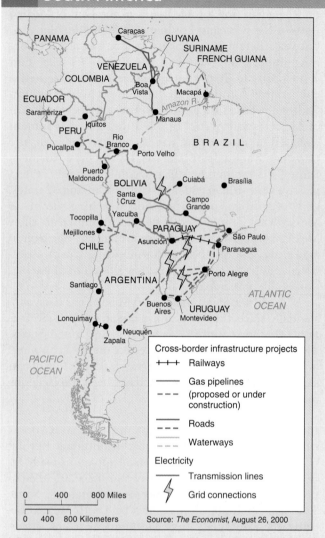

New Transport Routes Knitting South America

FIGURE 13-20 These roads, railroads, pipelines, and electrical grids are pulling together the population of South America, which map on the rear endpaper shows to be, for the most part, scattered around the fringes of the vast continent.

the Southern African Development Coordination Conference (SADCC), are economic. The purpose of each international organization is simply to provide a framework for consultation to reach agreement whenever possible in areas of mutual concern. In some cases, however, the members of these organizations can agree on very little and have even gone to war with one another. The African Union, for example (called the Organization of African Unity before 2002), and the Arab League serve few functions, although both are negotiating trade agreements. Many observers, however, fear that the proliferation of bilateral and regional trade blocs undermines the work of the World Trade Organization (discussed in Chapter 12).

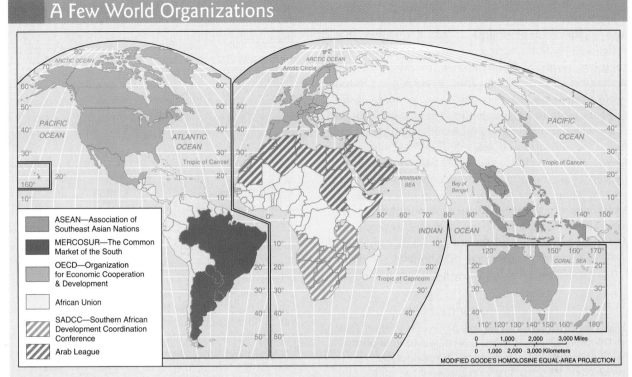

A Few World Organizations

ASEAN—Association of Southeast Asian Nations

MERCOSUR—The Common Market of the South

OECD—Organization for Economic Cooperation & Development

African Union

SADCC—Southern African Development Coordination Conference

Arab League

MODIFIED GOODE'S HOMOLOSINE EQUAL-AREA PROJECTION

FIGURE 13-21 These are just a few of the many international organizations that provide frameworks for international cooperation.

GLOBAL GOVERNMENT

The Dominican priest Francisco de Vitoria (1486–1546) earned recognition as "The Father of International Law" by arguing that each people has the right to its own ruler. Therefore, de Vitoria argued, relations among nations would have to recognize mutual rights. Subsequent centuries, however, have shown only token respect for this principle.

Discussions of world order are generally based on principles of state sovereignty and self-determination, but as the number of states has multiplied, the number of actual and possible conflicts among states has grown. Causes include aggressive ambitions, as with Iraq in the early 1990s; border disputes or rival claims on territory, as between Ethiopia and Eritrea; and domestic crises or policies that have effects abroad, causing other states to threaten to intervene, as with the former Yugoslavia. The necessity for common action in areas of common concern, such as the oceans or the global environment, is another pressure for global government. International society as a whole, however, has neither the centralized government, judicial system, or police that characterize a state,

> International society as a whole, however, has neither the centralized government, judicial system, or police that characterize a state, nor does it even have a consensus on what constitutes an international crime.

nor does it even have a consensus on what constitutes an international crime.

The first lasting international political structure was the League of Nations. The League was founded by the Allies after World War I, but several nations, including the United States, refused to join. The League provided a forum for discussion of world problems, and it carried out many humanitarian projects. It exercised little real power, however, and was unable to resolve the disputes that eventually led to World War II.

The United Nations

In 1942, 26 states that were allied against the Axis powers joined in a Declaration by the United Nations, pledging to continue their united war effort. A meeting of Allies in San Francisco in 1945 drew up a Charter for a United Nations organization to continue in existence after the war, and by the end of 1945 the new organization had 51 members. As more colonies received independence and joined the United Nations, membership rose to 191 by 2003—virtually all Earth's sovereign states.

The U.N. General Assembly serves as a rudimentary legislature of a world government, but it has no power of enforcement (Figure 13-22). The power that the United Nations does exercise is vested in the 15-member Security Council. Five members of the Council (the United States, China, France, the United Kingdom, and Russia—originally the U.S.S.R.) hold their seats permanently, and the remaining 10 members are elected by the General Assembly for two-year terms. This allocation of seats reflects the allocation of world power in 1945, but recent discussions have suggested that granting additional seats to, perhaps, Germany, Japan, India, and Brazil might more correctly reflect the present world balance of population and power. The U.N. Secretariat, with the Secretary General at its head, handles all administrative functions.

The Security Council has occasionally voted for the United Nations to field armies. United Nations

Cyprus

FIGURE 13-23 Since 1974 Cyprus has been divided into Greek and Turkish portions, with little mingling between the two groups. The Turkish sector has declared itself to be the Turkish Republic of Northern Cyprus, but only Turkey recognizes it as an independent country. The current members of the EU hoped that these sections could be reconciled before the Greek portion joins in 2004.

FIGURE 13-22 United Nations headquarters in New York City. The United Nations found a permanent home on the east side of Manhattan Island, where its headquarters buildings were designed by an international committee of architects, including Le Corbusier. The high slab of the Secretariat dominates the group, with the domed General Assembly to the north. The land legally is not part of the United States of America. (Rafael Macia/Photo Researchers, Inc.)

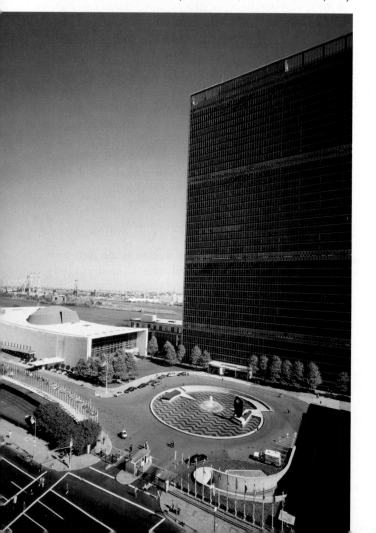

forces fought in Korea from 1950 to 1953. In 1990, a Security Council resolution gave the United States and its allies sanction to force Iraq to retreat from Kuwait, and on September 28, 2001, the United States obtained a Security Council resolution obliging U.N. members to cooperate in combating terrorism. In a greater number of cases, the United Nations has sent observer troops to patrol world trouble spots, and these incidents have been increasing. Examples include Kashmir; the Golan Heights; Lebanon; Cambodia; Yugoslavia; the Mediterranean island of Cyprus (Figure 13-23), split by feuding Greeks and Turks; and Haiti in 1993. U.N. troops have attempted both *peacekeeping* and *peacemaking* efforts. Peacekeeping involves patrolling a situation in which combatants have agreed to a peace. Peacemaking, however, is trying to impose a cease-fire or peace on combatants, and United Nations forces have had less success in these efforts.

In many cases the United Nations has assumed responsibility for monitoring the freedom of national elections. This idea that national governments can be legitimized by international standards represents a reversal of international law, which presumes the priority of national governments.

The International Court of Justice in The Hague is another branch of the United Nations that on the surface works like a branch of government. The litigant states, however, must choose to come before it, and the court has no power to enforce its decisions.

Perhaps the greatest accomplishment of the United Nations is when nothing happens; in other words, when countries are at peace. States that are at least talking to each other are seldom making war on each other. In September 2000 the United Nations hosted the greatest meeting of heads of state ever—more than 150 came together. U.S. Secretary of State Madeleine Albright pointedly attended a speech by Iranian President Mohammad Khatami, and he attended hers. Two days later U.S. President Clinton found himself standing next to Cuban president Fidel Castro and shook hands and spoke a few words. This was the first time since Castro came to power in 1959 that a sitting U.S. president recognized him. Such small civilities mark steps toward reconciliation between countries officially "not talking to each other."

U.N. special agencies   The United Nations has more than 40 specialized agencies. Some facilitate communications among member states: the Universal Postal Union, the International Civil Aviation Organization, the World Meteorological Organization, and the Intergovernmental Maritime Consultative Organization. The U.N. International Telecommunications Union, for example, has allocated frequencies on the radio spectrum, thus allowing international expansion of mobile telephones, satellite computer transmissions, and digital radio broadcasting. Other U.N. agencies encourage international cooperation toward specific humanitarian goals: the Food and Agriculture Organization; the World Tourist Organization; the United Nations Educational, Scientific, and Cultural Organization (UNESCO, Figure 13-24); the International Atomic Energy Agency; the International Bank for Reconstruction and Development; and the International Monetary Fund. The U.N. Conference on Trade and Development (UNCTAD) works to assist the economic development and trade of the poor countries. UNCTAD has, for example, developed a standardized computer customs collection system that is in use in over 50 countries. In 2000 the United Nations sponsored the Millennium World Peace Summit of Religious and Spiritual Leaders; about 800 attended from around the world.

The United Nations is also helping define principles for multinational business policies. In 2000, 50 leading multinational corporations joined labor associations and nongovernmental organizations in a pact that commits all to support human rights, eliminate child labor, allow free trade unions, and refrain from polluting. This was a declaration of principles, not a legal code, but it sets guidelines for international standards in an area of growing international concern.

Nongovernmental Organizations and Terrorist Organizations

Today states are no longer the only players on the international stage. Initiatives are increasingly taken by **nongovernmental organizations (NGOs).** These are organized international interest groups that cross state boundaries and put pressure on the existing governments of states. They include multinational corporations (p. 533), religious and charitable organizations, investors able to move their money almost instantly from one country to another and thus destabilize national economies (p. 538), but also drug cartels (p. 538), Mafia, and terrorists. The backlashes against biotechnology (p. 343) and globalization (p. 540) are led by NGOs. (It is ironic that the worldwide struggle against globalization is one of the world's most globalized activities.) Fewer activities are either purely international or purely domestic anymore.

The global civil society is unmanageable and hard to study. A world stage with millions of private empowered actors means a world of unlimited vulnerability. It is paradoxical that this is especially frightening for the United States, because the United States has done so much to destroy borders and walls, to shape a world market, and to promote freedom of communications,

FIGURE 13-24 Yellowstone National Park. Yellowstone Park was the first World Heritage Site. UNESCO supervises the Convention Concerning the Protection of the World Cultural and Natural Heritage. Signatories to the Convention—virtually all of the world's countries today—have agreed that certain areas have such unique worldwide value that they should be treated as part of the heritage of all humanity. Designated sites today range from Sinharaja Forest Reserve in Sri Lanka to the ancient city of Hatra, Iraq, Canterbury Cathedral in the United Kingdom, Dinosaur Provincial Park in Alberta, Canada, and Bolivia's historic capital of Sucre. Yellowstone National Park was the world's first national park and its 100th anniversary in 1972 inspired the global convention. (Ray Mathis/Corbis/Stock Market)

information, and movement. This modern reality demonstrates again what Henry David Thoreau and other nineteenth-century critics of communications—starting especially with the telegraph (p. 245)—argued: Accelerated and increased communication among peoples is not sufficient to guarantee good will or even understanding of one another.

Few governments interfere with the work of the charitable NGOs, such as, for example, Doctors without Borders, which provides emergency medical help in all situations around the globe. Other NGOs, however, have more militant or aggressive purposes. Some have demonstrated the willingness—and the ability—to use violence in the pursuit of their ends. The Earth Liberation Front, for example, founded in England in 1993, has engaged in acts of violence against people or activities it deems are harming the environment. In 2003, the Front claimed "credit" for destroying sport utility vehicles at dealerships across California.

Today even the ability to make war is no longer confined to state governments. Many world terrorist groups, such as Al Qaeda, are not recognized governments, nor do they represent the governments of recognized states. In much of Africa and parts of central Asia, centralized state power has effectively collapsed, and wars are fought by irregular armies commanded by political and religious organizations, often clan-based and prone to savage internecine conflicts. Rich societies cannot be insulated from the repercussions of collapsing states and the new forms of war. The attacks on the United States on September 11, 2001, proved that the ragged, irregular armies of the world's most collapsed state—Afghanistan—can reach to the heart of its richest and most powerful state.

Governments' difficulties in fighting NGOs were illustrated by the United States' first efforts to strike back at Al Qaeda. On Friday, September 14, 2001, the U.S. Congress gave President Bush the right to "use all necessary and appropriate force against those nations, organizations, or persons he determines planned, authorized, committed, or aided the terrorist attacks." The United States has fought several military engagements since World War II—in Korea, Vietnam, Nicaragua, Panama, and still more countries—but Congress has passed no declarations of war since entering World War II. This new resolution reminded many observers of the 1964 Gulf of Tonkin Resolution that had authorized U.S. attacks on Vietnam. In U.S. and international law, a declaration of war would define new sets of circumstances regarding presidential prerogatives, federal spending, civil rights, and many other domestic and international matters. In this case, however, it is unclear against what government or organization the U.S. could legally declare war. Resolutions such as the Tonkin Gulf Resolution and this new resolution create confusing legal situations in, for example, the treatment of terrorists captured abroad or even at home.

To carry on a war against terrorism, we must distinguish types of terrorists. Some have limited missions and do not see the United States as their principal enemy. In Sri Lanka (p. 455), Northern Ireland (p. 547), Chechnya, Palestine, and in the Basque provinces (p. 454), the terrorists see themselves as freedom fighters. The United States probably will not act against these.

Not all states in which terrorists operate are their willing accomplices. Some states are too weak to exert control (e.g., Lebanon), and others are insufficiently vigilant (even the United States itself). A determined project to rid the world of all rogues and terrorists would be interpreted around the world as imperialism.

Some observers fear that authoritarian rulers in some countries may use the generally approved war against terrorism as an excuse to repress any dissidents in their own countries. The United States often defends international stability, even when many peoples resent or struggle against present circumstances or political structuring. The map of states from Morocco to China, for example, was not drawn by the indigenous peoples, but by European imperialist powers, and the governments of each of these states, when each first gained independence, were installed by imperialists (see p. 452). In some cases, the original governments or even forms of government were overthrown by the populations. The United States, as leader of the Western alliance, did not always welcome what may have been genuine democratic revolutions. The original revolutions may have been democratic even if the replacement rulers turned out to be autocratic themselves, as in Libya, Iran, Iraq, Egypt, and Syria. In other countries, autocratic kings, emirs, and sheiks are still supported by the United States. America cooperates with these regimes and even defends them contrary to America's own values, so the local peoples see American forces not as symbols of freedom but as the support of a corrupt order they seek to replace.

America must, however, respond to the groups who have declared war on America. Some of these are religious fanatics who think the United States or all of the West is a machine for cultural subversion, political domination, and economic subjection. This kind of conviction feeds on experiences of despair and humiliation, and these issues can be understood and addressed.

Is National Sovereignty Inviolable?

National sovereignty means that international borders are inviolable. No matter how monstrous any regime may be, no matter how much it persecutes its own people or oppresses minorities, no outside state or international agency has the right to interfere. This is true even when internal persecution triggers international flows of refugees. The U.N. Charter insists, "Nothing contained in this present Charter shall authorize the United Nations to intervene in matters which are essentially within the domestic jurisdiction of any state." This is the legacy of the Treaty of Westphalia (see page 447).

Disarray in several states, however, and the formation of powerful NGOs may have signaled the beginnings of a change in international law. The first war against Iraq in 1991 did not overthrow Iraq's government, but the United Nations did subsequently protect some Iraqis from their own government (see the discussion of Iraq on page 578). In 1992 the Security Council voted to send a U.S.-led force to Somalia, which was wracked by civil war, to guarantee the distribution of food-relief supplies. The purpose of that mission was defined as "humanitarian," but the mission expanded to restoring central civil government to Somalia. Several states contributed troops (notably Canada and Pakistan), but Somalia has not achieved peace.

In 1999 U.N. Secretary General Kofi Annan, reviewing cases of civil war and slaughter in Africa, argued, "Nothing in the Charter precludes a recognition that there are rights beyond borders," and referred to a "developing international norm in favor of intervention to protect civilians from wholesale slaughter. . . ." The Secretary General has pressed the United Nations formally to establish the principle that massive and systematic violations of human rights must be stopped by international intervention. Many world leaders, however, agree with President Abdelaziz Bouteflika of Algeria, who argued that "interference can only occur with the consent of the state concerned." Many countries fear that U.N. intervention would serve as a cloak for Western interference. Russia's Foreign Minister Igor Ivanov said bluntly, "Human rights are no reason to interfere in the internal affairs of a state." China's President Jiang Zemin emphasized, "Dialogue and cooperation in the field of human rights must be conducted on the basis of respect for state sovereignty. . . . So long as there are boundaries between states, and people live in their respective countries, to maintain national independence and safeguard sovereignty will be the supreme interests of each government and people." Global debate over the "absoluteness" of state sovereignty will probably intensify.

The globalization of justice   After World War II, the victorious Allies tried leading German and Japanese officials for "crimes against humanity" and for "genocide." Since then, international law has affirmed that these crimes are subject to universal jurisdiction. They transcend the province of the state where the crimes occurred, and any state may prosecute. Few states, however, have been willing to do so. Therefore, in 1993 the United Nations created a War Crimes Tribunal to try Serbian leaders for "crimes against humanity" that had occurred during the fighting that accompanied the breakup of Yugoslavia. Other U.N. tribunals have investigated African wars, and another is trying the leaders of the Khmer Rouge that killed millions in Cambodia during its rule 1975–79.

In 1998 the United Nations voted to create a permanent International War Crimes and Genocide Tribunal, and two years later delegates from over 100 nations agreed on a catalog of acts that will constitute international crimes. The required 60 signatory nations ratified the treaty by 2002, so 21 Justices were sworn in and the Court began operating in March 2003. The Court must defer to national courts; its prosecutors can issue indictments only when national courts are unwilling or unable to deal with atrocities such as war crimes, crimes against humanity, and genocide. The United States originally signed the treaty to establish the court, but it never ratified the treaty, and in 2002 the U.S. government repudiated even having signed the treaty. American officials remember that in 1967 an International War Crimes Tribunal sitting in Stockholm found the United States guilty of terrorism and genocide in Vietnam, and with more troops engaged in more actions in more countries all around the world than any other country has, the United States fears that groundless charges would frequently be brought against U.S. officers and officials.

While the legal framework for international prosecution is rising, the judicial systems of individual countries have brought charges against foreign individuals. One famous case has been that of General Augusto Pinochet. General Pinochet, with the assistance of U.S. intelligence forces, led a military coup d'etat that overthrew a democratically elected government in Chile on September 11, 1973. In the following years of his dictatorship, many Chileans and foreigners in Chile were murdered; one opponent of the regime, along with several U.S. citizens, was killed by a bomb in Washington, D.C., in 1976. (In 2000 the U.S. Central Intelligence Agency released documents revealing that the Chilean responsible for the murders, General Manuel Contreras, was a paid CIA informant.) Pinochet surrendered power to an elected government in 1990. He demanded, however, and received, lifetime immunity from prosecution by the Chilean government for any crimes he may have committed while in power. In 1998, however, while visiting England, he was arrested on the order of a Spanish magistrate who wanted him extradited to Spain for crimes against Chileans and Spaniards in Chile. For 503 days the world awaited Britain's response. Finally, in March 2000, Britain sent Pinochet home to Chile, arguing that Pinochet's health precluded his facing charges in Spain; and Britain refused to bring charges itself. Upon Pinochet's return to Chile, the Supreme Court there stripped him of his immunity from prosecution, and a trial began. Thus, the courts of individual countries have been extending their jurisdiction to the citizens of other countries across international borders.

Regional Focus on

AFGHANISTAN, THE TALIBAN, AND AL QAEDA

In 1979 the Soviet Union invaded Afghanistan because Soviet rulers feared that Afghans were encouraging political unrest in the nearby Islamic Soviet republics. Afghanistan is an undeveloped, landlocked state in the middle of Asia that came into existence in the nineteenth century as a buffer between the imperial power of Britain in India to the south, and that of the Russian Empire spreading its power from the north. The population includes many ethnic groups, clans, and tribes that are quite independent of one another. Few observers would argue that there is an "Afghan nation."

In the late 1980s, Osama bin Laden, a native of Saudi Arabia, joined thousands of other Arabs who went to Afghanistan to help the Afghans in their *jihad,* or holy war, against Soviet occupation. Most of these Arabs were followers of Wahabism, the fundamentalist Islamic sect that is the official religion of Saudi Arabia (see p. 296). Bin Laden founded a group, Al Qaeda ("the foundation" in Arabic), which allied with an indigenous Wahabist Afghan group called the Taliban (meaning "student"), the members of which were largely Pathan. These united fighters called themselves *mujahedeen,* or "holy warriors." Arabs remained the backbone of Al Qaeda.

The United States assisted the Taliban and Al Qaeda in the war against the Soviet Union, although the exact degree of collaboration is a U.S. government secret. The Afghans were armed and financed with help from the United States, the government of Saudi Arabia, Libya's dictator Muammar al-Qaddafi, and Pakistan. The U.S. Central Intelligence Agency and the Pakistani intelligence service arranged deliveries of supplies, and agents of bin Laden also purchased arms in the United States.

The Soviet Union failed to subdue Afghanistan completely, but it did install a Communist government before it withdrew in 1989, after suffering some 15,000 fatalities. When this Communist government fell in 1992, the United States withdrew its support for Afghans—military, political, food, and other aid. The country fell into a many-sided civil war. Many Afghans of all backgrounds came to feel that America had used the Afghan people as pawns in the Cold War and then abandoned them.

The Taliban, together with its Al Qaeda allies, eventually gained control of about 90 percent of Afghanistan. They imposed upon the people the most severe restrictions of life and rights, following extreme fundamentalist teachings. The foreigners constituted a virtual army of occupation in Afghanistan, supporting the Taliban and exporting extremism. The Taliban repudiated religious toleration spectacularly in March 2001, when it destroyed colossal statues of Buddha carved into rock cliffs in Bamyan province some 2,000 years ago. These sacred Buddhist statues counted among the greatest of humankind's monuments.

The "Axis of Evil"

In January 2002 President George W. Bush labeled Iran, Iraq, and North Korea as an "Axis of evil, arming to threaten the peace of the world." The term "axis" was coined by Italian dictator Benito Mussolini to describe the World War II alliance of Italy, Germany, Romania, Bulgaria, and Japan. The three countries named in 2002, however, were not allied. Iran and Iraq were bitter enemies.

Iran   Iran is an important potential ally in the war against international terrorism, although alliance with Iran seems repellent to many Americans. Iran took 52 Americans hostage in 1979 and kept them for 444 days, and in November 2001 the confiscated former U.S. embassy in Tehran was converted into an anti-American museum. Iran's leaders refer to the United States as "The Great Satan" (see p. 259). The

United States insists that Iranian support of several Palestinian groups constitutes support for international terrorism.

Nevertheless, Iran is one of the Middle East's most populous countries, with 70 million people, and it is strategically located, bordering several powerful or volatile states. Iran, as a Shiite nation, was a natural enemy of the Taliban Wahabists in Afghanistan, and during the years of fighting in Afghanistan, Iran absorbed millions of refugees. Iran has made several friendly gestures toward the United States: After the attacks on the United States in 2001, Iran observed moments of silence for the American victims, and when U.S. attacks on Afghanistan began, Iran said that it would rescue any American military personnel in distress in its territory.

Iranians' major grievance against the United States remains the fact that the U.S. Central Intelligence Agency engineered the removal of a democratic

The Taliban failed to win international diplomatic recognition. Saudi Arabia recognized the Taliban government, and Pakistan provided it with considerable assistance, but most other countries and organizations, including the United States and the United Nations, recognized another government, called the Northern Alliance. This group, dominated by minority Uzbeks and Tajiks, held onto only about 10 percent of Afghanistan, in the Northeast.

The Taliban and Al Qaeda, emboldened by their triumph over the Soviet Union, then began a struggle against the United States and other countries that they believed were preventing the creation of true Islamic government throughout the Middle East. Bin Laden was a man of enormous wealth, and he used his wealth to finance several terrorist attacks. Through the 1990s, Al Qaeda was "credited" with bombings in Saudi Arabia, Kenya, and Tanzania, and with the ramming of the naval destroyer *U.S.S. Cole* by suicide bombers in the harbor of Yemen in October 2000. New York's World Trade Center Towers were bombed in 1993 by a group of terrorists that may have been affiliated with bin Laden. At least 15 of the 19 men identified as carrying out the attacks on the United States on September 11, 2001, were Saudi Arabs.

Many Islamic governments in other countries had supported the Taliban, and these governments eventually learned that they had been "playing with fire," for the extremists would eventually determine that many of these governments were themselves insufficiently religious, and the extremists would turn on them (see page 302). Pakistan supported extremists both in Pakistan and in Afghanistan, partly in order to recruit and train guerrillas committed to fight Indian rule in Kashmir (p. 308). Afghanistan became an impoverished nursery for zealots and an exporter of governmental instability to many countries. Al Qaeda formed alliances with Islamic Jihad, the Egyptian extremist group that had assassinated Egypt's President Anwar Sadat in 1981 (p. 303), and other international terrorist groups.

Russia, China, and India have long identified these groups as dangerous sources of terrorist activity. Islamic fundamentalist volunteers have appeared in the Chinese province of Xinjiang, trying to detach it as an independent Islamic republic (p. 473), as well as in the Russian provinces of Chechnya and Daghestan (p. 552), seeking to carve out independent Islamic states there.

From October until December 2001, a U.S.-led coalition invaded Afghanistan and overthrew the Taliban regime, although Osama bin Laden was not captured. Allied forces established a new government, which has moved to adopt a new constitution. This attempt at nation-building, however, continues to challenge the allies. At the end of 2003, regional warlords still ruled without interference from the capital city, Kabul, the Taliban had reformed and showed new strength, and exports of drugs increased.

government there in 1953 and replaced it with the tyrannical Shah, who was himself toppled by the religious revolution in 1979. As noted in Chapter 7, Iran is today a theocracy against which the United States encourages internal rebellion.

Other incidents in U.S.-Iranian relations still rankle Iran. In September, 2001, Iran's leader Ayatollah Khamenei criticized U.S. support of Israel (see p. 290), and then he commented, "[Americans] divide terrorism into good and bad. In the skies of the Persian Gulf they down an Iranian airliner with hundreds of passengers on board, but then they give the commander of the warship an award." In 1988 the *U.S.S. Vincennes* did shoot down an unarmed passenger liner. The U.S. government claimed it was an accident and awarded the ship's commander a medal for "exceptional conduct." The Ayatollah continued, "Evidence shows that the American government intends to repeat in Central Asia what it did in the Persian Gulf. They intend to come and establish themselves under the pretext of a lack of security here." In 2002 the United States built 13 new military bases in the nine Muslim countries from Bulgaria to Tajikistan, and many Iranians believe that America intends to establish a permanent presence in Central Asia and thus capture regional oil supplies. After the United States conquered and occupied Iraq in 2003, Iran was surrounded by U.S. troops in Iraq and in Afghanistan, and Iranians expressed greater concern about the continuing U.S. presence in the region.

Iran is developing nuclear installations. The government insists that these are only to generate electricity, but the United States insists that Iran is developing nuclear weapons. U.N. inspectors regularly visit Iranian nuclear sites, and if ever the International Atomic Energy Association votes to declare Iran in violation of the Nuclear Non-Proliferation Treaty, to which Iran is a signatory, the U.N. Security Council could then apply sanctions against Iran, even a war.

Focus on

IS THE UNITED STATES THE WORLD'S POLICEMAN?

Recent U.S. diplomatic initiatives have stimulated consideration of America's evolving view of its own role in world peacekeeping.

Early in U.S. history, statesmen stated an *isolationist* diplomatic position, that is, one that turns inward and is concerned only with its own domestic affairs. President George Washington's farewell message (1797) emphasized, "It is our true policy to steer clear of permanent alliances with any portion of the foreign world," but America could have "temporary alliances for extraordinary emergencies." President Thomas Jefferson's inaugural address (1801) similarly promised "no entangling alliances." On July 4, 1821, then-Secretary of State John Quincy Adams spoke against sending the U.S. Navy to interfere in the affairs of Spain's former colonies Colombia and Chile: "America does not go abroad in search of monsters to destroy. She is the well-wisher to the freedom and independence of all. She is the champion only of her own."

The United States took a more active role in 1823 with the proclamation of the Monroe Doctrine. It stated that whereas the United States would not interfere with existing European colonies in the Western Hemisphere, it would allow no new colonial claims. Thus, the United States defined itself as the "defender" of the Western Hemisphere.

When President William McKinley asked Congress to declare war on Spain in 1898, he injected a new consideration into U.S. foreign policy: a concern for human rights elsewhere. Referring to alleged Spanish mistreatment of Cubans, McKinley said: "It is no answer to say this is all in another country, and is therefore none of our business. It is especially our duty, for it is right at our door." The Spanish-American War involved fighting not only in the Caribbean, but around the world in the Spanish colony of the Philippines. It was this war that provoked

satirical writer Ambrose Bierce's wry observation, "War is God's way of teaching Americans geography." After defeating Spain, the United States took four more years to conquer the Philippines in order to hold it as its own colony (until 1946). The Philippines provides a humbling perspective on nation building, because although U.S. troops finally left the Philippines in 1990, they returned in 2002 to help hunt guerrillas tied to Al Qaeda; and democracy remains chronically insecure in the Philippines.

McKinley's successor President Theodore Roosevelt proclaimed the 1904 "Roosevelt Corollary" to the Monroe Doctrine, according to which "chronic wrongdoing . . . [may] . . . require intervention by some civilized nation, and in the Western Hemisphere the adherence of the United States to the Monroe Doctrine may force the United States, however reluctantly, in flagrant cases of wrongdoing or impotence, to the exercise of an international police power." This is a clear proactive stance.

After World War I, President Woodrow Wilson (who served 1913–21) repudiated isolationism and announced U.S. readiness to join in a permanent association of nations working for peace. His inability to compromise with Congress, however, led to the failure of the United States to join the League of Nations. After World War II, the United States helped found the United Nations, and the 1947 Truman Doctrine (President Harry Truman served 1945–53) pledged U.S. assistance to any country anywhere fighting a takeover by Communists. It was to be presumed that communism emanated exclusively from the U.S.S.R., and was, therefore, a foreign attack upon an independent nation. This policy came to be called one of "containing" communism during the Cold War. The degree to which individual Communist movements around the world were in fact initiatives of the U.S.S.R. or else were homegrown nationalist movements will long be debated.

President Jimmy Carter (who served 1977–81) accepted the necessity of accepting as our allies in the

Iraq   Saddam Hussein was an officer in the Iraqi army who became Iraq's president and virtual dictator in 1979. The next year he launched a war against neighboring Iran, which was then a U.S. adversary, and in 1983 U.S. President Ronald Reagan sent Donald Rumsfeld (later U.S. Secretary of Defense in the George W. Bush administration) to affirm U.S. support for Iraq. The amount and type of military support the U.S. gave Iraq

remains a secret. Iraq and Iran fought until both were exhausted in 1988.

In 1990 Iraq invaded Kuwait, claiming (with considerable justification) that the territory of Kuwait was in fact historically Iraqi, and that existence of the Kuwaiti state was an illegitimate relic of colonial rule. In 1991 the United States and several allies, fighting under U.N. auspices, forced Iraq to retreat from Kuwait, but

Cold War some countries that were by no means truly "free," but part of his legacy was renewed concern with the protection and encouragement of human rights in other countries.

The attacks on the United States on September 11, 2001, changed many Americans' isolationist world view and ended American willingness to respect the sovereignty of states that harbor terrorists. On September 20, 2001, President George W. Bush told Congress, "We will . . . drive [terrorists] from place to place, until there is no refuge or no rest. And we will pursue nations that provide aid or safe haven to terrorism. Every nation, in every region, now has a decision to make. Either you are with us, or you are with the terrorists. From this day forward, any nation that continues to harbor or support terrorism will be regarded by the United States as a hostile regime." Shortly thereafter, the president emphasized: "Every nation has a choice to make. In this conflict, there is no neutral ground. If any government sponsors the outlaws and killers of innocents, they have become outlaws and murderers themselves. And they will take that lonely path at their own peril." Many observers feel that with these words, President Bush repudiated even the 1648 Treaty of Westphalia (see Chapter 11).

The world has shrunk, and the United States no longer stands relatively safe and isolated behind ocean barriers as in George Washington's day. President Bush argued that if other countries cannot control terrorists who operate from within their borders, then the United States does not have to recognize the territorial sovereignty of those countries, but has the right preemptively to destroy enemies of America. If we accept that preemptive strikes against enemies may ever be necessary, then secrecy may be needed; how can such an issue be debated in a democracy? Who should make the decision to attack first, and to whom should the action be explained or justified? Could it even be necessary for the government to withhold information or sources in order to protect the sources? Questions such

as these prompted the young Congressman Abraham Lincoln to write in 1848, when he was protesting the U.S. War against Mexico, "Allow the President to invade a neighboring nation whenever he shall deem it necessary to repel an invasion . . . and you allow him to make war at pleasure. If today he should choose to say he thinks it necessary to invade Canada to prevent the British from invading us, how could you stop him? You may say to him, 'I see no probability of the British invading us,' but he would say, 'Be silent: I see it if you don't.' Lincoln's own later experience as president, however, convinced him of the necessity for presidential secrets. Questions such as these have always plagued diplomats and political theoreticians.

The debate continues within the United States. In 2002, Patrick Buchanan, senior advisor to three Republican presidents and the Reform Party's presidential candidate in 2000, wrote, "What happened on September 11, 2001, was a direct consequence of an interventionist U.S. policy in an Islamic world where no threat to our vital interests justifies our massive involvement. We are a republic, not an empire. And until we restore the foreign policy urged upon us by our Founding Fathers—of staying out of other nations' quarrels—we shall know no end of war and no security or peace in our own homeland." Today, however, in contradiction to the isolationist tradition in American foreign policy, perhaps the United States must "go abroad in search of monsters to destroy" in its own defense.

Another official U.S. government document, "The National Security Strategy of the United States of America," released in October 2002, emphasizes that "Our forces will be strong enough to dissuade potential adversaries from pursuing military build-up in hopes of surpassing, or equaling, the power of the United States." This document suggests to many readers that the United States has defined for itself a role as preeminent global policeman.

they neither conquered Iraq nor brought down Hussein's government. The United Nations did, however, establish "secure zones" within which the Iraqi government was prevented from attacking its own citizens who had cooperated with the U.N. allies (Figure 13-25). Shiites were protected south of the 33rd parallel, and the Kurds enjoyed a sanctuary north of the 36th parallel; there, through the following years, the Kurds

created a virtual Kurdish free state. This was much to the annoyance of the Turks, who only in 2002 lifted a 15-year state of emergency over Turkey's Southeast, where the Turks feared Kurdish separatism (see p. 453).

The 1991 war devastated Iraq's economy and infrastructure, and the United Nations subsequently allowed Iraq to export oil only under international supervision and only in exchange for food and medicine.

Protected Zones in Iraq

Occupied by Allied Forces, February 27–May 6, 1991

Area controlled by Kurdish nationalists since March 1991

Area established as a "safe haven" for Kurdish refugees by Allied Forces since May 1991

FIGURE 13-25 The United Nations war against Iraq in 1991 did not topple Iraq's government, but by the time of the second war 12 years later the United Nations was still protecting certain Iraqis from their own government.

Saddam Hussein awarded those supplies to his loyal soldiers rather than to Iraqi children, and the deaths of thousands of Iraqi children brought Iraq considerable world sympathy. Much wrath was directed at the U.N. sanctions rather than, perhaps more properly, at Saddam Hussein.

The United Nations also insisted that Iraq allow U.N. monitors to guarantee that Iraq was not producing nuclear, chemical, or biological weapons of mass destruction (WMDs). After Iraq drove out those inspectors in 1998 and repeatedly frustrated efforts to ensure that Iraq had no WMDs, the United States invaded Iraq again in 2003. This time acting without U.N. sanction, the United States was joined by few allies (notably the United Kingdom). It was argued that Iraq had WMDs, that it assisted international terrorist organizations, and that Saddam Hussein was persecuting his own people. Neither of the first two of these arguments had been confirmed by the autumn of 2003. The Second War brought down the Hussein government within weeks, but American efforts to establish a new government in Iraq met considerable difficulty. Few nations were willing to help rebuilding and nation-building efforts without U.N. sanction, and, as of autumn 2003, the situation in Iraq remained challenging. The headquarters

of U.N. aid missions in Iraq was blown up with considerable loss of life in August 2003. Many observers had feared that a U.S. invasion of Iraq would not quash terrorism, but, rather, stimulate terrorist activities around the globe. Indeed, within weeks of the U.S. overthrow of the Hussein government, terrorists set off bombs in Saudi Arabia, Indonesia, Morocco, and elsewhere.

Korea After the 1950–53 war in which United Nations forces defended South Korea against North Korea, South Korea prospered while North Korea stagnated as a Communist dictatorship under first Kim Il Sung (who ruled 1948–94) and then his son Kim Jong Il. North Korea has remained uniformly hostile to the United States, perhaps partly as a way of binding its own people, and it has created the most paranoid, totalitarian state in history while millions of its people starved to death (Figure 13-26). As late as 2003, the United States maintained 37,000 troops in South Korea in case of another North Korean invasion. Nevertheless, as South Korea prospered, it began to resent the American presence and domination and to yearn for reconnection with the North (see Figure 11-10). In fact, South Korea can hardly absorb the thousands of refugees it receives each year; trying to unite and harmonize its economy, culture, and politics with North Korea would be immeasurably harder even than it was for Germany to achieve unification.

North Korea almost certainly has biological weaponry, nuclear capability, and the capability to deliver these weapons by missile to it neighbors and possibly even to the United States. It has provided weaponry to many global terrorist organizations and hostile states. The United States and North Korea agreed in 1994 that, in exchange for regularization of diplomatic relations, trade, and aid, North Korea would end its

FIGURE 13-26 "Ruthless Punishment to U.S. Imperialism." Posters such as these are found everywhere in North Korea, exhorting the people to sacrifice in "defense" of their homeland, although the attack on the U.S. Capitol here is clearly offensive. (AP/Wide World Pictures)

nuclear programs. Neither side, however, fulfilled the terms of that agreement, and negotiations have continued fitfully. China has long defended and aided North Korea (providing up to 80 percent of its food and oil at concessionary prices), and it does not want the United States to attack. China also fears continuing mass migration of desperately poor North Koreans, who early in the twenty-first century already virtually overran China's contiguous regions. Japan fears North Korean nuclear arms, and many states fear that Japan may be provoked to develop its own nuclear capability. Japan disarmed after World War II, and, despite Japan's possible contribution of peacekeeping forces in Iraq, Japan has little military capability.

Human Rights

Global government presumes global acceptance of fundamentals of a political culture, which rests in turn on fundamentals of human culture. Two hundred years ago Thomas Jefferson could confidently write that certain truths regarding human rights were "self-evident," despite the fact that he himself owned slaves.

The U.N. General Assembly adopted a Universal Declaration of Human Rights very early, in 1948. At that time, however, most U.N. members were Western countries, and they were still holding most African and Asian societies as colonies. As these societies have achieved political independence, they have begun to challenge Western ideas of human rights. They deny that any human rights are in fact universal and "self-evident." Human rights, they insist, may vary from culture to culture.

Two questions have arisen in international organizations. First, what is the balance between the rights of individuals and those of the whole society? Even Western countries disagree on this issue. Canada, for example, is more authoritarian than the United States. Canada's equivalent to the United States's Bill of Rights is the Charter of Rights and Freedoms, passed only in 1982. It reflects a commitment to gender equality and social equality far beyond that of U.S. documents, but its commitment to civil liberties is arguably less. It guarantees "freedom of thought, belief, opinion, and expression," but these are subject "to such reasonable limits prescribed by law as can be demonstrably justified in a free and democratic society."

The second question is the balance between political rights and what some people call "economic rights." In 1997 President of China Jiang Zemin insisted, "As a developing country of 1.2 billion people, China's very reality determines that the right of subsistence and development is the most fundamental and most important human right in China. Before adequate food and clothing is insured for the people, the enjoyment of other rights would be out of the question."

At a 1993 U.N. Conference on Human Rights, many countries were concerned that Western countries would use their definitions of human rights as excuses for withholding aid or other economic benefits from developing nations that may not accept the same definitions. Singapore's Foreign Secretary Kishore Mahbubani insisted, "Too much stress on individual rights over the rights of the community will retard progress. . . . In the future, these agreements will assert the rights of society over the rights of individuals." Indonesia's foreign minister stated flatly that "Human rights exist as a function of history, culture, value systems, geography, and phases of development." Several Asian countries jointly offered a statement that fairness and justice should be measured against "regional particularities and various historical, cultural, and religious backgrounds." An African statement demanded that observers must take account of "the historical and cultural realities of each nation." Not all cultures recognize the equality of women with men, and the assembled nations could not even agree on a condemnation of torture. They insisted, however, that "the right to development is inalienable." The United States hesitates to accept a statement that development is a "right," because that could lead to demands that the rich countries redistribute their wealth among the poor countries.

After many days of debate a final document was adopted that says that while differences of "historical, cultural and religious background must be borne in mind, it is the duty of states, regardless of their political, economic and cultural systems, to promote and protect all human rights and fundamental freedoms." In other words, the states agreed to disagree.

Many observers feel that in January 2003 the United Nations embarrassed itself by awarding the chairmanship of its Commission on Human Rights to Libya, where opponents of the dictator Moammar Gadhafi are routinely tortured to death, whose agents have planted terrorist bombs throughout Europe, whose police have committed racist attacks on black African refugees, and who forbids visits by the U.N.'s own human rights investigators. The chairmanship is not awarded by merit but by rotation. In 2003 it was Africa's turn to hold the chairmanship, and the African Union chose Libya. The choice is usually made by consensus, but this time the United States pushed for a recorded vote. Thirty-three votes went to Libya, 17 countries abstained, and three countries (the United States, Canada, and Guatemala) voted no.

Jurisdiction over Earth's Open Spaces

One of the most important issues requiring international agreement is the adjudication of the rights that individual states have in Earth's open spaces. Open spaces include the Arctic, Antarctica, and the world's seas.

The Arctic and Antarctica Eight countries have territory north of the Arctic Circle: the United States, Canada, Denmark, Finland, Iceland, Norway, Sweden, and Russia. In 1996 these countries formed an Arctic Council pledged to cooperate in monitoring Arctic pollution and in protecting the region's plant and animal life.

Canada claims dominion over the uninhabited Arctic islands north of Canada all the way to the North Pole, but the United States has challenged Canada's claims. Concern is rising over this international dispute for two reasons. One is that the United States continues to attempt to realize a passage for shipping from Alaska's North Slope across the Arctic to the U.S. East Coast, and stronger ice-breakers and the melting of Arctic ice by global warming make the realization of such a passage more realistic. Second, the development of larger airplanes with increasing ranges of flight, plus the end of the Cold War, increase the feasibility of transpolar passenger flights from North America to Asian locations. Canada insists on the right to regulate flights over its territory.

Antarctica has never had any permanent inhabitants, but seven states claim overlapping sovereignty. Australia, New Zealand, Chile, and Argentina, the world's most southerly states, claim sovereignty on the basis of proximity. The claims of Britain, Norway, and France are based on explorations (Figure 13-27).

FIGURE 13-27 The United States does not recognize any of these overlapping claims to Antarctica.

Neither the United States nor Russia, which support most of the scientific research there, make any territorial claims, nor do they recognize other nations' claims. Forty states are parties to a 1959 Antarctic Treaty that resolves that "Antarctica shall continue forever to be used exclusively for peaceful purposes and shall not become the scene or object of international discord." The treaty prohibits military installations, but it opens the continent to all nations for scientific research and study. The treaty also prohibits signatories from harming the Antarctic environment.

Despite the terms of this treaty, however, some countries, led by Malaysia, demand that exploration should be undertaken for Antarctic resources and that any resources found should be exploited for the good of all nations. Other nations, led by France and Australia, insist that Antarctica should be set aside as a wilderness preserve.

Territorial waters   The world's states have divided up all the land on Earth (except Antarctica), but international agreement negotiates the question of how far out to sea the territorial claim of a country can reach.

The Dutch jurist Hugo Grotius (1583–1645) first argued that the world's seas are open space, *mare liberum*, which no political unit can claim. Each coastal state, however, has claimed coastal waters for defensive purposes. A later Dutch jurist, Cornelius van Bynkershoek, accepted this in his book *De Domina Maris* (1702). The limit of sovereignty was set at three nautical miles from shore, and this was generally accepted for more than 200 years. All distances at sea are measured in international nautical miles of one minute of latitude—1,852 meters (6,076 feet). Territorial sea is measured from the low-tide mark, and disputes between states arising from irregular coastlines or islands are resolved by individual negotiations. A 1958 International Conference on the Law of the Sea set rules for ascertaining the seaward extension of land borders (Figure 13-28).

The United States accepted a 3-mile limit in 1793, but in 1945 President Truman broke that international covenant and claimed sole right to the riches of the continental shelf up to 200 nautical miles out. A **continental shelf** is an area of relatively shallow water that surrounds most continents before the continental slope drops more sharply to the deep-sea floor. The 1958 Sea Law Conference agreed on a water depth of 200 meters (656 feet) as the definition of the outer edge of a continental shelf (Figure 13-29).

The Truman Proclamation claimed shelf mineral rights, but it did not claim control over fishing or shipping in the seas over the shelf beyond the three-mile territorial limit. It nevertheless triggered extended claims to fishing rights by other nations. In 1976 the United States extended its own claims to exclusive fishing rights up to 200 nautical miles out. As noted in Chapter 8, about 90 percent of the world's marine fish harvest is caught within 200 miles of the coasts.

A Disputed U.S.-Canadian Border

FIGURE 13-28 The United States and Canada dispute the seaward extension of their border here in the Arctic, and the United States has auctioned off the rights to explore for oil in the disputed tracts of the Beaufort Sea.

The 1982 United Nations Convention on the Law of the Sea In 1982 the United Nations proposed a new Law of the Sea Treaty. Sixty countries ratified it by 1993, so it came into force among the ratifying parties in 1994. The Law of the Sea Treaty authorizes each coastal state to claim a 200-mile exclusive economic zone (EEZ), in which it controls both mining and fishing rights. Therefore, possession of even a small island in the ocean grants a zone of 326,000 square kilometers (126,000 square miles) of sea around that island. This might help explain why France retains many small island colonies. France claims a full 10.4 percent of the world total EEZ. The United States claims 8.4 percent, New Zealand 5.8 percent, and Indonesia 4.7 percent. No other state claims more than 4 percent. The potential for discoveries of oil in the world's seas has triggered international scrambles to claim even tiny uninhabited outcrops of rock. For example, China, Japan, and South Korea lay conflicting claims to areas of the East China Sea. Vietnam, China, Taiwan, the Philippines, Thailand, and Malaysia extend overlapping claims in the South China Sea.

Another clause of the 1982 treaty guarantees ships **innocent passage** through the waters of one state on the way to another. The extension of countries' territorial waters, however, means that countries now claim many of the world's narrow waterways. They remain open in times of peace, but they can be closed in time of war (Figure 13-30).

The United States at first refused to sign the Sea Law Treaty in protest against provisions calling for internationalization of seabed mineral resources and the creation of an International Seabed Authority to control mining. The United States insisted on rights for free enterprise. Several of the treaty's other provisions, however, did serve U.S. interests, so U.S. President Reagan announced that the United States would regard all but the seabed provisions as law, even though

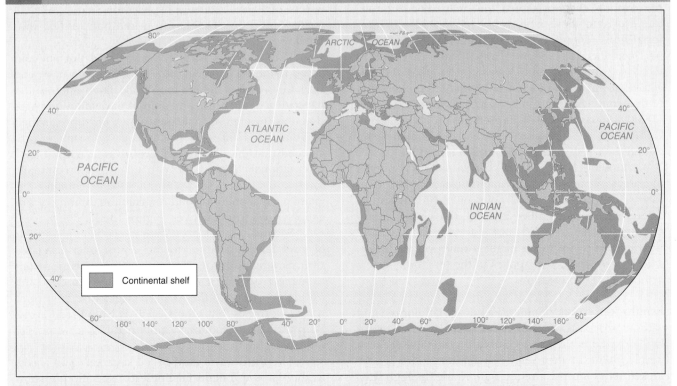

Continental Shelves

FIGURE 13-29 The various continental shelves extend out from as little as a few hundred yards to more than 600 miles.

Landlocked States and Strategic Straits

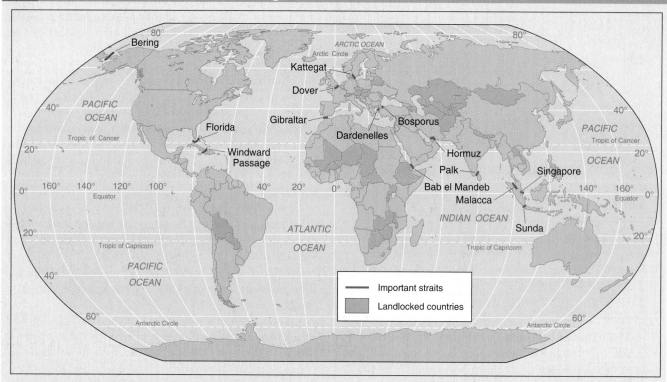

FIGURE 13-30 This map shows the world's landlocked states and some of the world's most important narrow sea lanes. As states have extended their seaward claims, many of these narrow waterways have been claimed by the adjacent states. A terrorist act in one of these narrow straits could disrupt the world economy. The Straits of Malacca, for example, is at points only 2.5 km (1.5 miles) wide, but it carries one-third of world trade.

the country had not signed the treaty. For example, one provision allows nations to declare a 12-mile territorial limit. Therefore, in accordance with the treaty, on December 28, 1988, the United States extended its territorial limit to 12 miles. Given the length of the U.S. coastline, this seaward extension enlarged the territory of the United States by some 478,000 square kilometers (185,000 square miles). Compromises on disputed provisions were eventually reached, and the United States signed the treaty in 1994 and ratified it in 1996.

Another clause of the 1982 treaty allows countries to claim some powers in a "contiguous zone" of 12 miles further out beyond the 12 miles of territorial waters. Therefore, on September 2, 1999, the United States announced that it would start enforcing U.S. law and boarding ships up to 24 nautical miles off the coast. The U.S. proclamation applies to the territorial waters around all U.S. possessions.

Landlocked states Some 42 states are landlocked and without seacoasts. Several other states are not totally landlocked, but their own coasts are unsuitable for port development, so they rely on neighbors' ports. Landlocked states must secure the right to use the high seas, the right of innocent passage through the territorial

waters of coastal states, port facilities along suitable coasts, and transit facilities from the port to their own territory.

Landlocked or partially landlocked states may gain access to the sea in one of three ways. First, any navigable river that reaches the sea may be declared open to the navigation of all states. Freedom of navigation on rivers that flow through several countries was first proclaimed by France in 1792. The revolutionary government proclaimed that the freedom of rivers was a "natural law." International commissions regulate navigation on many international rivers, and often these same commissions guard against pollution and regulate the drawing of irrigation waters from the river (Figure 13-31).

Second, a landlocked state may obtain a corridor of land reaching either to the sea or to a navigable river. Several countries have long, thin extensions out to seaports (Figure 13-32). Some of these, such as the Congo's corridor to the Atlantic Ocean, are important transport routes, but others, such as Namibia's Caprivi Strip to the international Zambezi River, serve no significant traffic function.

The third way a landlocked state can gain access to the sea is to obtain facilities at a specific port plus freedom of transit along a route to that port. Coastal

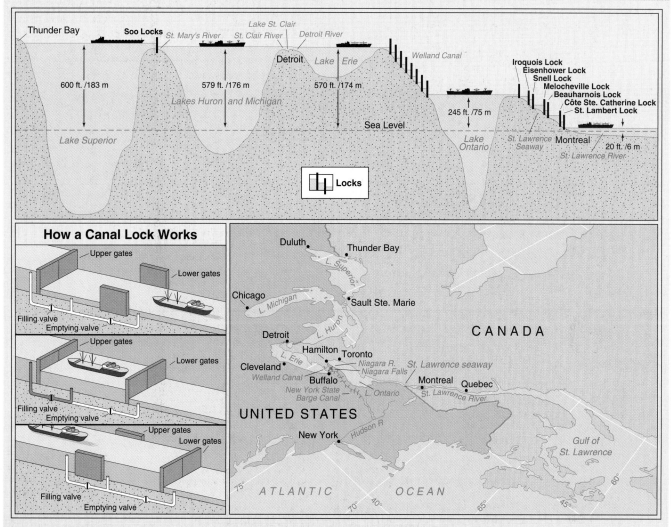

The Saint Lawrence Seaway

How a Canal Lock Works

FIGURE 13-31 Completed jointly by the United States and Canada in 1959, the Saint Lawrence Seaway is an example of international cooperation on an international waterway. It allows ocean vessels to reach from the Atlantic Ocean into the Great Lakes; the joint project included the construction of hydroelectric power plants.

states have signed international conventions promising to assist the movement of goods across their territories from landlocked states without levying discriminatory tolls, taxes, or freight charges. Chile, for example, helped build a railroad connecting La Paz, Bolivia's principal city, to the Chilean port of Arica, and Chile guarantees free transit. Argentina grants Bolivia a free zone at the Argentine city of Rosario on the Paraná River, and Peru gives Bolivia a free trade zone in the port of Ilo. In 1993 Ethiopia joined the ranks of the landlocked states when its coastal province of Eritrea gained independence. Eritrea promised to assist Ethiopian import and export trade, but in fact most of Ethiopia has long relied on transit via the port of Djibouti (look back at Figure 12-27). The 1998 treaty between Ecuador and Peru guaranteed Ecuador navigation rights in Peruvian ports on the Amazon River.

Coastal states can profit by granting transit rights. The coastal state's railroads have a captive customer. Coastal states' ports gain extra business and opportunity to serve as break-of-bulk points for processing imported or exported raw materials (see Chapter 10). The fact that a landlocked state misses these opportunities can hinder its economic development. Landlocked states also lose out in the apportionment of the resources of the continental shelves, exclusive economic zones, and fishing opportunities.

The high seas All nations agree that beyond territorial waters are high seas, where all nations should enjoy equal rights. Several problems, however, threaten any area where rights are not specifically assigned. These include depletion of resources (such as overfishing) and pollution. International conventions regulate each of these

Precious Access to the Sea

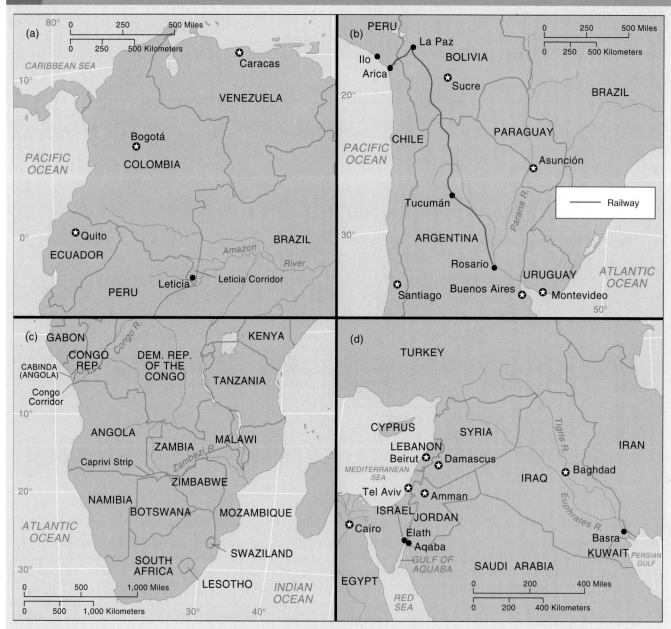

FIGURE 13-32 Several countries that would otherwise be landlocked have narrow corridors of territory reaching to provide access to the sea or to a navigable river. Alternatively, neighboring countries grant the landlocked state special railroad access to ports, as in (b). Sucre is the constitutional capital of Bolivia, although La Paz is the seat of government.

potential abuses of the high seas. The 1972 London Dumping Convention, for example, signed by 70 countries, originally banned the dumping of radioactive and other "highly dangerous" wastes at sea, but it was extended to ban dumping of all forms of industrial wastes in 1995. The enforcement of international agreements such as these is uncertain, and their effect is limited.

Airspace   The question of how far up a nation's sovereignty extends is as complex as the question of how far

out to sea it extends. Ships enjoy innocent passage through territorial waters, but airplanes have never been granted innocent passage to fly over countries. Airlines must negotiate that right, and states commonly prescribe narrow air corridors, the altitudes at which aircraft must fly, and even the hours of the day the passages are open. U.S. President Dwight Eisenhower first proposed an "Open Skies" treaty in 1955, and a treaty was finally realized in 1992. The treaty establishes conditions for unarmed observation flights over the entire territory of participants. In accordance with the treaty, a Russian Air Force jet spent five days in August 1997 flying over the United States taking photographs of U.S. military installations. This astonished anyone who grew up during the Cold War.

Diplomats cannot agree on the altitude at which national airspace ends. One possible definition of national airspace would limit it to the lowest altitude at which artificial unpowered satellites can be put into orbit at least once around Earth. That ranges between 113 and 161 kilometers (70 and 100 miles). The U.N. Office for Outer Space Affairs does try to formulate recommendations for designing and flying space vehicles to reduce the amount of debris they produce and their chances of colliding with one another.

No country wants to be the first to militarize space, but suggestions have been made that the eventual parceling-out of space is almost inevitable, as a measure of national security. In October 1997 the United States successfully targeted a laser beam on an orbiting satellite. Many countries objected that this was the first step to the militarization of space. Countries have so far refrained from extending territorial claims to the Moon or planets, but several have insisted that they reserve the right to do so (Figure 13-33). Our study of geography, therefore, is so far restricted to Earth.

FIGURE 13-33 Moon exploration. When U.S. astronauts visited the Moon, they planted U.S. flags. The United States denies that the gesture was intended as a claim. (NASA Headquarters)

Focus on

REGULATION OF WHALING

Whales are an excellent example of a commonly owned natural resource that can only be managed by international negotiation and agreement. Whales have been hunted for thousands of years, but the development of large-scale commercial whaling in the eighteenth and nineteenth centuries reduced many species populations to dangerously low levels. By the 1920s whalers began to realize that some species had been overfished, and in 1946 the International Whaling Commission (IWC) was formed to regulate the industry. In its early years the IWC's conservation achievements were modest at best, but by the 1980s widespread support for whale conservation had developed in much of the world. In 1986 the IWC declared a worldwide moratorium on whaling except for purposes of scientific research. The moratorium also allowed continued whaling by indigenous Arctic peoples in northern Russia and Alaska. Member nations retain their sovereignty over their whaling fleets,

however, and the IWC cannot enforce its restrictions. Norway and Japan continue to harvest whales, though fewer than were killed before the IWC was formed. In late 1997 the IWC agreed to ban whaling completely on the high seas but to allow commercial whaling in coastal waters. An agreement of this type effectively limits whaling to those areas where national sovereignty can be exercised. If it holds, it will allow more effective management of whaling and reduce the likelihood of overhunting.

International regulation of whaling contrasts with the lack of controls over harvesting other marine species. As described in Chapter 8, there is no effective worldwide regulation of fishing, and only in a few cases (most notably the European Union's Common Fisheries Policy) have nations been able to agree on regulation of fisheries. Even close friends like the United States and Canada have bitter disputes over shared fisheries, and the consequence has been worldwide devastation of fish populations similar to that experienced prior to regulation of whaling.

PROTECTING THE GLOBAL ENVIRONMENT

Just as the world has become more integrated and interdependent in recent decades, so has it become apparent that we share common concerns about environmental quality. While local environmental quality issues such as urban air pollution remain important—and are growing in importance in industrializing regions—global environmental issues have risen in importance. International governmental and nongovernmental organizations are increasingly involved in finding ways to manage concerns about the envioronment in a changing world.

Changing patterns of economic activities at the global scale mean changing patterns of resource consumption and pollution outputs. While most of the world's industrial activity and most output of industrial pollutants are still concentrated in the rich countries, a significant shift in pollution output to the poorer countries is occurring. Two factors are driving this shift. First, heavy manufacturing, and the pollution that accompanies it, is shifting from rich countries to poor ones. Second, population is growing faster in poor countries than in rich ones, so consumer-generated pollution such as sewage, solid waste, and automobile emissions are growing rapidly in poor countries. Unfortunately, the poor countries are less able to afford pollution-control systems, so their populations increasingly suffer the negative effects of living with pollution.

Energy consumption The poor countries are using rapidly rising amounts of natural resources, whereas the use of natural resources in the rich countries is growing slowly. The rich countries were provoked to decrease their natural resource use and to increase the energy efficiency of their economies by the rise in energy prices in the 1970s. The most obvious example of this increase in energy efficiency is the change in automobiles between the early 1970s and the 1990s. Modern autos are much more fuel efficient: A typical mid-1990s five-passenger sedan travels at least 50 percent farther on a liter of gasoline than a comparable-sized car built 20 years earlier. At the same time, since the 1980s Americans have been buying larger vehicles, especially minivans and sport utility vehicles, that are less efficient than the sedans and station wagons that were more common earlier. This has caused a reduction in our overall vehicle fuel efficiency.

Air pollution-control regulations are another factor in this increased efficiency. Using the auto as an example again, one of the easiest ways to reduce emissions is to improve the efficiency of fuel use. The replacement of carburetors and mechanical ignition points with computer-controlled fuel injection and electronic ignition both reduces pollution output and improves fuel efficiency. Similar changes that have occurred throughout all industries have reduced both the amount of resources used and the pollution generated.

At the same time that the rich countries were increasing their efficiency of fossil-fuel use, they were also experiencing a rapid economic transition away from energy- and pollution-intensive heavy industries toward light-industry and service-based economies. Steel manufacturing and related heavy-equipment industries were especially hard hit by the oil price increases of the 1970s. The steel industry in the United States, Canada, and much of Europe was aging, and many mills used less efficient, older technologies. These mills also faced increased competition from other regions. Developing-world countries built their own steel industries, and global production capacity was greater than demand. The domestic automobile industry in the United States and Canada faced increasing competition from Japan. When fuel prices rose, consumers demanded smaller foreign-made cars, not the gas guzzlers that the North American plants were producing. Increases in energy costs caused a general slowdown in manufacturing. The combined effects of these factors contributed to the decline of heavy industry in the northeastern United States and adjacent Canadian cities, in northern Britain, and to a lesser extent, in Germany. Along with this came a significant drop in fossil-fuel use, which helped to reduce pollution emissions in these regions (Figure 13-34).

The decline of the traditional heavy industrial regions of the developed world was accompanied by the growth of manufacturing in rapidly industrializing regions of east Asia and Latin America. Much of this manufacturing, however, is labor- and capital- rather than energy-intensive, so it has made only a minor contribution to increased energy consumption in these regions. The increase in wealth that this manufacturing brings, especially in urban regions, is much more significant. This added wealth raises demand for energy-consuming machines, especially automobiles. The number of autos in use in China, Malaysia, Brazil, and other nations with rapid economic growth rates is rising, so petroleum use and air pollution also rise.

Energy efficiency trends It takes energy to generate income. Some ways of generating income take more energy than others, however. Making steel uses large amounts of fuel per dollar earned, whereas making televisions requires much less. Typically, manufacturing takes much more energy than providing services. Also, some ways of making a given product are more efficient than others. Modern factories or office buildings usually use much less energy per unit of output or office space than old ones. We can see changes in energy efficiency by examining the relationship between how much economic activity changes in relation to changes in energy use.

The linkage between economic and environmental trends over the past two decades can be seen in

(a)

(b)

FIGURE 13-34 Changing locations of heavy industry. (a) An abandoned steel mill in Volkingen, Germany, and (b) a steel mill in Benxi, Liaoning Province, China. The locus of heavy industry in the 1950s and 1960s was in Europe, North America and Japan, but today much of this industry has shifted to developing nations, and along with it, much energy consumption and pollutant emissions. ([a] Haackenberg/Masterfile Corporation; [b] Michael Wolf/Aurora & Quanta Productions Inc.)

trends of economic efficiency of energy use. The amount of energy used to produced a dollar of GDP (also known as *energy intensity*) is shown for a few representative countries for the period 1980–2001 in Table 13-1. In this period, the GDP of the United States grew at an average rate of 3.1 percent per year (adjusted for inflation), while energy use grew at only 1.0 percent per year. In 2001 a unit of energy use produced about

50 percent more income in the United States than it did in 1980. Most of the wealthy industrial powers experienced a similar pattern. Income grew faster than energy use. In contrast, in many poorer countries, energy use grew faster than income. Thailand, for example, experienced economic growth averaging 5.9 percent per year (in constant U.S. dollars), but energy use grew at an average of 8.6 percent per year. In some poor countries, notably China, economic growth exceeded growth in energy use. Still, the general trend over that time period is for faster growth of energy use in the poorer world than in the wealthy world.

Energy use is an imperfect measure of pollution, but it is a useful one. In general, the more fossil fuels we use, the more pollution we produce. Certainly this is true for carbon dioxide emissions. For other pollutants, the correspondence is not so close, and pollution depends a lot on how the energy is used. Nevertheless, to the extent that energy use can be equated with pollution output, both energy use and pollution are increasing rapidly in the poor world, especially in those countries with high economic growth rates.

Development, Pollution, and the Quality of Life

Both monetary wealth and environmental conditions contribute to our quality of life. They contribute in different ways, however. Monetary income provides us with the ability to purchase sufficient quantities of good-quality food and electric or gas stoves for cooking, routine medical checkups, and safe water delivered

TABLE 13-1 Energy Use per Dollar of GDP (Constant 1995 Dollars)

	ENERGY USE (THOUSAND BTUS PER DOLLAR)		Percent Change
	1980	2001	
China	106	36	−66
United States	16	11	−34
Canada	23	17	−25
France	7	6	−20
Japan	5	4	−16
Zimbabwe	39	34	−12
Mexico	17	16	−5
India	26	25	−2
Korea, South	11	13	12
Nigeria	6	8	32
Turkey	11	15	33
Malaysia	13	20	54
Thailand	10	17	69

Source: Energy Information Administration

to our homes. It also makes available automobiles, telephones, computers, and large, warm houses and vacations at the shore. Good environmental quality gives us clean air to breathe, fish that are free from disease-carrying bacteria and toxins, and soil that produces good crops without excessive inputs of fertilizer. It also gives us diversity of wildlife to observe and enjoy, parks for relaxing, and the satisfaction of knowing that future generations will have at least some of the same opportunities to enjoy these things as we do.

> Both monetary wealth and environmental conditions contribute to our quality of life.

Economic activity and environmental quality both contribute things that make life materially and observably better in ways that virtually everyone agrees are important, especially health-related factors such as occurrence of disease, infant mortality, or longevity. Wealth helps prevent disease by paying for educational expenses and salaries of medical personnel, as well as the offices, laboratories, medicines, and supplies they need to do their jobs well. A clean environment also helps to prevent diseases such as chronic lung problems that occur in areas with bad air pollution. Wealth helps prevent infant mortality and improves longevity by making possible balanced diets that include sufficient protein and fresh vegetables regardless of season. A clean environment that yields drinking water that is free from chemical and biological contaminants also helps reduce infant mortality and lengthen lifespans.

While everyone needs good health and sufficient food supplies, the value of other things may vary considerably from one individual to another, or one culture to another. Clean water in a lake may be especially important to one who likes to swim in lakes or to one who likes to eat fish. But for someone who spends every day in a city and prefers movies and basketball games for entertainment, clean lakes may not be very important. Wealth and environmental quality provide amenities that are valued differently by different cultures. Wealth brings us televisions, on which we can watch comedy or sports or the news. Such entertainment is a central part of some cultures today, but it would be hard to argue on objective grounds that this makes us better people or that it makes our lives materially better than alternative activities such as conversation, participant sports, or reading. A wilderness park can be an exhilarating place to hike, learn, or simply contemplate the beauty of nature, but who is to say that this is a more meaningful experience than attending an opera? The value of many of the goods we produce through extracting resources and polluting the environment depends solely on individual or cultural attitudes.

Balancing development and environmental preservation If there were simple and effective ways to increase both wealth and environmental quality

simultaneously, then the process of development would be a straightforward one. Many of the paths to greater monetary wealth, however, reduce environmental quality, and improving environmental quality often has significant costs.

In the wealthy countries of North America and Western Europe, industrialization generated a standard of living and level of technology that were high enough so that people could give cleaner air a high priority, and these countries could afford to make the needed investments in pollution control. Wealth created the pollution, as consumers bought the things that were made in the smoke-belching factories, but wealth also provided the impetus for eliminating the pollution.

In many poorer countries the situation is quite different. The contrast in wealth between rich and poor countries is enormous. People in poor countries are anxious to raise their standards of living, and rapid industrial development is seen as the quickest way to gain that wealth. Governments thus channel any available funds toward industrial development and actively seek foreign investment in new plants. Environmental protection may be a concern, but it is certainly a secondary one, so environmental regulations that would slow development are few and weak.

The industrial growth that is taking place in poorer countries is heavily concentrated in urban areas, where the physical, political, and economic infrastructure necessary to support it is most available. Seeking employment opportunities, people migrate to these cities from rural areas, swelling urban populations. The resulting urban growth is rapid and uncontrolled, and governments find it difficult to provide the basic services that urban dwellers in rich countries expect, such as safe drinking water, sewage collection and treatment, and solid-waste removal.

The result is that some of the worst air and water pollution in the world occurs not in the rich countries, which consume most of the world's resources, but in the poor ones. Many newly industrializing cities, such as Mexico City, Beijing, Cairo, Yangon, and São Paolo, have severe air-pollution problems. The story for water pollution is no better. Only about 70 percent of the urban population of Asia is served by sewage collection; in Africa the figure is only 53 percent. Similarly, "safe" drinking water reaches only about 80 percent of urban dwellers in Asia and 70 percent in Africa.

In rural areas of poor countries air quality is usually better than in urban areas, but the problem is not eliminated. This is because indoor air pollution is perhaps even more a problem in rural areas than in urban ones. Clean energy sources like electricity or gas are rarely available in rural areas, and most people use wood, dung, or other biomass fuels for cooking and heating. These fuels are burned in simple stoves or fireplaces, often poorly ventilated. Therefore, chronic lung diseases are common, especially among women and children.

Can we trade wealth for environmental quality?   It is estimated that at least 5 million people, mostly children, die each year in Asia, Africa, and Latin America from diarrhea caused by contaminated drinking water. A similar number is estimated to die from chronic respiratory problems caused by poor air quality. These numbers are very crude estimates, but they certainly highlight some key issues regarding wealth, environmental quality, and the quality of life. Why do these deaths occur? Are they caused by poverty or pollution? How can they be prevented?

One way to eliminate these deaths would be to increase wealth. A small amount of money in the hands of a mother in Mumbai would allow her to buy kerosene to boil drinking water for her children, or it might buy simple rehydration therapies that would allow them to survive the diarrhea. The availability of cheap electricity or gas in Beijing might allow people to cook on electric stoves or heat their homes with gas instead of burning coal, and thus reduce the severe particulate air pollution of that city (Figure 13-35). If, however, that increase in wealth were achieved through an intensification of coal-fueled industry with few emission controls, as it is in

China and India, aren't we just trading one problem for another, with no net gain?

There are no simple answers. Clearly there is a need to increase wealth, especially in the poorer countries, while avoiding the environmental destruction that all too often accompanies that development. As discussed in Chapter 12, such an approach is usually called sustainable development. It has become the stated goal of virtually every governmental, nongovernmental, and international agency involved in promoting economic development.

Sustainable development as a concept makes sense, but making it a reality is quite another matter. Many involved in promoting sustainability agree that the key is to assign realistic economic values to environmental attributes, and to make sure these values are considered in the development process. This is similar to calculating wealth in terms of gross sustainable product, as was discussed in Chapter 12, but at the project level. For example, when a natural resource such as land covered in forest is converted to another use, say, agriculture, we normally think only of the positive value of the agricultural land created, and perhaps subtract the value of the timber production it replaced. But what of the value of the clean water that came from the forest and that has been replaced by streams polluted with agricultural runoff? What of the value of the biodiversity

FIGURE 13-35 This map shows concentrations of particulate matter in urban areas. Particulate matter is a major cause of health problems such as asthma, chronic bronchitis, and acute respiratory problems. (Redrawn from The World Bank Atlas, 2003)

that may have been lost in the process? Planning projects and measuring their achievements in these terms may help to ensure that the values in the present environment are given due consideration when we plan modifications in the name of economic development.

The central tensions over globalization of economic activity and the liberalization of trade rules under the World Trade Organization (WTO) focus on the equitable distribution of costs and benefits of production. Those who argue that industrial production should not be shifted from wealthy nations to poor ones without adequate environmental protections feel that the benefits of wealth generated by manufacturing jobs are less than the costs endured by those living near the factories. From their perspective, the costs of production are being transferred from rich countries to poor ones while the benefits, in the form of profits and consumer goods, remain in rich countries. Those who argue that free trade is good for poor countries obviously see the benefits of accepting those factories as outweighing their environmental costs.

These tensions were prominent at the World Summit on Sustainable Development, held at Johannesburg, South Africa, in 2002. The summit focused on improving people's lives and conserving natural resources in a world that is experiencing increasing demands for food, water, shelter, sanitation, health services, and economic security. The conference was attended by leaders from around the world; the United States was represented by Secretery of State Colin Powell. When Powell spoke he stated that "the United States is taking action to meet environmental challenges, including global climate change," but a skeptical audience responded with jeers and heckling. Clearly many outside the United States feel that the most powerful and influential nation in the world bears a responsibility to address these issues more effectively.

International Equity in Environmental Management

Measuring the value of environmental quality and natural resources is difficult enough in itself, but it is even more difficult when we recognize that people in different countries value resources differently. Many North Americans and Europeans place a high value on tropical rainforest biodiversity and argue that deforestation should be halted in order to protect that diversity, even if it means curtailing some economic benefits of deforestation. The governments of most tropical nations, however, are much less concerned about biodiversity and more concerned with the economic gains to be made by exploiting the forest. For environmental issues of only regional or national concern, such issues are usually resolved at that level. But when we consider resources of global importance, international negotiations must resolve conflicts derived from differences in

valuing resources. In this section we will discuss some examples of major environmental issues that have been the subjects of global-scale negotiations or that are likely to become so in the near future.

Ozone depletion and the Montreal Protocol   As we learned in Chapter 2, several trace gases in the atmosphere are important in absorbing radiation. One of these is ozone, a form of oxygen with three atoms per molecule instead of the usual two. Ozone is relatively abundant in the stratosphere, a layer of the atmosphere located between about 15,000 and 55,000 meters (50,000 and 180,000 feet) above sea level. Stratospheric ozone absorbs significant amounts of incoming solar radiation, especially in the ultraviolet wavelengths. Ultraviolet radiation is harmful to many things at Earth's surface, but our greatest concern is that it is a cause of skin cancer.

As early as the 1960s scientists began to raise concerns about the potential impacts of air pollution on ozone concentrations in the stratosphere. One of the substances that was particularly worrisome is a class of chemicals called chlorofluorocarbons (CFCs). CFCs are human-made chemicals that are used in a wide variety of applications, including refrigeration and air-conditioning and many industrial processes.

The usefulness of CFCs stems from the fact that they are very stable and do not react with other substances readily. But this very stability allows them to remain in the atmosphere for a long time after they are released—long enough to reach the stratosphere. There they are broken down by the intense solar radiation, and the chlorine atoms that are released when the CFCs break down enter reactions that contribute to a loss of ozone in the stratosphere.

In the 1970s there were strong theoretical reasons to believe that CFCs would harm stratospheric ozone, and these were sufficient to cause most nations to ban the use of CFCs as propellants in aerosol spray cans. It was not until the mid-1980s, however, that we accumulated clear evidence that damage to the ozone shield was in fact occurring, especially in high latitudes, and that this damage was attributable to CFCs.

When this became known, international action was obviously called for. CFCs were being manufactured and used in many countries, mostly the wealthy industrialized ones. In 1987 an international agreement was reached that has since been ratified by over 125 countries. The agreement, called the Montreal Protocol, called for signatory nations to freeze CFC production at 1989 levels and then cut production 50 percent by 1999. Developing nations were allowed to increase CFC production for 10 more years, and they were not required to phase out CFCs until 2010. As additional evidence confirming damage to the ozone layer accumulated, the agreement was strengthened. Amendments in 1990 and 1992 accelerated the phaseout, resulting in a complete cessation of CFC production in rich

Annual Global CFC Production, 1931-2001

FIGURE 13-36 Chlorofluorocarbon production. Production of chlorofluorocarbons peaked in 1988 and declined dramatically thereafter as a result of an international agreement to phase out these ozone-destroying chemicals. (Data from Worldwatch Institute)

countries (which accounted for about two-thirds of all production) in 1996. CFC emissions by the rich countries dropped faster than expected, and by 1997 levels of chlorine in the stratosphere stabilized and began to decline (Figure 13-36). Production has risen more rapidly than expected in the poorer countries, however, and there are concerns that this increase may significantly offset the decreases achieved in the rich countries.

Despite these concerns, the Montreal Protocol is generally regarded as a success story in international environmental management. One of the factors in this success was that the principal nations involved had similar stakes in the outcome. (1) The countries for whom skin cancer represents the greatest health problem are generally those in which (a) other health problems, such as those associated with basic sanitation, are less critical, and (b) populations are mainly in high latitudes. It is also noteworthy that skin cancer is most common among fair-skinned people. Thus the wealthy high-latitude nations of Europe and North America are among those with the greatest concern about the negative impacts of ozone destruction. (2) The industrial powers of North America were the major producers and consumers of CFCs. These same wealthy nations were the ones with the greatest chemical research and development capabilities and were thus most likely to be able to develop substitute chemicals quickly. (3) Finally, the agreement established different provisions for poor nations than for rich ones. The two-tiered agreement recognized the greater need of poor countries to continue producing these useful chemicals, and perhaps, a lower level of concern about their environmental impacts. Thus the countries that bore the greatest burden of complying with the terms of the treaty were also the ones that benefited the most from its enactment.

Global warming and CO$_2$ controls   Among the international meetings that have been sponsored by the United Nations, two major international conferences stand out: the Conference on the Human Environment held in Stockholm in 1972 and the World Conference on Environment and Development (Earth Summit) held 20 years later, in 1992, in Rio de Janeiro. In both cases the contrasting agendas of rich and poor nations have been major barriers to international agreements and action to protect the global environment, although international friction was less in 1992 than in 1972.

At the 1992 Earth Summit in Rio de Janeiro, an international agreement was signed that committed most countries to undertake serious efforts to reduce global emissions of carbon dioxide and other greenhouse gases. That agreement did not require any country to reduce emissions. Rather, it was a promise to talk about reducing emissions, with the expectation that significant action would be agreed to within five years. Nonetheless, it marked the beginning of an ongoing effort to tackle this difficult international environmental issue.

In 1995 negotiators met in Berlin, and they took some small steps toward controlling CO$_2$. The conference resulted in the Berlin Mandate, which called upon the most industrialized nations to control emissions voluntarily through setting specific emission-reduction targets. It soon became clear that voluntary controls were not going to work for many of the largest emitters. The United States, in particular, is not prepared to shoulder the burden of significantly controlling fossil-fuel use. The European Union, however, committed itself to reducing carbon emissions to a level 15 percent below 1990 emissions by the year 2010. This commitment reflects many Europeans' genuine concern about the consequences of global warming. It was also facilitated by a widespread conversion from coal to natural gas and other cleaner sources of energy that had been under way for some time.

Who would be the winners and who would be the losers if carbon emissions were effectively controlled? We don't know, exactly, and that is part of the problem. The most vocal groups opposing carbon controls are those who benefit most from fossil-fuel use. These include the United States (the world's foremost oil-consuming nation), major oil-producing nations, and the oil industry. Arguments in favor of emissions controls come from environmentalists, but also from a coalition of insurance companies and small island nations. These interests are most threatened by sea-level rise. The insurance industry in general has become highly vulnerable to major losses caused by hurricanes and other storms, as discussed in Chapter 2. Such disasters are most likely in nations that have long coastlines and valuable investments such as resorts along those coasts. Hence small island nations like Jamaica, Trinidad, Samoa, and the Seychelles think that preventing global warming is very much in their interest.

Focus on
MALARIA AND DDT

In May of 2003 President Bush made a pledge of $15 billion toward fighting infectious diseases, and a major focus of that pledge was AIDS in Africa. While AIDS is indeed the most serious health issue in Africa, malaria remains a serious problem. Malaria kills about 1 million people per year in Africa, mostly children. Children who survive malaria often suffer from learning impairments or brain damage, and the disease is also a leading cause of perinatal death, low birth weight, and maternal anemia.

While several treatments for malaria are available after an individual has been infected, there is only one way to prevent infection and that is through preventing mosquito bites. And the most effective agent for controlling the mosquitoes is DDT, a pesticide invented in 1937 (its inventor won the Nobel Prize for his work) and used with great success ever since. While DDT has been banned in most developed nations, it is still widely used for mosquito control in many relatively poor tropical nations, where it is considered an essential tool in the battle against malaria.

As discussed in Chapter 4, DDT is a persistent pesticide—one that remains in the environment for a relatively long period of time. DDT is harmless to humans in the low concentrations in which it is normally found, but it is passed up the food chain, reaching lethal concentrations in some carnivores. DDT is one of a group of similar toxic chemicals known as persistent organic pollutants, or POPs. POPs can evaporate in relatively warm conditions, such as are common in the tropics, and be carried on the wind to distant locations where they can be redeposited on Earth's surface. Through this mechanism such chemicals have been redistributed throughout the globe. They are found in even the most remote places, including Arctic regions

where they can accumulate because it is too cold for them to evaporate again.

People who live in Arctic regions depend largely on meat for their food, and they are traditionally hunters of seals, whales, and other carnivorous animals that, by their position at the top of the food chain tend to have high concentrations of POPs. The problem is worsened by the fact that these people have diets that are high in fat, and POPs are most concentrated in animal fat. POPs have been found in mothers' breast milk among the Inuit of northern Canada at concentrations nine times higher than among women in southern Canada. Inuit people are now being warned to reduce their consumption of foods that are staples in their culture, particularly seal.

Health officials thus face a dilemma. The most effective available tool for fighting malaria in the tropics is also one that is poisoning the food upon which arctic people depend. The environmental rules banning DDT that are viewed by North Americans as essential may be wholly unacceptable to many tropical nations. Donor agencies in wealthy nations are reluctant to fund malaria-control programs that include the use of DDT. In 1995, under international pressure, South Africa eliminated DDT from its malaria programs. Mosquitoes quickly became resistant to the substitute chemical, and the number of malaria cases rose dramatically. In 2001, 90 countries signed a treaty to eliminate POPs from the environment, and in that treaty an exemption was made for the use of DDT in poorer tropical nations. South Africa resumed the use of DDT in 2001, and the number of cases quickly dropped to near their 1995 level. It remains to be seen whether the new U.S. initiative to improve health in Africa and elsewhere will be effective at controlling malaria in ways that do not cause DDT to continue to diffuse around the globe.

Negotiations on reducing CO_2 emissions culminated in a conference at Kyoto, Japan, in December 1997. Under the pressure of the five-year deadline for agreement set in 1992, delegates struggled to hammer out an agreement that would bridge the enormous gap between countries that wanted major reductions in CO_2 output and countries that wanted no restrictions at all.

U.S. representatives went to the conference insisting they would agree only to reduce emissions to 1990

levels, and no lower, by 2010. By the year 2000 U.S. emissions were expected to be 13 percent higher than they were in 1990, so a reduction to 1990 levels by 2010 would be a substantial cut. The United States also demanded meaningful commitments to CO_2 controls on the part of poorer, less-industrialized countries that are expected to produce most of the growth in emissions in the next few decades. A third key U.S. requirement was that countries with high emissions be able to buy credit for reducing emissions by paying for emission reductions in other countries. It would probably be cheaper to pay Brazil to reduce deforestation, for example, than to curtail use of coal in the United States.

For their part, European representatives insisted on a 15 percent reduction in emissions below 1990 levels by 2010. Such a target seems highly ambitious, but Europe had already made significant progress in the early 1990s as a result economic changes there resulting from the collapse of the Soviet Union, combined with efforts to reduce acid rain. The small island nations also demanded substantial cuts in emissions. The less-industrialized bloc of countries refused to commit themselves to any reductions. They insisted that the industrialized countries that had caused the problem were primarily responsible for controlling it within their own borders.

With such enormous differences among the countries represented, compromise on many issues was difficult or impossible. The conference did reach an agreement, though an incomplete one. The United States agreed to reduce emissions to 7 percent below 1990 levels by 2010. It was also agreed that the treaty would not take effect unless it were ratified by countries whose emissions totaled 55 percent of global emissions. That effectively gave the United States a veto over the treaty, because it contributes 22 percent of total emissions. Without U.S. participation ratification would be nearly impossible.

Some issues, notably enforcement and the terms under which emission reductions could be traded internationally, were not resolved. The conferees set a deadline of late 1998 for agreement on these issues, but that deadline was ultimately extended to late 2000. Meetings were held in 1998 and 1999, and some issues regarding provisions for trading emissions were worked out at those meetings.

What was to be the final round of negotiations took place in The Hague, Netherlands, in 2000. One of the key remaining issues was whether or not countries would be able to take credit for carbon that was stored, or sequestered, in biomass through activities such as reforestation. The position of the United States, and other industrialized nations including Australia and Japan, was that such carbon storages should count against emissions. This was particularly important to the United States, because forest growth on abandoned farmland is storing a substantial amount of carbon there. The U.S. Senate had made it clear that U.S. ratification of the treaty was highly unlikely even if significant concessions to the U.S. position were made; nonetheless, there was a desire to reach an agreement so that a step toward emissions control, however small, would be made.

In 2000 the scientific evidence of the dangers of global warming became even stronger. The International Panel on Climate Change revised upward its predictions of future warming. A U.S. report documented the increased likelihood of weather-related disasters in a warmed climate, as described in Chapter 2. Then, in November 2000, just weeks before the final conference, a series of massive storms hit Great Britain, causing widespread flooding. The British government warned that this was a sign of things to come, and that more should be done to prepare for future disasters.

At the Hague conference, negotiations focused on the differences between the U.S. position (also supported by Australia, Canada, and Japan) and the European position. Europe insisted that carbon sequestration in forests not be allowed to count against emissions, and that real emissions reductions should be made. The United States, on the other hand, demanded that credits for sequestration in forests be allowed. Britain worked hard to forge a compromise between the U.S. and Europe, but in the end the parties failed to reach an agreement and negotiations broke down. Finally, in 2001, negotiations resumed. This time agreement was finally reached after Japan won concessions on accounting for carbon sequestration in forests. The result was seen by many as watered-down and ineffective, but an important step nonetheless. The U.S. did not participate in the negotiations in the final round, and is not expected to join the treaty any time soon, if ever. The Bush administration has since acknowledged that global warming is probably occurring and is probably caused by CO_2 emissions, but argues that the solution to the problem lies in adapting to climate change rather than in trying to prevent it.

Perhaps one reason why negotiations over CO_2 have proved so difficult, while those over CFCs were relatively easy, is that with CFCs the nations that would bear the costs of reducing pollution were also those that had much to lose (in the form of increased cancer rates). With CO_2, however, the United States would have to make significant changes in its habits; yet most U.S. citizens see little hazard in global warming. Europeans and small island nations, on the other hand, perceive a major threat. Europe, as a major emitter of CO_2, is the only bloc of nations that both emits large amounts of CO_2 and perceives a significant benefit from controlling those emissions. Unfortunately the problem is a global one, and Europe acting on its own will not be enough to control global warming.

CONCLUSION: CRITICAL ISSUES FOR THE FUTURE

As we consider the many consequences of global integration and greater cultural and economic interaction between rich and poor nations, the greatest challenges will be those arising from differences in the value systems, both economic and cultural, of the people involved. When Brazilian coffee growers and Chicago commodity traders are linked electronically so they are only seconds away, and if their governments are both members of trade pacts and other international agreements, then they become part of the same political-economic system. That joint membership requires that they share certain values and agree to operate by the same set of rules. The political gulf between the United States and China, on the other hand, is much larger. These two nations are the two largest emitters of carbon dioxide in the world, and it will be more difficult for them to reach agreement on controlling global warming. The enormous economic gaps between rich and poor, both between and increasingly within countries, only multiply the challenges posed by increased global interaction. Will economic liberalization be the key to economic development in the poorer nations of Africa, Asia, and Latin America? Will this economic development be rapid enough to reduce the disparities in national income between these countries and the rich countries? Or will it perpetuate

and strengthen the inequities that result from the greater ability of rich countries to capture value added? And will economic development benefit the lower economic strata of the societies it affects, or only those already relatively well off? These questions remain to be answered. Finally, as we grapple with the many social and political tensions created by global change and integration, will we have the opportunity to protect what remains of the natural environment? Or will we instead be so focused on economic development alone that the idea of sustainable development will remain more an idea than a reality?

These questions cannot be answered today. In searching for the answers, however, it is clear that we need to consider a wide range of topics simultaneously: climate, resources, culture, religion, politics, and economics. Such a wide-ranging viewpoint is open to students of any subject, but geography especially encourages such broad-minded analysis.

Chapter Review

SUMMARY

Through the twentieth century, the empires of the European powers and the United States have fragmented. The U.S.S.R.—the Russian Empire in a new name—collapsed only in 1991. New economic blocs are appearing, most notably the European Union and the North American Free Trade Area, but also MERCOSUR in South America and the Association of Southeast Asian Nations. Other international groups seek cultural or political goals.

The United Nations was created after World War II. Today it serves as a forum for international negotiation and coordination of peacekeeping activities, and as a means of organizing social programs, economic development programs, and other international cooperative undertakings. The Law of the Sea Treaty governs activities in the oceans.

Nongovernmental organizations have assumed a leading role on the world stage. Many of these are charitable, but

others are guilty of international terrorist acts. Some of these are based in collapsing states, but others may be actively encouraged by hostile states. Under such conditions, the United States and other powers are newly questioning the degree to which any state's territorial sovereignty must be respected absolutely.

Economic growth in the twentieth century has caused dramatic increases in energy use and pollution output, threatening the global environment. International negotiations focus on balancing the need for economic development with concerns for environmental protection. Rich countries focus their attention on pollution control, but poor nations are concerned with their need for immediate economic development. International agreements have been reached on ozone-destroying substances, but efforts to control predicted global warming are likely to be much more difficult.

KEY TERMS

bloc p. 546
common market p. 557
continental shelf p. 582
customs union p. 557
European Union (EU) p. 557
free trade area p. 557

innocent passage p. 583
international organization p. 546
local content requirement p. 567
maquiladora p. 567
Marshall Plan p. 557

nongovernmental organizations (NGOs) p. 573
North Atlantic Treaty Organization (NATO) p. 556
supranational organization p. 546

QUESTIONS FOR REVIEW AND DISCUSSION

1. What small remnants of the British Empire present political problems today?

2. How was the Russian Empire restructured after World War I?

3. Describe the economic bonds holding together the U.S.S.R. until 1991.

4. What independent countries today were ever ruled by the United States?

5. Name the members of the European Union. Describe the government of the European Union.

6. What activities and results of the European Union prove that it is more than a free trade area or customs union, but a full common market?

7. Is NAFTA a free trade area, customs union, or common market?

8. What are the major branches of the United Nations and what are the activities of each?

9. What are territorial waters and what is an EEZ? Give several reasons why these delineations are disputed.

THINKING GEOGRAPHICALLY

1. Investigate and analyze the assignment of seats in the European Parliament. Does it represent population? Economic power?

2. Chapter 5 told how the Canadian government's effort to preserve Native American cultures sabotages the goal of achieving a high standard of living for all Canadians. At a global level, international diplomacy generally observes a principle of nonintervention in sovereign states, and this principle often preserves misgovernment and unnecessary suffering. On what grounds could the principle of nonintervention be overturned?

3. List factors of European cultural unity and disunity. As the European Union expands, how will various factors boost or retard cooperation among the members and economic development?

4. Most Canadian citizens know more about the United States than most U.S. citizens know about Canada. Why?

5. Do you think that each country should have an equal vote in the U.N. General Assembly, or would you recommend a weighted system of voting—by population, wealth, or territory?

6. Investigate a few of the projects of the specialized agencies of the United Nations, such as UNICEF, the FAO, or WHO.

7. French television commentator Christine Ockrent has said, "The only true pan-European culture is American culture." What did she mean by that? Can we say that the only true global culture is American culture?

8. How has physical geography encouraged or discouraged the formation of multinational bonds in Europe, in the CIS states, or in North America? Are there obvious complementarities in products? Do transportation or settlement patterns act as centripetal forces?

9. What impact will the European Union have on the role of the United States in world affairs?

10. What might happen to Antarctica if a practical way is found for extracting its resources?

SUGGESTIONS FOR FURTHER LEARNING

Berezin, Mabel, Martin Schain, and Roland Axtmann. *Re-Mapping Europe: Territory, Membership and Identity in a Transnational Age.* Baltimore: Johns Hopkins, 2004.

Canadian Journal of Regional Science. Vol. 13:2/3. Special double issue devoted to the spatial impact of the U.S.-Canadian free trade agreement. Québec: Université du Québec.

Chaturvedi, Sanjay. *The Polar Regions: A Political Geography.* New York: John Wiley & Sons, 1997.

Chinn, Jeff, and Robert Kaiser. *Russians as the New Minority: Ethnicity and Nationalism in the Soviet Successor States.* Boulder, CO: Westview, 1996.

Churchill, R. R., and A. V. Lower. *The Law of the Sea,* 2nd ed. New York: Juris Publishing, Inc., 1999.

Collier, George. *Basta! Land and the Zapatista Rebellion in Chiapas.* Oakland, CA: Institute for Food and Development, 1996.

Combating Terrorism: Axis of Evil, Multilateral Containment or Unilateral Confrontation. Hearings before the Subcommittee on National Security. Washington, D.C.: U.S. Government Printing Office, 2003.

Dawisha, Karen, and Bruce Parrott. *Russia and the New States of Eurasia.* New York: Cambridge University Press, 1994.

Demko, George, and William B. Wood, eds. *Reordering the World: Geopolitical Perspectives on the Twenty-first Century.* Boulder, CO: Westview, 1994.

Earney, Fillmore. *Marine Mineral Resources.* New York: Routledge, 1990.

Forsythe, David P., Roger Coate, and Thomas G. Weiss. *The United Nations and Changing World Politics.* New York: Westview Publishing, 2000.

Freeland, Chrystia. *The Sale of the Century: Russia's Wild Ride from Communism to Capitalism.* New York: Crown Business, 2000.

Funck, Bernard, ed. *European Integration, Regional Policy, and Growth.* Washington, D.C.: The World Bank, 2003.

Gleditsch, N. P. "Armed conflict and the environment: A critique of the literature." *Journal of Peace and Conflict Resolution* 35, no. 3 (1998): 381–400.

Hakim, Peter, ed. *The Future of North American Integration: Beyond NAFTA.* Washington, D.C.: The Brookings Institution, 2002.

Huntington, Samuel. *The Clash of Civilizations and the Remaking of World Order.* New York: Simon & Schuster, 1996.

Ingram, Helen, Nancy Laney, and David Gillilan. *Divided Waters: Bridging the U.S.-Mexican Border.* Tucson: University of Arizona Press, 1997.

Klotz, F. G. *America on the Ice: Antarctic Policy Issues.* Washington, D.C.: National Defense University Press, 1990.

Langewiesche, William, "Anarchy at Sea," *The Atlantic,* 292, no. 2 (September 2003): 50–81.

The Law of the Sea. New York: United Nations, 1983.

MacArthur, John R. *The Selling of "Free Trade" NAFTA, Washington, and the Subversion of American Democracy.* New York: Hill & Wang, 2000.

Macmillan, Margaret. *Paris 1919: Six Months That Changed the World.* New York: Random House, 2001.

McCormick, John, *Understanding the European Union: A Concise Introduction,* 2nd ed. New York: Palgrave Macmillan, 2002.

Nations of the Earth Report (a three-volume study of the Earth environment country by country). Geneva: United Nations, 1992.

Palmer, Mark. *Breaking the Real Axis of Evil: How to Oust the World's Last Dictators by 2025.* Boston: Rowman & Littlefield, 2003.

Polman, Linda. *We Did Nothing: Why the Truth Doesn't Always Come Out When the UN Goes In.* New York: Viking, 2003.

Prescott, John Robert Victor. *Maritime Political Boundaries of the World.* New York: Methuen, 1986.

Salonius, P. "Population growth in the United States and Canada: a role for scientists." *Conservation Biology* 13, no. 6 (1999): 1518–1519.

Shaw, Denis J. B. *Russia in the Modern World: A New Geography.* London: Institute of British Geographers Studies in Geography, 1999.

We the Peoples: The Role of the United Nations in the Twenty-first Century (report of the Secretary General). United Nations Document A/54/2000.

Wolf, A. T. "Conflict and cooperation along international waterways." *Water Policy* 1, no. 2 (1998): 251–265.

Young, Oran, George Demko, and Kila Parti Ramakrishna, eds. *Global Environmental Change and International Governance.* Keene, NH: University Press of New England, 1997.

WEB WORK

The United Nations home page will take you to a tremendous variety of economic, political, cultural, and social organizations:

http://www.un.org/

The UN High Commissioner for Human Rights is at

http://www.unhchr.ch/

Information about the European Union can be found by starting at its home page at

http://europa.eu.int/

Europe Magazine is at

http://www.eurunion.org/

The International Criminal Court is at

http://www.un.org/law/icc/

Information on the law of the sea and ocean affairs is provided by the Council on Ocean Law at

http://www.oceanlaw.org/

A number of world organizations have rich Web sites. Among the good ones are NATO at

http://www.nato.int/

the Commonwealth of Nations at

http://www.thecommonwealth.org/

and the Organization for Economic Cooperation and Development at

http://www.oecd.org/

The U.S. Department of Commerce has a special page devoted to NAFTA affairs at

http://www.mac.doc.gov/nafta/

but Ralph Nader's "The Public Citizen" is critical of NAFTA:

http://www.citizen.org/

APPENDIX I
MAP SCALE AND PROJECTIONS

Phillip C. Muehrcke

Unaided, our human senses provide a limited view of our surroundings. To overcome those limitations, humankind has developed powerful vehicles of thought and communication, such as language, mathematics, and graphics. Each of those tools is based on elaborate rules, each has an information bias, and each may distort its message, often in subtle ways. Consequently, to use those aids effectively, we must understand their rules, biases, and distortions. The same is true for the special form of graphics we call maps: We must master the logic behind the mapping process before we can use maps effectively.

A fundamental issue in cartography, the science and art of making maps, is the vast difference between the size and geometry of what is being mapped—the real world, we will call it—and that of the map itself. Scale and projection are the basic cartographic concepts that help us understand that difference and its effects.

MAP SCALE

Our senses are dwarfed by the immensity of our planet; we can sense directly only our local surroundings. Thus, we cannot possibly look at our whole state or country at one time, even though we may be able to see the entire street where we live. Cartography helps us expand what we can see at one time by letting us view the scene from some distant vantage point. The greater the imaginary distance between that position and the object of our observation, the larger the area the map can cover but the smaller the features will appear on the map. That reduction is defined by the *map scale,* the ratio of the distance on the map to the distance on Earth. Map users need to know about map scale for two reasons: (1) so that they can convert measurements on a map into meaningful real-world measures and (2) so that they can know how abstract the cartographic representation is.

Real-World Measures

A map can provide a useful substitute for the real world for many analytical purposes. With the scale of a map, for instance, we can compute the actual size of its features (length, area, and volume). Such calculations are helped

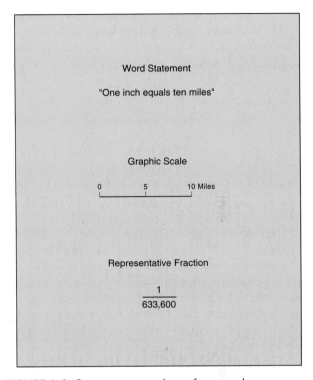

FIGURE A-1 Common expressions of map scale.

by three expressions of a map scale: a word statement, a graphic scale, and a representative fraction.

A *word statement* of a map scale compares X units on the map to Y units on Earth, often abbreviated "X unit to Y units."

For example, the expression "1 inch to 10 miles" means that 1 inch on the map represents 10 miles on Earth (Figure A-1). Because the map is always smaller than the area that has been mapped, the ground unit is always the larger number. Both units are expressed in meaningful terms, such as inches or centimeters and miles or kilometers. Word statements are not intended for precise calculations but give the map user a rough idea of size and distance.

A *graphic scale,* such as a bar graph, is concrete and therefore overcomes the need to visualize inches and miles that is associated with a word statement of scale (see Figure A-1). A graphic scale permits direct visual comparison of feature sizes and the distances between features. No ruler is required; any measuring aid will do.

It needs only to be compared with the scaled bar; if the length of 1 toothpick is equal to 2 miles on the ground and the map distance equals the length of 4 toothpicks, then the ground distance is 4 times 2, or 8 miles. Graphic scales are especially convenient in this age of copying machines, when we are more likely to be working with a copy than with the original map. If a map is reduced or enlarged as it is copied, the graphic scale will change in proportion to the change in the size of the map and thus will remain accurate.

The third form of a map scale is the *representative fraction* (RF). An RF defines the ratio between the distance on the map and the distance on Earth in fractional terms, such as $\frac{1}{633,600}$ (also written 1/633,600 or 1:633,600). The numerator of the fraction always refers to the distance on the map, and the denominator always refers to the distance on Earth. No units of measurement are given, but both numbers must be expressed in the same units. Because map distances are extremely small relative to the size of Earth, it makes sense to use small units, such as inches or centimeters. Thus, the RF 1:633,600 might be read as "1 inch on the map to 633,600 inches on Earth."

Herein lies a problem with the RF. Meaningful map-distance units imply a denominator so large that it is impossible to visualize. Thus, in practice, reading the map scale involves an additional step of converting the denominator to a meaningful ground measure, such as miles or kilometers. The unwieldy 633,600 becomes the more manageable 10 miles when divided by the number of inches in a mile (63,360).

On the plus side, the RF is good for calculations. In particular, the ground distance between points can be easily determined from a map with an RF. One simply multiplies the distance between the points on the map by the denominator of the RF. Thus, a distance of 5 inches on a map with an RF of 1/126,720 would signify a ground distance of 5 × 126,720, which equals 633,600. Because all units are inches and there are 63,360 inches in a mile, the ground distance is 633,600 ÷ 63,360, or 10 miles. Computation of area is equally straightforward with an RF. Computer manipulation and analysis of maps is based on the RF form of map scale.

Guides to Generalization

Scales also help map users visualize the nature of the symbolic relation between the map and the real world. It is convenient here to think of maps as falling into three broad scale categories (Figure A-2). (Do not be confused by the use of the words *large* and *small* in this context; just remember that the larger the denominator, the smaller the scale ratio and the larger the area that is shown on the map.) Scale ratios greater than 1:100,000, such as the 1:24,000 scale of U.S. Geological Survey topographic quadrangles, are large-scale maps. Although

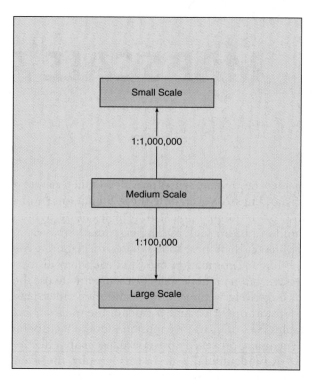

FIGURE A-2 The scale gradient can be divided into three broad categories.

those maps can cover only a local area, they can be drawn to rather rigid standards of accuracy. Thus, they are useful for a wide range of applications that require detailed and accurate maps, including zoning, navigation, and construction.

At the other extreme are maps with scale ratios of less than 1:1,000,000, such as maps of the world that are found in atlases. Those are small-scale maps. Because they cover large areas, the symbols on them must be highly abstract. They are therefore best suited to general reference or planning, when detail is not important. Medium- or intermediate-scale maps have scales between 1:100,000 and 1:1,000,000. They are good for regional reference and planning purposes.

Another important aspect of map scale is to give us some notion of geometric accuracy; the greater the expanse of the real world shown on a map, the less accurate the geometry of that map is. Figure A-3 shows why. If a curve is represented by straight line segments, short segments (X) are more similar to the curve than are long segments (Y). Similarly, if a plane is placed in contact with a sphere, the difference between the two surfaces is slight where they touch (A) but grows rapidly with increasing distance from the point of contact (B). In view of the large diameter and slight local curvature of Earth, distances will be well represented on large-scale maps (those with small denominators) but will be increasingly poorly represented at smaller scales. This close relationship between map scale and map geometry brings us to the topic of map projections.

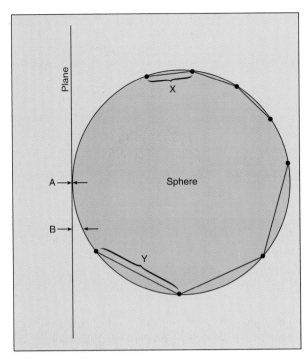

FIGURE A-3 Relationships between surfaces on the round Earth and a flat map.

MAP PROJECTIONS

The spherical surface of Earth is shown on flat maps by means of map projections. The process of "flattening" Earth is essentially a problem in geometry that has captured the attention of the best mathematical minds for centuries. Yet no one has ever found a perfect solution; there is no known way to avoid spatial distortion of one kind or another. Many map projections have been devised, but only a few have become standard. Because a single flat map cannot preserve all aspects of Earth's surface geometry, a mapmaker must be careful to match the projection with the task at hand. To map something that involves distance, for example, a projection should be used in which distance is not distorted. In addition, a map user should be able to recognize which aspects of a map's geometry are accurate and which are distortions caused by a particular projection process. Fortunately, that objective is not too difficult to achieve.

It is helpful to think of the creation of a projection as a two-step process (Figure A-4). First, the immense Earth is reduced to a small globe with a scale equal to that of the desired flat map. All spatial properties on the globe are true to those on Earth. Second, the globe is flattened. Since that cannot be done without distortion, it is accomplished in such a way that the resulting map exhibits certain desirable spatial properties.

Perspective Models

Early map projections were sometimes created with the aid of perspective methods, but that has changed. In

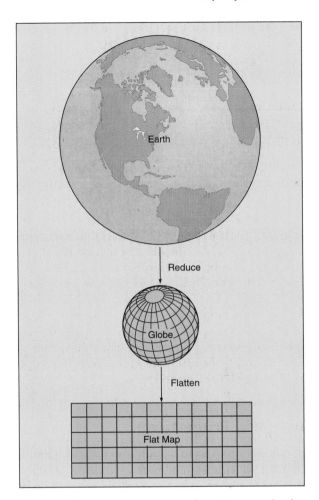

FIGURE A-4 The two-step process of creating a projection.

the modern electronic age, projections are normally developed by strictly mathematical means and are plotted out or displayed on computer-driven graphics devices. The concept of perspective is still useful in visualizing what map projections do, however. Thus, projection methods are often illustrated by using strategically located light sources to cast shadows on a projection surface from a latitude/longitude net inscribed on a transparent globe.

The success of the perspective approach depends on finding a projection surface that is flat or that can be flattened without distortion. The cone, cylinder, and plane possess those attributes and serve as models for three general classes of map projections: *conic, cylindrical,* and *planar* (or azimuthal). (Figure A-5) shows those three classes, as well as a fourth, a false cylindrical class with an oval shape. Although the *oval* class is not of perspective origin, it appears to combine properties of the cylindrical and planar classes (Figure A-6).

The relationship between the projection surface and the model at the point or line of contact is critical because distortion of spatial properties on the projection is symmetrical about, and increases with distance from, that point or line. That condition is illustrated for the cylindrical and planar classes of projections in Figure A-7. If

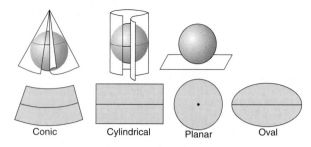

FIGURE A-5 General classes of map projections. (Courtesy of ACSM)

the point or line of contact is changed to some other position on the globe, the distortion pattern will be re-centered on the new position but will retain the same symmetrical form. Thus, centering a projection on the area of interest on Earth's surface can minimize the effects of projection distortion. And recognizing the general projection shape, associating it with a perspective model, and recalling the characteristic distortion pattern will provide the information necessary to compensate for projection distortion.

Preserved Properties

For a map projection to truthfully depict the geometry of Earth's surface, it would have to preserve the spatial attributes of *distance, direction, area, shape,* and *proximity.* That task can be readily accomplished on a globe, but it is not possible on a flat map. To preserve area, for example, a mapmaker must stretch or shear shapes; thus,

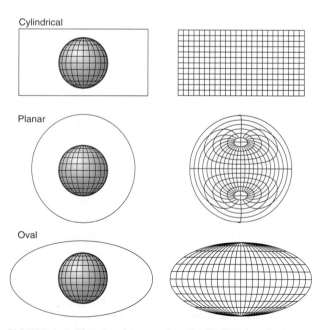

FIGURE A-6 The visual properties of cylindrical and planar projections combined in oval projections. (Courtesy of ACSMJ)

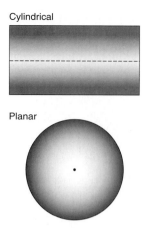

FIGURE A-7 Characteristic patterns of distortion for two projection classes. Here, darker shading implies greater distortion. (Courtesy of ACSM)

area and shape cannot be preserved on the same map. To depict both direction and distance from a point, area must be distorted. Similarly, to preserve area as well as direction from a point, distance has to be distorted. Because Earth's surface is continuous in all directions from every point, discontinuities that violate proximity relationships must occur on all map projections. The trick is to place those discontinuities where they will have the least impact on the spatial relationships in which the map user is interested.

We must be careful when we use spatial terms, because the properties they refer to can be confusing. The geometry of the familiar plane is very different from that of a sphere; yet, when we refer to a flat map, we are in fact making reference to the spherical Earth that was mapped. A shape-preserving projection, for example, is truthful to local shapes—such as the right-angle crossing of latitude and longitude lines—but does not preserve shapes at continental or global levels. A distance-preserving projection can preserve that property from one point on the map in all directions or from a number of points in several directions, but distance cannot be preserved in the general sense that area can be preserved. Direction can also be generally preserved from a single point or in several directions from a number of points but not from all points simultaneously. Thus, a shape-, distance-, or direction-preserving projection is truthful to those properties only in part.

Partial truths are not the only consequence of transforming a sphere into a flat surface. Some projections exploit that transformation by expressing traits that are of considerable value for specific applications. One of those is the famous shape-preserving *Mercator projection* (Figure A-8). That cylindrical projection was derived mathematically in the 1500s so that compass bearing (called rhumb lines) between any two points on Earth would plot as straight lines on the map. That trait let navigators plan, plot, and follow courses

FIGURE A-8 The useful Mercator projection, showing extreme area distortion in the higher latitudes. The white lines are great circles—the shortest route between two points on Earth's surface. (Courtesy of ACSM)

between origin and destination, but it was achieved at the expense of extreme areal distortion toward the margins of the projection (see Antarctica at the bottom of Figure A-8). Although the Mercator projection is admirably suited for its intended purpose, its widespread but inappropriate use for nonnavigational purposes has drawn a great deal of criticism.

The *gnomonic projection* is also useful for navigation. It is a planar projection with the valuable characteristic of showing the shortest (or great circle) route between any two points on Earth as straight lines. Long-distance navigators first plot the great circle course between origin and destination on a gnomonic projection (Figure A-9,

FIGURE A-9 A gnomonic projection (A) and a Mercator projection (B). (A rhumb line describes the course of a ship that crosses all meridians of longitude at the same angle.)

top). Next they transfer the straight line to a Mercator projection, where it normally appears as a curve (Figure A-9, bottom). Finally, using straight-line segments, they construct an approximation of that course on the Mercator projection. Navigating the shortest course between origin and destination then involves following the straight segments of the course and making directional corrections between segments. Like the Mercator projection, the specialized gnomonic projection distorts other spatial properties so severely that it should not be used for any purpose other than navigation or communications.

Projections Used in Textbooks

Although a map projection cannot be free of distortion, it can represent one or several spatial properties of Earth's surface accurately if other properties are sacrificed. The two projections used for world maps throughout this textbook illustrate that point well. Goode's homolosine projection, shown in (Figure A-10), belongs to the oval category and shows area accurately, although it gives the impression that Earth's surface has been torn, peeled, and flattened. The interruptions in Figure A-10 have been placed in the major oceans, giving continuity to the land masses. Ocean areas could be featured instead by placing the interruptions in the continents. Obviously, that type of interrupted projection severely distorts proximity relationships. Consequently, in different locations the properties of distance, direction, and shape are also distorted to varying degrees. The distortion pattern mimics that of cylindrical projections, with the equatorial zone the most faithfully represented (Figure A-11).

An alternative to special-property projections such as the equal-area Goode's homolosine is the compromise projection. In that case no special property is achieved at the expense of others, and distortion is rather evenly distributed among the various properties, instead of being focused on one or several properties. The *Robinson projection*, which is also used in this textbook, falls into that category (Figure A-12). Its oval projection has a global feel, somewhat like that of Goode's homolosine. But the Robinson projection shows the North Pole and the South Pole as lines that are slightly more than half the length of the equator, thus exaggerating distances and areas near the poles. Areas look larger than they really are in the high latitudes (near the poles) and smaller than they really are in the low latitudes (near the equator). In addition, not all latitude and longitude lines intersect at right angles, as they do on Earth, so we know that the Robinson projection does not preserve direction or shape either. However, it has fewer interruptions than the Goode's homolosine does, so it preserves proximity better. Overall, the Robinson projection does a good job of representing spatial relationships, especially in the low to middle latitudes and along the central meridian.

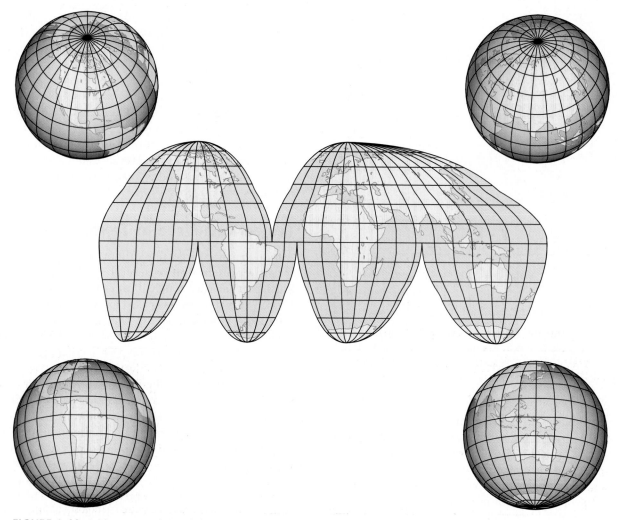

FIGURE A-10 An interrupted Goode's homolosine, an equal-area projection. (Courtesy of ACSM)

FIGURE A-11 The distortion pattern of the interrupted Goode's homolosine projection, which mimics that of cylindrical projections. As in Figure A-7, the darker shading implies greater distortion. (Courtesy of ACSM)

SCALE AND PROJECTIONS IN MODERN GEOGRAPHY

Computers have drastically changed the way in which maps are made and used. In the preelectronic age, maps were so laborious, time-consuming, and expensive to make that relatively few were created. Frustrated, geographers and other scientists often found themselves trying to use maps for purposes not intended by the map designer. But today anyone with access to computer mapping facilities can create projections in a flash. Thus, projections will be increasingly tailored to specific needs, and more and more scientists will do their own mapping rather than have someone else guess what they want in a map.

Computer mapping creates opportunities that go far beyond the construction of projections, of course. Once maps and related geographical data are entered into computers, many types of analyses can be carried out involving map scales and projections. Distances, areas, and volumes can be computed; searches can be conducted; information from different maps can be combined; optimal routes can be selected; facilities can be allocated to the most suitable sites; and so on. The term used to describe such processes is *geographical information system,* or GIS (Figure A-13). Within a GIS, projections provide the mechanism for linking data from different sources, and scale provides the basis for size calculations of all sorts. Mastery of both projection and scale becomes the user's responsibility because the map user is also the mapmaker. Now more than ever, effective geography depends on knowledge of the close association between scale and projection.

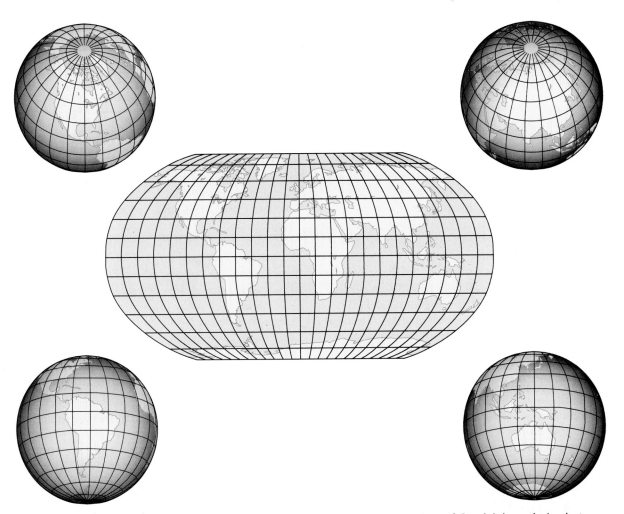

FIGURE A-12 The compromise Robinson projection, which avoids the interruptions of Goode's homolosine but preserves no special properties. (Courtesy of ACSM)

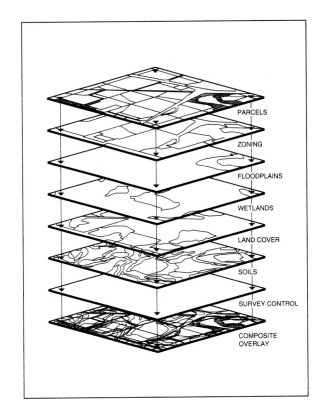

PARCELS

ZONING

FLOODPLAINS

WETLANDS

LAND COVER

SOILS

SURVEY CONTROL

COMPOSITE OVERLAY

FIGURE A-13 Within a GIS, environmental data attached to a common terrestrial reference system, such as latitude/longitude, can be stacked in layers for spatial comparison and analysis.

Climate classification is a challenging intellectual and scientific enterprise. Climates vary in many different ways, the most significant of which are temperature, precipitation, and seasonal variations in these variables. Each of these dimensions of variation should be portrayed by a useful classification system.

In addition to having multiple criteria that must be considered, none of these variables falls into simple categories, like male and female. Temperature and precipitation vary through wide ranges, with gradual transitions from place to place. Climate classification is a little like naming colors. Most people would agree on what is meant by "blue," and what is meant by "green." Finding agreement on the difference between bluish green and greenish blue (not to mention names like "turquoise" or "teal") is not so simple.

Wladimir Köppen (1846–1940) was a German climatologist who undertook this difficult task. Köppen recognized that there is a strong correlation between climate and vegetation types, and he reasoned that if climate types corresponded to significant vegetation types then the classification would be more useful and meaningful. The correlation between vegetation and climate is also useful in mapping climates in areas where few weather data are available but where vegetation can be mapped. In this book we use a modified version of Köppen's system, summarized in this appendix.

Köppen's approach was to identify five basic climate types and name them A, B, C, D, and E, in rough order of their latitudinal location on Earth (A in the tropics; E at the poles). Four of these are humid climate types and one, B, is an arid type. Each is divided into subtypes, with the second letter providing information about rainfall, and the third indicating temperature, except for polar climates in which only two letters are used and the second letter pertains to temperature. All climates are distinguished on the basis of mean monthly data.

For the humid climates, the second letter of the classification (f, s, m, or w) is used to describe the seasonal distribution of rainfall. In midlatitude climates,

f indicates evenly distributed rainfall, s indicates summer dry, and w indicates winter dry. In tropical (A) climates, the letter m is used to indicate a monsoonal precipitation regime, and the summer (or high-sun) dry type, s, does not exist. The third letter (a, b, c, or d) is used to distinguish warmer and cooler variations of the subtropical and midlatitude (C and D) climates, with a being the warmest and d the coolest.

For dry climates (B), the second letter (W or S) distinguishes desert from semiarid climates. The specific criteria for distinguishing these two from humid climates and from each other are based on mathematical formulas comparing temperature and rainfall. This is because plant water needs are temperature dependent. In a cool climate less rainfall is needed to support vegetation than in a warm climate. This is also the reason for differences in rules for identifying "winter dry" and "summer dry" climates.

Some of the climates distinguished by Köppen do not relate clearly to vegetation differences. In eastern North America, for example, the transition from Dfa to Dfb climates passes through the middle of the broadleaf deciduous forest region; neither does the line distinguish any major agricultural regions. However, the boundary between Dfb and Dfc corresponds relatively closely to the northern limit of commercial agriculture in this area.

In this book we have simplified Köppen's scheme somewhat, especially in Fig. 2-32 and the section describing climate types. We chose to identify 11 different major climate types that correspond reasonably well to significant vegetation differences. These types are named humid tropical, seasonally humid tropical, desert, semiarid, humid subtropical, marine west coast, Mediterranean, humid continental, subarctic, tundra, and icecap and ice sheets. Each of these corresponds to one to four subtypes of a Köppen major type. Our simplifications are designed to retain distinctions that relate clearly to natural vegetation and agricultural regions, while eliminating those that divide major vegetation regions.

| LETTER | | | | | | |
First	Second	Third	Derivation	Description	Definition	Climate types as used in this book
A			Alphabetical	Tropical humid climates	Average temperature of each month above 18° C (64° F)	
	f		German: *feucht* ("moist")	No dry season	Average rainfall of each month at least 6 cm (2.5 in.)	Humid tropical (Af)
	m		Monsoon	Monsoonal; short dry season compensated by heavy rains in other months	1 to 3 months with average rainfall less than 6 cm (2.5 in.)	Humid tropical (Am)
	w		Winter dry	Dry season in low-sun season	3 to 6 months with average rainfall less than 6 cm (2.5 in.)	Seasonally humid tropical (Aw)
B			Alphabetical	Dry climates; potential evaporation exceeds precipitation	Dry climates are determined by a set of formulas relating temperature and precipitation. Different formulas are used depending on whether precipitation is concentrated in the summer, winter, or neither. If the wettest winter month has three or more times the rainfall of the driest summer month, the climate is considered to have a winter concentration of rainfall. If the wettest summer month has 10 or more times the rainfall of the driest winter month, the climate is considered to have a summer concentration of rainfall. If neither of these is true, rainfall is considered evenly distributed.	
	W		German: *wuste* ("desert")	Arid climates	If precipitation is concentrated in the high-sun season, average annual precipitation (cm) less than mean annual temperature (°C). If precipitation is concentrated in the low-sun season, average annual precipitation (cm) less than mean annual temperature (°C) plus 14. If precipitation is evenly distributed through the year, average annual precipitation (cm) less than mean annual temperature (°C) plus 7.	Desert
	S		Steppe, or semiarid	Semiarid climates	If precipitation is concentrated in the high-sun season, average annual precipitation (cm) less than 2 times mean annual temperature (°C). If precipitation is concentrated in low-sun season, average annual precipitation (cm) less than 2 times mean annual temperature (°C) plus 14. If precipitation is evenly distributed through the year, average annual precipitation (cm) less than 2 times mean annual temperature (°C) plus 7.	Semiarid

608 Appendix II   The Köppen Climate Classification System

First	Second	Third	Derivation	Description	Definition	Climate types as used in this book
		h	German: *heiss* ("hot")	Low-latitude dry climate	Average annual temperature above 18° C (64° F)	Hot desert (BWh) or hot semiarid (steppe) climate (BSk)
		k	German: *kalt* ("cold")	Midlatitude dry climate	Average annual temperature below 18° C (64° F)	Cool desert (BWk) or cool semiarid (steppe) climate (BSk)
C			Alphabetical	Midlatitude humid climates	Average temperature of coldest month between 18° C (64° F) and −3° C (27° F)	
	s		Summer dry	Dry summer	Driest summer month has less than one-third the precipitation of the wettest winter month	Mediterranean (Csa, Csb)
	w		Winter dry	Dry winter	Driest winter month has less than one-tenth the precipitation of the wettest summer month	Humid subtropical (Cwa)
	f		German: *feucht* ("moist")	No dry season	Does not fit either s or w above	Marine west coast (Cfb, Cfc)
		a	Alphabetical	Hot summers	Average temperature of warmest month above 22° C (72° F)	Humid subtropical (Cfa)
		b	Alphabetical	Warm summers	Average temperature of warmest month below 22° C (72° F); at least 4 months with average temperature above 10° C (50° F)	Marine west coast (Cfb, Cfc)
		c	Alphabetical	Cool summers	Average temperature of warmest month below 22° C (72° F); fewer than 4 months with average temperature above 10° C (50° F)	
D			Alphabetical	Humid midlatitude climates with cold winters	4 to 8 months with average temperatures above 10° C (50° F)	Humid continental (Dfa, Dfb, Dwa, Dwb); Subarctic (Dfc, Dfd, Dwc, Dwd)
	Second and third letters same as in C climates					
		d	Alphabetical	Very cold winters	Average temperature of coldest month below −38° C (−36° F)	
E			Alphabetical	Polar climates	No month with average temperature above 10° C (50° F)	
	T		Tundra	Tundra climates	At least one month with average temperature above 0° C (32° F)	Tundra (ET)
	F		Frost	Ice-cap climates	No month with average temperature above 0° C (32° F)	Ice-cap and ice sheets (EF)

APPENDIX III
THE WORLD TODAY: POPULATION, ECONOMY, ENVIRONMENT

	Total Population (millions— 2001)	Average Annual Population Growth Rate % (1980– 2001)	Total Fertility Rate (births per woman— 2001)	Life Expectancy at Birth (years— 2001)	Infant Mortality Rate (per 1,000 live births— 2001)	Gross National Income per Capita ($—2001)	Carbon Dioxide Emissions (per capita metric tons— 1999)	Commercial Energy Use per Capita (kg of oil equivalent— 2000)	Freshwater Resources (total renewable resources per capita cu. m— 2000)
Afghanistan	27.2	2.6	6.8	43	165	—	0.0	—	2,448
Albania	3.2	0.8	2.2	74	23	1,340	0.5	521	13,593
Algeria	30.8	2.4	2.9	71	39	1,650	3.0	956	471
Angola	13.5	3.1	6.6	47	154	500	0.8	584	14,009
Argentina	37.5	1.4	2.5	74	16	6,940	3.8	1,660	24,276
Armenia	3.8	1.0	1.4	74	31	570	0.8	542	2,787
Australia	19.4	1.3	1.8	79	6	19,900	18.2	5,744	25,649
Austria	8.1	0.4	1.3	78	5	23,940	7.6	3,524	10,357
Azerbaijan	8.1	1.3	2.1	65	77	650	4.2	1,454	3,615
Bangladesh	133.3	2.1	3.0	62	51	360	0.2	142	9,238
Belarus	10.0	0.2	1.3	68	17	1,290	5.7	2,432	5,797
Belgium	10.3	0.2	1.6	78	5	23,850	10.2	5,776	1,561
Benin	6.4	3.0	5.4	53	94	380	0.2	377	4,114
Bolivia	8.5	2.2	3.8	63	60	950	1.4	592	37,305
Bosnia and Herzegovina	4.1	0.0	1.6	74	15	1,240	1.2	1,096	9,429
Botswana	1.7	3.0	3.9	39	80	3,100	2.4	—	8,776
Brazil	172.4	1.7	2.2	68	31	3,070	1.8	1,077	43,022
Bulgaria	8.0	−0.5	1.3	72	14	1,650	5.1	2,299	2,595
Burkina Faso	11.6	2.4	6.4	44	104	220	0.1	—	1,286
Burundi	6.9	2.5	5.9	42	114	100	0.0	—	529
Cambodia	12.3	2.8	3.9	54	97	270	0.1	—	39,613
Cameroon	15.2	2.6	4.7	49	96	580	0.3	427	18,352
Canada	31.1	1.1	1.5	79	5	21,930	14.4	8,156	94,314
Central African Republic	3.8	2.3	4.7	43	115	260	0.1	—	37,934
Chad	7.9	2.7	6.3	48	117	200	0.0	—	5,589
Chile	15.4	1.5	2.1	76	10	4,590	4.2	1,604	58,115
China	1,271.8	1.2	1.9	70	31	890	2.3	905	2,241
Hong Kong, China	6.7	1.4	0.9	80	3	25,330	6.2	2,319	—
Colombia	43.0	2.0	2.5	72	19	1,890	1.5	681	49,930
Congo, Democratic Rep.	52.4	3.2	6.1	45	129	80	0.0	292	23,809
Congo, Republic	3.1	3.0	5.9	51	81	640	0.8	296	275,679
Costa Rica	3.9	2.5	2.4	78	9	4,060	1.6	861	29,501
Côte d'Ivoire	16.4	3.3	4.7	46	102	630	0.8	433	4,790
Croatia	4.4	−0.2	1.4	74	7	4,550	4.8	1,775	16,301

	Total Population (millions— 2001)	Average Annual Population Growth Rate % (1980– 2001)	Total Fertility Rate (births per woman— 2001)	Life Expectancy at Birth (years— 2001)	Infant Mortality Rate (per 1,000 live births— 2001)	Gross National Income per Capita ($—2001)	Carbon Dioxide Emissions (per capita metric tons— 1999)	Commercial Energy Use per Capita (kg of oil equivalent— 2000)	Freshwater Resources (total renewable resources per capita cu. m— 2000)
Cuba	11.2	0.7	1.6	77	7	—	2.3	1,180	3,405
Czech Republic	10.2	0.0	1.2	75	4	5,310	10.6	3,931	1,382
Denmark	5.4	0.2	1.8	77	4	30,600	9.3	3,643	1,124
Dominican Republic	8.5	1.9	2.7	67	41	2,230	2.8	932	2,508
Ecuador	12.9	2.3	2.9	70	24	1,080	1.9	647	34,161
Egypt, Arab Republic	65.2	2.2	3.2	68	35	1,530	2.0	726	1,071
El Salvador	6.4	1.6	3.0	70	33	2,040	0.9	651	2,836
Eritrea	4.2	2.7	5.4	51	72	160	0.1	—	2,148
Estonia	1.4	−0.4	1.2	71	11	3,870	11.7	3,303	9,346
Ethiopia	65.8	2.7	5.6	42	116	100	0.1	291	1,711
Finland	5.2	0.4	1.7	78	4	23,780	11.3	6,409	21,268
France	59.2	0.4	1.9	79	4	22,730	6.1	4,366	3,218
Gabon	1.3	2.9	4.1	53	60	3,160	3.0	1,271	133,333
Gambia, The	1.3	3.5	4.9	53	91	320	0.2	—	6,140
Georgia	5.3	0.2	1.1	73	24	590	1.0	533	12,395
Germany	82.3	0.2	1.4	78	4	23,560	9.7	4,131	2,165
Ghana	19.7	2.9	4.1	56	57	290	0.3	400	2,756
Greece	10.6	0.4	1.3	78	5	11,430	8.2	2,635	6,913
Guatemala	11.7	2.6	4.4	65	43	1,680	0.9	628	9,591
Guinea	7.6	2.5	5.1	46	109	410	0.2	—	30,479
Guinea-Bissau	1.2	2.3	5.7	45	130	160	0.2	—	22,519
Haiti	8.1	2.0	4.3	52	79	480	0.2	256	1,633
Honduras	6.6	2.9	4.1	66	31	900	0.8	469	14,945
Hungary	10.2	−0.2	1.3	72	8	4,830	5.6	2,448	11,855
India	1,032.4	1.9	3.0	63	67	460	1.1	494	1,878
Indonesia	209.0	1.6	2.4	66	33	690	1.2	706	13,759
Iran, Islamic Republic	64.5	2.4	2.6	69	35	1,680	4.8	1,771	2,018
Iraq	23.8	2.9	4.2	62	107	—	3.3	1,190	4,776
Ireland	3.8	0.6	1.9	77	6	22,850	10.8	3,854	13,706
Israel	6.4	2.4	2.8	79	6	16,750	10.0	3,241	273
Italy	57.9	0.1	1.2	79	4	19,390	7.3	2,974	3,281
Jamaica	2.6	0.9	2.4	76	17	2,800	4.0	1,524	3,653
Japan	127.0	0.4	1.4	81	3	35,610	9.1	4,136	3,389
Jordan	5.0	4.0	3.6	72	27	1,750	3.1	1,061	143
Kazakhstan	14.9	0.0	1.8	63	81	1,350	7.4	2,594	7,278
Kenya	30.7	2.9	4.3	46	78	350	0.3	515	1,004
Korea, Democratic Rep.	22.4	1.3	2.1	61	42	—	9.4	2,071	3,462
Korea, Republic	47.3	1.0	1.4	74	5	9,460	8.4	4,119	1,485
Kuwait	2.0	1.9	2.6	77	9	18,270	24.9	10,529	0
Kyrgyz Republic	5.0	1.5	2.5	66	52	280	1.0	497	9,461
Lao PDR	5.4	2.5	4.9	54	87	300	0.1	—	63,175
Latvia	2.4	−0.4	1.2	70	17	3,230	2.8	1,541	14,924
Lebanon	4.4	1.8	2.3	71	28	4,010	4.0	1,169	1,109
Lesotho	2.1	2.0	4.3	43	91	530	—	—	2,555
Liberia	3.2	2.6	5.9	47	157	140	0.1	—	74,121
Libya	5.4	2.7	3.4	72	16	—	8.3	3,107	113

	Total Population (millions— 2001)	Average Annual Population Growth Rate % (1980– 2001)	Total Fertility Rate (births per woman— 2001)	Life Expectancy at Birth (years— 2001)	Infant Mortality Rate (per 1,000 live births— 2001)	Gross National Income per Capita ($—2001)	Carbon Dioxide Emissions (per capita metric tons— 1999)	Commercial Energy Use per Capita (kg of oil equivalent— 2000)	Freshwater Resources (total renewable resources per capita cu. m— 2000)
Lithuania	3.5	0.1	1.3	73	8	3,350	3.8	2,032	7,102
Macedonia, FYR	2.0	0.4	1.8	73	22	1,690	5.6	—	3,151
Madagascar	16.0	2.8	5.3	55	84	260	0.1	—	21,710
Malawi	10.5	2.5	6.2	38	114	160	0.1	—	1,668
Malaysia	23.8	2.6	2.9	73	8	3,330	5.4	2,126	24,925
Mali	11.1	2.5	6.2	41	141	230	0.0	—	9,225
Mauritania	2.7	2.7	4.5	51	120	360	1.2	—	4,278
Mauritius	1.2	1.0	2.0	72	17	3,830	2.1	—	1,853
Mexico	99.4	1.8	2.5	73	24	5,530	3.9	1,567	4,675
Moldova	4.3	0.3	1.4	67	27	400	1.5	671	2,735
Mongolia	2.4	1.8	2.5	65	61	400	3.2	—	14,512
Morocco	29.2	1.9	2.8	68	39	1,190	1.3	359	1,010
Mozambique	18.1	1.9	5.1	42	125	210	0.1	403	11,870
Myanmar	48.3	1.7	2.9	57	77	—	0.2	262	21,898
Namibia	1.8	2.8	4.9	44	55	1,960	0.1	587	25,896
Nepal	23.6	2.3	4.2	59	66	250	0.1	343	9,122
Netherlands	16.0	0.6	1.7	78	5	24,330	8.5	4,762	5,716
New Zealand	3.8	1.0	2.0	78	6	13,250	8.1	4,864	85,361
Nicaragua	5.2	2.8	3.5	69	36	—	0.8	542	37,409
Niger	11.2	3.3	7.2	46	156	180	0.1	—	3,000
Nigeria	129.9	2.9	5.2	46	110	290	0.3	710	2,206
Norway	4.5	0.5	1.8	79	4	35,630	8.7	5,704	87,508
Oman	2.5	3.9	4.1	74	12	—	8.5	4,046	415
Pakistan	141.5	2.6	4.6	63	84	420	0.7	463	1,610
Panama	2.9	1.9	2.5	75	19	3,260	2.9	892	51,647
Papua New Guinea	5.3	2.5	4.4	57	70	580	0.5	—	156,140
Paraguay	5.6	2.8	3.9	71	26	1,350	0.8	715	17,103
Peru	26.3	2.0	2.7	70	30	1,980	1.2	489	67,852
Philippines	78.3	2.3	3.3	70	29	1,030	1.0	554	6,251
Poland	38.6	0.4	1.3	74	8	4,230	8.1	2,328	1,594
Portugal	10.0	0.1	1.5	76	5	10,900	6.0	2,459	7,294
Puerto Rico	3.8	0.9	1.9	76	10	10,950	2.7	—	—
Romania	22.4	0.0	1.3	70	19	1,720	3.6	1,619	9,463
Russian Federation	144.8	0.2	1.2	66	18	1,750	9.8	4,218	30,904
Rwanda	8.7	2.5	5.8	40	96	220	0.1	—	611
Saudi Arabia	21.4	3.9	5.4	73	23	8,460	11.7	5,081	116
Senegal	9.8	2.7	5.0	52	79	490	0.4	324	4,134
Sierra Leone	5.1	2.2	5.7	37	182	140	0.1	—	31,803
Singapore	4.1	2.6	1.4	78	3	21,500	13.7	6,120	—
Slovak Republic	5.4	0.4	1.3	73	8	3,760	7.2	3,234	15,293
Slovenia	2.0	0.2	1.2	76	4	9,760	7.3	3,288	9,402
Somalia	9.1	1.6	7.0	47	133	—	0.0	—	1,789
South Africa	43.2	2.1	2.8	47	56	2,820	7.9	2,514	1,168
Spain	41.1	0.5	1.2	78	4	14,300	6.8	3,084	2,753
Sri Lanka	18.7	1.2	2.1	73	17	880	0.5	437	2,708
Sudan	31.7	2.4	4.5	58	65	340	0.1	521	4,792

	Total Population (millions— 2001)	Average Annual Population Growth Rate % (1980– 2001)	Total Fertility Rate (births per woman— 2001)	Life Expectancy at Birth (years— 2001)	Infant Mortality Rate (per 1,000 live births— 2001)	Gross National Income per Capita ($—2001)	Carbon Dioxide Emissions (per capita metric tons— 1999)	Commercial Energy Use per Capita (kg of oil equivalent— 2000)	Freshwater Resources (total renewable resources per capita cu. m— 2000)
Swaziland	1.1	3.0	4.3	45	106	1,300	0.4	—	4,306
Sweden	8.9	0.3	1.6	80	3	25,400	5.3	5,354	20,656
Switzerland	7.2	0.6	1.4	80	5	38,330	5.7	3,704	7,437
Syrian Arab Republic	16.6	3.1	3.6	70	23	1,040	3.4	1,137	2,761
Tajikistan	6.2	2.2	3.0	67	91	180	0.8	470	12,853
Tanzania (mainland only)	34.4	2.9	5.2	44	104	270	0.1	457	2,701
Thailand	61.2	1.3	1.8	69	24	1,940	3.3	1,212	6,750
Togo	4.7	2.9	5.0	49	79	270	0.3	338	2,651
Trinidad and Tobago	1.3	0.9	1.8	72	17	5,960	19.4	6,660	2,921
Tunisia	9.7	2.0	2.1	72	21	2,070	1.8	825	481
Turkey	66.2	1.9	2.3	70	36	2,530	3.1	1,181	3,593
Turkmenistan	5.4	3.1	2.3	65	69	950	6.4	2,627	11,523
Uganda	22.8	2.7	6.1	43	79	260	0.1	—	2,972
Ukraine	49.1	−0.1	1.2	68	17	720	7.5	2,820	2,820
United Arab Emirates	3.0	5.0	3.1	75	8	—	31.3	10,175	69
United Kingdom	58.8	0.2	1.7	77	6	25,120	9.2	3,962	2,503
United States	285.3	1.1	2.1	78	7	34,280	19.7	8,148	9,985
Uruguay	3.4	0.7	2.2	74	14	5,710	2.0	923	39,856
Uzbekistan	25.1	2.2	2.5	67	52	550	4.8	2,027	4,623
Venezuela, RB	24.6	2.3	2.8	74	19	4,760	5.3	2,452	29,892
Vietnam	79.5	1.9	2.2	69	30	410	0.6	471	11,350
West Bank and Gaza	3.1	—	5.0	72	21	1,350	—	—	—
Yemen, Republic	18.0	3.6	6.1	57	79	450	1.1	201	234
Yugoslavia, Fed. Republic (today's Serbia & Montenegro)	10.7	0.4	1.7	73	17	930	3.7	1,289	17,674
Zambia	10.3	2.8	5.2	37	112	320	0.2	619	11,498
Zimbabwe	12.8	2.8	3.7	39	76	480	1.4	809	1,117

Source: The United Nations

GLOSSARY

A

acid deposition The deposition of acidic substances on the ground, primarily as a result of sulfur and nitrogen oxide pollution of the atmosphere

actual evapotranspiration (ACTET) The amount of water evaporated and/or transpired in a given environment

acculturation The process of adopting some aspect of another culture

adiabatic cooling Cooling of air as a result of expansion of rising air; *adiabatic* means "without heat being involved"

advection Horizontal movements of air or substances by wind or ocean currents

African Diaspora The migration of black peoples out of Africa

agglomeration The bringing of people and activities together in one place for greater convenience

Agricultural Revolution The application of science and technology to agriculture, resulting in greatly increased yields and releasing workers

alluvial fan A fan-shaped deposit of sediment formed where a stream emerges from a narrow canyon onto a wider valley floor

alpine glacier A glacier occupying a valley in a mountainous area. The movement of an alpine glacier is primarily governed by the underlying topography

angle of incidence The angle at which solar radiation strikes a particular place at a point in time

apartheid A policy of racial segregation enforced in South Africa 1948–1993

aquaculture Herding or domesticating aquatic animals and farming aquatic plants

arithmetic density The number of people per unit of area

artifact A material object of culture; literally, "a thing made by skill"

asylum The safety a country grants refugees

atmosphere A thin layer of gases surrounding Earth to an altitude of less than 480 kilometers (300 miles)

autumnal equinox September 22 or 23, one of two dates when at noon the perpendicular rays of the Sun strike the equator (the Sun is directly overhead along the equator)

B

basic sector The part of a city's economy that is producing exports

beach A deposit of wave-carried sediment along a shoreline, on which waves break

behavioral geography The study of how we perceive our environment and of how our thoughts and perception influence our behavior

biodiversity The amount of variety of living things in a given environment

biogeochemical cycle Recycling process that moves essential substances such as carbon, nitrogen, and other nutrients between the biosphere and other components of Earth

biogeochemical oxygen demand The amount of dissolved oxygen in a water body that is consumed by decay of organic pollutants added to the water

biomagnification The tendency for substances that accumulate in body tissues to increase in concentration as they are passed to higher levels in a food chain

biomass The dry mass of living or formerly living matter in a given environment

biome A large grouping of ecosystems characterized by particular plant and animal types

biosphere All living organisms on Earth

biotechnology New techniques for modifying biological organisms and their physiological processes for applied purposes

bloc A group of countries that presents a common policy when dealing with other countries

boreal forest An evergreen needleleaf forest characteristic of cold continental climates

brain drain The emigration of a country's best-educated people and most skilled workers

broadleaf deciduous forest A forest with broadleaved trees that lose their leaves in the winter; characteristic of humid midlatitude environments

brownfields Abandoned industrial facilities in central cities, many of which are today being redeveloped

C

capital-intensive activity An activity in which a large amount of capital is invested per worker

capitalist system One in which the state defers to private enterprise and a stock market raises and allocates capital

carbon cycle The movement of carbon among the atmosphere, hydrosphere, biosphere, and lithosphere as a result of processes such as photosynthesis, respiration, sedimentation, weathering, and fossil-fuel combustion

carbon dioxide (CO_2) A trace gas found in the atmosphere; a major contributor to the greenhouse effect

carbon monoxide A pollutant formed by incomplete combustion of fossil fuels

carnivore An animal whose primary food supply is other animals

carrying capacity The ability of a given region to support human population indefinitely with a given technology and quality of life

cartel An organization formed to control the market for a particular commodity, usually by restricting supply

cartogram A maplike image designed to convey the magnitude of something rather than exact spatial locations

cartography Mapmaking

caste A group in the rigid social hierarchy of Hinduism

central business district (CBD) The traditional core of the city, concentrating office buildings and retail shops

central place theory A model of the distribution of cities across an isotropic plain

centrifugal forces Forces that tend to pull states apart

centripetal forces Forces which bind a state together

chemical weathering The breakdown of rocks or minerals through chemical reactions at Earth's surface

city A concentrated nonagricultural human settlement

climate The totality of weather conditions over a period of several decades or more

climax community The end point of community succession

cloning The production of identical organisms by asexual reproduction from a single cell of a preexisting organism

cognate A word that clearly looks or sounds like one in another language to which it is related historically

cognitive behavioralism The theory that people react to their environment as they perceive it

cold front The boundary formed when a cold air mass advances against a warmer one

collapsed state A state that has proven incapable of providing its citizens with either economic development or even peace and security

commercial agriculture Raising food or fiber to sell

Commercial Revolution The tremendous expansion of global trade between about 1650 and 1750

common market A customs union within which common laws create similar conditions of production

composite cone volcano A volcano formed by a mixture of lava eruptions and more explosive ash eruptions

concentration The distribution of a phenomenon within a given area

condensation Water changing from a gas state (vapor) to a liquid or solid state

conformal map A map that distorts size but preserves shapes

congregation Residential clustering by choice

consolidated metropolitan statistical area (CMSA) Two or more contiguous MSAs

contiguous diffusion Diffusion that occurs from one place directly to a neighboring place

continental glacier A thick glacier hundreds to thousands of kilometers across, large enough to be only partly guided by underlying topography

continental shelf An area of relatively shallow water that surrounds most continents, landward of the place where continental slope drops more sharply to the deep seafloor

convection Circulation in a fluid caused by temperature-induced density differences, such as the rising of warm air in the atmosphere

convergent plate boundary A boundary between tectonic plates in which the two plates move toward one another, destroying or thickening the crust

core area The historic homeland of a nation or area of greatest settlement

Coriolis effect The tendency of an object moving across Earth's surface to be deflected from its apparent path as a result of Earth's rotation

corporativism Government by representatives of interest groups

creole A pidgin that has become a mother tongue

crude birth rate The annual number of live births per 1,000 people

crude death rate The annual number of deaths per 1,000 people

cultural diffusion The spreading of cultural attributes

cultural ecology The study of the ways societies adapt to environments

cultural geography The study of the geography of human cultures

cultural imperialism The substitution of one set of cultural traditions for another, either by force or by degrading those who fail to acculturate and rewarding those who do

cultural landscape A landscape that reveals the many ways people modify their local environment

cultural mosaic A phrase describing Canada and suggesting that various cultural groups retain individuality there

cultural subnationalism The splitting of a state population among several local primary allegiances

culture A bundle of attributes of shared behavior or belief. These may include virtually anything about the way a people lives

culture realm The region throughout which a culture prevails

culture region The region throughout which a culture prevails

customs union A free trade area that enforces a common external tariff

cyberspace The extension of reality through global electronic means of communication

cyclone Large low-pressure areas in which winds converge in a counter-clockwise swirl (in the Northern Hemisphere, clockwise in the Southern Hemisphere)

D

delta A deposit of sediment formed where a river enters a lake or ocean

demographic transition model A model that describes the historical experience of population growth in the countries that are today rich

demography The analysis of a population in terms of specific characteristics, such as age or income levels

density The frequency of occurrence of a phenomenon in relation to its geographic area

dependency ratio The ratio of the combined population less than 15 years old and adult population more than 65 years old to the population of those between 15 and 64 years of age

desert A vegetation type with sparsely distributed plants, specifically adapted for moisture gathering and moisture retention

desert climate A climate with low precipitation and temperatures warm enough to cause potential evapotranspiration to be substantially higher than precipitation

desert pavement The stony surface of a desert soil formed by selective removal of fine particles by surface erosion

desertification The process of a region's soil and vegetation cover becoming more desertlike as a result of

human land use, usually by overgrazing or cultivation

dialect A minor variation within a language

diffusion The process of an item or feature spreading through time

diffusionism The theory that aspects of civilization were developed in very few places and then diffused from those places

diminishing returns A condition that exists when, in adding equal amounts of one factor of production, such as fertilizer or labor, each successive application yields a smaller increase in production than the application just preceding

discharge The quantity of water flowing past a point on a stream per unit time

dissolved oxygen Oxygen found in dissolved form in water. It is essential for aquatic animals, and depleted by pollution

distance The extent of space between two objects or places; it can be measured absolutely, in terms of miles or kilometers, or in terms of other units, such as time or cost to cross

distance decay The diminution of the presence or impact of any cultural attribute away from its hearth area

distribution The position, placement, or arrangement of a phenomena

divergent plate boundary A boundary between tectonic plates in which the two plates move away from each other, and new crust is created between them

domestication The process of adapting plants and animals to intimate association with humankind, to the advantage of humankind

double cropping Harvesting two crops from each field per year

doubling time The number of years it would take any country's population to double at present rates of increase

downstream activities Economic activities away from raw materials and closer to goods' ultimate consumers

drainage basin The geographical area that contributes runoff to a particular stream, defined with respect to a specific location along that stream

drainage density The total length of streams in a drainage basin divided by the drainage area

dune An accumulation of windblown sand, shaped by the wind

E

earthquake A sudden release of energy within Earth, producing a shaking of the crust

ecology The scientific study of ecosystems

economic development The process of progressively increasing the value of goods and services that a place is able to produce to enjoy or export

economic geography The study of how various peoples make their living and what they trade

economies of scale The economic factors determining that as the number of units of a good produced increases, the production cost per unit generally falls

ecosystem An interrelated collection of plants and animals and the physical environment with which they interact

ecotourism Travel to see distinctive examples of scenery, unusual natural environments, or wildlife

ejido A Mexican form of land tenure in which a peasant community collectively owns a piece of land and the natural resources and houses on it

electoral geography The study of voting districts and voting patterns

electronic highway Electronic methods of communication, such as e-mail, facsimile machines, and computer modems

El Niño A circulation change in the eastern tropical Pacific Ocean, from westward flow to eastward flow, that occurs every few years

emigration Movement away from a place

enculturation (socialization) Teaching youngsters a society's values and traditions, its political and social culture

endogenic processes Forces within Earth that affect its surface, such as plate tectonics, volcanic eruptions, and earthquakes

environmental determinism The simplistic belief that human events can be explained entirely as the result of the effects of the physical environment

epicenter The location on Earth's surface immediately above the focus of an earthquake

epidemiological transition The shift within a country of the principal causes of death from infectious to degenerative diseases

epidemiology The study of the incidence, distribution, and control of disease

equal-area map A map projection that preserves size but distorts shape

equator Earth's imaginary midline everywhere equidistant between the poles

ethnic group A dubious term suggesting that a particular cultural group is in the minority or "not normal" for a particular place and time

ethnocentrism A tendency to judge foreign cultures by the standards and practices of one's own, and usually to judge them unfavorably

etymology The study of the origin and history of words

European Union (EU) A bloc of European countries enjoying free trade and committed to eventual full political union

evapotranspiration The transfer of water from Earth's surface to the atmosphere by either evaporation, transpiration, or both

evolutionism The theory that a culture's sources of change were embedded in the culture from the beginning, so the course of development was internally determined

exclusive economic zone (EEZ) A 200-mile zone within which a coastal state controls both mining and fishing rights

exogenic processes Forces originating in the atmosphere that, aided by gravity, shape Earth's surface. Erosion by running water, glaciers, wind, and waves are examples

export-led economic growth A national economic policy of welcoming foreign investment to build factories that will manufacture goods for international markets

external economies The range of goods and services in a city that can be rented or hired temporarily

external refugees People who seek refuge in foreign countries

externality An exchange of a good or service that takes place without agreement of the parties in a market. Pollution is generally considered an externality

exurbs The settlements that make up the outermost ring of expanding metropolitan areas

F

failed state A state that has proven incapable of providing its citizens with either economic development or even peace and security

fault A fracture in Earth's crust along which displacement of rocks has occurred

federal government A form of government in which a central government shares power with subunits

felt needs Things people have become convinced that they need

fertility rate The number of children born per year per 1,000 females

First World During the Cold War, all reasonably well-to-do, non-Communist countries

fishery A concentration of aquatic species suitable for commercial harvesting

floodplain A low-lying surface adjacent to a stream channel and formed by materials deposited by the stream

focus (of an earthquake) The location in Earth where motion originates in an earthquake

folk culture A culture that is handed down and preserves traditions

food chain The sequential consumption of food in an ecosystem, beginning with green plants, followed by herbivores and carnivores and ending with decomposers

footloose activities Activities that can move or relocate freely

foreign direct investment (FDI) Investment by foreigners in wholly owned enterprises that are operated by the foreigner

formal region A region defined by essential uniformity in one or more physical or cultural features

fossil fuel A source of chemical energy stored in formerly living plant and animal tissue. Coal, oil, and natural gas are fossil fuels

free trade area An international territory having no internal tariffs, but its members are free to set their own tariffs on trade with the rest of the world

friction of distance The effort, time, or cost necessary to move or transport items

front A boundary between warm air and cold air

frontier An undeveloped region that may offer potential for settlement

functional region A region defined by interaction among places, such as trade and communication

fundamentalism The strictest adherence to traditional religious beliefs

G

gene splicing (or recombinant DNA) The joining of the genes of two organisms to produce recombinant (recombined) genetic material

genetic engineering The manipulation of species' genetic material through selective breeding or recombinant DNA

genetically modified (GM) Organisms and food products that have had their genetic makeup permanently changed as a result of gene splicing, that is, the recombining of the DNA from different organisms

gentrification The occupation and restoration of select urban residential neighborhoods by successful urban white-collar workers

geographic information systems (GIS) Computer hardware and software systems used to organize, store, analyze, and display geographic information

geography The study of the interaction of all physical and human phenomena at individual places and of how interactions among places form patterns and organize space

geomorphology The study of the shape of Earth's surface and the processes that modify it

geopolitics A pseudo-science studying "the natural and necessary trend towards [national] expansion as a means of self-preservation;" today often loosely applied to studies of military strategies

gerrymandering The drawing of voting district lines in ways that include or exclude specific groups of voters, so that one group gains an unfair advantage

glacier A large mass of flowing, perennial ice

globalization Increased interaction among people and societies at the global scale, including both economic and cultural phenomena

global positioning system (GPS) A navigational tool consisting of a fleet of satellites orbiting Earth, broadcasting digital codes, and a portable receiver that can "hear" those codes and determine its location

global warming A general increase in temperatures over a period of at least several decades believed to be caused primarily by increased levels of carbon dioxide in Earth's atmosphere

grade A condition in which a stream's ability to transport sediment is just balanced by the amount of sediment delivered to it

green revolution An intensive effort, starting about 1950, to develop new grain varieties and associated agronomic systems and to establish them in developing countries. It focused on certain crops (wheat, rice) and certain techniques (breeding for responses to fertilizer inputs). It was driven largely by private foundations

greenhouse effect Atmospheric warming that results from the passage of incoming shortwave energy and the capture of outgoing longwave energy

greenhouse gases Trace substances in the atmosphere that contribute to the greenhouse effect; water vapor, carbon dioxide, ozone, methane, and chlorofluorocarbons are important examples

Greenwich Mean Time (GMT) The time at the prime meridian or 0° longitude at Greenwich, England

gross domestic product (GDP) The total value of all goods and services produced within a country

gross national income (GNI) A country's GDP plus any income residents receive from foreign investments, minus any money paid out of the country to foreign investors

groundwater Water beneath Earth's surface at a depth where rocks and/or soils are saturated with water

gyre A circular ocean current beneath a subtropical high-pressure cell

H

hearth The place where a distinctive culture originated

herbivore An animal whose primary food supply is plants

hierarchical diffusion Diffusion that occurs downward or upward in a hierarchy of organization; when mapped, it shows up as a network of spots

hinterland The region to which any city provides services and upon which it draws for its needs

historical consciousness A people's consciousness of past events insofar as that consciousness influences their present behavior

historical geography The study of the geography of the past and how geographic distributions have changed

historical materialism The belief that technology has historically increased humankind's control over the environment and improved material welfare, and that this improvement prompts other historical events and movements

horizon A layer in the soil with distinctive characteristics derived from soil-forming processes

human development index (HDI) A statistic that combines statistics of life expectancy, school enrollment, literacy, and income to compare the quality of life around the world

human geography The study of the geography of human groups and activities

humid continental climate A climate characterized by cold winters and warm summers, with moderate levels of precipitation

humid subtropical climate A climate with cool winters, hot summers, and moderately high levels of precipitation

humid tropical climate A climate with high temperatures and high rainfall amounts all the year

hunter-gatherers People who live on what they can hunt or harvest from Earth

hurricane An intense tropical cyclone that develops over warm ocean areas in the tropics and subtropics, primarily during the warm season. Hurricanes in the Pacific Ocean are called typhoons, and in the Indian Ocean they are called cyclones

hydrocarbon A chemical substance composed of carbon and hydrogen; hydrocarbons in the atmosphere contribute to the formation of photochemical smog

hydroelectric power Electricity generated by water passing through turbines at a dam

hydrologic cycle The movement of water from the atmosphere to Earth's surface, across that surface, and back to the atmosphere

hydrosphere The water realm of Earth's surface, including the oceans, surface waters on land (lakes, streams, rivers), groundwater in soil and rock, water vapor in the atmosphere, and ice in glaciers

I

ice-cap climate A climate with very cold temperatures all the year, and with summer temperatures rarely above freezing

iconography A state's set of symbols, including a flag and anthem

igneous rocks Rocks formed by crystallization of a magma

immigration Movement into a place

import-substitution economic growth A national economic policy of protecting domestic infant industries

incorporation The process of defining a city territory and establishing a government

indigenous peoples As defined by the U.N., "descendants of the original inhabitants of a land who were subjugated by another people coming after them"

indirect rule The imperialist use of native rulers as intermediaries between the imperialists and the people

Industrial Revolution The evolution first in Europe between about 1750 and 1850 from an agricultural and commercial society to an industrial society relying on inanimate power and complex machinery

industrial society A society with a significant share of its output from the secondary sector

inertia The force that keeps things stable or fixed in place

infant industries Newly developing industries that probably cannot compete with imports

infant mortality rate The number of infants per thousand who die before reaching one year of age

infiltration capacity The maximum amount of water that can soak into a soil per unit time

informal or underground sector The economic activities that do not appear in official accounts

infrastructure Fixed assets in place, such as buildings, dams, and roads

innocent passage The internationally guaranteed right of the ships of one state to pass through the territorial waters of another on the way to a third

insolation The amount of solar energy intercepted by a particular area of Earth

internal economies Those goods and services that a large company can provide for itself

International Date Line Imaginary line on Earth's surface where, by international agreement, travelers traveling eastward subtract one calendar day, and travelers traveling westward add one calendar day. The line generally follows the 180° meridian, but it deviates for political and economic convenience

international organization An organization that coordinates activities among two or more countries

intertropical convergence zone (ITCZ) A low-pressure zone between the Tropic of Cancer and the Tropic of Capricorn where surface winds converge

irredenta Territory that one state claims from another

isogloss A line around places where speakers use a linguistic feature in the same way

isostatic adjustment A vertical movement of Earth's crust caused by loading or unloading of the buoyant crust

isotropic plain A theoretical perfectly flat surface with absolutely no variations across it

K

karst An assemblage of landforms found in areas of intense subsurface chemical weathering, often including features such as caves and underground drainage

kleptocracy Government by thieves or theft

L

labor-intensive activity An activity that employs a high ratio of workers to invested capital

laissez-faire capitalism A capitalist system that minimizes the government's role in the economy

landform A characteristic shape of the land surface, such as a hill, valley, or floodplain

language A set of words, plus their pronunciation and methods of combining them, that is used and understood to communicate within a group of people

language family The languages that are related by descent from a common protolanguage

large-scale map Map that shows a given area in a large space

latent heat Heat stored in water when it melts or evaporates, not detectable by people; *latent* means "hidden"

latent heat exchange The exchange of energy necessary to change water from one of its states to another—solid, liquid, or gaseous

latitude The location of a place measured as angular distance north and south of the equator

lava Magma that reaches Earth's surface and erupts

liberation theology The belief in putting the problems of overcoming poverty at the heart of Christian theology

life expectancy The average number of years that a newborn baby within a given population can expect to live

lingua franca A second language held in common for international discourse

liquidity The quality of being readily convertible into cash

lithosphere The solid Earth, composed of rocks and sediments overlying them

Little Ice Age The period between about 1500 to 1750, when climates on Earth were especially cool

local content requirement A definition of the percentage of the total value of a good entering one country that must have been added in the second country for that product to qualify as a product of the second country

location The place where a thing is; can be defined absolutely or relatively

loess An accumulation of windblown silt

longitude The location of a place measured as angular distance east and west from the prime meridian

longshore current A current in the surf zone along a shoreline, parallel to the shore

longshore transport Sediment transport by a longshore current

longwave energy Energy radiated by Earth in wavelengths of about 5.0 to 30.0 microns

M

magma Molten rock beneath Earth's surface

Malthusian theory The pessimistic argument that population increases will always outpace increases in food production, causing cycles of war, famine, and disease; articulated by Thomas Malthus (1766–1834)

mantle The portion of Earth above the core and below the crust

map A two-dimensional (flat) representation of some portion of Earth's surface

maquiladora A factory in Mexico specializing in assembling items for export to the U.S. market

marine terrace A nearly level surface along a shoreline, elevated above present sea level, formed by coastal erosion at a time when sea level at the location was higher than at present

marine west coast climate A climate with moderately cool winters, moderately warm summers, and moderate to high rainfall all year

market-oriented manufacturing Manufacturing that locates close to the market either because the processing increases

the perishability of the product or because the processing adds bulk or weight to the product

Marshall Plan A plan of financial aid for the economic rehabilitation of Europe after World War II, named for U.S. Secretary of State George C. Marshall

mass movement Downslope movement of rock and soil at Earth's surface, driven mainly by the force of gravity acting on those materials

material-oriented manufacturing Manufacturing that locates close to the source of the raw material either because the raw material is heavy or bulky or because the raw material is perishable

meandering The tendency of flowing water to follow a sinuous course with alternating right- and left-hand bends

mechanical weathering The breakdown of rocks into smaller particles caused by application of physical or mechanical forces

Mediterranean climate A climate with warm, dry summers and cool, moist winters

melting pot Name given to the United States in a 1914 novel of that title, suggesting that ethnic and racial differences among immigrants melt together to form one culture

meltwater channel A river channel carved by water from a melting glacier

mental map The ideas that we have about places, whether our ideas are true or false

meridians Imaginary lines extending from pole to pole and crossing all parallels at right angles

meritocracy A society in which the most capable people can rise to the top on merit alone

metamorphic rocks Rocks formed by modification of other rock types, usually by heat and/or pressure

metropolitan statistical area (MSA) According to the U.S. Census Bureau, "an integrated economic and social unit with a recognized large population nucleus"

microclimate Local climatic conditions, as in cities, that differ from those of surrounding areas

midlatitude cyclone A storm characterized by a center of low pressure in the midlatitudes usually associated with a warm front and a cold front

midlatitude low-pressure zones Regions of low pressure and air converging from the subtropical and polar high-pressure zones

migration chain A network of social and communication linkages that attracts migrants to follow others

model An idealized, simplified representation of reality

monoculture The specialized production of one crop

monotheism Belief in the existence of only one god

monsoon circulation Seasonal reversal of pressure and wind, most prominent in Asia, in which winter winds from the continental interior produce dry winters, and summer winds blowing inland from the oceans produce wet summers

moraine An accumulation of rock and sediment deposited by a glacier, usually in or near the melting area

multinational corporation An enterprise that produces and markets goods in several countries

multinational state A state containing several nations

multiplier effect The fact that jobs in a city's basic sector multiply jobs in the nonbasic sector

N

nation A group of people who want to have their own government and rule themselves

nationalized resources and enterprises Resources and enterprises that are owned by the government of a country in the name of the people

national self-determination The idea of the nation-state as defended by U.S. President Woodrow Wilson after World War I

nation-state A state ruling over a territory containing all the people of a nation and no others

natural landscape A landscape without evidence of human activity

natural population increase (decrease) The difference between the number of births and the number of deaths

natural resource Something that is useful and that exists independent of human activity

network hypothesis The theory that central city unemployment is caused by a lack of social networks

nitrogen oxide (NO$_x$) A compound of nitrogen and oxygen; a component of air pollution

nonbasic sector The part of a city's economy serving the needs of the city itself

nongovernmental organizations (NGOs) Organized international interest groups that cross state boundaries and put pressure on the existing governments of states

nonrenewable resource A resource that is either not being produced by nature or is produced at rates much slower than it is used by humans

North Atlantic Treaty Organization (NATO) A military bloc founded in 1949 and in 2003 including Belgium, Denmark, France, Great Britain, Iceland, Italy, Luxembourg, the Netherlands, Norway, Portugal, Canada, the United States, Greece, Turkey, Germany, Spain, Poland, the Czech Republic, and Hungary; seven more countries will join in 2004: Latvia, Lithuania, Estonia, Bulgaria, Slovakia, Slovenia, and Romania

O

official language The language in which legal documents are kept in a country

oligarchy An elite privileged clique

omnivore An animal that feeds on both plants and other animals

opportunity cost A return sacrificed by leaving capital invested in one form or activity rather than another

orographic precipitation Precipitation caused when air is forced by regional circulation patterns to rise above a mountain

orthography The study of writing, or a system of writing

outwash plain An accumulation of sand and gravel carried by meltwater streams from a glacier, usually deposited immediately beyond the terminal moraine from the glacier

overland flow Water flowing across the soil surface on a hillslope, usually resulting from precipitation falling faster than the ground can absorb it

Overseas Chinese Chinese migrants who often remain linked across international borders by bonds of family, clan, and common home province

oxygen cycle The movement of oxygen between the biosphere and atmosphere through photosynthesis and oxydation or respiration

ozone (O$_3$) A gas composed of molecules with three oxygen atoms; a highly corrosive gas at ground level, but in the upper atmosphere essential to protecting life on Earth by absorbing ultraviolet radiation

P

parallels Lines connecting all points of the same latitude

parent material Mineral matter such as rocks or transported sediments from which soil is formed

particulate A small solid particle in the air; a component of air pollution

pastoral nomadism A style of life that does not have fixed residences, but drives flocks from place to place to find grazing lands and water

pathogen A disease-causing organism

pattern The arrangement of objects within an area

permafrost Soil or rock with a temperature below 0° C all year

photochemical smog A mixture of air pollutants including oxidants such as ozone, formed by interaction of sunlight and pollutants such as nitrogen oxides and hydrocarbons

photosynthesis A chemical reaction that occurs in green plants in which carbon dioxide and water are converted to carbohydrates and oxygen

photovoltaic cell A device that converts light to electricity

physical geography The study of the characteristics of the physical environment

physiological density The density of population per unit of arable land

plate tectonics theory The movement of large, continent-sized slabs of Earth's crust relative to one another

Pleistocene Epoch A period of geologic time consisting of the first part of the Quaternary Period beginning about 3 million years ago and ending about 12,000 years ago

polar front A boundary between cold polar air and warm subtropical air circling the globe in the midlatitudes

polar high-pressure zones Regions of high pressure and descending air near the North and South poles

political community A willingness to join together and form a government to solve common problems

political culture The set of unwritten ways in which written rules are interpreted and actually enforced

political economy The study of individual countries' organization and regulation of their economies

political geography The study of the interaction between political processes and the distributions of all other activities and transformations of the landscape

pollution A human-caused increase in the amount of a substance in the environment

pollution prevention A strategy for reducing pollution that focuses on reducing the amount of pollutants created, rather than on removing them from waste streams

polyculture The raising of a variety of crops

polyglot state A country that grants legal equality to two or more languages

polytheism The worship of many gods

popular culture The culture of people who embrace innovation and conform to changing norms

population geography The study of the distribution of humankind across Earth

population projection A forecast of the future population, assuming that current trends remain the same or else change in defined ways

population pyramid A graphic device that shows the shares of a nation's population by age groups

possibilism The theory that the physical environment itself will neither suggest nor determine what people will attempt, but it may limit what people can profitably achieve

postindustrial society A society with the bulk of its economic activity in the tertiary sector

potential evapotranspiration (POTET) The maximum amount of water that could be evaporated from a moist surface and/or transpired by plants if it were available

potential resource Something that is not useful today but may become so in the foreseeable future

prairie A vegetation type characterized by dense grass up to 2 meters high, found in midlatitude semiarid climates

preindustrial society A society with the bulk of its economic activity in the primary sector

prime meridian The meridian passing through the Royal Observatory in Greenwich, England, from which longitude is measured

primary sector That part of the economy that extracts resources directly from Earth, including agriculture, fishing, forestry, and mining

primate city A large city concentrating a national population or national political, intellectual, or economic life

privatize To give or sell government assets to private individuals or investors

projection A method of portraying Earth or any portion of it on a flat map

proselytize To try to convert others to your religious beliefs

protolanguage The ancestor that is common to any group of several of today's languages

proxemics The study of how people perceive and use space

pull factors Considerations that attract people to new destinations

push factors Anything that makes a person want to leave a place and seek a better life elsewhere

Q

Quaternary Period The period of geologic time encompassing approximately the last 3 million years

R

race Groups of people variously defined by relatively minor biological differences within the human species

racism The false belief in the inherent superiority of one race over another and the linking of human ability, potential, and behavior to racial inheritance

radiation Energy in the form of electromagnetic waves that radiate in all directions

recombinant DNA (or gene splicing) The joining of the genes of two organisms to produce recombinant (recombined) genetic material

refugee As defined by the 1951 Geneva Convention, someone with "a well-founded fear of being persecuted in his country of origin for reasons of race, religion, nationality, membership of a particular social group or political opinion"

region A territory that exhibits a certain uniformity

regional geography An inventory analysis of all characteristics of any individual place

relative humidity The *actual* water content of the air expressed as a percentage of how much water the air *could* hold at a given temperature

relative location The location of a place relative to other places

relocation diffusion Diffusion from one widely separated point to another

remote sensing The acquisition of data about Earth's surface from a satellite orbiting the planet or from high-flying aircraft

renewable resource Something that is produced by nature at rates similar to those at which it is consumed by humans

replacement rate A total fertility rate of 2.1, which stabilizes a population

respiration A chemical reaction that occurs in plants and animals in which carbohydrates and oxygen are combined, releasing water, carbon dioxide, and heat

runoff Flow of water from the land, either on the soil surface or in streams

S

sanitary landfill A site at which solid waste is deposited and covered with layers of earth

saturation vapor pressure The maximum water vapor that air can hold

savanna A vegetation type characterized by grasses and scattered trees, characteristic of seasonally dry tropical climates

sawah A flooded rice field (in Indonesian)

scale A quantitative statement of the relative sizes of an object on a map and in reality

scientific revolution in agriculture The continuing application of science to agriculture

seafloor spreading Creation of new oceanic crust where two tectonic plates are diverging on the seafloor

sea level The general elevation of the sea surface, averaging out variations caused by waves, storms, and tides

seasonally humid tropical climate A climate with warm temperatures all the year, a season with high rainfall, and a pronounced dry season

Second World During the Cold War, the Soviet bloc dominated by totalitarian Communist governments

secondary sector That part of the economy that transforms raw materials into manufactured goods

sectoral evolution A shift in the concentration of activity from an economy's primary sector to its secondary and tertiary sectors

secularism A lifestyle or policy that deliberately ignores or excludes religious considerations

sediment transport The movement of rock particles by surface erosion

sedimentary rocks Rocks formed through accumulation of many small rock fragments at Earth's surface

segregation Residential clustering as the result of discrimination

seismic waves Vibrations or shock waves originating at the focus of an earthquake and transmitted through Earth

seismograph A device for recording movements of Earth's crust, such as earthquakes

semiarid climate A climate with precipitation slightly less than potential evapotranspiration for most of the year

sensible heat Heat detectable by sense of touch, or with a thermometer

service sector That part of the economy that services the primary and secondary sectors

shamanism A belief in the power of mediums (shamans) who characteristically go into autohypnotic trances, during which they are thought to be in communion with the spirit world

Sharia Islamic teachings that are often incorporated into civil law in Islamic countries

shield The ancient core of a continent

shield volcano A volcano with relatively gentle slopes formed by eruption of relatively fluid lavas

shortwave energy Radiant energy emitted by the Sun in wavelengths about 0.2 to 5.0 microns

sial Crust formed of relatively less dense minerals, dominated by silicon and aluminum (for SIlicon-ALuminum)

sima Crust formed of relatively dense minerals, dominated by silicon and magnesium (for SIlicon-MAgnesium)

site The characteristics of the absolute location of a place

situation The characteristics of the relative location of a place

small-scale map A map that shows a given area in a very small space

social Darwinism The theory of British sociologist Herbert Spencer that "Nature's law" calls for "the survival of the fittest," even among cultures and whole peoples

socialization (enculturation) Teaching youngsters a society's values and traditions, its political and social culture

soil A dynamic, porous layer of mineral and organic matter at Earth's surface

soil creep Slow downslope movement of soil caused by many individual, near-random particle movements such as those caused by burrowing animals or freeze and thaw

soil fertility The ability of a soil to support plant growth through storing and supplying water, air, and nutrients

soil order A major category in the U.S. soil classification system

sojourner A migrant who intends to stay in a new location only long enough to save capital to return home to a higher standard of living

solar energy Radiant energy from the Sun

spatial mismatch hypothesis The hypothesis that central city unemployment is caused by the removal of job opportunity to the suburbs and the concentration of the poor in the central city

speech community A group of people who speak together

standard language Any language that is spoken and written according to formal rules of diction and grammar

state An independent political unit that claims exclusive jurisdiction over a defined territory and over all of the people and activities within it

steppe A vegetation type characterized by relatively short, sparse grasses, found in midlatitude semiarid climates

storm surge An area of elevated sea level in the center of a hurricane that may be several meters high, and which does most of the damage when a hurricane comes ashore

structural landform A landform whose major characteristics are derived from endogenic processes, or by erosional exposure of rock structures

subarctic climate A high-latitude climate characterized by brief, cool summers and long, cold winters

subsistence agriculture Raising food only for oneself, not to sell

substitutability The degree to which one commodity can be substituted for another in various uses

subtropical high-pressure (STH) zones Regions of high pressure and descending air at about 25° north and south latitudes

succession A process of ecological change in which organisms modify their immediate environments in ways that allow other species to establish themselves and dominate

sulfur oxide (SO_x) An air pollutant consisting of compounds of sulfur and oxygen, derived mainly from combustion of coal and oil

summer solstice For places in the Northern Hemisphere, June 20 or 21, the date when at noon the Sun is directly overhead along the parallel of 23.5° north latitude; for places in the Southern Hemisphere, December 21 or 22, the date when at noon the Sun is directly overhead at places along the parallel of 23.5° south latitude

superimposed boundaries Boundaries imperialists drew to establish colonies over natives' territorial demarcations

supply and demand The interplay of buyers and sellers of a commodity in the marketplace

supranational organization An organization that exercises power over countries

surface erosion Downslope movement of rock and soil at Earth's surface, driven mainly by air, water, or ice moving across the surface

sustainable development Economic development that increases wealth but avoids environmental destruction

sustained yield A way of managing a renewable natural resource such that harvest can continue indefinitely

swidden Slash-and-burn clearing and cultivation

syncretic religions Religions that combine Christianity with traditional practices

system An interdependent group of items that interact in a regular way to form a unified whole

systematic geography The study of universal laws or principles that apply to all places; topics may be as diverse as the geography of soils (pedology), of life forms (biogeography), of politics (political geography), of economic activities (economic geography), and of cities (urban geography)

T

tap route A transport network that consists of lines penetrating into the country from its ports

tectonic plate A large, continent-sized piece of Earth's crust that moves in relation to other pieces

telecommuting Working at home at a computer terminal connected to an office

temperature inversion A layer in the atmosphere in which relatively warm air lies above cooler air

terminal moraine An accumulation of rock and sediment at the downstream end of a glacier

territoriality A drive—possibly biological—to lay claim to territory and defend it against members of one's own species

terrorism Violent acts intended to frighten and to intimidate for political ends a civilian population beyond their immediate victims

tertiary sector That part of the economy that services the primary and secondary sector

theocracy A form of government where a church rules directly

Third World During the Cold War, the countries seen as characterized by underdevelopment, overpopulation, irrationality, religion, and political chaos

threshold principle The belief that a nation must have some minimal population and territory to merit self-determination

topical geography The study of universal laws or principles that apply to all places; topics may be as diverse as the geography of soils (pedology), of life-forms (biogeography), of politics (political geography), of economic activities (economic geography), and of cities (urban geography)

topography Surface relief

toponymy The study of place names

tornado A rapidly rotating column of air usually associated with a thunderstorm, often having winds in excess of 300 km/hour (185 miles/hour)

total fertility rate The average number of children that would be born to each woman in a given society if, during her childbearing years (15–49), she bore children at the current year's rate for women of that age

toxic substance A pollutant that can be harmful even at very low concentrations

trade wind Prevailing wind in subtropical and tropical latitudes that blows toward the Intertropical Convergence Zone, typically from the northeast in the Northern Hemisphere and from the southeast in the Southern Hemisphere

transform plate boundary A boundary between tectonic plates in which the two plates pass one another in a direction parallel to the plate boundary

transnational corporation An enterprise that produces and markets goods in several countries

transpiration The use of water by plants, normally drawing it from the soil via roots, evaporating it in the leaves and releasing it to the atmosphere

trophic level A position in the food chain relative to other organisms, such as producer, herbivore, or carnivore

Tropic of Cancer The parallel of 23.5° north latitude

Tropic of Capricorn The parallel of 23.5° south latitude

tropical rain forest Broadleaf evergreen vegetation characteristic of humid tropical environments

tsunami An extremely long wave created by an underwater earthquake; it may travel hundreds of kilometers per hour

tundra A low, slow-growing vegetation type found in high-latitude and high-altitude conditions in which snow covers the ground most of the year

tundra climate A climate characterized by long, very cold winters and short, cool summers. Summer temperatures are only a few degrees above freezing

typhoon The name applied to a hurricane in the Pacific Ocean

U

undocumented, irregular, or **illegal immigrants** People who cross borders without completing legal papers

unitary government A form of government in which the balance of power lies with the central government

untouchable One of a group of people considered so low that their status is below the formal structure of the Hindu caste system

urban enterprise zones Areas within which governments create generous conditions for enterprises to encourage the creation of jobs

urban geography The geographic study of cities

urban heat island A dome of heat over a city created by urban activities and conditions

urbanization The process of concentrating people in cities

V

value added by manufacturing The difference between the value of a raw material and the value of a product manufactured from that raw material

vernacular region A region defined by widespread popular conception of its existence

vernal (spring) equinox March 20 or 21 (in the northern hemisphere) or September 22 or 23 (in the southern hemisphere), when at noon the perpendicular rays of the Sun strike the equator (the Sun is directly overhead along the equator)

volcano A vent in Earth's surface where lava emerges

W

warm front A boundary formed when a warm air mass advances against a cooler one

water budget An accounting of the amounts of precipitation, evapotranspiration, soil moisture storage, and runoff at a given place

wavelength The distance between successive waves of radiant energy, or of successive waves on the surface of a water body

weather Patterns of atmospheric circulation, temperature, and precipitation over short time periods such as hours to days

weathering The chemical and/or mechanical breakdown of rocks into smaller particles at Earth's surface

winter solstice For places in the Southern Hemisphere, June 20 or 21, the date when at noon the Sun is directly overhead at places along the parallel of 23.5° north latitude; for places in the Northern Hemisphere, December 21 or 22, the date when at noon the Sun is directly overhead at places along the parallel of 23.5° south latitude

workers' remittances Money that migrant workers send home from elsewhere

X

xerophyte A plant adapted to living in arid conditions

Z

zero population growth A stabilized total population

Zionism The belief that the Jews should have a homeland of their own

zoning Restricting or prescribing the use to which parcels of land may be put

INDEX

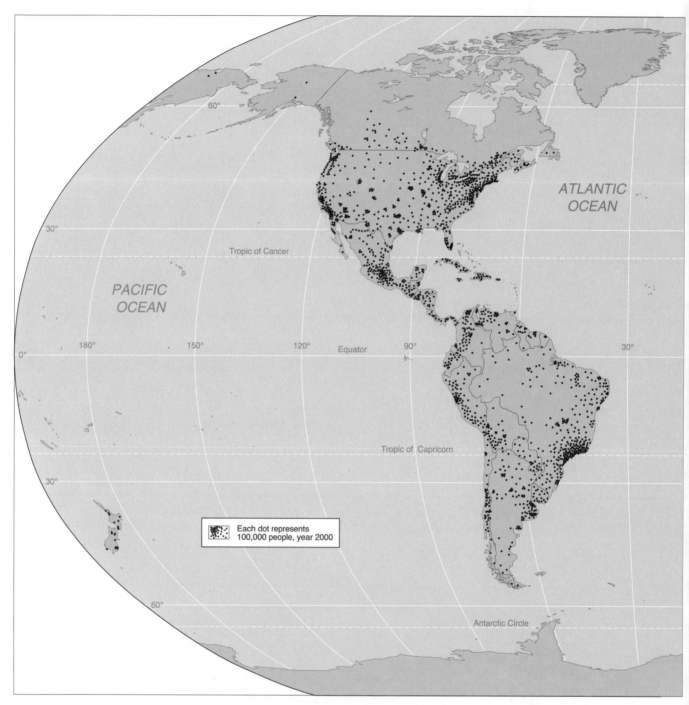

Dot maps like this cannot be exact, because dots cover other dots in dense areas, and the locations of some dots must be generalized. Nevertheless, the map conveys the pattern of concentration.